Library of
Davidson College

THE TECHNOLOGICAL LEVEL OF SOVIET INDUSTRY

THE TECHNOLOGICAL LEVEL
OF SOVIET INDUSTRY

edited by
RONALD AMANN, JULIAN COOPER
AND R. W. DAVIES
(WITH THE ASSISTANCE OF HUGH JENKINS)

Centre for Russian and East European Studies
University of Birmingham

YALE UNIVERSITY PRESS
NEW HAVEN AND LONDON
1977

Copyright © 1977 by Yale University

All rights reserved

This book may not be reproduced, in whole or in part, in
any form (except by reviewers for the public press),
without written permission from the publishers

Designed by John Nicoll and set in Monophoto Times

Filmset and printed in Great Britain by
BAS Printers Limited, Over Wallop, Hampshire

Published in Great Britain, Europe, Africa, the Middle East,
India and South East Asia by Yale University Press, Ltd., London

Distributed in Latin America by Kaiman & Polon, Inc., New York City;
in Australasia by Book & Film Services, Artarmon, N.S.W., Australia;
in Japan by Harper & Row, Publishers, Tokyo

Library of Congress Cataloging in Publication Data

Main entry under title:

The Technological level of Soviet industry.

   Includes index.
   1. Technology—Russia. I. Amann, Ronald,
1943–   II. Cooper, Julian, 1945–
III. Davies, Robert William.
J26.R9T44   338.4'7   77-76298
ISBN 0-300-02076-7

Preface

The essays in the present volume, prepared by an inter-university group of scholars based on the Centre for Russian and East European Studies, University of Birmingham, attempt to assess the comparative technological level of Soviet industry and the changes which have taken place in it over the past twenty years. The problem is approached by means of case-studies: in the first two chapters our approach is explained and an attempt is made to bring together the results of the individual studies.

The present volume is intentionally confined to presenting as systematically as possible the evidence which we have assembled about Soviet technological level; we do not assess the *efficiency* of Soviet technological performance or attempt to *explain* the complex patterns of technological level which appear to exist. In a second volume, which is now being prepared, we will examine the innovation process in the Soviet Union as exemplified in the various industries we have studied.

The project was undertaken as part of the Centre's research programme in Soviet science and technology; it arose from our general study of science and industry completed in 1969 and published in *Science Policy in the USSR* (OECD, Paris), which led us to the conclusion that a substantial improvement of our knowledge of the Soviet research and development system required a very thorough investigation of particular industries. Soviet data are extremely scattered, and assembling them from a variety of sources, especially newspapers and technological journals, is necessarily time-consuming. We were therefore fortunate in obtaining a four-year grant from the British Social Science Research Council, which enabled Dr Cooper to be employed on the project full-time as a Research Fellow for three years (1972–5) and Dr Amann, seconded from his normal duties as a lecturer in the Centre, for two years (1973–5). Professor Davies devoted a substantial part of his personal research time to the project. The project also provided for a full-time Research Associate—this post was held by Mr Kocourek in 1973–4 and Mr Hugh Jenkins in 1974–5. The other contributors undertook their work as part of their normal activities as university lecturers, or in their spare time; and we are grateful to the Universities of Aston, Brunel, Edinburgh and Loughborough, to the Electricity Research Council and to Bailey Meters Ltd for facilitating their collaboration.

These studies were feasible only as a result of close collaboration between different disciplines in the social sciences and in technology: the three editors are a political scientist, an economist and an economic historian, and they were advised in designing the project by Dr T. J. Grayson, lecturer in production engineering in the Centre, whose illness unfortunately prevented him from contributing his own case-study. The other authors include two political scientists, two economists, three engineers and a chemist. While each author in the team is responsible for his own chapter, both the project as a whole and each case-study were strongly influenced by debate among ourselves; all the authors met at informal conferences on several occasions in the course of 1973–5, and our drafts were discussed in detail both at these conferences and in meetings between individual authors and the three editors.

The authors were in close contact throughout the project with two other research groups engaged in similar work: the Science Policy Research Unit of the University of Sussex and the Osteuropa-Institut in Munich. Much valuable advice was received in discussions with Professor Freeman and his colleagues at Sussex on the basis of their vast knowledge of Western industry. We greatly benefited from acquaintance with the well-known studies of Soviet oxygen steel, computers and other technologies by Dr Heinrich Vogel, Dr Jiří Sláma and their colleagues, and from wide-ranging exchanges of opinion with them at two meetings in Birmingham and in the course of a study visit to Munich by Dr Amann, financed by the Deutsche Forschungsgemeinschaft. Dr Philip Hanson of our Birmingham Centre, whose work on the role of technology transfer in the Soviet economy is complementary to our own, contributed valuable comments on several of our case-studies and on the project as a whole.

Authors of particular chapters wish to acknowledge help and support as follows:

Dr R. Amann wishes to thank Professor David Granick (University of Wisconsin), Mr M. R. Freeman (British Sulphur Corporation), Dr Keith Norris (Brunel University) and Dr J. R. Thomas (National Science Foundation, Washington, DC) for their extensive comments and advice. The following colleagues at Birmingham University also provided invaluable help on particular parts of the work: Mr Neville Brown, Mr Henry Scott, Professor J. R. Harris, Mr Peter Kneen and Mr Stuart Burchell.

Dr J. M. Cooper is indebted to Dr B. W. Rooks (Department of Mechanical Engineering, Birmingham University) for his assistance in clarifying some technical problems in the development of NC machine tools and also wishes to thank Mr A. E. De Barr (Director of the Machine Tool Industry Research Association) and the staff of MTIRA, who generously provided Western machine tool statistics.

Mr M. E. Cave is grateful for information and advice from David Bicket,

Jim Hackett and Paul Stoneman. He has relied heavily on an earlier article on Soviet computer technology by Professor Richard Judy of Toronto University, who was also kind enough to send him an unpublished account of his most recent work.

Mr M. J. Berry wishes to acknowledge the valuable help which he received from the Machine Tool Industry Research Association (MTIRA) and the Machine Tool Trades Association (MTTA). The Leverhulme Trust also provided him with generous financial support at an earlier stage in his work.

Mr D. J. Holloway would like to acknowledge help from R. M. Ogorkiewicz (Imperial College, London University) and Joe Thompson (Science Policy Research Unit, Sussex University) concerning tank technology; Milton Leitenberg (Cornell University) and Peter Nailor (Lancaster University) provided helpful comments on the section of his work concerned with ICBMs.

Mr M. Kocourek is greatly indebted to Professor G. A. Tokaty (City University, London), who read the first draft of his chapter and offered invaluable technical advice on such important problems as rocket thrust.

Dr Z. A. Siemaszko wishes to thank the management of Bailey Meters and Controls Ltd for their indispensable cooperation while he was engaged in preparing his chapter.

Mr W. G. Allinson received useful advice from Dr Hugh Duffy (Electricity Research Council), the Delle Alsthom Company (France) and Dr I. F. Elliot, for which he wishes to express his gratitude.

In order to pull together the various contributions to this volume a great deal of time-consuming effort went on behind the scenes. In this connection, the editors are most grateful to Mr Hugh Jenkins. Both in the collection of data from the Soviet press and in the editorial assistance which he gave in the final stages of the work he observed exacting and professional standards; his good-natured but just exposure of each contributor's failure to observe stylistic rules will long be remembered. Our Centre secretary, Miss Anthea Roth, was also a tower of strength, managing to preserve her equanimity as a rapid succession of large and untidy manuscripts landed with a thud on her desk. We would also like to express collective gratitude to each of our wives and children for their forbearance.

In view of the large number of individuals and organisations consulted during the course of our research it is particularly important to stress that all remaining errors and omissions are, of course, the responsibility of the editors and authors and not of those whom we have approached for comments and advice.

December 1976

RA
JMC
RWD

List of Contents

List of Tables   xix
List of Text Figures   xxix
Glossary   xxxi

CHAPTER 1   RONALD AMANN

Some approaches to the comparative assessment of Soviet technology: its level and rate of development   1

    Aggregated economic assessments of technology: general methodological problems   2

    Some aggregate economic estimates of Soviet technology and its rate of development: results and limitations   8

    Disaggregated measures of the level and rate of technological development: general approaches   23

       Inputs   24
          Higher education and scientific manpower
          Expenditure on research and development
       Outputs   24
          Research performance
          Technical innovation
          Diffusion of technology
          Composition of foreign trade

Some disaggregated measures of the level of technology in the USSR ... 26
Our approach and its limitations ... 30

CHAPTER 2 R. W. DAVIES

The technological level of Soviet industry: an overview ... 35

Introduction ... 35
Research ... 38
Prototype and first commercial production ... 39
Diffusion of new technology ... 47
Quality ... 58
The 'research-production cycle' ... 59
The role of foreign technology in Soviet industry ... 63
Conclusion ... 66

CHAPTER 3 JULIAN COOPER

Iron and steel ... 83

Introduction ... 83
Ironmaking ... 84
    The preparation of raw materials ... 85
    The scale of blast furnaces ... 86
    Intensification of the smelting process ... 89
    Mechanisation and automation ... 91
    Direct reduction of iron ore ... 91
Steelmaking ... 92
    Open-hearth steel ... 93
    Oxygen converter steel ... 96
    Electric arc furnace steel ... 100
Continuous casting of steel ... 101
Alloy and quality steel ... 104
    Refining steel with synthetic slag ... 105
    Vacuum degassing of steel ... 106
    Electro-slag remelting ... 106

    Vacuum arc remelting   107
    Vacuum induction remelting   107
    Other processes   107
Rolled metal products   108
    Tube and pipe making   112
    Rolling mills   112
Automation of processes in the iron and steel industry   118
Conclusion   118

CHAPTER 4   M. J. BERRY AND JULIAN COOPER

Machine tools   121

Machine tools: stock and production   122
    Numbers and distribution of stocks by industry   122
    Age pattern of machine tool stocks   125
    Weight and value of machine tools produced   126
    Balance between metalcutting and metalforming machine tools   129
    Composition of machine tool stocks by type   132
    Reasons for differences between machine tool stocks   139
        The significance of the differences in stocks   143
    Changes in relative technological levels   145
Soviet imports and exports of metalcutting machine tools as an indicator of technological level   146
    The historical background   146
    The pattern of Soviet imports—the case of the United Kingdom   152
    Soviet export performance in the United Kingdom market   157
Numerically controlled machine tools   160
    The early years—from research to industrial production   161
    Some problems of Soviet NC development before 1968   163
    Soviet NC development since 1968   167
    Types of NC machine tools produced   174

The NC machine tool stock	175
Machining centres	179
Control systems	181
Types of control systems	184
Programming of NC machine tools	186
Some new paths of development of numerical control	188
Some aspects of the use of NC machine tools	191
The problems of technological level	193
The path of technological development	193
Innovation performance	194
The technical qualities of NC machine tools	196
The utilisation of NC machines	197
Conclusion	198

CHAPTER 5 W. G. ALLINSON

High voltage electric power transmission — 199

Introduction	199
Power systems and their development in the USSR, 1917–mid 1950s	199
Basic features of HVAC and HVDC	202
HVAC developments up to 1960—the introduction of 400–500 kV	204
HVAC in the 1960s—the move to 700 kV and above	207
HVDC	215
Recent trends in HVAC	220
Summary and conclusions	222

CHAPTER 6 RONALD AMANN

The chemical industry: its level of modernity and technological sophistication — 227

Some distinctive features of the chemical industry in industrialised societies	227
Summary	239
The diffusion of technology	240
Main phases in the development of chemical technology	240

Phase one: the inorganic chemical industry	240
Phase two: the early organic chemical industry	241
Phase three: the synthetic materials industry	242
'Generations' of chemicals?	243
The pattern of output of the Soviet chemical industry	245
Traditional inorganic chemicals	246
Fertilisers	248
Heavy organic chemicals	249
Synthetic materials	252
Some general perspectives	257
The pattern of Soviet foreign trade in chemicals and chemical plant	258
Trade in chemicals and related products	259
Trade in chemical plant and equipment	262

Technological development — 265

Patents	265
Problems of interpreting patents statistics	265
The Soviet patents system	266
Soviet inventive activity	268
The first commercial production of synthetic materials	271
Plastics and synthetic resins	272
Artificial and synthetic fibres	276

The level of research — 285

Outstanding contributions to chemical science	288
The objectivity of Nobel Prizes as a measure of outstanding contributions to chemistry	289
Outstanding contributions to chemistry in the twentieth century	290
Citation of Soviet chemical research	291

Conclusions — 296

Appendix 6A: Statistical data — 298

Appendix 6B: An analysis of the level of Soviet polymer research by means of citations. (by ARTHUR HOLT) — 320

CHAPTER 7 Z. A. SIEMASZKO

Industrial process control — 328

Introduction — 328

The essentials of control	329
Signal and supply media: pneumatic and electronic	334
Unified signals	334
Intrinsic safety	335
Amplifying elements	336
The mechanism of innovation in process control	337

Analogue control up to 1970 — 338

Early history (until *c.* 1950)	338
The 'Great Debate' of the 1950s	341
The intermediate stage (AUS, EAUS, MZTA)	342
The state system of instruments (GSP)	344
Diversion to pneumatics (USEPPA)	347
Minor systems	349
Analogue equipment in current production: summary	349

Equipment for the 1970s and 1980s (GSP and 'all-regime' RP2) — 352

The new GSP3 system	352
Physical construction of GSP3 (system UTK)	355
The power industry's new system (all-regime RP2)	357
Success indicators of Soviet systems currently under development	359

Computer control — 359

Earlier attempts	359
Process control in ASU (ASUTP)	362

Critical review of the technology — 364

Development-production cycle	364
UK-Germany-USA comparison	366
Soviet lags with respect to the UK	366
Analogue control in current production	366
Analogue control currently undergoing development	370
Computer control	370
Soviet leads with respect to the UK	370
Analogue equipment in current production	370
Analogue equipment being developed at present	371
Computer control	371
Production	371
Summary of lags and leads	371

CHAPTER 8 MARTIN CAVE

Computer technology — 377

Introduction — 377
Problems of comparing technological level — 379
Soviet computer systems since 1968 — 381
Second generation computers — 381
Third generation computers, other than the ES series — 383
The Edinaya Sistema (ES EVM) of third generation computers — 384
Peripherals — 386
Time-sharing — 387
Software — 388
The diffusion of computers — 391
The availability of peripherals — 396
Summary and conclusions — 397

CHAPTER 9 DAVID HOLLOWAY

Military technology — 407

Introduction — 407
Earlier studies of Soviet military technology — 407
Design philosophy — 414
Conclusion — 415
Medium tanks — 416
Soviet tank development — 417
Soviet and Western tanks 1950–75 — 421
Protection — 424
Firepower — 427
Mobility — 435
Design philosophy — 438
Diffusion — 440
Conclusions — 441
Intercontinental ballistic missiles — 446
Soviet rocket research to 1945 — 448
The development of Soviet nuclear weapons — 451
The Soviet missile programme 1945–57 — 455

Soviet and United States ICBM programmes, 1957–75	459
Assessment of the level of technology	468
Yield	468
Accuracy	470
Throwweight	474
Multiple warheads	475
Propellants	477
Penetration aids	478
Missile silos	479
Countersilo lethality	480
Conclusion	483
Final conclusions	488

CHAPTER 10 MILAN KOCOUREK

Rocketry: level of technology in launch vehicles and manned space capsules — 490

Introduction	490
Comparative history of Soviet rocketry	491
Soviet rocketry 1928–45	492
Soviet rocketry after the Second World War	497
Soviet space policy	499
Soviet space technology	499
Launch vehicle building	500
Rocket thrust and design philosophy	500
Development of launch vehicles	501
Exploitation of launch vehicles	507
Spacecraft building	510
Design philosophy	512
Engineering	514
Conclusions	517

CHAPTER 11 M. J. BERRY AND M. R. HILL

Technological level and quality of machine tools and passenger cars — 523

Introduction	523

Western and Soviet assessments of the technological level of Soviet machine tools with special reference to quality 523
Western comments 524
Soviet comments 527
Conclusions 530
The quality and reliability of Soviet general-purpose machine tools 530
The role of Soviet state standards for machine tools 532
The legal status of Soviet state standards 532
Soviet organisation and planning of state standards 532
Soviet state standards for machine tools 533
Dimensional capacity 534
Machine accuracy 535
Accuracy of Soviet machine tools produced before 1970 536
Accuracy of Soviet machine tools produced since 1970 539
Other aspects of machine tool specifications 545
Speed and feed range 545
Noise levels 546
Vibration 546
Control systems 547
Machine reliability and performance in service 547
Passenger cars 550
Product mix since 1970 550
Technical characteristics of Soviet passenger cars 551

Appendix 11A: Assessments of the accuracy of Soviet-made horizontal knee and column milling machines and centre lathes 556
Horizontal knee and column milling machines 556
Centre lathes 557

Appendix 11B: Assessment of the accuracy of the major design elements of the 6M82 milling machine 558
Slideways 558
Spindle assembly 560

Appendix 11C: Soviet-built machine tools in use in Britain 561
    Radial drilling machines 561
    Knee-type milling machines 562
    Surface grinding machines 563

Index of Names 564

Subject Index 566

List of Tables

Table 1.1	Some macro-economic measurements relating to the level and rate of 'technical' progress in the USSR	16
Table 1.2	Areas and modes of assessment	32
Table 2.1	Research development and innovation in numerically controlled machine tools	41
Table 2.2	Plastic materials: lag in introduction in USSR	43
Table 2.3	Chemical fibres: lag in introduction of first commercial production in USSR	43
Table 2.4	Prototypes or first commercial production of computers	44
Table 2.5	First commercial production of electronic controllers	44
Table 2.6	Maximum thrust power of launch vehicles	45
Table 2.7	Some Boretsky indicators of 'technological change' in Soviet industry between 1940 and 1960 compared with the United States	49
Table 2.8	Comparative indicators of Soviet technological change, 1955–73	50
Table 2.9	Rate of growth of industrial production: annual average	53
Table 2.10	The rate of diffusion of new technology in the USSR and other countries	55
Table 2.11	Production of artificial and synthetic fibres in 'comparable periods': USA and USSR	57
Table 2A.1	Statistical Appendix (containing raw data for summary Table 2.8, above)	67

Table 3.1	Output of pig iron and blast furnace ferro-alloys	85
Table 3.2	The structure of iron produced	85
Table 3.3	The average proportion of sinter and pellets in the ore component of the blast furnace burden	87
Table 3.4	Indicators of work of blast furnaces of different volume	87
Table 3.5	Average and maximum capacity of blast furnaces	88
Table 3.6	Average coke consumption per tonne of pig iron produced	90
Table 3.7	Average daily output per cubic metre of furnace volume	91
Table 3.8	Output of crude steel	93
Table 3.9	The structure of steel making processes	94
Table 3.10	The output of steel smelted in open hearth furnaces	94
Table 3.11	Year of introduction of first and second oxygen convertors and their capacity	96
Table 3.12	The diffusion of oxygen steelmaking	98
Table 3.13	The average and maximum capacity of oxygen convertors	99
Table 3.14	The output of electric steel	101
Table 3.15	Year of introduction of first experimental and industrial continuous casting installations and their capacity	102
Table 3.16	The production of steel by continuous casting	103
Table 3.17	The output of alloy steel	105
Table 3.18	Output of finished rolled steel	108
Table 3.19	The structure of production of finished steel	109
Table 3.20	Output of selected end products	110
Table 3.21	Structure of consumption of rolled metal, tubes and general metal products (*metizy*) in the Soviet Union and USA in 1965	111
Table 3.22	Output of steel tubes	113
Table 3.23	Welded tubes as a proportion of total tube output	113
Table 3.24	The technical level of rolled metal production	114
Table 4.1	Total number of machine tools in the Soviet Union and the United States	123

LIST OF TABLES

Table 4.2	Output of machine tools for selected years	124
Table 4.2a	Annual average output	124
Table 4.3	Machine tool stock in engineering and metalworking	125
Table 4.4	Distribution of machine tools in the Soviet economy	126
Table 4.5	Proportion of machine tools in basic and auxiliary production in the Soviet engineering industry	127
Table 4.6	Breakdown of machine tool stock by age	128
Table 4.7	Age pattern of Soviet machine tools in the economy as a whole and in engineering and metalworking	128
Table 4.8	Output by value and unit value for selected years	129
Table 4.9	Comparative output of metalcutting machine tools by value for selected years for main rouble-dollar conversion rates	130
Table 4.10	Comparative data on low value machine tools produced in the Soviet Union and United States	131
Table 4.11	The division between metalcutting and metalforming machines in the machine tool stock of the United States and the Soviet Union	131
Table 4.12	The division of the Soviet machine tool stock under 10 years old between metalforming and metalcutting	132
Table 4.13	Proportion of metalforming and metalcutting machine tools in the machine tool stock of basic production shops of Soviet engineering factories	133
Table 4.14	Breakdown of machine tool stock by type of machine tool	134
Table 4.15	Comparison of breakdown of metalcutting machine tool types under 5 and 10 years old at most recent census	136
Table 4.16	Breakdown of turning group of machine tools for USA and Soviet Union	136
Table 4.17	Breakdown of 'grinding group and all other' machine tools for Soviet Union and United States	137

Table 4.18	Cutting off, tapping and threading machines	138
Table 4.19	Ultrasonic and electrical machinery units	138
Table 4.20	Other metalcutting machine tools	139
Table 4.21	Breakdown of metalforming machines by type: Soviet Union and United States	140
Table 4.21a	Soviet stock of metalforming equipment by type as a proportion of American stock	141
Table 4.22	Summary of main differences in machine tool stock of Soviet Union and United States	142
Table 4.23	Soviet imports and exports of metalcutting machine tools, 1955 to present	147
Table 4.24	Distribution of Soviet trade in machine tools among leading capitalist and socialist countries, 1955 to present (units and value)	148
Table 4.25	Distribution of Soviet trade in machine tools among leading capitalist and socialist countries (%)	149
Table 4.26	Distribution of Soviet trade in machine tools by type of country, 1967 to present	150
Table 4.27	Value of Soviet trade and balance of trade	151
Table 4.28	Unit value of machine tools imported and exported by USSR, 1955 to present	151
Table 4.29	Main types of metalworking machinery imported into USSR from the United Kingdom	154
Table 4.30	Soviet demand for machine tools and planned output for 1965 according to the 1959–65 seven-year plan, and actual output in 1965	155
Table 4.31	Unit value and value per cwt of British metalcutting and metalforming machine tool imports in 1968 (by country)	158
Table 4.32	British imports of metalcutting machine tools from the Soviet Union, 1967–9	159
Table 4.33	Average weight, unit value and value per cwt of types of metalcutting machine tools imported to UK from Soviet Union in 1967–9 compared with average weight, unit value and value per cwt of all machine tools of these types imported to UK in 1968	160
Table 4.34	NC machine tools: stages of the innovation process	165

LIST OF TABLES

Table 4.35	Planned and actual output of metalcutting machine tools with NC	169
Table 4.36	Annual output of NC machine tools	172
Table 4.37	NC machine tools as a proportion of total machine tool output	172
Table 4.38	Types of NC machine tools produced	175
Table 4.39	The stock of NC machine tools	176
Table 4.40	NC machine tools as a proportion of the total machine tool stock	177
Table 4.41	The stock of NC machine tools in the machine tool and tooling industry	178
Table 4.42	The structure of the NC machine tool stock in the machine tool and tooling industry	178
Table 4.43	Type of control systems fitted to NC machine tools	185
Table 4.44	Rate of completion of stages of the innovation process	194
Table 4.45	Leads and lags: the innovation process and major developments	195
Table 5.1	Main events in the development of 700 kV and above	210
Table 5.2	Characteristics of existing 500 kV lines	211
Table 5.3	Lengths of lines at various voltages	212
Table 5.4	Details of first and some later HVDC transmissions brought into operation	217
Table 5.5	Relative costs of fuel transport and HVDC transmission	218
Table 5.6	Relative costs of AC and DC: Urals to West Siberia	218
Table 5A.1	Lengths of HVAC transmission lines, 1959–70	225
Table 5A.2	Total power station capacity	226
Table 5A.3	Ratio of line lengths to installed capacity	226
Table 6.1	Change in the chemical intensity (*khimkoemkost'*) of 'gross social product', 1959–66	228
Table 6.2	Capital intensity and labour productivity in the Soviet chemical industry	231
Table 6.3	Structure of capital investment in the Soviet chemical industry	232
Table 6.4	Relative structure of expenditure on the manufacture of industrial products	233

Table 6.5	Percentage of all employees classified as administrative technical or clerical in British industry in 1971	234
Table 6.6	The ratio of relative growth of the chemical industry (*koeffitsient operezheniya*)	235
Table 6.7	Research intensiveness of British and US industries	237
Table 6.8	Share of various countries in world output of chemicals	246
Table 6.9	Proportion of petrochemicals used in the manufacture of synthetic materials	251
Table 6.10	Structure of the raw material base for organic synthesis in 1970	251
Table 6.11	Output of polycondensation plastics per tonne of polymerised plastics	252
Table 6.12	The proportion of synthetics in total output of man-made fibres	254
Table 6.13	The structure of Soviet manmade fibre production	255
Table 6.14	Weight of synthetic materials as a percentage of weight of inorganic chemicals	259
Table 6.15	British trade in chemicals in 1971	260
Table 6.16	Balance of trade as a proportion of domestic production	262
Table 6.17	Grants of patents in chemistry, 1968–72	269
Table 6.18	Patents granted to Soviet institutes and organisations by the London Patent Office and those granted to some major Western chemical companies, 1971–2	271
Table 6.19	First commercial production of various plastic materials	278
Table 6.20	Soviet purchases of plastics plant and equipment	280
Table 6.21	First commercial production of various chemical fibres	286
Table 6.22	Soviet acquisition of manmade fibre plant from abroad	287
Table 6.23	'League table' of Nobel Prizes in chemistry, 1901–72	291
Table 6.24	The 'effectiveness' of Soviet scientific contributions in chemistry	295

LIST OF TABLES

Table 6A.1	Comparative growth rates of key industrial products in selected countries, 1960–70	298
Table 6A.2	Production of soda ash in selected countries	300
Table 6A.3	Production of sulphuric acid in selected countries	301
Table 6A.4	Output of superphosphates	302
Table 6A.5	Output of nitrogenous fertilisers	303
Table 6A.6	Output of ethyl alcohol in selected countries	304
Table 6A.7	Comparative aggregate growth rates of organic chemicals	305
Table 6A.8	Output of key organic chemicals	306
Table 6A.9	Output of plastics and synthetic resins	307
Table 6A.10	Output of key plastic materials	308
Table 6A.11	Output of chemical fibres	309
Table 6A.12	Output of synthetic rubber	310
Table 6A.13	Soviet trade with the world in chemicals and related products	311
Table 6A.14	Soviet trade with the world in 'chemical products'	312
Table 6A.15	Soviet trade in 'plastics and materials for production of plastics' with individual countries	313
Table 6A.16	Soviet trade in manmade fibres (artificial and synthetic) with selected countries	315
Table 6A.17	Soviet trade in 'miscellaneous chemicals' with individual countries	316
Table 6A.18	Soviet imports of chemical equipment	318
Table 6A.19	Grants of patents to nationals and foreigners, 1968–72	319
Table 6B.1	Bibliographic citations according to origin of publication cited	326
Table 6B.2	Results weighted for population of Soviet Union vs West	327
Table 7.1	Amplifying elements for control systems	337
Table 7.2	Characteristics of electronic proportional-plus-integral controllers in 1960s	344
Table 7.3	Soviet aggregated control complexes	354
Table 7.4	Technological characteristics of GSP3 and all regime RP2	358

Table 7.5	The rate of development of selected Soviet and British control systems	365
Table 7.6	Technological level of the British control and instrumentation industry compared with those of FRG and USA	367
Table 8.1	Soviet production of computational equipment	392
Table 8.2	Stock of computers held in various countries	393
Table 8.3	Soviet lag in computer stock	394
Table 8.4	Date of first production of comparable Soviet and American computers	401
Table 8.5	Soviet lag in entering successive generations of computers	401
Table 8.6	Characteristics of disc units	401
Table 8.7	Characteristics of line printers	401
Table 8A.1	Characteristics of Soviet computers, 1953–73	404
Table 8A.2	Some advanced American computers, 1946–73	406
Table 8A.3	Some British computers, 1960–73	406
Table 9.1	Results of US Department of Defense comparison of Soviet and American military technologies in 1972	411
Table 9.2	Production of tanks during the Second World War	422
Table 9.3	Soviet tanks of the 1930s	423
Table 9.4	Basic combat features of Soviet and German tanks, 1943–5	424
Table 9.4a	T-50, T-60 and T-72 tanks	424
Table 9.5	Combat features of battle tanks, c. 1950	425
Table 9.6	Combat features of battle tanks, c. 1960	426
Table 9.7	Combat features of battle tanks, c. 1970	428
Table 9.8	Armour thickness on selected MBTs	429
Table 9.9	Chronology of Soviet medium tank development since 1945	430
Table 9.10	Firepower and ammunition characteristics of Soviet and Western tanks	432
Table 9.11	Tank engines	444
Table 9.12	Tanks, 1950–70	445
Table 9.13	Summary of results in Table 9.12	446
Table 9.14	Structure of strategic forces, 1965–75	448
Table 9.15	Classification of military ballistic missiles by range	449

Table 9.16	Steps in the development of nuclear weapons	454
Table 9.17	Nuclear explosions, 1945–74 (announced and presumed)	456
Table 9.18	US land-based ballistic missiles	460
Table 9.19	Soviet land-based ballistic missiles	462
Table 9.20	Deployment of US ICBMs	465
Table 9.21	Development of Soviet ICBMs	466
Table 9.22	US strategic missile engines	467
Table 9.23	Soviet strategic missile engines, 1957–62	467
Table 9.24	Megatonnage deliverable by ICBM forces	470
Table 9.25	Accuracy of Soviet and US ICBMs	473
Table 9.26	Missile payload options	474
Table 9.27	Throwweights of selected Soviet and United States ICBMs	476
Table 9.28	Total throwweight of Soviet and US ICBM forces	477
Table 9.29	Number of independently targetable warheads on ICBMs	478
Table 9.30	Total K required to destroy each Soviet silo with probabilities of 97 and 90 per cent	481
Table 9.31	K values of Soviet and US ICBMs	482
Table 9.32	K value of US ICBM force	483
Table 9.33	K value of the Soviet ICBM force	484
Table 9.34	Landmarks in the development of ICBMs	487
Table 10.1	Pioneers in the theory of reactive motion and first launchings of their LPRs	491
Table 10.2	Development of major Soviet liquid propelled rocket engines, 1929–45	494
Table 10.3	Dvelopment of thrust and payload capability of Soviet and US launch vehicles	502
Table 10.4	Some characteristics of major Soviet launch vehicles	504
Table 10.5	Some characteristics of major US launch vehicles	508
Table 10.6	Soviet and US launches by types of launch vehicles, 1957–73	511
Table 10.7	Soviet and US manned spaceflight data, 1961–73	518
Table 10.8	Major space 'firsts', 1957–73	521

Table 11.1	Some foreign comments on Soviet machine tools from recent issues of *Stankoimport Review*	524
Table 11.2	Some Western views on Soviet machine tools	525
Table 11.3	Percentage of output on level of best foreign machine tools	528
Table 11.4	Soviet views of technological level of Soviet machine tools	531
Table 11.5	Condensed specifications for selected Soviet and British machines	537
Table 11.6	Geometric alignment accuracy tests and tolerances for knee and column milling machines having overall dimensional specifications listed in Table 11.5 above	540
Table 11.7	Geometric accuracy alignment tests and tolerances for centre lathes	542
Table 11.8	Accuracy requirements of samples of components processed on centre lathes by the Soviet and British engineering industries	544
Table 11.9	Tolerance requirements for Soviet turned shafts	544
Table 11.10	Estimated comparative accuracy requirements of components processed on 400 mm swing centre lathes in the Soviet and British engineering industries	544
Table 11.11	Comparative age structure of Soviet machine tool state standards	545
Table 11.12	Technical characteristics of Soviet passenger car product range (1971)	552
Table 11.13	Comparative technical data relating to the Moskvich 412 and passenger cars produced by Western European firms	553
Table 11.14	Frequency of detected faults on nine models of Moskvich 412	555
Table 11B.1	Comparison of accuracy of slideways	559
Table 11B.2	Comparison of bearing spindle and housing tolerances	560

List of Text Figures

Fig 3.1	Oxygen steel as a proportion of total steel output	97
Fig 4.1	NC machine tools: stages of the innovation process	166
Fig 4.2	Annual output of NC machine tools	173
Fig 5.1	Graph of maximum transmission voltages (HVAC)	208
Fig 5.2	System linking lines for European USSR	213
Fig 5.3	Relative costs of 765 and 345 kV	214
Fig 6.1	Comparative growth rates of key industrial products, 1960–70	247
Fig 6.2	Comparative growth rates of selected sectors of the chemical industry, 1960–70	258
Fig 6.3	Location of known outstanding scientific talent in chemistry, 1901–72	292
Fig 7.1	Fundamentals of process control	330
Fig 7.2	Front of a Soviet electronic proportional-plus-integral controller, type VTI, launched in 1951	340
Fig 7.3	Front of a British electronic proportional-integral-derivative controller, with control station, launched by Evershed & Vignoles in 1949	340
Fig 7.4	Electronic proportional-plus-integral controllers of 1960s	346
Fig 7.5	Miniature pneumatic controllers	350
Fig 7.6	Soviet control systems for the second half of the 1970s	360

Fig 7.7	A British general-purpose control system for the second half of the 1970s	361
Fig 7.8	A very old circular-chart recorder, originating from 1930s, still in full production in 1968	372
Fig 7.9	A modern Soviet circular-chart recorder	373
Fig 7.10	Antiquated heavy electrical Soviet control panel instruments	374
Fig 7.11	Instrument panel of pneumatic USEPPA-START system	374
Fig 7.12	Stepping motors	375
Fig 7.13	An extreme example of delayed innovation	376
Fig 7.14	The outer case of an American recorder of the 1930s used over 30 years later	376
Fig 8.1	Stock of computers per million inhabitants	395
Fig 8.2	Comparison of maximum operating speeds of British, American and Soviet computers, 1949–73	398
Fig 8.3	Comparative performance of best American and best Soviet computers	399
Fig 8.4	Comparative performance of best British and best Soviet computers	400
Fig 9.1	Development of power-to-weight ratios	436
Fig 9.2	Tank range, 1945–75	437
Fig 9.3	Tank speeds (max), 1945–75	437
Fig 9.4	Nominal ground pressure, 1945–75	438
Fig 9.5	Development of gun calibre, 1945–75	443
Fig 9.6	Rounds of ammunition carried, 1945–75	445
Fig 9.7	Classification of military rockets	450
Fig 9.8	Development of Soviet and US ICBMs	464
Fig 9.9	Megatonnage deliverable by ICBM forces	471
Fig 9.10	Number of independently targetable warheads on US and Soviet ICBMs	479
Fig 9.11	K values of Soviet and US ICBM forces	485
Fig 10.1A	Number of Soviet and US launches compared	512
Fig 10.1B	Weight launched into space by Soviet Union and United States, 1957–73	512

Glossary

ASU (*Avtomatizirovannaya Sistema Upravleniya*): Automated System of Management and Control.

ASUTP (*Avtomatizirovannaya Sistema Upravleniya Teknologicheskimi Protsessami*): Automated Process Control System; according to official definition, a system based on control computers.

ENIMS (*Eksperimental'nyi Nauchnyi Institut Metallotrezhushchikh Stankov*): Experimental Scientific Institute for Metalcutting Machine Tools.

Fondoemkost': capital intensity; unit value of fixed capital per unit value of output.

Fondootdacha: rate of return on capital.

Giproiv (*Gosudarstvennyi institut po proektirovaniyu predpriyatii iskusstvennogo volokna*): State Project Institute for Artificial Fibre Enterprises.

Gipromez (*Gosudarstvennyi soyuznyi institut po proektirovaniyu metallurgicheskikh zavodov*): State Union Project Institute for Metallurgical Factories.

Glavtyazhstankoprom (*Glavnoe Upravlenie Tyazheloi Stankostroitel'noi Promyshlennosti*): Chief Administration for the Heavy Machine Tool Industry.

GOELRO (*Gosudarstvennaya Kommissiya po Elektrifikatsii Rossii*): State Commission for the Electrification of Russia.

Gosplan (*Gosudarstvennaya Planovaya Kommissiya*) (*SSSR*): (USSR) State Planning Commission; chief planning body of the Soviet Union. There also exists a State Planning Commission for each Union Republic.

Gosstandart (*Gosudarstvennyi Komitet Standartov*): State Committee for Standards.

GOST (*Gosudarstvennyi Standart*): State Standard.

IAT (*Institut Avtomatiki i Telemekhaniki*): Institute of Automation and Remote Control.

ITR (*Inzhenerno-Tekhnicheskie Rabotniki*): Engineering and technical personnel, with higher or secondary technical education engaged in technical management of industry or transport.

Khimkoemkost': Chemical intensity; expenditure on chemical products per unit value of output in industry, construction and agriculture.

Koeffitsient operezheniya: Rate of relative growth. Measure of the extent to which the growth rate of a sector of the economy exceeds that of others.

Kompleksnaya avtomatizatsiya: Complex or comprehensive automation; a term denoting the automation of an entire production process, as opposed to individual elements thereof.

Mosstankin: Moscow Machine Tool and Tooling Institute.

Nauchnyi rabotnik: Scientist; in Soviet usage this term includes, in additon to the personnel of research establishments, the teaching staff of higher education establishments.

Nauka: Science, including social sciences and humanities; also section in state budget including most general expenditure on R and D, excluding that within the enterprise, and capital investment on new research establishments and extensions to existing ones.

NIITEKHIM (*Nauchno-issledovatel'skii institut tekhniko-ekonomicheskikh issledovanii po khimii*): Research Institute for Technical and Economic Studies in Chemistry.

NIITeplopribor (*Nauchno-issledovatel'skii institut teploenergeticheskogo priborostroeniya*): Research Institute for Thermal Power Instrument Building.

Ob"edinenie: Association; an economic unit consisting of a number of enterprises, sometimes incorporating research institutes.

Oblast': Region; administrative sub-division of a Union Republic.

Osnovnye proizvodstvennye fondy: Fixed capital.

OST (*Otraslevoi Standart*): Branch Standard.

Oveshchestvlennyi trud: Embodied labour; past labour embodied in machinery, fuel, raw materials, etc, used in production.

Promyshlennyi masshtab: Full-scale industrial production, as opposed to experimental production, production of prototypes and trial batches.

Rabochii: Manual worker, as distinct from *rabotnik*, who may be engaged in either mental or manual labour.

Sharaga: an R and D project staffed by imprisoned scientists, as depicted in A. Solzhenitsyn's *The First Circle*.

SKB (*Spetsial'noe Konstructorskoe Byuro*): Special Design Bureau. A common term for important design bureaux, for example in the machine tool industry. Also sometimes used for design bureaux at major enterprises.

Sovnarkhoz (*Sovet narodnogo khozyaistva*): Economic Council. These organisations were re-established in 1957 after having existed for some time after the Revolution, thus creating a regional, rather than branch, system of economic administration. Each council was responsible for a region, or number of regions. They were abolished in 1965 when the previous Ministerial system was restored.

Tipazh: Range of product types and sizes.

Tipovoi zavod: Factory constructed according to a standard design.

Trudoemkost': Labour intensiveness.

TsNIICherMet (*Tsentral'nyi nauchno-issledovatel'skii institut chernoi metallurgii imeni I. P. Bardina*): I. P. Bardin Central Research Institute for Ferrous Metallurgy.

TsNIIKA (*Gosudarstvennyi vsesoyuznyi Tsentral'nyi Nauchno-Issledovatel'skii Institut Kompleksnoi Avtomatizatsii*): State All-Union Central Research Institute for Complex Automation.

TsSU (*Tsentral'noe Statisticheskoe Upravlenie*): Central Statistical Administration.

VINITI (*Vsesoyuznyi Institut Nauchnoi i Tekhnicheskoi Informatsii*): All-Union Institute for Scientific and Technical Information.

VNIIV (*Vsesoyuznyi Nauchno-Issledovatel'skii Institut Iskusstvennogo Volokna*): All-Union Research Institute for Artificial Fibre.

Zhivoi trud: Live labour; labour actually involved in current production.

1. Some approaches to the comparative assessment of Soviet technology: its level and rate of development

Each of the contributions in this volume attempts to evaluate the comparative level of Soviet technology in a key sector of the economy. It was felt that a substantial number of sectors should be examined in order to obtain a balanced picture. For reasons discussed below, the conceptual and practical difficulties involved in measuring technological 'lead' and 'lag' are considerable and the results of this exercise should be treated with caution. But such an assessment needs to be made in order to provide a context within which problems of innovation and research policy can be discussed meaningfully. Many of the basic studies of Soviet science and technology which have appeared in the West during the past decade have analysed the special problems that stem from the operation of an 'administered economy' and have attempted to measure the research and development effort in the USSR.[1] Unfortunately, it is virtually impossible to decide whether the 'capabilities' and 'problems' identified by these studies are significant ones without recourse to precise indices of the comparative *performance* of the nation or its constituent branches of industry. This is a perennial problem for the researcher as he ploughs his way through the Soviet press and economic journals: how far are the critical comments of officials and academics that he reads objective and how far are they coloured by insufficient knowledge of Western practice, by the desire to exhort workpeople to greater efforts, and by an environment which strongly encourages specific criticisms of a practical kind? It was largely in response to this problem that the present study was conceived; namely, to supplement the existing body of work at a crucial point by the provision of hard evidence.

[1] See, for example, N. De Witt, *Education and Professional Employment in the USSR*, Washington DC (1961); C. Freeman and A. Young, *The Research and Development Effort in Western Europe, North America and the Soviet Union*, OECD, Paris (1965) (the Soviet figures in this comparison were calculated by G. R. Barker, R. W. Davies and R. Fakiolas); A. Korol, *Soviet Research and Development; its Organisation, Personnel and Funds*, Cambridge, Mass. (1965); R. Amann, M. J. Berry and R. W. Davies, Part V, *Science Policy in the USSR*, OECD, Paris (1969).

What do we understand by the term 'technology'? First, it refers to the ranges of equipment, mechanisms and processes which transform raw materials into products or services that may or may not have existed before. The extent to which technology is effective is determined by the quality, output, novelty and profitability of the final product. In a broader sense, 'technology' also includes the skills of the workforce: first, the skill to operate the equipment and bring it to peak performance; second, a practical familiarity and theoretical understanding of the field ('know-how') which enables the nation or firm to adapt, improve upon and extend the existing range of technological hardware. 'High (or advanced) technology' and 'technological level' are both inherently comparative concepts. The former refers to technical and scientific *sophistication* and is often used when comparing *different* technologies. The concept of 'technological level', on the other hand, is usually used when comparing the *achievement* of a nation or firm within the *same* broad area of technology, which may not necessarily be a sophisticated one. In practice, of course, the two concepts are closely related because an important aspect of the 'technological level' of an industry is the relative proportion of 'high' and 'low' technology within it. The 'rate of technological development' refers to the rapidity with which the technology of a firm or nation is raised from a given level to a higher and more effective level. As we shall see below, this general definition of technology, adopted for the present study, differs in most cases from the broader concept of 'technical progress', which is used widely by economists. Our definition has no greater merit. It is simply more consistent with our particular interests and intentions.

For the purposes of empirical enquiry, the meaning of 'technology' is much more elusive than would be suggested by the general definition outlined above. In some ways the concept of 'technology' resembles the concepts of 'heat', 'light', 'weight' or 'speed'; the phenomena can be described in a general way but they acquire precise meaning only through measurement. Unfortunately, although there is a considerable body of literature on the measurement of the level of technology and its rate of development, no general consensus of opinion has formed as to how these measurements should be carried out. In the remainder of this chapter, the strengths and weaknesses of some alternative approaches are discussed from the point of view of their methodology and, where appropriate, their application to the USSR. Finally, the approaches used in the various parts of this volume are elaborated and their limitations are considered.

Aggregated economic assessments of technology: general methodological problems

Direct assessments of technological level can be obtained by comparing the operating characteristics of particular processes or the quality of their end-products. In the case of machinery, for example, this might involve the comparison of such indices as reliability, power/weight ratio or operating speed. But as Komzin has pointed out,[2] these indices are numerous and tend to be specific to particular items of equipment. In cases where they are a common feature, they differ in importance depending upon the exact purpose of the equipment; moreover, the task is made even more difficult by the proliferation of technical standards and tolerances, which are often not strictly comparable. Thus, in the absence of general engineering criteria, direct assessments of technological level are necessarily confined to a very limited range of equivalent machines. It would be dangerous to base any general conclusions on individual studies of this kind unless the technology involved could be shown to be a 'key' one, which decisively influenced the overall level of a branch of industry or the economy as a whole.

As an alternative to this confusing mass of detailed engineering data, the evaluation of which demands highly specialised knowledge, a neater approach is to attempt to isolate the level of technology and its rate of change by means of macro-economic methods. Probably the most simple criterion that might be used for this purpose is labour productivity, on the grounds that the level of labour productivity at any given point in time or its growth over a given period of time reflects the accumulated level of technology or its rate of development, respectively. Unfortunately, this is a very partial and incomplete indicator because it fails to differentiate the strictly technological factors from other important influences on productivity, such as the amount of capital per worker, economies of scale, and the degree of utilisation of productive capacity.[3] A more satisfactory approach is to apply some measure of total factor productivity. Assuming that output is a function of the main productive factors, capital and labour, increases in output that cannot be accounted for by increases in factor inputs are attributed to a residual combination of influences often equated with 'technical progress'. These relationships are commonly expressed in the form of a production function. For the most part, economists have been interested in using total factor productivity as a measure of technical *change* (or 'dynamic efficiency') over time, though in some cases, as we shall see below, it has been used as a non-dynamic measure of the *level* of technology (or 'static efficiency') in different economies. From a general methodological point of view the use of production functions represents a distinct improvement upon a simple measure of labour productivity, but a number of critical problems remain, notably in the calculation of the input variables and in deciding what the measure of 'technical progress' that is produced really means; Professor R. Nelson has referred to this as 'a measure of our ignorance'![4]

The American economist R. Solow[5] was one of the first to use this approach. For his main variables Solow took time-series data on GNP at constant prices, labour man-hours (unadjusted for quality and composition) and aggregate

---

[2] B. I. Komzin in *Sovremennaya nauchno-tekhnicheskaya revolyutsiya v razvitykh kapitalisticheskikh stranakh* (1971), p. 215.
[3] E. Mansfield, *Industrial Research and Technological Innovation*, London (1968), p. 3.
[4] R. Nelson, *American Economic Review*, September 1964.
[5] R. Solow, *Review of Economics and Statistics*, August 1957, pp. 312–20.

capital stock. He assumed that returns to scale were constant (i.e., that if inputs were increased by a given percentage, output would increase in the same proportion) and made two further assumptions about the nature of technical progress that were to be seriously challenged in later writings: first, that technical progress inhered in the knowledge and skills of the workforce and affected the productivity of old and new machinery alike (i.e., it was 'disembodied');[6] second, that technical progress was exogenously determined and, hence, not influenced by any of the input variables. Solow concluded that from 1909–49 87·5 per cent of the increase of per capita output in the United States was due to 'technical progress'.[7]

Common sense might suggest this was rather a substantial contribution for 'technical progress' to make, especially in view of its disembodied nature as defined by Solow. Economists subsequently argued that the residual was a 'catch-all' category, which included many non-technological factors. Mansfield,[8] for example, points out that the residual can include the effects of economies of scale, improved allocation of resources, changes in product mix, increases in the level of education and improved health and nutrition; these are lumped together with the effects of R and D and innovation. If we take the USSR as a particular example, a Solow-type residual would fail to discriminate between the effects on productivity of innovation and the diffusion of new technology, on the one hand, and such factors as: increased effort and motivation due to party campaigns, greater specialisation and amalgamation of industrial enterprises, and the impact of economic reforms (for example, the use of the profitability indicator, the imposition of a capital charge and the extension of the Shchekino experiment, which were introduced in order to improve factor utilisation). In some ways, the Soviet concept of *intensive development*, which encompasses such influences as improved factor allocation and utilisation and better education as well as the accelerated introduction into the economy of the latest developments of science and technology, is a less ambiguous term for the residual than 'technical progress'. But it is precisely the composition as well as the size of the residual that one wants to know. It seems probable that the relative importance of its component parts will change during the course of industrial development even though the overall impact of the residual on productivity could remain much the same throughout. For example, at a relatively early stage of industrial development the growth of managerial efficiency and the absorption of foreign technology will tend to be at least as important as indigenous innovation. The need for innovation in order to maintain the rate of economic growth will only become critical when all other avenues have been explored. Thus, to compare the residuals of two

---

[6] This understanding of 'technology' corresponds, for example, with that of C. Kennedy and A. P. Thirlwall (*Economic Journal*, March 1972, p. 12), who define it as 'useful knowledge pertaining to the art of production'. We, on the other hand, would wish to consider the actual hardware that results from and in a sense 'embodies' this knowledge as well as the knowledge itself.

[7] The residual in this case was the difference between the rate of growth of per capita output and rate of growth of capital stock per man.

[8] E. Mansfield, *Science*, Vol. 175, 4 February 1972, p. 477.

different countries at varying phases of economic development is insufficiently informative so far as the nature of technical development is concerned.

Later writers on this theme tried to refine Solow's original approach by adjusting the input variables in order to pull irrelevant factors out of the residual. Denison,[9] for example, adjusts the labour input to take account of the level of education and the age and sex structure of the working population. He concludes that from 1950–62, 'advances in knowledge' (his name for the residual) accounted for approximately one third of output growth in the United States.[10] An even more dramatic result was obtained by Jorgenson and Griliches,[11] who made an unprecedented number of quality adjustments to the capital and labour inputs on the grounds that real product is a function of real factor inputs (i.e., scarce resources that have alternative uses) and that costless sources of productivity play a relatively minor role. Thus, according to this approach, residuals are largely phoney and arise from the failure to price inputs properly. Jorgenson and Griliches concluded that 96·7 per cent of the growth of output in the United States from 1945–65 could be accounted for by the rate of growth of inputs. In order to achieve this result the authors were obliged to make a number of tenuous quality adjustments, but the ghost of 'disembodied technology' was well and truly laid. Unfortunately, this approach does not lead to any clear conclusion about the contribution of 'technical progress' to productivity, however it may be defined. Though this is partly a matter of classification,[12] the work of Jorgenson and Griliches does raise some very real problems, which tend to undermine the use of production functions for the purposes described above.

The crux of the problem is that the nature of inputs and the relationships between them are extremely complex. A particular difficulty arises in the measurement of capital stock. This is achieved by aggregating past investments with an allowance for depreciation (usually only a very rough valuation of existing stock relative to the cost of replacement). These estimates do not take into account technological improvements which are 'embodied' in new plant and in a sense represent 'more capital'. To make an aggregate quality adjustment for this, one would have to know something about the rate of technical change but since it is the whole object of the exercise to find this out, to impute the approximate rate of technical change at this stage would be to predetermine the final result in an important way. Moreover, according to Norris and Vaizey, estimates of capital stock in Britain have a reliability of plus or minus 10–20 per cent or worse.[13] The lack of accurate estimates is particularly

---

[9] E. Denison, *Journal of Business*, April 1962, pp. 109–21; *Why Growth Rates Differ*, Washington DC (1967).

[10] E. Denison, *op. cit.* (1967), p. 314. Denison's figures for capital inputs do not take technical change into account though he acknowledges that this is an important consideration that would have influenced his final result. His costing of education is also criticised, especially on the grounds that education imposes the substantial opportunity cost of non-productive labour for several years.

[11] D. Jorgenson and Z. Griliches, *Review of Economic Studies*, July 1967, pp. 249–83.

[12] Since output, in an accounting sense, must equal income, the authors could almost be described as eliminating residuals by definition.

[13] K. Norris and J. Vaizey, *The Economics of Research and Technology*, London (1973), p. 151.

regrettable in view of the arguments of some economists that technical change is not exogenous but is to a large extent a function of capital accumulation. Kaldor,[14] for example, has posited a two-way relationship between investment and technical change; increases in capital per man depend upon technical change (e.g., new techniques in automation) while, conversely, most new techniques require more capital per man. Hence, it is impossible to separate out increases in output due to technical change and those due to increases in capital per man.[15] Another important phenomenon, which emphasises the dependence of technical change on capital formation, has been observed by Arrow.[16] Arrow argues that technical change is part of a learning process which occurs in the course of mastering new plant and equipment.[17] This process of 'learning by doing' is subject to diminishing returns but is stimulated by successive injections of new capacity. But how can the impact of this important factor be handled in aggregated calculations? Statistics on the rate of capital replacement would certainly give some indication but other variables such as *novelty of design* and *structure of renewal* are likely to have an important effect and these are less easy to measure.[18]

Another group of criticisms can be directed at the measurement of the output variable. Leaving aside for the moment the problem of devising appropriate exchange rates for measuring the GNPs of different countries on a comparative basis, it is questionable whether GNP itself adequately captures the full effect of technical change. The products of the research-intensive space and defence sectors, for example, are usually valued at cost, while the improvements in the quality of products stemming from 'spin-off' also tend to be under-priced relative to the substantial burden of R and D expenditure, which has been largely carried out beyond the confines of the individual civilian firm. In general, measurements of output do not take into account the effect on consumer welfare of increased choice arising from the introduction of new products; thus, in the case of the USSR, it might be argued that the tendency to maximise the output of a relatively narrow range of simple products militates against choice and quality and consequently exaggerates

[14] N. Kaldor, *Economic Journal*, December 1957, pp. 591–624; N. Kaldor and J. Mirlees, *Review of Economic Studies*, June 1962, pp. 174–92.

[15] A similar two-way relationship can be observed in relation to labour inputs. Denison maintained that technical progress (or 'advances in knowledge') was influenced by the level of education. But it is also the case that due to cross-fertilisation of knowledge, the level of technical progress also influences the level of education. See E. Mansfield, *op. cit.* (1972), p. 478, and R. Nelson, *American Economic Review*, September 1964, p. 591.

[16] K. Arrow, *Review of Economic Studies,* June 1962, pp. 155–73.

[17] The standard examples used for demonstrating this phenomenon are (a) the observation that the number of man-hours needed to produce an airframe is a decreasing function of the number of airframes of the same type previously produced (see C. E. Ferguson, *The Neoclassical Theory of Production and Distribution*, Cambridge (1969), p. 293, quoted by K. Norris and J. Vaizey, *op. cit.*) and (b) that in the Herndal iron works in Sweden no new investment occurred for 15 years, yet output per man increased by two per cent per year (quoted by S. Gomulka, *Inventive Activity and the Stages of Economic Growth*, Aarhus (1971), pp. 36–7).

[18] For example, from the point of view of the economy as a whole, capital accumulation embodying bold new designs, which is channelled into new branches of industry such as plastics, is likely to have a relatively high impact on technical change; due to the fact that the opportunities for adaptation and change are great, it is possible that the learning process will escalate more quickly than in more traditional sectors.

performance. For these and other reasons Mansfield takes the view that, '... the current state of the art in the area [i.e., production functions] is not strong enough to permit very accurate estimates of the contribution of R and D to economic growth [in the United States].'[19]

If the effect of research and development is what one is primarily interested in, a different approach from the one discussed above would be to dispense altogether with the calculation of residuals and to introduce the R and D effort as an explicit input. This view has been advanced by Mansfield. He is mainly interested in discovering the contribution of R and D to economic growth among firms in a particular country; however, if a close relationship between R and D and productivity could be firmly established as a general phenomenon, it might be possible to draw conclusions about technological level and rate of technical change in different countries by comparing their R and D efforts. Most of the work in this area has been confined to industrial firms and the evidence is conflicting. In a study of the performance of selected chemical and pharmaceutical firms in the United States from 1947–57, J. Minasian found that there was a high correlation between R and D expenditure and total factor productivity;[20] similar results have been obtained by Mansfield.[21] In the case of British industry, on the other hand, J. Sargent found no significant correlation between qualified scientific manpower (QSE) and total factor productivity between 1955–64.[22] Some of these conflicting results may, perhaps, be explained by the problems involved in correlating R and D effort and productivity. First, there is the possibility of spurious correlations due to the fact that progressive managements tend to devote more resources to R and D, but the high productivity of their firms could be due to their managerial skills rather than to R and D activity.[23] Second, apart from the general difficulties of measuring output referred to above, micro-studies are beset with special difficulties in this respect. The effects of R and D carry beyond the firm, industry or nation and are only partially reflected in productivity at a given level; thus the impact of R and D is invariably underestimated and is only marginally compensated by revenue obtained from the sale of licenses.

Some of the problems referred to above are less crucial (but still present) in correlations of R and D expenditure and productivity for the economy as a whole, which are more relevant to the purpose of this volume. Unfortunately, at this level another series of even more difficult problems comes into play. On the input side, it is a very complex task to obtain internationally comparable estimates of national spending on research and development because one has to separate this element out from routine activity in each set of national statistics; moreover, in the case of countries like the USSR one cannot obtain

---

[19] E. Mansfield, *op. cit.* (1972), p. 478.

[20] J. Minasian, in (Ed.) R. Nelson, *The Rate and Direction of Inventive Activity*, Princeton, New Jersey (1962).

[21] E. Mansfield, *Industrial Research and Technological Innovation*, New York (1968), pp. 74–80.

[22] J. Sargent, *The Distribution of Scientific Manpower*, Paper read to the International Economics Association Conference, 1971, cited by K. Norris and J. Vaizey, *op. cit.* (1973), p. 132.

[23] E. Mansfield, *op. cit.* (1972), p. 479.

any reliable figures for the substantial R and D spending in the aerospace and military sectors; nor is it possible to price R and D in terms of the official rouble exchange rate, which is notoriously misleading. But even when reasonably accurate estimates of national R and D efforts have been available (as they were for 1964, for example)[24] there appears to be a negative correlation between the proportion of GNP devoted to R and D in a given year and the rate of economic growth during the subsequent period. Admittedly, if the military and aerospace sectors are discounted and, more specifically, if one focuses on R and D resources channelled into research intensive industries[25] a more positive picture emerges, but some ambiguities still remain. It is clear that the *efficiency* of the R and D process varies between countries to a considerable extent. This in turn is a function of how the specific organisational and economic mechanisms in each country operate.[26] In view of these limitations, estimates of the R and D effort cannot be regarded as satisfactory indicators of the relative levels of technology in different countries.

Most of the general problems referred to above apply in one way or another to the various attempts by economists to measure the rate of 'technical progress' in the USSR. These limitations are inherent in the methodology of aggregate assessments and will be assumed in subsequent discussion. However, in relation to the Soviet Union there are special problems involved in the measurement and interpretation of data. It is to these questions that we now turn.

Some aggregate economic estimates of Soviet technology and its rate of development: results and limitations

In recent years, several Soviet economists have become interested in estimates of technological level in order to determine the 'progressiveness' of particular industrial processes and branches of industry. By these means it is hoped to achieve an optimal pattern of resource allocation for capital investment. Like their Western counterparts, these Soviet economists consider that such estimates require a relatively sophisticated approach. Komzin, for example, criticises the convention of using 'labour productivity' and 'recoupment' (*fondootdacha*) as criteria for estimating 'technical progress' in the USSR. Taken on its own, 'labour productivity' is obviously misleading because it conceals capital costs. Under the conditions of the 'scientific and technical revolution' the role of physical labour is declining relative to capital investment and it is fundamentally incorrect to regard basic production funds as 'gifts of nature'. The 'recoupment' indicator is also inadequate for assessing technical

---

[24] As one of its contributions to the International Statistical Year the OECD calculated comparable estimates of the R and D effort for several of its member countries. See *The OECD Observer*, October 1967, No. 30, pp. 32–7.

[25] For these calculations see *The OECD Observer*, April 1968, No. 33, pp. 18–28.

[26] In order to illustrate this point, it is interesting to compare the results of the following studies, which reveal substantial differences between how the R and D process is handled in different countries: *Reviews of National Science Policy: United States*, OECD, Paris (1968) and *Science Policy in the USSR*, OECD, Paris (1969).

level because it is influenced by increases in the intensiveness of utilisation; moreover, the application of 'recoupment' to international comparisons tends to make relatively primitive economies with small amounts of capital per worker appear 'progressive', and if it was used as a criterion for investment decisions it would favour alternatives with a relatively high component of physical labour. According to Komzin, a measure of total factor productivity is needed, which in Soviet terminology includes the effects of physical labour and intellectual labour (*zhivoi trud*), embodied in new plant and machinery (*oveshchestvlennyi trud*).[27] Komzin is anxious to establish that the 'formal coincidence' between his approach and that of some Western economists does not signify a common purpose. Whereas, in his view, Western aggregate measurements (especially the Cobb-Douglas production function) were created as instruments to justify the allocation of national income between labour and capital in the capitalist system, in the USSR these measurements are used to compare the effectiveness of different investment projects and branches of industry.[28]

Soviet interest in these questions has undoubtedly been prompted by a downward trend in the rate of economic growth in the USSR during the 1960s (with some fluctuations). Whereas Soviet GNP grew by 15 per cent in 1950–1, it grew by only 7 per cent in 1968–9.[29] This decline has been to some extent precipitated by the drying-up of 'extensive' factors of production. The labour pool can no longer be supplemented by the mass migration of population from the countryside into the towns on a scale comparable with that of the 1930s, although considerable internal migration does still take place. The increased expectations of the population since the death of Stalin, combined with more accurate knowledge of living standards in Western countries, impose constraints on capital accumulation relative to consumption. Thus, it is recognised that a major role in future Soviet economic development needs to be played by 'intensive factors' (a concept which includes both the development of new technology and other sources of increased efficiency). In the 1960s, according to V. A. Trapeznikov, 'intensive factors' were responsible for one third of total economic growth in the Soviet Union. Ya. Kotkovskii has argued that if all known technology was applied and diffused in the USSR, national income could be increased by 50 per cent without having to employ more capital or labour.[30] The works of some East European scholars, though their definitions of 'intensive factors' vary considerably, reveal that although the intensiveness of Soviet economic growth has been increasing during the 1960s

---

[27] B. I. Komzin, *op. cit.* (1971), p. 217. Komzin's approach is based on that expounded by V. A. Trapeznikov, *Avtomatika i Telemekhanika*, 1969, No. 1, pp. 5–24. Similar work has been done by I. G. Kurakov, *Effektivnost' nauchno-tekhnicheskogo progressa* (1969), pp. 85–111, and *Voprosy Filosofii*, 1966, No. 5, pp. 3–13; 1966, No. 10, pp. 3–14 (cited by Komzin).

[28] *Ibid.*, p. 233.

[29] John Whalley (London School of Economics), *Some Recent Growth Accounting Exercises for the USSR and Eastern Europe*, Paper presented at the University of Birmingham, 4 December 1973.

[30] V. A. Trapeznikov, *op. cit.*; Ya. Kotkovskii, *Voprosy Ekonomiki*, 1967, No. 4, pp. 74–5. These sources are cited by J. Wilczynski, *Technology in Comecon*, London (1974), p. 14.

it is still well behind the economies of Czechoslovakia and the GDR in this respect.[31]

Despite general agreement about the progressive character of science and technology in broad terms, it is interesting that there are sharp clashes of opinion among economists and planners in the USSR as to their precise impact on economic growth. This appears to be part of a continuing debate about the allocation of resources between R and D and capital investment, which first surfaced in the Soviet press and economic journals in the middle of the 1960s and focuses typically on the rate of return yielded by one rouble of expenditure on 'science'. In this connection, a very critical survey by L. Glyazer is worthy of note, not least because it echoes many of the doubts about estimates of 'technical progress' expressed by Western economists.[32] According to Glyazer, the Soviet economic literature is scattered with widely conflicting estimates of the annual economic effect of one rouble spent on science; these range from a 'realistic' 28–31 kopeks at one extreme to 3–5 roubles at the other. He concludes that these variations are not the result of minor errors but of serious shortcomings in the methodology. Again, the main culprit is the residual. Glyazer refers to this picturesquely as a *chulan dlya khlama* (literally, 'lumber room' or 'glory hole') and argues along lines familiar to Western economists that it is a serious mistake to attribute the whole of the growth of the *chulan* to the influence of science. Thus

> the measurement of this or that [economic] process produces convincing results only when it corresponds to the objective content of the given process. If the methodological approaches which are used squeeze real phenomena into the framework of an artificially constructed scheme, then the results of this exercise are far removed from reality.[33]

One of the most celebrated Western attempts to compare Soviet economic performance with the economic performances of Western countries, on the basis of 1960 data, is that of Abram Bergson.[34] This study is primarily concerned with the comparative efficiency of these economies in terms of total factor productivity, but this sheds some light on the level and rate of technological development, which is a substantial component of 'efficiency'. Bergson distinguishes between *static efficiency* (what a country can produce at a point in time, given the supply of labour, capital and technical knowledge) and *dynamic efficiency* (the capacity to improve performance over time due to the creation and utilisation of new technical possibilities). The two concepts are closely connected in the sense that static efficiency at a given point in time is the

---

[31] *Ibid.*, pp. 10–11. (The works referred to are: J. Kleer, *Wzrost intensywny w krajach socjalistycznych* (Intensive Growth in Socialist Countries), Warsaw (1972), p. 101; K. Falas, *Kozgazdasagi szemle* (Economic Review), Budapest, 1970, No. 9, p. 1017.)
[32] L. Glyazer, *Voprosy Ekonomiki*, 1971, No. 9, pp. 61–76.
[33] *Ibid.*, p. 76.
[34] Abram Bergson, *Planning and Productivity under Soviet Socialism*, New York (1968) (based on the Benjamin F. Fairless Lectures given by the author at Carnegie-Mellon University, November 1967).

result of past dynamic efficiency and at the same time creates the surplus needed for future dynamic efficiency. For his measure of static efficiency Bergson selected output per unit of composite labour and capital in terms of both national prices and US dollars; dynamic efficiency is correspondingly revealed in the rate of growth of this index over time. On the basis of these calculations it was alleged that the static efficiency of the USSR in 1960 was between one third and one half of the United States' level and well behind the leading West European countries with the exception of Italy.[35] There tended to be an inverse relationship between static and dynamic efficiency. Bergson explained this by the fact that countries with relatively low static efficiency have a lot of catching up to do; this catching up can be facilitated by various forms of technology transfer compared to the leading countries, which depend substantially on the more time-consuming and costly business of innovation. Thus, although the dynamic efficiency of the USSR from 1950–62 was superior to that of the United States and Great Britain, it was smaller than that of the other West European countries in the sample and, significantly, it was markedly inferior to Italy, which had a roughly equivalent level of static efficiency.[36]

Bergson's work raises a number of interesting problems relating to the measurement of the main variables. For his labour input he uses 'employment' (adjusted for educational level and sex distribution)[37] rather than the 'total potential labour force'. But if one were to regard 'hidden unemployment' in the USSR as a conscious and socially desirable policy compared with the policies of explicit unemployment pursued in Western countries, and if the Soviet labour input figure was adjusted to take this fact into account, it would have a substantial effect on productivity. There is also the problem of relative factor endowment. Bergson argues that the better quality of agricultural land in Western countries is offset by the more extensive area of cultivation in the USSR; but from the viewpoint of productivity, the cultivation of such an extensive and less fertile area requires substantially greater expenditures on transportation, labour and agricultural chemicals and it is doubtful whether agricultural factor productivity *could* equal that of Western countries in these circumstances. If it did it would imply relatively *greater* efficiency. A further point is that since the end of the Second World War a large number of American subsidiaries have been created in West European countries, especially in technology-intensive branches of industry. If this phenomenon were taken into account it would certainly diminish the efficiency of the West European countries relative to both the United States and the Soviet Union (unless, of course, it was argued that the toleration or encouragement of American subsidiaries by European governments was an aspect of efficient economic management!). Finally, as Hanson has pointed out,[38] there are technical difficulties involved in international comparisons which include the USSR and

[35] *Ibid.*, p. 23.
[36] *Ibid.*, p. 56.
[37] This is based on earnings differentials in the USA and, as the author admits, is not exact.
[38] P. Hanson, *Soviet Studies*, January 1973, pp. 340–1.

these represent potential sources of inaccuracy. In particular, calculations of static efficiency are highly sensitive to the imputed weights used to aggregate the labour and capital inputs and, by manipulation, it is even possible to produce the result that the USSR has a higher level of static efficiency than the USA. The possibility of obtaining satisfactory weights is inhibited by the fact that the prices of Soviet goods and services do not express accurately their scarcity value.

Bergson's conclusions about the relative efficiency of the Soviet economy are by no means cut and dried and can be interpreted in different ways. The author himself is aware of the fact that what he may be measuring is not static efficiency but, rather, relative levels of economic development.[39] This problem arises because output per man is higher in industry than in agriculture. Thus, the static efficiency of countries with a large agrarian sector, such as the USSR, is likely to be relatively low. Although Bergson acknowledges this problem it does not seem to inhibit him from drawing more or less firm conclusions.[40] Problems of developmental perspective also touch Bergson's argument at other points. In order to establish a comparable base for measures of relative efficiency it is assumed that the USSR has had plenty of time since 1928 to assimilate and apply the stock of technical knowledge available to Western countries; its failure to do so is thus a reflection of its inefficiency. But, ultimately, efficiency must be judged in relation to the concrete set of circumstances existing in different countries at any given point in time; these circumstances are not part of efficiency but represent the basis upon which efficiency during a future period must be achieved. Thus, it might be argued that the USSR in 1928 was not a 'blank sheet of paper' upon which technical knowledge could be inscribed as easily as in Western countries at a comparable stage; extensive war damage, the flight of key specialists and the lack of experience in applying the results of research, due to the pre-revolutionary separation of Russian science from foreign-owned factories, were important obstacles. This argument can be extended. If we are concerned to measure efficiency during the post-war period should the 'Stalinist heritage' of economic structures and attitudes be looked upon as an integral part of that efficiency/inefficiency or an aberration which is exogenous and should be adjusted for? A final point to be mentioned is whether efficiency, in a strictly economic sense, is definitive. Economic development takes place within a broader political and social framework in which there is a trade-off between economic objectives, political objectives and social objectives. In response to its past isolation and present desire to play a major role in world affairs the USSR has devoted considerable resources to prestige projects, defence and foreign aid. From the standpoint of our own values we may deplore this pattern of priorities, but is it less *efficient*? These limitations are not intended as

---

[39] Abram Bergson, *op. cit.* (1968), p. 65.

[40] This point is made by Philip Hanson (*Soviet Studies*, October 1971, pp. 296–301) in the course of a spirited exchange with Bergson (*Soviet Studies*, October 1971, pp. 282–95). Much of this discussion is taken up by the question of how far Bergson's results shed light on the efficiency of *socialism* as distinct from the specific features of the Soviet economy.

criticisms which are specific to Bergson's work. They will be present to some extent in all purely *economic* estimates of efficiency.

In 1970, Weitzman published a pioneering article in which he made explicit use of production function techniques to analyse the main sources of Soviet economic growth since the end of the Second World War.[41] From 1950–69, and especially after the late 1950s, a decline could be observed in the rate of economic growth in the USSR, without a proportional diminution in the growth of factor inputs. This result remained the same despite the use of alternative factor weights. Moorsteen and Powell[42] have suggested that Soviet economic growth rose rapidly after 1953 as the system freed itself of some of the worst rigidities of the Stalin era, but after an initial burst of productivity the USSR was unable to sustain this rate of growth by means of further increases in efficiency and 'technical progress' (the 'residual'). Weitzman, on the other hand, finds no steep decline in efficiency and concludes that the rate of 'technical progress' in the USSR during the post-war period has proceeded at about 2 per cent per annum, which is 'respectable if not spectacular by world standards'. Approximately 15–20 per cent of average increases in output can be attributed to this residual. Thus, in Weitzman's opinion, 'The Soviet post-war record looks very much like a classical model of economic growth with high rates of capital accumulation serving as the prime mover'.[43]

For our purposes, Weitzman's work contains most of the problems associated with aggregate production functions, but the author is aware of these and does not claim that he has isolated the rate of technical development precisely. He also disarms his critics by acknowledging a number of other limitations, which are inevitable in view of the non-availability of data. First, labour inputs are not adjusted for quality on the grounds that Soviet wage rates do not reflect marginal productivity but are often determined by social criteria. Second, due to lack of hard evidence, the official capital stock figures are not adjusted for degree of utilisation, rates of replacement or quality. Finally, Weitzman accepts that a rapid rate of introduction of new products at high prices may tend to inflate economic growth artificially. Brubaker, however, takes the view that Weitzman's estimates of capital productivity are 'incredibly' high in the light of the considerable amount of anecdotal evidence that exists about the inefficiency of Soviet capital investment. These sources of inefficiency include the systematic understatement of real costs due to the lack of an explicit interest rate throughout most of the period, which encouraged hoarding and a bias towards capital intensive projects. Thus, Brubaker argues that Weitzman's implied estimate that the average gross rate of return to capital in Soviet industry during the 1950s and 1960s ranged between 20–40 per

---

[41] Martin L. Weitzman, *American Economic Review*, September 1970, pp. 676–92. In this study a CES production function was used in preference to Cobb-Douglas. The author argues that this function 'fitted' Soviet data more satisfactorily than American data; this was due to relatively more unpredictable nature of US growth (business cycles, variations in capacity utilisation, greater role of technical change).

[42] R. H. Moorsteen and R. P. Powell, *Two Supplements to the Soviet Capital Stock*, New Haven (1968), p. 9 (cited by Weitzman, p. 678).

[43] Martin L. Weitzman, *op. cit.* (1970), p. 685.

cent is 'remarkable to say the least' and, moreover, is inconsistent with his general conclusion that diminishing returns are the main source of deceleration in the rate of growth of Soviet industrial output after 1965.[44]

An approach very similar to that of Weitzman has been used recently by Padma Desai to compare the rates of 'technical progress' in seven Soviet industries.[45] The outcome of these calculations is that Soviet heavy industry (chemicals, machine building and electric power) appears to be subject to more rapid technical change than light industry and the food industry, with construction materials and ferrous metals occupying an intermediate position. This comparison is less interesting than one between the same branches of industry in different countries. It is fairly well established that branches of industry such as chemicals, for example, have rapid rates of 'technical progress' because of relatively greater *opportunities* for innovation and other forms of technical development; but the extent to which different economies are able to exploit these opportunities is a much more significant question, which will perhaps be answered in future work.[46]

Before moving on to consider other forms of economic estimate relating to the level and rate of development of Soviet technology, it is important to note the general point that the estimates referred to above are decisively influenced by how the production function is specified in each case. In particular, the more common Cobb-Douglas function used by Bergson and others (see Table 1.1) and the CES function used by Weitzman and Desai rest on different assumptions about the processes of economic change and raise serious difficulties of interpretation when they are applied to the Soviet economy. In the Cobb-Douglas approach, the weights attached to labour and capital inputs are fixed because it is assumed that the elasticity of substitution between capital and labour is equal to one. The CES approach, on the other hand, allows the elasticity of substitution to be of any value (including unity), provided that it is constant over time. Thus, due to the high rate of capital accumulation in the post-war Soviet economy, combined with a relatively lower growth rate of labour inputs, estimates using Cobb-Douglas tend to arrive at a residual that is declining over time; this suggests that the rate of technical change may have been slowing down since the 1950s. Conversely, Weitzman found that the rapid growth of capital relative to labour, combined with a relatively low elasticity of substitution of capital for labour, produced a sort of diminishing returns effect that tended to slow Soviet growth while the residual was approximately constant. A fundamental difficulty here is in establishing firm grounds for

---

[44] E. R. Brubaker, *American Economic Review*, Vol. LXII, 1972, pp. 675–8. Weitzman's reply appears in the same issue.

[45] Padma Desai, *Technical Change, Factor Elasticity of Substitution and Returns to Scale in Branches of Soviet Industry in the Post-war Period*, manuscript of article prepared at the Institute of International Studies, University of California, Berkeley, April 1974. As distinct from Weitzman, Desai uses, alternately, both gross output and gross value-added as her dependent variables and explicitly considers raw material inputs.

[46] Desai might also have mentioned that the relatively high rate of technical progress in the Soviet chemical industry has been assisted by massive imports of plant and equipment during the 1960s.

preferring one approach to any other.[47] Table 1.1 attempts to summarise, in a systematic way, the results of some major Soviet and Western studies of the Soviet economy which use production functions. The considerable variations, in both the results and the methods employed, make it awkward to generalise. However, the tentative outline that seems to emerge is that the *rate* of technological development in the USSR during the post-war period has been superior to that of the United States and roughly on a par with that of other major Western countries whereas the cumulative *level* of technology continues to be inferior to all major Western countries, particularly the United States. All this, of course, is subject to the important reservations noted above about the extent to which the conventional economic concept of 'technical progress' (measured by the residual) equates with the definition of the level and rate of technological development, which was adopted at the beginning of this chapter.

Is it possible to obtain an aggregate assessment of technical change in the USSR and elsewhere while avoiding the problems associated with production functions and calculations of total factor productivity? A daring approach to this problem has been suggested by Michael Boretsky, who has attempted to 'deresidualise' the concept of technical progress.[48] This is a substantial and intricate piece of work that needs to be examined in more detail than is possible here. Boretsky proceeds on the assumption that technical progress is 'embodied' in new capital stock, energy, sources or materials.[49] 'Technical progress' is equated with the impact of the introduction and diffusion of key technologies in each of these three areas on the reduction of labour costs.[50] Thus a representative sample of key technologies can be used to construct an index of technical progress, weighted according to the relative contribution of each item to the reduction of labour costs in the economy. This is a more direct and ingenious way of isolating that portion of the 'residual' in the approaches discussed earlier, which represents technological development in the strict sense (i.e., the innovation and diffusion of new technology). In practice, the examples used in the sample are not key innovations so much as *key indicators*, which in Boretsky's opinion subsume a whole range of 'satellite innovations' (for example, use of fertilisers per unit of cropland, which reflects the greater use of chemicals and consumption of electric power, which reflects the greater application of 'electrotechnology'). Having constructed his weighted index, Boretsky applies it to the appropriate

[47] The points contained in this paragraph lie outside the repertoire of a simple political scientist such as myself. I am very grateful to Dr P. Hanson for bringing them to my attention. Any errors remaining are entirely due to my imperfect understanding.

[48] Michael Boretsky, *New Directions in the Soviet Economy*, Joint Economic Committee of US Congress, Washington DC (1966), p. 143.

[49] Boretsky accepts the importance of such factors as: increasing skill, 'learning by doing', improvements in management, inter-industry shifts in factors, economies of scale and changes in the conditions of demand, but he sees these as being *fundamentally dependent* on 'embodied' technological development (*ibid.*, p. 145). This argument cleverly incorporates the objections of Arrow and Kaldor to the approaches discussed earlier.

[50] Labour costs are used by Boretsky as a proxy measure of total costs on the grounds (a) that the economies resulting from technical change are mainly made up of labour cost savings and (b) that capital savings, if any, are in a fixed proportion to labour savings (*ibid.*, p. 148).

Table 1.1 *Some macro-economic measurements relating to the level and rate of 'technical progress' in the USSR*

Author(s) and date of publication	Years to which estimates refer	Purpose(s) of measurement	Technique(s) used.	Results
F. Seton (1958)[5]	1950–55	Sources of Soviet economic growth.	Cobb-Douglas production function—'dynamic version' where elasticities do not add up to unity.	Continuing high rates of Soviet economic growth 1950–55 are mainly the result of increasing efficiency (technological, administrative or both) rather than capital accumulation or labour influx. Technical progress proceeded at over 6% per annum.[12]
A. Bergson (1963)[3]	1950–58	Dynamic efficiency (growth in total factor productivity) of Soviet economy compared with selected Western countries.	Production function with the following variables: NNP valued at 1937 prices and 'given year' prices, labour, reproducible fixed capital and livestock herds.	Annual average rate of growth of total factor productivity in USSR = 2·7% (unadjusted for intersectoral shifts in employment or changes in inventories and assuming a 20% annual net return for capital, inventories and livestock). Equivalent figure for USA = 1·7%.
B. Balassa (1964)[2]	1950–58	Dynamic efficiency.	As above but output measured in terms of GNP; capital weighted in terms of share of GNP including an allowance for depreciation to make it consistent with MIT estimates for Western countries.[4]	Average annual rate of growth of total factor productivity for economy as a whole: Japan = 4·1% (1951–59); FRG = 3·4% (1950–59); USSR = 3·3% (1950–58); USA = 2·6% (1948–57); UK = 0·7% (1949–59). Soviet performance can be largely explained by rapid accumulation and re-allocation of labour from agriculture to industry.
J. S. Berliner (1964)[1]	1960	Static efficiency (total factor productivity) of Soviet and American economies.	Kendrick-type production function (homogeneous of degree one: $P = \alpha L + \beta K$), and Cobb-Douglas ($P = \beta L^{\alpha} K^{1-\alpha}$). No adjustment made for relative qualities of land, labour and capital.	'Relative efficiency of the USSR ranges from 36–39% of USA if the USA used Soviet inputs and if outputs were valued at Soviet prices, to 87–98% of the USA if the USSR used American inputs and if outputs were valued at American prices.' Due to absence of qualitative adjustments, the estimates measure relative efficiency of total economies rather than efficiency of pure systems of economic organisation.
S. H. Cohn (1966)[6]	1950–58 and 1958–64	Sources of Soviet economic growth.	'Simplified' Cobb-Douglas production function with GNP as the output variable.	Annual average rates of increase: L K Residual Production 1950–58 1·2 8·3 3·7 7·1 1958–64 0·6 9·4 2·0 5·3

Author	Period	Topic	Method	Results
B. N. Mikhalevskii and Yu. P. Solov'ev (1966)[7]	1951–63	Sources of Soviet economic growth.	Production function of the type specified by Dhrymes and Kurz.[8] Measure of output excludes non-productive sectors (c.f. Cohn). Inputs include working capital, consumer durables, agricultural land and mineral resources.	Annual average rates of increase 1951–63: $L+K$ Residual Production 4·8 2·2 7·0 This estimate of 'dynamic efficiency' corresponds closely to Cohn's estimate for a similar period; as with Cohn's estimate, the rate of growth of inputs declined less rapidly than factor productivity. The authors further suggest that 1·5% of the residual (2·2%) is accounted for by 'embodied and disembodied technical progress'.
A. Tolkachev (1966)[9]	1950–64	Sources of Soviet economic growth.	Cobb-Douglas production function: $Q = 1·186 \times X^{0·155} \times Y^{0·845}$ where Q = net output X = man hours Y = production funds.	The equation reveals that a 1% increase in production funds leads to 0·845% growth in net output whereas a 1% growth in man hours leads to only a 0·155% growth (N.B. The 1·186 is not the 'residual'; it is an adjustive constant).
A. Bergson (1968)[10]	(a) 1960 (b) 1950–62	(a) Static efficiency. (b) Dynamic efficiency: Soviet economy compared with major Western countries.	Implied Cobb-Douglas production function. Employment variable adjusted for educational level and sex distribution.	(a) Static efficiency of USSR economy in 1960 was between $\frac{1}{3}$ and $\frac{1}{2}$ that of USA and well behind leading West European countries with the exception of Italy. (b) Dynamic efficiency of USSR economy 1950–62 superior to that of USA and Britain but inferior to that of other West European countries especially Italy.
M. L. Weitzman (1970)[11]	1950–69	Sources of Soviet economic growth.	CES, Hicks neutral production function.	Soviet decline in rate of economic growth throughout the period especially after late 1950s. This is explained primarily by diminishing returns to capital rather than a declining 'residual'; Weitzman estimates that 'technical change' has occurred at a rate of 2% per annum throughout the period, which is 'respectable' by world standards. This result is challenged by E. R. Brubaker (1971).[13]

Author	Year	Topic	Method	Findings
E. R. Brubaker (1972)[15]	1960	Static efficiency of *particular* sectors of selected West European countries and USSR.	Systematic comparison of data on *output per man hour* compiled by Bergson (1968) *op. cit.*, Cohn (1966) *op. cit.* and Maddison (1965).[16]	There are wide variations in the results, especially between estimates using American price weights. The general picture is that in 1960 the Soviet economy, overall, appears to have been 'unambiguously less efficient' than West Germany, France and Britain and on a par with Italy. This aggregate shortfall is primarily due to the low output per man hour in commerce and other services; other sectors, such as industry, agriculture and transport and communications perform relatively well according to this criterion.
A. Bergson (1973)[14]	1970–75–80	Future sources of Soviet economic growth.	Cobb-Douglas production function.	The 1970–75 plan envisages an annual average rate of growth of material income of 6·7%. This can only be obtained by considerable increases in productivity, given that even in official figures there is a planned decline in the rate of growth of capital investment (6·7% for 1970–75 c/p 7·5% for 1967–70). Employment is most unlikely to grow at an increasing rate. Thus a new model of economic growth is implied, founded on sharp increases in efficiency and technical progress. Assuming a continued decline in the rate of growth of capital stock, in order to fulfil the plan the average annual rate of increase of total factor productivity would need to be 3·0% (c.f. actual rate of 1·7% in 1950–58 and 0·7% in 1958–67).
J. Hocke, O. Kyn and H. J. Wagener (1973)[17]	1950–70	Growth of factor productivity of Soviet *industry* (i.e., dynamic efficiency).	Six variants of Cobb-Douglas production function (three assuming constant returns to scale and three allowing for increasing or decreasing returns to scale).	Average rate of 'technological change' in Soviet industry was between 3–5·5%. 'This is neither extremely low nor extremely high . . . [and] . . . indicates a good overall performance.' Rate of change may have been greater at the beginning of the period (4–7·3%) than at the end (3–3·7%); thus there has been an average annual decline in the rate of technological change of 0·09%–0·18%.
P. Desai (1974)[18]	1955–71	Comparative rates of 'technical change' in seven branches of Soviet industry.	Cobb-Douglas and CES production function with constant and variable returns to scale with the assumption of Hicks neutrality. 'Best fits' were obtained where gross output was the dependent variable and raw materials were explicitly considered (c/p Weitzman).	Soviet heavy industry appears to be more progressive technologically than light industry. Rates of 'technical change' for the period, by branches of heavy industry, are: Chemical and mining-chemical 7·68% Machinery and metalworking 4·89% Construction materials 2·80–3·36% Electric power 1·69–3·15% Ferrous metals 1·74%

Sources:
1 J. S. Berliner, *American Economic Review*, May 1964, pp. 480–9.
2 B. Balassa, *ibid.*, pp. 490–505.
3 A. Bergson, in (Eds.). A. Bergson and S. Kuznets, *Economic Trends in the Soviet Union*, Cambridge, Mass. (1963), pp. 1–37.
4 E. D. Domar, *Review of Economics and Statistics*, February 1964.
5 F. Seton, *American Economic Review*, May 1959, pp. 1–14.
6 S. H. Cohn, in *New Directions in the Soviet Economy*, Joint Committee of US Congress, Washington DC (1966), pp. 99–132.
7 B. N. Mikhalevskii and Yu. P. Solov'ev, *Ekonomika i Matematicheskie Metody*, 1966, No. 6, pp. 823–40. Comment by F. Denton in *Soviet Studies*, 1967–68, No. 4, pp. 501–9.
8 P. Dhrymes and M. Kurz, *Econometrica*, 1964, No. 3, p. 32.
9 A. Tolkachev, *Planovoe Khozyaistvo*, 1966, No. 6, pp. 1–9.
10 A. Bergson, *Planning and Productivity under Soviet Socialism*, New York (1968).
11 A. Bergson, *American Economic Review*, September 1970, pp. 676–92.
12 See also P. J. D. Wiles, *The Prediction of Communist Economic Performance*, Cambridge (1971), p. 337.
13 E. R. Brubaker, *American Economic Review*, Vol. LXII, 1972, pp. 675–8.
14 A. Bergson, *Problems of Communism*, March–April 1974, pp. 1–9.
15 E. R. Brubaker, *Soviet Studies*, 1971–1972, No. 3, pp. 435–49.
16 A. Maddison, *Banca Nazionale Del Lavoro Quarterly Review*, March 1965.
17 J. Hocke, O. Kyn and H. J. Wagener, *Forschungsbericht 1973*, Osteuropa-Institut, Munich, pp. 1–44.
18 P. Desai, *Technical Change, Factor Elasticity of Substitution and Returns to Scale in Branches of Soviet Industry in the Postwar Period*, manuscript of article prepared at Institute of International Studies, University of California, Berkeley, April 1974.

Soviet and American statistics of production or consumption in order to calculate an aggregate measure of the level and rate of technical progress. He concludes that in 1962 the level of technical progress in the USSR was 25 years behind that of the United States. Furthermore, according to these calculations, the rate of technical change in the Soviet Union from 1940–62 was roughly 60 per cent of the American rate (depending on how you measure it), which implies that the USSR has actually tended to fall behind during the post-war period. To those who claim that the USSR is catching up due to cheap technology transfer and relatively rapid rates of capital accumulation Boretsky replies bluntly that 'this is not true'.[51]

Boretsky's argument seems to rest on two crucial pillars: (a) very accurate measurements of labour savings and (b) the careful selection of a representative sample of key innovations. Let us examine the second of these points first. It can be argued that the innovation and diffusion of new technology is not revealed clearly in quantitative increases in his general indicators. Changes in the composition of these indicators are more significant. Thus, from the point of view of technical progress, the growth of concentrated compound fertilisers is more important than the overall output of fertilisers; the output of particular varieties of thermoplasts is more important than total output of plastics and synthetic resins; the production of acrylic and polyester fibres is more significant than the total output of non-cellulose fibres, and so forth. Global output statistics, expressed in non-value terms, fail to pick up those significant trends; for example, one of the most interesting aspects of the American plastics industry during the 1960s has been the rapid growth of small-tonnage/high value plastics and resins. Unfortunately, at this level of detail, Soviet statistics are difficult to obtain and even if they were available it is extremely doubtful whether they could be matched up with information about labour savings. In the event of such data being available, it seems probable that the level of technical progress in the USSR would be made to appear even more backward, using Boretsky's approach. A second major problem with Boretsky's indicators is that many of them are closely connected with geographical factors and patterns of natural resources and are thus not entirely a reflection of 'technical progress'. The mode of power generation and transmission, for example, is dependent upon the availability of water power and transmission distances; the choice of steelmaking technology is to some extent related to the comparative cost of different sources of energy; the mode of transportation is a function of distance and topography as well as technical capabilities; the choice of hydrocarbon fuel is tied in with the comparative cost and availability of coal, oil and natural gas in addition to the technological sophistication of refineries and of the organic sector of the chemical industry. If one compared the USSR with any major West European country, many of those indicators would be inappropriate. This methodology cannot be regarded as being generally applicable. Whether it is applicable to a comparison between the USA and the USSR hangs on an evaluation of these factors, which does not seem to be present in Boretsky's analysis. On purely

[51] *Ibid.*, p. 150.

intuitive grounds, it is difficult to disagree with Boretsky's view that the USSR is not likely to be at a substantial disadvantage in regard to any of the natural resources referred to above; impressions are, nevertheless, potential sources of inaccuracy.

The second important element in Boretsky's argument concerns the measurement of labour savings. For this purpose he uses the estimates of Soviet project-planning organisations. Admittedly these estimates tend to measure 'potential increases rather than actual',[52] but it is argued that such biases will be constant between indicators and will therefore not affect the weighted index. Boretsky may be too trusting here. In the USSR, the estimates of project-planning organisations have been widely criticised in the economic literature; such estimates are often wildly inaccurate and, given the casual basis upon which they are sometimes advanced, there are no good grounds for supposing that they are equally inaccurate in all cases. This seems to be a very shaky foundation on which to erect such an elaborate statistical analysis.

Finally, some criticism of Boretsky's computational methods has been made by Joseph Berliner.[53] Berliner suggests that an index based on per unit magnitudes does not have any economic meaning. Using Boretsky's figures, he calculates *total labour savings* for only three of the innovations in the sample (plastics, manmade fibres and data processing); in Berliner's view, the comparison of these totals for the USA and the USSR would give a more meaningful measure of relative technical progress in the two countries. The surprising result of these calculations is that over the 22 year period average annual labour savings in the USSR, arising from the diffusion of these three innovations, are approximately one million man years. This is a very substantial figure and if the other items in Boretsky's sample were included, not to mention innovations outside the sample, they would produce a total of labour saving far in excess of the most extravagant Soviet claims. Berliner considers that 'it becomes increasingly difficult to take a serious view of a method that implies such results'.[54]

Despite differences in approach, all the studies reviewed above seem to agree that some kind of technology gap exists between the USSR and Western countries, especially between the USSR and the United States. Is this gap likely to be a permanent one? If not, the above assessments are less important because they highlight transitory phenomena rather than the permanent inferiority of Soviet economic structures and adaptive mechanisms. Opinions on this question differ. R. V. Burks,[55] for example, has argued that the curve of technological development in Western countries is exponential and, since the USSR is heavily dependent on borrowing technology from abroad, the Soviet economy will never catch up and may become progressively more obsolete. In a similar vein, Academician Sakharov has drawn an analogy between two skiers

[52] *Ibid.*, p. 183. Boretsky equates the accuracy of estimates of Soviet project organisations with those of Arthur D. Little in the USA. This may be an optimistic view!
[53] J. S. Berliner, *The ASTE Bulletin*, Autumn 1971, pp. 18–24.
[54] *Ibid.*, p. 23.
[55] R. V. Burks, *Technological Innovation and Political Change in Communist Europe*, Santa Monica, Cal. (1969).

in a downhill race; the 'red' skier can not equal the sophisticated techniques of the leading 'striped' skier and is unable to catch him up, despite the fact that he has the advantage of running in his tracks.[56] A more recent monograph by Stanislaw Gomulka has added some new insights into this problem.[57] Gomulka begins with a schematic analysis of the major factors influencing the level of technology, which is 'embodied' in successive vintages of capital. These factors are:

1. The diffusion of free scientific and technical information, forced by the extent of the technology gap.[58]
2. New technology 'embodied' in imported plant and equipment.
3. 'Learning by doing'.
4. Indigenous innovation carried out by the domestic R and D establishment.

By means of an abstract model of industrialisation in a particular country it is possible to hypothesise the interplay between these factors and capital and labour inputs, which would characterise a particular phase of development. On this basis, Gomulka identifies a 'hat-shaped curve' in the rate of growth of labour productivity. In the long term, labour productivity in all countries will settle down to a low but steady rate of increase because the technological lags between countries will have been overcome and further economic expansion will depend primarily upon population growth. Thus, the strength of diffusion declines as the gap closes, while future development is constrained by the relative scarcity of capital and labour resources. Gomulka predicts that:

> The relative technology gap between the United States, on the one hand, and Western Europe, the Soviet Union and Japan on the other, is gradually narrowing. All these countries are becoming one technologically unified and leading area. The unification will probably be completed as early as 1990 in the case of Japan and about 2010–30 in the case of the USSR and Western Europe. Then, rapid growth will probably disappear over the area.[59]

Gomulka's argument is pitched at a very high level of generality and the richness of the concepts is not matched at this stage of his work with a corresponding abundance of relevant data. His predictions are therefore highly speculative. It is possible to imagine alternative scenarios in which the labour and capital constraints are overcome (or substantially postponed) by means of ingenious 'factor stretching' and 'resource extending' innovations (for example: solar energy, robots, re-cycling and exploitation of hitherto inaccessible sources of raw materials). Success in this would probably come to countries with a R and D process which is geared to indigenous innovation. Gomulka does not say anything about the R and D process in different countries, yet its efficiency and mode of operation affects all his main variables.

---

[56] A. D. Sakharov, *Progress, Coexistence and Intellectual Freedom*, London (1968), pp. 65–6.
[57] S. Gomulka, *Inventive Activity, Diffusion and the Stages of Economic Growth*, Economics Institute, Aarhus University, Monograph 24, 1971.
[58] Gomulka assumes that this process is costless. For a different view, see C. Freeman in (Eds.) D. O. Edge and J. N. Wolfe, *Meaning and Control*, London (1973), pp. 231–40.
[59] S. Gomulka, *op. cit.* (1971), p. 58.

It seems probable that a country with an independent capability to develop and introduce new technology will be relatively better equipped to fight off the diminishing returns of technical progress; such a country may come to realise that advanced technology is a valuable national asset and may wish to restrict the use of this knowledge in order to maintain a lead over its rivals.[60] When one's time-scale extends into the twenty-first century there is no limit to the alternatives that one can imagine. However, Gomulka's work is a thought-provoking answer to those who assume that because Soviet technology lags behind American technology at present, the USSR is perpetually condemned to a position of inferiority unless it initiates radical changes in its economic structures and attitudes.

Disaggregated measures of the level and rate of technological development: general approaches

Many of the attempts to estimate and compare the level of technical progress in different countries have been carried out using more disaggregated methods than those discussed above. It is at this level of analysis that one encounters some of the more polemical studies of the 'technology gap'. At the end of the 1960s, for example, Servan-Schreiber wrote an influential book in which he alerted European governments to the dangers of economic penetration by American firms, based on their superior technology, know-how and managerial methods. This trend was regarded as a '. . . strange phenomenon, so dangerous, so massive in its size and power, it [was] hypnotizing and overwhelming'.[61] Contrary to Gaullist protectionism, the author believed that the American challenge could only be fought off by arming European industry with similarly effective weapons; this required a concentration on technological growth points and a complete revitalisation of traditional economic structures and attitudes. If these steps were not taken he predicted that by 1980 the third largest power in the world after the United States and the Soviet Union, would be American industry in Europe.[62] A broadly similar view was advanced by Academician Sakharov and his colleagues in a letter to the Central Committee of the Soviet Communist Party.[63] These three scholars argued that the initial impetus of industrialisation in the USSR had been lost. The Soviet Union was now falling behind the advanced capitalist countries and in key areas like computer technology it virtually 'lived in another age'; these gaps could be closed only by the elimination of administrative mediocrity and by a substantial degree of liberalisation, which would transform the whole structure of economic management and would release creative energies repressed by the existing system of controls. In view of the fears (and admiration) expressed above, it is extremely ironic that by the early 1970s the United States itself should be going through a period of self-doubt. In a

---

[60] This point has been made forcefully by A. C. Sutton, whose work is discussed below.
[61] J. J. Servan-Schreiber, *The American Challenge*, London (1968), p. xiii.
[62] *Ibid.*, p. 11.
[63] A. D. Sakharov, V. F. Turchin and R. A. Medvedev, *Le Monde*, 11 April 1970.

balanced appraisal of the situation, Harvey Brooks acknowledged that the West European countries and Japan were encroaching upon the American share of international trade, even in high technology goods.[64] But this was to be expected. Earlier perceptions of American superiority, both in the United States and abroad, were in his view exaggerated. Instead, the industrialised nations of the world were approaching some kind of technological saturation and the United States, as the most advanced nation with the highest per capita GNP, had reached this plateau ahead of its rivals. Brooks concluded that 'the scientific system is increasingly international, so that the very concept of national superiority in science and technology is obsolescent'.[65]

The conflicting views expressed above can only be reconciled by more thorough investigation. How might this be carried out? The main kinds of disaggregated criteria used in assessments of technological level are systematically expounded in a massive study conducted by the OECD in the late 1960s.[66] As distinct from aggregate economic assessments, which strive for analytical elegance, these studies depend to a large extent on the accumulation of very substantial quantities of data by teams of researchers. The following areas of investigation are thought to be especially revealing.

INPUTS

HIGHER EDUCATION AND SCIENTIFIC MANPOWER

The conventional measures used here are: absolute and per capita resources devoted to education; numbers of graduates, especially in key occupations such as engineering and R and D; the organisational distribution of highly qualified specialists between universities, independent institutes and manufacturing industry.

EXPENDITURE ON RESEARCH AND DEVELOPMENT

The conventional measures used here are: expenditure in absolute terms and as a percentage of GNP; the allocation of expenditure between the defence/space sectors and manufacturing industry (and within the latter, the resources devoted to science-based industries); the allocation of expenditures according to sector of performance.

OUTPUTS

RESEARCH PERFORMANCE[67]

The level of research is an important factor to assess because it is the breeding ground of discoveries, which may later find their way into industrial practice in the form of new products and processes. The main problem here is that of 'objective' assessment. It is difficult to see how this can depend, ultimately, on

---

[64] Harvey Brooks, *Harvard Business Review*, Vol. 50, May–June 1972, No. 3, pp. 110–18.

[65] *Ibid.*, p. 118. Of course, the USSR lies largely outside the network of relations in multinational firms and institutions that Brooks refers to.

[66] *Gaps in Technology: Analytical Report*, OECD, Paris (1970). In addition to this summary volume there are detailed sector reports: *Scientific Instruments* (1968); *Electronic Components* (1968); *Plastics* (1969); *Pharmaceuticals* (1969); *Non-Ferrous Metals* (1969).

[67] This element is not explicitly considered in the OECD series of studies.

anything other than the expert opinion of the scientific community itself. For this reason, the number of times that a country has succeeded in winning an international award (for example, Nobel Prizes) or the degree to which the work of its scientists is cited in foreign journals are often used as appropriate criteria, although these are both subject to acute problems of interpretation.[68]

TECHNICAL INNOVATION

In this group of indicators the intention is to evaluate success in transforming research findings into industrial applications. Two related aspects can be considered. The first aspect concerns the level of inventiveness in devising laboratory scale prototypes or experimental batches of new products and can be approached through an examination of the award of patents and international trade in licenses and know-how. The second aspect is concerned with the first commercial production of new and important products (or the first operation of a new process on an industrial scale) in different countries. This approach, which has been used by both Freeman[69] and Hufbauer,[70] provides a reasonably firm basis for comparing technological leads and lags.

DIFFUSION OF TECHNOLOGY

Diffusion is the final phase in the cycle. It concerns the extent to which the new products or processes become incorporated within the economy as a whole and indicates how far industry is able to replicate the original development, scale-up to large-tonnage output and manufacture products at a competitive price and acceptable quality. In their pioneering study, Nabseth and Ray look upon diffusion as having the most decisive influence on the rate and level of technical progress.[71]

COMPOSITION OF FOREIGN TRADE

Countries which are superior in terms of research, innovation and diffusion will tend to have a relatively large proportion of high technology goods in their exports. Posner[72] has argued that countries in which an innovation first occurs retain a manufacturing advantage for up to 10–15 years after first commercial production and it is only in the long term that a country with lower wages can compete with this. However, there is an important practical problem in using trade statistics. Once a new process has become standardised and is running smoothly in the leading country, subsidiaries will begin to be formed in foreign countries.[73] This may go some way towards explaining the so-called 'Leontief paradox', namely, that the ratio of capital:labour inputs in US exports is lower

---

[68] See, for example, D. de S. Price, *Little Science: Big Science*, New York (1963).
[69] C. Freeman, *National Institute Economic Review*, November 1963, No. 26, pp. 22–63.
[70] G. C. Hufbauer, *Synthetic Materials and the Theory of International Trade*, London (1966).
[71] L. Nabseth and G. F. Ray, *The Diffusion of New Industrial Processes*, Cambridge (1974); the interim report on this study was written by G. F. Ray and published in *National Institute Economic Review*, May 1969, No. 48, pp. 40–83.
[72] M. V. Posner, *Oxford Economic Papers*, October 1961, No. 3, pp. 323–41.
[73] R. Vernon, *Quarterly Journal of Economics*, May 1966, pp. 190–207.

than in the ratio of US production displaced by imports.[74] Assessments based on trade statistics which are not adjusted to take this phenomenon into account will tend to inflate the performance of West European countries relative to the United States (the source of subsidiaries) and to the Soviet Union (which has no US subsidiaries).

The forms of measurement described above are subject to a number of difficulties over and above the detailed problems of interpretation, which are encountered in using patents, citations and trade statistics or in establishing precisely when first commercial production of a product began in a particular country. These methods do not produce a neat and precise answer to the problem of comparing the levels of technology in different countries. On the contrary, the resulting data are often relatively chaotic and inconsistent. Since one is not dealing with a single residual but with numerous criteria it is not clear how these criteria should be ranked in order of importance. This dilemma has the advantage of not giving a false impression of precision but lays bare the fact that in the last analysis the researcher is exercising his own subjective judgement in arriving at a general view. If this conclusion is based only on incomplete data (for example, data on only a few phases in the research-production cycle or data confined to relatively few branches of industry) it must be treated with a good deal of caution. Thus, in order to achieve higher levels of confidence, disaggregated measures of technical progress require the collection of more and more data.

Some disaggregated measures of the level of technology in the USSR

Not all the measures expounded by the OECD in their study of technology gaps between member countries have as yet been applied to the USSR. There is still scope for more solid information in this area. However, one element which has received attention from Western specialists is the magnitude of the Soviet research and development effort. As has been pointed out earlier, international comparisons of research and development efforts are beset with problems of definition and by the forms in which statistics are collected in different countries. A pioneering attempt to include the USSR in an international comparison of this kind, based on the OECD 'Frascati definition'[75] was published in 1965.[76] A more refined and extensive version of this early attempt

---

[74] W. Leontief, *Proceedings of the American Philosophical Society*, Vol. XCVII, 1953, pp. 332–49; *Review of Economics and Statistics*, Vol. XXXVIII, November 1956, pp. 386–407. An interesting recent article by Steven Rosefielde (*American Economic Review*, 1975, Vol. LXIV, No. 4, pp. 670–81) demonstrates that in general Soviet trade *with the world* tends to be biased towards capital intensive exports and labour intensive imports; the trends over time reveal, surprisingly, that even Soviet exports to the *West* have become more capital intensive than imports from this source.

[75] *Proposed Standard Practice for Surveys of Research and Development*, OECD, Paris (1963).

[76] C. Freeman and A. Young, *The Research and Development Effort in Western Europe, North America and the Soviet Union*, OECD, Paris (1965).

was prepared by R. W. Davies and M. J. Berry in 1969.[77] The authors concluded that the total number of personnel engaged in R and D in the USSR in 1966 (excluding the social sciences and humanities) was somewhere between 1,655,000 and 2,291,000; the number of graduates employed in R and D was between 476,000 and 670,000. In both cases, this pool of manpower was substantially larger in absolute terms than that of any other country, including the United States.[78] More recent figures calculated by R. A. Lewis[79] suggest that the decline in the rate of increase of scientific manpower in the USSR during the 1960s as a whole was relatively slight compared to the USA, where, since 1969, there has been an absolute decline; in mid-1971, 3·5 per cent of all R and D scientists in the United States were unemployed. Even more striking is the fact that the rate of growth of Soviet spending on R and D actually increased during the late 1960s whereas in the United States R and D expenditure (at constant prices) reached a plateau.[80] Clearly, the USSR now deploys a larger R and D effort than any other country. This reflects the great faith that Soviet leaders appear to retain in the social and economic benefits to be derived from scientific activity.

Are the results of scientific activity in the USSR commensurate with the great effort that is put into it? As a first attempt at answering this question it is interesting to compare the views of Soviet spokesmen and Western visitors about the relative levels of technology in various fields in the USSR. A sample of Soviet assessments has been collected from comments made at plenary sessions of the Soviet Communist Party's Central Committee.[81] It is difficult to place the technologies and the indicators that are mentioned in any clear order of importance. However, the general picture that seems to emerge is that in relation to Western countries Soviet leaders believe themselves to be 'equal or in the lead' in the priority areas of space exploration, aircraft, military equipment, and in some of the more traditional sectors of industry, such as blast furnaces, rolling mills and welding equipment; on the other hand, they acknowledge that the USSR is 'lagging behind' in chemicals, electronics, light industry and construction materials. A roughly comparable set of assessments, confined to Soviet civilian technology, has been compiled by A. C. Sutton.[82] These assessments are based on the views of delegations of Western engineers who have visited the USSR and are somewhat more critical of Soviet achievements. With regard to blast furnaces and chemicals, the views of the Western specialists are broadly in line with those of Central Committee

[77] R. W. Davies and M. J. Berry in *Science Policy in the USSR*, OECD, Paris (1969), pp. 501–34.

[78] Berry and Davies were sceptical about comparing R and D *expenditures* on the grounds (a) that it was not possible to distinguish Soviet spending on R and D from 'science' (*nauka*) in the statistics, and (b) that it was not possible to work out an accurate research-rouble exchange rate. However, they tentatively suggested that Soviet expenditure on R and D in 1965 was approximately $14·8–$20·7 milliard (based on the Freeman-Young exchange rate: $2·5–3·5 = 1 rouble).

[79] The figures are analysed in R. Amann, *La Recherche*, December 1972, pp. 1027–34.

[80] *Funds and Manpower in the United States*, Washington DC (1972) p. 5.

[81] R. W. Davies and M. J. Berry, *op. cit.*, pp. 496–500.

[82] A. C. Sutton, *Western Technology and Soviet Economic Development 1945 to 1965*, Hoover Institute (1973), p. 379.

members, but the delegations were less impressed with Soviet metallurgical technology and identified 20–30 year lags in steel rolling, ore benefication and the manufacture of large-diameter pipes.

Systematic studies of the level of innovation and the diffusion of new technology in the USSR are few and far between. Work on innovation is virtually non-existent. In the case of diffusion, there are a number of relevant studies of Soviet computers and machine tools.[83] Apart from these specific studies some other interesting results have been obtained by a group of researchers at the Osteuropa Institute in Munich. By comparing the industrial structure of the USSR and Western Germany between 1950–70 it was found that relatively large increases in key branches such as machine building and chemicals broadly corresponded to an earlier phase in the development of German industry, though with some lags. Given what appears to be an identical strategy of modernisation, the 'lagging share in the progressive branches may to some extent be interpreted as an overall technological lag in the common process of industrial development.'[84] The Munich group has also carried out studies of the diffusion of computer technology and oxygen steel technology in the USSR.[85]

In the area of foreign trade, one of the more novel exercises carried out by the Munich group has been to examine the unit value of Soviet exports compared with Western exports for key equivalent types of products. This was justified on the grounds that products with a high value relative to weight or numbers embody more sophisticated technologies. Thus, a study of machine tools sold on the 'less discriminating' Yugoslavian market showed that Soviet products had a lower unit value; a high proportion of Soviet exports were simple general-purpose machine tools. In synthetic dyes, Soviet exports in 1970 realised $1·69 per kg compared with an average of $3·4 per kg received by Western countries. For major types of chemical process equipment, such as pumps, centrifuges and filters, a similar tendency could be observed.[86]

A more pointed question about Soviet achievements in developing sophisticated new products and processes, which does not receive a clear answer from the collection of Soviet statistics or from standard Soviet accounts of industrial development, is how far the USSR has been dependent on Western technology. This important theme has been explored by A. C. Sutton in his three-volume study,[87] which is based largely on Western archives,

[83] Richard Judy, 'The case of computer technology' in (Ed.) S. Wasowski, *East–West Trade and the Technology Gap* (1970), pp. 43–71; Michael Boretsky in *Dimensions of Soviet Economic Power*, Joint Economic Committee of US Congress, Washington DC (1963); *Science Policy in the USSR*, OECD, Paris (1969), pp. 491–3. These technologies are discussed at length below by Martin Cave and Julian Cooper and their main findings are summarized by R. W. Davies in the following chapter. It is not appropriate, therefore, to discuss them at this point.

[84] J. Sláma in (Ed.) C. Watrin, *Struktur und stabilitatspolitische Probleme in alternativen Wirtschaftssystemen*, Berlin (1974).

[85] J. Sláma and H. Vogel, *Osteuropa-Wirtschaft*, 1974, No. 2; J. Sláma and H. Vogel, *Jahrbücher für Nationalökonomie und Statistik*, Vol. CLXXXVII, 1973, No. 3. These studies are reviewed in chapter 2 below.

[86] J. Sláma and H. Vogel in (Ed.) Z. M. Fallenbuchl, *Economic Development in the Soviet Union and Eastern Europe*, Vol. 1, New York (1974), pp. 197–220.

[87] A. C. Sutton, *Western Technology and Soviet Economic Development 1917 to 1930*, Hoover Institute (1968); Vol. 2, ... *1930 to 1945* (1971); Vol. 3, ... *1945 to 1965* (1973).

memoirs and reports of visiting delegations to the USSR. The impressive quantity of evidence that is presented seriously undermines some notions that were popular in the West after the launching of the first Sputnik in 1957. Sutton attempts to demonstrate that this was an isolated success that by no means reflected the general level of Soviet technological accomplishment at the time, or since. He argues that:

> No fundamental industrial innovation of Soviet origin has been identified in the Soviet Union between 1917–65. . . . Soviet innovation has consisted, in substance, in adapting those made outside the USSR or using those made by Western firms specifically for the Soviet Union and for Soviet industrial conditions and factor patterns.[88]

This often involves scaling-up, which gives the impression of indigenous achievement. According to Sutton, the USSR has had plenty of time since 1928 to catch up with the West and its failure to do so is the result of systemic weaknesses, not of science, but of the capability to transform the results of science into new products and processes. The sharp political conclusion to be drawn from this is that innovation is a valuable 'natural resource' of Western systems which has been acquired cheaply by the USSR since 1917 and has been an indispensable factor in promoting Soviet industrial development and the growing political influence of the USSR in world affairs. This technological transfer was permitted because the pressures for trade in the West were greater than any other consideration.[89]

Sutton's assessments of the level of technology in particular Soviet industries will be reviewed, and in some cases criticised, in the various contributions to this volume. However, his general case, that the USSR is dependent on Western technology to a degree which gives Western governments powerful means of political and economic leverage, is disputed by Philip Hanson.[90] Hanson distinguishes between 'negotiable technology transfer' (plant imports, sales of licenses and know-how agreements) and 'non-negotiable technology transfer' (information obtained through journals, visits of delegations and industrial espionage). It is only over the former, especially over imports of industrial plant, that Western governments could conceivably exercise sure control. Here, Soviet imports of Western machinery in recent years have not represented more than four per cent of total domestic investment in plant and equipment, a figure which is below that of the OECD countries.[91] By comparing the percentage of total machinery imports for a particular industrial sector with the corresponding share of that sector in total domestic investment it is possible to arrive at the approximate conclusion that Soviet dependence on foreign equipment has been highest in the chemical industry

---

[88] *Ibid.* (1973), p. xxv.
[89] *Ibid.*, p. 418.
[90] P. Hanson, *External Influences on the Soviet Economy since the mid-1950s: the Import of Western Technology*, CREES Discussion Paper, RC/B 7, Birmingham (1974).
[91] *Ibid.*, p. 13.

and in shipping.[92] This information is useful but leaves some loose ends. Basically, the problem centres on what 'technology' is and how it is valued. The Soviet electric power industry provides an interesting example of this. It is extremely expensive to create a system for the high voltage transmission of electric current. This involves large expenditures on excavation, pylons, cables, power stations, and so forth. However, the technological linch-pins of this system are the circuit breakers, which in the Soviet case were at one important point in time imported initially from France: these may represent a relatively small proportion of total investment but without them the system would not work. Thus, dependence on these 'key' technologies (possibly priced in terms of their value to other *Western* countries) is not highlighted in global statistics of machinery imports. The import dependence of the OECD countries may indeed seem to be greater than that of the USSR but it may not in practice be so. Whether it is or not will depend upon the extent to which the Western countries would be able to produce the imported machinery themselves if they wished to, and thus on how far these imports are simply the result of comparative cost advantages arising from national specialisation. If the latter is the case, this form of interdependence could be quite different from the position of a country like the USSR, which may have been compelled to abandon its autarchic aspirations in order to import *key* and *basic* items of industrial plant. The answer to this question can emerge from studies of the innovation process in the USSR and elsewhere, combined with a very detailed breakdown of machinery imports for particular industries.

Our approach and its limitations

The authors of the case studies of technological level undertaken at Birmingham have all chosen to use disaggregated forms of assessment. We accept the important limitation that disaggregated estimates are essentially 'clusters of insights' that are difficult to rank in order of importance. But this approach can be justified on a number of grounds. First, it does not appear that aggregated approaches, for example through the use of production functions, produce results which command general agreement. It is true that a subjective element creeps into the interpretation of disaggregated criteria, but in comparisons of total factor productivity or production functions these assumptions, judgements and approximations are implicit in the various stages of the calculation and, indeed, are often inevitable in view of the poverty of the statistical data. Second, since our future aim is to try to understand how the research-production cycle works in particular Soviet industries, aggregate estimates of technical level are much less useful than more specific measures of research, development, and diffusion; the latter provide a quantitative indication of strengths and weaknesses, which can be explored in future work.

The main ground that we cover is laid out schematically in Table 1.2 and the results of this work are analysed in detail by R. W. Davies in the following

[92] Unfortunately, Hanson could find no convenient way of working out a precise figure for the degree of dependence of particular industries on imported equipment for their investment plans.

chapter. A tick is placed in the appropriate column of Table 1.2 only if a substantial part of the case study is devoted to a particular mode of assessment. In most cases all these modes are considered, but not in detail. Often this is because the data are not available or because the measure is not relevant to the particular industry or technology involved; in other cases, we did not have the expertise or the time available to squeeze out inaccessible information. It can be seen from the table that measures of innovation and diffusion are common to nearly all the case studies. The gaps in column (5) are to some extent deliberate. Much of this evidence has been collected by Sutton and we could therefore spend our time more efficiently by looking for evidence of domestic innovation in Soviet sources in order to evaluate Sutton's conclusions. Transfer of space and military technology, if it has occurred, is of a kind that for obvious reasons cannot be investigated by us. The gaps in columns (1) and (7) are mainly due to lack of expertise and of accessible information that could be processed in a simple form. The lack of data on patterns of foreign trade (column (6)) is primarily due to the fact that this form of assessment is not relevant for all the case studies; in some cases (for example, space technology, military equipment and, to some extent, computers) trade has been inhibited by strategic embargoes; in other cases, either the volume of trade is very small or it cannot be broken down in such a way as to reveal anything relevant about technological level. Thus, following Nabseth and Ray, 'We have adapted the analysis to the nature of the problem and the quality of the data rather than insisting that a particular methodology should be followed at all costs.'[93]

Before concluding this chapter it is necessary to be more self-critical about the significance of results produced by the methods described above. The *manifestations* of technological level and change that are identified and measured by these approaches are in some ways more ambiguous than those that are the concern of aggregate economic assessments or of engineering assessments. While aggregate economic estimates are concerned with overall increases in productivity and engineering assessments compare the operating characteristics of equivalent items of equipment, our approach largely focuses on *comparative patterns* of technical development and diffusion. There is a danger of ethnocentricity in this. What is to be the standard of comparison? It is generally accepted, for example, that the United States exhibits all the characteristics of a technologically advanced society in their most mature form. It is tempting to carry this assumption a stage further and to *define* advanced technology as that pattern which prevails in the United States; the level of Soviet technology in particular industries could thus be assessed in terms of how far it matched the American pattern. This may be true in many cases but it could be misleading in others. The approach raises two particular difficulties. First, it leads one to ignore or diminish the significance of indigenous technology, where that technology is not diffused widely in Western countries. The paradigm case here might be that of deciding whether modern anaesthetic techniques are more 'advanced' than acupuncture; a similar problem crops up with some Soviet manmade fibres (for example,

[93] L. Nabseth and G. F. Ray, *op. cit.* (1974), p. 15.

Table 1.2 *Areas and modes of assessment*

Field	Author	Level of research (1)	Comparative dates of first commercial production of key technologies (2)	Rate of diffusion of key technologies (3)	Overall pattern of diffusion in branches of industry/technology (4)	Technology transfer (including import of key processes) (5)	Patterns of foreign trade in products (6)	Comparative parameters of key items of equipment (7)
High voltage electric power transmission	W. G. Allinson							
Chemical industry:								
General assessment	R. Amann	√	√	√	√			
Polymer research	A. Holt	√	√	√	√	√	√	
Computer technology	M. Cave		√	√	√			√
Iron and steel industry	J. Cooper		√	√	√			
Machine tools:								
General assessment	M. J. Berry				√		√	√
Numerically controlled	J. Cooper		√	√	√			
General purpose	M. R. Hill							√
Passenger cars	M. R. Hill				√			√
Military technology	D. Holloway		√	√	√			√
Rocketry	M. Kocourek		√	√	√			√
Process control instruments	Z. A. Siemaszko		√	√				√

'Enant'/Nylon-7), which apparently have good technical characteristics but are not produced in large quantities in Western countries. A second objection to the above approach is that it assumes that all countries are striving towards a uniform pattern of development and does not take into account the economic appropriateness of particular patterns of technology, which are influenced by factor endowment and domestic consumption. If we take the chemical industry as an example, the optimality of the pattern of output at a given point in time is, to some extent, dependent upon the relative costs of inputs and the ability of other industries to use sophisticated chemical products. The fact that the Soviet Union produces large quantities of fertilisers and simple inorganic chemicals relative to modern synthetic materials could be seen in relation both to the substantial deposits of minerals in the USSR and to the conservatism of Soviet manufacturing industry in substituting synthetic materials for traditional materials. Under these circumstances it is difficult to judge how far the pattern of innovation and diffusion in the USSR is determined by technological backwardness and how far by conscious economic choice.[94] It is true that one of the explicit aims of Soviet economic policy is self-sufficiency and a desire to catch up with and eventually overtake the United States in production capacity and technology. But even if Soviet pronouncements were accepted at face value, it seems unlikely that these objectives would be reflected in total emulation. Thus, to some extent, it is necessary to look at the Soviet pattern of technical development and diffusion within its *own* economic context.

Having obtained estimates of technological level, subject to the limitations described above, do these results cast any light on the relative efficiency of the Soviet R and D system? This is obviously the question that one most wants to ask, yet it is much more difficult to answer than might appear at first sight. How this question is answered will depend on how far the USSR is thought to be at an equivalent point to Western countries in terms of its phase of economic development, because it is only when the phases of development are equivalent that one can talk about inherent strengths and weaknesses. What criteria could be used to define the phase of development? Obviously, one could not use the 'level of technology' for this purpose because it would produce a tautologous argument in which it would be impossible to demonstrate that the USSR had any systemic weaknesses that would not be eradicated in the natural course of development.[95] Possibly, indices such as the proportion of the working population engaged in manufacturing industry, together with some per capita output figures for a sample of basic industries, might be used as appropriate criteria. This would provide a more meaningful perspective of development but it would still be far from accurate because it would conceal some important institutional factors. The process whereby the rapid expansion of these basic industries in the USSR took place, tended to impose an opportunity cost; it led

---

[94] See discussion of Bergson's work, above.

[95] I.e., this argument would assert that the level of technology in the USSR was inferior because of its relatively earlier phase of economic development (defined in terms of level of technology).

to a skewed pattern of development, within which the R and D establishment became accustomed to absorbing foreign technology rather than initiating independent innovation and it accommodated itself to an economic and administrative framework appropriate for forced industrialisation. What might now appear as an approximately equivalent phase of development to Western countries in terms of production and employment indices is not so in terms of the way that economic institutions operate in the USSR. The Soviet Union is still, to some extent, a prisoner of its past. Thus, to make comparative judgement about R and D efficiency (as opposed to straightforward measures of 'technological level') between countries with radically different patterns and time scales of economic development raises issues that are almost unanswerable and certainly unmeasurable.

Is a superior level of technology, as it is measured in the various chapters of this volume, necessarily a good thing? Since the end of the 1960s there has been a noticeable shift in attitudes on this question, which is reflected in a recent report published by the OECD.[96] The report argues that due to the declining quality of life in Western countries and the inadequate development of the services sector relative to the manufacturing sector, 'economic growth, *per se*, is no longer a sufficient overall objective.'[97] More active governmental direction is needed within the guidelines of 'a more comprehensive and socially oriented framework than in the past'.[98] If it is accepted that qualitative factors are important it implies that conventional aggregated and disaggregated measures of technical level and development give an incomplete picture. The Soviet economic planning system, in theory at least, possesses both the outlook and the controls that are necessary to maintain a satisfactory balance between economic growth and the quality of life. This does not mean, in practice, that environmental and other problems do not exist in the USSR. They clearly do; but the relative position would need to be weighed in the balance. This consideration lies outside the scope of the present study.

---

[96] *Science, Growth and Society: A New Perspective*, OECD, Paris (1971).
[97] *Ibid.*, p. 89.
[98] *Ibid.*, p. 92.

2. The technological level of Soviet industry: an overview

Introduction

Our studies of the technological level of Soviet industry were undertaken with two major objectives. First, we hoped to improve our knowledge of the comparative international position of the Soviet economy. The slackening over the past 20 years of the rate of growth of Soviet national income and particularly of Soviet industrial production is well established. But assessments of the interrelationship between economic and technological development have been less successful. Attempts to determine the role of 'technical progress' and 'technological change' in Soviet economic growth have produced uncertain and inconsistent results (see pp. 8–23 above), and comparisons of Soviet and Western factor productivity have rested on this shaky foundation. In view of these uncertainties we concluded that it would be useful to examine technological level as such, in order to establish whether in these terms Soviet industry has in the past 20 years been moving towards or overtaking its technologically more advanced rivals. Has the slackening in the rate of growth of Soviet industry been accompanied by, and perhaps caused by, an improvement in its comparative technological level, or has the gap between Soviet and Western technological levels widened?

Second, we hoped to find out the particular respects in which Soviet technology was advanced or backward as compared with that of Western countries, to establish some patterns of variation between the levels of Soviet and Western technology.

For both these objectives it was obviously important that we should select industries and technologies which were representative, at least in the sense that they reflected the different processes at work in the Soviet economy. We therefore included both high priority and apparently successful industries (iron and steel, electric power, space, weapons), and apparently unsuccessful industries which have received some degree of priority (chemicals, computers). We shall eventually include one or two low priority (e.g., consumer) industries, omitted so far through lack of time. We included traditional industries with a relatively low R and D content (iron and steel, electric power), intermediate industries (chemicals, machine tools) and science-based industries (weapons, space, computers). We examined both whole branches of industry and, where sufficient expertise was available, groups of products and even individual products (numerically controlled machine tools, process control instruments, tanks).

The development of a new technology involves an elaborate cycle of activities from research to commercial production; and the technological level of an economy in respect of a particular industry or group of products depends on the extent of the diffusion of the more advanced technology. In the present project we have attempted to establish, wherever possible, the Soviet comparative position for each industry or group of products at each of the main stages of what is known in the USSR as the 'research-production cycle', as shown below.

1. Research.
2. Experimental development to the prototype or pilot plant stage.
3. First commercial production (i.e., for general sale or use)—i.e., innovation in the sense in which the term is used by economists.
4. Diffusion: here we were concerned with the general technological level of (a) the production processes of the industry itself; (b) the products it was currently manufacturing; (c) the stock of the products of the industry in the economy at large.

We were equally concerned, of course, to establish changes over time, and initially hoped to assess each stage of R/D/innovation/diffusion for a series of bench-mark years at five year intervals: 1950, 1955, 1960, 1965, 1970 and 1975 (or the latest year before 1975 for which information was available).

It is hardly surprising that it rarely proved possible to keep to this neat programme. Sometimes it was inappropriate for a particular technology; more often, available information was inadequate. In particular, we usually lacked sufficient information to distinguish reliably between stages 2 and 3; and for most industries we did not have time to make an adequate examination of stage 1. On the other hand, several authors decided it would be misleading to confine themselves strictly to the period since 1950, and in the case of the industries for which they were responsible (e.g., space rocketry, chemicals) the study is taken back in outline to the inter-war period. For all the industries and products, we have some information over time about stages 2 and 3 combined, and about stage 4, and have therefore been able to make an assessment of the changes in the comparative Soviet level in respect of both innovation and diffusion. In view of the rapidity of technological change in both the Soviet Union and the West, it is important to note that the last year for which comparative information was available when our research was completed was almost invariably 1973. In the present survey chapter, the findings of similar studies carried out by other scholars (for example, by H. Vogel and J. Sláma, and by the Rand Corporation) have been considered together with our own results; they are generally consistent with them. The information in the present chapter has been derived from the case studies published in this volume, except where otherwise stated.

Wherever possible, comparisons have been made with five major industrialised countries: USA, UK, France, FRG and Japan. We have thus included both long established industrial countries like the UK and recently industrialised countries like Japan (whose industrialisation took place at roughly the same time as that of Russia and the Soviet Union). These countries suffered different degrees of devastation during the Second World War, though all of them suffered less than the Soviet Union. For individual products and processes, we have usually restricted comparisons to the United States or the UK, simply because of lack of expertise and time. We have tried to resist the temptation to assess Soviet technology solely in terms of the *best* achievement of capitalist countries, but may not always have succeeded.

The problem of what criteria to use to assess 'technological level' proved less agonising in practice than in principle. We have lacked resources and information to attempt anything like the multiple criteria which have been successfully used to assess the performance of US and Soviet turbine engines,[1] and have normally been restricted to two or three performance criteria for each product or process. But to our surprise technologists generally agreed among themselves that a certain product or process was unambiguously 'technologically advanced' at a particular date, and that one or two specified performance criteria for that product or process were a satisfactory indicator of its technological level. But such criteria do not take sufficiently into account the *quality* of Soviet products, in terms, for instance, of their reliability or durability. In order to assess the weaknesses of Soviet industry in this respect, and the success of the heroic efforts of the Soviet authorities in recent years to overcome them, specific comparisons were undertaken between certain Soviet machine tools and passenger cars and their Western equivalents (chapter 11). These costly efforts to improve quality might be a further factor contributing towards the slackening in growth rates.[2]

In most of our case studies, some attempt has also been made to assess the contribution of foreign technology to the development of Soviet industry. We obviously considered the quantitative importance of foreign machinery or know-how for each industry or group of products, where the information was available, and it has also been important, for reasons adduced in chapter 1 above, to consider how far foreign technology was crucial to a particular Soviet development. Our excursions into this subject were usually undertaken as background information rather than being serious independent research efforts.

A failure or a success of Soviet industry in developing or diffusing a particular product or process does not, of course, necessarily indicate inefficiency or efficiency. Thus in armaments, technological progressiveness does not always imply military effectiveness; yet all that has been undertaken in the present project are 'side-by-side' comparisons of technological level

---

[1] A. J. Alexander and J. R. Nelson, *Measuring Technological Change: Aircraft Turbine Engines*, Rand R-1017-ARPA/PR, Santa Monica, May 1972.

[2] In certain industries, notably engineering, improvements in quality were, however, immediately and even disproportionately reflected in price and hence in output series (see, for example, A. S. Becker in *Soviet Studies*, Vol. XXVI, 1974, pp. 363–79).

rather than 'face-to-face' comparisons of effectiveness. Technologically advanced does not, of course, always mean economically most effective. The availability of natural and human resources in a particular economy, and its degree of development, will all influence the effectiveness of a particular technology. The continued production of steel by open hearth furnaces *might* be economically justifiable if plentiful scrap were available at low cost and the open hearth process were very efficient. Automation or computerisation at a high capital cost per unit of output would not be justified if unskilled or clerical labour were available sufficiently cheaply. Nor can the inclusion in the official plan of particular kinds of advanced technology be taken to prove that they are economically effective; *per contra*, failure to achieve a particular government programme does not in itself demonstrate economic inefficiency. But some important new technologies are both capital- and labour-saving, and in such cases failure to introduce them may well indicate a weakness in the economic mechanism.

The material assembled in the present volume frequently confronts us with such issues, turning largely on the extent to which technological 'lag' is due to 'systemic weakness' rather than to economic or national circumstance. Tantalisingly, perhaps, we decided that at this stage of our work (though this Ranke-esque self-denying ordinance has been somewhat unevenly observed by the contributors) we would not bewail, justify or explain the technological level, or rather patterns of technological levels, attained by the USSR, but rather confine ourselves to assessing what these patterns are. We intend to turn our attention to an explanation of these patterns of innovation and diffusion in a further volume.

Research

Most assessments of particular branches of research have concluded that the Soviet contribution to knowledge is less substantial than that of the United States. In terms of publications, the USSR was responsible for 15·6 per cent of physics abstracts in 1961 and 20·7 per cent of chemistry abstracts in 1965, as compared with 31·6 and 28·5 per cent for the United States.[3] Kapitsa guessed very roughly in 1966 that the Soviet Union produced only half the number of scientific papers produced in the United States, although it had nearly the same number of scientists.[4] Similar conclusions have been reached in studies concerned with the quality as well as the quantity of output of particular branches of research. On the basis of Soviet publications one author concludes that in semiconductor thin film research the USSR is 'generally a few years behind the United States, with the exception of some narrow aspects of materials in which it is ahead'.[5] According to another survey of publications, in

[3] D. J. Price, in *Journeys in Science: Small Steps, Great Strides*, 1967; at this time the United States population was approximately five sixths that of the USSR.

[4] *Komsomol'skaya pravda*, 20 January 1966; Kapitsa reported that according to American studies the USA produced one third and the USSR one sixth of the world's 'scientific output'.

[5] G. Rudins, *Soviet Research in Semiconductor Thin Films: a Survey*, Rand R-1181-ARPA, Santa Monica, February 1973, p. v; the author does not, however, explain how he reached this conclusion.

1971 Soviet chemical laser research was frequently 'heavily bogged down in theoretical analyses, while their eyes are turned on the West, eagerly awaiting experimental developments'. The author reported, however, 'steady progress with a prospect for sharper focussing on truly important problems in the near-term future'.[6] In some important fields, moreover, Soviet research is reported to be in advance of the West: thus the Soviet research effort on high current electron beams is 'more intensive ... than in the West', and has made a 'number of significant advances'.[7] In one field directly relevant to our own project, chemical fibres, several foreign specialists have praised the quality of Soviet research (see p. 284 below).

Our present project has not primarily been concerned with Soviet research as such, but the small amount of evidence we have assembled on the whole fits in with these assessments. In a pilot study, the comparative level of polymer chemistry was examined by a citations count (see Table 6B.2, p. 327 below). It was found that in 1973 23·9 per cent of citations in the main Soviet polymer journal were from Western publications, while only 7·3 per cent of citations in an equivalent Western journal were from Soviet publications (these figures have been adjusted to take account of differences in population). East European journals also rated the West more highly in applied and considerably more highly in fundamental polymer research. The relative level of Soviet polymer science has, however, improved since 1960 by this method of reckoning. A wider analysis of citations undertaken in the Soviet Union (summarised on pp. 293–5 below) reaches similar results, but attributes the low level of Soviet citations in the West to the isolation of the USSR from the main information streams rather than to the lower level of its research. In chapter 6 below, the small number of Nobel prizes received by Soviet scientists (a half of one prize in the period 1901–72 in the case of chemistry) is adduced as evidence that at the highest level of science there have also been relatively few outstanding Soviet contributions (pp. 290–91).

All this does not, however, entitle us to do much more than draw the very general conclusion that Soviet scientific research tends to be smaller in quantity (in terms of equivalent populations) and lower in quality than in the United States and some other Western countries. No definite conclusions can be drawn about the relative level of fundamental and applied research within each branch; in chemical lasers, fundamental research is reported to be more advanced; in polymer chemistry, applied research (see p. 325 below). No systematic attempts have yet been made to examine in comparative terms the progress of Soviet research over time.

Prototype and first commercial production

The iron and steel industry (chapter 3 below) is correctly described as 'traditional', both because it is a long established industry and because the cost

---

[6] Y. Ksander, *Soviet Chemical Laser Research*, Rand R-921/1-ARPA, Santa Monica, November 1971, p. ix.
[7] S. Kassel and C. D. Hendricks, *Soviet Research and Development of High-Power Gap Switches*, Rand R-1333-ARPA, Santa Monica, January 1974, pp. 1, 6–7, 30.

of research and development is a relatively small proportion of total costs. This does not, of course, mean that R and D in the industry is not important to its technology; and in the USSR there has been a long tradition of native iron and steel R and D, dating back to the beginning of this century. In the USSR the industry has always, with the exception of a brief period under Khrushchev, been given very high priority, and the high level of its technology has long been a justifiable source of Soviet pride. As early as 1930 the new iron and steel complex at Magnitogorsk was intended to be technologically more advanced than the most modern United States plant. This, then, is an industry which might be expected to be in the forefront of world technology.

In the manufacture of iron and crude steel by the long established processes the USSR has successfully developed very large blast and open hearth furnaces (here size is generally agreed to be a reasonable indicator of technological level, as it can bring about reduced costs and improved product quality). In iron smelting, the largest Soviet blast furnaces were consistently the largest in the world until 1970, and during the past 20 years the USSR has pioneered a number of improvements (including higher top gas pressure, substitution of natural gas for coke, evaporative cooling). Although in 1969–73 it temporarily lost its leading position owing to the construction of larger furnaces in Japan, its 5,000 m^3 furnace completed at the end of 1974 is again the largest in the world. (See pp. 88–91 below.)

In steelmaking, although the Soviet industry has lagged in scrap preparation methods, it has remained generally in a leading position in traditional methods of production: the 900-tonne open hearth furnaces completed by 1965 are the largest in the world (see p. 93 below). In the past 20 years, however, in the major Western industries the traditional methods of steel-making in which the USSR has excelled have been successfully superseded by continuous casting of steel and oxygen steelmaking. At the stage of experimental development and pilot plant, the Soviet industry was successful in pioneering the continuous casting of steel more or less simultaneously with the Austrians: the first Austrian plant was installed in 1952, the first Soviet plant in 1953 or 1955. The USSR is now building, with German assistance, the largest continuous casting plant in the world. The oxygen steel process was first developed in Austria in 1952; the first Soviet plant went into operation in 1956, two years after the United States, which like the USSR had a well established and reasonably efficient traditional industry. Although between 1960 and 1970 the maximum size of new oxygen convertors was consistently lower than in most other industrialised countries, the size of the new 300-tonne convertor introduced in 1973 was exceeded only in the FRG. (See pp. 96–9, 102 below.) In summary, the USSR, ever since the 1940s, has been generally in a leading position in scaling-up traditional large-scale iron and steel processes; and it developed the new processes at approximately the same time as the United States, West Germany and Japan, though later than the Austrians.

The Soviet record has been much less impressive in the other major branches of steelmaking. While it was a pioneer in electroslag remelting, it tended to neglect other processes for the manufacture of alloy and high quality steel; and in most processes for the rolling of steel on which information is available the Soviet lag is considerable, though it may be diminishing (see pp. 104–8, 112–17 below).

The machine tool industry offers a second example of a more traditional industry in which the introduction of new technology has been given high priority by the Soviet authorities over a long period. We at first intended to compare systematically the characteristics of new Soviet machine tools with the equivalent new machines in Western countries over the period 1950–70: this proved, however, to be too complex and time-consuming a task to be undertaken within the framework of our present project. We accordingly confined ourselves, as far as experimental development and innovation are concerned, to numerically controlled machine tools. Partly through lack of information, we were not able to examine the technological parameters of this group of machines in detail. But sufficient information has been available to show a continuing lag behind the United States and the other major countries, which grew in the 1960s and has been somewhat diminished only in 1971–73 (see Table 2.1). The lead time between research and first prototype was substantially longer (eight years) than in other countries, perhaps because of the lack of open availability of United States technology, though (unexpectedly in terms of the usual stereotype of the Soviet system) the period of seven years between first prototype and commercial production was no longer than in the United Kingdom, and somewhat shorter than in Japan. Soviet failure to keep up with some recent major developments appears to have been due primarily to the weakness of the control systems; this in turn was a result of the backwardness not of the machine tool industry itself, but of the science-based electronics industry (see p. 164 below) (this also proved a crucial obstacle in several other Soviet industries).

The electric power industry provides our third example of a traditional high priority industry in which the Soviet Union has a long research and production

Table 2.1 *Research, development and innovation in numerically controlled machine tools*

	Reached by USSR in	USSR (+ in advance; − behind) in relation to:			
		USA	UK	Japan	FRG
Start of research	1949	−2	−1	+4	+6
First prototype	1958	−6	−2	=	=
Start of industrial production[a]	1965	−8	−2	+1	−1
First machining centre	1971	−12	(−10)	−5	−10
First third generation control system	1973	−7	(−5)	(−5)	(−5)
First use of computer for control	1973	−6	(−4)	−5	(−4)

Source: see Table 4.45 below.

Note: ( ) – estimate. a – 50 units or more per annum.

experience: in this industry we were able to undertake a case study of power transmission, of particular economic significance in the USSR (chapter 5 below). In power transmission, the higher the transmission voltage, the lower the losses on the line, as the voltage carried by a particular type of line can be taken as an approximate indication of technological level (assuming that long-distance transmission is economically justified). The three main types of power transmission, in chronological order, are high-voltage alternating current (HVAC), high-voltage direct current (HVDC) and, very recently, ultra-high-voltage alternating current (UHVAC) (the term 'ultra-high' is taken to refer to voltages above 1,000 kV). The Soviet Union has been notably successful in developing HVAC and moved into a leading position in world technology in the middle and late 1950s. It introduced a 400 kV line in 1956, four years later than the Swedes but many years in advance of the Americans; in 1959 it was the first country to introduce a 500 kV line; it introduced an experimental 750 kV line in 1967, two years after the first Canadian but two years before the first United States' line of similar voltages (see Fig 5.1 below). In two major recent developments the Soviet Union has also on the whole retained a leading position. First, much effort was devoted in the 1960s, in Sweden, the USSR and elsewhere, to the development of HVDC lines and the associated equipment for AC–DC conversion. A 200 kV DC line was constructed in Germany before the Second World War; in 1951, the USSR was the second country to construct a line of this capacity. Ten years later, the USSR took the lead: in 1962–5 it completed a second 800 kV (i.e. ± 400 kV) line; an equal capacity was not attained elsewhere until the USA completed a line of the same capacity in 1970. A 1,500 kV HVDC Soviet line has been projected for some years and equipment for its first stage is now in production; a 2,200 kV line is in preparation. But a 900 kV line is expected to be completed in Canada in 1976, and a 1,066 kV line in Mozambique in 1975 (see Table 5.4 below). The latter line is of particular significance, as it will use thyristor convertors, now favoured against mercury arc convertors by all Western companies except GEC: in this the Soviet Union lags by up to six years (see p. 220 below). Second, UHVAC transmission is being developed both in the USSR and in four Western countries: the USSR plans a 1,150 kV line, and lines of similar voltages are planned in the West, but relative progress cannot yet be assessed (see pp. 221–2 below).

In the chemical industry, an intermediate industry in the sense that it has a higher proportion of R and D in total costs than in the traditional industries, but less than in the high technology industries, a substantial effort has long been devoted to research, but the industry itself was, according to all Soviet accounts, relatively neglected until about 1958: since then it has been afforded a much higher priority. The evidence assembled in chapter 6 below about pilot plants and first commercial production reveals no diminution whatsoever in the lag behind the more advanced countries in the past 20 years. In the case of plastic materials (see Table 2.2), a *diminishing* lag in the last years of Stalin appears to have been replaced by an *increasing* lag since the mid-1950s; and in chemical fibres (see Table 2.3) no improvement can be observed. We must hasten to add that this may well be misleading. Some of the plastics introduced

Table 2.2 *Plastic materials: lag in introduction in USSR (years)*

	Plastics introduced in USSR:					
	Pre-1941		1947–52		Post-1952	
	Mean	Median	Mean	Median	Mean	Median
1. From *first* introduction in West	21·3	15·5	10·8	10	13·3	13
2. From *last* introduction in established industrialised countries[a]	11·5	6	−2·0	−4	5·3	6
3. From *last* introduction in recently industrialised countries[b]	−1·5	0	−2·6	−4	3·3	1

− indicates Soviet lead.

Source: Calculated from Table 6.19 below.

Notes: a – USA, Germany (or FRG), UK, France.   b – Italy and Japan.

Table 2.3 *Chemical fibres: lag in introduction of first commercial production in USSR*

	Pre-1941 Mean	1948–54 Mean	Post-1954 Mean
1. From *first* introduction in West	27·5	20·8[a]	19·8+[b] (n=5)
2. From *last* introduction in established industrialised countries	5·5	14[a]	6 (n=2)
3. From *last* introduction in recently industrialised countries	5 (n=2)	8·5[a] (n=4)	8·5+[c] (n=4)

Source: Calculated from Table 6.21 below.

Notes:
a – biased by case of acetate continuous filament; mean of other three cases is (1) 12; (2) 8·3; (3) 5·7 years.
b – three of the five items were not produced in the USSR by 1970.
c – two of the four items were not produced in the USSR by 1970.

in 1947–52 were produced only on a small scale, while plastics introduced since 1952 are much more technologically and commercially significant than those introduced earlier. Nevertheless, there is clearly a continuing and substantial lag. The Soviet Union alone in this entire group of countries has never been the original innovator of a major plastic material or chemical fibre (though the last British success was in 1937; the last French in 1941); and since 1958 for each new product it has been the last in the entire group to begin commercial production (see Tables 6.19 and 6.21 below).

Our findings for the science-based industries which have been established since the Second World War were similar. Chapter 8, which brings Richard Judy's valuable work on *computers* up to date, indicates that the lag of the

USSR has not been reduced since the 1950s in respect of computer hardware (see Table 2.4), though a substantial improvement has occurred in software, from a previous low level (see pp. 388–91 below). Computers have received a considerable degree of priority; and it is perhaps not surprising that for *industrial process control instruments* (chapter 7), where the priority is lower, the lag is even greater. One original Soviet innovation was the USEPPA pneumatic system, in 1964 the most advanced pneumatic system in the world. This provides an interesting example of an innovation resulting from industrial demand in circumstances of technical backwardness: it was developed as a substitute for electronic systems in view of the slow Soviet progress with semiconductor devices. But in 1972 USEPPA was superseded by a West German pneumatic controller. No new Soviet system, pneumatic or electronic, has been fully developed since 1964, whereas in Britain new systems are produced on average once in two years. If the earlier lead has now been lost in the case of pneumatic controllers, in electronic controllers the gap has widened, at least in relation to the United Kingdom (see Table 2.5). The continuing lag in

Table 2.4 *Prototypes or first commercial production of computers*

(a) *Lag of best Soviet computer behind best US computer (in years)*[a]

1955	4
1960	9
1965	14
1967	4
1970	7
1973	10

Source: See Fig 8.3 below.

Note: a – criterion: operations per second. BESM-6 has been included (see Fig 8.3 below).

(b) *Date of introduction of Soviet computer compared with US equivalent*

	US equivalent	Soviet model	Lag (in years)
Ural 4	1955	1962	7
BESM 6	1962	1966	4
Nairi 1	1960	1964	4
ES Series	1965–6	1972–3	6–8

Source: See Table 8.4 below.

Table 2.5 *First commercial production of electronic controllers*

	USSR	UK	Soviet lag (in years)
First electronic controller	1951	1949	2
First fully transistorised controller	none by 1974	1959	15+

both computers and control instruments is in turn largely due to the backwardness of the Soviet electronics industry.

As examples of 'high technology' industry, in which research and development amounts to a high proportion (usually 10 per cent or more) of the total cost of output, we examined civilian space rockets and ballistic missiles; both of these were given a very high priority by the Soviet authorities, the former for prestige and the latter for military reasons.

In the case of *space rockets* (chapter 10), thrust power has (with some qualifications) been taken as an indicator of the technological level of the launch vehicle. The USSR was clearly in the lead from the launching of Vostok in 1957, followed by Proton, until 1967 when the United States launched Saturn V. Saturn V has more than twice the thrust power of Proton, and nearly eight times the thrust power of Vostok (see Table 2.6). In the eight years 1967–75 no new Soviet rocket has been launched. Lead has given way to lag. In manned spacecraft, too, the Soviet Union has lost the leading position which it temporarily obtained in 1961. Apollo, first launched in 1968, is clearly a more advanced craft than Soyuz, first launched in 1967; and Salyut, the world's first orbital station, launched in 1971, is only one quarter the weight of Skylab, launched two years later, and is in many respects less sophisticated (see Table 10.7 and pp. 516–17 below). Both the Soviet Union and the United States have, however, secured important space 'firsts' in the whole period from 1957–73 (see Table 10.8 below). In any case, we are here, of course, comparing Soviet accomplishments with the best achievements of the rest of the world: no other power has attempted a space programme which has the slightest pretension to rival those of the United States and the USSR.

Other studies of Soviet high technology generally support these conclusions.

The level of Soviet weapons development is perhaps crucial to our assessment of the level of Soviet technology as a whole. This is, of course, an ultra-high-priority area. The extent of this priority is illustrated by the relation of the military to the civilian space programme: in chapter 10 below, it is shown that in the USSR military aspects predominate over civilian aspects of space research to an extent which has undoubtedly hindered the success of the civilian programme (see pp. 517–20 below). It is also commonly assumed that

Table 2.6 *Maximum thrust power of launch vehicles (in thousand kg)*

	USSR	United States	
1957	510	—	
1958	↓	179	
1959	600	↓	Soviet lead
1964	650	240	
1965	1800	1336	
1967–	↓	4048 ↓	United States lead

Source: See Tables 10.4 and 10.5 below.

the Soviet Union has been uniformly successful by international standards in the experimental development of weapons, while being less successful in technology which is not directly military; if this were true, the general lag in all civilian development might well be explained by the concentration of R and D on weapons development. It is therefore interesting to note that the studies undertaken in chapter 10 below have, with some exceptions, unexpectedly revealed a pattern not dissimilar to that in civilian industry.[8] In the case of medium tanks, the war-time T-34 was generally recognised to be superior to medium tanks produced in other countries, with the possible exception of the German Panther. The present situation of the USSR in relation to the United States, is, however, roughly as follows (based on Tables 9.12 and 9.13 below):

	USSR:USA
Protection	
Armour	equal
Nuclear, Biological, Chemical Protection	lead
Firepower (calibre)	equal after lag
Firecontrol	
Rangefinding	lag
Stabilisation	lead
Infra-red	equal after lag
Mobility	
Speed	lead
Cruising range	lag after lead
Power-weight ratio	lag
Nominal ground pressure	lag after lead

The tentative conclusion drawn in chapter 9 is that Soviet tank technology lost its wartime lead by 1950, regained it between 1950 and 1960, and then 'between 1960 and 1970 lost its marked superiority to a new generation of Western tanks' (see pp. 441–6 below).

The position was in some respects similar in the case of Intercontinental Ballistic Missiles (ICBMs). In the 1950s several striking examples occurred of Soviet 'firsts'. While, as in the case of the chemical industry, it is unclear whether these successes conclusively demonstrate that Soviet technology was more advanced at that time, they were certainly spectacular achievements (see pp. 486–7 below). In the 1960s, however, on available evidence the Soviet rate of innovation lagged behind that of the United States (see Table 9.34 below). The USSR has been unable to eliminate the United States' technological lead in several important areas in which more sophisticated technology is required. The accuracy of United States' ICBMs increased ten-fold in 11 years (1959–70), while in a period of 13 years (1962–75) the accuracy of Soviet ICBMs has increased only four- to six-fold, and has continuously lagged

[8] It should be emphasised that a side-by-side rather than a face-to-face comparison has been undertaken; the study therefore does not tell us about military effectiveness.

behind the United States in absolute terms (see Table 9.25 below); the Soviet lag is greater in multiple warhead technology than in the earlier types of ICBMs. The Soviet lag in these respects has, however, been compensated by the massive use of simpler technologies. In terms of yield and throwweight available per ICBM, the Soviet Union has been consistently ahead since the mid-1960s, and the SS-18, first tested in 1972, may, owing to its huge yield, be far more lethal than any other ICBM (see p. 468 below).

A Rand Corporation analysis of 28 Soviet turbine engines also deserves mention here. It concludes that after a period, up to about 1955, in which Soviet engines were on the whole technically more advanced (though with a shorter endurance) than those of the United States, American engines have been consistently superior to those of the USSR, and that the gap gradually widened between 1955 and 1971, the concluding date of the analysis. Soviet avionics also lags behind that of the United States, and the lag continued in the 1960s.[9] The situation in naval warfare, however, appears to vary considerably from technology to technology.[10]

The comparative level of Soviet experimental development and innovation thus varies considerably between the different technologies we have studied. There is a marked contrast between the traditional industries and the science-based or high technology industries. In traditional industries, as illustrated by iron and steel, machine tools and electric power transmission, the Soviet Union generally occupied a leading position by the middle 1950s (though it was backward in certain branches of technology such as steel rolling) and that position has on the whole not been lost in the period of 20 years covered by our study. In the science-based industries, from the long established but rapidly developing intermediate chemical industry to the new industries such as control instruments and computers, the considerable Soviet lag in the middle 1950s (particularly great in relation to the rapid technological advances taking place in these industries elsewhere) has not been reduced in the past 20 years: in the chemical and control instruments industries it may even have increased. There is also some evidence of a relative decline in the Soviet technological level in high priority high technologies: for instance, in launch vehicles, spacecraft and even certain weapons, such as military turbine engines.

Diffusion of New Technology

The only published attempt to examine at a general level the comparative change in Soviet and United States technology is Boretsky's study of the years 1940–62, published in 1966.[11] Although Boretsky's quantitative assessment of the rate of Soviet technological change was unsuccessful, for reasons adduced by Berliner and others (see pp. 20–22 above), a number of his individual indicators in our opinion provide very useful pointers to comparative changes

[9] R. Perry, *Comparisons of Soviet and US Technology*, Rand R-827-PR, Santa Monica (1973), pp. 24–34, 39; this study is discussed further in chapter 9, p. 410 below.
[10] See chapter 9, pp. 412–13 below.
[11] M. Boretsky, in *New Directions in the Soviet Economy*, Joint Economic Committee of the US Congress, part II-A, Washington DC (1966), pp. 133–256.

in the technological levels of the two countries. In Table 2.7, we summarise Boretsky's conclusions for those 7 of his 15 groups of indicators which in our opinion are unambiguously acceptable. If these indicators are representative, they confirm Boretsky's general conclusion that what he terms 'the rate of technological change' was faster in the United States than in the USSR during this period. In 1940–55, the United States outpaced the Soviet Union in four of the seven groups of indicators and equalled it in two others, being outpaced by the USSR only in one group. This was, of course, the period of the Second World War and its aftermath, in which the United States suffered no destruction of industrial capacity, while the USSR was devastated. The Soviet reconstruction of civilian industry in 1946–50, and further development in 1951–5, was to a considerable extent based on pre-war technology.[12] This reconstruction was extremely rapid: in 1940–55, even on the somewhat conservative estimates of Warren Nutter, the output of Soviet industry, in spite of war devastation, increased by 5·1 per cent per annum as compared with 5·2 per cent in the United States; in the last years of the period, 1950–5, after Soviet output had reached its pre-war level, it expanded much more rapidly than that of the United States.[13] There was thus a dichotomy between technological development and industrial growth when considered in relation to the United States.

From the middle 1950s, in terms of Boretsky-type indicators we begin to enter a new era. In the last seven years of Boretsky's study, 1955–62, which are also the first period of our own work, a definite if small improvement took place, as the second column of Table 2.7 shows: the relative Soviet performance in synthetic resins and plastics and in the installation of telephones in the economy improved, though the United States substantially improved its position in power turbine technology.

We have extended our calculation of Boretsky-type indicators to 1973, and as is shown in Table 2.8, the relatively more rapid Soviet technological advance on the whole continued.[14] In 1955–73 as a whole, although the lag in iron and steel technology continued and a lag developed in electric power transmission, the Soviet Union developed more rapidly than the United States in terms of electric power per worker and in both indicators for the composition of machine tool stock, as well as in plastics and synthetic resins and the number of telephones in the economy. Our comparison with three other major industrialised countries for the same period shows that in terms of these indicators technological advance was more rapid in the USSR than in the United Kingdom, less rapid than in the FRG, and much less rapid than in Japan. Comparing the USSR with the four countries taken together, the proportion of cases in which Soviet technology is developing more rapidly than or at the same rate as in the other countries rose in 1960–5 as compared with

[12] *Ibid.*, p. 149.
[13] G. Warren Nutter, in *Comparisons of the United States and Soviet Economies*, Part I, Washington DC (1959), p. 105.
[14] Table 2.8 includes all the indicators from Table 2.7, with the exception noted, and certain other indicators for which information has been available.

Table 2.7 *Some Boretsky indicators of 'technological change' in Soviet industry between 1940 and 1962 compared with the United States*

	Indicator Rate of growth of:	1940–55	Period 1955–62	1940–62
2	Consumption of electric power per production worker	=	=	=
4a	Maximum capacity of steam turbines for electricity production	+	−	=
b	Length of HVAC transmission lines (over 400 kV)	n.a.	+	+
5	Proportion of aluminium and magnesium in total basic metal consumption	−	−	−
6	Percentage of steel output by electric arc or O_2 process	−	−	−
7b	Percentage of metalforming machine tools in total stock of metalworking machine tools	−	−	−
c	Output of NC machine tools	nil	−	−
8a	Output of synthetic resins and plastics*	−	=	−
9a	Output of chemical fibres*	+	+	+
14a	Number of telephones in the economy	−	+	−
	Total** +	1	2	1
	−	4	3	4
	=	2	2	2
	Grand total	7	7	7

+ more rapid Soviet progress; − more rapid United States progress; = approximately equal progress.

General notes:
Other indicators suggested by Boretsky have been omitted for the following reasons (numbered as in Boretsky's table, *op. cit.*, pp. 156–9).
1. Percentage distribution of fuel consumed: we are not sure that a rise in the proportion of natural gas and oil in total fuel, or a decline in wood and peat, clearly indicates technological advance. Boretsky's case rests on the smaller resources required to extract natural gas and oil, but assumes infinite resource availability.
3. Mechanical power per production worker in industry: US data unreliable.
4. Changes in percentage of installed electricity generator by type: Boretsky himself reaches no definite conclusion about whether an increase in the proportion of hydropower as against thermal power represents technical advance.
10. Ratio of engineering and technical personnel to production workers in industry: too indirect an indicator of technological change.
11. Use of computers and punched card systems: no data available about change over time in this period.
12. Percentage distribution of freight transport by type: decline in proportion of railways not a clear indicator of technological advance, and only indirectly relevant to our present study of Soviet industry.
13. Percentage distribution of inter-city passenger transport by type: as for 12.
14. Number of telephones in the economy: included by us as providing information about the scale of the diffusion of the products of the telecommunications industry, but perhaps not strictly relevant.
15. Agriculture: use of machinery: not directly relevant to our present study.

Notes to Table:
* indicators 8 and 9 have been grouped as they are both part of the same industry.
** where one group has two indicators, the convention has been adopted that a + and a − are aggregated to an =, and that a + and an + are aggregated to a +, a − and an = to a −.

Table 2.8 *Comparative indicators of Soviet technological change, 1953–1973*

Rate of growth of:	1955–1960 USA	JPN	UK	FRG	1960–1965 USA	JPN	UK	FRG	1965–1970 USA	JPN	UK	FRG	1970–1973 USA	JPN	UK	FRG	1955–1973 USA	JPN	UK	FRG
1 Electricity consumed per person employed in industry and construction	+	–	–	+	+	+	+	+	=	–	–	–	–	–^{a}	–^{a}	–	–	+^{a}	–^{a}	–	+^{a}
2a AC transmission lines of 300 kV and above as % of total AC lines	+		+	+	+	–	–	=	–	–	–	–					–^{b}		–^{b}	–^{b}
b Output of nuclear power stations as % of total electricity output					+	–	+	–	–	+	+	–	–^{a}	+^{a}	+	+	–^{ac}	–^{ac}	+^{c}	–^{c}
3a O$_2$ steel as % of total steel output					–	–	–	–	+	+	+	+	+	+	–	+	–^{d}	–^{d}	–^{d}	–^{d}
b Continuously cast steel as % of total steel output					–	–	–	–	–	–	+	–	+	+	–	–	–^{f}	–^{e}	+^{e}	–^{e}
4a Metal-forming machine tools as % of total stock in machine-building and metal-working					–^{g}		+^{g}		+^{h}		–^{h}		+^{h}		=^{h}		+^{j}		+^{j}	
b Metalcutting machine tools 10 years old or less as % of total stock in machine building and metalworking																	+^{jk}		+^{jk}	
c NC machines as % of total m.c. machine tool output	+	–	=	=	–				+			+	+			+	+		+^{e}	+^{e}
5a Output of plastics and synthetic resins per capita	+	–	+	–	+	+	+	+	+	–	+	=	+	+	=	=	+	–^{m}	+	–
b Output of chemical fibres per capita	=	–	=	+	=	+	+	+	=	–	+	+	–	+	+	=	+	–	+	+
c Output of synthetic fibres per capita	=	–	–	–	+	+	+	+	–	–	–	–	=	+	+	=	+	–	=	–
d Output of synthetic rubber per capita	+	+	–	–	+	–	–	–	+	–	–	+	=^{a}	–^{a}	=^{a}	=^{a}	+^{b}	–^{b}	–^{b}	–^{b}
6 Number of telephones per capita	=	–	=	–	–	–	–	=	+	–	–	+	=^{a}	–^{a}	=^{a}	=^{a}	+^{a}	–^{a}	+^{a}	–^{a}
Number of groups in which USSR more rapid	3	0	1	2	4	2	4	1	3	1	3	2	2	3	3	2	4	1	3	2
Number of groups in which other country more rapid	0	3	1	2	1	4	1	2	1	4	2	3	2	2	2	2	2	5	1	4
Number of groups in which rate approximately equal	1	0	2	0	1	0	1	2	2	1	1	1	2	1	1	2	0	0	2	0
Total number of groups	4	3	4	4	6	6	6	5	6	6	6	6	6	6	6	6	6	6	6	6

+ USSR more rapid; – other country more rapid; = approximately equal development ± 1 percentage point

General Notes:
Two indicators in Table 2.7 (Maximum capacity of turbines for electric power generation; Proportion of aluminium and magnesium in total consumption of basic metals) have been omitted owing to the lack of adequate information.
The following indicators have been added: Output of nuclear power stations as a proportion of total electricity output; Metalcutting machine tools 10 years old or less as proportion of total stock of machine tools in machine building and metalworking; Numerically controlled machine tools as a proportion of total metalcutting machine tool output; Output of synthetic fibres; Output of synthetic rubber. Figures for increases in output (lines 5a, 5b, 5c and 5d) have been calculated in terms of increases in output per capita rather than the increases in total output included in Boretsky and in Table 2.7.

Sources:
See Appendix Table 2A.1

Notes:
a – to 1972.
b – 1960–70.
c – from 1965; these years have been selected because in 1960 the proportion in the USSR was very low and no nuclear power was produced in Japan or FRG.
d – from 1960.
e – from 1965.
f – 1965–70 (other years not available).
g – 1962–66 in USSR; 1963–68 in USA; 1961–66 in UK.
h – 1966–72 in USSR; 1968–73 in USA; 1966–71 in UK.
j – 1962–72 in USSR; 1963–73 in USA; 1961–71 in UK.
k – Soviet figures are for total stock.
m – to 1970.

1955–60, fell in 1965–70 and rose in 1970–3.[15] It may therefore be tentatively suggested that some decline in the relative Soviet position in the later 1960s has been followed by an improvement in 1970–3.

These relative improvements in technology coincided with a decline in the rate of growth of Soviet industrial production. Industrial production continued to grow more rapidly than in almost all other major industrialised countries, and as a result of the slackening of the rate of growth of United States industry since the middle 1960s the Soviet rate of growth has continued to be higher than that of the United States. But the gap between Soviet growth rates and those of several other countries has certainly narrowed. In general it might be argued on this basis that the relative improvement in Soviet technology and relative slackening of its industrial growth rate has produced a certain convergence in the economic behaviour of the Soviet Union and other countries. Notable exceptions are Japan, which has outpaced the Soviet Union and all other countries both in technology and in production, and the United Kingdom, which has lagged badly behind the other major countries, including the Soviet Union, in both respects.

In spite of this 'convergence', the Soviet pattern of diffusion of new products and processes is substantially different from that in the industrialised capitalist countries. This is already apparent in the statistics for industrial production. During the past 20 years, some major traditional industries continued to expand rapidly in the Soviet Union, while in the other countries their rate of growth slowed down considerably. Thus, throughout the period from 1950 to 1973, Soviet production of crude steel per capita expanded much more rapidly than that of the United States and the United Kingdom,[16] and the output of cotton yarn per capita steadily expanded although it fell in the other major countries: Soviet output of cotton yarn per capita is now higher than that of Japan, the UK and the FRG and is rapidly approaching that of the United States (see Appendix Table 2A.1).[17] This reflects a more conservative pattern of industrial production in the Soviet Union: plastics and artificial or synthetic fibres have replaced steel and cotton much more slowly than elsewhere.

A similar pattern has appeared within each of the industries we have studied. In Table 2.10, the rates of diffusion of new technologies within industries are compared, using criteria similar to those developed by Nabseth and Ray in their study of the capitalist countries. In the steel industry, production by traditional methods continued to increase in the USSR even after the introduction of oxygen smelting and continuous casting, while in the other industrialised countries the new processes have tended to drive out the old. In

---

[15] The summary figures are as follows:

	1955–60	1960–5	1965–70	1970–3
USSR: more rapid	6	11	9	10
USSR: equal	3	4	5	6
USSR: less rapid	6	8	10	8
Total number of groups	15	23	24	24

[16] In Japan, however, per capita output increased so rapidly that by 1973 it was 208 per cent of the Soviet level, as compared with only 38 per cent in 1950.

[17] In Japan, the output of cotton yarn per capita rose rapidly in the 1950s but has declined since the middle 1960s.

Table 2.9 *Rate of growth of industrial production: annual average* (in %)

	1951–60	1961–5	1966–70	1971–4
USSR: official	11·89	8·6	8·5	7·5[1]
Greenslade-Robertson[a]	9·8[2]	6·9[2]	6·8[2]	5·6[2,b]
USA	3·9	6·3	3·6	4·0[1]
FRG	9·5	5·7	6·0	2·9[1]
France	6·1	5·2	6·4	5·3[1]
UK	3·1	3·5	2·1	1·5[1]
Japan	14·3[3]	11·6[3]	15·4[3]	6·0[1]

Sources:
Calculated from *Narodnoe khozyaistvo SSSR v 1972 godu* (1973), p. 91, except in the following cases:
1 Calculated from *UN Monthly Bulletin of Statistics*, July 1975, pp. 24–32.
2 *Soviet Economic Prospects for the Seventies*, Washington DC (1973), pp. 271, 278.
3 Calculated from *Economic Statistics Annual—Japan*, Tokyo (1965), p. 223; (1974), p. 217.

Notes:
a – this index is calculated in 1968 weights; unlike the Soviet official index, which is for gross industrial production of all industries including defence industries, it is for the value of final production and covers civilian production only.
b – preliminary, 1971–2 only.

terms of the Nabseth-Ray criterion for measuring diffusion of new technology, it took the USSR 16 years (1956–72) before the oxygen process amounted to 20 per cent of all production, as compared with between 2 and 12 years in the other major countries.[18] As is shown in Table 2.10(a), each stage of diffusion of oxygen steel (from first installation to 5 per cent and from 5–20 per cent) was slower than in the other four countries. J. Sláma and H. Vogel have shown that this result was contrary to the plans of the Soviet authorities.[19] A similar pattern may be observed in the relationship between the production of artificial fibres and the new synthetic fibres. Although the total production of chemical fibres expanded more rapidly in the USSR than in other countries between 1955 and 1973 (see Table 2.8), the Soviet rate of diffusion of synthetic fibres was distinctly slower; in particular, it took 11 years for synthetic fibres to increase from 10 to 33 per cent of total production, as compared with 5–8 years in other countries (see Table 2.10(c)). It is shown in Table 2.11 that in 'comparable' years, when Soviet and United States synthetic fibre production were similar, production of the longer established artificial fibres continued to grow much more rapidly in the Soviet Union than in the United States at a similar period (United States output of artificial fibres eventually levelled out at about 600,000 tonnes). While there are some exceptions (see Table 2.10(g)), this slower pattern of diffusion is characteristic of the chemical industry generally.

[18] For the other countries see (Eds.) L. Nabseth and G. F. Ray, *The Diffusion of New Industrial Processes*, Cambridge (1974), p. 17.

[19] See their papers in *Forschungsbericht 1974*, Osteuropa Institut, Munich, 1975, pp. 129–56, where there is an interesting discussion of what they term the 'conservative' Soviet pattern of diffusion, and in *Jahrbücher für Nationalökonomie und Statistik*, Vol. CLXXXVII, 1973, pp. 245–61.

A similar pattern with some variations appears in some of the sub-branches of industry examined in this volume. In electric power transmission, although the USSR was ahead of the USA, the UK and FRG in the central development of HVAC lines at 300 kV and above in the late 1950s, by 1970 the proportion of 300+ kV lines (measured in km) was slightly lower than in the United States and FRG and much lower than in the UK (see Table 2.10(f)).

Certain exceptions may be found to the general conservative pattern. The production of numerically controlled machine tools, which lagged behind that of other industrialised nations in the 1960s, has risen extremely rapidly since 1970, greatly outstripping the rate of increase elsewhere; and here the rate of diffusion has since the late 1960s been more rapid than in other countries (see Table 2.10(g)). Group Technology, a system for identifying and bringing together new components into families in order to achieve economies of scale, pioneered in the USSR, was diffused very rapidly in Soviet engineering factories between 1953 and 1965, certainly much more rapidly than in any other country. Since the abolition of the system of regional economic councils (*sovnarkhozy*) in 1965, however, the diffusion of Group Technology has slowed down considerably, while continuing to grow substantially in other major countries.[20]

Space rockets are also in part an exception to the general rule. A comparison of the number of launchings by the Soviet Union and the United States provides some indication of the 'output' of this sophisticated activity. Annual Soviet launches, lower than those of the United States until 1967, have exceeded them by increasing amounts since that year; the cumulative total of Soviet launches has also exceeded that of the United States since 1972. In terms of the weight launched, the Soviet total has always exceeded that of the United States: in the six years 1968–73 the USSR launched an average of 355 tonnes annually compared with a United States figure of 190 tonnes (calculated from Tables 10.5–10.7 and Fig 10.1 below). This is because the most frequently used Soviet rocket, Vostok (61 per cent of all launches) has several times the thrust power of Atlas and Thor, the most frequently used US rockets (72 per cent of all launches); Saturn V, with six times the thrust power of Vostok, designed specially for flights to the moon, was used on only 13 occasions. In these terms, then, the Soviet Union has overtaken the United States in space activity. But, in manned spaceflight, the most complex space accomplishment, the number of US flights has exceeded, and the weight of spacecraft launched greatly exceeded, those of the Soviet Union. Some important reservations about the use of number of launches or weight launched as an indicator are, however, suggested in chapter 10.

In the pattern of distribution of launches between rockets of different generations and thrust powers, the normal differences in diffusion of new technology between the United States and the Soviet Union are not quite so marked. Both countries follow a 'conservative' pattern in the sense that they have continued to use, until the present, rockets developed in the late 1950s.

---

[20] It is hoped that T. J. Grayson's study of Group Technology will be included in our second volume.

Table 2.10 *The rate of diffusion of new technology in the USSR and other countries*

(a) *Oxygen steelmaking*

	First industrial installation year	Year output of O₂ steel as proportion of total output reached:		Number of years between:		
		5%	20%	First indl. instn. and 5%	5% and 20%	First indl. instn. and 20%
USSR	1956	1966	1972	10	6	16
USA	1954 −2	1962 −4	1966 −6	8	4	12
Japan	1957 +1	1960 −6	1962 −10	3	2	5
UK	1960 +4	1963 −3	1965 −7	3	2	5
FRG	1955 −1	1962 −4	1966 −6	7	4	11

Sources: See pp. 96–8 below.

(b) *Continuous casting of steel*

	First industrial installation year	Year output of continuously cast steel as proportion of total output reached:		Number of years between:		
		1.5%	5%	First indl. instn. and 1.5%	1.5% and 5%	First indl. instn. and 5%
USSR	1955	1966	1972	11	6	17
USA	1962 +7	1967 +1	1969 −3	5	2	7
Japan	1960 +5	(1966) =	1970 −2	6	4	10
UK	1958 +3	1966 =	1974 +2	8	8+	16
FRG	1954 −1	1965 −1	1968 −4	11	3	14

Sources: See pp. 102–3 below. ( ) – estimate.

(c) *Synthetic fibres*

	Year first produced commercially (nylon)	Year output of synthetic fibres as a proportion of total chemical fibres output reached:			Number of years between:		
		10%	20%	33%	First prodn. and 10%	10% and 33%	First prodn. and 33%
USSR	1948	1962	1966	1973	14	11	25
USA	1938 −10	1951 −11	1954 −12	1959 −14	13	8	21
Japan	1942 −6	1958 −4	1960 −6	1963 −10	16	5	21
UK	1941 −7	1957 −5	1960 −6	1964 −9	16	7	23
FRG	1941 −7	1958 −4	1961 −5	1964 −9	17	6	23

Sources: See pp. 286, 309 below, and R. Amann, *The Soviet Chemical Industry; its Level of Modernity and Technological Sophistication*, CREES Discussion Paper RC/C11, Birmingham (1974), pp. 98–100.

(d) *Polyolefins*

	Year first produced commercially		Year output of polyolefins as a proportion of total plastics output reached: 15%		Number of years between first production and 15%
USSR	(1953)a		1970³		17
USA	1941	−12	(1956)²	−14	15
Japan	1954	+1	1963¹	−7	9
UK	1937	−16	(1955)	−15	18
FRG	1944	−9	1965¹	−5	21

Sources:
1 Calculated from *The Chemical Industry 1969/70*—supplement, OECD, Paris, p. 33.
2 Estimate based on known output of low density polyethylene (*Studies in petrochemicals*, UN, Vol. 1 (1966), p. 472).
3 Derived from appendix Tables 6A.9 and 6A.10, below.

Note: a – estimate—polyethylene first produced in 'early 1950s' (see p. 278 below).

(e) *HVAC transmission lines*

	Year first 300 kV line¹		Year first 500 kV line¹		Year first 750 kV line¹		Year lines over 300 kV as a proportion of total lines (over 100 kV) reached:²			
							5%		10%	
USSR	1956		1959		1967		1960		1970	
USA	1954	−2	1965	+6	1969	+2	1966	+6	1970	=
UK	1963	+7	−	+	−	+	1966	+6	1970	=
FRG	(1955)	−1	−	+	−	+	1971	+11	−	+

Sources:
1 See pp. 208–9 below; UK and FRG—*The Situation and Prospects of Europe's Electric Power Supply Industry*, UN, Geneva, various years.
2 See p. 225 below, and *Survey of Electric Power Equipment*, OECD, Paris, various years.

(f) *Nuclear power*

	Year first commercial power station		Year output of nuclear power as a proportion of total electricity output reached:						Number of years between:		
			0.5%		1.0%		2.0%		First stn. and 0.5%	0.5 and 2.0%	First stn. and 2.0%
USSR	1954		1971		1973		(1975)		17	4	21
USA	1957	+3	1967	−4	1970	−3	1971	−4	10	4	14
Japan			1970	−1	1970	−3	1971	−4		1	
UK	1956	+2	1959	−12	1959	−14	1962	−13	3	3	6
FRG	1961	+7	1967	−4	1969	−4	1970	−5	6	3	9

Sources:
Proportion of electricity produced by nuclear power stations calculated from *UN Statistical Yearbook*, various years.
Year first commercial power station—UN, *The electric power situation in Europe and its future prospects, 1963–65*, New York (1967), pp. 71–2.

(g) *NC machine tools*

	Year first prototype	Year output of NC mc. ts. reached:		Number of years between:		
		50 units a year	1% of total mc. t. output	Prototype and 50 units	50 units and 1%	Prototype and 1%
USSR	1958	1965	1971	7	6	13
USA	1952 −6	1957 −8	1965 −6	5	8	13
Japan	1958 =	1966 +1	1973 +2	8	7	15
UK	1956 −2	1963 −2	1968 −3	7	5	12
FRG	1958 =	1964 −1	— ++	6	9+	15+

Sources: See pp. 172–3, 194–5 below.

Table 2.11. *Production of artificial and synthetic fibres in 'comparable' periods: USA and USSR (in thousand tonnes)*

	USSR			USA	
	Artificial fibres	Synthetic fibres		Artificial fibres	Synthetic fibres
1965	330	77	1951	587	77
1966	362	96	1952	515	96
1967	395	116	1953	543	112
1968	424	130	1954	492	129
1969	441	142	1955	572	172
1970	456	167	1956	521	182
1971	473	203	1957	517	234
1972	507	239			
1973	543	287	1960	466	307

Sources:
1965–71: See R. Amann, *The Soviet Chemical Industry, its Level of Modernity and Technological Sophistication,* CREES Discussion Paper RC/C 11, Birmingham (1974), pp. 98–100.
1972–3: *Narodnoe khozyaistvo SSSR v 1973 godu* (1974), p. 274.

Nevertheless, the Soviet pattern was more conservative than that of the United States: it continued to manufacture Vostok (developed in 1954–9) unabated throughout the period, launches by Vostok reaching a maximum in 1973, while in the United States the use of Atlas and Thor, also developed in the 1954–9 period, fell after 1967 (see Table 10.6 below).

The conservative pattern of production also predominates in the examples of weapons production studied in chapter 9 below. Production of previous models of tanks did not cease when new models were introduced: the T-34/85 continued to be produced after the T-54 had been introduced, and it is reported that the T-55 is still being produced, although its successor the T-62 was introduced over 10 years ago (see p. 441 below).

In spite of some exceptions, it may safely be concluded that the rate of diffusion of new products and processes in terms of their share in total output is lower in the USSR than in the other industrialised countries. This is evidently partly because existing production capacity is withdrawn from use much more slowly than in the West. Existing steel plants and textile plants have continued to operate in the USSR when they would have been withdrawn from service elsewhere. This is a frequent pattern. Soviet machine tools have been withdrawn from service more slowly than in other countries (see pp. 41–2 below), and the relative youth of the stock is explained by its rapid expansion. The Soviet military continue to retain in service ICBMs and other weapons from generations which have been withdrawn in the United States (see pp. 488–9 below). It would be worth investigating whether within the capital equipment industry, and the engineering industry generally, existing capacity for the manufacture of long established plant and machinery continues in operation when in other countries it would have been withdrawn from service and replaced by new capacity capable of manufacturing new plant and machinery for the production of new products. Such continued production of less technologically advanced plant must have taken place in order to bring about the continued expansion in the capacity of the cotton textile industry, and of capacity for the manufacture of artificial fibres, or of open hearth steel, at a stage when these products or industries would have been withdrawn from production in other countries.

This low rate of capital retirement in Soviet industry generally might be economically defensible in certain cases in view of the continued availability throughout this period of adequate supplies of fresh labour. It is interesting to note that, as would be expected if the cost of labour relative to capital increased, the rate of obsolescence significantly increased in the Soviet Union between the middle 1950s and the late 1960s.[21] But technological conservatism may often be a result of the economic mechanism, which provides little encouragement to the diffusion of new technology. In the control instruments industry, for example, it would be difficult to find an adequate economic justification for the continued production of five major and six minor systems, some of which are antiquated even by the standards of the most advanced Soviet production, which has itself been basically unchanged for 10 years.

Quality

Our account so far has tacitly assumed that the quality of similar processes and products is the same in the USSR as in the other major countries. A large body of Soviet literature testifies that this is not the case at least in civilian industry; and this commonly accepted view has been confirmed by our specific studies of, for example, control instruments and machine tools. In chapter 11 below, it is shown that with a group of Soviet general-purpose machine tools

[21] According to one estimate, the implicit length of service of capital resources fell from 31 to 22 years between the five year periods 1956–60 and 1966–70; from 1967–70, however, the annual rate of capital retirement declined from 3·0 to 1·8 per cent (G. Fink and J. Sláma, in *Forschungsbericht 1973*, Osteuropa Institut, Munich (1974), pp. 160, 162).

their accuracy, as indicated by the 1954–6 Soviet standards in force until recently, is lower than that of their British equivalents (see Tables 11.6 and 11.7 below), and, in the case of two out of three types studied, their reliability is substantially lower than that of their British equivalents, even in the case of Soviet export models (see Appendix 11.C, pp. 561–3 below). Similar results were obtained for passenger cars (see pp. 551–6 below).

In both these industries, however, there are signs of recent improvements. Soviet standards for machine tools introduced since 1970, for example, provide for production of a proportion of Soviet machine tools of 'improved precision' which would have a *higher* accuracy than their standard British equivalents (see Tables 11.6 and 11.7, and pp. 539–45 below). There is a great deal of evidence that very substantial efforts have been devoted to the improvement of quality in both capital goods' and consumer goods' industries;[22] and the resulting improvements, usually not taken into account in our indicators of technological level, should be borne in mind in comparing Soviet and Western industry.

The 'research-production cycle'

Western studies of Soviet technological development, including previous studies by the present authors, have commonly assumed that research is more advanced than experimental development, that development is more advanced than innovation, and that the Soviet economy is least advanced in the diffusion of new technology. As a new product or process proceeds through the various stages from research to production, it has been argued, difficulties increase. High priority sectors, especially military production, have, however, been treated as a major exception to this general pattern:

> Aviation, rocketry, space exploration and atomic energy are four clear examples of science-based industries in which the USSR has achieved many outstanding successes. These are 'priority' areas; ... they work on a different plane from the normal run of Soviet industry.[23]

Our studies of the traditional industries, including high priority civilian industries, have on the whole confirmed this generally accepted stereotype. Here we review briefly the evidence that has been collected. Although insufficient evidence has been assembled for a general assessment of the relationship between research and the rest of the 'research-production cycle', it is interesting to note that research related to chemical fibres, including synthetic fibres, is reported to be of a high standard,[24] while diffusion has lagged considerably. It can be stated more firmly that the Soviet lag was often less marked in experimental development (prototype or pilot process) and innovation (first commercial production) than in diffusion, as emerges clearly

[22] See, for example, M. C. Spechler's article in *The ACES Bulletin*, Vol. XVII, 1975, No. 1, pp. 62–87.
[23] R. Amann, M. J. Berry and R. W. Davies, in E. Zaleski *et al.*, *Science Policy in the USSR*, OECD, Paris (1969), p. 435; for the view summarised above, see Part V of this book, *passim*.
[24] Our pilot study of polymer chemistry, however, produced less favourable results.

from Table 2.10. Thus in oxygen steelmaking, a lag of only two years behind the USA in introducing the first industrial installation became a lag of six years in terms of the diffusion of the new process, and in continuous casting of steel a lead of seven years became a lag of three. HVAC transmission lines did not quite follow this pattern (see Table 2.10(e) and 5.1 below). The USSR by the 1960s had established a clear lead over the United States in the technology of its most advanced transmission line (as measured in terms of voltage), and, unlike the situation in the steel industry and in nuclear power, the Soviet lead was also increased in the first stage of diffusion;[25] in the second stage of diffusion,[26] however, the United States has caught up with the USSR. In this group of traditional industries, the Soviet Union maintained its relative position only in the case of numerically controlled (NC) machine tools: a lag of six years behind the United States in manufacturing the first prototype in 1958 remained a lag of six years when output of NC machines in the Soviet Union reached one per cent of its total metalcutting machine tool output in 1971 (see Table 2.10(g)).

The relatively poor Soviet performance in diffusion of new technology was particularly marked in the chemical industry. Thus the first commercial production of 11 major plastic materials which have been introduced into the Soviet Union since 1940 took place on average 11 years later than in the United States; but by 1972, the Soviet Union was some 19 years behind the United States in per capita production of plastics and synthetic resins (see Tables 6.19 and 6A.9 below). A similar though smaller imbalance may be observed in chemical fibres. The first commercial production of six major synthetic fibres introduced since 1948 took place on average at most 15 years later than the United States;[27] but in 1972 the Soviet Union was over 17 years behind the United States in per capita production and 15 years behind it in terms of the proportion of synthetic fibres in total production of chemical fibres (see Tables 6.21 below and source cited in Table 2.10(c) above).

For other science-based and high technology products adequate information is not available. For control instruments and weapons we lack even total production figures which can be reliably compared with other countries. Each of the three cases for which we have information exhibits a different pattern. Nuclear power followed the pattern of the traditional industries. A three year Soviet lead over the United States in starting up the first commercial power station in 1954 had become a lag of at least four years by the time both countries had increased nuclear power to two per cent of total electricity output (see Table 2.10(f)). In the second case, computers, the only figure available is for their total stock, normally estimated to have been 6,000 in 1970, equal to the United States' production in 1962 (see p. 393 below). This eight year lag is roughly equal to the estimated Soviet lag in production of the latest type of computer as compared with the USA (see Table 2.4). No breakdown of production between different types of computer is available for

[25] On the criterion that lines of 300 kV and above have reached 5 per cent of total lines over 100 kV in km.
[26] On the criterion that lines of 300 kV and above amount to 10 per cent.
[27] The lag was only 11 years if the odd case of acetate continuous filament, which was not produced in the USSR until 35 years after production began in the United States, is omitted.

any period. In the third case, space rockets, the crucial problem for the Soviet Union has not been its ability to produce rockets in large numbers but its failure to complete the experimental development of a rocket as powerful as Saturn V; the lag has been at the development and innovation stages.

Thus in the traditional industries the diffusion of new technology in the Soviet Union is slower than experimental development and innovation, in comparison with the major capitalist countries. In the science-based industries, the situation is more varied, but in some important cases, such as nuclear power, diffusion has certainly been slow.

An important qualification should, however, be made. Partly in our information and partly in reality, the line is often blurred between what we here termed 'experimental development' and 'first commercial production'. In certain cases, however, firm evidence is available enabling us to distinguish between them. With NC machine tools, while the period between research and development was longer than elsewhere, the period between first prototype and commercial production was no longer than in the UK and shorter than in Japan (see Table 2.1). With continuous casting of steel, in which the USSR was the innovator, information is available about the dates on which the first experimental plant and the first industrial plant were established, both in the USSR and in other major countries (see Table 3.15 below). In the USSR it took 11 years, from 1944 to 1955, to move from experimental to industrial production, and this is equal to or better than FRG (1943 to 1954), UK (1946 to 1958) and USA (1946 to 1952). But the diffusion of continuous casting after the successful innovation of 1955 was particularly slow. Thus on this evidence the hold-up did not occur at either the development or the innovation stage.

But even here the possibility is not excluded of some cost or quality weakness in the full-scale production process. Thus even in the case of continuous casting, where the Soviet process has been patented in 30 countries and introduced into a number of them (including Japan), the Soviet Union is now employing Demag and Mannesmann, two leading FRG firms, to supply and install complete installations (see p. 104 below). Similarly diffusion of the oxygen steel process rapidly increased after substantial Soviet purchases of Austrian plant in 1964 and 1965.[28] The difficulty of determining when industrial production was fully under way is particularly acute in the chemical industry. Thus in the case of at least 7 of the 11 major plastics introduced into production in the USSR since 1936, large imports of plant took place in the late 1950s and 1960s after production had already been established on an industrial or semi-industrial scale in the USSR (see Tables 6.19 and 6.20 and p. 276 below). It may also be significant that the USSR has proved able to develop specialised synthetic fibres, some of them not available in the West, and produce them in small quantities (see pp. 284–5 below). Evidently the transition from experimental to genuine full-scale production (i.e., from development to innovation) may often be as difficult for the Soviet economy as the transition from innovation to diffusion.

Thus our results on the whole seem to confirm that Soviet research and

[28] See pp. 98–9 below; and also Table 4 in J. Sláma and H. Vogel, *op. cit.* (1973), p. 259.

experimental development were relatively more successful than the later stages of the research-production cycle. In some cases, however, the lower technological level appeared to be due to insufficiently flexible attention to alternative or more recent processes at the R and D stages. Thus alternatives to their own SKB process for the manufacture of synthetic rubber were almost ignored (see p. 256 below), and processes for the manufacture of alloy and quality steel other than electric-slag remelting received inadequate attention (see pp. 104–17 below). In iron and steel, as in many other major industries, research and development are concentrated into national units under ministerial control, a common technological policy is enforced at a national level, and complaints are frequently found that experimentation is restricted by the monopoly position of the major R and D organisations (see, for example, A. Bek's fictionalised documentary, *Novoe naznachenie*, Frankfurt (1971)). Synthetic rubber may have been in a similar position.

A further conclusion of several of our studies has been that at all stages of the research-production cycle Soviet industry displays a particular aptitude for improving and scaling up existing processes, rather than for bringing basically new processes and products into full-scale operation. Thus the scaling up of blast furnaces and open hearth furnaces has been particularly successful. The space programme has been dominated by what has been called an 'add-on' philosophy: it has relied heavily and for a long period on the Vostok launch vehicle, and spacecraft have been developed not by designing a new craft for each mission but by building on to a standard craft. In the United States, the technological differences between the Atlas, Thor, Titan and Saturn rockets, and the Mercury, Gemini and Apollo spacecraft are very much greater than in the Soviet Union (see pp. 514–17 below). While the contrast between the evolution of Soviet and United States tanks is less clear-cut (see pp. 438–9 below), a Rand study of aircraft design procedures reports that the Soviet Union does not make the development of a new aircraft depend on successful new developments in airframe, engine and avionics; it is more prepared than the United States to incorporate existing technology into a new airframe. The Rand study commends the efficiency of these and related Soviet R and D arrangements, which enable more aircraft to be taken to the test and production stages than in the United States, with smaller teams of designers.[29] Whether or not the economical conservatism of the Soviet approach should be preferred to the expensive innovationary dynamism of the United States must remain an open question. The Rand study agrees that the propulsion and avionics systems in Soviet aircraft are inferior to those of the United States; and our chapter on the Soviet space programme similarly concluded that the Soviet pattern of innovation and philosophy of design were responsible for the loss of tempo and of technological lead in relation to the United States (see p.

---

[29] R. Perry, *Comparisons of Soviet and US Technology*, Rand R-827-PR, Santa Monica (1973), pp. 11–19. Other aspects of his interesting analysis will be considered in our next volume: he praises the Soviet arrangements in which the design team is responsible for development of the prototype in its own experimental factory; in the United States development and production are normally passed to industrial contractors, and this makes for the impermanence of design teams and reduces the number of prototypes from which the choice can be made.

522 below). The Soviet approach may itself be partly a result or a reflection of the existing technological level: reliance on Vostok was due as much to lack of success in producing a rocket with a greater thrust power than Saturn V as to keenness in principle on economical standardisation. All these are issues to which we shall return in our second volume: for the moment it is interesting to note that an 'add-on' approach is seen by several studies to be a characteristic of Soviet technological development as distinct from the 'jumpiness' of the United States development. This evidence contradicts some previous assumptions.[30] There may be a variety of patterns in Soviet industry. While in some important industries, as we have seen, insufficient attention was paid to entirely new products and processes, in others, where the management of R and D was less centralised, the number of varieties of the new technology developed *and put into production* has tended to be surprisingly high: control instruments and NC machine tools are outstanding examples. The extent of this proliferation of alternatives as compared with the situation in other industrialised countries clearly deserves further investigation.

The role of foreign technology in Soviet industry

We did not examine the influence of foreign technology systematically; and this section merely summarises information scattered through the book. Quantitatively, imports were not large: imports of Western machinery amounted in 1970 to about 4 per cent of all Soviet investments in capital equipment and all imports of machinery, including those from Eastern Europe, to something of the order of 12 per cent. This is only about the same proportion of machinery imports in total capital equipment as in the case of the United States.[31] But foreign technology is much more significant in Soviet development than this global comparison suggests. The role of machinery imports was much greater in certain industries, and in any case the quantity of imports is often an inadequate guide to their importance in technological development.[32]

The industries we studied may be divided into at least three groups. First, those with a strong indigenous technology in which imports of machinery are

---

[30] See, for example, D. Granick, *Soviet Metal-Fabricating and Economic Development*, Madison, Milwaukee and London (1967), pp. 245–51, which concludes that 'improvement in the state of the arts in the Soviet Union should proceed in a less continuous fashion than in the United States'; according to Granick, in space technology development work in the United States 'aims at a steady expansion of the existing borders of knowledge', in contrast to the Soviet reliance on 'a series of theoretic break-throughs'.

[31] P. Hanson, *External Influences on the Soviet Economy since the mid-1950s: the Import of Western Technology*, CREES Discussion Paper, RC/B 7 (1974), pp. 13, 16–17, 19, 36; the figure of 12 per cent is suggested by the present author, on the assumption that the 'real' value of machine imports by the USSR from Eastern Europe should be assumed to be about 50 per cent higher than its foreign trade value. See also P. Hanson in *Soviet Economy in a New Perspective*, Joint Economic Committee of the US Congress, Washington DC (1976), pp. 786–811.

[32] A fairly speculative estimate by D. W. Green and H. S. Levine suggests that the marginal productivity of imported Western capital goods in the USSR was much higher than that of domestic Soviet capital (see their papers to NATO Economic Directorate Colloquium, 1976, and to National Science Foundation Workshop on Soviet Science and Technology, November 1976).

of minor importance. These include, of course, space rocketry, weapons, and nuclear power, as well as certain well established civilian industries, such as electric power. In these industries, imports are extremely small; the electric power industry received in 1970 only 0·1 per cent of all machinery imported from Western countries although its share in investment was as much as 6·1 per cent in 1971.[33] This does not, of course, mean that Soviet technology in these industries is not influenced by the West: purchases of Western equipment on a one-off basis are undertaken whenever possible, the Western technical press is diligently analysed, and espionage is not unknown. In the case of HVAC transmission lines, the purchase in 1957 of air-blast circuit breakers from the French company Delle Asthom was of considerable importance to the successful introduction of the 500 kV line. But the technology was rapidly assimilated: the Soviet Union appears now to produce its own 500 kV circuit breakers, and not to have imported circuit breakers for its 750 kV line, while Canada and the USA were still dependent on Delle Asthom (see pp. 206, 214 below).

Second, the iron and steel industry provides a case of an industry which has a well established indigenous technology and development capability, and has been responsible for pioneering important innovations, but in which foreign technology has nevertheless been of substantial importance in certain respects. Only 1·7 per cent of all machinery imported from the West went to this industry in 1970 although its share of investment in 1971 was 3·7 per cent.[34] But, as we have seen, foreign equipment and know-how appears to have been important even in the diffusion of new processes in which the USSR is itself a leading innovator. Imports of Austrian equipment took place at one stage in the diffusion of the oxygen process, and extensive German assistance is also being employed in the construction of a continuous casting plant (see pp. 98–9, 104 below). Elsewhere in the industry new processes are being introduced by large-scale agreements with foreign firms: a German consortium is to construct a very large integrated steel works with a plant for the direct reduction of iron ore with a capacity of 2·5 million tonnes a year, by far the largest in the world (see p. 92 below). And while the vast majority of equipment in most of the industry is manufactured in the Soviet Union, the rolling of steel, the most backward sector of the industry, has been heavily dependent on imports, which increased from 5·2 per cent of equipment in 1950 to 44·0 per cent in 1965, falling to 39·2 per cent in 1970 (most of this equipment was imported not from the West but from Eastern Europe) (see p. 92 below).

The Soviet machine tool industry also seems to come into this intermediate category. A fairly high proportion of Soviet machine tools (20 per cent by value) has been imported in recent years. These imports have been concentrated on sophisticated machines which the Soviet industry manufactures either at insufficiently high quality or in insufficient numbers. The imports have thus assisted the Soviet industry to cope with the higher requirements imposed by its consumers (see pp. 152–7 below). At the same time imports

[33] Hanson, *op. cit.* (1974), p. 22.
[34] Hanson, *op. cit.* (1974), p. 19.

from abroad and technological agreements with foreign firms have helped the Soviet machine tool industry to improve its product: thus for NC machines a number of agreements with foreign control systems firms have been signed since 1969 (see p. 184 below).

In a third group of industries, Soviet dependence on foreign technology is very high. In the case of computers, Richard Judy concluded in the late 1960s that 'technology in the Soviet Union is virtually entirely imported from the West'.[35] The Soviet Union possesses a substantial indigenous computer manufacturing facility, which it has developed without direct foreign assistance; but according to Judy virtually all innovations have first occurred in the West and been transferred to the USSR either through examination of the Western technical literature or through importing foreign computers. Other specialists have argued, however, that the complexities of computer technology are such that the Soviet Union could not have constructed, say, a third generation computer without a large development and innovation effort of its own (see p. 403 below). A study of the Soviet scheme to introduce a computerised system for management of the economy, to appear in our second volume, may shed further light on this question.

The chemical industry is the outstanding example of Soviet dependence on foreign technology. Since 1965, imports have accounted for one third of all chemical plant and equipment though within this overall figure the share of Eastern European countries has steadily increased (see p. 262 below). Plant for all the major plastics and synthetic fibres has been imported, largely from Western countries: in the case of plastics, imports generally occurred after production had been established in the USSR, so that they were evidently important in the diffusion of the new processes; in the case of polymerised fibres, process plant imported from abroad was responsible for first commercial production in the USSR (see pp. 276, 284 below).

Soviet reliance on foreign technology has on the whole tended to increase in the course of the past 20 years. The share of Western machinery imports in Soviet domestic equipment has risen steadily but not dramatically since the mid-1950s.[36] But at the same time Soviet exports of machinery have also increased substantially, though these have almost invariably been of a relatively unsophisticated kind. In certain sectors, however, the increase has been very substantial in both imports and exports. For machine tools, for example, imports are now 10 times as high as they were in 1955 and exports are as much as 18 times as high; in 1974 imports amounted to 212 million roubles and exports to 120 million roubles.[37] To what extent this is an increase in 'interdependence' characteristic of all advanced industrial countries rather than an increase in 'dependence' cannot be decided without a more detailed

---

[35] R. Judy, in (Ed.) S. Wasowski, *East–West Trade and the Technology Gap*, New York (1970), p. 63.

[36] Hanson, *op. cit.* (1974), p. 17.

[37] In the case of machine tools, the average value of each machine tool imported rose from 9,004 roubles in 1955 to 18,570 in 1974, while the value of each machine tool exported rose only from about 7,250 to 7,380 roubles (calculated from Table 4.24 below). This would appear to indicate that the sophistication of imports increased more than the sophistication of exports.

examination of the pattern of Soviet trade and technological borrowing in different industries as compared with that of other countries.

Conclusion

In most of the technologies we have studied there is no evidence of a substantial diminution of the technological gap between the USSR and the West in the past 15–20 years, either at the prototype/commercial application stages or in the diffusion of advanced technology. Four reservations should be made. First, we have covered only a limited number of technologies. It is not clear whether our present selection is likely to have been to the favour or disfavour of the USSR. The bias has been towards technologies in which the USSR is normally believed to be in a strong position: we have not yet completed a study of a consumer goods industry. It is possible that Soviet technology has improved more rapidly in the previously most backward industries, but otherwise our selection is likely to be biased in favour of the USSR. Second, and this will have led us to underestimate the technological level of the USSR, we are likely to lack knowledge of the most recent achievements in the USSR to a greater extent than in the case of Western countries. Third, there is some evidence that the position has improved in the past four or five years in certain respects: NC machine tools provide an important recent example of rapid diffusion, and quality of production in a number of industries has recently begun to improve. Finally, the recent depression in capitalist countries may well be inhibiting diffusion of new technology in the West and thus providing the Soviet Union with a better opportunity to catch up.

Appendix Table 2A.1 *(Prepared by Julian Cooper. It forms the basis of summary tables 2.7 and 2.8 in the text.)*

1. Electric power

	1940	1950	1955	1960	1965	1970	1973
A. Production of electricity (gross, million kWh)							
USSR	48309[1]	91226[1]	170225[1]	292274[1]	506672[1]	740925[1]	914653[2]
USA	179907[3]	409828[4]	666066[4]	891581[4]	1226795[6]	1739796[7]	2073070[9]
Japan	35473[3]	46300[4]	65200[4]	115500[4]	192159[8]	359549[8]	470088[10]
UK	29976[3]	67642[5]	95230[5]	138748[5]	196495[8]	249193[8]	282128[11]
FRG	62964[3a]	46717[5]	80484[5]	119028[5]	172340[8]	242612[8]	298995[11]
B. Per capita production (kWh)							
USSR	252	507	868	1363	2194	3052	3662
USA	1362	2691	4014	4934	6314	8492	9819
Japan	497	559	733	1239	1943	3446	4339
UK	639	1336	1860	2642	3627	4497	5044
FRG	1109	931	1536	2147	2940	4000	4825
C. USSR per capita production as a percentage of:							
USA	18·5	18·8	21·6	27·6	34·7	35·9	37·3
Japan	50·7	90·7	118·4	110·0	112·9	88·6	84·3
UK	39·4	37·9	46·7	51·6	60·5	67·9	72·6
FRG	22·7	54·5	56·5	63·5	74·6	76·3	75·9
D. Index of growth per capita production (1960 = 100)							
USSR	18	37	64	100	161	224	269
USA	28	55	81	100	128	172	199
Japan	40	45	59	100	157	278	350
UK	24	51	70	100	137	170	191
FRG	52	43	72	100	137	186	225

Sources:
1 *Narodnoe khozyaistvo SSSR za 1922–72 gody*, p. 158.
2 *Narodnoe khozyaistvo SSSR v 1973 godu*, p. 254.
3 *UN Statistical Yearbook* (1949–50), pp. 280–4.
4 OECD, *Basic Statistics of Energy, 1950–64*, pp. 201–21, 252.
5 *Ibid.*, supplementary issue, pp. 110, 324.
6 OECD, *Statistics of Energy, 1953–67*, p. 135.
7 *Ibid., 1958–72*, p. 124.
8 *UN Statistical Yearbook* (1973), pp. 360–8.
9 OECD, *Statistics of Energy, 1959–73*, p. 184.
10 *UN Monthly Bulletin of Statistics*, March 1975, p. 99.
11 Eurostat, *Energy Statistics, Quarterly Bulletin*, 1974, No. 3, p. 96.

Note:
a – Pre-war boundaries.

Production of electricity by nuclear power stations

	1955	1960	1965	1970	1973
1. Capacity of nuclear power stations[a] (mW, end of year)					
USSR	5[1]	105[2]	310[5]	900[6]	3200[6]
USA	—	297[2]	926[3]	6493[4]	21070[8]
Japan	—	—	13[7]	1336[7]	1836[9]
UK	—	296[2]	2835[3]	3427[4]	4281[4]
FRG	—	—	15[3]	890[4]	2200[4]
Capacity of nuclear power stations as a proportion of total electricity generating capacity (per cent)					
USSR	0·01	0·1	0·3	0·5	1·6
USA	—	0·2	0·3	1·8	4·8
Japan	—	—	0·03	2·0	3·6
UK	—	0·9	6·2	5·5	5·9
FRG	—	—	0·04	1·9	4·1

Sources:
1. UNECE, *Annual Bulletin of Electric Energy Statistics for Europe* (1959), p. 26.
2. *Ibid.* (1964), pp. 34–5.
3. *Ibid.* (1968), p. 17.
4. *Ibid.* (1973), pp. 10–15.
5. Calculated from *Gosudarstvennyi pyatiletnii plan razvitiya narodnogo khozyaistva SSSR na 1971–1975 gody* (1972), pp. 2–7.
6. *Elektricheskie Stantsii*, 1973, No. 12, pp. 2–7.
7. *UN Statistical Yearbook* (1973), p. 359.
8. *Referativny Sbornik, Ekonomika Promyshlennosti*, 1975, Item 2V74.
9. *Ibid.*, 1974, Item 9V64.

Note:
a – maximum net capacity of plants in continuous operation; definition of capacity for USSR 1965–1974 not known.

Production of electricity by nuclear power stations (*continued*)

	1960	1965	1970	1972	1973
2. Output of electricity from nuclear power stations ('000 kWh, gross)					
USSR	100[1]	1647[2a]	3200[3]	7300[4]	11700[3]
USA	554[5]	3913[6]	23324[7]	57814[7]	89100[9]
Japan	—	36[8]	4581[8]	9480[8]	12800[9]
UK	2224[10]	16338[8]	26012[8]	29378[11]	27996[11]
FRG	—	117[8]	6030[8]	9137[11]	11755[11]
Output per 10,000 people (kWh)					
USSR	5	70	132	295	468
USA	31	201	1138	2768	4235
Japan	—	4	439	886	1181
UK	424	3016	4694	5265	5006
FRG	—	20	994	1482	1897
USSR output per 10,000 people as per cent of:					
USA	16·1	34·8	11·6	10·7	11·1
Japan	—	17·0	30·1	33·3	39·6
UK	1·2	2·3	2·8	5·6	9·3
FRG	—	3·0	13·3	19·9	24·7
Output of electricty from nuclear power stations as a proportion of total electricty output (per cent)					
USSR	0·03	0·31	0·43	0·85	1·28
USA	0·06	0·32	1·34	2·94	4·30
Japan	—	0·02	1·27	2·21	2·72
UK	1·60	8·32	10·44	11·14	9·92
FRG	—	0·07	2·49	3·33	3·93

Sources:
1 *Soviet Studies*, 1974, No. 4, p. 576.
2 UNECE, *Annual Bulletin of Electric Energy Statistics for Europe* (1969), p. 31.
3 *Elektricheskie Stantsii*, 1973, No. 12, pp. 2–7 (approximate).
4 *Ekonomicheskaya Gazeta*, 1973, No. 4, p. 2 (approximate).
5 OECD, *Basic Statistics of Energy, 1950–1964*, p. 253.
6 OECD, *Statistics of Energy, 1953–1967*, p. 135.
7 *Ibid., 1958–1972*, p. 124.
8 *UN Statistical Yearbook* (1973), p. 368.
9 OECD, *Statistics of Energy, 1959–1973*, pp. 184, 202.
10 OECD, *Basic Statistics of Energy, 1950–1964*, supplementary issue, p. 325.
11 UNECE, *Annual Bulletin of Electric Energy Statistics for Europe* (1973), pp. 21, 31–2.

Note:
a – 1966.

High voltage electricity transmission lines

Length of AC transmission lines by voltage (km)

	1955	%	1960	%	1965	%	1970	%	1972	%
USSR										
100–299 kV	35032[1a]	100	98700[2]	94·7	175100	92·0	241300[2]	89·8	273900[3]	89·3
300–400 kV			3100	3·0	7110	3·7	13950	5·2	16300	5·3
500 kV	—		2400	2·3	8170	4·3	13140	4·9	15700	5·1
750 kV	—		—		—		160	0·1	800	0·3
USA										
100–299 kV	n.a.		197967[4b]	97·6	276102[5]	96·6	342846[7]	89·8	n.a.	
300–400 kV	n.a.		4970	2·4	9199	3·2	26425	6·9	n.a.	
500 kV	—		—		476	0·2	10885	2·9	n.a.	
735–765 kV	—		—		—		1668	0·4	n.a.	
UK										
100–299 kV	16475[6c]	100	20963[4b]	100	30486[5]	96·5	40292[7]	89·7	41583[8]	88·8
300–400 kV	—		—		1107	3·5	4621	10·3	5241	11·2
500 kV and over	—		—		—		—		—	
FRG										
100–299 kV	26252[5c]	99·0	28010[4b]	97·9	39369[5]	96·8	47318[7]	95·3	48518[8]	94·3
300–400 kV	258	1·0	598	2·1	1296	3·2	2340	4·7	2940	5·7
500 kV and over	—		—		—		—		—	

Sources:
1 *Promyshlennost' SSSR* (1964), p. 236.
2 A. S. Pavlenko and A. M. Nekrasov (Eds.), *Energetika SSSR v 1971–1975 godakh* (1972), p. 186.
3 *Narodnoe khozyaistvo SSSR v 1973 godu*, p. 257.
4 UN, *The Situation and Prospects of Europe's Electric Power Supply Industry in 1960–61*, p. 74.
5 UN, *The Electric Power Situation in Europe and its Future Prospects* (1967), p. 73.
6 UN, *Developments in the Situation of Europe's Electric Power Supply Industry during the Post-War Period* (1959), p. 108.
7 Approximate: calculated from 1965 totals updated by reported (or estimated) annual additions reported in OECD, *Survey of Electric Power Equipment*—21st, pp. 32–3, 73; 22nd, pp. 31–2, 67; 24th, pp. 25–7, 72; 25th, pp. 24–5, 81. Note—in so far as the withdrawal of lines from service is not taken into account, the proportion of lines of lower voltages is probably overstated.
8 Approximate: calculated as in source 7 above; 1971 and 1972 additions from OECD, *25th Survey of Electric Power Equipment*, pp. 24–5.

Notes:
a – 110–330 kV.
b – 1959.
c – 1956.

Electricity consumed per worker in industry and construction

	1938	1950	1955	1960	1965	1970	1972
Electricity consumed per person employed in industry and construction (kWh)							
USSR	§+1986[1a]	§3638[1a]	§5598[1b]	7796[1a]	10905[2a]	13047[2a]	13451[2c]
USA	4725[3d]	*10236[3d]	15820[4e]	20326[5f]	24395[6g]	29369[7g]	32509[7g]
Japan	n.a.	*2716[8d]	3938[8e]	5937[8f]	8008[9g]	12836[10g]	14707[10g]
UK	2055[3d]	*3303[3d]	3886[4e]	5662[5f]	6830[6g]	8669[7g]	9701[7g]
FRG	n.a.	†*4956[3d]	5230[4e]	5958[5f]	7728[6g]	10130[7g]	11378[7g]
USSR consumption as a percentage of:							
USA	42.0	35.5	35.4	38.4	44.7	44.4	41.4
Japan	n.a.	133.9	142.2	131.3	136.2	101.6	91.5
UK	96.6	110.1	144.1	137.7	159.7	150.5	138.7
FRG	n.a.	73.4	107.0	130.8	141.1	128.8	118.2

Sources:
Electricity consumed in industry and construction:
1 *Narodnoe khozyaistvo SSSR v 1965 godu*, p. 162.
2 *Statisticheskii ezhegodnik stran-chlenov SEV* (1974), pp. 54–5.
3 UNECE, *Annual Bulletin of Electric Energy Statistics for Europe* (1956), p. 39 (USA estimated).
4 *Ibid.* (1959), pp. 64–5.
5 *Ibid.*, (1964), p. 89.
6 *Ibid.* (1968), pp. 31–4.
7 *Ibid* (1973), pp. 41–2.
8 OECD, *Basic Statistics of Energy, 1950–1964*, pp. 201–21.
9 OECD, *Statistics of Energy, 1953–1967*, p. 127.
10 *Ibid., 1958–1962*, pp. 131–5.

Number employed in industry and construction:
a *Narodnoe khozyaistvo SSSR za 1922–1972 gody*, pp. 346–7.
b *Trud v SSSR* (1968), p. 25.
c *Narodnoe khozyaistvo SSSR v 1973 godu*, pp. 574–5.
d ILO, *Yearbook of Labour Statistics* (1953), pp. 71–7.
e *Ibid.* (1960), pp. 81–91.
f *Ibid.* (1966), pp. 271–84.
g *Ibid.* (1974), pp. 278–92.

Notes:
+ – 1937.
* – 1951.
† – excluding Saar.
§ – without account of electricity consumed in construction sector—about 4 per cent of industry-plus-construction consumption per person employed in 1960.

The iron and steel industry

	1940	1950	1955	1960	1965	1970	1973
1. Production of crude steel (million tonnes)							
USSR	18·3[1]	27·3[1]	45·3[1]	65·3[1]	91·0[1]	115·9[1]	131·5[2]
USA	60·8[3]	87·8[3]	106·2[3]	90·1[4]	119·0[6]	119·1[7]	136·5[8]
Japan	6·9[3]	4·8[3]	9·4[3]	22·1[5]	41·2[6]	93·3[7]	119·3[8]
UK	13·2[3]	16·6[3]	20·1[3]	24·7[4]	27·4[6]	28·3[7]	26·6[8]
FRG	21·5[3]	14·0[3]	24·5[3]	34·1[4]	36·8[6]	45·0[7]	49·5[8]
Per capita output (kg)							
USSR	95	152	231	305	394	477	527
USA	460	577	640	499	612	581	649
Japan	97	58	106	237	417	894	1101
UK	273	328	393	470	506	511	474
FRG	308	279	468	615	628	742	799
USSR per capita output as a percentage of:							
USA	20·7	26·3	36·1	61·1	64·4	82·1	81·2
Japan	97·9	262·1	217·9	128·7	94·5	53·3	47·9
UK	34·8	46·3	58·8	64·9	77·9	93·3	110·7
FRG	30·8	54·5	49·4	49·6	62·7	64·3	66·0

Sources:
1 *Narodnoe khozyaistvo SSSR za 1922–1972 gody*, p. 165.
2 *Narodnoe khozyaistvo SSSR v 1973 godu*, p. 264.
3 *UN Statistical Yearbook* (1959), pp. 237–8.
4 *UN Quarterly Bulletin of Steel Statistics for Europe,* Vol. XIV, 1963; No. 1, pp. A–2, 20, 22.
5 *Ibid.*, Vol XV, 1964, No. 1, p. A–28.
6 *Ibid.*, Vol. XIX, 1968, No. 1, pp. A–9, 22, 26, 28.
7 *Ibid.*, Vol. XXIII, 1972, No. 1, pp. A–6, 24, 26, 28.
8 *Ibid.*, Vol. XXV, 1974, No. 2, pp. 12, 19, 21.

Iron and steel (continued)

	1960	1965	1970	1973
2. Production of steel by oxygen converter (million tonnes)				
USSR	2·5[1]	3·8[3]	19·9[5]	28·1[6]
USA	3·0[1]	20·7[4]	57·5[5]	75·6[6]
Japan	2·6[2]	22·6[4]	73·9[5]	96·1[6]
UK	0·4[1]	5·6[4]	9·1[5]	12·6[6]
FRG	0·9[1]	7·0[4]	24·1[5]	33·6[6]
Per capita output (kg)				
USSR	12	16	82	113
USA	17	107	281	359
Japan	28	229	708	887
UK	8	103	164	225
FRG	16	119	397	548
USSR per capita as a percentage of:				
USA	70·6	15·0	29·2	31·5
Japan	42·9	7·9	11·6	12·7
UK	150·0	17·5	50·0	50·2
FRG	75·0	15·1	20·7	20·6
Oxygen steel as a proportion of total steel output (per cent)				
USSR	3·8	4·2	17·3	21·4
USA	3·4	17·4	47·0	55·4
Japan	11·9	55·0	79·1	80·5
UK	1·7	20·5	32·1	47·3
FRG	2·7	19·1	55·8	67·8

Sources:
1 *UN Quarterly Bulletin of Steel Statistics for Europe*, Vol. XIV, 1963, No. 1, pp. A–8, 20–3.
2 *Ibid.*, Vol XV, 1964, No. 1, p. A–28.
3 UN, *The European Steel Market in 1967*, p. 58.
4 *UN Quarterly Bulletin of Steel Statistics for Europe*, Vol. XIX, 1968, No. 1, pp. A–9, 22–8.
5 *Ibid.*, Vol. XXIII, 1972, No. 1, pp. A–6, 22–8.
6 *Ibid.*, Vol. XXV, 1974, No. 2, pp. 12, 18–21.

Iron and steel (continued)

	1965	1970	1972	1973
3. Output of continuously cast steel (million tonnes)				
USSR	1·2[1]	5·0[2]	6·5[3]	7·0[4]
USA	1·0[5]	13·7[5]	n.a.	n.a.
Japan	0·4[6]	5·2[7]	16·5[8]	24·7[8]
UK	0·4[9]	0·5[10]	0·6[8]	0·8[8]
FRG	0·8[9]	3·7[10]	6·1[8]	8·1[8]
Per capita output (kg)				
USSR	5	21	26	28
USA	5	67	n.a.	n.a.
Japan	4	50	154	228
UK	7	9	11	14
FRG	14	61	99	131
USSR per capita as a percentage of:				
USA	100	31	n.a.	n.a.
Japan	125	42	17	12
UK	71	233	236	200
FRG	36	34	26	21
Continuously cast steel as a proportion of total steel output (per cent)				
USSR	1·3	4·3	5·5	5·3
USA	0·8	11·5	n.a.	n.a.
Japan	1·0	5·6	17·0	20·7
UK	1·4	1·8	1·8	3·0
FRG	2·1	8·3	13·9	16·3

Sources:
1 *Steel International,* September 1966, p. 126.
2 UN, *The Steel Market in 1971,* p. 82.
3 UN, *The Steel Market in 1971,* p. 92.
4 *Ibid.*, 1973, p. 97.
5 L. Nabseth and G. Ray, *The Diffusion of New Industrial Processes* (1974), p. 241.
6 *Referativnyi Sbornik, Ekonomika Promyshlennosti,* 1972, Item 3V113.
7 UN, *The Steel Market in 1972,* p. 62.
8 OECD, *The Iron and Steel Industry in 1973 and Trends in 1974,* p. 26.
9 OECD, *The Iron and Steel Industry in 1966 and Trends in 1967,* Table 3.
10 OECD, *The Iron and Steel Industry in 1971 and Trends in 1972,* p. 36.

The engineering industry: the machine tool stock

	1958	1962	1966	1973
1. Total stock in the economy ('000 units)				
USSR				
Metalcutting machines	1916[1]	2442[2]	2797[3]	4080[3]
Metalforming machines	394[1]	497[2]	616[3]	948[3]
Proportion of forming machines (%)	17.1	16.9	18.0	18.9
USA				
Metalcutting machines	2204[4]	2537[5a]	2620[6b]	2923[7]
Metalforming machines	683[4]	816[5a]	850[6b]	837[7]
Proportion of forming machines (%)	23.7	24.3	24.5	25.3

2. Stock in the machine building and metal working industry

	1962	1966	1972
USSR			
Metalcutting machines	1360[2]	1622[8]	2411[9]
Metalforming machines	263[8]	339[8]	523[9]
Proportion of forming machines (%)	16.2	17.3	17.8
Proportion of cutting machines 10 years old or less (%)	55	n.a.	60[c]
Number of machines per 1,000 employed in the MCMW industry[11]	180	188	230

	1963	1968	1973
USA			
Metalcutting machines	2137[5]	2175[6]	2362[7]
Metalforming machines	671[5]	695[6]	703[7]
Proportion of forming machines (%)	23.9	24.2	22.9
Proportion of cutting machines 10 years old or less (%)	36	37	33
Number of machines per 1,000 employed in the MCMW industry	352[5]	287[6]	279[7]

	1961	1966	1971
UK			
Metalcutting machines	1035[10]	968[10]	728[10]
Metalforming machines	199[10]	172[10]	129[10]
Proportion of forming machines (%)	16.1	15.1	15.0
Proportion of cutting machines 10 years old or less (%)	41	37	42
Number of machines per 1,000 employed in the MCMW industry[12]	321	286	226

Sources:
1 *Narodnoe khozyaistvo SSSR v 1965 godu*, p. 65.
2 D. M. Pal'terovich, *Planirovanie potrebnosti v oborudovanii* (1972), p. 122.
3 *Narodnoe khozyaistvo SSSR v 1972 godu*, p. 68.
4 *American Machinist*, 17 November 1958.
5 *Ibid.*, 10 June 1963.
6 *American Machinist*, 10th inventory.
7 *American Machinist*, 29 October 1973.
8 Estimated from source 3 above, on the basis of the proportions of cutting and forming machines in the total stock of MCMW in 1962 and 1972.
9 Calculated from source 3 above, using the proportions in MCMW given by L. Kostin, *Proizvoditel'nost' truda i tekhnicheskii progress* (1974), p. 142.
10 *Metalworking Production Survey of Machine Tool and Production Equipment in Britain* (1973).
11 All workers and employees in MCMW: 1972—*Statisticheskii ezhegodnik stran-chlenov SEV* (1974), p. 133; 1966—*ibid.* (1971), p. 124; 1962—calculated using index—ILO, *Yearbook of Labour Statistics* (1968), p. 361.
12 Number employed in MCMW—*Department of Employment Gazette*, December 1972, pp. 1186–7.

Notes:
a – 1963.
b – 1968.
c – total stock.

The engineering industry: numerically-controlled machine tools

	1960	1965	1970	1973
Output of NC machine tools (units)				
USSR	16[1]	49[1]	1687[1]	3788[2]
USA	402[3]	2094[3]	1819[3]	2286[3]
Japan	—	39[4]	1651[4]	2765[4]
UK	n.a.	(225)[5]	600[5]	467[5]
FRG	n.a.	162[6]	762[6]	992[6]
Output per million people (units)				
USSR	0·07	0·21	6·95	15·17
USA	2·22	10·78	8·88	10·87
Japan	—	0·39	15·82	25·52
UK	n.a.	4·15	10·83	8·35
FRG	n.a.	2·76	12·56	16·01
USSR output per million people as a percentage of:				
USA	3·2	2·0	78·3	139·6
Japan	—	53·8	43·9	59·4
UK	n.a.	5·1	64·2	181·7
FRG	n.a.	7·6	55·3	94·8
Output of NC machines as a proportion of total metalcutting machine tool output (%)				
USSR	0·01	0·03	0·83	1·77
USA	0·31	1·14	1·01	0·91
Japan	—	0·04	0·64	1·30
UK	n.a.	0·11	0·46	0·63
FRG	n.a.	0·36	1·03	0·86

Sources:
1 *Narodnoe khozyaistvo SSSR v 1972 godu*, pp. 172–3.
2 *Ibid., 1973*, pp. 216–17.
3 US Department of Commerce, *Current Industrial Reports*, Series MQ-35W.
4 Japan Machine Tool Trade Association, *Japan's Machine Tool Industry*.
5 UK Department of Trade, *Business Monitor*, Series PM 332 and PQ 332.
6 Data of Verein Deutscher Maschinenbau Anstalten.

Notes:
( ) – estimate.
n.a. – not available.

The chemical industry: production of plastics and synthetic resins

	1940	1950	1955	1960	1965	1970	1973
Output of plastics and synthetic resins ('000 tonnes)							
USSR	11[1]	67[1]	160[1]	312[1]	803[1]	1673[1]	2320[2]
USA	n.a.	1034[3]	1762[3]	2850[4]	4009[5]	8712[5]	9905[8]
Japan	n.a.	19[3]	109[3]	737[4]	1609[5]	5154[5]	n.a.
UK	16[6]	158[7]	310[7]	588[7]	957[5]	1440[5]	1976[9]
FRG	n.a.	84[7]	362[7]	964[7]	1999[5]	4170[5]	6408[9]
Output per capita (kg)							
USSR	0·06	0·37	0·82	1·46	3·48	6·89	9·29
USA	n.a.	6·79	10·62	15·77	25·27	42·52	47·08
Japan	n.a.	0·23	1·23	7·91	16·27	49·40	n.a.
UK	0·33	3·12	6·05	11·20	17·66	26·15	35·33
FRG	n.a.	1·67	6·91	17·39	34·10	68·75	103·40
USSR per capita as a percentage of:							
USA	7·3	5·4	7·7	9·3	13·8	16·2	19·7
Japan	n.a.	160·9	66·7	18·5	21·4	13·9	n.a.
UK	18·2	11·9	13·6	13·0	19·7	26·3	26·3
FRG	n.a.	22·2	11·9	8·4	10·2	10·0	9·0

Sources:
1 *Narodnoe khozyaistvo SSSR za 1922–1972 gody*, p. 192.
2 *Narodnoe khozyaistvo SSSR v 1973 godu*, p. 273.
3 *The Chemical Industry in the European Member Countries of OECD* (1964), p. 51.
4 *UN Statistical Yearbook* (1966), p. 283.
5 *Ibid.* (1973), p. 272.
6 *Annual Abstract of Statistics, 1938–1948*, p. 150.
7 OECD, *Industrial Statistics, 1900–1962*, p. 155.
8 *Survey of Current Business,* March 1975, p. S–26.
9 *UN Monthly Bulletin of Statistics,* March 1975, p. 72.

n.a. – not available.

The chemical industry: production of chemical fibres

	1940	1950	1955	1960	1965	1970	1973
1. Output of chemical fibres ('000 tonnes)							
USSR	11[1]	24[1]	111[1]	211[1]	407[1]	623[1]	830[2]
USA	214[3]	626[4]	744[4]	773[5]	1500[6]	2250[6]	3062[7]
Japan	228[3]	116[4]	348[4]	552[5]	879[6]	1550[6]	1851[7]
UK	77[3]	172[4]	232[4]	269[5]	391[6]	599[6]	731[8]
FRG	308[3]	162[4]	217[4]	281[5]	471[6]	723[6]	979[8]
Per capita output (kg)							
USSR	0·06	0·13	0·57	0·98	1·76	2·57	3·32
USA	1·62	4·10	4·48	4·28	7·72	10·98	16·47
Japan	3·19	1·40	3·91	5·92	8·89	14·86	17·08
UK	1·60	3·40	4·53	5·12	7·22	10·81	13·07
FRG	4·41	3·23	4·14	5·07	8·03	11·92	15·80
USSR per capita output as a percentage of:							
USA	3·7	3·2	12·7	22·9	22·8	23·4	20·2
Japan	1·9	9·3	14·6	16·6	19·8	17·3	19·4
UK	3·8	3·8	12·6	19·1	24·4	23·8	25·4
FRG	1·4	4·0	13·8	19·3	21·9	21·6	21·0

Sources:
1 *Narodnoe khozyaistvo SSSR za 1922–1972 gody*, p. 172.
2 *Narodnoe khozyaistvo SSSR v 1973 godu*, p. 274.
3 *UN Statistical Yearbook* (1948), pp. 202–4.
4 *Ibid.* (1959), pp. 207–9.
5 *Ibid.* (1966), pp. 263–7.
6 *Ibid.* (1973), pp. 246–50.
7 *Survey of Current Business*, March 1975, p. S–39.
8 *UN Monthly Bulletin of Statistics*, March 1975, pp. 59–63.

The chemical industry: production of chemical fibres (continued)

	1950	1955	1960	1965	1970	1973
2. Output of synthetic fibres ('000 tonnes)						
USSR	1[1]	9[1]	15[1]	77[1]	167[1]	287[2]
USA	56[3]	172[3]	307[3]	807[4]	1627[4]	2862[7]
Japan	1[5]	16[5]	118[6]	380[4]	1028[4]	1308[8]
UK	4[3]	18[3]	61[3]	148[4]	337[4]	454[8]
FRG	1[3]	12[3]	52[3]	179[4]	497[4]	810[8]
Per capita output (kg)						
USSR	0·01	0·05	0·07	0·43	0·69	1·15
USA	0·37	1·04	1·70	4·15	7·94	13·60
Japan	0·01	0·18	1·26	3·84	9·85	12·07
UK	0·08	0·35	1·16	2·73	6·08	8·11
FRG	0·02	0·23	0·94	3·05	8·19	13·07
USSR per capita as a percentage of:						
USA	2·7	4·8	4·1	10·4	8·7	8·5
Japan	100·0	27·8	5·6	11·2	7·0	9·5
UK	12·5	14·3	6·0	15·8	11·3	14·2
FRG	50·2	21·7	7·4	14·1	8·4	8·8
Synthetic fibres as a proportion of total chemical fibres output (per cent)						
USSR	4·2	8·1	7·1	18·9	26·8	34·6
USA	8·9	23·1	39·7	53·8	72·3	82·6
Japan	0·9	4·6	21·4	43·2	66·3	70·7
UK	2·3	7·8	22·7	37·9	56·3	62·1
FRG	0·6	5·5	18·5	38·0	68·7	82·7

Sources:
1 *Narodnoe khozyaistvo za 1922–1972 gody*, p. 172.
2 *Narodnoe khozyaistvo SSSR v 1973 godu*, p. 274.
3 OECD, *Industrial Statistics, 1900–1962*, p. 139.
4 UN, *The Growth of World Industry* (1972), Vol. 2, pp. 307–17.
5 *UN Statistical Yearbook* (1959), p. 207.
6 *Ibid.* (1966), p. 263.
7 *Survey of Current Business*, March 1975, p. S–39.
8 *UN Monthly Bulletin of Statistics*, March 1975, pp. 59–63.

Cotton yarn

	1940	1950	1955	1960	1965	1970	1973
Output of cotton yarn ('000 tonnes)							
USSR	650[1]	663[1]	1038[2]	1169[2]	1292[3]	1435[4]	1535[5]
USA	1140[6a]	1823[7]	1711[8]	1644[8]	1958[9]	1618[5]	1592[5]
Japan	415[1]	238[1]	418[10]	536[2]	566[9]	526[9]	555[5]
UK	541[6b]	433[1]	339[1]	271[8]	220[9]	159[9]	115[5]
FRG	319[6a]	282[1]	373[8]	350[8]	295[9]	239[9]	215[5]
Per capita output (kg)							
USSR	3·4	3·7	5·3	5·5	5·6	5·9	6·2
USA	8·8	12·0	10·3	9·1	10·1	7·9	7·6
Japan	5·8	2·9	4·7	5·8	5·8	5·0	5·1
UK	11·3	8·6	6·6	5·2	4·1	2·9	2·1
FRG	4·7	5·6	7·1	6·3	5·0	3·9	3·5
USSR per capita as a percentage of:							
USA	38·6	30·8	51·5	60·4	55·4	74·7	81·6
Japan	58·6	127·6	112·8	94·8	98·2	118·0	121·6
UK	30·4	43·0	80·3	105·8	136·6	203·4	295·2
FRG	63·0	66·1	74·6	87·3	112·0	151·3	177·1

Sources:
1 *UN Statistical Yearbook* (1956), p. 261.
2 *Ibid.* (1963), pp. 245–6.
3 *Ibid.* (1970), pp. 258–9.
4 *Narodnoe khozyaistvo SSSR za 1922–1972*, p. 198.
5 *Cotton—World Statistics, Quarterly Bulletin*, April 1975, p. 36.
6 OECD, *Industrial Statistics, 1900–1962*, p. 133.
7 *The Textile Industry in Europe* (1958), p. 75.
8 *The Textile Industry, 1961–1962*, Table 36.
9 *The Textile Industry in OECD Countries, 1972–73*, p. 40.
10 *UN Statistical Yearbook* (1959), p. 199.

Notes:
a – 1938.
b – 1939.

The chemical industry: production of synthetic rubber

	1940	1950	1955	1960	1965	1970	1973
Output of synthetic rubber ('000 tonnes)							
USSR	100[1]	184[2]	235[2]	439[2]	667[2]	878[2]	n.a.
USA	3[3]	484[4]	986[4]	1459[4]	1842[6]	2232[6]	2585[7]
Japan	—	—	—	19[5]	161[6]	698[6]	972[9]
UK	—	—	—	92[4]	174[6]	306[6]	354[8]
FRG	41[3]	—	11[4]	81[4]	173[6]	321[6]	397[9]
Per capita output (kg)							
USSR	0·5	1·0	1·2	2·1	2·9	3·6	n.a.
USA	0·02	3·2	5·9	8·1	9·5	10·9	12·3
Japan	—	—	—	0·2	1·6	6·7	9·0
UK	—	—	—	1·8	3·2	5·5	6·3
FRG	0·6	—	0·2	1·5	3·0	5·3	6·4
USSR per capita output as percentage of:							
USA	2500	31·3	20·3	25·9	30·5	33·0	n.a.
Japan	—	—	—	1050	181	53·7	n.a.
UK	—	—	—	117	90·6	65·5	n.a.
FRG	83·3	—	600	140	96·7	67·9	n.a.

Sources:
1 V. I. Kas'yanenko, *Zavoevanie ekonomicheskoi nezavisimosti SSSR, 1917–1940 gody* (1972), p. 219.
2 Estimate—R. Amann.
3 *UN Statistical Yearbook* (1949–50), p. 235.
4 OECD, *Industrial Statistics, 1900–1962*, p. 165.
5 *UN Statistical Yearbook* (1966), p. 274.
6 UN, *The Growth of World Industry, 1963–1972*, Vol. 2, p. 306.
7 *Current Survey of Business*, March 1975, p. S–37.
8 *Monthly Digest of Statistics*, April 1975, p. 92.
9 *UN Monthly Bulletin of Statistics*, March 1975, p. 65.

Communications

	1940	1950	1955	1960	1965	1970	1972
Number of telephones in use in the economy ('000 units)							
USSR	1729[1]	2313[1]	3190[1]	4301[1]	6399[1]	10987[1]	13199[2]
USA	19450[3a]	43004[4]	56243[5]	74342[6]	93656[7]	120218[8]	131108[8]
Japan	1131[3a]	1664[4]	3123[5]	5526[6]	13999[7]	26233[8]	34021[8]
UK	3019[3a]	5376[4]	6830[5]	8208[6]	10621[7]	14967[8]	17572[8]
FRG	3624[3]	2393[4]	4040[5]	5994[6]	8802[7]	13835[8]	16521[8]
Number of telephones per thousand people (units)							
USSR	9	13	16	20	28	45	53
USA	166	282	339	411	482	587	628
Japan	16	20	35	59	142	251	318
UK	64	106	133	156	196	268	315
FRG	52	48	77	108	150	228	268
USSR number per thousand people as a percentage of:							
USA	5·4	4·6	4·7	4·9	5·8	7·7	8·4
Japan	56·3	65·0	45·7	33·9	19·7	17·9	16·7
UK	14·1	12·3	12·0	12·8	14·3	16·8	16·8
FRG	17·3	27·1	20·8	18·5	18·7	19·7	19·8

Sources:
1 *Transport i svyaz' SSSR* (1972), p. 282.
2 *Narodnoe khozyaistvo SSSR v 1973 godu*, p. 531.
3 *UN Statistical Yearbook* (1951), p. 358.
4 *Ibid.* (1954), pp. 334–5.
5 *Ibid.* (1959), pp. 367–9.
6 *Ibid.* (1965), pp. 488–90.
7 *Ibid.* (1970), pp. 501–2.
8 *Ibid.* (1973), pp. 495–8.

Note:
a – 1937.

3. Iron and steel

Introduction

In 1913 Imperial Russia produced 4·3 million tonnes of crude steel, an output attained by the USA over 20 years earlier. During the civil war following the Revolution, output declined to a low point of only 150 thousand tonnes in 1919, but from then on the industry was gradually restored so that the pre-war output level had again been reached by the beginning of the first five year plan in 1929. During the three pre-1941 five year plans the foundations of the modern Soviet iron and steel industry were laid. The great Ural-Kuznetsk combine was built, leading to a substantial increase in iron and steel output obtained in the Urals and beyond; major new works in the Ukraine and European Russia included Azovstal', Krivoi Rog and the Novo-Tula factory. In the pre-war years the Soviet industry assimilated the production of a wide range of the quality steels and ferro-alloys required for the modern engineering and defence industries. In 1940 the output of steel reached 18·3 million tonnes, 3·7 times the 1929 level and 13 per cent of world steel output in that year. During the Second World War much capacity in the Ukraine was destroyed, but production to the east of the Urals was expanded. Pre-war steel output was reached again in 1948 and doubled by 1963. By 1965 the Soviet industry accounted for one fifth of total world steel production and in 1971 output at 120·6 million tonnes exceeded the US level for the first time.

Apart from some small-scale production at engineering factories, iron and steel production in the USSR is administered by the Ministry of Ferrous Metallurgy. This ministry is responsible for all stages of metallurgical production from the extraction and preparation of iron ore to the making of finished metal products, but not the making of equipment for the industry. In January 1973 the iron and steel industry included 9·8 per cent of all industrial fixed capital ('basic funds') in the USSR and employed a total of 1·75 million people.[1] The industry is responsible for about five per cent of total annual

---

[1] *Ekonomika i Organizatsiya Promyshlennogo Proizvodstva*, 1974, No. 4, pp. 71–2.

industrial production, and about nine per cent of total annual industrial investment,[2] indicating the relatively high capital intensity of iron and steel production. Over 30 research organisations employ a staff of 18,500 persons including 9,000 scientists; about 100 million roubles a year is spent on scientific research work.[3]

The basic technology of iron and steel making has not changed fundamentally for many years, but it has undergone progressive improvements, especially since the war. Iron ore is treated in a blast furnace to make iron which then serves as the basic material for steelmaking in a converter or furnace. The molten steel is cast into ingots and then, after reheating, rolled into various shapes as required or, in the case of the new technology of continuous casting, the molten metal is directly cast into billets or slabs for subsequent rolling, eliminating the need for primary rolling and reheating. Each of the stages of the iron and steelmaking process as such is considered in detail below: the preliminary stage, the mining and treatment of iron ore, is only briefly mentioned, as it has a technology quite different from that of the main metallurgical processes.

Ironmaking

In 1970 the Soviet Union displaced the United States as the leading iron producing country, although during the 1960s the fastest rate of growth of output was achieved by the Japanese industry (see Table 3.1). Output per head of population remains, however, lower than in all the other industrialised countries except the United Kingdom; in 1973 it was 384 kg in the USSR, compared with 436 kg in the USA, 829 kg in Japan, 594 in Germany and 305 in the United Kingdom.

The main product of the ironmaking industry in all countries is conversion pig iron for steelmaking. In addition, a certain proportion of output takes the form of foundry pig iron used directly by customers, mainly in engineering; in recent years the proportion of foundry iron has fallen in all major producing countries as parts made from cast iron have been replaced by pressings, forgings and welded items made from rolled steel. In the Soviet Union the proportion of foundry iron is higher than in other countries, reflecting the relatively greater role of heavy engineering and the greater reliance on metalcutting techniques. Since the Second World War the smelting of ferro-alloys in blast furnaces has been replaced by the use of electric furnaces. (See Table 3.2.)

Although the basic technology of ironmaking in blast furnaces has not undergone any qualitative changes for several decades, there has been quite significant technical progress in the form of an accumulation of partial refinements, particularly in the last 10 to 15 years. The main directions of technical change are considered under the following headings: preparation of raw materials, scale of blast furnaces, intensification of the smelting process, and mechanisation and automation of processes.

[2] P. A. Shiryaev, *Kapital'noe stroitel'stvo v chernoi metallurgii* (1973), p. 53.
[3] *Ekonomika i Organizatsiya Promyshlennogo Proizvodstva*, 1974, No. 4, p. 71.

Table 3.1 *Output of pig iron and blast furnace ferro-alloys (thousand tonnes)*

	1950	1955	1960	1965	1970	1973
USSR	19175	33310	46757	66184	85933	95933
USA	57667	68623	59358	80612	83295	91813
Japan	2233	5217	11896	27502	68048	90007
FRG	9473	16482	25739	26990	33627	36828
UK	9633	12470	15763	17740	17339	17032

Source: *Quarterly Bulletin of Steel Statistics for Europe*, various years.

Table 3.2 *The structure of iron produced (% of total by weight)*

| | 1960[1] | | | 1965[1] | | | 1973[2] | | |
	S	F	A	S	F	A	S	F	A
USSR	82·3	14·9	2·8	84·0	13·6	2·4	89·0	9·1	1·0
USA	92·2	6·6	1·2	93·8	5·4	0·8	97·2	2·4[a]	0·4
Japan	90·5	9·5	—	94·7	5·3	—	97·6	2·4	—
FRG	91·3	6·4	2·3	91·5	6·5	2·0	94·3	5·0[a]	0·7
UK	86·5	12·2	1·3	90·6	8·3	1·1	92·9	6·0	1·1

S – steelmaking iron; F – foundry iron; A – blast furnace ferro-alloys.

Sources:
1 *Osnovnye napravleniya nauchno-tekhnicheskogo progressa* (1971), pp. 82–3.
2 Calculated from *Quarterly Bulletin of Steel Statistics for Europe*, 1974, No. 2, pp. 26–30.

Note: a – includes some blast furnace ferro-alloys.

THE PREPARATION OF RAW MATERIALS

The Soviet Union is the world's largest producer of iron ore, a position which it has occupied since the late 1950s. While other major producing countries rely quite heavily on imported ores (Japan almost completely), the USSR is a major exporter. As high-grade deposits have been depleted in recent years the iron content of mined ore has steadily declined, from 44·5 per cent in 1960 to 37·3 per cent in 1970, but the development of ore concentration and benefication processes has provided a gradual improvement in the iron content of merchantable ores, which has risen from 54·3 per cent in 1960 to 58·8 per cent in 1970.[4]

The iron content of merchantable ores is nevertheless lower than in other major producing countries: e.g., in 1970 the iron content was 61·9 per cent in the USA, 63·3 per cent in Japan and 64·3 per cent in Canada.[5]

The pre-treatment of iron ore has been extensively developed by the Soviet industry and since 1954 the USSR has been the world's largest producer of

[4] D. I. Popov, *Povyshenie effektivnosti metallurgicheskogo proizvodstva* (1972), p. 88.
[5] *Ekonomika i Organizatsiya Promyshlennogo Proizvodstva*, 1974, No. 4, p. 86.

agglomerates, mainly in the form of sinter. Sintering is a means of raising the quality of ores by removing impurities and ensuring a homogeneous material input into the furnace. Furthermore, it provides a means of using finely powdered ore which would otherwise be wasted. The use of sinter in the furnace improves the quality of iron and also raises the efficiency of smelting because some processes previously performed within the furnace are completed in advance outside the furnace. The provision of a high quality, homogeneous input is also a basic precondition for improving the efficiency of smelting by other means to be discussed below. One important path of improving the quality of ores is the making of self-fluxing iron ore agglomerates by pre-treating the ore with limestone. The use of self-fluxing agglomerates leads to fuel savings and raises efficiency. In the Soviet Union the use of self-fluxing agglomerates has reached a very high level. In 1950 15 per cent of the output of agglomerates was of this type, in 1970 almost 100 per cent, compared with under 40 per cent in the USA in 1969, and 95 per cent in Japan in 1968.[6]

One of the most economical forms of agglomeration is pelletisation, a process which ensures a high degree of homogeneity of both the chemical composition and physical shape of the material input into the furnace. In the United States the production of pellets reached a large scale during the 1950s and by 1969 output had reached 53 million tonnes, or 53 per cent of the total production of agglomerates.[7] In the USSR the making of pellets began in 1964 and output grew from 0·3 million tonnes in 1965[8] to 21·5 million tonnes in 1973, when it represented 12·8 per cent of total agglomerate production.[9] In Japan the output of pellets reached 6·4 million tonnes in 1973, or 5·7 per cent of total agglomerate output.[10]

The high level of production of agglomerates in the Soviet Union is matched by an unusually high proportion of agglomerates in the total iron ore component of the blast furnace burden as Table 3.3 illustrates.

THE SCALE OF BLAST FURNACES

One of the notable features of the development of the ironmaking industry during the last 10 years has been the steady increase in the maximum volume of blast furnaces. Soviet specialists claim that increased scale results in reduced fuel consumption, greatly improved labour productivity and reduced capital investment per tonne of iron produced. On the other hand, the output per cubic metre of furnace volume has a tendency to decline with size. On the whole, cost savings from increased size are not large, as Table 3.4 shows.

As Table 3.5 shows, both the average and maximum capacity of Soviet blast furnaces have been high by international standards.

The Soviet ironmaking industry is more highly concentrated than that of the

---

[6] (Ed.) N. P. Bannyi, *Ekonomika chernoi metallurgii SSSR* (1971), pp. 96–7.
[7] A. Sutulov, *The Soviet Challenge in Base Metals*, Utah (1971), p. 88.
[8] D. I. Popov and N. I. Mityaev, *Ispol'zovanie osnovnykh fondov metallurgicheskikh predpriyatii* (1972), p. 82.
[9] *Quarterly Bulletin of Steel Statistics for Europe*, 1974, No. 2, p. 18.
[10] *Ibid.*, p. 21.

USA: in 1970 furnaces having a useful volume of over 1,500 cubic metres accounted for 53·6 per cent of the total useful volume of all blast furnaces, compared with only 23·8 per cent in the USA in 1967; while units of 100 cubic metres or less accounted for 15·3 per cent of total capacity and 36·2 per cent respectively.[11] However, the role of large furnaces is now greater in Japan than

Table 3.3 *The average proportion of sinter and pellets in the ore component of the blast furnace burden (per cent)*

	1956	1960[1]	1965	1970	1973
USSR	61·6[1]	73·0	87·0[1]	92·6[1]	94·4[1]
USA	27·5[1,a]	52·5	63·9[2]	70·0[2,c]	n.a.
Japan	44·0[3,b]	46·5	62·6[1]	75·9[1]	79·8[1]
FRG	29·3[1]	41·0	67·4[1]	70·1[1]	73·5[1]
UK	28·1[1]	49·6	71·4[1]	71·2[1]	66·0[1]

Sources:
1 *The European Steel Market in 19..*, various years.
2 British Steel Corporation, *Statistical Handbook*, London (1969).
3 *Principal Factors Affecting Labour Productivity: Trends in the Iron and Steel Industry*, UNECE, NY (1969), p. 21.

Notes:
a – 1957.
b – 1955.
c – 1969.

Table 3.4 *Indicators of work of blast furnaces of different volume*

Indicator	Furnace useful volume (m³)						
	1033	1386	1719	2000	2700	3200	5000
Annual output (mt)[1]	1·0	1·3	1·4	1·6	2·1	2·5	4·0
Coefficient of utilsn. of useful volume (m³/t per day)[1]	0·362	0·381	0·427	0·438	0·448	0·452	0·455
Expenditure of dry coke per t of iron[1]	412	410	406	403	398	395	380
Annual output per worker of blast furnace shop[1] ('000 t)	10·2	12·2	12·2	12·7	14·9	16·9	24·5
Capital investment per t annual output (index)[1]	100	98	96	94	89	84	60
Cost per t of iron (index)[2]	100	99·7	99·2	99·1	98·3	97·1	96·9

Sources:
1 P. A. Shiryaev, *Kapital'noe stroitel'stvo v chernoi metallurgii* (1973), p. 27.
2 (Ed.) N. P. Bannyi, *Ekonomika chernoi metallurgii SSSR* (1971), p. 421.

[11] R. S. Livshits, *Effektivnost' kontsentratsii proizvodstva v promyshlennosti SSSR* (1971), p. 75

in the USSR. At the beginning of 1971 the USSR had 16 furnaces of 2,000 cubic metres useful volume or more out of a total of 132 blast furnaces, whereas Japan in 1969 had 18 such furnaces out of a total of 61 furnaces;[12] at the end of 1973 11 of the 15 largest blast furnaces in the world were Japanese, but only one Soviet (the thirteenth).[13] Furthermore, whereas the annual output per furnace in operation in the early 1960s was approximately the same in the USSR, USA and Japan, by 1969 the average output per furnace in the USSR was 618,000 tonnes, in the USA 516,000 tonnes and in Japan 960,000 tonnes,

Table 3.5 *Average and maximum capacity of blast furnaces (useful volume—cubic metres, year end)*

	1940	1950	1960	1965	1970	1972	1974
USSR:							
average	589[1]	639[1]	904[1]	1082[1,b]	1135[2]	n.a.	n.a.
maximum	1300[1]	1370[1]	2002[1]	2300[2]	2700[2]	3200[3]	5000[4]
USA:							
average	714[5,a]	805[5,e]	966[5]	1087[5,b]	n.a.	n.a.	n.a.
maximum	n.a.	1500[6,e]	n.a.	2150[7]	2520[8]	2520[8]	2520[8]
Japan:							
average	n.a.	n.a.	n.a.	1181[9,c]	1260[10]	2070[11]	n.a.
maximum	n.a.	n.a.	1947[12,d]	2142[7]	3363[13]	4363[13]	4617[13]

Sources:
1. (Ed.) N. P. Bannyi, *Ekonomika chernoi metallurgii SSSR* (1971), p. 421.
2. D. I. Popov, *op. cit.*, pp. 67–8.
3. *Journal of Metals*, December 1973, p. 4.
4. *Pravda*, 3 January 1975.
5. R. S. Livshits, *Effektivnost' kontsentratsii proizvodstva v promyshlennosti SSSR* (1971), p. 76.
6. *Advances in Steel Technology in 1956*, UNECE, Geneva.
7. (Eds.) A. S. Tolkachev, I. M. Denisenko, *Osnovnye napravleniya nauchno-tekhnicheskogo progressa* (1971), p. 252.
8. *Journal of Metals*, February 1972, p. 19. (No subsequent reports of larger furnaces in 1973 or 1974.)
9. *Principal Factors Affecting Labour Productivity: Trends in the Iron and Steel Industry*, UNECE, NY (1969), p. 100.
10. P. A. Shiryaev, *op. cit.*, p. 16.
11. *Journal of Metals*, October 1972, p. 4.
12. *Mirovaya Ekonomika i Mezhdunarodnye Otnosheniya*, 1973, No. 6, p. 53.
13. *Journal of Metals*, December 1973, p. 4.

Notes:
a – 1938.
b – 1967.
c – 1966.
d – 1962.
e – 1945.

[12] I. A. Zaitsev, *Nauchno-tekhnicheskii progress i effektivnost' proizvodstva*, Kiev (1973), pp. 51–2.
[13] *Journal of Metals*, December 1973, p. 4.

rising to over 1·1 million tonnes in 1972, when the Soviet level reached 694,000 tonnes.[14]

A feature of the Soviet ironmaking industry has been the slow rate of withdrawal of old and small units. The average age of units withdrawn from use in the period 1959 to 1970 was about 70 years, and in the early 1970s there were 26 blast furnaces in operation having a useful volume of less than 500 cubic metres. These old and small units tend to have a very low productivity of labour and high repair costs.

The average age of blast furnaces also shows a tendency to rise, from 20·7 years in 1959 to 25 years in 1971. At the beginning of 1971 47 out of a total of 132 blast furnaces were fully depreciated (over 25 years old); these units accounted for about a fifth of total output.[15] In the USSR old units are still kept in regular production, but in the USA such units tend to be maintained as a reserve for use only at times of high demand; thus in 1969 there were 228 blast furnaces for the production of pig iron and blast furnace ferro-alloys, but of these 59 were idle (in 1965 85 out of 236).[16] Very high levels of capacity utilisation are claimed for the Soviet ironmaking industry: 90·5 per cent utilisation in 1966, rising to 95·5 per cent in 1970.[17]

INTENSIFICATION OF THE SMELTING PROCESS

Gradual improvement of the pre-treatment of iron ore has enabled an increase in the efficiency of blast furnace operation, particularly in the form of the reduced consumption of coke, the primary fuel of ironmaking. One important means of intensifying the smelting process has been the increase in the pressure of the top gas of the furnace. This method was known before the war, but was first widely used in the Soviet Union in the 1950s following experiments which began in 1940.[18] By 1955, 46 furnaces were working with raised pressure compared with 13 in the USA.[19]

In 1965 102 out of 129 blast furnaces in the USSR were using raised top gas pressure,[20] but in 1968 in the USA only 30 out of 173 active furnaces,[21] and in Japan 22 out of 58 furnaces.[22] Another means of raising efficiency pioneered by the Soviet industry is the use of a blast of high temperature. In 1970 about 60 per cent of furnaces worked with a blast of 1,000°C or more.[23] This method has been widely applied by the Japanese industry, the majority of furnaces in Japan working with blast temperatures of 1,100–1,200°C.[24]

[14] 1969—*Referativniy Sbornik, Ekonomika Promyshlennosti*, 1972, Item 3V113; except USSR—calculated from known output data and number of furnaces; 1972—number of furnaces, *Iron and Steel Engineer*, January 1972, p. D-27.

[15] D. I. Popov and N. I. Mityaev, *Ispol'zovanie osnovnykh fondov metallurgicheskikh predpriyatii* (1972), pp. 37–9.

[16] *British Steel Corporation—Statistical Handbook, 1969*, London (USA Section) p. 3.

[17] P. A. Shiryaev, *Kapital'noe stroitel'stvo v chernoi metallurgii* (1973), p. 16.

[18] L. N. Roitburd, *Ocherki ekonomiki chernoi metallurgii* (1960), p. 465.

[19] P. Belan and I. Denisenko, *Perspektivy razvitiya chernoi metallurgii SSSR* (1962), p. 80.

[20] D. I. Popov, *op. cit.*, p. 95.

[21] N. E. Sidorov, *Tekhnicheskii progress i snizhenie energoemkosti produktsii chernoi metallurgii*, Kiev (1974), p. 15.

[22] V. A. Minenko, *Proizvoditel'nost' truda v chernoi metallurgii SSSR* (1973), p. 20.

[23] D. I. Popov, *op. cit.*, p. 100.

[24] N. E. Sidorov, *op. cit.*, p. 16.

Table 3.6 *Average coke consumption per tonne of pig iron produced (kg)*

	1957	1960	1965	1967	1968	1969	1970	1971	1972	1973
USSR	817	724	586	561	588	581	575	566	564	558
USA	842	770	650	639	633	636	658	646	619	595
Japan	718	619	507	501	504	496	478	451	442	438
FRG	954	834	672	604	579	564	559	521	487	495
UK	947	825	680	657	656	650	625	610	580	543

Source: The European Steel Market in 19.., UNECE, various years.

In all ironmaking countries since the war there has been a search for cheaper substitutes for coke as a fuel for smelting. The Soviet industry has placed particular emphasis on the use of natural gas, first used in world practice in the USSR in 1957.[25] In 1970 86 per cent of all iron produced in the USSR was smelted with the use of natural gas which was applied at 73 per cent of all furnaces.[26] In 1968 about half the furnaces in operation in the USA used natural gas, a smaller proportion than in the USSR where two thirds of all furnaces were using natural gas in 1965.[27] Natural gas has been found to be particularly effective if combined with oxygen; oxygen injection was tried in the Soviet Union before the war, but because of certain technical problems was not used to any extent until the 1960s, when it was diffused quite rapidly: in 1960 13 furnaces worked with oxygen enriched blast, in 1965 56 furnaces and in 1970 75 furnaces accounting for 56 per cent of total iron output.[28] Oxygen is used to a greater extent than in other countries. In 1969 33 cubic metres of oxygen were used per tonne of iron smelted, compared with only 4·6 cubic metres in the USA, 4·2 in Japan and 6·5 in Germany.[29] However, it appears that in other countries much greater use is made of alternative forms of blast enrichment, notably heavy oil and fine coal, little used in the USSR. In summary—while the Soviet Union pioneered the general use of a number of methods of raising the efficiency of the smelting process, other countries have in recent years been applying these methods, and others, on an ever wider scale. The Japanese industry has been especially active: in 1973 it was reported that nearly all its blast furnaces now work with high top pressure, with injection of heavy oil as a fuel and the use of a high temperature, oxygen enriched blast.[30]

One major consequence of this heightened efficiency has been the very significant reduction in coke usage which has occurred in all producing countries since the late 1950s. The rate of decline of coke expended per tonne of iron smelted has slowed down in the USSR in recent years, whereas in other countries it has tended to rise after a period of slower decline in the mid-1960s.

[25] *Steel International*, September 1966, p. 124.
[26] D. I. Popov, *op. cit.*, p. 95.
[27] N. E. Sidorov, *op. cit.*, p. 15; D. I. Popov, *op. cit.*, p. 95.
[28] D. I. Popov, *op. cit.*, p. 95.
[29] N. E. Sidorov, *op. cit.*, pp. 15–16.
[30] *Journal of Metals*, October 1973, p. 6.

Table 3.7 *Average daily output per cubic metre of furnace volume (tonnes)*

	1961	1963	1965	1967	1969	1971	1973	Max. 1973
USSR	1·37[1]	1·42[1]	1·51[1]	1·59[1]	1·65[1]	1·69[1]	1·75[2]	2·17[2]
Japan	1·16[3]	1·25[3]	1·42[3]	1·64[3]	1·85[3]	1·84[3]	2·03[4]	2·52[4]

Sources:
1 *Narodnoe khozyaistvo SSSR za 1922–72 gody* (1972), p. 168.
2 *The Steel Market in 1973*, UNECE, NY (1974), p. 96.
3 *Journal of Metals*, October 1972, p. 4.
4 *Ibid.*, August 1974, p. 5.

As a result, the absolute level of coke expended per tonne of iron is now substantially greater than in Japan and Germany (see Table 3.6).

An indication of the rate of improvement of blast furnace performance is also provided by the average daily output per cubic metre of furnace volume shown in Table 3.7.

This clearly reveals the rapid progress of the Japanese industry in recent years.

One major Soviet technical innovation, a system of evaporative cooling for blast furnaces, has been widely introduced abroad on a licence basis.[31]

MECHANISATION AND AUTOMATION

In 1974 a Soviet expert acknowledged that blast furnaces were not as well provided with modern equipment as those of Japan.[32]. It is also admitted that there remains a substantial amount of non-mechanised work in Soviet ironmaking: the proportion of manual workers comprises as much as about 45 per cent of the labour force.[33] The conveyer system of feeding the charge to the top of the furnace was being introduced in the Soviet Union at furnaces newly built during the ninth five year plan, whereas 10 furnaces were so equipped in Japan in 1968.[34] The use of computers for process control also appears to be at a fairly early stage of development.[35]

DIRECT REDUCTION OF IRON ORE

In recent years great efforts have been made to develop an alternative to the traditional blast furnace process of ironmaking. These efforts have been prompted partly by fear of a future shortage of suitable coking coals and partly by the need to create viable iron and steelmaking units of a relatively small size for developing countries and for special purposes in developed countries. A number of different processes have been developed but to date they have only been economic in small-scale use. In the *direct reduction* process, pre-treated

---

[31] *Literaturnaya Gazeta*, 22 November 1972, p. 10; *Soviet News*, 8 May 1973.
[32] N. E. Sidorov, *op. cit.*, p. 17.
[33] V. A. Minenko, *op. cit.*, p. 22.
[34] N. E. Sidorov, *op. cit.*, p. 17.
[35] D. I. Popov, *op. cit.*, pp. 194, 203–306.

iron ore of high iron content is reduced to 'sponge iron' using gas or a non-coke solid fuel and this sponge iron, together with scrap, is converted to steel in an electric arc furnace. Substantial cost savings are claimed, construction times are greatly reduced and labour productivity can achieve very high levels, superior to those attained in large enterprises.

The direct reduction process enables the establishment of viable 'mini-factories' and a number of these are now in operation: in the USA there are about 50 with a 50,000–500,000 tonnes per year unit capacity; over half of these have been constructed since 1967. World production of directly reduced iron now amounts to about six million tonnes a year.[36] In the Soviet Union some early experience was obtained through the transfer of German plants to the USSR as reparations after the war.[37] It is claimed that one of the methods of direct reduction now employed in the USA and patented there in 1967 very closely resembles a process (the 'KShS' process) patented in the USSR in 1959 and tested under laboratory conditions.[38] At present there are apparently no 'mini-factories' in the USSR, but recently the idea of building such works on the basis of direct reduction has been promoted as a means of meeting the iron and steel needs of the Soviet far east and eastern Siberia.[39]

The last two or three years have seen a number of practical measures for the creation of direct reduction capacity in the USSR. In late 1974 it was reported that construction was beginning on the first Soviet electro-metallurgical complex using direct reduction of highly concentrated ores produced by the Lebedinsk mining and concentration works near Kursk in the Ukraine.[40]

Furthermore the Soviet industry has reached agreement with a consortium of German firms, including Krupp, for the construction of a very large integrated steel works near Kursk. This complex will incorporate a direct reduction plant of 2·5 million tonnes a year capacity (much larger than any such plant now in existence), electric arc furnaces and continuous casting installations. The first stage of this project will be built from 1977.[41]

Steelmaking

Throughout the 1950s and 1960s the Soviet Union was second only to the USA in its steel production. In 1971 the USSR displaced the USA as the world's leading steel producer, but temporarily lost this lead again in 1973. A notable feature of the last decade has been the extremely rapid rise of the Japanese steel industry. The growth of output of the Soviet industry has been very stable compared with that of other countries, especially that of the US industry, which has found it necessary to retain considerable reserve capacity. Output

---

[36] *Ekonomika i Organizatsiya Promyshlennogo Proizvodstva*, 1974, No. 4, p. 87.
[37] A. C. Sutton, *Western Technology and Soviet Economic Development, 1945–65*, Stanford, Cal., (1973), pp. 123–4.
[38] *Ekonomika i Organizatsiya Promyshlennogo Proizvodstva*, 1974, No. 4, p. 88.
[39] *Ibid.*, pp. 81, 85–92.
[40] *Iron and Steel International*, December 1974, p. 435.
[41] *Ibid.*, June 1974, p. 163; *Press Bulletin, Moscow Narodny Bank*, 2 January 1975, p. 5.

Table 3.8 *Output of crude steel (thousand tonnes)*

	1950	1955	1960	1965	1970	1973[2]
USSR	27329[1]	45272[1]	65294[1]	91021[1]	115889[1]	131481
USA	87848[3]	106173[3]	90068[2]	18985[2]	119140[2]	136462
Japan	4839[3]	9408[3]	22138[2]	41161[2]	93322[2]	119322
FRG	14019[2]	24502[2]	34100[2]	36821[2]	45041[2]	49521
UK	16554[2]	20108[2]	24695[2]	27440[2]	28316[2]	26649

Sources:
1 *Narodnoe khozyaistvo SSSR za 1922–72 gody* (1972), p. 165.
2 *Quarterly Bulletin of Steel Statistics for Europe*, various years.
3 *UN Statistical Yearbook*, NY (1955) and (1962).

per capita, however, is still less than that in the other leading producing countries.[42]

The Soviet share of world steel output rose steadily from 14·4 per cent in 1950 to 19·9 per cent in 1965, but, primarily owing to the expansion of the Japanese industry, has declined slightly in recent years to 19·6 per cent in 1970 and 19·2 per cent in 1974.[43]

Until about 1960, the Siemens-Martin open hearth steelmaking process was for several decades responsible for 85–90 per cent of total crude steel output in all major steel producing countries, except those using the basic Bessemer (Thomas) process for ores of high phosphorus content. Since 1960 the oxygen converter has rapidly displaced the open hearth furnace as the primary process, notably in Japan, Germany and the USA. In a number of countries the proportion of steel produced in electric arc furnaces has also risen. The changing role of different processes is shown in Table 3.9.

OPEN HEARTH STEEL

In all the main steel-producing countries, with the exception of the USSR, the output of open hearth steel has declined in both relative and absolute terms since the early 1960s. In the USSR the absolute output of open hearth steel continues to rise:

Whereas the construction of open hearth furnaces ceased in Japan in the late 1950s and the USA by the mid-1960s, the Soviet industry was introducing new capacity as late as 1970. A distinguishing feature of Soviet open hearth furnaces is their high average and maximum capacity. In the USA the largest furnace has a capacity of 550 tonnes,[44] but the Soviet industry has several of 900 tonnes, the world's largest open hearth installations.

[42] See Table 2A.1, p. 72.
[43] Calculated from: *British Steel Corporation, Statistical Handbook 1969; 1974;* London *Financial Times*, 22 January 1975; *Pravda*, 25 January 1975.
[44] (Eds.) L. M. Gatovskii *et al.*, *Voprosy optimal'nogo razmera predpriyatii v promyshlennosti SSSR* (1968), pp. 245–6. (In 1965 there were six Soviet open hearth furnaces of 850–900 tonnes capacity, when the largest American furnace was 550 tonnes.)

Table 3.9 *The structure of steelmaking processes (proportion of total steel output by type of process—per cent[b])*

	1940	1950	1955	1960	1965	1970	1973
USSR							
Open hearth furnace	84.8[1]	89.9[1]	88.1[1]	84.4[2]	84.0[2]	72.5[2]	67.8[2]
Oxygen converter	—	—	—	3.8	4.4	17.3	21.4
Electric furnace	5.9	6.3	7.5	8.9	9.4	9.2	10.0
USA							
Open hearth furnace	91.9[3]	89.1[3]	90.0[3]	87.0[2]	71.7[2]	37.3[2]	26.4[2]
Oxygen converter	—	—	—	3.4	17.4	47.0	55.4
Electric furnace	2.5	6.2	6.9	8.4	10.5	15.7	18.2
Japan							
Open hearth furnace	80.8[3]	80.4[3]	83.1[3]	68.0[3]	24.7[3]	4.1[2]	1.6[2]
Oxygen converter	—	—	—	11.9	55.0	49.1	80.5
Electric furnace	15.8	15.6	12.6	20.2	20.3	16.8	17.9
FRG							
Open hearth furnace	51.1[2,a]	51.0[2]	49.1[2]	47.2[2]	42.9[2]	26.3[2]	18.3[2]
Oxygen converter	—	—	—	2.7	19.1	55.8	67.8
Thomas converter	44.7	46.0	46.8	43.7	29.4	8.1	3.5
Electric furnace	3.5	2.4	4.0	6.4	8.5	9.8	10.4
UK							
Open hearth furnace	88.2[3]	87.7[2]	87.2[2]	84.5[2]	63.7[2]	47.2[2]	31.7[2]
Oxygen converter	—	—	—	1.7	20.5	32.1	47.3
Electric furnace	3.4	4.5	5.6	6.9	13.1	19.5	19.9

Calculated from:
1 (Ed.) N. P. Bannyi, *Ekonomika chernoi metallurgii SSSR* (1971), p. 434.
2 *Quarterly Bulletin of Steel Statistics for Europe*, various years.
3 *British Iron and Steel Federation, Statistical Handbook*, London, various years.

Notes:
a – 1938.
b – percentages may not sum to 100.0 because of the exclusion of 'other processes'.

Table 3.10 *The output of steel smelted in open hearth furnaces (thousand tonnes)*

	1960	1965	1970	1974
USSR	55109[2]	76530[2]	84052[1]	90717[1]
USA[1]	78351	85451	45531	32203
Japan[1]	15045	10166	3855	1553
FRG[1]	16087	15805	11819	9287
UK[1]	20863	17490	13374	6189

Sources:
1 *Quarterly Bulletin of Steel Statistics for Europe*, various years.
2 *British Iron and Steel Federation, Statistical Handbook*, London (1965).

The average capacity of Soviet furnaces rose from 154 tonnes in 1956[45] to 250 tonnes in 1970,[46] compared with 185 tonnes in the USA in 1967.[47] Furthermore, the Soviet open hearth capacity is more highly concentrated: units of 500 tonnes or more account for 20 per cent of total capacity, compared with only 0·9 per cent in the USA; while units of less than 200 tonnes account for 24·7 per cent and 50·2 per cent respectively.[48] There are nevertheless many small open hearth furnaces in the USSR. In 1971 52·5 of the total number of furnaces had a capacity of under 200 tonnes (including 26 per cent less than 100 tonnes), compared with 10·9 per cent with a capacity over 500 tonnes.[49] In the West part of the remaining open hearth capacity is now being used for smelting alloy metals.

The Soviet industry has widely adopted measures for intensifying the operation of open hearth furnaces. One of the main methods has been the use of oxygen enrichment. The use of oxygen was first tried experimentally before the war in the Soviet Union, but not applied on a regular basis until the 1950s. By 1970 oxygen was used in the smelting of 77 per cent of Soviet open hearth steel.[50] Oxygen enrichment has been widely applied in other countries but in most cases to a lesser degree than in the USSR, where 45·5 cubic metres of oxygen were used per tonne of open hearth steel, as compared with 1968 levels of 30·8 cubic metres for the USA, 34·4 for Germany, 31·3 for the United Kingdom. Japan is again a major exception, using as much as 61 cubic metres per tonne.[51] Natural gas has also been used in Soviet open hearth furnaces since 1957, and by 1970 almost 60 per cent of open hearth steel was smelted with its use.[52]

Since 1965, when an existing open hearth furnace was converted, Soviet specialists have also been developing and introducing a new form of furnace having two baths instead of the single bath of the traditional open hearth process. The use of two baths makes fuller use of the heat generated, speeds up the smelting process and in effect transforms the open hearth method into a semi-continuous process. It is also claimed that fuel consumption is greatly reduced and that the refractories have a longer life. Economic indicators are claimed to be much improved, labour productivity being estimated at 20–30 per cent higher than for the usual open hearth furnace. Several furnaces of this new type are already in operation in the Magnitogorsk combine and the Cherepovetsk factory; more are now being built by reconstructing existing open hearth units. While this new form of furnace represents an advance on the traditional steelmaking process, it is acknowledged that economic performance is inferior to that of converters.[53]

---

[45] *Advances in Steel Technology in 1956*, UNECE, Geneva, p. 9.
[46] I. A. Zaitsev, *Nauchno-tekhnicheskii progress i effektivnost' proizvodstva chernykh metallov*, Kiev (1973), p. 53.
[47] R. S. Livshits, *Effektivnost' kontsentratsii proizvodstva v promyshlennosti SSSR* (1971), p. 76.
[48] *Ibid.*, p. 75.
[49] A. G. Livshits, *Sebestoimost' stali i puti ee snizheniya* (1974), p. 43.
[50] N. E. Sidorov, *op. cit.*, p. 18.
[51] D. I. Popov, *op. cit.*, p. 105.
[52] *Narodnoe khozyaistvo SSSR za 1922–72 gody* (1972), p. 112.
[53] D. I. Popov and N. I. Mityaev, *op. cit.*, pp. 87–8; *Pravda*, 9 June 1973.

A measure of the gradual improvement in open hearth technology achieved in the USSR in recent years is provided by the average daily yield of steel per square metre of furnace area. This indicator rose from 5·36 tonnes in 1950 to 7·69 in 1960, 9·16 in 1970 and 10·15 in 1974.[54]

Soviet open hearth steelmaking is clearly at a high technological level but the level of some inputs is not up to the best foreign standards. In a recent work a Soviet specialist observed that 'The open hearth production of the Ukraine and the USSR as a whole is the most developed in the world by its capacity, technical equipment and the refinement of its technology. Its indicators would be still higher if the open hearth furnaces were secured better prepared scrap and iron ore deoxidising agents, and the quality of refractories were better. In these questions our production lags behind the indicators of the technically developed foreign countries.'[55]

OXYGEN CONVERTER STEEL[56]

One of the most significant innovations in the steel industry since the war has been oxygen steelmaking, first introduced in Austria in 1952. As Table 3.11 shows the Soviet Union introduced this process relatively early.

The diffusion of oxygen steelmaking is illustrated in Table 3.12 and Fig 3.1. Initially Soviet introduction of the L-D oxygen process was at a rate

Table 3.11 *Year of introduction of first and second oxygen converters and their capacity*

	Year of introduction		Capacity of first converter (t)
	First	Second	
USSR	1956[a]	1958[b]	25
USA	1954	1957	50
Japan	1957	1959	50
FRG	1955	1957	60
UK	1960	1962	30
France	1960	1961	20
Austria	1952	1953	30
Canada	1954	1958	40

Source: *Comparison of Steel-making Processes*, UNECE, New York (1962), pp. 78–82.

Notes:
a – Petrovskii factory, Dnepropetrovsk.
b – Lenin factory, Krivoi Rog.

[54] *Narodnoe khozyaistvo SSSR za 1922–72 gody* (1972), p. 168 (for 1950, 1960 and 1970). *Ekon. Gaz.*, 1975, No. 4, p. 2 (for 1974).
[55] N. E. Sidorov, *op. cit.*, p. 19. The relatively low technical level of scrap preparation is also acknowledged by D. I. Popov and N. I. Mityaev, *op. cit.*, pp. 90–1.
[56] For a more detailed study of the development of oxygen steel in the USSR, see J. Sláma and H. Vogel, in *Jahrbücher fur Nationalökonomie*, Vol. 187, 1972-3, pp. 245–61.

comparable to that of other major producing countries, but during the years 1959 to 1964, when rapid diffusion began in many countries, progress in the Soviet Union was slight. Since 1964 the rate of diffusion has increased but at a lower rate than in other countries. At the end of 1973 there were 32 oxygen converters in use in the USSR, compared with 92 in Japan, 74 in the USA, 36 in Germany and 23 in Britain.[57]

Not only has the rate of diffusion of oxygen steelmaking been relatively slower in the Soviet Union than in other leading steel-producing countries, but the scale of converters introduced has also been somewhat smaller, as shown in Table 3.13. During the current ninth five year plan period a number of converters built during the 1960s in the USSR are being reconstructed in order to increase their capacity from 100–130 tonnes to 145–150 tonnes. At the same

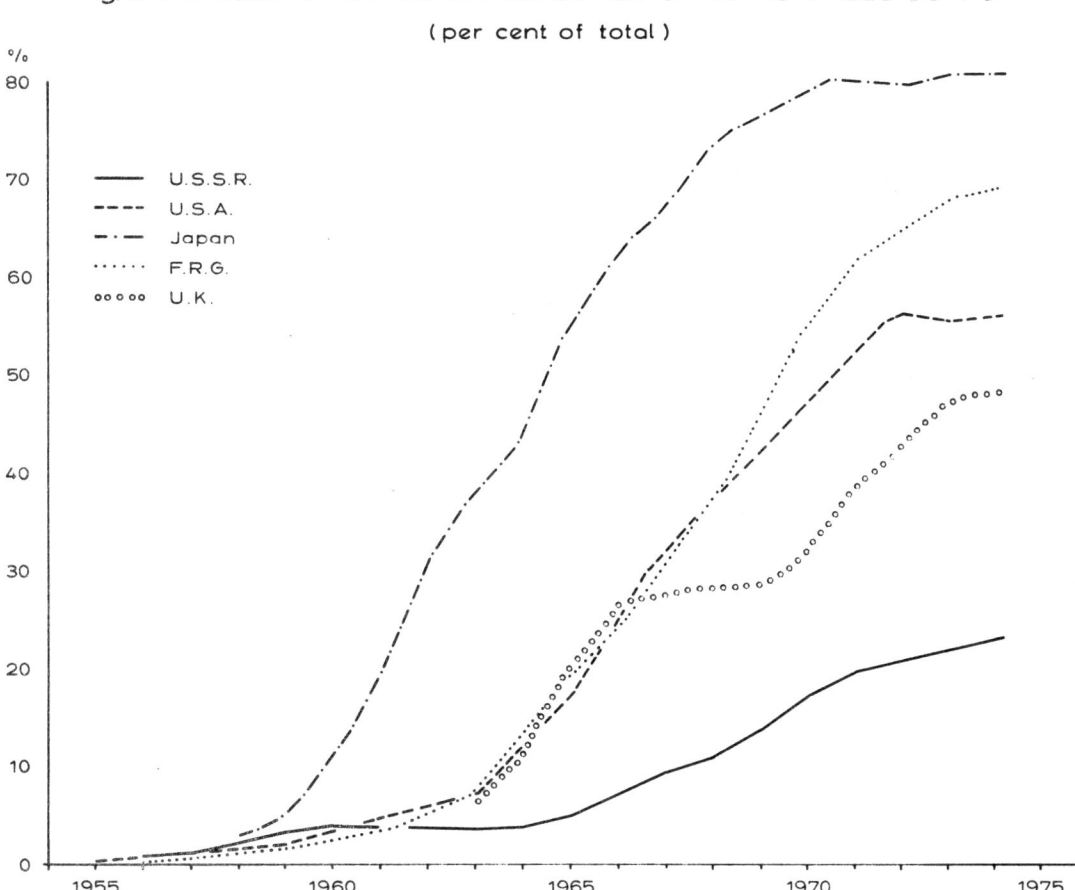

Fig. 3·1 OXYGEN STEEL AS A PROPORTION OF TOTAL STEEL OUTPUT (per cent of total)

[57] Calculated from *Stahl und Eisen*, 1974, No. 3, pp. 108–10.

Table 3.12 *The diffusion of oxygen steelmaking*

Year	USSR a	USSR b	USA a	USA b	JAPAN a	JAPAN b	FRG a	FRG b	UK a	UK b
1955	—	—	0.3[3]	0.3[3]	—	—	n.a.	n.a.	—	—
1956	0.07[2]	—	0.4[3]	0.4[3]	—	—	n.a.	n.a.	—	—
1957	0.5[2]	1.0[2]	0.6[3]	0.5[3]	n.a.	n.a.	0.01[1]	—	—	—
1958	1.2[2]	2.1[2]	1.2[3]	1.6[3]	0.4[1]	3.0[1]	0.3[1]	1.3[1]	—	—
1959	1.9[2]	3.2[2]	1.7[3]	2.0[3]	0.8[1]	4.4[1]	0.5[1]	1.9[1]	—	—
1960	2.5[1]	3.8[1]	3.0[3]	3.4[3]	2.6[1]	11.9[1]	0.9[1]	2.5[1]	n.a.	n.a.
1961[1]	2.5	3.5	3.6	4.9	5.4	19.0	1.2	3.6	n.a.	n.a.
1962[1]	2.7	3.5	5.0	5.6	8.4	30.6	1.7	5.2	n.a.	n.a.
1963[1]	2.7	3.4	7.7	7.8	12.0	38.2	2.4	7.8	1.5	6.7
1964[1]	3.3	3.8	14.0	12.8	17.6	44.2	5.2	14.0	3.0	11.4
1965[1]	4.1	4.5	20.7	17.4	22.6	55.0	7.0	19.1	5.6	20.2
1966[1]	6.4	6.6	30.8	25.3	29.9	62.6	8.7	24.5	6.5	26.1
1967[1]	9.4	9.2	37.6	32.6	41.7	67.2	11.6	31.5	6.7	27.6
1968[1]	11.3	10.6	44.1	37.1	49.3	73.7	15.3	37.1	7.4	28.1
1969[1]	15.2	13.8	54.6	42.6	63.2	76.9	20.8	46.0	7.4	28.2
1970[1]	19.9	17.2	57.5	47.1	73.9	79.1	24.1	55.8	9.1	32.1
1971[1]	23.2	19.2	58.0	53.2	70.8	80.0	24.9	61.8	9.4	38.7
1972[1]	25.8	20.5	67.6	56.0	77.0	79.4	28.2	64.7	10.8	42.6
1973[1]	28.1	21.4	75.6	55.4	96.1	80.5	33.6	67.8	12.6	47.3
1974[1]	30.8	22.7	74.0	56.0	94.7	80.8	36.6	68.8	10.8	48.1

a – output of oxygen steel (million tonnes); b – oxygen steel as a proportion of total crude steel output (per cent); — – oxygen steel not produced; n.a. – no data.

Sources:
1 *Quarterly Bulletin of Steel Statistics for Europe*, various years.
2 *Tekhniko-ekonomicheskie issledovaniya v chernoi metallurgii* (1963), p. 44.
3 *Statistical Abstract of the United States*, US Bureau of the Census, Washington DC (1962).

time they are being thoroughly modernised and equipped with automatic control systems.[58] New oxygen converters now under construction are all of 300 or 320 tonnes capacity.[59]

Considerable economies of scale are claimed for oxygen steelmaking: according to Gipromez calculations the transition from converter shops with 130 tonne units to shops with 300 tonne units will give a reduction of investment per unit output of 14 per cent, a rise in labour productivity of 20 per cent and a reduction in the cost of steel of 1.3 per cent.[60]

In 1963 an agreement was concluded for the supply of Austrian equipment for the construction of the Novo-Lipetsk oxygen converter shop and in 1964–6 23,800 tonnes of equipment worth 26.3 million roubles were

[58] A. G. Livshits, *op. cit.*, p. 41; P. A. Shiryaev, *op. cit.*, p. 28.
[59] *Stahl und Eisen*, 1974, No. 3, pp. 109–10.
[60] P. A. Shiryaev, *op. cit.*, pp. 29–30.

imported.[61] This suggests that there may have been technical problems in the early years of oxygen steel development in the USSR.

The existence of a very efficient open hearth technology and the availability of relatively low-cost scrap, which tends to make the traditional method of steelmaking more economic, are two factors which appear to have influenced the rate of introduction of the oxygen process.

The L-D oxygen process cannot use scrap as a material input to the same extent to which it can be used with the open hearth process, but if the proportion of scrap used in the latter process is increased, in order to absorb the scrap freed by the transition to the new process, beyond a certain point costs will rise. One authoritative Soviet work published in 1971 considered that in view of the economies of scrap use the optimal relationship between oxygen and open hearth steel in Soviet conditions would be 40:60, i.e., open hearth steel would continue to play a major role for many years.[62]

Until recently oxygen converters have been top blown, i.e., the oxygen is injected into the converter from the top. In 1967 in Germany the world's first commercial bottom blown oxygen unit went into operation. This process, known in the USA as the 'Q-BOP', offers a number of advantages over the conventional oxygen technique, as a higher proportion of scrap can be smelted

Table 3.13 *The average and maximum capacity of oxygen converters (tonnes)*

		1960	1965	1970	1973
USSR	Average	50	75	107	112
	Maximum	50	100	245	300
USA	Average	95	167	180	181
	Maximum	110	270	300	300
Japan	Average	82	100	131	144
	Maximum	245	245	245	300
FRG	Average	99	114	156	154
	Maximum	180	250	370	370
UK	Average	n.a.	108	134	148
	Maximum	n.a.	150	300	300

Source:
Calculated from *Stahl und Eisen*, 1974, No. 3, pp. 108–10, on the basis of converters in operation in 1973. The table therefore somewhat overstates the average capacities for 1960 and 1965, in so far as smaller units were withdrawn or enlarged before 1973.

[61] Yu. N. Kapelinskii, *Torgovlya SSSR s kapitalisticheskimi stranami* (1970), p. 126.
[62] (Ed.) N. P. Bannyi, *Ekonomika chernoi metallurgii SSSR* (1971), pp. 446–7.

and fuel economy is improved. By March 1973 there were two units in operation in the USA, including the largest then in use (200 tonnes), two in Germany, five in France, two in Belgium and single examples in a number of other countries. A very rapid rate of diffusion is predicted. One great advantage of this process is that it can be obtained by converting existing open hearth installations. One variant, the 'SIP' (submerged injection process), can use up to 60 per cent scrap and is particularly well suited to conversion from open hearth capacity. As Western observers have noted, this new technique could offer the possibility of rapidly converting the Soviet open hearth units into highly efficient oxygen installations at relatively low capital cost.[63] No evidence is so far available of the introduction of these new processes in the Soviet industry.

ELECTRIC ARC FURNACE STEEL

Electric steelmaking is extremely flexible: capacity can be introduced and withdrawn quickly according to changes in demand, and a range of steels of all qualities can be smelted. Before the Second World War, electric arc furnaces were primarily used to make alloy steels. In recent years, however, electric steelmaking has become a viable process for making ordinary carbon steel. The price of electricity has fallen in relation to other fuels, and the development of the oxygen steelmaking process has led to a decline in the demand for scrap, which can be used in a high proportion in electric arc furnaces. In consequence, the proportion of electric steel in the total steel output of major producing countries has gradually risen throughout the 1950s and 1960s, except in Japan, where the oxygen process has predominated. During the 1950s the share of electric arc steel in the Soviet Union was rather higher than in all other producing countries except Japan, but the other countries have pushed ahead more rapidly since 1960: the process is still restricted to the smelting of alloy steels in the Soviet Union.

Moreover, although it is acknowledged that there are economies of scale in electric steel production, the average and maximum size of Soviet furnaces has tended to be rather lower than in other major producing countries. Thus, in 1965 the maximum capacity was 100 tonnes,[64] compared with 180 tonnes[65] in the USA; and the average was 26 tonnes[66] compared with 36 tonnes in the USA in 1960 and 54 tonnes in 1970.[67] Since 1969 the Soviet industry has introduced a number of 200-tonne electric furnaces, but the maximum capacity in the USA is now 400 tonnes.[68]

The extent to which the limited use of the electric arc process is justified in Soviet conditions is a matter of dispute in the USSR itself. According to one estimate, a reduction in the cost of electricity of 50–60 per cent would be

---

[63] *Journal of Metals*, March 1973, pp. 34–40; *Iron and Steel International*, October 1973, pp. 440–8.
[64] A. S. Osintsev, *Ekonomika chernoi metallurgii SSSR* (1969), p. 252.
[65] I. M. Denisenko, *Mezhotraslevye svyazi struktury chernoi metallurgii* (1969), p. 287.
[66] *Ibid.*, p. 311.
[67] *Referativnyi Sbornik—Ekonomika Promyshlennosti*, 1973, No. 6, Item V110.
[68] *Journal of Metals*, August 1973, p. 17.

Table 3.14 *The output of electric steel*

Year	USSR		USA		JAPAN		FRG		UK	
	a	b	a	b	a	b	a	b	a	b
1950[1]	1·7	6·3	5·5	6·2	0·8	15·6	0·3	2·4	0·8	4·5
1955[2]	3·4	7·5	7·2	6·9	1·2	12·6	1·0	4·0	1·1	5·6
1960[2]	5·8	8·9	7·6	8·4	4·5	20·3	2·2	6·4	1·7	6·9
1965[2]	8·6	9·4	12·5	10·5	8·4	20·3	3·1	8·5	3·5	13·1
1970[2]	10·7	9·2	19·2	15·7	15·6	16·8	4·4	9·8	5·5	19·5
1974[2]	13·5	9·9	25·8	19·6	20·9	17·9	5·7	10·8	5·3	23·5

a – output (million tonnes); b – electric steel as a proportion of total (per cent).

Sources:
1 (Ed.) N. P. Bannyi; *Ekonomika chernoi metallurgii SSSR* (1971), p. 469.
2 *Quarterly Bulletin of Steel Statistics for Europe*, various years; except for 1950.

required before the process could compete with the open hearth method for the production of ordinary steel;[69] more recently, another Soviet specialist has argued that this view does not take full account of the real costs involved in both cases, and that it would now be expedient to produce ordinary carbon and low alloy steel on a regular basis in electric arc furnaces.[70]

Continuous casting of steel

One of the most significant developments in the steel industry since the war has been the replacement of the traditional method of casting in ingot moulds by continuous casting. Whereas the old processes required the reheating of the ingot for primary rolling, continuous casting directly provides semi-finished billets or slabs suitable for subsequent hot working operations.

The elimination of the ingot stage and primary rolling reduces capital and space requirements and the process gives a product of superior quality, as steel cast continuously possesses greater homogeneity than does steel cast by the traditional method.

Continuous casting has a long history,[71] but its industrial application dates from the Second World War. In intermediate forms of technology, transitional to fully continuous casting, the Soviet industry has a particularly strong claim to the role of pioneer. In 1938 a conveyor type mould machine was introduced at the Bezhitsk works, and in 1951 a vertical, semi-continuous unit was introduced at the 'Krasnyi Oktyabr' works.[72] Table 3.15 refers to fully continuous casting installations.

[69] (Ed.) N. P. Bannyi, *Ekonomika chernoi metallurgii SSSR* (1971), p. 474.
[70] *Steel in the USSR*, April 1974, pp. 293–6.
[71] *Economic Aspects of Continuous Casting*, UNECE, New York (1968), pp. 1–9.
[72] *Continuous Casting of Steel in the USSR*, OECD, Paris (1964), pp. 34–46.

Table 3.15 *Year of introduction of first experimental and industrial continuous casting installations and their capacity*[a] *(tonnes)*

	First experimental/pilot plant		First industrial plant	
USSR	1944	(1·5)	1955	(50)
USA	1946	(6·0)	1962	(20)
Japan	1955	—	1960	(30)
FRG	1943	(1·5)	1954	(35)
UK	1946	(2·5)	1958	(7·5)
Canada			1953	(30/50)
France	1950	(1·5)	1953	(1·5)
Italy			1958	(30/75)
Austria	1947	(7·5)	1952	(10)

Source:
Compiled from *Economic Aspects of Continuous Casting*, UNECE, New York (1968), pp. 26–46; (Eds.) D. L. McBride and T. E. Dancy, *Continuous Casting*, New York (1962), pp. 6 and 16–23; *Advances in Steel Technology* in 1956, UNECE, Geneva, pp. 95–102. This table has been compiled from a number of sources which in some cases provide conflicting information. There are problems of definition of 'experimental' and 'industrial' plants, as some of the former were later transformed into production installations.

Note: a – ladle capacity.

It will be observed that the Soviet Union was one of the first countries to develop and introduce continuous casting. Initial work was undertaken by the research institute TsNIICherMet; in 1953 an experimental industrial-scale installation began operation at the Novo-Tula works, and the first fully industrial installation was completed in 1955 at the 'Krasnoe Sormovo' works. The Soviet industry was the first to use continuous casting in conjunction with oxygen converter steelmaking.[73] In 1967, at the Novo-Lipetsk works, it was the first to develop a steelmaking plant relying on continuous casting alone; the first American works of this kind began operation in 1968–9.[74]

The Soviet type of continuous casting installation has been patented in about 30 countries and also built in a number of them, including Italy, Finland, the GDR and Japan.[75]. Three of the major continuous casting installations in Japan in 1972 were of the Soviet type.[76]

As shown in Table 3.16 the rate of growth of output of continuously cast steel in the USSR compared quite favourably with that of other leading countries until about 1967. During the seventies output in both absolute and relative terms has fallen below the levels achieved by the USA and Japan by a large margin, and in 1973 fell below the German level in absolute terms for the first time. British output in absolute terms lags substantially behind that of the Soviet Union, but reached the Soviet proportion of total steel output in 1974.

[73] B. P. Bel'gol'skii and V. T. Fadeev, *Ekonomiya metalla v prokatnom proizvodstve* (1972), p. 200.
[74] D. I. Popov, *op. cit.*, p. 126; *Iron Age*, 31 May 1973, p. 45.
[75] *Steel International*, September 1966, p. 126.
[76] *Journal of Metals*, December 1972, p. 4.

Table 3.16 *The production of steel by continuous casting (output in million tonnes and as a percentage of total steel output)*

		1965	1966	1967	1968	1969	1970	1971	1972	1973	1974
USSR	Output	1·2[1]	1·7[2]	n.a.	3·9[3]	n.a.	5·0[4]	5·9[4]	6·5[4]	7·0[4]	7·4[4]
	%	1·3	1·8	n.a.	3·7	n.a.	4·3	4·9	5·5	5·3	5·4
USA	Output	1·0[5]	1·6[5]	3·7[5]	5·8[5]	7·8[5]	13·7[5]	n.a.	n.a.	23[6,a]	n.a.
	%	0·8	1·3	3·2	4·9	6·1	11·5	n.a.	n.a.	n.a.	n.a.
Japan	Output	0·4[7]	n.a.	n.a.	n.a.	3·3[8]	5·2[8]	9·9[8]	16·5[8]	24·7[8]	29·4[8]
	%	1·0	n.a.	n.a.	n.a.	4·0	5·6	11·2	17·0	20·7	25·1
FRG	Output	0·8[8]	0·8[8]	1·4[8]	2·2[8]	3·3[8]	3·7[8]	4·1[8]	6·1[8]	8·1[8]	10·3[8]
	%	2·1	2·4	3·7	5·3	7·3	8·3	10·2	13·9	16·3	19·4
UK	Output	0·4[8]	0·4[8]	0·4[8]	0·4[8]	0·5[8]	0·5[8]	0·4[8]	0·5[8]	0·8[8]	1·1[8]
	%	1·4	1·6	1·8	1·7	1·8	1·8	1·7	1·8	3·0	5·5

Sources:
1 *Steel International*, September 1966, p. 126.
2 *Principal Factors Affecting Labour Productivity: trends in the Iron and Steel Industry*, UNECE, New York (1969), p. 163.
3 K. I. Klimenko and E. V. Petrova, *Ekonomicheskaya effektivnost' tekhnicheskogo progressa v tyazheloi promyshlennosti SSSR* (1971), p. 31.
4 *The Steel Industry in 19.. and Trends in 19..*, OECD, Paris, various years.
5 L. Nabseth and G. Ray, *The Diffusion of New Industrial Processes*, Cambridge (1974), p. 241, (output calculated).
6 *Journal of Metals*, October 1973, p. 40.
7 *Referativnyi Sbornik, Ekonomika Promyshlennosti*, 1972, No. 3, Item V113.
8 *The Iron and Steel Market in 19.. and Trends in 19..*, various years.

Notes:
a – capacity.

Some reduction in the rate of introduction of continuous casting capacity in the Soviet Union is acknowledged by Soviet writers and there are indications that there have been major differences of opinion in the steel industry as to the relative merits of the new process compared with the traditional casting and rolling technique.[77] One factor which may explain the rate of growth in recent years is the relative emphasis on different types of steelmaking processes in the USSR compared with Japan and the USA. In Japan priority has been given to oxygen steel and the oxygen converter is the most suitable installation for continuous casting; in the USA the great majority of continuous casting installations have been attached to electric arc furnaces, which are also well suited to the process. In the USSR, however, most of the installations of the 1960s were at open hearth steelmaking units. It is therefore possible that the slower development in the last five years is related to the slow rate of introduction of oxygen steelmaking capacity. At the present time the largest continuous casting plant in the world is being built at the Novo-Lipetsk works. This plant will have two oxygen converters of 300 tonnes each and five continuous casting installations. Total output will be four million tonnes a

[77] *Pravda*, 23 October 1972.

year. This works, the first stage of which came into use in December 1974, is being built with West German assistance. The Demag firm is supplying and building complete continuous casting installations and in this work is cooperating with the Mannesmann company, one of the world's leading specialists in continuous casting.[78]

Alloy and quality steel

One of the most significant developments in the iron and steel industry in recent years has been the increasing emphasis on the production of alloy and quality steels. Such steels have for a long time been used in the making of tools, bearings and armaments, but with the rapid development of the aerospace, chemical, atomic and electronics industries the demand for special alloy steels has correspondingly increased, as is shown in Table 3.17.

The Soviet industry has paid particular attention in recent years to the production of low alloy steels.

The output of these steels rose from 5·8 million tonnes (5·8 per cent of total steel output) in 1960 to 11·9 million tonnes (10·3 per cent) in 1970.[79] Production of rolled metal from alloy and low alloy steel has also correspondingly increased, from 3·9 and 4·4 million tonnes respectively in 1965 to 6·9 and 7·7 million tonnes in 1970,[80] and according to the ninth five year plan output of low alloy rolled metal in 1975 should reach 15·5 million tonnes.[81] However, the heat treatment of alloy and low alloy steels which greatly raises their effectiveness is poorly developed in the USSR. Heat treatment was applied to only 1·7 million tonnes of rolled metal in 1965 and 3·4 millions in 1970.[82] A Soviet specialist acknowledged a significant lag with respect to heat treatment in a work published in 1969; in the United States almost all low alloy steel is used in a treated form, in the USSR it is usually used without treatment.[83]

Information about the production of stainless steel in the USSR is sparse, but according to one recent work such steel represents 9·9 per cent of all alloy steel output, i.e., 1·3 million tonnes in 1970, compared with 1·2 million tonnes in the USA (1·4 in 1963 and 1972), and 1·6 million tonnes in Japan.[84]

Quite a large proportion of Soviet alloy steel production takes place at enterprises outside the Ministry of Ferrous Metallurgy: in 1970 over one third of alloy steel output (4·6 million tonnes) and 8·5 per cent of low alloy steel output (1·0 million tonnes).[85]

[78] *Sots. Ind.*, 14 February 1974; *Journal of Metals*, December 1973, p. 8.
[79] Calculated from D. I. Popov, *Povyshenie effektivnosti metallurgicheskogo proizvodstva* (1972), p. 37.
[80] *Ibid.*, p. 41.
[81] N. F. Sklokin and A. A. Brodov, *Osnovnye napravleniya i ekonomika uluchsheniya kachestva chernykh metallov* (1974), p. 28.
[82] D. I. Popov, op. cit., p. 41.
[83] I. M. Denisenko, *Mezhotraslevye svyazi i struktura chernoi metallurgii* (1969), pp. 94–5.
[84] N. F. Sklokin and A. A. Brodov, op. cit., p. 22; *Sheet Metal Industries*, 1974, No. 5, p. 226.
[85] Calculated from B. P. Bel'gol'skii and V. T. Fadeev, *Ekonomiya metalla v prokatnom proizvodstve* (1972), p. 244 and D. I. Popov, op. cit., p. 37.

Table 3.17 *The output of alloy steel (million tonnes and as a percentage of total crude steel output)*

	1960		1965		1970	
	Output	% total	Output	% total	Output	% total
USSR	7·0[1]	10·8	9·9[1]	10·9	13·4[1]	11·6
USA	7·5[2]	8·3	13·4[2]	11·3	12·8[3]	10·8
Japan	1·9[2]	8·4	3·9[2]	9·4	n.a.	n.a.

Sources:
1 Output calculated from D. I. Popov, *op. cit.*, p. 37.
2 *Principal Factors Affecting Labour Productivity: Trends in the Iron and Steel Industry*, UNECE, NY (1969), p. 23.
3 *The Iron and Steel Industry in 1971 and Trends in 1972*, OECD, Paris (1973), p. 45. The USA output given by this source (11·6 m t) excludes stainless steel (1·2 m t in 1970—*Sheet Metal Industries*, 1974, No. 5, p. 226).

As in other producing countries, special steels in the USSR are made in electric arc furnaces and small open hearth furnaces of 50–100 tonnes capacity.[86]

Since the Second World War a number of new processes have been developed for raising the quality of steel:

REFINING STEEL WITH SYNTHETIC SLAG

Refining steel in a ladle with liquid synthetic slag is a means of purifying and raising the homogeneity of steel produced by any steelmaking process. The use of this process can raise the quality of steel made by open hearth or oxygen converter methods to the level obtained in an electric arc furnace. Soviet specialists consider this to be one of the best means of improving the quality of steels in wide use; the product has been found to be especially well suited to welded structures in the Soviet North.[87] The use of liquid lime-alumina slag in a ladle was pioneered by Soviet research undertaken at TsNIICherMet from 1959. It was first used on an industrial scale at the Chelyabinsk metallurgical factory in 1963,[88] and is now employed at a number of leading enterprises. Output of steel refined by this method has risen from 0·3 million tonnes in 1965[89] to 0·8 million tonnes in 1970, with a 1975 aim of 2·5 million tonnes,[90] rising to 5 million tonnes by 1980.[91] The diffusion of this process has not been as rapid as envisaged in the eighth five year plan period: a capacity of 3 million tonnes in 1970 was originally foreseen.[92]

[86] *Steel in the USSR*, April 1973, p. 933.
[87] N. F. Sklokin and A. A. Brodov, *op. cit.*, p. 9.
[88] V. V. Lempitskii, I. N. Golikov, N. F. Sklokin, *Progressivnye sposoby povysheniya kachestva stali* (1968), p. 51.
[89] I. M. Denisenko, *op. cit.*, p. 300.
[90] N. F. Sklokin and A. A. Brodov, *op. cit.*, p. 11.
[91] *Steel in the USSR*, April 1973, p. 933.
[92] V. V. Lempitskii *et al.*, *op. cit.*, p. 60.

VACUUM DEGASSING OF STEEL

This process is employed for obtaining high quality steel by eliminating gases, notably hydrogen, which if not removed may lead to the formation of internal cracks.

Vacuum treatment has been employed in the USSR since about 1954.[93] By 1968 there were 18 industrial installations compared with 59 in the USA, 39 in Japan and 22 in Britain.[94] In 1965 200,000 tonnes of vacuum treated steel were produced in the USSR,[95] and in 1970 capacity reached 1·5 million tonnes[96] (as compared with the five year plan target of 2 million tonnes)[97] and output is planned to reach 3 million tonnes in 1975.[98] Although Soviet metallurgical enterprises have now gained experience of the main techniques of vacuum degassing, one specialist writing in 1973 acknowledged that 'in this field . . . Soviet works are still failing to meet the requirements of the various consumer sectors and are still a long way behind the standards obtained in some capitalist countries.'[99] Considerable new capacity for vacuum degassing is now planned.[100]

ELECTRO-SLAG REMELTING

The above mentioned processes involve treating the molten steel before casting. There are a number of alternative techniques for raising the quality of steels which involve the remelting of the metal under conditions securing the elimination of impurities and gases. The main process employed in the Soviet Union is electro-slag remelting. This technique was developed independently in both the USA and the USSR. The first reference to it was made in a Russian work published in 1893; pioneering work was undertaken by Hopkins in the USA during the 1930s but development was slow, partly because of the existence of the competitive vacuum arc refining process.

Soviet specialists appear to have been unaware of the American work when they began research in 1948.[101] Research and development was undertaken by the Paton Welding Institute and was prompted by the need to meet the demands of the aerospace industry at a time when the vacuum arc process had not been developed. A pilot plant was in operation in 1958 and by the end of that year two industrial-scale plants had been commissioned. In Britain research began in 1961 and a commercial plant was in operation in 1964. Japan and France both acquired licences for the Soviet process and introduced their first plants in 1963 and 1965 respectively.[102] By September 1966 the Soviet electro-slag remelting process had been patented in more than 30 countries,

---

[93] *Continuous Casting of Steel in the USSR*, OECD, Paris (1964), p. 54.
[94] V. V. Lempitskii *et al.*, *op. cit.*, p. 61.
[95] I. M. Denisenko, *op. cit.*, p. 299.
[96] B. P. Bel'gol'skii and V. T. Fadeev, *op. cit.*, p. 244.
[97] I. M. Denisenko, *op. cit.*, p. 317.
[98] P. A. Shiryaev, *Kapital'noe stroitel'stvo v chernoi metallurgii* (1973), p. 31.
[99] *Steel in the USSR*, April 1973, p. 933.
[100] N. I. Sheftel', *Uluchshenie kachestva i sortamenta prokata* (1973), p. 16.
[101] W. E. Duckworth and G. Hoyle, *Electro-slag Refining*, London (1969), pp. 1–3.
[102] *Ibid.*, pp. 4–6.

including the USA,[103] where a number of firms have taken out licences.[104] The process has the merits of simplicity and low cost compared with alternative technologies: according to Soviet calculations capital investment per tonne of steel refined by the electro-slag process is one fifth of that required for the vacuum arc process, and one tenth of that for the vacuum induction process.[105] Output of electro-slag refined steel rose by 50 per cent during the years 1966–70 and was planned to rise by 150 per cent during the years of the ninth five year plan.[106]

VACUUM ARC REMELTING

The vacuum arc remelting process is widely employed in the USA for the making of special, high quality steels for many industries including aerospace and armaments production. It is particularly well suited to the refining of heat resistant steels and alloys.

In the USSR this process is used at a number of factories for refining steel for bearings, and various heat resistant steels and alloys, but the electro-slag process remains the primary technique.[107] During the ninth five year plan period it was intended to double the output of vacuum arc remelted steel.[108]

VACUUM INDUCTION REMELTING

In the USA and other countries this process has been developed alongside vacuum arc remelting, but while there are a number of installations in the USSR, the level of use appears to be substantially lower than in the USA.[109] In 1973 it was reported that the USSR was buying a vacuum-induction furnace of nine tonnes per hour capacity from an American firm.[110]

OTHER PROCESSES

Electron-beam remelting secures very highly refined metals. It is widely employed in the USA for making alloy steels and as early as 1963 there were 10 industrial installations. The process has also been developed in Japan, West Germany and Britain. In the USSR laboratory and industrial-scale research began in 1965, but there appears to be little industrial use of the technique at present.[111] Laboratory research into *plasma remelting* is also being undertaken.[112] This technique was first developed by the Paton Institute in Kiev.[113]

In summary, it appears that in this sphere of very rapidly developing technology of vital importance for the aerospace, armaments, electronics and

---

[103] *Steel International*, September 1966, p. 126.
[104] *Pravda*, 14 June 1973.
[105] N. F. Sklokin and A. A. Brodov, *op. cit.*, p. 14.
[106] *Gosudarstvennyi pyatiletnii plan razvitiya narodnogo khozyaistva na 1971–1975 gody* (1972), p. 112.
[107] N. F. Sklokin and A. A. Brodov, *op. cit.*, p. 13.
[108] P. A. Shiryaev, *op. cit.*, p. 30.
[109] V. V. Lempitskii, *op. cit.*, p. 46.
[110] *Press Bulletin, Moscow Narodny Bank*, 5 September 1973, p. 7.
[111] P. A. Shiryaev, *op. cit.*, p. 30; N. I. Sheftel', *op. cit.*, pp. 36–7.
[112] N. F. Sklokin and A. A. Brodov, *op. cit.*, p. 14.
[113] *Pravda*, 25 June 1974.

chemical industries, the Soviet Union initially concentrated its efforts on one process, electro-slag remelting, and achieved success in its development. Other processes widely employed in the West have until recently received less attention, and the Soviet industry is now striving to overcome this lag.

Rolled metal products

Consideration of the structure of rolled steel products in the Soviet Union and other major producing countries reveals significant differences which must be taken into account in any assessment of the overall technical level of the Soviet steel industry. The most striking difference is the high proportion of sectional steel in the Soviet output and the correspondingly low proportion of flat steel, in particular sheet metal. Total output of finished rolled steel in the USSR is now at about the same level as that of the United States, but a feature of the last decade has been the very rapid increase of output in Japan: in 1973 the Japanese output exceeded that of the USSR for the first time (see Table 3.18). In per capita terms, production in the USSR is, of course, still substantially lower than in the USA, Japan and FRG.

Table 3.18 *Output of finished rolled steel (thousand tonnes)*

	1960[1]	1965	1970[1]	1973[1]
USSR	44806	61650[2]	82142	93118
USA	64546	89528[1]	82373	101090
Japan	17200	33421[1]	76032	100404
FRG	25841	28476[1]	35974	40226

Sources:
1 *Quarterly Bulletin of Steel Statistics for Europe*, various years.
2 *Statisticheskii ezhegodnik stran-chlenov SEV* (1973), p. 84.

The structure of production of finished steel is shown in Table 3.19 and the output of a number of basic end products is shown in Table 3.20. The latter table reveals the relatively low level of production of a number of progressive forms of flat steel, notably cold-rolled sheet and cold-reduced strip. While the rate of growth of output of these products has exceeded that of the USA since 1960, it has been lower than the rates achieved in Japan and Germany.

A major explanation of these structural differences is clearly the nature of the demand in each country for finished steel, particularly by the engineering industry. Table 3.21 indicates approximately the structure of rolled metal consumption in the Soviet Union and the USA by branch of industry and sector of the economy for the year 1965.

The smaller proportion of sheet steel in the Soviet output of finished metal is clearly related to the relative size of the motor industry and the production

Table 3.19 The structure of production of finished steel (per cent of total)*

Category of finished steel	USSR 1960	USSR 1966[f]	USSR 1973	USA 1960	USA 1965	USA 1973	Japan 1960	Japan 1965	Japan 1973	FRG 1960	FRG 1965	FRG 1973
Sections:												
Rail track materials	7·1	4·9	4·0	1·2	1·0	1·2	2·5	1·6	0·6	2·6	1·5	1·3
Heavy sections	30·7	38·8	38·6	7·4	7·0	6·4	4·7	6·5	8·7	7·5	8·0	6·2
Light sections	9·3			14·9	15·1	16·3	26·6	24·2	18·3	23·4	19·7	15·9
Wire rods	7·9	8·6	8·6	5·5	5·1	4·7	1·7	9·1	7·6	8·4	10·1	10·7
Semis; solids and ingots for tubes	8·7	7·8	7·4	3·6	4·3	3·5	3·5	3·0	2·5	3·6	4·3	6·3
All sections	63·7	59·6	58·6	32·6	32·5	32·1	46·0	44·0	37·7	48·9	46·8	40·4
Flat:												
Strip	6·2	8·6	9·6	10·3	5·5	7·7	5·5	3·7	2·4	9·2	7·9	7·6
Heavy plates[a]	15·2[d]	12·9[d]	12·4[d]	8·6	11·6	8·7	15·6	15·6	17·0	13·9	13·0	12·9
Medium plates[b]	1·6[e]	5·1[e]	5·8[e]	45·5			4·2	2·4	1·8	4·2	6·0	10·8
Sheets[c]	10·2	11·5	11·8		47·6	47·8	28·4	33·7	41·0	12·5	15·2	20·8
All flat	33·2	38·1	39·6	64·4	64·7	64·2	53·7	55·4	62·2	39·8	42·1	52·1
Other (incl. semis for sale)	3·1	2·3	1·8	3·0	2·8	3·7	0·3	0·2	0·1	11·3	11·1	7·5
Total	100·0	100·0	100·0	100·0	100·0	100·0	100·0	100·0	100·0	100·0	100·0	100·0

Source:
Quarterly Bulletin of Steel Statistics for Europe, various years.

Notes:
a – thickness 4·75 mm plus.
b – thickness 3·00 to 4·75 mm.
c – thickness less than 3·00 mm (hot rolled).
d – 5 mm plus.
e – 3·0 to 5·0 mm.
f – 1965 not available.

*The above table does not take into account a number of minor differences in the definition of certain categories for each country: see original source.

Table 3.20 *Output of selected end products (thousand tonnes)*

End product		USSR	USA[1]	Japan[1]	FRG[1]
Cold-rolled sheet	1960	1533[1]	13123	2793	1841
	1965	3862[2,a]	18809	5749	3697
	1970	5178[1]	12927	14166	6718
	1973	5813[1]	18486	18966	8238
1973 as a % of 1960		379	141	679	447
Cold-reduced strip	1960	n.a.	1203	159	1130
	1965	112[2,a]	2207	222	1133
	1970	119[1]	1045	377	1553
	1973	320[1]	1400	556	1866
1973 as a % of 1960			116	350	165
Tin plate	1960	312[1]	4958	450	404
	1965	466[2,a]	4936	876	558
	1970	501[1]	6018	1428	817
	1973	561[1]	6088	1906	863
1973 as a % of 1960		180	121	424	214
Galvanised sheet	1960	237[1]	2773	938	241
	1965	332[2,a]	4469	1430	417
	1970	508[1]	4359	3878	1004
	1973	1007[1]	6247	5375	1472
1973 as a % of 1960		425	225	805	611
Welded tubes	1960	2539[1]	4110	773	606
	1965	4739[2]	4796	2414	1002
	1970	7042[1]	4199	5999	1812
	1973	8311[1]	4711	7549	2378
1973 as a % of 1960		327	115	977	392

Sources:
1 *Quarterly Bulletin of Steel Statistics for Europe*, various years.
2 *Statisticheskii ezhegodnik stran-chlenov SEV* (1973), p. 85.

Note: a – 1966.

of domestic appliances and cans in the Soviet Union as compared with the USA, Japan and Germany. According to Soviet calculations, in the USA as much as about 40 per cent of all metal products are consumed in the production of consumer goods as against less than 8 per cent in the USSR.[114] There is also evidence, however, that in all branches of engineering and metal working in the Soviet Union, not merely those working for the consumer

[114] I. M. Denisenko, *Mezhotraslevye svyazi i struktura chernoi metallurgii* (1969), p. 90.

Table 3.21 *Structure of consumption of rolled metal, tubes, and general metal products (metizy) in the Soviet Union and USA in 1965*

Branch	Metal consumed (million tonnes)		Proportion of total metal consumed (%)	
	USSR	USA	USSR	USA
Automobile industry	3·6	25·7	6·5	27·9
Agricultural machinery and tractors	5·5	2·9	9·5	3·3
Electrical engineering	2·1	4·4	3·7	4·8
Rail transport engineering	1·6	3·0	2·8	3·3
General and 'other' engineering	8·4	8·5	14·4	9·2
Production of cans	1·0	9·6	1·7	10·4
Production of domestic appliances and trade equipment	1·6	6·2	2·8	6·7
Other branches of engineering and metalworking	6·1	1·1	10·1	1·2
Total engineering and metal working industries	29·8	61·4	51·0	66·6
Mining and timber industries	1·2	0·5	2·1	0·6
Oil and gas industries	0·9	2·0	1·5	2·1
Other branches of industry	4·3	0·4	7·4	0·4
Total industry	36·2	64·3	62·0	69·7
Construction	18·7	26·8	31·9	29·1
Agriculture	0·3	0·3	0·5	0·3
Rail transport	2·3	0·8	3·9	0·9
Other branches	1·0	—	—	—
All	58·5	92·2	100·0	100·0

Source: I. M. Denisenko, *Mezhotraslevye svyazi i struktura chernoi metallurgii* (1969), p. 87.

industries, a relatively greater proportion of sectional as opposed to sheet steel is consumed.[115] Several factors have contributed to the predominance of sectional steel. A higher proportion of steel is consumed in repair work and a greater reliance is placed on metalcutting techniques at the expense of stamping and welding. Many types of Soviet machines are heavier than their Western equivalents, demanding greater use of heavier sections of steel.[116] The system of performance indicators tends to encourage the production of heavier profiles of rolled metal.[117] Thus both structural factors, due to the nature of demand, and systemic factors, springing from the nature of the planning system, have resulted in a high level of production of sectional steel.

[115] *Ibid.*, p. 91.
[116] *Ibid.*, pp. 89–95.
[117] *Pravda*, 5 August 1973.

TUBE AND PIPE MAKING

In 1962 the Soviet Union became the world's leading tube producer by volume of output. The growth of output is shown in Table 3.22.

Recent years have seen a shift from the traditional seamless tube making process to the more progressive welding of tubes, with an increasing use of electrical welding methods; in this respect, developments have been more rapid in the USSR than in the USA.

The very rapid development of the Soviet oil and gas industry in the east of the country and the intensive construction of pipelines have given rise to a demand for large-diameter tubes which the domestic industry has been unable to satisfy. Imports of tubes as a proportion of domestic output have risen from 8·5 per cent in 1965 to 10·5 per cent in 1970 and 13·9 per cent in 1973.[118] In 1965 less than a quarter of the imports by weight took the form of large-diameter welded tubes, but in 1973 the proportion had risen to 64 per cent.[119] In 1974 the Italian state steel concern 'Finsider' concluded the largest ever contract for the supply of steel to the USSR. This provided for the supply of 500,000 tonnes of large-diameter welded tubes for each of the five years 1975 to 1979 to a total value of 1,500 million dollars.[120] New capacity for the production of tubes of larger diameter (1,220, 1,420 and 1,620 mm) is being introduced during the current five year plan at the Khartsyzsk tube works.[121]

ROLLING MILLS

The great diversity of types of rolling mills and their products makes comparison of their technical level difficult. This section is therefore restricted to an account of the views of Soviet specialists on the comparative technical level of rolling mills in Soviet industry. In 1970 360 mills were in use for hot rolling: of these 269 were at metallurgical enterprises; the remaining 91 were at engineering factories. There were also 25 cold rolling mills.

It is generally acknowledged that in recent years the technical level of rolling has been lower than that of iron and steelmaking.[122] Very old and antiquated mills have continued to be used alongside modern equipment. The rolling mills at engineering factories are generally old and of a low technical level and account for less than five per cent of total annual output.[123] In 1965 120 of all hot rolling mills were over 30 years old,[124] and there were more than 40 mills dating from the last century.[125] Costs at these old mills are substantially higher and labour productivity is very low. It has been estimated that the withdrawal of 75 of the most obsolete rolling mills would free not less than 20,000 workers

---

[118] Calculated from *Vneshnyaya torgovlya SSSR za 19. . g.*, various years.
[119] *Ibid.*
[120] *Press Bulletin, Moscow Narodny Bank*, 2 October 1974, p. 4.
[121] P. A. Shiryaev, *Kapital'noe stroitel'stvo v chernoi metallurgii* (1973), p. 36.
[122] A. F. Mets, *Effektivnost' novoi tekhniki i tekhnologii v prokatnom proizvodstve* (1968), p. 8; I. M. Denisenko, *op. cit.*, p. 288.
[123] B. P. Bel'gol'skii and V. T. Fadeev, *Ekonomiya metalla v prokatnom proizvodstva* (1972), p. 205.
[124] *Ibid.*, p. 195.
[125] D. Ya. Komarova and D. I. Popov, *Trudoemkost' proizvodstva metalloproduktsii*, 1968, p. 96.

Table 3.22 *Output of steel tubes (million tonnes)*

	1960[1]	1965	1970[1]	1973[1]
USSR	5805	9014[2]	12434	14369
USA	6398	8111[1]	7056	8285
Japan	1227	3146[1]	7672	9560
FRG	2033	2598[1]	3622	4408

Sources:
1 *Quarterly Bulletin of Steel Statistics for Europe*, various years.
2 *Statisticheskii ezhegodnik stran-chlenov SEV* (1973), p. 85.

Table 3.23 *Welded tubes as a proportion of total tube output (per cent of total by weight)*

	1955	1960	1965	1970	1973
USSR	33·9[1]	43·8[1]	52·7[1]	57·1[1]	57·8[2]
inc. electro-welded	16·3[1]	30·4[1]	39·3	46·1	n.a.
USA	61·7[3]	64·3[2]	59·1[1]	59·5[1]	56·9[2]
inc. electro-welded	29·6[3]	36·8[2]	34·7[2]	44·6[2,a]	n.a.
Japan	n.a.	63·0[2]	78·8[2]	78·2[2]	79·0[2]
FRG	22·8[3]	29·8[2]	38·6[2]	50·0[2]	53·9[2]

Sources:
1 D. I. Popov, *op. cit.*, p. 134.
2 *Quarterly Bulletin of Steel Statistics for Europe*, various years.
3 Calculated from *British Iron and Steel Federation, Statistical Handbook*, London, various years.
Note: a – 1968.

and produce significant cost savings for the branch.[126] But withdrawal has been impossible in view of the shortage of rolled metal experienced for many years, in circumstances in which both the volume of investment in rolling and the capacity of the metallurgical engineering industry are inadequate.[127]

It is usual to classify rolling mills into 'modern' and old types, modern units being mechanised and having continuous or semi-continuous operation. Table 3.24 shows the growing role of modern mills in the output of all types of rolled products, but also the continued existence of many old mills.

Soviet rolling mills for the production of large and heavy sections, which, as has been shown, predominate in Soviet output, are said to be up to foreign standards, but mills for sheet metal and some of the newer progressive rolled metal products tend to be relatively less advanced. Thus roughing mills for blooms and slabs, and mills for large sections and rails, are said to be of a

---

[126] D. I. Popov and N. I. Mityaev, *Ispol'zovanie osnovnykh fondov metallurgicheskikh predpriyatii* (1972), p. 41.
[127] A. F. Mets, *op. cit.*, p. 8.

Table 3.24 *The technical level of rolled metal production*

Type of rolling mill	1958			1965			1970a		
	No. all	Inc. modern no.	modern % prodn.	No. all	Inc. modern no.	modern % prodn.	No. all	Inc. modern no.	modern % prodn.
rail	6	3	76·5	5	3	73·2	5	3	76·3
Large section	22	3	48·4	24	5	57·7	25	7	77·5
Medium section	32	3	32·2	34	5	35·3	36	10	65·6
Light section	33	4	38·6	42	11	58·7	43	16	72·5
Strip	2	2	100·0	3	3	100·0	3	3	100·0
Universal	4	1	59·2	4	1	60·0	3	1	71·2
Wire rod	16	3	43·0	21	8	57·4	22	10	65·6
Sheet and plate	101	9	59·3	105	13	72·2	85	15	82·5
Tyre	3	1	72·2	3	1	69·0	3	1	71·0
Other	4	2	100·0	5	5	100·0	6	6	100·0
Total	225	34	54·8	240	47	61·3	234	74	76·5

Source: I. M. Denisenko, *op. cit.*, p. 290. N.B.: some mills excluded: total 1970 was 269.

Note: a – planned: as seen in 1969.

technical level comparable to those in use abroad.[128] Soviet mills of these types tend to have a higher productivity (output per mill per year) than those employed elsewhere: they are usually operated on a round-the-clock basis, roll metal of larger cross-section and tend to be more specialised.

This is possible because the range of shapes and sizes rolled in the Soviet Union is significantly lower than normal in other countries.[129] According to one recent estimate there are four times as many shapes-sizes of rolled metal in the USA as in the USSR.[130] Mills for medium sections, wire rod and some light sections are considered to lag behind foreign levels.[131] In the mid-1960s the speed of rolling on most modern wire rod mills in the USSR was no higher than 30 metres per second, as compared with 37 metres on foreign mills of this type.[132] According to a leading Soviet specialist, 'In the USSR the quality of the surface finish of rolled metal, its dimensional accuracy, homogeneity and workability, have still not achieved the level of the best world standards, especially for wire rod and light sectional steel.'[133] But some improvement has evidently taken place in recent years: the mills for wire rod and light sections installed since the mid-1960s are considered to be up to the best foreign standards.[134]

[128] A. F. Mets, *op. cit.*, pp. 13–14; B. P. Bel'gol'skii and V. T. Fadeev, *op. cit.*, p. 205.
[129] A. F. Mets, *op. cit.*, p. 9.
[130] *Ekonomika i Organizatsiya Promyshlennogo Proizvodstva*, 1974, No. 4, p. 74.
[131] A. F. Mets, *op. cit.*, p. 20; I. M. Denisenko, *op. cit.*, p. 291.
[132] A. F. Mets, *op. cit.*, p. 20.
[133] *Ibid.*
[134] D. I. Popov and N. I. Mityaev, *op. cit.*, p. 71.

Until recently one of the most backward sectors of rolling has been the making of flat products, in particular sheet steel. Table 3.24 indicates the large number of old type mills in use alongside a relatively small number of mills of the continuous and semi-continuous types. In 1965 there were 7 continuous and semi-continuous modern wide strip mills for rolling sheet steel in the USSR, compared with 42 in the USA, 11 in Japan and 5 in the United Kingdom.[135] By 1971 there were 9 modern wide strip mills in the USSR. But 53 old type 'in train' mills and pack mills continue to be used: output per hour at one of these pack mills, which involve considerable hand work, is about one tonne, compared with 600 tonnes at the modern '2000' mill of the Novo-Lipetsk factory.[136]

Labour productivity at these 53 old mills is one ninth of the level at the modern mills and cost per tonne of sheet steel almost six times greater. These 53 mills have an aggregate capacity of only two million tonnes and it is estimated that their replacement by one modern continuous mill would provide a higher level of output while freeing 9,000 workers.[137] (In the USA the number of old type sheet mills was reduced from 46 in 1954 to 25 in 1960 and those remaining in use in the 1960s were used for producing special steel sheet.)[138] Before 1958 the rolling of flat metal was generally undertaken on imported equipment. From the late 1950s new Soviet wide strip mills were designed on the basis of the best foreign technology of the time, but it took so long to manufacture the equipment and construct the rolling shops (about 10 years) that by the time they were commissioned they were found to be of a lower technical level than new Western mills of a similar type: the '1700' mill of the Il'ich works and the '2500' mill of the Magnitogorsk combine were slower, less powerful and less highly automated than British and American equivalents.[139]

As shown in Table 3.20 the production of cold rolled sheet steel has been substantially below that in other major steel-producing countries. In 1972 it was admitted that the shortage of cold rolled steel was restricting the production of a number of very important types of industrial products.[140] Cold rolling is essential for securing high quality thin sheet, but in order to secure the necessary accuracy, automatic regulation of the sheet thickness is needed. In the 1960s the level of automation of Soviet cold rolling mills was inadequate, in particular the automatic regulation of thickness.[141] Securing accurate and regular thickness of sheet metal appears to have been a widespread problem. Writing in 1968, one Soviet specialist observed that 'many rolling mills producing finished rolled metal do so with tolerances exceeding the standards acceptable for industrially developed countries by 50 to 200 per cent.'[142] Not

---

[135] A. F. Mets, *op. cit.*, p. 169.
[136] D. I. Popov and N. I. Mityaev, *op. cit.*, p. 41.
[137] *Ibid.*, pp. 42–3.
[138] I. M. Denisenko, *op. cit.*, p. 291.
[139] A. F. Mets, *op. cit.*, p. 23.
[140] B. P. Bel'gol'skii and V. T. Fadeev, *op. cit.*, p. 231.
[141] A. F. Mets, *op. cit.*, p. 25.
[142] *Ibid.*, p. 71. See also V. V. Lempitskii, I. N. Golikov, N. F. Sklokin, *Progressivnye sposoby povysheniya kachestva stali* (1968), pp. 112–13, for evidence that the thickness of Soviet sheet metal is more variable than usual in the West.

only has the thickness of Soviet sheet steel been more variable than usual in the West, but the average thickness also appears to be greater; this is partly a consequence of the underdevelopment of cold rolling. Soviet industry is at present much concerned with this problem and has achieved some success; between 1966 and 1969 the average thickness of sheet steel produced was reduced from 2·14 mm to 1·86 mm. Progress is hindered, however, by the continuation of the practice of assessing the performance of enterprises not by the area of sheet metal produced but by its weight.[143] The problem of producing high quality sheet steel now appears to be being tackled under pressure of demand from the rapidly growing motor industry. A work published in 1972 observed that significant changes had taken place in the rolling of sheet steel, previously one of the most backward sectors. A number of modern new mills have been introduced, e.g., the '1700' mill of the Karaganda combine and the '2000' mill of the Novo-Lipetsk works, both possessing a high level of automation and other features, such as rigid stands, securing high productivity and accuracy.[144] More new sheet mills are being introduced during the current five year plan period, including the '2000' wide continuous strip mill at the Cherepovets factory and the '3600' mill for plate at the Azovstal works. The automation of the Cherepovets mill is being undertaken by the German Siemens firm.[145] At the same time, some older mills are being reconstructed and fitted with automatic control equipment.[146]

The production of tinplate and galvanised steel are two further sectors which until recently have lagged behind the practice of other major producing countries. Tinplate requires fine rolling, best carried out on a continuous cold rolling mill of high accuracy. Soviet shortcomings are shown by the fact that in the late 1960s the average thickness of tinplate was 0·27 mm, whereas half the tinplate produced in the USA had a thickness of 0·06–0·09 mm.[147] In the USA in 1954 there were 54 lines for the electrolytic coating of tinplate, a progressive process which permits considerable savings of lead compared with the traditional hot coating process, but the Soviet Union had only one installation of Austrian manufacture.[148] By 1962 the electrolytic method accounted for 96 per cent of the American output, and 78 per cent of the British, as against only 8 per cent of the Soviet output; the proportion in the USSR increased rapidly to about 30 per cent by 1970.[149] Production of tinplate was planned to rise to 750,000 tonnes by 1970 according to the eighth five year plan,[150] but in fact only reached just over 500,000 tonnes. Since 1969 imports of tinplate have risen sharply, averaging about 16 per cent of domestic output in 1969–73.[151] The technology of galvanising steel has also lagged behind usual Western practice.

[143] D. I. Popov, *Povyshenie effektivnosti metallurgicheskogo proizvodstva* (1972), pp. 50–1.
[144] D. I. Popov and N. I. Mityaev, *op. cit.*, pp. 71–2.
[145] *Iron and Steel International*, August 1973, p. 295.
[146] N. F. Sklokin and A. A. Brodov, *Osnovnye napravleniya i ekonomika uluchsheniya kachestva chernykh metallov* (1974), p. 37.
[147] I. M. Denisenko, *Mezhotraslevye svyazi i struktura chernoi metallurgii* (1969), p. 95.
[148] A. F. Mets, *op. cit.*, p. 25.
[149] I. M. Denisenko, *op. cit.*, pp. 144, 285; N. F. Sklokin and A. A. Brodov, *op. cit.*, p. 67.
[150] A. F. Mets, *op. cit.*, p. 26.
[151] *Vneshnyaya torgovlya SSSR za 19. . g.*, 1969–73 inclusive.

The first continuous line for zinc coating was introduced in 1964 in the USSR, using Austrian equipment; at that time there were already 50 such lines in the USA and 21 in Japan.[152] Other forms of anti-corrosion flat steel were also less developed during the 1960s. In 1964 550,000 tonnes of plastic coated sheet and strip were produced in the USA, which had 12 installations for plastic coating, compared with 9 in Britain and 7 in Japan. At this time no plastic coated steel was produced in the USSR on an industrial scale,[153] and even now it is produced on a much smaller scale than in other leading countries.

Special steels for use in the electrical engineering industry have to be of highly pure metal and be accurately thin-rolled with a good surface finish in order to reduce power losses to a minimum. In the mid-1960s Soviet produced electro-technical steel was of lower quality than that imported from abroad; the poor quality resulted in substantial losses. In other countries almost all transformer steel was produced by cold rolling at a thickness of under 0·35 mm, but in the USSR only half was cold rolled, at a thickness of 0·5 mm.[154] Until recently all Soviet dynamo steel was produced by hot rolling. During the current five year plan, however, improvements have taken place. Plant for rolling electrical steels of the required thickness and quality, including shops for cold rolling transformer and dynamo steels, has been installed at the Verkh-Isetsk and Cherepovets metallurgical factories.[155] Electrical and other equipment for cold rolling mills for electro-technical steel is being supplied by a French firm.[156]

Rolling is the only sector of the Soviet iron and steel industry to rely to any significant extent on imported equipment. Explaining the relatively lower level of rolling in comparison to iron and steel making, Mets, writing in 1968, referred to the inadequacy of the capacity of the metallurgical equipment industry and also observed that in recent years the quality of the production of rolling mills has been low.[157]

The proportion of imported equipment in total annual consumption was as follows: 5·2 per cent in 1950, 21·6 per cent in 1960, 44·0 per cent in 1965 and 39·2 per cent in 1970.[158] Most imported rolling mill equipment came from Czechoslovakia and the GDR, but more recently the Federal German Republic has also been a major supplier.

This survey indicates that in the post-war years rolling has been one of the most backward sectors of the Soviet iron and steel industry. During the 1950s and early 1960s its relative technical level deteriorated but since about 1965, notably during the early 1970s, the lag behind the rolling technology of other leading iron and steel producing countries has been reduced.

---

[152] A. F. Mets, *op. cit.*, p. 26.
[153] *Ibid.*, pp. 63–5.
[154] *Ibid.*, pp. 61–2, D. Ya. Komarova and D. I. Popov, *op. cit.*, p. 136.
[155] P. A. Shiryaev, *op. cit.*, pp. 32–3.
[156] *Press Bulletin, Moscow Narodny Bank*, 23 May 1973, p. 5.
[157] A. F. Mets, *op. cit.*, pp. 9, 14.
[158] *Narodnoe khozyaistvo SSSR za 1922–72 gody* (1972), p. 495.

Automation of processes in the iron and steel industry

It is difficult to make any assessment of the comparative level of automation of processes in the iron and steel industry because the appropriate information is sparse. Visitors to the Soviet Union in the mid-1950s were impressed by the degree to which automatic control instruments were employed in the Soviet industry.[159] A British delegation in late 1964 noted that the Soviet industry was ahead of Britain in the automation of blast furnaces and sinter plants, almost level in steelmaking, but behind in the automation of rolling mills and finishing processes.[160] But since the electronic computer was developed for process control the Soviet industry has not apparently attained the rate of progress in automation which has been achieved in the West. In 1972 it was estimated that there were over 100 electronic computers installed in the industry under the Ministry of Ferrous Metallurgy, including 63 at enterprises. Of the total, 38 per cent were universal machines of medium size (Minsk, Ural, etc.), 22 per cent small computers (Mir, Nairi, etc.) and 40 per cent were for control purposes.[161]

At this time there were 34 process control computers in industrial use at enterprises, including 8 for blast furnace production, 5 for oxygen converters, and the rest for various types of rolling mill.[162] The Japanese iron and steel industry possessed 480 computers early in 1974, including 200 for systems of production control.[163] There is no doubt that the number of computers in use in the United States, while lower than in Japan, is also substantially higher than in the Soviet Union. The lag in Soviet automation is obviously related to the continuing lag in the diffusion of computer technology generally.

Evidence that automation in the Soviet iron and steel industry is not up to the level attained in other major producing countries was provided by a resolution of the Central Committee of the CPSU in August 1972, 'On the course of work on the automation of technological processes in ferrous metallurgy.' This expressed dissatisfaction with the rate of introduction of automatic processes, noting that many of the aggregates introduced during the years of the eighth five year plan had not been equipped with computerised systems of comprehensive automation and that many auxiliary processes remained unmechanised.[164]

Conclusion

The iron and steel industry provides an interesting example of a well established industry which has experienced rapid technical change since the Second World War and, in particular, since the early 1960s. The basic oxygen process, continuous casting, the use of the electric arc furnace for ordinary crude steelmaking, electro-slag remelting and a number of other refining

---

[159] M. G. Clark, *The Economics of Soviet Steel*, Cambridge, Mass. (1956), pp. 261–2.
[160] *The Russian Iron and Steel Industry—BISRA report of November/December 1964 delegation*, London (1966), p. 79.
[161] D. I. Popov, *op. cit.*, p. 193.
[162] *Ibid.*, p. 194.
[163] *Referativnyi Zhurnal, Metallurgiya*, 1974, No. 8, Item V29.
[164] *Pravda*, 30 August 1972.

processes are all innovations of the last 20 to 25 years, and these new developments have been accompanied by a gradual improvement of the traditional ironmaking technology.

The last decade has seen the rapid introduction of computerised process control and the mechanisation and automation of auxiliary processes traditionally performed by manual labour. These process innovations have made possible the satisfaction of, and been promoted by, new demands posed by technologically sophisticated industries which have developed very rapidly since the beginning of the 1960s.

The performance of the Soviet industry has been uneven. The rapid rate of growth of output of iron and steel since the war has put the Soviet Union ahead of the USA by volume of production. In respect of the traditional processes, notably blast furnace technology and open hearth steel production, it has on the whole performed well and established a number of technical leads; in recent years, however, the Japanese industry has provided a powerful challenge, notably in relation to blast furnace technology. Soviet industry showed an early awareness of the significance of the two major post-war innovations, oxygen steel and continuous casting; in the latter case it played a prominent role in the development of the process. But the rate of diffusion of oxygen steel has been much slower than in the other major producing countries, at least during the 1960s, when diffusion rates were very rapid in Japan, the USA and Germany. An early lead was lost in the course of a five to six year period from about 1959 to 1965, when the oxygen process maintained an almost constant share of total steel output. Whether this actually represents an economic as well as a technological 'lag' requires further analysis of the costs and benefits of the oxygen process in relation to the Soviet open hearth process. The scale and technology of open hearth steel making in the USSR may have been such as to deprive the new process of the marked economic superiority it appears to have shown in the industrially advanced capitalist countries.

The rate of diffusion of continuous casting has also been much lower than in the USA, Japan and the FRG since about 1967. The merits of this process have also been disputed in the Soviet Union, albeit less convincingly: its introduction is in any case now seen as a high priority task. In the West the electric arc furnace has been developed since the war for the making of ordinary steel and this has led to a steady increase in the scale of installations. In the Soviet Union this shift has not yet taken place and today the volume of Soviet electric arc furnaces is on average lower than in other major producing countries.

The electro-slag refining process is a successful Soviet innovation widely used abroad; concentration on this process appears to have been accompanied by a lag in other refining processes. Soviet industry has particularly lagged in the production of high quality thin sheet steel, essential for the mass production of light cars and domestic appliances. There are clearly fundamental structural reasons for this lag: Soviet rolling mill plant was primarily developed for the production of sectional steel and thick plate required by heavy engineering and construction. Lags are also acknowledged in the technology of making tinplate and other types of coated metal, and in particular plastic coated steels which have been introduced on a rapidly increasing scale during recent years in the other major steel-producing countries. The Soviet iron and steel industry also lags in computerised process control. In all these cases a substantial effort is now being made to catch up.

4. Machine tools*

The machine tool is the primary equipment of the engineering industry. The technological level of the machine tool industry and its products is therefore an important determinant of the technological level of a substantial sector of industry in all modern economies.

Since the creation of the Soviet machine tool industry in the early 1930s its progress has continued interrupted only by war until today the Soviet Union has the largest stock of machine tools in the world. Over 10 years have now elapsed since the pioneering studies by Boretsky and Daukas in this field;[1] and with the recent publication of censuses of the machine tool stock of the United States and the Soviet Union an attempt may now be made to reassess the standing of the Soviet Union in comparison with other industrialised countries, and in particular with the United States.[2]

This chapter is devoted to a study of various aspects of the technological level of the Soviet machine tool industry. The first section examines the main comparative statistical material which is available on the stock, production and utilisation of machine tools and discusses the main differences which emerge. In the second section the evidence on technological level provided by Soviet foreign trade in machine tools is analysed. A third and final section is devoted to a detailed study of Soviet numerically-controlled machine tools; numerical control represents one of the most significant technical advances in machine tool technology in the post-war years. Further evidence on the technological level of Soviet machine tools is presented in chapter 11, which focuses on problems of quality and reliability.

---

* M. J. Berry wrote pp. 122–60, and Julian Cooper pp. 160–98.
[1] M. Boretsky and A. Daukas, in *Dimensions of Soviet Economic Power*, Joint Economic Committee of US Congress, Washington DC (1962), pp. 69–143 and pp. 165–80 respectively.
[2] In addition to census data two recent studies by a Soviet writer have proved particularly valuable: D. M. Palterovich, *Park proizvodstvennogo oborudovaniya* (1970) and *Planirovanie potrebnosti v oborudovanii* (1972).

Machine tools: stock and production

NUMBERS AND DISTRIBUTION OF STOCKS BY INDUSTRY

The latest published census figures show that the Soviet Union now has by far the largest stock of machine tools of any country in the world, with a total in 1972 of around five million; the United States in 1973 had less than four million. In this total figure the Soviet Union has about one million more metalcutting machine tools than the US but only slightly more metalforming machines (see Table 4.1).

The background to this remarkable situation is that for some years at least the Soviet output has exceeded the American for metalcutting machine tools but has yet to overtake the American production of metalforming machines. In practice, however, Table 4.2 does not give a true picture because of fluctuations in American output—thus output for the years 1966–9 was over 225,000 in each year and the average output over periods of years in Table 4.2a gives a better impression. As one Soviet specialist has put it, in terms of capacity 'the United States machine tool industry is considerably more powerful'[3] (*znachitel'no moshchnee*).

Although the increase in production has slowed down in recent years it shows no sign of coming to a stop and the machine tool stock continues to expand. In the distribution of this large machine tool stock in the economy, however, a different pattern emerges from that found in the United States. First, the United States uses a far greater proportion of its machine tools in the engineering and metalworking sector (see Table 4.3). Thus although the Soviet stock of metalcutting machine tools in 1972 was about a million more than that of the United States, the total number used in engineering and metalworking was about the same. In the case of metalforming machines, where total stocks are about equal, the Soviet Union has only about three quarters of the American stock in engineering and metalworking.

As Table 4.4 shows, the remaining machine tools are located in a wide variety of industries outside engineering and metalworking and also in other sectors of the economy. The proportion outside engineering and metalworking increased between 1958 and 1962, and, although it had fallen by 1972, it had increased considerably in absolute size from around a million to about a million and a half for metalcutting machines and from around 250,000 to nearly 380,000 for metalforming machines.[4] These machines are mainly used for repair work and in particular for the manufacture of spare parts.

Repair work is also a problem in the engineering and metalworking industry itself and is estimated to have accounted for about 8–10 per cent of the stock of metalcutting machine tools in the engineering industry in 1958.[5] In addition a further 13–15 per cent are used for the manufacture of tooling and fixtures at engineering factories in the absence of sufficient supplies from specialised

---

[3] See P. M. Penkov, *Puti razvitiya stankostroeniya i novye metalloobrabatyvayushchie stanki*, Kiev (1962), p. 1.
[4] These figures must be considered approximate because of the revised stock figures published recently for the 1962 census.
[5] S. A. Kheinman, *op. cit.*, p. 170.

factories.[6] Thus almost a quarter of the stock in engineering and metalworking was used outside the main production shops. In the United States it is probable that almost all machine tools in the engineering sector would be used in basic production. It is likely that there has been little change in the USSR since 1958: a recent work gave the following breakdown for metalcutting machine tools in engineering without specifying the year: basic production—57 per cent; tooling shops—14 per cent; repair shops—8 per cent; other—21 per cent—giving a total of 43 per cent outside basic production.[7] On the other hand, there is some evidence that the proportion of machine tools outside basic production may be lower in the recent period, as is indicated by the results of the regular two-yearly surveys by the Central Statistical Administration (TsSU) of the utilisation of machine tools given in Table 4.5. These surveys cover about two thirds of the total number of metalcutting machine tools in engineering and metalworking and about half of the metalforming equipment.

Table 4.1 *Total number of machine tools in the Soviet Union and the United States*

	Soviet Union[2]			United States[1]		
	metal-cutting	metal-forming	total	metal-cutting	metal-forming	total
1958	1916	394	2310	2204	683	2887
1962	2227	468	2695			
1963				2537	816	3353
1966	2797	610	3407			
1968				2620	850	3470
1972	3925	901	4826			
1973				2923	887	3810

Sources:
1 *American Machinist. The Eighth, Ninth, Tenth and Eleventh Inventories of Metalworking Equipment.* Unless otherwise stated all American data are taken or calculated from these sources.
2 1958—*Promyshlennost' SSSR* (1964), p. 78. 1962, 1966, 1972—*Narodnoe khozyaistvo SSSR v 1972 godu* (1973), p. 68. All the original figures for these years have been revised in this source. The following statistical handbook, *Narodnoe khozyaistvo SSSR v 1973 godu* (1974), p. 61, published revised figures for 1972 as follows: metalcutting—4,137,000; metalforming—934,000, giving a total of 5,071,000, but earlier figures have not been revised so this does not give a consistent series. A short paper has been prepared discussing the Soviet statistics and the problem of comparability. This can be obtained on application to M. J. Berry at CREES, University of Birmingham, England.

Note:
All figures are based on census data except the 1966 figures for the Soviet Union which are official estimates.

[6] *Ibid.*
[7] V. K. Fal'tsman, *Prognozirovanie potrebnosti v oborudovanii* (1970), p. 79.

Table 4.2 *Output of machine tools for selected years (× 1000)*

	Metalcutting		Metalforming	
	USSR[1]	USA	USSR[1]	USA
1940	58·4	110[2]	4·7	n.a.
1945	38·4	103[2]	2·9	n.a.
1950	70·6	41·5[2]	7·7	n.a.
1955	117·1	n.a.	17·1	n.a.
1960	115·9	133·5[3]	29·9	49·6[3]
1965	186·1	184·2[3]	34·6	66·3[3]
1970	202·3	188·5[4]	41·3	63·9[4]
1971	207·2	157·0[4]	42·3	54·8[4]
1972	210·0	207·3[4]	44·0	57·5[4]
1973	211·1	251·5[4]	46·5	55·0[4]
1974	225·0	272·6[4]	49·0	61·5[4]
1975	232·0		50·5	

Sources:
1 *Narodnoe khozyaistvo SSSR v 1973 godu* (1974), pp. 218–19; *Pravda*, 25 January 1975; 1 February 1976.
2 H. D. Wagoner, *The US Machine Tool Industry, 1900–1950*, Cambridge, Mass. (1968), p. 363.
3 *Statistical Abstract of the United States 1971*, US Bureau of the Census, Washington DC (1971), p. 716.
4 *Economic Handbook of the Machine Tool Industry 1975–6*, National Machine Tool Builders Association, McLean, Virginia (1975), pp. 73, 89.

Note:
US metalcutting machine tool output has been subject to fluctuation. Comparative annual average output is therefore shown in Table 4.2a below.

Table 4.2a *Annual average output*

	Metalcutting	
	USSR[1]	USA[c]
1940	58·4	110[2]
1941–5	32·6	199·4[2]
1946–50	58·1	52·2[2]
1951–5	91·4	86·4[2a]
1956–60	139·4	138·5[3b]
1961–5	179·0	148·2[3]
1966–70	199·7	224·6[3]
1971–4	213·3	222·1[3]

Sources:
1 As for Table 4.2, and M. J. Berry, *Research, Development and Innovation in the Soviet Machine Tool Industry*, unpublished research report, Appendix B.
2 H. D. Wagoner, *op. cit.*, p. 363.
3 *Economic Handbook of the Machine Tool Industry, 1975–6*, NMTBA, McLean, Virginia (1975), p. 73.

Notes:
a – 1951 to 1953.
b – 1959 to 1960.
c – output by number for 1954–8 is not given in the sources used.

AGE PATTERN OF MACHINE TOOL STOCKS

The age pattern of the machine tool stock in the USSR presents a remarkable contrast to that of the United States. In the past 15 years the high levels of output have led to a steady increase in the proportion of machines under 10 years old. In 1958 the Soviet Union had a greater proportion of machines over 20 years old than the United States; by 1973 the Soviet proportion was half that of the United States (see Table 4.6). In absolute terms, however, there was a considerable increase in the number of machines in use over 20 years old in both the Soviet Union and, more particularly, the United States which now has around one million tools over 20 years old.[8]

One curious feature which the limited age data for the Soviet Union does bring out is that the stock in the engineering sector tends to be slightly older than the stock in the economy as a whole, at least in the 1958 and 1962 censuses. It is not clear what the current situation is. In the case of the United States for the years for which comparable statistics are available (1958 and 1963) the stock in the engineering and metalworking sector tends to be slightly younger than the figures for the whole economy.

Table 4.3 *Machine tool stock in engineering and metalworking*

	Metalcutting				Metalforming			
	USSR		USA[4]		USSR		USA[4]	
	No. ×1000	% of total stock	No. ×1000	% of total stock	No. ×1000	% of total stock	No. ×1000	% of total stock
1958	1115[1]	38[1]	1707	73	247[1]	63[1]	511	75
1962	1360[2,a]	56[2,a]			263[2,a]	53[2,a]		
1963			2137	84			671	82
1966	—	—			—	—		
1968			2175	83			695	82
1972	2394[3]	61[3]			523[3]	58[3]		
1973			2362	81			703	79

Sources:
1 S. A. Kheinman, *Organizatsiya proizvodstva i proizvoditel'nost' truda* (1961); p. 169. See also (Ed.) L. M. Gatovskii, *Problemy ekonomicheskogo stimulirovaniya nauchno-tekhnicheskago progressa* (1967), p. 96.
2 D. M. Palterovich, *op. cit.* (1972), p. 122.
3 B. A. Kostin, *Proizvoditel'nost' truda i tekhnicheskii progress* (1974), p. 142.
4 As for Table 4.1.

Note:
a – 1962 figures for the Soviet Union are based on the original figures for that year *not* the revised figures given in Table 4.1.

---

[8] Calculated from Tables 4.1 and 4.5 on the assumption that the age pattern is the same for the economy as a whole as for engineering and metalworking to give a minmum figure.

WEIGHT AND VALUE OF MACHINE TOOLS PRODUCED

Although the number and age of the machine tool stock are frequently taken as the main criteria for comparing the strengths of the industry of two countries, some writers have pointed out the deficiencies of this system.[9] Thus Daukas argued that the large number of recently made machines in the Soviet stock was misleading because this did not mean that they were modern machines, and Palterovich has argued that it is more realistic to use weight and value to make comparisons. Although the data are not available to make full use of Palterovich's criteria it is possible to show the significance of these criteria and in particular of value.

In the case of weight we know that the average weight of a Soviet metalcutting machine tool is much higher than that of an American machine

Table 4.4 *Distribution of machine tools in the Soviet economy*

Branch	1958[1] Metal-cutting	1958[1] Metal-forming	1962[2a] Metal-cutting		1962[2a] Metal-forming	
Industry	75	78	71·6		76·3	
incl.						
Engineering	48	40	} 56·8		} 58·2	
Metalworking	10	23				
Ferrous metallurgy			2·27		5·2	
Non-ferrous metallurgy			1·0		1·1	
Chemical			1·3		1·8	
Woodworking			1·0		0·9	
Building materials	} 17	} 15	1·1	} 14·8	2·1	} 18·1
Light			1·7		1·3	
Food			1·5		0·6	
Timber			1·0		0·4	
Other			4·0		4·7	
Construction			4·2		9·7	
Transport	} 25	} 22	4·0	} 28·4	3·2	} 23·7
Agriculture (not *kolkhozy*)			2·6		2·3	
Other branches			17·6		8·5	
Total	100	100	100		100	

Sources:
1 S. A. Kheinman, *op. cit.*, p. 169.
2 D. M. Palterovich, *op. cit.*, p. 105.

Note:
a – the breakdown for 1962 refers to the original figures published for that year, *not* the revised totals given in Table 4.1.

[9] See, for example, A. Daukas, *op. cit.*, p. 177, and D. M. Palterovich, *op. cit.* (1970), pp. 46–52.

Table 4.5 *Proportion of machine tools in basic and auxiliary production in the Soviet engineering industry*

	Metalcutting				Metalforming			
	Basic production		Auxiliary production		Basic production		Auxiliary production	
	No. ×1000	%	No. ×1000	%	No. ×1000	%	No. ×1000	%
1967[1]	787·9	66	399·1	34	179·2	84	23·7	16
1969[2]	863·0	67	430·0	33	195·8	89	25·1	11
1971[3]	933·2	67	459·8	33	214·1	88	28·6	12
1973[4]	1015	67	497	33	239·6	88	31·5	12

Sources:
1 *Vestnik Statistiki*, 1968, No. 7, pp. 89–90.
2 *Ibid.*, 1970, No. 8, pp. 89–90.
3 *Ibid.*, 1972, No. 3, p. 94.
4 *Ibid.*, 1974, No. 5, p. 88.

tool—about 3,200 kg for a Soviet machine tool as opposed to about 1,200 kg for an American.[10] Although this refers to current production it can be used to make a crude comparison of the weight of the respective metalcutting machine tool stocks. This gives a Soviet total of around 12·5 million tonnes and a United States figure of around 3·5 million tonnes. In part the heavier Soviet machine tools can be explained by the continued emphasis on heavy industry and the relative weakness of light industry.

Following the recent publication of value figures for Soviet machine tool output some rough comparisons can be made of the value of output in the Soviet Union and the United States. Here again the Soviet figures show a steady rate of increase which has continued despite a levelling off in numbers produced (see Table 4.8).

The chief problem in assessing the comparative value of output is the conversion rate to be used. A number of possible figures have been suggested by both Western and Soviet specialists and Table 4.9 shows the comparative output in dollars which emerges from using some of these conversion rates. The table shows that for several of the years cited Soviet output was well ahead of the US output and since 1970, according to Boretsky's rate of exchange, has been 250–400 per cent larger, and according to Palterovich's rate has been 75–300 per cent larger. These results must be treated cautiously and much more work is required before any firm conclusions can be reached.

By far the most interesting point which the question of value brings out is the composition of output in value terms. As Daukas has pointed out, the United States produces a large number of low value machines which have few parallels in the Soviet product mix.[11] These, in his view, tend to hinder comparisons and

[10] D. M. Palterovich, *op. cit.* (1970), p. 309.
[11] A. Daukas, *op. cit.*, p. 168–9.

distort the United States output statistics in units. Some Soviet specialists have argued in recent years that production of small machines should be increased in the USSR; and one of them quotes striking comparative data in support of this case (see Table 4.10). Even allowing for a wide margin of error in price comparisons these figures bring out very clearly a major difference in the product mix of the two countries, which must also be reflected in their machine tool stocks. Thus Daukas argues that for a true comparison of production one

Table 4.6 *Breakdown of machine tool stock by age (in years as % of total)*

	Country	Metalcutting <10	10–20	>20	Metalforming <10	10–20	>20	Total[a] <10	10–20	>20
1958	USSR[2]	49	30	21	53	27	20	50	29	21
1958	USA[1]	39	43	18	39	38	23	39	42	19
1962	USSR[3]	54	21	25	59	20	21	58	20	22
1963	USA[1]	35	44	21	36	40	24	35	43	22
1966	USSR	—	—	—	—	—	—	—	—	—
1968	USA[1]	37	41	22	32	41	26	36	41	23
1972	USSR[3]	60	26	14	63	24	13	61	25	14
1973[b]	USA[1]	33	38	29	31	40	29	33	49	28

Sources:
1 As for Table 4.1.
2 *Narodnoe khozyaistvo SSSR v 1962 godu* (1963), p. 55.
3 As for Table 4.1.

Notes:
a – the breakdown for the Soviet Union has been calculated.
b – these figures cover engineering and metalworking only, not the whole economy.

Table 4.7 *Age pattern of Soviet machine tools in the economy as a whole and in engineering and metalworking (in years as % of total)*[1a]

	Machine	Total stock <10	10–20	>20	Engineering and metalworking <10	10–20	>20
1958	Metalcutting	49	30	21	45	34	21
	Metalforming	53	27	20	45	31	24
1962	Metalcutting	57	20	23	55	19	26
	Metalforming	62	19	19	55	18	27

Source:
1 *Narodnoe khozyaistvo SSSR v 1962 godu* (1963), p. 55.

Note:
a – these breakdowns are based on original data (1962) *not* revised figures.

Table 4.8 *Output by value and unit value for selected years*

	Metalcutting				Metalforming			
	USSR[1a]		USA[2b]		USSR[1a]		USA[2b]	
	Total value m r	Unit value ×1000 r	Total value m $	Unit value ×1000 $	Total value m r	Unit value ×1000	Total value m $	Unit value ×1000 $
1940	67·8	1·2	440	4·0	6·5	1·4	n.a.	n.a.
1945	55·5	1·4	424	4·1	2·8	1·0	n.a.	n.a.
1950	95	1·3	316	7·6	1·6	2·1	n.a.	n.a.
1955	230	2·0	666	n.a.	6·7	3·9	195	n.a.
1960	419	2·7	539	4·0	12·9	4·3	238	4·8
1965	638	3·4	1048	5·7	16·1	4·6	410	6·2
1970	978	4·8	1098	5·8	24·6	6·0	487	7·4
1971	1053	5·1	714	4·5	27·3	6·5	372	6·5
1972	1142	5·4	802	4·0	29·7	6·8	396	6·7
1973	1256	5·9	1188	4·8	32·7	7·0	476	9·0
1974	1385	6·2	1514	5·6	36·9	7·5	580	9·4

Sources:
1 *Narodnoe khozyaistvo SSSR v 1973 godu* (1974), pp. 218–19.
2 As for Table 4.2. For 1950, the figure for metalcutting from the *Statistical Abstract of the United States*, US Bureau of the Census, Washington DC (1954) p. 854, has been preferred to Wagoner's figure of 306.

Notes:
a – Soviet Union data based on 1967 prices.
b – United States figures are current prices.

should compare only the more valuable machines,[12] that is 91 per cent of Soviet output with only 28 per cent of American output in units according to Table 4.10. This would give the Soviet Union a much greater lead in terms of numbers of machines although their relative position in value terms would remain about the same. A similar situation probably exists in metalforming machinery, although comparative Soviet data are not available.[13] Here again this disparity can in part be explained by the Soviet emphasis on heavy industry and the relative neglect of light industry.

BALANCE BETWEEN METALCUTTING AND METALFORMING MACHINE TOOLS

The proportion of metalforming machine tools in the Soviet economy is lower than in the United States (see Table 4.11). This criterion was used by Boretsky to indicate the technological backwardness of the USSR[14] and in recent years

[12] *Ibid.*

[13] In the American output of metalforming machines about 60 per cent of the output in terms of units accounted for only about 3 per cent of the value (D. M. Palterovich, *op. cit.* (1970), p. 309).

[14] M. Boretsky, in *New Directions in the Soviet Economy*, Joint Economic Committee of US Congress, Washington DC (1966), pp. 171–2.

the Soviet Union has taken some steps to increase this proportion. This is shown by the data for the Soviet machine tool stock under 10 years old (see Table 4.12). Surprisingly this trend is much less pronounced among the machine tools in basic production shops covered by the quite large biennial surveys by TsSU in recent years (see Table 4.13).

Table 4.9 *Comparative output of metalcutting machine tools by value for selected years for main rouble–dollar conversion rates (in million dollars)*

	USA	Soviet output			
	output[1]	Official rate[3]	American Machinist[2a]	Boretsky[b] 1 r=$2·75[1]	Palterovich[c] 1 r=$2[1]
1940	440	75		186	136
1945	424	62		153	111
1950	316	105		261	190
1955	666	255		633	460
1960	539	465		1152	838
1965	1048	708	662	1755	1276
1970	1098	1086	803	2690	1966
1971	714	1169	860	2896	2106
1972	802	1265	1029	3141	2284
1973	1188	1402	1274	3454	2512
1974	1514	1538	1343	3812	2772

Sources:
1 Figures taken or calculated from Table 4.8.
2 1965—*American Machinist*, 16 January 1967, p. 131.
 1970—*Ibid.*, 24 January 1972, p. 82.
 1971—*Ibid.*, 22 January 1973, p. 77.
 1972—*Ibid.*, 21 January 1974.
 1973 and 1974—*Ibid.*, 15 January 1975, p. 69.
3 Official rate—1 rouble=$1·1. Given in *Economic Handbook of the Machine Tool Industry 1975–6*, McLean, Virginia, p. 173. Soviet rouble prices are converted at 1967 exchange rate.

Notes:
a – *American Machinist*: this has been included since its annual review of world output is often used for international comparisons. The US figures given in the first column do not always correspond precisely to the data given in *American Machinist* for the US but the reason for this is not known. The procedure for arriving at a Soviet output figure in dollars is not completely clear. The 1965 figure is said to be converted at the 'editor's estimate of a realistic exchange rate rather than at the official rate' (*American Machinist*, 16 January 1967, p. 131). 1970 figures are said to be converted at 'commercial rates of 1972' (*Ibid.*, 24 January 1972, p. 82). Subsequent figures are converted at Soviet official rates (see sources for 1971–4 quoted above).
b – Boretsky in his 1962 study used this exchange rate based on the 1955 rouble (see M. Boretsky, *op. cit.*, pp. 117–19). In a debate with Professor Alec Nove in *Survival* (July–August 1972, pp. 169–77) he explains that this ratio was based on 644 observations of all types of machinery and related products, produced in all types of scales and subject to all types of Soviet government priorities' (*Ibid.*, p. 169). The ratio was used to convert 1955 roubles into 1964 dollars. Its use here is a little arbitrary but it was felt that it did provide some guide to relative output value.
c – This conversion rate is based on work done by Palterovich in NIIMash and refers specifically to machine tools. See D. M. Palterovich, *op. cit.* (1970), pp. 150–1. The method of calculation is explained in D. M. Palterovich, *Metodicheskie voprosy mezhdunarodnykh ekonomicheskikh sravnenii po otraslyam mashinostroeniya* (1965), pp. 14–18.

Table 4.10 Comparative data on low value machine tools produced in the Soviet Union and United States (year not given)

	Roubles	USSR % Number	USSR % Value	USA % Number	USA % Value
Metalcutting machine tools	<500	9·3	0·4	72	3·6
	>500	90·6	99·6	28	96·4
Incl. drilling machines	<500	47·1	8·7	87	9
	>500	52·9	91·3	13	91
grinding machines	<500	11·6	0·78	68·8	2·7
	>500	88·4	99·2	31·2	97·3
lathes	<500	9·38	0·04	31·8	2·9
	>500	99·62	99·96	68·2	96·1

Source:
(Ed.) M. A. Vilenskii, *Ekonomicheskie problemy povysheniya promyshlennoi produktsii* (1969), p. 80.

Note:
No source is given for these data. The US figures given here for metalcutting machine tools are similar to those quoted in D. M. Palterovich, *op. cit.* (1970), p. 307, for machines over and under 1,000 dollars in value.

Table 4.11 The division between metalcutting and metalforming machines in the machine tool stock of the United States and the Soviet Union (%)

	Total stock USSR mc	Total stock USSR mf	Total stock USA mc	Total stock USA mf	Engineering and metalworking USSR mc	Engineering and metalworking USSR mf	Engineering and metalworking USA mc	Engineering and metalworking USA mf
1958	82·9	17·1	76·3	23·7	81·8	18·2	77·0	23·0
1962	82·6	17·4			83·8	16·2		
1963			75·7	24·3			76·1	23·9
1966	82·1	17·9			n.a.	n.a.		
1968			75·5	24·5			75·8	24·2
1972	81·3	18·7			82·1	17·9		
1973			76·8	23·2			77·1	22·9

Source:
Calculated from Table 4.1.

The ninth five year plan (1971–5) proposed an expansion of the output of metalforming equipment from 16·9 per cent of total machine tool output in 1970 to 20·6 per cent in 1975.[15] It should be pointed out that the breakdown of the United Kingdom stock in engineering and metalworking shows a lower proportion of metalforming than in both the USA and the USSR—about 15 per cent in 1971.[16]

COMPOSITION OF MACHINE TOOL STOCKS BY TYPE

In considering the technological level of Soviet industry, the breakdown of the machine tool stock between the various types of machine tools is of course more important than these general indicators.[17] This is a very complicated and confused area in which there are few published Soviet statistics and the problem of comparability of US and Soviet types is a severe one. The most detailed Soviet statistics for both metalcutting and metalforming machine tools are those given by Palterovich.[18] These are particularly valuable since they are presented in a comparative table with other countries and he has clearly taken some trouble to make the breakdowns comparable. Unfortunately, the information published relating to the 1972 census is not so detailed. A further complication is that while Palterovich gives a detailed breakdown for 1962 for engineering and metalworking which is the coverage of the American breakdown, the 1972 Soviet census covers 'economic accounting enterprises and organisations', a rather vague term which has not been used before.[19]

The groups which show the greatest divergence—'lathes' and 'grinding and all others' will be looked at in greater detail below. The other types of machine

Table 4.12 *The division of the Soviet machine tool stock under 10 years old between metalforming and metalcutting*

Age	Total stock		Stock in economic accounting organisations	
	Metalcutting	Metalforming	Metalcutting	Metalforming
Up to 5 yrs	80	20	79	21
6–10 yrs	82	18	81	19

Source:
Narodnoe khozyaistvo SSSR v 1972 godu (1973), p. 68.

---

[15] *Gosudarstvennyi pyatiletnii plan razvitiya narodnogo khozyaistva SSSR na 1971–5 gody* (1972), pp. 128, 347.
[16] See *Metalworking Production, Third Survey of machine tools and production equipment in Britain*, London (1971), p. 2.
[17] This excludes numerically controlled machine tools which are dealt with in a later section of this chapter.
[18] D. M. Palterovich, *op. cit.* (1970), and *op. cit.* (1972), p. 127.
[19] See Table 4.1.

Table 4.13 *Proportion of metalforming and metalcutting machine tools in the machine tool stock of basic production shops of Soviet engineering factories (survey data) (%)*

Type	1967	1969	1971	1973
Metalcutting	81·5	81·5	81·3	80·9
Metalforming	18·5	18·5	18·7	19·1

Source:
Calculated from data given in Table 4.5.

tools appear to fall into two groups—those of which the Soviet Union has a smaller proportion in total than the USA and probably a larger proportion in engineering and metalworking (broaching, milling and gear cutting machines); and those of which the Soviet Union has a larger proportion in total and probably the same or a lower proportion in engineering and metalworking (boring, drilling and planing and shaping machines).[20]

This general picture is largely confirmed in the case of machines less than 10 years old (Table 4.15).[21] The major discrepancies occur in the 'lathe' group and the 'grinding and all other' group.

In the lathe group (see Table 4.16) the Soviet proportion of lathes properly so-called is almost double that of the United States. The proportions of turrets and automatics, on the other hand, are broadly similar for the years 1958 and 1962, the only years for which comparable data are available (for the engineering sector).

The group of 'grinding and all other' machines is by far the most complicated and has been broken down into four separate subgroups in Table 4.17 (the methods used are explained in the notes to the table). The proportion of *grinding and polishing* machines in the Soviet stock is still lower than in the United States but has caught up considerably in recent years.[22] For *tool and cutter grinders* the Soviet Union had a larger proportion in 1962 in engineering, but this may have declined since then. In *bench and floor grinding and sharpening machines* the Soviet Union is likely to still have a larger proportion than the United States whose statistics show a steady decline. These machines are concerned with the initial machining of castings and the American figures reflect a shift to more precision casting.

Finally, among the '*other machines*' covered under this heading the most striking difference is found in 'cutting-off machines' (see Table 4.18). This is one of the few types of machine for which the United States shows a steady increase over the four censuses; the American proportion in 1962 was over four times the Soviet proportion. The United States also had a larger proportion of threading and tapping machines in 1962.

[20] This must clearly be tentative in the absence of a recent breakdown for engineering and is based on the TsSU survey data.

[21] The one exception is gear cutting and finishing machines, where the Soviet proportion is higher.

[22] It is possible that the 17·7 per cent for 1972 is comparable with the other entries in the table but this cannot be finally proved, and does seem rather unlikely.

Table 4.14 *Breakdown of machine tool stock by type of machine tool: (a) Soviet Union*

Type of machine tool	1958[1] census Engineering and metalworking	1962[2] census Engineering and metalworking	1962[2] census Total stock excluding kolkhozy	TsSU survey: main production shops 1967[3]	TsSU survey: main production shops 1969[4]	1972[5]: Stock in economic accounting organisations
Lathes	33·6	33·4	33·3	37·9	37·5	29·1
Boring	2·0	2·0	2·4	2·2	2·3	2·6
Drilling	17·3	18·0	22·3	17·2	17·4	22·8
Planing and shaping	4·1	3·4	4·0	1·4	1·3	2·9
Broaching	0·5	0·6	0·4	0·8	0·8	0·4
Gear cutting and finishing	2·8	3·0	2·0	4·4	4·4	2·2
Milling	12·7	13·2	10·2	14·2	14·2	10·8
Grinding and all others	27·0	26·4	25·4	22·1	22·0	31·2
Total	100·0	100·0	100·0	100·0	100·0	100·0

Sources:
1 Yu. D. Matevosov, *Ekonomicheskie problemy razvitiya stankostroitel'noi promyshlennosti v Zakavkaz'e*, Erevan (1968), p. 50. The source does not state that this covers engineering and metalworking but data for 1962 in same table correspond to those under this heading in Palterovich's table.
2 D. M. Palterovich, *op. cit.* (1972), p. 127. Based on original census data.
3 *Vestnik Statistiki*, 1968, No. 7, p. 89.
4 *Vestnik Statistiki*, 1970, No. 8, p. 89.
 Both these breakdowns are based on surveys of over three quarters of a million machine tools in over 2,500 engineering factories. Percentages calculated.
5 *Narodnoe khozyaistvo SSSR v 1972 godu* (1973), p. 69. Percentages calculated. This breakdown covers machine tools in 'economic accounting enterprises and organisations'. These account for about 90 per cent of total stock.

Breakdown of machine tool stock by type of machine tool: (b) USA

Type of machine tool	1958	1963	1968	1973
Lathes (turning)	21·5[a]	22·0[a]	21·0	20·9
Boring	2·6[a]	2·6[a]	2·3	2·3
Drilling	20·9	20·8	19·3	19·3
Planing and shaping	2·8	2·4	1·8	1·5
Broaching	0·7	0·7	0·6	0·6
Gear cutting and finishing	2·5	2·4	1·6	1·8
Milling	10·3	11·0	12·2	12·2
Grinding and all others	38·5	38·0	41·2	41·5
Total	100·0	100·0	100·0	100·0

Source:
As for USA in Table 4.1. Percentages calculated.

Note:
a – these two figures have been adjusted to transfer 'vertical boring mills' to the turning group for comparative purposes following Palterovich. *American Machinist* made the change for the 1968 census.

(c) Divergences between Soviet and American stock

Type of machine	Engineering and metalworking		Total stock excl. kolkhozy	TsSU surveys		Economic accounting organ- isations
	1958	1962	1962	1967	1969	1972
Lathes	12·1	11·4	11·3	16·9	16·5	8·2
Boring	−0·6	−0·6	−0·2	−0·1	0·0	0·3
Drilling	−2·6	−2·8	1·5	−2·1	−1·9	3·5
Planing and shaping	1·3	1·0	1·6	−0·4	−0·5	1·4
Broaching	−0·2	−0·1	−0·3	0·2	0·2	−0·2
Gear cutting and finishing	0·3	0·6	−0·4	2·8	2·8	0·4
Milling	2·4	2·2	−0·8	2·0	2·0	−1·4
Grinding and all others	−11·5	−11·6	−12·6	−19·1	−19·2	−10·3

Source:
Calculated from Table 4.14(a) and (b).

Note:
Soviet data are compared to closest US data, i.e., 1958 to 1958, 1962 to 1963, 1967 and 1969 to 1968 and 1972 to 1973. All positive figures indicate Soviet Union has a greater number and a minus sign indicates that the Soviet Union has less than the United States.

Breakdown of metalcutting machine tool stock by type of machine tool: (d) Soviet stock as a percentage of American stock

Type of machine	Engineering and metalworking		Total stock excl. kolkhozy	TsSU surveys		Economic accounting organ- isations
	1958	1962	1962	1967	1969	1972
Lathes	156	152	151	180·5	179	139
Boring	77	77	92	96	100	113
Drilling	83	87	207	89	90	118
Planing and shaping	146	142	167	78	72	193
Broaching	71	86	57	133	133	67
Gear cutting	112	125	83	275	275	122
Milling	123	120	93	116	116	89
Grinding and all others	70	69	67	54	53	75

Source:
Calculated from Table 4.14(a) and (b).

Table 4.15 *Comparison of breakdown of metalcutting machine tool types under 5 and 10 years old at most recent census*

	USSR[2]				USA[1]	
	Under 5 years		Under 10 years		Under 10 years	
Machine	No. (× 1000)	%	No. (× 1000)	%	No. (× 1000)	%
Turning	304·8	26·5	580·8	27·9	149·9	18·9
Boring	33·5	2·9	60·2	2·9	16·5	2·1
Drilling	282·9	24·6	483·8	23·2	133·5	16·8
Milling	112·2	9·7	219·5	10·5	109·0	13·7
Broaching	2·9	0·3	5·7	0·3	3·5	0·4
Planing and shaping	20·9	1·8	40·6	1·9	10·1	1·3
Grinding and all others	373·8	32·5	652·6	31·3	369·9	46·8
Total	1151·6	100	2082·6	100	795·1	100

Sources:
1 *American Machinist*, 11th inventory.
2 Calculated from percentage data given in *Narodnoe khozyaistvo SSSR v 1972 godu* (1973), p. 69.

Table 4.16 *Breakdown of turning group of machine tools for USA and Soviet Union*

(a) *Soviet Union*[1]

Type of machine tool	1958 census Engineering and metalworking	1962 census Engineering and metalworking	Total stock excl. kolkhozy	TsSU survey of main production shops 1967	1969	1972 census in econ. acctng. organisns.
Lathes	24·0	24·1	26·9	20·5	20·3	
Vertical boring mills			0·5	1·0	1·0	
Turrets	5·0	4·7	3·1	6·7	6·4	
Automatic and semi-automatic	4·2	4·6	2·8	9·7	9·8	
Total % (of total)	33·2	33·4	33·3	37·9	37·5	29·1

Source:
As for Table 4.14 (a).

(b) *United States*[1]

Type of machine tool	1958	1963	1968	1973
Lathes	11·1	11·6	11·9	11·9
Vertical boring mills	1·1	1·1	0·9	0·8
Turrets	4·4	4·2	3·6	3·1
Automatics	5·0	5·1	4·6	5·1
Total % (of total)	21·5	22·0	21·0	20·9

Source:
1 As for Table 4.14 (b)

Table 4.17 *Breakdown of 'grinding group and all other' of machine tools for Soviet Union and United States*

	USSR					USA		
Type of machine	Engineering and metalworking		Total stock excl. *kolkhozy*	Stock in econ. acctng. organisns.		Engineering and metalworking		
	1958	1962	1962	1972	1958	1963	1968	1973
Grinding and polishing	11·3	10·6	8·1	17·7	17·2	18·1	20·1	19·6
Tool and cutter grinding	3·4	3·0	4·3	n.a.	3·9	2·8	2·9	2·7
Bench and floor grinding and sharpening	n.a.	7·4	8·6	n.a.	6·1	5·4	4·6	4·5
Other (non-grinding)	n.a.	5·4	4·4	n.a.	11·3	11·7	13·6	14·7
Total (% of total)	n.a.	26·4	25·4	n.a.	38·5	38·0	41·2	41·5

Source:
As for Table 4.14.

Note:
This is a difficult group and apart from Palterovich's 1962 data the Soviet statistics lack sufficient detail for a full comparison. On analysing Palterovich's table and the adjustments he has made, the following would appear to emerge. The 'grinding group of machine tools' in the US covers grinding machines, honing and lapping machines and polishing and buffing machines and the *total* of these three types equals the *total* for the following three Soviet types—grinding and polishing machines, tool and cutter grinding (*obdirochno-shlifoval'nye*) and sharpening (*tochil'nye*) machines. In order to adapt the American figures to the Soviet definitions (according to Palterovich's methodology), it is necessary to deduct from grinding machines 'tool and cutter' grinding machines and 'bench floor and snag grinding machines' (which Palterovich identifies with *obdirochno-shlifoval'nye, tochil'nye*). It is then necessary to add to the balance 'honing and lapping' machines and 'polishing and buffing' machines.

Table 4.19 deals with a small subgroup in the 'other machines' category: the modern machines concerned with electrical and ultrasonic machinery units. Some of the early work in this field was done in the Soviet Union; it seems clear that it has about the same number of these machines as the United States.

Table 4.20 covers the last remaining subgroups of 'other machines'. As far as is known, the Soviet figures mainly cover unit head machines and machines in automatic lines.

The Soviet Union is evidently trying to bring the composition of its stock closer to the American pattern. The changes in the breakdown of output of metalcutting machine tools in the Ministry for the Machine Tool and Tooling Industry between 1965 and the planned output for 1975 show that the largest changes occur mainly in those types which differ most from the United States. Thus the proportion of lathes is to fall by some five per cent, while broaching, grinding and cutting-off machines are to increase their share of output.[23]

The information on metalforming machines in Soviet sources is again rather limited but a number of striking differences emerge[24] (see Table 4.21). The

---

[23] See *Ekonomicheskie problemy razvitiya mashinostroeniya* (1975), p. 53.

[24] The following is based closely on the comments made by Palterovich to his table which has been used in Table 4.21 (D. M. Palterovich, *op. cit.* (1970), pp. 166–9).

Soviet Union has far fewer of the type of machines which are not involved directly in the manufacture of forgings and hot stampings—riveting machines, wire and metalforming and, to a lesser extent, bending and forming machines. Palterovich quotes the example of profile bending machines of which the United States had 14,600 in 1963 compared with 868 in the Soviet Union in 1962. One result of this is that although the Soviet Union in 1962 had about 61 per cent of the total American stock of metalforming machines it produced about twice as many forgings and hot stampings. Among presses the Soviet Union had rather more hydraulic presses, rather less pneumatic and hydropneumatic presses. According to Palterovich forging equipment made up 13·9 per cent of Soviet output in 1960–7 and only 1·3 per cent of American output; and much of the Soviet output was made up of hammers while the United States produced more advanced equipment. This is brought out in the stock figures: the share of hammers in the Soviet stock is some seven times that of the United States.

Table 4.18 *Cutting-off, tapping and threading machines (%)*

Type of machine	USA (engineering and metalworking)				USSR—1962	
	1958	1963	1968	1973	Engineering and metalworking	Total stock excl. *kolkhozy*
Cutting-off	5·6	7·0	8·0	8·3	1·7	1·6
Threading and tapping	2·6	2·2	2·3	2·2	1·6	1·4

Source:
As for Table 4.14.

Table 4.19 *Ultrasonic and electrical machinery units* (in units)

Type of machine	USA Engineering and metalworking				USSR (× 1000)		
					1962		1972
	1958	1963	1968	1973	Engineering and metal-working	Total exc. *kolkhozy*	Econ. acctng. organisns.
Ultrasonic	178	548	444	n.a.	2·0	2·4	1·3
Electrical discharge	578	2385	6250	9145			8·4
Electro-chemical	n.a.	n.a.	887	1183	n.a.	n.a.	1·9
Electrolytic grinders	n.a.	710	751	848	n.a.	n.a.	n.a.

Source:
As for table 4.14. USSR 1962 data calculated from percentage data in original table.

Note:
Units are used because of small numbers involved.

Table 4.20 Other[a] metalcutting machine tools (%)

Type of machine	USA (engineering and metalworking)				USSR—1962	
	1958	1963	1968	1973	Engineering and metalworking	Total stock excl. *kolkhozy*
Multi-function	n.a.	n.a.	0·4	0·6	⎫	⎫
Special way-type	0·7	0·7	0·6	0·6	⎪	⎪
Automatic assembly	0·2	0·3	0·6	0·7	⎬ 1·9	⎬ 1·3
Contour sawing and filing	1·0	1·2	1·1	1·0	⎪	⎪
Other	1·3[b]	0·2	0·3	0·7	⎭	⎭
Total	3·2	2·4	3·0	3·6	1·9	1·3

Source:
As for Table 4.14. Palterovich compares 1963 American data with 1962 Soviet data. His US total of 2·3 is reasonably close to the 1963 figure given above and suggests that the right composition of types has been used.

Notes:
a – according to the Soviet source this includes 'unit head and transfer machines'. No comparable later data are available.
b – includes 0·4 'pipe cutting and threading machines', afterwards apparently excluded from census.

Although the available information summarised above makes the general situation considerably clearer much still remains uncertain. First, the headings used in the published Soviet data are very broad, and the far more detailed American breakdown by types cannot be used to identify the comparative position in relation to particular types of advanced machine tools. Second, machine tools are not classified according to their precision. Third, the proportion of low value simple machines in the American stock cannot be precisely ascertained.

With these provisos the major differences between the Soviet and American stocks of machine tools are summarised in Table 4.22. The most striking feature to emerge from the comparison of stocks is that the Soviet Union has over one million more metalcutting machine tools than the United States overall, and in the case of those under 10 years old has almost 1,300,000 more, or two and a half times as many. The significance of this is underlined by the fact that the American stock is likely to include a significant proportion of machine tools smaller and simpler in design than those in the Soviet stock.

REASONS FOR DIFFERENCES BETWEEN MACHINE TOOL STOCKS

Some of the reasons for this difference have already been indicated: first, a large number of machine tools are used outside the engineering and metalworking sector; secondly, a high proportion are not used in the main production shops. The first reason is a consequence of the small amount of spare parts produced in the Soviet Union.[25] This situation has arisen because it has been more profitable to use all the parts produced to make up finished machines. Repair was treated as of secondary importance, and very little attempt was made to centralise it and organise it more efficiently. The high

Table 4.21 *Breakdown of metalforming machines by type: Soviet Union and United States (%)*

Type of machine	USA[1,a]				USSR[2] 1962		1972
	1958	Engineering and metalworking 1963	1968	1973	Total excl. *kolkhozy*	Engineering and metal-working	Total econ. accntng. organisns.
Presses:							
mechanical	47·1	43·1	42·4	41·4	39·6	50·6	28·6
hydraulic	5·9	7·2	8·0	8·7	12·0	9·8	22·4
pneumatic and hydropneumatic	2·0	2·1	2·0	2·2	0·9	1·4	
Forging equip.:							
hammers	1·9	1·6	1·1	0·9	} 14·3	} 10·3	6·4
forging machines	2·1	3·1	3·2	3·0			0·5
incl. hot and cold headers	n.a.	2·0	1·9	1·5	1·8	2·3	
Pressed forging and automatics[b]	n.a.	n.a.	n.a.	n.a.	n.a.	n.a.	4·7
Punching and shearing	13·6	11·9	13·6	14·4	15·2	12·9	13·6
Bending and forming	12·8	14·0	13·6	14·0	13·4	9·4	15·4
Riveting machines	9·0	9·1	8·3	7·9	0·4	1·4	} 8·4
Thread rolling machines	n.a.	1·5	1·4	1·2	1·8	2·4	
Wire and metal ribbon formers	3·3	3·0	3·2	3·7	1·4	1·1	
Other	2·3	3·4	3·1	2·5	0·5	0·7	
Total	100·0	100·0	100·0	100·0	100·0	100·0	100·0

Sources:
1 As for Table 4.1. Percentages calculated. D. M. Palterovich, *op. cit.* (1970), p. 167.
2 *Narodnoe khozyaistvo SSSR v 1972 godu* (1973), p. 70.

Notes:
a – the American data has been adjusted according to Palterovich's methodology. Analysis of his table which gives the 1968 US data shows that all his entries correspond to the published US statistics except the entries 'punching and shearing machines' and 'other'. Palterovich gives 13·6 per cent and 3·3 per cent whereas the US published breakdown gives proportions of 11·8 and 4·9. It would appear that 'roll-flowing' and 'roll-forming' machines under 'other' have been transferred to the other group though this is not completely certain. This gives 13·6 and 3·1 respectively; and the discrepancy in the proportion of other machines can perhaps be explained by Palterovich taking this as a residual since otherwise, because of rounding, the column does not add up to 100·0.
b – this is a category used only in the 1972 Soviet census and has no equivalent in the American statistics. It presumably includes machines previously found under both 'presses' and 'forging equipment' and clearly distorts the figures in these groups.

[25] According to one source in the machine tool industry in the Soviet Union spare parts account for 1·5 to 2 per cent of output while in the United States they account for 12 per cent. *Planovoe Khozyaistvo*, 1968, No. 11, pp. 24–5. Other engineering industries are probably broadly similar.

proportion of machine tools outside main production shops in the Soviet engineering industry is largely a result of the organisation of production which has emerged in the industry. Many factories have a largely closed cycle of production and are specialised by end product, most components being made by the factory itself. This was largely due to the supply system. Tooling also became a low priority and inefficient area as production was largely decentralised.[26] Serious efforts have been made in the past 10 years to improve the level of specialisation in tooling.

In spite of the size of the machine tool stock, few existing machines are replaced: the bulk of the substantial and increasing production of machine tools goes to increase the size of the stock.[27] A rough indication of this is that between the 1962 and 1972 censuses the stock increased by around 1,700,000

Table 4.21a *Soviet stock of metalforming equipment by type as a proportion of American stock*

	1962[a]		1972[a]
Type of machine	Total excl. *kolkhozy*	Engineering and metalworking	Total in econ. acctng. organisns.
Presses:			
mechanical	92	117	69
hydraulic	167	136	257
pneumatic and hydropneumatic	43	67	
Forging equipment:			
hammers	} 304	} 219	711
forging equipment			17
incl. hot and cold headers	90	115	
Pressed forging automatics	—	—	—[b]
Punching and shearing	128	108	94
Bending and forming	96	67	110
Riveting machines	10	15	
Thread rolling machines	120	160	} 55
Wire and metal ribbon forming	47	37	
Other	15	21	

Source:
Calculated from Table 4.21.

Notes:
a – 1962 is compared with US 1963, 1972 with US 1973.
b – see Table 4.21, note b.

[26] See, for example, *Ekon. Gaz.*, 17 May 1961, pp. 2–3.
[27] By way of contrast the metalcutting machine tool stock in the United Kingdom has declined by almost 30 per cent between 1961 and 1971. See *Metalworking Production, Third Survey of Machine Tools and Production Equipment in Britain*, London (1971), p. 21. About two thirds of United States machine tool production is used to replace existing stock.

machines while output from 1962 to 1971 inclusive totalled just over 1,900,000, imports and exports being roughly in balance.[28] The ninth five year plan provided that 25–30 per cent of metalcutting machine tools should go to engineering factories to replace existing stock.[29]

The low level of replacement for machine tools as compared with the United States is clearly related to the level of demand. Machine tools are still in very

Table 4.22 *Summary of main differences in machine tool stock of Soviet Union and United States*

Criteria	Soviet Union	United States
1. *Numbers*		
Total stock	Larger from *c.* 1968	
incl. metalcutting	Larger from *c.* 1965	
metalforming	Larger from *c.* 1971	
Stock in engineering and metalworking (total stock)		Larger
incl. metalcutting	Larger from *c.* 1971	
metalforming		Larger
Stock outside engineering	Larger	
Stock in engineering used outside basic production	Larger	
2. *Age*	More machines under 10 years old	
Weight	Heavier	
Value	Greater from *c.* 1955–60	More low value machines
Pattern of output of machine tool and tooling industry	More machine tools	More tooling and fixtures
3. *Composition*		
Proportion of metaforming		Larger
Types: metalcutting	More lathes	More grinding and polishing machines
	More bench and floor grinders	More cutting-off machines
	More in total stock: boring, drilling, planing and shaping	More in engineering: boring, drilling, planing and shaping
	More in engineering: broaching, milling, gearcutting	
metalforming	More hammers	More riveting machines
		More wire and metal ribbon forming machines

[28] The low renewal rate shown by these figures is probably somewhat exaggerated, however, since it can be calculated that the number of machine tools under 10 years old in 1972 was 2,355,000 (60 per cent of the total stock—see Tables 4.1 and 4.5)—which supports the view that the 1972 stock figures include some machines not covered by the production figures.

[29] *Gosudarstvennyi pyatiletnii plan razvitiya narodnogo khozyaistva SSSR na 1971–75 gody* (1972), p. 128.

short supply in the Soviet Union and demand far exceeds supply.[30] A number of factors is involved here. First, the demand comes from all the sectors in which machine tools are used, and as we have seen these tend to be wider than in the US. The pressure of demand is aggravated by the common Soviet complaint of overordering. In satisfying demand, a major role is played in the United States by secondhand machines,[31] while in the Soviet Union the secondhand market is almost non-existent.[32] A consequence of this is that new machine tools are allocated to factories in all parts of the economy, even to those uses for which secondhand machines would be suitable. Lastly, the large stock of machine tools is needed in the Soviet Union because of the low general level of utilisation. This is reflected in part by the average shift use[33] which has been declining in the Soviet Union in recent years.[34] Adequate comparative data are lacking but it seems likely that the Soviet Union has a lower shift coefficient than the United States.[35] The level of utilisation is also affected by such factors as stoppages during shifts, and the proportion of machines not in use because they are not installed or are undergoing maintenance or repair work.[36] Much depends, moreover, on the effectiveness with which the machine is used. In the Soviet Union a large proportion of machines is used in one-off and small batch production, and this tends to result in a low machining time.[37] On the other hand, it is necessary to take into account that in the United States the fluctuating level of demand means that production capacity, including machine tools, is often not fully utilised.

THE SIGNIFICANCE OF THE DIFFERENCES IN STOCKS

The significance of the differing composition of machine tool production and stock in the two countries is much more difficult to assess. The differing pattern partly reflects the technological capabilities of the machine tool industries themselves. But it is also itself a result of the pattern of demand, and hence of the technological requirements, and technological level, of the engineering industries and the other major consumers of machine tools.

In this respect it is useful to distinguish three aspects of the demand for machine tools: what can be called 'non-progressive' demand; 'progressive' demand; and 'official' or 'recognised' demand.

Demand is high for 'non-progressive' universal machines with a wide range

---

[30] Demand for metalcutting machine tools as a percentage of output was: 1962—176 per cent, 1967—203 per cent, 1968—208 per cent (V. K. Fal'tsman, *Prognozirovanie potrebnosti v oborudovanii* (1970), p. 12).

[31] In 1960–2 sales of secondhand machines averaged around 90,000 per annum or about 50 per cent of output (see *American Machinist*, 10 June 1963, p. 529).

[32] Some attempt was made in the 1965 reform to encourage factories to get rid of surplus equipment but as far as can be judged these sales as a proportion of total output remain very small.

[33] That is, the average number of shifts worked per machine tool in 24 hours.

[34] See V. K. Fal'tsman, *op. cit.*, pp. 12–13.

[35] See K. I. Klimenko, *Ekonomicheskie problemy tekhnicheskogo progressa v mashinostroenii SSSR* (1965), p. 58.

[36] All these areas have been criticised in Soviet sources.

[37] See, for example, *Stankoinstrumental'naya Promyshlennost'*, 1961, No. 5, p. 6.

of speeds, feeds, etc., which can be used for a variety of tasks.[38] At the same time these are the machines which the Soviet machine tool industry likes to make because they give greater benefit to the producer. On the other hand 'progressive' demand is represented by those factories and industries which want advanced machine tools—the ballbearing industry and the motor industry are two good examples. These machines are less popular with the machine tool industry and where possible it often tends to try to avoid producing them.[39]

Gosplan has the task of striking a balance between these two conflicting aspects of demand, and the pattern of demand which it has 'recognised' for inclusion in the state plan becomes the 'official' demand. The technological conservatism of the industry is countered by the deliberate decisions of the government and Gosplan which, via the state plan, attempt to impose their own policy on output and hence on the composition of the stock, largely guided, it would seem, by the experience of capitalist countries.

This policy is sometimes expressed gradually but sudden campaigns can also produce startling changes in a short period. This can be illustrated by the example of numerically controlled machine tools which is described below in the final section of this chapter. An equally striking example is provided by the rapid increase in the production of precision machine tools. After remaining at 7,000–8,000 in the period 1956–9, output expanded rapidly, in particular following a government decree of 13 January 1960,[40] and by 1962 had reached over 20,000.[41]

In 1958 the demand for precision machine tools in 1965 had been estimated at 16,000[42] but by 1963 the planned output target for 1965 had been increased to 39,700[43] and although this target was not reached the 1965 output was probably some four times the 1958 output.[44] Production has continued to expand but at a slower pace and 1970 output was probably about one and a half times the 1965 figure.

Sometimes the state plan is used to counter the combined conservatism of both the producer and the consumer. Palterovich cites unsuccessful attempts to change the technological structure of machine tool output by a sharp reduction in the number of lathes produced and an increase in the number of grinding and boring machines. The change in output failed to alter the demand structure: a shortage of lathes resulted, while there was no increase in demand

---

[38] Thus, in the aggregated demand for the *sovnarkhozy* for the seven year plan period, 80 per cent was for universal machines (*Ekon. Gaz.*, 14 February 1961).

[39] This is obviously a very brief and simplified account. This whole question is discussed in more detail in the present writer's *Research, Development and Innovation in the Soviet Machine Tool Industry*, unpublished research report, University of Birmingham (1974).

[40] See *Stankoinstrumental'naya Promyshlennost'*, 1961, No. 5, pp. 5, 7.

[41] *Narodnoe khozyaistvo SSSR v 1962 godu* (1963), pp. 232–3.

[42] See P. M. Pen'kov, *Razvitie tipazha i struktura vypuska metallorezhushchikh stankov i komplektuyushchikh prinadlezhnostei k nim v 1958 i 1959–65 gody* (1958), pp. 9–10.

[43] *Metallorezhushchie i derevo-obrabatyvayushchie stanki, avtomaticheskie linii*, 1963, No. 2, p. 1.

[44] Based on percentage figure given in (Ed.) N. M. Oznobin et al., *Sovershenstvovanie struktury promyshlennogo proizvodstva* (1968), p. 136.

for the more progressive machines.[45] This can in part be explained by economic factors and there is evidence of some resistance on the part of user factories to acquiring the more advanced types: these are often much more expensive but only slightly more productive than existing machines.

The more important part of the explanation lies in the relationship between the present machine tool stock and the needs of the Soviet economy. Palterovich has pointed out: 'Sometimes in the literature the ratio between different types of metalworking equipment is looked upon as a cause determining the level of technology and the organisation of production in engineering. For example, the high proportion of metalcutting and the loss of large amounts of metal in swarf is interpreted as a consequence of the insufficiently high proportion of metalforming equipment. In reality the technological structure of metalworking machinery not only determines the character of the technology but also itself depends on the level of technology, organisation and culture of production which has been achieved.'[46] A not dissimilar view is expressed in a study of innovation in the American machine tool industry.[47] This pointed out that the industry 'has been traditionally dominated by its customers', and 'the industry has relied for direction in development work on what the customer wanted'. It concluded that 'in part the industry's conservatism as regards innovation has been a function of the conservatism of its users, of their endemic lack of interest in innovation in manufacturing processes'. Both of these writers, then, are arguing that the size and shape of the machine tool stock are essentially a symptom of the technological level of industry as a whole, rather than a cause. Thus, in the Soviet context, because of low quality castings it would probably be unwise to reduce the output of equipment used for machining these and increase the proportion of finishing machines, although this represents the American pattern.

Thus there are many differences in the demand for machine tools between the USA and the USSR and to argue that the composition of the Soviet stock should be made more similar to the composition of the American stock, unless the pattern of industrial production was generally similar, would be unreasonable. It seems unlikely that this will be the case in the near future, despite the moves made in the Soviet Union to expand light industry and other consumer orientated industries. Moreover, in comparing the two stocks the inefficient use of machine tools in Soviet industry would have to be set against the cyclical pattern of production and the frequently under-utilised capacity and potential in American industry.

CHANGES IN RELATIVE TECHNOLOGICAL LEVELS

With the evidence so far available it is difficult to draw any firm general conclusions about changes in the relative technological levels of the machine

---

[45] D. M. Palterovich, *Planirovanie potrebnosti v oborudovanii* (1972), pp. 117–18.
[46] D. M. Palterovich, *Park proizvodstvennogo oborudovaniya* (1970), p. 169.
[47] In *Patterns and Problems of Technical Innovation in American Industry*, Report to National Science Foundation, September 1963, Arthur D. Little Inc., p. 111.

tool industries of the two countries in the past 20 years. As we have seen, there has been a tendency for the Soviet Union to move towards an American pattern of stock and output but there are still considerable divergences, in part reflecting the differences in the use of the machine tool stock and its distribution in the economy. Thus the Soviet Union still has a much larger proportion of machines for the more basic coarser operations, e.g., lathes and hammers, and a smaller proportion of machines concerned with finishing operations and the more complex metalforming operations. A similar problem emerges in assessing changes in the relative levels of complexity and sophistication of machine tools produced. The output and stock data fail to give any clear indication and the situation is also complicated by the significant proportion of relatively simple low value machines in the American statistics. In one of the few areas where comparative figures are available, ultrasonic and electrical machining techniques, the two countries appear to have kept roughly in step.

We have therefore undertaken three further studies which may reveal more clearly changes in relative technological level which may have occurred. The first study looks at Soviet imports and exports as an indicator of technological level. The second deals with one of the frontier areas of development—numerically controlled machine tools, which is perhaps one innovation in recent years which is capable of forcing a transformation of the pattern of production in industry by changing the patterns of thought about machine tools which have been built up over decades. The third study appears in chapter 11 below, which deals with the quality of production; one of our case studies is a product group on which a great deal of Soviet effort has been concentrated—universal machine tools, and we have also assembled in this chapter some Soviet and Western opinions on Soviet machine tools.

Soviet imports and exports of metalcutting machine tools as an indicator of technological level

THE HISTORICAL BACKGROUND

In the machine tool industries of leading Western industrialised countries there has evolved over the years some degree of specialisation of output so that no one country produces every possible kind of machine tool and the major producers of machine tools are also major importers. In the case of the Soviet Union the situation is rather different, since self-sufficiency was from the beginning a major element of Soviet policy. As a result the range of types and sizes in production continues to expand rapidly, although in recent years there has been some attempt to rationalise output within the framework of Comecon.

Imports reached a high level in the early 1930s but declined steadily thereafter, except for the war period, and in the period 1947–54 averaged only 1,400 per annum—well below exports. From the mid-1950s, however, Soviet imports of machine tools began to expand (see Table 4.23). Initially the main role was played by the other East European countries but in recent years, however, Soviet purchases in the West have increased dramatically in value terms and in 1973 accounted for almost 50 per cent of imports (by value) (see Table 4.25a). Soviet exports also have increased considerably (although the picture given in Table 4.23 is slightly distorted as 1955 was well below previous years) and in 1974 totalled 57 per cent of the value of imports. Here again, the main East European countries play the major role although in 1955 China and Mongolia were the main recipients. Surprisingly, exports to developing countries have declined in recent years while exports to Comecon have increased considerably and even exports to developed capitalist countries have increased slightly (see Table 4.26).

The balance of trade in value terms has been against the Soviet Union since 1953 and the deficit has steadily increased until in 1974 it was over 90 million roubles,[48] most of which is accounted for by trade with the main capitalist countries (see Table 4.27). The Soviet Union has a net surplus only in its trade with 'other countries'.

Table 4.23 *Soviet imports and exports of metalcutting machine tools, 1955 to present*

	1955	1960	1965	1970	1973	1974
Imports: No. ($\times 1000$)	2·4	7·6	6·5	9·2	12·9	11·4
Value ($\times 1000$ r)	21852	56666	83178	135045	252962	211703
Exports: No. ($\times 1000$)	0·9	2·1	5·6	12·2	14·0	16·3
Value ($\times 1000$ r)	6522	12893	39507	78792	95155	120267

Source:
Vneshnyaya torgovlya SSSR: statisticheskii sbornik for appropriate years.

Note:
Data in this and following tables refer only to metalcutting machine tools (100 in the Soviet foreign trade code). They do not include automatic (or transfer) lines of machine tools (10401). It is also possible that other machine tools are included under other headings, e.g., 'Equipment for car factories' (10514). Both of these have been published (as separate entries) only recently. Imports of automatic lines in value terms were as follows:

	1970	1971	1972	1973	1974
Value of automatic lines imported ($\times 1000$ r)	28733	11049	38401	58382	34925

See also B. L. Kostinsky, *Description and Analysis of Soviet Foreign Trade Statistics, Foreign Economic Report No. 5*, US Department of Commerce, July 1974, pp. 15–39, especially for a discussion of the problem of Soviet foreign trade statistics referring to complete factories. This shows that metalcutting machine tools were an important component of exported complete plants.

[48] This would be even higher if automatic lines of machine tools were included. See note to Table 4.1.

Table 4.24 Distribution of Soviet trade in machine tools among leading capitalist[a] and socialist[b] countries, 1955 to present[1] (in units and value × 1000r)

	1955		1960		1965		1970		1973		1974	
	No.	Value	No.	Value	No.	Value	No.	Value	No.	Value	No.	Value
(a) *Imports*												
Main socialist	1850	19360	5686	42154	5442	64133	5779	74379	7518	109218	6115	101242
Main capitalist	556	2450	1816	13362	1021	18053	1708	53307	3718	128176	2478	87148
Other[d]	21	42	109	1150	40	992	1700[c]	7359	1700[c]	15568	2800[c]	23313
Total	2427	21852	7611	56666	6503	83178	9200[c]	135045	12900	252962	11400	211703
(b) *Exports*												
Main socialist	198	2994	1005	7852	2391	25323	7255	58268	8751	71822	11052	94157
Main capitalist	—	—	40	225	354	1542	697	5941	1090	7811	1190	8458
Other[d]	700[c]	3528	1100	4816	2900	12648	4200	14583	4200[c]	16022	4000[c]	17652
Total	900[c]	6522	2100[c]	12893	5600[c]	39507	12200[c]	78792	14000[c]	95155	16300[c]	120267

Source:
Taken from *Vneshnyaya Torgovlya SSSR: statisticheskii obzor* for appropriate years.

Notes:
a – capitalist countries—United Kingdom, Italy, United States, FRG, France, Switzerland, Japan.
b – socialist countries—Bulgaria, Hungary, GDR, Poland, Romania, Czechoslovakia.
c – rounded to nearest 1000.
d – calculated by residual.

Table 4.25 Distribution of Soviet trade in machine tools among leading capitalist[a] and socialist[a] countries (%)

	1955		1960		1965		1970		1973		1974	
	No.	Value	No.	Value	No.	Value	No.	Value	No.	Value	No.	Value
(a) *Imports*												
Main socialist	(76)	88·6	(75)	74·4	(84)	77·1	(63)	55·1	(58)	43·2	(54)	47·8
Main capitalist	(23)	11·2	(24)	23·6	(16)	21·7	(19)	39·5	(29)	50·7	(22)	41·2
Other[b]	(1)	0·2	(1)	2·0	(1)	1·2	(18)	5·4	(13)	6·2	(25)	11·0
Total	100	100	100	100	100	100	100	100	100	100	100	100
(b) *Exports*												
Main socialist	(22)	45·9	(48)	60·9	(43)	64·1	(59)	74·0	(63)	75·0	(68)	78·3
Main capitalist	—	—	(2)	1·7	(6)	3·9	(6)	7·5	(8)	8·2	(7)	7·0
Other[b]	(78)	54·1	(50)	37·4	(51)	32·0	(35)	18·5	(30)	16·3	(25)	14·7
Total	100	100	100	100	100	100	100	100	100	100	100	100

( ) Percentages in brackets are approximate because of rounding.

Notes:
a – see notes to Table 4.24.
b – by residual.

Table 4.26 *Distribution of Soviet trade in machine tools by type of country, 1967 to present (in units ($\times 1000$))*

Year	Imports					Exports					
	Total	Socialist All	Comecon	Capitalist Developed	Capitalist Developing	Total	Socialist All	Comecon	Capitalist Developed	Capitalist Developing
1967[1]	6·0	4·7	4·5	1·3	—	9·1	4·6	4·1	1·7	2·7
1968[2]	7·8	5·5	4·8	2·3	—	10·5	5·8	5·1	1·9	2·8
1969[2]	7·9	6·3	5·3	1·6	—	12·4	7·2	6·5	2·0	3·3
1970[3]	9·2	7·4	5·8	1·8	—	12·2	7·9	7·3	1·9	2·4
1971[3]	11·9	9·8	8·7	2·1	—	11·9	7·7	6·8	1·9	2·3
1972[4]	16·1	12·7	10·5	3·4	—	11·4	7·5		2·0	1·9
1973[4]	12·8	9·0	7·5	3·8	—	14·0	9·4	9·0	2·6	2·0
1974[5]	10·9a	8·8	6·1	2·6	—	16·3	12·1	11·7	2·6	1·6

Sources:
1 *Vneshnyaya Torgovlya*, 1969, No. 10, p. 52 and No. 11, p. 50.
2 *Ibid.*, 1970, No. 9, pp. 50, 55.
3 *Ibid.*, 1972, No. 8, p. 54, and No. 9, p. 55.
4 *Ibid.*, 1974, No. 6, p. 51, and No. 7, p. 54.
5 *Ibid.*, 1975, No. 6, p. 51, and No. 7, p. 53.

Note:
a – does not correspond with total in annual foreign trade statistical handbook which gives 11,400, used in other tables (*Vneshnyaya Torgovlya SSSR za 1974 god: statisticheskii obzor* (1975), p. 39).

Table 4.27 *Value of Soviet trade and balance of trade (× 1000 r)*

	1955	1960	1965	1970	1973	1974
Total						
Imports	21852	56666	83178	135045	252962	211703
Exports	6522	12893	39507	78792	95155	120267
Balance	−15330	−43773	−43671	−56253	−157807	−91436
Main socialist countries[a]						
Imports	19360	42154	64133	74379	109218	101242
Exports	2994	7852	25323	58268	71322	94157
Balance	−16366	−34302	−38810	−16111	−37896	−7085
Main capitalist countries[a]						
Imports	2450	13362	18053	53307	128176	87148
Exports	—	225	1542	5941	7811	8458
Balance	−2450	−13137	−16511	−47466	−120365	−78690
Other countries						
Imports	42	1150	992	7359	15568	23313
Exports	3528	4816	12642	14583	16022	17652
Balance	+3486	+3666	+11650	+7224	+454	−5661

Source:
Vneshnyaya torgovlya: statisticheskii obzor data for appropriate years.

Note:
a – see note to Table 4.24.

Table 4.28 *Unit value of machine tools imported and exported by USSR, 1955 to present*

	Imports			Exports		
	All[b]	Main capitalist[a] countries	Main socialist[a] countries	All[b]	Main capitalist[a] countries	Main socialist[a] countries
1955	9·0	4·4	10·5	7·2	—	15·1
1960	7·4	7·4	7·4	6·1	5·6	7·8
1965	12·8	17·7	11·8	7·1	4·4	10·6
1970	14·7	31·2	12·9	6·5	8·5	8·0
1973	19·6	34·5	14·5	6·8	7·2	8·2
1974	18·6	35·2	16·6	7·4	7·1	8·5

Source:
Calculated from Table 4.24.

Notes:
a – see notes to Table 4.24.
b – approximate because of rounding of total numbers to nearest 100.

The main feature which emerges from an examination of total trade statistics is the imbalance between imports and exports in recent years in value terms, although in number terms exports are higher. This means that the unit value of machines imported is much higher than those exported (approximately double—see Table 4.25). This feature can be brought out more clearly by considering only a number of leading capitalist countries. Exports from these countries are mainly expensive machines, about five times the unit value of machines they import from the USSR (see Table 4.28). This does suggest that the Soviet Union has a long way to go before it can achieve a balanced trade with the advanced capitalist countries. On the other hand, it is perhaps necessary to point out that the Soviet machines exported are likely to be cheaper than equivalent Western machines; there is also probably still some resistance to buying from the Soviet Union because it has only recently become a major exporter of machine tools and has yet to fully establish itself.

THE PATTERN OF SOVIET IMPORTS—THE CASE OF THE UNITED KINGDOM

One crucial question in relation to the sharp rise in Soviet imports in recent years is the reasoning behind it—are machines imported because the Soviet Union cannot make them, or is it because the Soviet Union cannot make *enough* of them at the present time (for example, for the crash programme of expanding the motor industry) and the extension in capacity is not justified, or, alternatively, are machines imported in order to be copied, as Sutton, for example, suggests?[49]

The last possibility seems least likely because of the numbers of machines involved—since for copying purposes only one or two machines would presumably be needed. It is clear that more information is needed to discuss the first two possibilities. In the absence of a detailed breakdown of all Soviet imports it is useful to examine the breakdown of British exports to the USSR given in British sources (see Table 4.29). This brings out very clearly the major role played by grinding machines in Soviet imports—in the period 1965–72 they accounted for about 44 per cent by value of total Soviet imports from Britain. Table 4.29 also brings out very clearly the considerable fluctuations in the volume of imports among the other types of machine tools listed. These fluctuations seem to depend entirely on the policy of the Soviet Union since efforts to explain them in terms of the British machine tool industry have been unsuccessful.[50]

It is interesting to compare the pattern of recent UK exports to the USSR with the requirements for Soviet industry included in the so-called Bulganin-Khrushchev £1,000 million 'shopping list' which was discussed at the time of

---

[49] See A. C. Sutton, *Western Technology and Soviet Economic Development, 1945–65*, Stanford, Cal. (1973), pp. 303–17.

[50] See G. S. Nash, *An Analysis of British Exports to the Socialist Countries of Eastern Europe*, Final Year Project Report for Chemistry/Management Degree, Loughborough University of Technology (1975) especially pp. 30–8. I am grateful for the opportunity to see this unpublished research report.

their visit to Britain in 1956. This included the following metalcutting machines:

Transfer machines	£27–36 million
Special and multiple head machines	500 units
Grinding machines:	
planetary	40
gear cutters and shavers	50
optical profile	40*
thread	40*
broach and spline	80
Gear cutters (up to 160 inch dia.)	10*
Gear planers (up to 64 inch dia.)	25*
Profile millers: multispindle	25*
Universal pattern milling machines	100
Jig borers	45*
Relieving lathes	47
Thread milling machines	20**
Auto-lathes 4 and 6 spindle	40

* complete embargo.
**partially embargoed, part subject to quantitative control.

The shopping list also included over 300 metalforming machine tools and over 1,000 items of foundry equipment.[51] With the exception of milling machines, jigboring machines and relieving lathes, all the listed machine tools were prominent in subsequent trade, although it is difficult to judge the role played by special machines since these are not separated out in the available statistics. In the case of jigboring machines it is clear that Soviet output was expanded in the 1960s and it was recently stated that existing capacity for their production was not being fully used.[52] Overall, judging by the UK statistics, it would appear that the pattern of Soviet requirements has changed little since 1956 in these key areas although in the case of the other types imports from Britain have been low.

Another indication of the unsatisfied demand in the Soviet Union for many types of machine tools can be found by comparing the demand for machine tools in 1965 as calculated for the seven year plan (1959–65) with the planned output (see Table 4.30). At the time the plan was drawn up there appeared to be no intention to increase imports and a balance of exports over imports of 56,000 units was envisaged.[53] In practice, however, there was a deficit of about 23,500 units.

[51] See *Metalworking Production*, 16 November 1956, pp. 185–7.
[52] See *Pravda*, 12 May 1972.
[53] P. M. Pen'kov, *Razvitie tipazha i struktura vypuska metallorezhushchikh stankov i komplektuyuschchikh prinadlezhostei k nim v 1958 i 1959–65 gody* (1958), p. 52.

There appear to have been some subsequent changes in the plan, judging by the large increase in output of broaching machines and tool and cutter grinders compared with the original plan. Table 4.30 does, however, bring out very clearly the important shortfalls in the production of those machines which form the bulk of Soviet imports from the UK—thus, for grinding machines demand was satisfied by only 39·5 per cent, automatic lathes by only 47 per cent and gear cutting machines by only 58 per cent. It seems clear from this that Soviet imports reflect a shortage of capacity in the Soviet Union, particularly for the more sophisticated machines. Production could not be expanded fast enough to satisfy demand.

Similarly it is interesting to relate the pattern of Soviet imports to the comparison between the stocks of machine tools in the Soviet Union and the United States given in the first section of this chapter.[54] Unfortunately, because of the large groupings of machine tools used it is difficult to draw any firm conclusions except in the case of grinding machines, of which the US has had a much higher proportion until very recently. Both countries appear to have

Table 4.29 *Main types of metalworking machinery imported into USSR from the United Kingdom in value terms* (× £1000)

	Type of machine tools						
	All unit construction and transfer	All gear making	All grinding, honing and lapping	Automatic lathes	Screwing and threading and tapping	Other[a]	Total[a]
1960[1]	—	92	104	—	—	180	376
1965[1]	—	98	1535	—	—	3	1636
1966[1]	—	117	910	—	67	309	1403
1967[1]	—	—	362	—	—	28	390
1968[1]	124	1054	4101	16	933	352	6580
1969[1]	1349	876	5540	501	1388	1816	11470
1970[1]	903	352	4170	1190	573	1729	8917
1971[2]	1000	266	1367	1135	261	2093	6122
1972[2]	248	770	1943	3049	1624	1216	8850

Sources:
Calculated from figures in UK trade in metalworking machinery with the Soviet Union.
1 Taken from materials prepared for the Machine Tool Trade Association by the journal *Machinery and Production Engineering*.
2 Taken from HM Customs and Excise *Annual Statement of Overseas Trade* (the data in this and the above source for the most part appear to be approximately the same, although for 1968 there are some major differences).

Note:
a – figures refer to both metalcutting and metalforming machines. 'Other' covers all other types of machine tools and also all metalforming machines which in some years made an important contribution to total exports.

[54] See Tables 4.14–20.

about the same proportion of gearmaking machines and automatic lathes while the United States has a larger proportion of screwing, threading and lapping machines and probably a larger proportion of unit construction and transfer machines.

Finally, it is necessary to consider whether imports are influenced primarily by questions of quality. Evidence on this point is not completely convincing. The list of machine tools given above for which the Soviet Union is behind the rest of the world includes some of those which have been imported in large numbers from Britain but also others which have not been imported on a large scale. It seems necessary to distinguish two aspects of this question: first, machines required for those areas such as light industry and other consumer orientated industries—e.g., the car industry—which until recently were not given high priority in the Soviet Union, and, second, machinery for major long-established Soviet industries.

Table 4.30 *Soviet demand for machine tools and planned output for 1965 according to the 1959–65 seven year plan and actual output in 1965*

Type of machine tool	No. (× 1000)			%	
	Demand[1]	Planned output[1]	Actual output[2]	Planned output/ Demand	Actual output/ Demand
Lathes and vertical lathes	47.4	41.7	} 55.5	88.0	} 85.7	94.7
Turrets	11.2	8.5		75.9	
Lathes—autos and semi-autos	10.0	7.1	4.7	71	47
Boring	5.1	3.4	3.2	66.7	62.7
Drilling	35.0	28.8	28.3	82.3	80.9
Planers and slotters	5.0	4.2	4.3	84	86
Broachers	1.9	0.8	1.5	42.1	78.9
Milling	30.0	25.6	22.3	85.3	74.3
Gear cutting	6.0	4.9	3.5	81.7	58.3
Grinding	31.1	16.7	12.3	53.7	39.5
Tool and cutter grinding	9.3	4.8	11.0	51.6	118.3
Special, specialised and unit head	38.0	38.0	26.1	100	68.7
Sharpening	10.0	11.3	} 13.6	113	} 77.5	} 68.0
Sawing	3.0	1.5		50	
Thread cutting	3.5	2.1		60	
Others	3.5	0.6		17.1	
Total	250.0	200.0	186.1	80.0	74.4

Sources:
1 P. M. Pen'kov, *Razvitie tipazha i struktura vypuska metallorezhushchikh stankov i komplektuyushchikh prinadlezhnostei k nim v 1958 i 1959–65 gody* (1958), p. 60.
2 (Eds.) N. M. Oznobin et al., *Sovershenstvovanie struktury proizvodstva* (1968), p. 136.

It is probably reasonable in the first case that the Soviet Union should be dependent on imports for a time at least as, for example, in the programme of expansion of the car industry. This was clearly an area in which the Soviet Union was backward in the early 1960s and at the 1958 machine tool conference a Gosplan specialist blamed the machine tool industry for this state of affairs since 'they have not made the right machine tools' and the ZIL factory needed '2,200 special machines of a class which is hardly made in the machine tool industry at present'.[55]

An example of another industry for which the Soviet Union is apparently unable to provide satisfactory machinery is the ballbearing industry. There have been many complaints about the poor quality of the machine tools produced for this industry.[56] To some extent a vicious circle operates and one machine tool specialist complained 'we ask them [the ballbearing industry] for precision ballbearings for machine tools and they say give us the precision machine tools and then we'll make the precision ballbearings'.[57] He continued 'this had now gone on for two or three years and as yet there were no precision machine tools and no ballbearings. We suggested that they bought from abroad the machine tools to make the ballbearings and they suggested we bought the ballbearings to make the machine tools.'[58]

The problems the Soviet Union was having in this area were highlighted by its attempt in 1960 to buy from the United States 45 grinding machines used in the manufacture of miniature ballbearings used almost exclusively for the military or aerospace industries.[59] This case is particularly interesting because of the two totally different assessments of Soviet machine tool technology which emerged from it, which serves to emphasise the problems in this field. The US Department of Commerce had allowed the deal to go ahead because their 'technical people consider that a Russian machine is being designed to compare with anything produced in the West at this time'.[60] It was also claimed that the 'USSR is reported by US intelligence to be producing complete precision miniature bearings of quality equal to the highest quality specifications of the ballbearing industry despite the fact that the Soviet Union is not in possession of any Bryant model B Centalign grinders'.[61] In contrast the Department of Defense spokesman reckoned the Bryant machine 'superior to anything the Russians could produce themselves and to what they could buy abroad'.[62] The machine the Department of Commerce said was up to US

[55] *Otraslevoe soveshchanie po stankostroeniyu, Minsk iyul' 1958: vystupleniya uchastnikov soveshchaniya (sokrashchennaya stenogramma)* (1958), p. 47.

[56] See, for example, *Plenum tsentral'nogo komiteta kommunisticheskoi partii Sovetskogo Soyuza 24–29 iyunya 1959 goda stenograficheskii otchet* (1959), pp. 360–2; *Ekon. Gaz.*, 20 August 1960, p. 2; 20 October 1960, p. 2; 27 November 1960, p. 1; 25 August 1962, p. 18; 9 April 1962, p. 17.

[57] P. M. Pen'kov, *Puti razvitiya stankostroeniya i novye metalloobrabatyvayushchie stanki*, Kiev (1962), p. 3.

[58] *Ibid.*

[59] See *Proposed Shipment of Miniature Ball Bearing Machines to Russia*, US Congress, Committee of the Judiciary, United States Senate, Washington DC (1961).

[60] *Ibid.*, p. 7.
[61] *Ibid.*, p. 37.
[62] *Ibid.*, p. 87.

standards (the LZ-10) was considered to be of a type already 'obsoleted by American industry'.[63] J. Gwyer, a Library of Congress specialist, also claimed that 'copying equipment of the nature of the precision machine tool enters into a new realm' and reckoned it would take at least five years to produce a Soviet equivalent.[64]

On the whole, therefore, the evidence suggests that some Soviet imports at least are motivated by a desire to obtain a machine tool which is technologically superior to anything that the Soviet Union can produce at the time. Some imports are also made of machines which are in particularly short supply in the USSR and may not represent a technologically superior design although they may have other advantages such as better quality of manufacture and operator aids not found on the equivalent Soviet machine.

SOVIET EXPORT PERFORMANCE IN THE UNITED KINGDOM MARKET

The last point which must briefly be considered in using Soviet trade data to evaluate the quality and technological level of Soviet machine tools is the performance of the Soviet industry in the export field. As has been shown above (Tables 4.25 and 4.26) an increasing volume of Soviet exports go to the other socialist countries and the main capitalist countries account for well under 10 per cent of exports by value. These exports have a much lower unit value than Soviet imports from these countries (see Table 4.28). Here again shortage of detailed data makes it difficult to analyse the composition of Soviet exports but, in so far as one can judge, their exports to capitalist countries fall into two main groups: first, relatively cheap universal type machines, and, second, a small number of more sophisticated machines particularly of the heavier types.

These features are brought out by the data on Soviet exports to the United Kingdom given in Tables 4.31–33.[65] Table 4.31 shows that although the Soviet exports have a relatively high unit value they have a very low value per hundredweight and the lowest of all the countries exporting to the United Kingdom. Table 4.32 shows the composition by type of Soviet exports of metalcutting machine tools and brings out the major role of centre lathes and milling machines which account for about 60 per cent of the total number. Table 4.33 shows that these two groups have an average weight of 3 and $3\frac{1}{2}$ tons, respectively, while a number of other types imported have a much higher average weight—horizontal and vertical boring machines, vertical boring and turning mills, planing machines, etc. The main feature to emerge from Table 4.33 is, however, the low value per hundredweight of Soviet machine tools mentioned above. The low price to weight ratio of Soviet machine tools has already been examined by Sláma and Vogel in their study of the comparative kilogram/prices of exports of machine tools to Yugoslavia by

[63] *Ibid.*, p. 26.
[64] *Ibid.*, pp. 105–6.
[65] Imports from the Soviet Union account for only around one per cent of total British imports of metalcutting and metalforming machine tools.

the USSR and Western countries.[66] Their tentative conclusion was that the relatively low kilogram/prices of Soviet exports might in part be explained by the correspondingly lower technological level.[67] Even allowing for the heavier weight of Soviet machine tools and their probable lower price in relation to similar Western machines, it does seem reasonable to follow Sláma and Vogel in ascribing to a lower technological level the divergences in value per hundredweight of imports from the Soviet Union compared with total imports into the United Kingdom.

The United Kingdom statistics also show a small number of higher priced heavier machines imported and this feature is supported by information on exports to other capitalist countries reported in recent issues of *Stankoimport Review*. Thus almost all the machine tools reported as purchased by the United States and West Germany are produced by factories of Glavtyazhstankoprom

Table 4.31 *Unit value and value per cwt of British metalcutting and metalforming machine tool imports in 1968 (by country)*

Country	Unit value (£)	Value per cwt (£)
Canada	1893	84
Soviet Union	2410	22
Sweden	1611	86
Denmark	410	45
Poland	1961	27
FRG	3878	73
GDR	2809	31
Netherlands	3236	89
Belgium	1689	59
France	1715	70
Switzerland	3205	110
Spain	638	29
Italy	1452	54
Austria	464	77
Czechoslovakia	2093	33
Japan	932	44
United States	3184	120
Miscellaneous	1102	44
Total	2499	74

Source:
As for Table 4.29. Calculated.

[66] J. Sláma and H. Vogel, in *Jahrbuch der Wirtschaft Osteuropas*, Munich (1971), pp. 443–84.
[67] For a discussion of some of the problems of this approach see R. Amann and J. Sláma, *The Organic Chemicals Industry of the USSR: a Case Study in the Measurement of Comparative Technological Sophistication by Means of Kilogram/prices*, unpublished working paper, University of Birmingham, 1975, pp. 3–11.

Table 4.32 *British imports of metalcutting machine tools from the Soviet Union, 1967–9*

Type of machine		No.	Weight (cwt)	Value (£)
Boring:	horizontal	13	3737	100151
	vertical	1	298	8007
	jig	1	710	40076
Drilling:	radial	12	1182	17687
	vertical	2	n.a.	26
	other	2	452	9145
Grinding:	cylindrical	30	2467	77067
	surface	37	2187	159130
	other	22	797	26204
Lathes:	bar auto	1	43	2001
	centre	145	8878	213468
	copying	1	34	3016
	other	8	601	13454
Vertical boring and turning machines		2	680	16834
Milling (not NC)		163	11461	347007
Shaping and slotting		5	602	10894
Planing		3	5081	65233
Sawing (incl. friction and abrasive)		2	416	4210
Other metalcutting		47	63	8840
Total		497	39689	1022450

Source:
As for Table 4.29.

Note:
Table includes only types imported from Soviet Union in this period. See source for full list of types.

(the chief administration for heavy machine tools).[68] Some of these have Western NC systems—e.g., Sinumeric 270, Boche and Olivetti systems and, on one sold to Sweden, a Saab system.[69] This last factor is perhaps emerging as an important selling point in Western markets.

All in all, therefore, it does appear that the Soviet Union has a long way to go before establishing itself as a major seller of sophisticated machine tools, although it has had some success with its universal machines. Surprisingly, it has failed to emerge as a major supplier to the developing countries, and its exports to these countries have declined in recent years. There are, however, signs that it could perhaps become an important supplier in the area of the heavier types of machine tools, a number of which have been sold to the advanced capitalist countries.

[68] See *Stankoimport Review*, No. 37, p. 25; No. 38, p. 28; No. 39, p. 28; No. 40, p. 31.
[69] *Ibid.*, No. 37, p. 24.

Table 4.33 *Average weight, unit value and value per cwt of types of metalcutting machine tools imported to UK from Soviet Union in 1967–9 compared with average weight, unit value and value per cwt of all machine tools of these types imported to UK in 1968*

Type of machine tool		Average weight (cwt)		Unit Value (£)		Value per cwt (£)	
		USSR 1967–9	Total 1968	USSR 1967–9	Total 1968	USSR 1967–9	Total 1968
Boring:	horizontal	288	244	7704	9608	27	39
	vertical	298	167	8007	10684	27	64
	jig	710	81	40076	7944	56	98
Drilling:	radial	99	45	1474	891	15	20
	vertical	n.a.	8	13	491	n.a.	59
	other	226	55	4573	4167	20	75
Grinding:	cylindrical	82	66	2569	5227	31	80
	surface	59	39	1598	2256	27	58
	other	36	9	1191	849	33	97
Lathes:	bar auto	43	29	2001	3092	47	107
	centre	61	40	1472	1551	24	39
	copying	34	106	3016	7811	89	72
	other	75	21	1682	1394	22	65
Vertical boring and turning mills		340	423	8417	2061	25	50
Milling: other (not NC)		70	59	2129	3355	30	56
Shaping and slotting		120	19	2179	750	18	40
Planing		1694	10	21744	447	13	47
Sawing (incl. friction and abrasive)		208	8	2105	305	10	39
Other metalcutting		1	6	188	644	140	107

Source:
As for Table 4.29. Calculated.

Numerically controlled machine tools*

The most significant development in machine tool technology since the Second World War has been the introduction of numerical control permitting the automation of small and medium batch production in the engineering industry.[70] Earlier forms of automatic or semi-automatic machine tools were generally uneconomic for such production because of the inflexibility of the predetermined work cycle: any change in the configuration of the workpiece necessitated a change or rearrangement of the mechanical elements of the machine. Numerical control

---

*M. J. Berry wrote pp. 122–60, and Julian Cooper pp. 160–98.
[70] The importance of this can be judged from the fact that items made in batches of up to 50 comprise three quarters of all products produced by the Soviet engineering industry (O. I. Volkov, *Ekonomicheskie aspekty vnedreniya avtomatizatsii* (1972), p. 99), while about two thirds of all Soviet engineering factories make their products on a small batch or individual basis (V. A. Letenko et al., *Ekonomika mashinostroitel'noi promyshlennosti* (1968), p. 11).

is a kind of automation in which the sequence of operations needed to machine a given workpiece is set up in advance by means of numerical data recorded in some way on tape or cards. When the tape or cards are run through the control unit, the working parts of the machine, such as tool holders, workpiece holders and slides, go through the sequence automatically.[71]

Some numerical control systems also use dial or plugboard input. Numerically controlled machine tools can be quickly changed over from one job to another and the programme of machining information required for any particular workpiece can easily be stored for repeated use. Numerical control must be distinguished from other, simpler, forms of sequence programme control in which certain elements of a work cycle can be pre-set by dials or plugboards.[72] This chapter refers to the numerical control (NC) of metalcutting machine tools, unless otherwise specified.

The high degree of adaptability to different workpieces and work cycles is the main advantage of NC. When compared with the universal types of machine tools which are generally employed in small and medium batch production, NC machines can be operated by less highly skilled workers, need fewer jigs and fixtures, require less floor space for an equivalent output, and secure more uniform quality, thereby reducing inspection requirements. Disadvantages, especially during the early stages of the introduction of NC, include the need for skilled maintenance workers, the high initial cost of most NC machines, the need for facilities for preparing programmes, and the difficulty of securing a satisfactory organisational framework in order to take full advantage of the possibilities offered by NC.[73]

THE EARLY YEARS—FROM RESEARCH TO INDUSTRIAL PRODUCTION

Initial research into numerical control was undertaken from 1947 on the initiative of the United States Air Force, which was seeking a new method of machining intricate components for aircraft. The first prototype NC machine tool suitable for industrial use was built at the Massachusetts Institute of Technology in 1952 and two years later the first two models built by a commercial firm were shown at the Chicago Machine Tool Exhibition. In Britain research into NC began in about 1950, and the first marketable prototype appeared in 1956. A prominent role was played by the electronics industry and, as in the USA, research was primarily directed towards finding

---

[71] UNIDO, *Regional Seminar on Machine Tools in Developing Countries of Europe and the Middle East*, New York (1972), p. 45.

[72] Soviet writings employ two different terms: 'numerical programme control' (*chislovoe programmnoe upravlenie*—ChPU, or sometimes *tsifrovoe PU*) and 'sequence programme control' (*tsiklovoe programmnoe upravlenie*), the latter covering simpler, non-numerical forms of control. The general term 'programme control' (*programmnoe upravlenie*) is often used to refer to both types.

[73] On the disadvantages and advantages of NC, see A. E. De Barr, *The Production Engineer*, 1972, No. 9, pp. 381–2.

new methods of machining for use in the aircraft industry. In both France and Germany research began in about 1955, and the first examples appeared in 1957 and 1958 respectively. At the 1959 Paris Exhibition both these countries showed machines with automatic tool selection; the French GSP machine was one of the very first machining centres.[74] In the case of Japan, research began in about 1953; the first prototype appeared in 1958 and in the following year two commercial firms and the Government Mechanical Laboratory presented new models.[75]

The early history of NC in the USSR is rather obscure. In 1949 the Institute of Physics of the Ukrainian Academy of Sciences began research into recording electrical processes on magnetic tape and this work led to the making of a control system with tape input. After initial tests with a lathe, in 1952 a new semi-automatic turret lathe was built for use with this system.[76] At the Moscow Machine Tool and Tooling Institute (Mosstankin) a lathe controlled by punched tape was developed by L. A. Gleizer;[77] and in about 1954–5 another development group created a milling machine with an open-loop, positional control system.[78] A variety of organisations were involved in research and development of NC: in early 1956 it was reported that work was being carried out in the USSR Academy of Sciences by the Institute of Machine Science and the Institute of Automation and Remote Control, and also by the machine tool industry research institute ENIMS, the Leningrad Polytechnical Institute, and other bodies.[79] At this stage, however, it appears that all the examples built were of an experimental, laboratory type, not in a form suitable for regular production and industrial use. The first examples of marketable prototypes appeared in 1958, and were shown at the Brussels World Fair of that year.

Four different Soviet NC models were shown at Brussels; the most important, the 6N13Pr milling machine developed by ENIMS, gained a Grand Prix. This machine, later built by the Gor'kii milling machine factory, was fitted with a three coordinate, open-loop continuous path control system with magnetic tape input and stepping motor drive.[80] The other machines were a punched tape controlled lathe, a small profile milling machine with punched tape controlled two-dimensional positioning of its table, and a punched card controlled horizontal boring machine, the 262PR, built by the Sverdlov factory, Leningrad, and apparently representing the first Soviet NC machine to be built by an industrial enterprise as opposed to a research institute.[81] Thus the Soviet Union was relatively well advanced in the initial stages of development of NC. The technical level of its NC machines was certainly well

---

[74] See p. 179 below.
[75] *Metalworking Production*, 2 October 1959, p. 1580. The development of NC is summarised in Table 4.34.
[76] A. A. Zvorykin *et al.*, *Istoriya tekhnika* (1962), p. 559.
[77] *Trudy Instituta Istorii Estestvoznaniya i Tekhniki*, Vol. 45 (1962), p. 42.
[78] A. A. Spiridonov, *Metallorezhushchie stanki s programmnym upravleniem* (1964), p. 39.
[79] *Mashinostroitel'*, 1956, No. 2–3, p. 58.
[80] Spiridonov, *op. cit.*, p. 71; *Metalworking Production*, 23 May 1958, p. 918.
[81] *Metalworking Production*, 23 May 1958, pp. 918–19; G. Borisov and S. Vasil'ev, *Stankostroitel'nyi im Sverdlova*, Leningrad (1962), pp. 259–60.

below the level attained in the USA by 1958, however,[82] and also below the level of French and German NC machines shown at the 1959 Paris Exhibition. Work on NC began somewhat later in other socialist countries. In the case of Czechoslovakia, research started in 1957 and several models were shown at the Brno Fair two years later,[83] and the German Democratic Republic built its first NC prototype in 1963.[84]

In the United States the transition from the first NC prototypes to regular industrial production was quite rapid and an impressive rate of growth of output was achieved between 1959 and 1966 when the number of units built per year increased from 203 to 2,939.[85] In Britain the rate of growth of output was slower; an output of 50 units per year was attained in 1963. This output was reached in Germany in the following year, the USSR in 1965 and France and Japan in 1966. Thus the transition in the USSR from a marketable prototype to industrial scale production compared quite favourably with the performance of West European machine tool industries, although the length of time from the start of research was much longer.[86] Performance was much less satisfactory if the USA is taken as a standard, however, and if the Soviet NC output is compared with the total machine tool output and the size of the stock. During the mid-1960s, progress in the USSR was relatively slow compared with other leading machine tool building countries and official concern was evident by 1968 when the government intervened to accelerate the rate of development.

SOME PROBLEMS OF SOVIET NC DEVELOPMENT BEFORE 1968

In considering the relatively slow development of Soviet NC machines before 1968, a number of factors need to be considered. First, numerical control of machine tools represents an amalgamation of two distinct technologies and branches of industry, namely the traditional branch of metalcutting machine tool technology and the relatively new, science-based branch of electronics. In capitalist countries a very prominent role in the development of NC has been played by the electronics industry itself, as supplier of control equipment. Thus such firms as General Electric (USA), Ferranti, Plessey, Siemens, Alcatel and Fujitsu, having considerable experience in the electronics field, have been active innovators and, according to the 1970 OECD report on NC machine tools, cooperation between machine tool builders and control systems manufacturers has, on the whole, been good.[87] In the USSR, at least in the early years, the situation appears to have been much less satisfactory; poor

---

[82] In 1958 the American firm of Kearney and Trecker built the first production line of three NC machines linked by transfer mechanisms for use by the Hughes Aircraft Company. (*Metalworking Production*, 14 June 1958, p. 1039.)

[83] *Stanki i Instrument*, 1960, No. 9, p. 7; *Metalworking Production*, 16 October 1959, pp. 1669–73.

[84] *Metalworking Production*, January 1973, p. 78.

[85] See Table 4.36, p. 172.

[86] This information is summarised in Table 4.34 and Fig 4.1.

[87] *NC Machine Tools—Their Introduction in the Engineering Industries*, OECD, Paris (1970), p. 51.

cooperation between control systems production enterprises and the machine tool industry was frequently reported. Thus the first Soviet attempt to build a machining centre at the Sverdlov factory, Leningrad, in the early 1960s foundered largely because the factory tried to make the control equipment itself without outside assistance.[88] Similarly, in 1971 the Kosior factory, Khar'kov, reported that its attempts to make an NC cylindrical grinding machine were being held up by the need to make its own control systems.[89]

Furthermore, the relative backwardness of the Soviet electronics industry itself, in particular those sectors concerned with computing technology, has probably exerted a negative influence on NC development. It is significant that in 1958 at the machine tool industry conference in Minsk the director of ENIMS, A. P. Vladzievskii, in outlining policy with regard to programme control, stated that the control systems would use the minimum amount of computer technology.[90] The early emphasis on simpler programme control systems requiring a smaller 'input' of electronics technology, and the stress on open-loop, as opposed to closed-loop, NC systems, both provide evidence that the Soviet machine tool industry sought to minimise its demands on the electronics industry.[91]

Second, problems of coordination of a more general kind also appear to have harmed Soviet NC development. A frequent subject of complaint has been the absence of a single coordinating central body for NC, which has resulted in the scattering of work among a variety of different organisations. In 1962, 20 or 30 organisations were said to be involved in work on programme control.[92] This dispersal of the initial research and development effort resulted in the creation of a variety of different systems of NC. This variety had some good results, for the competition between various organisations may have helped the process of selecting the most viable systems for production on an industrial scale. Thus V. V. Boitsov, Chairman of the State Committee for Standards, writing in 1971, while calling for the standardisation of NC systems, noted that the use of several different systems of programme control in the initial period of their development is justified and explicable. The search for the best methods and means of numerical programme control of machine tools is a complex task and demands the development, study and verification in practice of different methods and paths.[93] In this respect, the Soviet experience was analogous to that of Britain or other countries, where there was a proliferation of different systems in the early years of NC development.

---

[88] *Izvestiya*, 12 July 1972.

[89] *Sots. Ind.*, 26 December 1971.

[90] *Otraslevoe soveshchanie po stankostroeniyu. Minsk iyul' 1958: vystupleniya uchastnikov soveshchaniya (sokrashchennaya stenogramma)* (1958), pp. 3–6.

[91] See below, p. 186. In the early years the Soviet industry devoted some attention to non-numerical sequence control, which makes limited use of electronic technology. In all, about 40 machine tools with this type of technology were built, rising to about 100 in 1962. (*Narodnoe khozyaistvo SSSR v 1962 godu* (1963), pp. 232–3, and known output of genuine NC machines.)

[92] P. M. Pen'kov, *Puti razvitiya stankostroeniya i novye metalloobratyvayushchie stanki*, Kiev (1962), p. 7.

[93] V. V. Boitsov, *Mekhanizatsiya i avtomatizatsiya v melkoseriinom i seriinom proizvodstvakh* (second edition) (1971), p. 324.

Table 4.34 *NC machine tools: stages of the innovation process*

	Year of:			Annual output[7] of:	
Country	Start of research	First prototype[a]	Start of industrial production[6b]	500 units or more	1000 units or more
USSR	1949[1]	1958[4]	1965	1969	1970
USA	1947[2]	1952[2]	1957[c,e]	1961	1962
Japan	1953[3]	1958[3]	1966	1969	1970
FRG	1955[2]	1958[5]	1964	1969	d
UK	1950[2]	1956[2]	1963	1968	d
France	1955[2]	1957[5]	1966[c,e]	d	d

Sources:
1 See text, p. 162.
2 (Eds.) L. Nabseth and G. F. Ray, *The Diffusion of New Industrial Processes*, Cambridge (1974), p. 27.
3 *Referativnyi Zhurnal, Tekhnologiya Mashinostroeniya*, 1974, No. 2, Item A190.
4 See text, p. 162.
5 *National Institute Economic Review*, May 1969, p. 53.
6 See table 4.36, p. 172.
7 *Ibid.*

Notes:
a – marketable prototype, i.e., suitable for production and use in industry.
b – defined as year in which output reached 50 units a year.
c – approximate.
d – not yet achieved.
e – year is estimated on the basis of available output data.

A third factor which may have influenced the intensity of effort devoted to NC was a campaign in the late 1950s and early 1960s by a number of specialists for highly adaptable unit-construction machine tools, which would permit the automation of small and medium batch production on the basis of traditional machine tool technology.[94] This type of machine tool has a long history in the Soviet Union and it is highly probable that they have been quite widely employed in certain industries, notably aviation. This could certainly have reduced the pressure for NC.[95] In capitalist countries the demand for NC machine tools and the intensity of NC development has been closely related to the role of the aerospace industry in different countries; thus NC developed vigorously and early in the USA and early in Britain and France, but was

[94] V. V. Boitsov, *op. cit.* (first edition) (1962), passim; *Pravda*, 7 August 1958.
[95] In the 1930s the Central Institute of Labour (TsIT) undertook considerable research into unit-construction machines of great flexibility and built many examples at its experimental factory. This institute was absorbed into the aircraft industry in 1940. (See *TsIT i ego metody NOT* (1970), p. 54 and ch. 7.) The leading advocate of this path in the 1958–62 period was V. V. Boitsov, then director of the Scientific Research Institute for Technology and Organisation of Production, an institute with interests very similar to those of TsIT, and apparently attached to the defence industry.

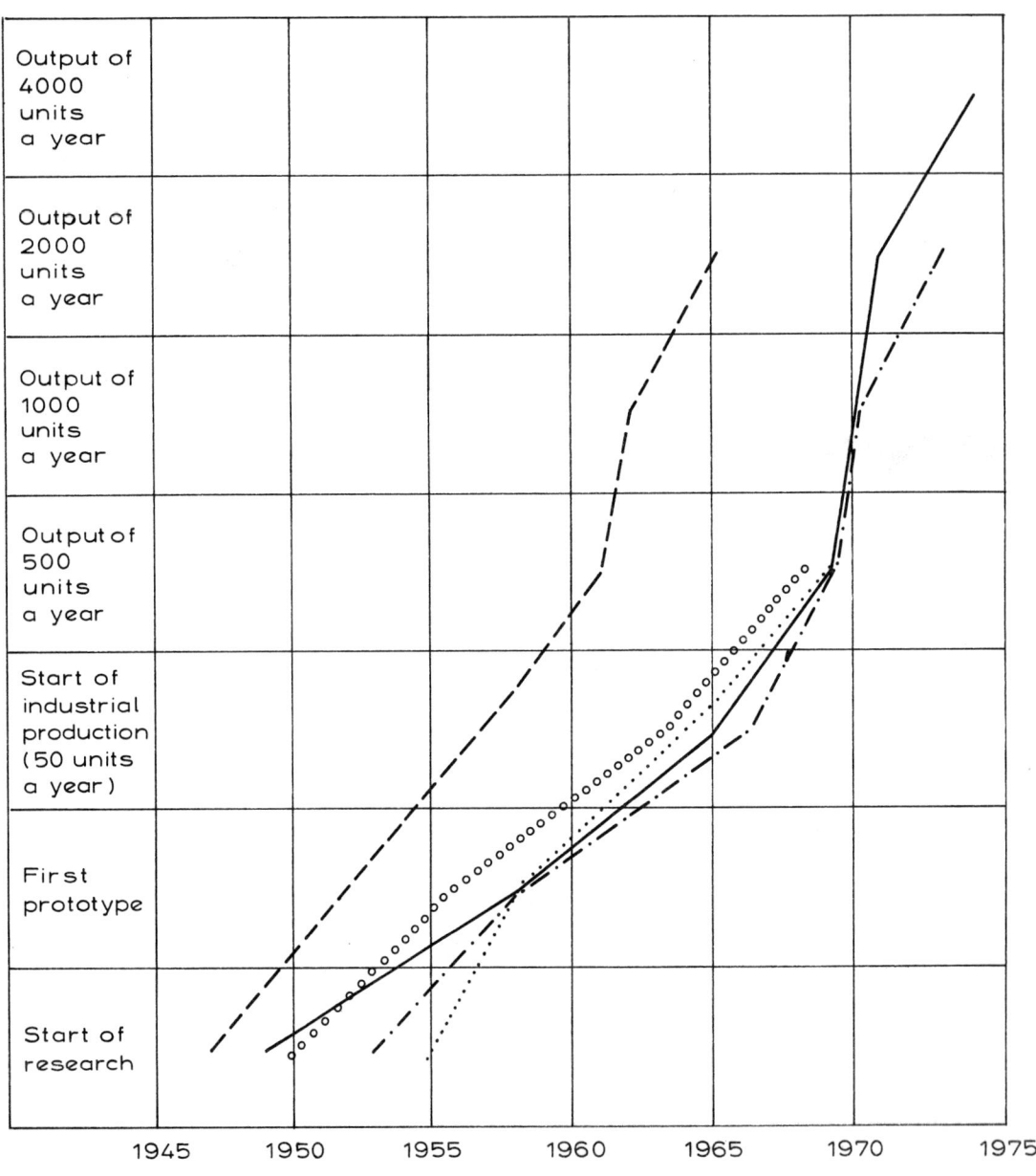

Fig. 4·1 NC MACHINE TOOLS: STAGES OF THE INNOVATION PROCESS

Source: Table 4.34

adopted later in Japan.⁹⁶ When the Soviet government promoted NC in 1968 the Ministry of the Aviation Industry was directly involved.⁹⁷

A fourth and final factor which must have influenced the pace of Soviet NC development is the policy of technical independence pursued by the Soviet industry. One Western observer has argued that the Soviet Union has made a 'false start' by attempting to develop its own NC systems instead of using an existing Western system.⁹⁸ This viewpoint is open to several objections, and it is uncertain whether there was a tried Western system of a type matched to Soviet needs available during the early 1960s. The type of control system widely adopted in the USSR is not that generally employed in the USA or Europe. Strategic considerations may have promoted a policy of self-reliance and, more generally, capitalist firms may have been unwilling, or unable, to enter into some form of licensing or cooperation agreement with the Soviet machine tool industry; NC machine tools have been covered by 'Co-Com' strategic embargo agreements between NATO countries. Sales of certain NC machines to the USSR have been banned: an American observer noted in 1971 that even though such restrictions were generally being relaxed, it was likely that few approvals would be granted for the direct sale of American NC machines to the USSR, especially those with continuous path control of the closed-loop type, or systems giving a high degree of precision.⁹⁹ Some US firms have sold less sophisticated NC equipment to the USSR through their European subsidiaries, including machining centres with positioning control.¹⁰⁰ Possibilities for cooperation with European and Japanese firms have increased in recent years; today, the Soviet Union has agreements with firms in a number of countries for cooperation in the field of control systems.¹⁰¹ British, German and Japanese machine tool building firms were able to draw directly on American NC practice from an early period and American control systems, often built by foreign subsidiaries, have been fitted to the machine tools of many countries. This transfer of technology involves some time lag, however, so that the Soviet industry has not been able to obtain the most advanced US technology through third countries.¹⁰²

SOVIET NC DEVELOPMENT SINCE 1968

A major turning point in the development of NC machine tools in the Soviet Union was a government decision of April 1968. This called on the Ministry of the Machine Tool and Tooling Industry and the Ministry of the Aviation

---

⁹⁶ *NC Machine Tools—Their Introduction in the Engineering Industries*, OECD, Paris (1970), pp. 42–4.
⁹⁷ See below, p. 168.
⁹⁸ A. W. Astrop, *Machinery and Production Engineering*, 27 January 1971, p. 123.
⁹⁹ *Iron Age*, 25 November 1971, p. 60.
¹⁰⁰ *Ibid.*, 1 January 1971, p. 36.
¹⁰¹ See below, p. 184.
¹⁰² In the course of a visit by US machine tool builders to the USSR in 1971 representatives of the Soviet industry are reported as having said that they considered that there was a lag of about two years in the adoption of the latest American technology by overseas machine tool building plants. This was offered as one reason for desiring direct contact with the US machine tool industry (*Iron Age*, 1 July 1971, p. 36).

Industry to try to significantly increase the output of NC machine tools in the years 1969–70. The machine tool industry was also obliged to provide users of NC machines with technical assistance with the preparation of programmes and with the training of operators and setters for NC equipment. The Ministry of Instrument Making, Means of Automation and Control Systems (Minpribor) was made responsible for both the development and the production of control systems and, together with the machine tool ministry and other organisations, was to present proposals to Gosplan for the development of NC machine tools and control systems during the 1971–5 period. The State Committee for Science and Technology, with the participation of relevant ministries, was to draw up and approve a plan of coordination of research and development for the creation of unified systems of NC for other types of equipment, such as metalforming, textile, woodworking and printing machinery.[103]

Serious results were not achieved for about two years. In December 1969 the industry was criticised for its poor performance in 'catching up lost time'.[104] In the spring of 1970 it was admitted that programme control was still 'a very backward area'.[105] But during these years the foundations for future growth were laid, and the subsequent increase in the output of NC machines is impressive, as is shown in Table 4.35. The increase in output during the last five years has been very rapid. In terms of units built per year the Soviet Union now occupies first place in the world, having overtaken the USA in 1971.[106] In this development non-specialised NC producers have been very important; and the secrecy surrounding this component of NC production, and the specific reference to the aviation industry in the 1968 decision, suggests that this industry is responsible for a large proportion of this production. These machines are presumably built for the aircraft industry itself and the factories of the machine tool ministry also supply NC machines to the aircraft industry;[107] it can therefore reasonably be assumed that this industry holds a substantial proportion of all Soviet NC machines. What types of NC machines are built outside the specialised machine tool industry is unknown, but in view of the priority of the aircraft industry and the complexity of its demands it is probable that they are of a high technological level and quality.

In 1974 over 30 enterprises of the machine tool ministry, about one third of the total number,[108] were building NC machine tools, but at the heart of the NC programme of the ministry during the ninth (1971–5) five year plan has been the large-batch production of 11 models at six enterprises using flow production methods: Krasnyi Proletarii in Moscow, the lathe factory in Ryazan, the milling machine factories in Gor'kii and L'vov, the drilling machine factory in Sterlitamak, and the Charentsavan boring machine factory

---

[103] *Izvestiya*, 26 April 1968.
[104] *Pravda*, 21 December 1969.
[105] *Stanki i Instrument*, 1970, No. 5, p. 2.
[106] See Table 4.36.
[107] *Sots. Ind.*, 8 May 1971.
[108] *Vestnik Mashinostroeniya*, 1972, No. 8, p. 6; *Mekhanizatsiya i Avtomatizatsiya Proizvodstva*, 1974, No. 8, p. 1.

Table 4.35 *Planned and actual output of metalcutting machine tools with NC*

Year	Total output (units) Planned	Total output (units) Actual	Enterprises of the machine tool machinery Planned	Enterprises of the machine tool machinery Actual	Enterprises outside the specialised industry Planned	Enterprises outside the specialised industry Actual	Proportion built by specialised industry[c] (per cent)
1965	—	49[1]	—	42[2]	—	7[a]	86
1969	—	520[1]	—	420[3,b]	—	100[a,b]	81[b]
1970	3·2 × 1969[4]	1687[1]	2·6 × 1969[4]	1100+[2]	—	587[a]	65
1971	2500[5]	2538[1]	—	} 3100[6]	—	} 2487[a]	56
1972	3050[7]	3049[1]	—		—		
1973	4200[7]	3783[1]	—	2000[9]	—	1800[a]	53
1974	4450[8]	4410[10]	2660[9]		1790[a]		

Sources:
1 As for Table 4.36.
2 *Vestnik Mashinostroeniya*, 1972, No. 8, p. 3.
3 Estimated on the basis of the known 1970 planned and actual output on the assumption that the 1970 plan was fulfilled.
4 *Stanki i Instrument*, 1970, No. 3, p. 2.
5 *Pravda*, 9 December 1970.
6 *Mekhanizatsiya i Avtomatizatsiya Proizvodstva*, 1973, No. 5, p. 2.
7 *Ekon. Gaz.*, 1973, No. 16, p. 4.
8 Calculated from *Stanki i Instrument*, 1974, No. 1, p. 2 — output of NC in units to increase by 17 per cent over 1973.
9 *Mekhanizatsiya i Avtomatizatsiya Proizvodstva*, 1974, No. 5, p. 3 (output 1971–1973 of 5,100 units).
10 *SSSR v tsifrakh* (1976), p. 89.

Notes:
a – residual.
b – estimate.
c – calculated.

in Armenia.[109] The most widely diffused models of NC machines in the USSR are those based on standard general-purpose models suitably modified to accept numerical control. The NC model forms part of a unified range together with ordinary machine tools, and in this way the cost of NC machines is reduced. Examples of the most widely diffused models of the years 1970 to 1974 are the 1K62F3[110] centre lathe of the Krasnyi Proletarii factory, the 1M63F3 engine lathe of the factory in Ryazan, the 6N13F3 milling machine of the factory in Gor'kii and the 2N55F2 radial drilling machine of the radial drilling machine factory, Odessa.[111] All these models are based on well established

[109] *Mekhanizatsiya i Avtomatizatsiya Proizvodstva*, 1974, No. 7, p. 2.
[110] Most NC machines built by the machine tool industry now have a model number of this form. The '1K62' refers to the basic model, the F indicates that it is fitted with NC, while the final number indicates the number of independently controlled machine functions (e.g., two axes positioning, plus tool selection). Sometimes the 'F' is preceded by the letter 'M' indicating that the machine has a magazine-type automatic tool selection, or 'R' indicating a turret-type tool head. Suffixes indicate the type of control system fitted, e.g., the 16K20F3S1 is fitted with the Kontur 2PT system; the 16K20F3S2 with the French Alcatel system; and the 16K20F3S5 with the new Soviet third generation N22-1M system.
[111] *Tekhnologiya i Organizatsiya Proizvodstva*, 1972, No. 3, p. 42.

general-purpose machines built in quantity. It is evidently this policy which has permitted the rapid build-up of output. In 1973 and 1974 enterprises have been switching from the above-mentioned basic models to new models of more modern design based on the latest general-purpose machines. Thus the Krasnyi Proletarii factory is now building the 16K20F3 centre lathe on a flow basis, the Ryazan factory has introduced the 1P752MF3 lathe, and the Gor'kii factory is building the 6R13F3 milling machine. These models also form the basis for a series of variants, some of which incorporate multi-tool magazines permitting automatic tool changing.[112] Some older models are now being withdrawn from production.[113] Parallel with this development there has been considerable attention devoted to the creation of more specialised NC equipment.

The transition to flow production of new more advanced models appears to have somewhat reduced the rate of growth of output in 1973 and 1974. The ambitious 1973 annual plan of 4,200 units was not fulfilled; the shortfall was 400 units. The 1974 target of 4,450 units (actual output: 4,410 units) was clearly substantially lower than it should have been if the original ninth five year plan target for 1975, 7,000 units, was to be met.[114] Nevertheless, the rate of growth of output has been impressive when compared with other countries in recent years. During the ninth five year plan period the number of NC models has also steadily increased. In 1970 a total of 30 models was in production at enterprises of the machine tool ministry; in 1972, 63 models; and, at the end of 1973, 92 models, compared with the annual plan of 100 models.[115]

Since 1968 the policy of the Soviet industry has been more outward-looking. Cooperation agreements have been entered into with French, German and Japanese firms in the field of control systems. Cooperation with CMEA has also increased. All the East European CMEA countries are now building NC machine tools, especially the GDR and Czechoslovakia. The GDR is actively selling its products in Western Europe through 'NC Centres' in Paris and Munich,[116] and exhibits at recent Leipzig Fairs have demonstrated that it is a leading country in NC technology, with probably the highest technical level of NC in the socialist countries. The development of NC machine tools forms part of the Comprehensive Programme of CMEA adopted at the twenty-fifth session in 1971, which outlined an ambitious programme of collaboration covering research, development and production of NC machines and control systems, including machining centres, to be carried out during the period 1971 to 1975. This work is to include the unification of NC systems and machine

[112] *Mekhanizatsiya i Avtomatizatsiya Proizvodstva*, 1974, No. 7, pp. 17–18; No. 8, p. 1; No. 11, pp. 1–2.
[113] *Stanki i Instrument*, 1974, No. 9, p. 2.
[114] *Gosudarstvennyi pyatiletnii plan razvitiya narodnogo khozyaistva SSSR na 1971–1975 gody* (1972), p. 126. However, according to many sources, e.g., *Stanki i Instrument*, 1972, No. 4, p. 4, output was to increase 3·5 times, apparently meaning 3·5 times the 1970 output, i.e. about 6,000 units in 1975. This suggests that the original target may have been revised. Of the total, the machine tool ministry enterprises were to produce 4,500 units in 1975 (*Vestnik Mashinostroeniya*, 1972, No. 8, p. 6). Actual output in 1975 was 5,532 units (*SSSR v tsifrakh* (1976), p. 89).
[115] *Ekon. Gaz.*, 1973, No. 16, p. 4; *Mekhanizatsiya i Avtomatizatsiya Proizvodstva*, 1974, No. 5, p. 55.
[116] *Metalworking Production*, September 1973, pp. 138–9.

tools and the development of a single programming language for use with NC machines, together with study of the possibility of setting up an international network for NC programming.[117] One aim of the programme is the full satisfaction of the needs of participant countries for components and systems for NC and ending of reliance on capitalist firms in this area.[118] Examples of work being carried out within this agreement are the joint designing of NC machines by ENIMS and the Hungarian machine tool research institute,[119] and the establishment of centres for testing NC machines built in the GDR at the Krasnyi Proletarii and Ordzhonikidze factories in Moscow.[120]

The upsurge of NC production in the USSR coincided with a downturn in the major capitalist producing countries in 1970 and 1971, followed by a strong recovery in 1973 and 1974, in both the USA and Japan. In the case of the USA the fall in output was caused both by general economic conditions and also changes in the structure of output, with a reduction in the proportion of relatively simple drilling machines and an increase in the proportion of NC lathes and machining centres. The downturn in output has promoted technical development as machine tool builders have attempted to expand sales by improving the performance of NC machines. Some of these developments are considered below.

The growth of NC machine tool output in major producing countries is shown in Table 4.36 and Fig 4.2. Soviet performance in the late 1960s was not as favourable as indicated by the absolute data provided in Table 4.36 because total machine tool output in the USSR is substantially greater than that of Britain and also larger than that of Germany. The position in relative terms is shown in Table 4.37.

Soviet NC output reached 1·95 per cent of total metalcutting machine tool output in 1974, and 2·38 per cent in 1975.[121] The impression from these figures that NC machine tool production is of very little significance is corrected by the figures for output in value terms. Unfortunately, no NC output statistics are available for the USSR in these terms. In the USA NC output represented 15 per cent of total output in 1962, rising to 27 per cent in 1968, but falling to about 20 per cent in 1970. In Japan the proportion was about 2·5 per cent in 1967, 10 per cent in 1970 and 12·5 per cent in 1972. In Germany the proportion has been rather less: 2·8 per cent in 1966, rising to 7·3 per cent in 1970. In Britain, NC output was 3·7 per cent of total output in 1966, rising to a peak of 13·3 per cent in 1971, but falling to 9·8 per cent in 1973.[122] Given that Soviet NC machines, like those of Japan, are predominantly of the open-loop control system type as opposed to the more expensive closed-loop type favoured in the USA and Britain, the proportion in the USSR during the last two to three years was probably 15 to 20 per cent.

---

[117] *Ekon. Gaz.*, 1971, No. 33, p. 12.
[118] *Ibid.*, 1971, No. 47, p. 20.
[119] *Pravda*, 19 May 1972.
[120] *Stanki i Instrument*, 1974, No. 1, p. 2; 1974, No. 9, p. 3.
[121] Calculated from *SSSR v tsifrakh* (1976), p. 89.
[122] These proportions are calculated from output data given in sources for Table 4.36.

Table 4.36 *Annual output of NC machine tools (units)*

Year	USSR	USA[5]	Japan[6]	FRG[6]	UK[7]	France[6]
1954						
1955						
1956		193				
1957						
1958						
1959		203				
1960	16[1]	402		15	553	
1961		518			(20)	170
1962		1049		15	(40)	
1963	620[2]	1220		38	(85)	
1964		1517		108	(165)	
1965	49[1]	2100	39	162	(225)	
1966	(90)[3]	2939	90	268	323	90
1967	150[3]	2957	129	429	458	92
1968		2917	388	490	558	154
1969	520[2]	2372	860	574	543	154
1970	1687[1]	1901	1651	762	600	147
1971	2538[1]	1248	1379	816	590	247
1972	3049[1]	1630	1350	765	381	230
1973	3783[3]	2286	2765	992	467	389

Sources:
1 *Narodnoe khozyaistvo SSSR v 1972 godu* (1973), pp. 172–3.
2 L. P. Miroshnikov, *Primenenie stankov s programmnym upravleniem*, Khar'kov (1972), p. 25.
3 (Eds.) A. S. Tolkachev and I. M. Denisenko, *Osnovnye napravleniya nauchno-tekhnicheskogo progressa* (1971), p. 132.
4 *SSSR v tsifrakh* (1974), p. 89.
5 US Department of Commerce, *Current Industrial reports, Series MQ-35W-Metalworking Machinery*.
6 Supplied by the Machine Tool Industry Research Association.
7 *Business Monitor*, Series P64 and PQ332; 1961–5 from chart in *BNCS News*, January 1973, p. 22.

Table 4.37 *NC machine tools as a proportion of total machine tool output (per cent of number produced)*

	1960	1965	1966	1967	1968	1969	1970	1971	1972	1973
USSR	0·01	0·03	0·05	0·08	n.a.	0·25	0·83	1·24	1·45	1·77
USA	0·31	1·14	1·21	1·25	1·29	1·03	1·01	0·80	0·81	0·91
Japan	n.a.	0·04	0·08	0·08	0·21	0·37	0·64	0·75	0·82	1·30
FRG	n.a.	0·11	0·20	0·41	0·40	0·40	0·46	0·50	0·51	0·63
UK	n.a.	0·36	0·53	0·80	1·02	0·96	1·03	1·29	0·87	0·86

Source:
Calculated from output data given in sources for Table 4.36.

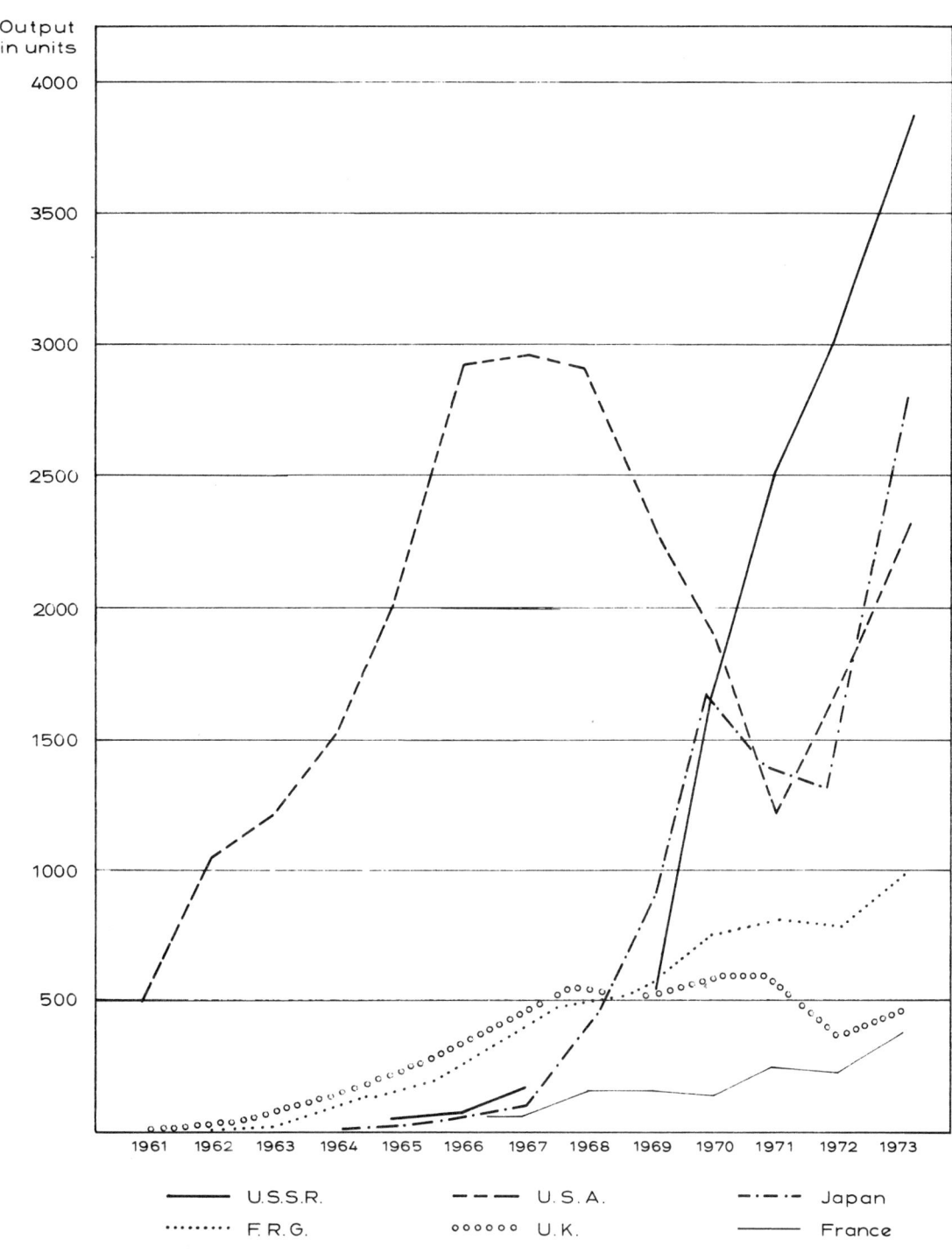

Fig. 4·2 ANNUAL OUTPUT OF NC MACHINE TOOLS (UNITS)

Source: Table 4·36

TYPES OF NC MACHINE TOOLS PRODUCED

Initial development of NC machine tools in the USA, Britain and Japan focused mainly on milling, drilling and boring machines. Development of milling machines with NC was promoted by the need to devise new means of machining complex parts of the type previously worked on copy-milling machines. In the USA 35·2 per cent of NC machines built in the years 1954 to 1958 were of the milling type, falling to 11·7 per cent in 1965. From the early 1960s in the USA and Britain there was strong emphasis on drilling and boring machines fitted with positioning control as opposed to the more complex contouring systems employed on milling machines. West Germany took a rather different path, as from an early date there was a marked interest in NC lathes. In Britain the first NC lathe was not built until 1965. In the USA the proportion of lathes rose steadily during the sixties and at the same time the proportion of drilling and boring machines fell as machining centres were built instead.

Although at first priority was probably given to milling machines in the Soviet Union as in other countries, the Soviet industry is exceptional in that attention was devoted to NC lathes from the very beginning. This early development of NC lathes may not have been significant. All the examples built up to about 1966 appear to have been fitted with positioning type control systems using punched cards as the programming input and such systems have limited technical possibilities. The first lathe fitted with contouring type control to be built on a regular basis was the 1K62PU built by the Krasnyi Proletarii factory from about 1966.

There are no statistics on the structure of Soviet NC output by type of machine tool, but it can be deduced approximately from indirect evidence. Milling machines form the largest group in the Soviet NC machine tool stock and are built on a batch basis. The second largest group is turning machines, also built on a batch basis. The drilling and boring group forms a small proportion of the total stock and these machines are built on a small batch basis.[123] Machining centres were not built until recently and must represent a very small percentage of total output. Certain other types are also being built with NC, including cylindrical grinding machines built by the Kosior factory, Khar'kov, and spark erosion machines built by a factory in Kirovakan. In 1971 at enterprises of the machine tool ministry there were 22 models of lathes, 18 models of milling machines and 13 models of drilling and boring machines in batch production.[124] In the 1971–5 five year plan emphasis has been shifting from milling machines to lathes and the drilling-boring group and, more recently, to machining centres. Of the planned product range (*tipazh*) to be built during the five years by the ministry's enterprises, 35 per cent of the models are planned to be of the turning group, 22 per cent of the milling type, 28 per cent of the drilling and boring group and 8 per cent of other types.[125] The

---

[123] *Tekhnologiya i Organizatsiya Proizvodstva*, 1972, No. 3, p. 42; *Vestnik Mashinostroeniya*, 1970, No. 3, p. 13.
[124] *Vestnik Mashinostroeniya*, 1972, No. 8, p. 4.
[125] *Mekhanizatsiya i Avtomatizatsiya Proizvodstva*, 1974, No. 7, p. 4.

structure of NC machines produced by the aircraft industry is not known, but this branch tends to employ a relatively high share of NC milling machines. This evidence suggests that milling machines and turning machines are now produced in about equal proportions, with drilling and boring machines representing a somewhat smaller share of the total Soviet output. Thus the structure in 1974 was probably of the form: milling machines 25 per cent, turning machines 35 per cent, drilling and boring machines 25 per cent, and other types, including machining centres, 5 per cent. The structure of types built in the leading producing countries is shown in Table 4.38. This output structure has a major influence on the type of control systems required: contouring systems are generally fitted to NC milling and turning machines, while simpler positioning systems are usually fitted to drilling and boring machines and to many machining centres based on boring machines. This question will be examined in more detail below.

Table 4.38 *Types of NC machine tools produced (per cent of total output in units)*

Type	USSR[1a] 1974	USA[2] 1966	USA[2] 1972	Japan[3] 1966	Japan[3] 1972	UK[4] 1966	UK[4] 1971	FRG 1966[5]	FRG 1970[6]	France[6] 1970
Turning	35·0	21·7	35·0	13·3	43·7	5·6	18·1	26·5	53·5	14·3
Milling	35·0	15·4	14·1	36·7	14·3	5·9	19·7	17·8	18·2	25·9
Drilling/Boring	25·0	56·1	12·4	32·2	13·5	80·8	42·5	52·3	16·8	42·1
Machining Centres	5·0	6·8	29·1	0·0	24·4	4·6	18·8	3·4	11·5	12·2
Other			9·4	17·8	4·1	3·1	0·9			5·5
Total	100·0	100·0	100·0	100·0	100·0	100·0	100·0	100·0	100·0	100·0

Sources:
1 Estimate, see text.
2 US Dept of Commerce, *Current Industrial Reports*, Series MQ-35W, *Metalworking Machinery*.
3 Supplied by MTIRA.
4 *Business Monitor*, Series P64 and PQ332.
5 *NC Machine Tools*, OECD, Paris (1970), p. 37.
6 *British Machine Tools in Europe*, NEDO, London (1972), p. 35.

Notes:
a – estimate.

THE NC MACHINE TOOL STOCK

The growth of the NC machine tool stock in a number of countries is shown in Table 4.39. In 1970 the Soviet stock was one tenth that of the USA, and it was also smaller than those of the United Kingdom, Japan and West Germany. Since 1970, the rapid rate of increase of output in the USSR, coupled with the downturn in output in the West, must have led to a substantial narrowing of the difference between Soviet and American stocks, and estimates suggest that the Soviet stock is now over twice those of Japan, West Germany and Britain.

In terms of the proportion of NC machines in the total metalcutting machine tool stock of the economy, however, the Soviet position is less favourable because of the relatively large total stock in the Soviet Union: an approximate indication is provided in Table 4.40. It is foreseen that by the end of 1975 the Soviet Union will have about 20,000 NC machine tools,[126] giving a proportion of the total stock of about 0·45–0·48 per cent. This is less than the level in the United States in 1968 and in the British engineering industry in 1971.

Information about the structure of the stock is virtually non-existent. In a number of countries the largest concentration of NC machine tools is found in the aerospace industry: 18·2 per cent of the American stock was located in this sector in 1967 (15·3 per cent in 1973),[127] and 19 per cent of the British[128] and 22 per cent of the French stock in 1971.[129] In Japan and Germany the proportion

Table 4.39 *The stock of NC machine tools (the total number of NC metalcutting machine tools installed in the economy—end of year[c])*

Year	USSR	USA	Japan	FRG	UK	France
1962	n.a.	2200[1]	50[1]	130[2]	170[2]	n.a.
1965	n.a.	7000[1]	100[1]	300[3a]	800[1]	200[1]
1968	400[4]	13700[5]	600[6]	1030[7]	2100[7]	630[6]
1970	2000[8]	20000[7]	3500[9]	1930[7]	3200[7]	1100[10]
1972	7600[11a]	24800[12]	5000[13b]	4200[12]	4200[14a]	1500[15]

Sources:
1. *Mirovaya Ekonomika i Mezhdunarodnye Otnosheniya*, 1968, No. 11, p. 123.
2. *Metalworking Production*, 30 May 1962, p. 10.
3. Estimated on the basis of known 1966 stock and production.
4. *Mashinostroitel'*, 1967, No. 10, p. 25.
5. *American Machinist*, 10th inventory, 1968, p. 48.
6. *The Production Engineer*, 1971, No. 5, p. 375.
7. L. Nabseth and G. F. Ray, *op. cit.* (1974), p. 31.
8. O. I. Volkov, *Ekonomicheskie aspekty vnedreniya avtomatizatsii* (1972), p. 100.
9. *Iron Age*, 25 November 1971, p. 60.
10. *British Machine Tools in Europe*, NEDO, London (1972), p. 40.
11. Estimated on the basis of known 1970 stock and 1971–2 production, but without account of imports, exports or scrapping.
12. *Werkstatt und Betrieb*, 1973, No. 8, p. 525.
13. *Ibid.*, 1973, No. 9, p. 691.
14. Estimated on the basis of known 1971 stock.
15. *Referativny Zhurnal, Tekhnologiya Mashinostroeniya*, 1974, No. 8, Item A4.

Notes:
a – estimate.
b – 1973.
c – stock statistics generally refer to the machine building and metalworking industries, but few NC machines are installed in other branches.

[126] *Voprosy Ekonomiki*, 1973, No. 11, p. 54 (estimate with account of imports and scrapping).
[127] *American Machinist*, 10th and 11th inventories.
[128] *Metalworking Production Survey*, London (1971).
[129] *Referativnyi Zhurnal, Tekhnologiya Mashinostroeniya*, 1972, No. 6, Item A143.

Table 4.40 *NC machine tools as a proportion of the total machine tool stock[a] (per cent of total stock in units)*

	USSR[1]			USA[2]			UK[3]
	All economy	Engineering industry		All economy	Engineering industry		Engineering industry
1968	0·01	0·01	1963	0·14	0·17	1966	0·10
1970	0·05	0·08	1968	0·50	0·63	1971	0·51
1973	0·27	0·45	1973	0·91	1·13		

Sources calculated from:
1 NC stock taken from table 4.39. 1973 figure is estimated.
   Total stock: *Narodnoe khozyaistvo SSSR v 1973 godu* (1974), p. 61. 1968—estimated. Engineering industry estimate—60 per cent of total stock, see A. A. Kostin, *Proizvoditel'nost' truda i tekhnicheskii progress* (1974), p. 143.
2 *American Machinist*, 9th, 10th and 11th inventories.
3 1966: NC stock—*National Institute Economic Review*, 1969, No. 48, p. 54.
   Total stock—*Metalworking Production Survey*, London (1966).
   1971: NC and total stock—*Metalworking Production Survey*, London (1971) (NC stock taken as 3,700).

Note:
a – approximate: the scope and definition of total stock is different in each case. Equivalent total stock data for Japan and FRG not available.

of the stock found in the aerospace sector is much smaller. In view of the role of the aviation industry in the production of NC machine tools in the Soviet Union, it seems highly probable that the aerospace sector possesses a large proportion of the total NC stock. According to Miroshnikov a 'significant part' of the NC output is allocated to enterprises of the aviation industry.[130]

One of the main users of NC machine tools in all industrially advanced countries is the machine tool industry itself, the high-precision individual and small-batch production associated with machine tool building being particularly well suited to the employment of NC equipment. In 1973 12·5 per cent of the total American NC stock was located in the machine tool and tooling industry;[131] in the United Kingdom in 1971 6 per cent of the total stock was located in the machine tool industry;[132] in France in 1971 the proportion was 15·7 per cent;[133] in Germany 21 per cent of all domestically produced NC machine tools built between 1966 and 1970 were acquired by the machine tool industry itself.[134] The situation is apparently similar in the USSR. Available information on the stock of NC machine tools in the Soviet machine tool and tooling industry is compared with US data in Table 4.41.

[130] L. P. Miroshnikov, *Primenenie stankov s programmnym upravleniem*, Khar'kov (1972), p. 25.
[131] *American Machinist*, 11th inventory.
[132] *Metalworking Production Survey*, London (1971).
[133] *Referativnyi Zhurnal, Tekhnologiya Mashinostroeniya* 1972, No. 6, Item A143.
[134] *British Machine Tools in Europe*, NEDO, London (1972), p. 41.

Table 4.41 *The stock of NC machine tools in the machine tool and tooling industry*

USSR[1]		USA[2]	
Year	Stock (units)	Year	Stock (units)
End 1970	360	1963	490
End 1972	857	1968	1778
End 1973	1030	Mid-1973	3324

Sources:
1 *Mekhanizatsiya i Avtomatizatsiya Proizvodstva*, 1974, No. 7, p. 12. (Note: the source gives a stock of 1,030 'at present', but the article apparently represents a conference report of early 1974.)
2 *American Machinist*, 9th, 10th and 11th inventories.

This suggests that about 10 per cent of the total Soviet NC stock is located in the machine tool and tooling industry, i.e., about the same proportion as in other industrially advanced countries. Plans for the growth of the NC stock of the machine tool industry have not been fulfilled. In 1973 an end-1975 stock of 3,000 units was foreseen[135] but more recently a stock of 2,000 units in 1976 was indicated.[136] The structure of the NC stocks in the machine tool industries of the USSR and the USA are very different (see Table 4.42). The Soviet industry uses a substantially higher proportion of NC lathes and a much smaller proportion of drilling and boring machines and of machining centres, which are also frequently of the drilling-boring type. The proportion of milling machines is relatively low considering the proportion in total NC output, but it is probable that they are used in large quantity in the aircraft industry.[137]

Table 4.42 *The structure of the NC machine tool stock in the machine tool and tooling industry (per cent of total stock)*

		Type of machine tool				
		Turning	Milling	Drilling-boring	Machining centres	Other types
USSR:[1]	End 1973	54	15	27	1	3
USA:[2]	1968	8	10	45	35	2
	Mid-1973	15	13	34	37	1

Sources:
1 *Mekhanizatsiya i Avtomatizatsiya Proizvodstva*, 1974, No. 7, p. 13.
2 *American Machinist*, 10th and 11th inventories.

[135] *Stanki i Instrument*, 1973, No. 6, p. 2.
[136] *Mekhanizatsiya i Avtomatizatsiya Proizvodstva*, 1974, No. 7, p. 12.
[137] Milling machines formed 36·4 per cent of the total NC stock in the American aerospace industry in 1973, compared with only 7·4 per cent turning machines and 19·8 per cent for drilling-boring machines. (*American Machinist*, 11th inventory, 1973.)

MACHINING CENTRES

Numerically controlled machine tools have a high degree of flexibility within the limits of a given operation—milling, turning, drilling, etc.—but when intricate components like casings and housings are machined, it is often desirable to carry out a number of different operations without transferring the workpiece from machine to machine. Various types of combined machine tools were built before the appearance of NC, but they were either uneconomic in small batch production or provided a very low level of automation. The development of NC made it possible to create flexible, automated, multi-operational machine tools. Several terms are employed to describe these machines—'multi-operation', 'multi-function', and 'machining centre'. In the Soviet Union two terms are generally employed—'multi-operational machine tools' (*mnogooperatsionnye stanki*) and, more commonly, 'machining centres' (*obrabatyvyayushchie tsentry* or OTs). The latter have been defined as 'multi-operational machine tools with NC and automatic tool change for the comprehensive machining of components.'[138]

The first NC machining centre was built by Kearney and Trecker in the USA in 1959.[139] This model, the Milwaukee Matic 11, was shown at the 1960 Chicago exhibition and by 1966 over 500 had been sold. In Europe the first example was also built in 1959, the French GSP tape-controlled drilling-boring-milling machine with a 54-tool magazine.[140] The first Japanese machining centre appeared in 1966.[141] In the USA peak output was 976 units in 1972,[142] and from then on annual output declined, reaching 475 units in 1972.[143] In Japan output rose from 77 units in 1968 to 338 in 1970 and 325 in 1972, but the proportion of these 'machining centres' having automatic tool change is very small—only 35 were so equipped in 1972.[144] In Britain a peak was attained in 1971 when 111 examples were built; in 1972 output fell to 61.[145] In mid-1973 8,851 machining centres with NC were in use in the American economy, about 30 per cent of the total NC stock.[146]

Proposals to build NC machining centres were made in the USSR as early as 1958, presumably as a direct response to the first American efforts and work began at the Sverdlov factory in Leningrad in 1961. By 1964 the first Soviet machining centre, with 30 tools, had been built. It was not a success, however, largely because the factory itself undertook all the development work, including the creation of the control equipment, and the primitive electronics rendered the machine unsatisfactory. The Leningrad sovnarkhoz decided that the project was not viable and withdrew the machining centre from the plan of

---

[138] *Mekhanizatsiya i Avtomatizatsiya Proizvodstva*, 1972, No. 10, p. 35. Note that the Soviet definition is more rigorous than that employed in some other countries, e.g., Japanese and American statistics for 'machining centres' include machines without automatic tool change.
[139] *Machinery* (US edition), March 1959, pp. 146–7.
[140] *Metalworking Production*, 18 September 1959, p. 1495.
[141] *Werkstatt und Betrieb*, 1971, No. 10, p. 763.
[142] *American Machinist*, 1970, No. 10, p. 57.
[143] *Metalworking Production*, September 1973, p. 127.
[144] *Ibid.*, p. 145.
[145] *Business Monitor*, Series P64 and PQ332.
[146] *Metalworking Production*, February 1974, p. 95.

new technology. This decision effectively delayed the development of machining centres in the Soviet Union by about five years: work resumed at the factory only in 1968. On this occasion organisation was much improved, and the factory cooperated with outside organisations, including a research institute in Novosibirsk which created the control system.[147] This particular machining centre appears to have been completed in 1971 and was shown at the Stanki-72 exhibition in the spring of 1972.

From the activity devoted to the creation of machining centres which developed from about 1969, it can be assumed that a decision must have been taken by the machine tool ministry to give priority to this area of work. In 1969 it was announced that it was planned to put four models of machining centres into production,[148] and in 1970 a number of statements were published to the effect that work was being carried out on centres with automatic tool changing.[149] The date of the 'first' example is not clear, but was probably 1971.[150] Three different machining centres were displayed at the Stanki-72 exhibition including the 100-tool Model '2B622F4' of the Sverdlov factory with combined milling, drilling and boring operations.[151] Advanced machining centres, especially those requiring control of four or more axes, need complex control systems, and this may have held back developments in the USSR. In recent years new control systems have been developed to meet this need, including the U52-2, Zig-Zag and Razmer 2M,[152] and the M-6010 small computer can also be used for this purpose.[153]

After a late start the Soviet machine tool industry is now devoting serious attention to machining centres, although the number actually built must still be quite small. In 1974 it was reported that of 50 new NC models entering batch production during the first three years of the 1971–5 five year plan, 13 were machining centres.[154] New models introduced in 1974 included a lathe machining centre with a 12-tool magazine, the 1P732F4, built by the Ordzhonikidze factory.[155] The Soviet industry has more than 10 years' foreign experience in this field to draw on. Recent Western opinion is more doubtful about the benefits of complex machining centres. Such machines allow the performance of a large number of machining operations at one setting of the workpiece, and thereby reduce time wasted in moving and fixing the workpiece, and machine utilisation is improved. But for much of the time these very expensive and complex machines are performing relatively simple

---

[147] *Izvestiya*, 12 July 1972.
[148] *Stanki i Instrument*, 1969, No. 3, p. 8.
[149] See, for example, *Vestnik Mashinostroeniya*, 1970, No. 5, p. 2.
[150] *Iron Age*, 25 November 1971, p. 60, refers to the existence of a prototype having 100 tools (this is presumably the Sverdlov factory model); *Stankostroenie SSSR*, Vyp. 6 (1971), p. 58, notes that ENIMS was working on the eight-tool model MA65F3 to the order of the aircraft industry. It is possible that machining centres were built at an earlier date outside the specialised machine tool industry, but no reference to one has been traced.
[151] *Mekhanizatsiya i Avtomatizatsiya Proizvodstva*, 1972, No. 10, p. 35; *Stanki i Instrument*, 1972, No. 4, p. 5.
[152] *Tekhnologiya i Organizatsiya Proizvodstva*, 1972, No. 3, p. 43.
[153] *Mekhanizatsiya i Avtomatizatsiya Proizvodstva*, 1974, No. 7, p. 10.
[154] *Ibid.*, p. 4.
[155] *Ibid.*, 1974, No. 9, p. 2.

operations like drilling, which would be more economical on a much cheaper machine. In the West, recent developments in NC technology are offering the possibility of replacing large complex machining centres of the type at present built by the Sverdlov factory.[156] Soviet specialists are not unaware of these problems: Academician B. Petrov has cautioned against excessive enthusiasm for machining centres and called for the development of alternative machining systems of greater potential.[157] The development of less complex machines with automatic tool changing is now being actively pursued by the Soviet industry: no less than 40–50 per cent of all NC machine tools planned to be built at enterprises of the machine tool ministry during 1974 were to feature automatic tool change mechanisms.[158]

CONTROL SYSTEMS

The main factory producing control systems for machine tools is the Leningrad electro-mechanical factory (LEMZ) of Minpribor. This factory is the primary supplier of Kontur and Koordinata systems widely fitted to Soviet NC machines, and more recently of the N-22 and N-33 systems based on integrated circuits. Other enterprises building control systems include the mathematical machines factory in Tomsk, a factory for small calculating machines in Smolensk, an electronic equipment factory in Astrakhan, and the Sibelektrotyazhmash factory.[159] Research and development for control systems is undertaken at ENIMS, the Kiev Institute of Automatics, the Scientific Research Institute for Technology and the Organisation of Production (NIAT) in Moscow, the special design bureau of machine tool building in Leningrad, the Novosibirsk Scientific Research, Project-Design and Technological Institute of Electric Drive Units (NIIKE), the design bureau of LEMZ, and also at a number of machine tool factories.[160] Responsibility for developing and building control systems was officially vested with Minpribor by the government decision of April 1968[161] and now it appears that most control systems are, in fact, built by the specialised control systems branch of Minpribor and not by the machine tool industry itself or other non-specialised organisations. But although Minpribor builds the control systems, it is itself dependent on the electronics industry for the supply of basic electronic components.[162]

In other countries producing NC machine tools in quantity, control systems are often made by specialised firms of the electronics industry. Thus in Japan the Fujitsu-Fanuc company produces 90 per cent of all Japanese control systems, only a very small proportion being built by the machine tool industry

---

[156] See A. E. De Barr, in *The Production Engineer*, September 1972, pp. 381–2.
[157] *Sots. Ind.*, 6 May 1972.
[158] *Mekhanizatsiya i Avtomatizatsiya Proizvodstva*, 1974, No. 7, pp. 5 and 11.
[159] *Sots. Ind.*, 18 May 1972; 21 September 1972; N. Acherkan, *Machine Tool Design*, Vol. 4 (1969), p. 281 (in English).
[160] A. A. Razi and A. I. Fedotov, *Chislovye sistemy programmnogo upravleniya mashinami i agregatami*, Leningrad (1973), passim.
[161] *Izvestiya*, 26 April 1968.
[162] *Vestnik Mashinostroeniya*, 1972, No. 4, p. 5.

itself.[163] In the USA the situation is rather different, because despite the fact that General Electric and Bendix supply a considerable proportion of control systems, machine tool firms tend to design and make their own systems. Such leading American firms as Cincinnati Milling Machines, Gidding and Lewis, Warner and Swasey, and Excello have specialised divisions producing control systems both for their own use and for sale. Similarly, in Britain the major producers are Cincinnati Milacron in the machine tool industry, and Plessey in the electronics industry. Furthermore, the machine tool firms have played an active role in control systems innovation: Cincinnati produced the first machine tool with a system based on integrated circuits.[164] This suggests that for machine tool firms which are very actively concerned with innovation in the NC field, a very close relationship between control systems development and the building of the actual machines has been found desirable. Even since the 1968 decision there is evidence that in the Soviet Union the development of control systems has not kept pace with the demands of the machine tool industry and that in the creation of new systems there has been inadequate coordination between systems builders and machine tool builders. Some factories still have to develop their own systems,[165] and some machines are fitted with systems of a type not fully matched with requirements. Early in 1972 some lathes were being fitted with control systems originally intended for milling machines, because systems for lathes were being produced in insufficient quantity.[166] At this time, a 'significant lag' in the design and production of numerical control systems by Minpribor was noted by two writers in the official journal of the Ministry.[167] Both in 1971 and in 1972 it was reported that control systems had been supplied to machine tool factories in a form unsuited to operation in shopfloor conditions.[168] The problem of designing control systems for industrial use is by no means unique to the USSR; elsewhere, shortcomings of this type have been a major source of poor reliability, especially in the early years of NC development.[169]

The first NC control systems used valves and relays. The technology, which was associated with radio communications, was well established and had advanced in response to military demands posed during the war. These systems, however, proved very unreliable. During the late 1950s and early 1960s in the West, and apparently somewhat later in the USSR, the second generation, based on the use of semi-conductors, was developed, and offered improved reliability at a lower cost. In the United States the first machine tool equipped with a third generation control system based on integrated circuits was produced in 1966.[170] Integrated circuits being about miniaturisation and

[163] *Werkstattstechnik*, 1973, No. 9, p. 535.
[164] E. K. Vasil'evskii, *Proizvodstvennyi apparat stankostroitel'noi promyshlennosti SShA* (1970), p. 54.
[165] See p. 164 above.
[166] *Tekhnologiya i Organizatsiya Proizvodstva*, 1972, No. 3, p. 42.
[167] *Pribory i Sistemy Upravleniya*, 1972, No. 3, p. 5.
[168] *Sots. Ind.*, 8 May 1971; 21 September 1972.
[169] *Metalworking Production*, March 1974, p. 71.
[170] The Cincinnati 'Acramatic IV' system—*Metalworking Production*, 18 October 1967, p. 70; *Iron Age*, 25 August 1966, p. 81.

much greater reliability, and by 1970 systems based on them were widely employed in the USA and Europe. In the USSR the first models of control systems based on integrated circuits were planned for 1972;[171] the Leningrad electro-mechanical factory planned to produce systems of the new type in 1973.[172] These third generation control systems, N-22 and N-33 are now in regular production in a number of variants and according to a Soviet assessment they conform to 'all modern technical demands'.[173] These new systems are now being fitted to the latest NC machine tools, e.g., the 6R13F3 milling machine of the Gor'kii factory and the 16K20F3 of the Krasnyi Proletarii works.

The multiplicity of organisations involved in NC development in the USSR in the early years, the inadequate control from the centre and the policy of independent design work in isolation from foreign practice, all resulted in a proliferation of programming media and coding systems. Until recently, the basic media employed were five channel punched tape and magnetic tapes of three different sizes.[174] In 1972 the Gor'kii factory complained that the machines it was building had 10 different programming systems working from magnetic and punched tape and 9 different machine languages, only one of which conformed to international standards.[175] In the USA and Western Europe the situation was similar in the early stages. Most European NC machines using punched tape were designed for five channel tape during the early years, whereas American and Japanese systems employed eight channel tape. Standardisation of input media and coding systems was, however, achieved more quickly in the West than in the Soviet Union. In Europe the switch to eight channel tape was made during the mid-1960s, but in the Soviet Union this transition has been taking place only in the 1970s.[176] American systems are designed for use with the standard EIA code[177] for coding information fed into the system, and in Europe the ISO[178] code is employed as well as the EIA, while many systems are adaptable for use with both. In the USSR a number of different codes are employed, but usually the BTsK-5 for punched tape. Efforts are now being made to achieve compatibility with Western systems: the ISO code can now be used on some Soviet systems by fitting a suitable module,[179] and the latest third generation systems are said to conform to the basic demands of ISO.[180] Until recently pressure to conform to international standards for control systems may have been rather less than in other countries, because exports have been limited and those machines which are exported tend to be fitted with control systems of foreign origin.

[171] *Vestnik Mashinostroeniya*, 1972, No. 4, p. 5.
[172] *BBC Summary of World Broadcasts*, Weekly Economic Review, 9 March 1973, SU/W715/A/10.
[173] *Mekhanizatsiya i Avtomatizatsiya Proizvodstva*, 1974, No. 11, p. 1.
[174] A. A. Razi and A. I. Fedotov, *op. cit.*, pp. 32–77.
[175] *Sots. Ind.*, 21 September 1972.
[176] New systems shown at the Stanki-72 exhibition were designed for eight channel tape—*Pribory i Sistemy Upravleniya*, 1972, No. 7, pp. 57–9.
[177] EIA—Electronic Industries Association.
[178] ISO—International Standards Organisation.
[179] E.g., on the STsM range—*Stanki i Instrument*, 1972, No. 5, p. 7.
[180] *Mekhanizatsiya i Avtomatizatsiya Proizvodstva*, 1974, No. 7, p. 8.

In developing NC control systems, the Soviet Union has designed and built its own systems rather than resorting to the use of existing designs. In recent years, however, there has been an increased willingness to cooperate with foreign control systems firms and agreements have been reached between the Soviet industry and firms in a number of countries. Since about 1969 as a result of cooperation with the French firm Alcatel, some Soviet designed machine tools are fitted with Alcatel systems.[181] In 1972, 52 such machines were to be produced jointly;[182] the agreement was still in force in 1975.[183] The Sverdlov factory, Leningrad, has had close contact with this French firm for five years and cooperation has involved an exchange of personnel between the French and Soviet enterprises.[184] In 1972 an agreement was signed between the Soviet machine tool industry and the leading Japanese control systems firm Fujitsu, makers of Fanuc systems.[185] Soviet NC machines sold on the West German market are now being fitted with Siemens control systems,[186] and those sold to Sweden are using Saab-built control units.[187] In early 1974 the possibility of cooperation between the British firm of Plessey and the Soviet industry was under discussion,[188] but this deal was not concluded.[189]

TYPES OF CONTROL SYSTEMS

There are two basic types of control systems, point-to-point (or positioning) and contouring.[190] Point-to-point type systems are simpler and are generally used for the control of machine tools performing drilling, boring, reaming or tapping operations. The more complex contouring systems are used for the control of milling, turning or grinding operations. It has been estimated that R and D costs for point-to-point type systems are half those for the contouring type.[191] Although some of the very first NC machine tools built in the USA and Britain were of the contouring type designed to meet the needs of the aircraft industry, from about 1961 the rapid growth in NC output in the USA was associated mainly with the simpler and cheaper point-to-point type control. The predominance of point-to-point control was also characteristic of other major producing countries during the early and mid-1960s, but more recently there has been a relative increase in the proportion of contouring systems, particularly in Japan and West Germany, a change associated with the

---

[181] *Soviet News*, 22 April 1969, p. 46.
[182] *Ibid.*, 27 June 1972.
[183] *Neva*, 1975, No. 1, p. 150.
[184] *Ibid.*
[185] *Electronics*, 8 May 1972, p. 55.
[186] *Press Bulletin, Moscow Narodny Bank*, 7 November 1973, pp. 7–8.
[187] *Ibid.*, 12 December 1973, p. 7.
[188] *Ibid.*, 6 February 1974, p. 1.
[189] *The Engineer*, Vol. 239, 1974, No. 6188, p. 34.
[190] Point-to-point, or positioning, systems accurately control the relative positions of the cutting tool and workpiece at discrete points, but not between points. Contouring, or continuous path, systems provide accurate control of the relative movement of the cutting tool and workpiece on a continuous basis and on more than one axis, permitting the machining of curvilinear surfaces.
[191] *NC Machine tools—their Introduction in the Engineering Industries*, OECD, Paris (1970), p. 51.

development of NC turning machines. The relative proportions of the two basic types of control systems in total NC output are shown in Table 4.43.

Initial Soviet NC development, following the USA and Britain, appears to have focused on contouring type control,[192] but in the early 1960s much attention was devoted to simple, point-to-point systems including types intended for retrofitting to lathes. In recent years, however, a distinct bias has appeared towards contouring type control for Soviet machines, a bias largely determined by the structure of NC output with its emphasis on milling and turning machines. In a recent survey of the main Soviet control systems, the majority are of the contouring type.[193]

Tabel 4.43 *Type of control systems fitted to NC machine tools (per cent of total NC output)*

	USA		UK		FRG		Japan	
	1966[1]	1970[2]	1966[1]	1970[3]	1966[1]	1970[3]	1966[1]	1970[3]
Point-to-point	80	57	94	82	83	52	73	n.a.
Contouring	20	43	6	18	17	48	27	n.a.

Sources calculated from:
1 *NC Machine Tools—their Introduction in the Engineering Industries*, OECD, Paris (1970), p. 37.
2 *Current Industrial Reports*, series MQ–35W.
3 *British Machine Tools in Europe*, NEDO, London (1972), p. 35.

Note: A few machines fitted with control systems of other types are omitted.

Control systems can be of the closed- or open-loop types. Closed-loop (or 'feedback') systems make it possible to overcome discrepancies between the situation specified by the programme and that actually existing on the machine tool, and are essential where great accuracy is required. Open-loop systems do not provide feedback and are employed in conjunction with electric or electro-hydraulic stepping motors, the angle of rotation of which is controlled by the application of a sequence of electrical pulses. These motors control the movement of the machine table. Open-loop systems, while not providing such a high level of precision, have the merit of greater simplicity, requiring less electronic circuitry. As a result, they tend to be substantially cheaper than closed-loop and are, therefore, suitable for fitting to batch produced, general-purpose NC models.

In the USA and Britain the great majority of systems are of the closed-loop type. In Germany both closed- and open-loop systems are built, but in Japan most machines are fitted with the Fanuc open-loop control systems and

---

[192] The Model 6N13Pr milling machine shown at the Brussels Fair in 1958 had open-loop continuous path control, with an electric stepping motor designed by ENIMS.
[193] A. A. Razi and A. I. Fedotov, *op. cit.*, pp. 32–58.

stepping motors. The Soviet Union from an early date developed open-loop systems with stepping motors and thus followed a similar path to that of Japan. In 1974 it was stated that 70 per cent of NC machines (this apparently refers to those built by the machine tool industry) were fitted with electro-hydraulic stepping motors built by the Gidroprivod factory, Shilute (Lithuania), and the experimental factory, Gidroavtomatika, Leningrad.[194] The widely-used Kontur and PRS series of systems and the new N-22 and N-33 models are all of the open-loop type, and are fitted to the basic batch produced NC machines. While the basic Soviet NC models are fitted with open-loop systems, more specialised models are supplied with closed-loop control, e.g., in the FSPU, STsP and StsM series.[195] Thus Soviet technical policy has been one of giving priority to the relatively more complex contouring systems, but at the same time emphasis has been placed on a simpler form of this type of control, less complex from the point of view of electronics, and suitable for use with batch produced general-purpose basic models. It is therefore probable that the average level of accuracy of Soviet NC machine tools is somewhat lower than that typical for British or American machines. One Soviet observer in reference to the basic Soviet NC models noted that closed-loop control 'significantly raises their reliability and the accuracy of machining'[196] and added that some of the open-loop contouring systems had limited technical possibilities and were used when accuracy requirements were not very high.[197] The need to develop closed-loop systems has been stressed on a number of occasions recently, but it appears that progress is being hampered by the lack of a suitable small, direct current motor with thyristor regulation: the electrical engineering industry at present does not make a motor of this type.[198]

PROGRAMMING OF NC MACHINE TOOLS

The programming of NC machine tools is a difficult and time-consuming operation. Writing in 1971, a prominent British specialist observed that the facilities available for preparing programmes were probably the least satisfactory aspect of the development of NC, and noted that on average it takes about 30 hours to prepare the input tape for one-hour's machining.[199] A Soviet writer estimated that 50 hours were required for preparing a programme of average complexity, while each machine tool in use in small or medium batch production needs on average 200 programmes.[200] Programmes can be prepared either by hand (aided by conventional desk calculating machines) or by computer. In the early years of NC almost all programming was done by hand, although manual programming is not very satisfactory for more complex machines, especially those with contouring control. Nevertheless, in

---

[194] *Mekhanizatsiya i Avtomatizatsiya Proizvodstva*, 1974, No. 7, p. 11.
[195] A. A. Razi and A. I. Fedotov, *op. cit.*, pp. 32–54.
[196] *Tekhnologiya i Organizatsiya Proizvodstva*, 1972, No. 3, p. 42.
[197] E.g., the 'PRS-3K' and 'Kontur 4MI-68' systems, widely fitted to milling machines and lathes (*ibid.*).
[198] *Mekhanizatsiya i Avtomatizatsiya Proizvodstva*, 1974, No. 7, pp. 3 and 19; No. 11, p. 2.
[199] A. E. De Barr, *The Production Engineer*, September 1972, p. 378.
[200] *Stanki i Instrument*, 1973, No. 8, p. 44.

1973 most NC machines in the Soviet Union were programmed manually[201] and this was also the case in Britain in 1971.[202] In the United States the use of computers as a programming aid was developed more vigorously; as early as 1966 it was reported that it was becoming a common practice in the metalworking industry.[203]

The need to develop computerised programmes was recognised at an early stage in the USSR. A leading representative of the machine tool ministry wrote in 1959 that the development of computing centres was essential for the introduction of programme controlled machine tools.[204] This is the policy now being pursued at the present time; the problem of programming is being tackled by the creation of a nationwide system of centralised programme preparation. The machine tool ministry is organising 10 computer centres on a zonal basis, 9 for metalcutting machines and one for metalforming; some of these have now been in operation for two or three years. Each centre is linked to the engineering factories in its zone by telephone, and also provides technical assistance to enterprises introducing NC machines. The service is provided on a contract basis. The 10 centres are being located at machine tool factories or research institutes: the Ryazan machine tool factory, for example, will act as a centre for the Moscow, Tula, Penza, Voronezh and Ryazan regions. A head (*golovnoi*) centre for preparing programmes (GTsPP) is being organised at ENIMS to coordinate the work of the whole system and develop special software. The zonal centres will at first use second generation computers (Minsk 32), but it is later intended to equip them with third generation models of the Ryad series (ES-1030 and ES-1050).[205] Programming is also being centralised in other countries, e.g., in Britain programme preparation is increasingly undertaken by commercial programming bureaux using large time-shared computers, the users being linked to the system by telephone.[206]

It is claimed that initial experience of the Soviet centralised system of NC programme preparation reveals that this path gives 'extremely positive results'.[207] In the autumn of 1974 about 200 enterprises were being served by the system, including 94 enterprises of the machine tool ministry. Considerable reductions in the cost of programme preparation are claimed: the average cost per programme was reduced from 200 to 60 roubles in 12 months.[208] In 1975 it is expected that over one million programmes will be prepared by the system, rising to eight million in 1977.[209] Similar programming centres have been organised for the aircraft industry.[210]

Computerised programming for NC requires suitable programming languages and considerable research has been undertaken in the West leading

[201] *Ibid.*
[202] A. E. De Barr, *op. cit.*, p. 378.
[203] *Iron Age*, 16 September 1966, p. 100.
[204] A. Vladzievskii, *Planovoe Khozyaistvo*, 1959, No. 3, p. 45.
[205] *Stanki i Instrument*, 1973, No. 8, pp. 44–5.
[206] *Metalworking Production*, March 1974, p. 77.
[207] *Mekhanizatsiya i Avtomatizatsiya Proizvodstva*, 1974, No. 7, p. 24.
[208] *Ibid.*, No. 12, p. 40.
[209] *Voprosy Ekonomiki*, 1974, No. 6, p. 115.
[210] *Ibid.*

to the creation of such part-programming languages as APT and ADAPT, both developed in the USA, and EXAPT created in Germany on the basis of APT. The USSR appears to be more backward in this area. There is now a cooperative CMEA research project directed towards the development of a standardised programming language.[211] The GDR machine tool building firm SAK has already developed a programming language called AUTOTECH, which is similar to the West German language, EXAPT.[212] In Czechoslovakia, a programming language called AUTOPROG has been developed on the basis of Group Technology principles.[213]

SOME NEW PATHS OF DEVELOPMENT OF NUMERICAL CONTROL

So far we have considered what could be broadly termed the first generation of NC machine tools. Characteristic features of these machines include reliance on special, 'hard-wired', control systems in which the configuration of the electronic circuitry predetermines the range of control functions that can be performed. Different types of control systems have to be used for different types of machine tools; the complexity of programming is a major obstacle to the widespread diffusion of NC technology; NC machines are regarded as individual units and as such have been incorporated into existing organisational and managerial structures. As a result, as De Barr has observed, 'Despite the rapid growth in the number of NC machine tools in use, the revolution in batch manufacturing methods that was expected when NC was first introduced has not materialised.'[214] As very often happens with new technical advances, for a period NC has been regarded as a variant of traditional machine tool technology, as a machine tool with a control system, rather than being seen as a new technical development in its own right.

Several recent developments have promoted a new approach to NC. Recent advances in computer technology brought about by the use of integrated circuits, have given rise to cheap and reliable miniature computers,[215] and time-sharing on large computers is now widespread. Computers are increasingly used in management and a systemic approach to production organisation has emerged. The inherent limitations of complex machining centres are increasingly realised, together with the possibility of greater flexibility and improved machine utilisation from the alternative approach of using interlinked separate machines. The new approach, at present at a very early stage, involves a convergence of machine tool and computing technologies, in contrast to the previous development of NC when these technologies evolved

---

[211] *Stanki i Instrument*, 1974, No. 1, p. 4.
[212] *Metalworking Production*, January 1973, p. 79.
[213] A. E. De Barr, *op. cit.*, p. 380.
[214] *Ibid.*, p. 382.
[215] The rapidity of this technical development is illustrated by the movement of computer prices in the USA. Computers of the late fifties and early sixties frequently sold for over $1 million. In 1961 the Digital Equipment Corporation introduced the PDP-1 which sold for $120,000. In 1963 the same firm introduced the first small computer, the PDP-5, selling for about $27,000, while in 1966 the very first mini-computer was introduced—the PDP-8 selling for about $19,000. By the end of 1972 mini-computers were being sold for only $3,500. (*Machinery*, September 1972, p. 54.)

in relative isolation. The radical difference of approach is summed up by the comment of a leading American specialist: 'In the future machine tools will be looked at as the output elements for computers.'[216]

Three separate but related paths of development can be distinguished: first, the use of general-purpose computers to control groups of NC machine tools (usually termed 'direct numerical control' (DNC)); second, the use of small general-purpose computers to replace the conventional control systems on individual NC machines (computer numerical control (CNC)); and, third, computer controlled, integrated, complex machining systems.

Direct numerical control. In DNC systems a central computer either feeds programmes directly to machine tools with conventional control systems, or, in more advanced forms, the computer both distributes the programmes and directly controls the machine tools, eliminating the individual control systems. The first example was in use in Japan in 1968[217] and DNC has been since developed strongly by the Japanese machine tool industry: by the end of 1972 26 systems had been installed.[218] While DNC was given great prominence in 1970–1, emphasis has since shifted more to CNC systems offering greater flexibility. No reference to a Soviet DNC system has been traced: it appears that more attention is being paid to integrated machining systems than to DNC as such.

Computer numerical control. The use of small and miniature computers to control machine tools instead of conventional hard-wired control systems offers many potential advantages. Programmes for machining different parts can be stored in the memory of the computer while a standard general-purpose computer can be employed for the control of many different types of machine tools simply by feeding in a suitable programme. The computer can be used for preparing and verifying programmes and can be employed to provide various forms of adaptive control, correcting for errors caused by tool wear or machine deflections. Owing to the memory facility the input programme which had to be read once per part with conventional NC only has to be read once for each batch of parts; this offers greater reliability because punched tape readers are one of the least reliable elements of conventional NC. The use of general-purpose computers of the type widely used in industry and commerce allows much greater economies of scale in production than is the case with control systems and should provide a cheaper means of control in the future.

There are disadvantages, however, especially in the short run. Computers are more complex than conventional control systems, and CNC puts considerable emphasis on the provision of suitable software. The development of CNC, both in the West and in the USSR, is to a large extent dependent on the level of the computer industry, and in particular, on the availability of

---

[216] James Geier (President of Cincinnati Milacron), *Control Engineering*, November 1970, p. 53.
[217] *Chartered Mechanical Engineer*, March 1974, p. 49.
[218] *Chartered Mechanical Engineer*, March 1974, p. 50.

cheap mini-computers. CNC originated in the United States in the late 1960s and was quickly taken up by the Japanese industry. In Europe the first CNC system was developed in Sweden.[219] In the Soviet Union CNC appears to be still at the experimental stage, but suitable small computers for use with machine tools are now in production. These computers, the M-6010 and M-6000, form part of the ASVT-M unified series of control computers, and are considered to be applicable as control units for machine tools, machining centres, transfer lines, industrial robots and complex machining systems.[220] It seems unlikely that CNC will be widely adopted until the production of mini-computers has been mastered on an industrial scale and at a cost competitive with that of conventional control systems. CNC machine tools have been built in the GDR and shown at the Leipzig Fairs of 1972 and 1973; mini-computers built by Zeiss were used for control.[221] This suggests that CNC is a possible future area of CMEA collaboration.

Integrated complex machining systems. A logical advance from conventional NC and DNC is the creation of production sections in which a group of NC machine tools, interlinked by transfer mechanisms, is controlled as an integrated system by a general-purpose computer, which not only controls the machines but also optimises the procedures of the entire system. Such systems offer the prospect of a radical new approach to batch production and eventually of completely automated shops and factories.

Pioneering work in the creation of computer controlled machining systems was undertaken in Britain by the Molins company in the form of their System 24, but the system was not fully developed and the individual machines were not directly computer controlled. The first integrated complex machining system in the world was built in the GDR and was demonstrated at the 1971 Leipzig Fair. This system, the ROTA-F-125NC, was designed for the small-batch machining of cylindrical components, and the control of the machine tools, workpiece handling and production scheduling were all undertaken by the central computer (an ICL 1905).[222] The first Soviet machining system, the AU-1 (*avtomatizirovannyi uchastok-1*), was built during 1972–3 and is now in experimental operation at the Stankokonstruktsiya factory. Designed by ENIMS, this system consists of 14 NC machine tools and a central computer, and is intended for machining a wide range of cylindrical rotating parts on a small batch basis. It is claimed that the system raises the productivity of labour of the machine tool operators by 13 times, reduces the number of required machines to one seventh and the required production area to one fifth. The computer prepares programmes and feeds them to the NC machines and also gives commands to a store of workpieces. The system is not yet fully automated

---

[219] *Metalworking Production*, December 1973, p. 83.
[220] *Mekhanizatsiya i Avtomatizatsiya Proizvodstva*, 1974, No. 7, pp. 8–10.
[221] *Metalworking Production*, September 1973, p. 139.
[222] *Metalworking Production*, December 1971, p. 79; September 1973, p. 139; A. E. De Barr, *op. cit.*, p. 385. Note: in the following year another system was shown, the ROTA-F2-200, controlled by a computer built in the GDR (R-40); *Metalworking Production*, September 1973, p. 139.

because the loading of the machine tools is still done by hand, although in future it is intended to perform this operation with the use of industrial robots.[223] Work is now being undertaken on a second large integrated machining complex, AP-1 (*avtomatizirovannoe proizvodstvo-1*), for machining more than 100 different casings of medium size for Moscow machine tool factories. A central computer will provide a high degree of automation of both technological processes and production control.[224] When the first AU-1 was designed, only second-generation computers were available and this hindered development work, but now suitable third-generation process control computers are available (the M-4030, M-6000 and M-6010).[225] Further development of machining systems requires the creation of what Soviet writers term 'third generation' NC machine tools, machines with provision for automated loading and unloading of workpieces, automatic programmed setting up, and adaptive control.[226] At present the Soviet Union is quite well advanced in this potentially important area of NC development.

SOME ASPECTS OF THE USE OF NC MACHINE TOOLS

In the early years of NC machine tool development in the Soviet Union, as in other countries, many problems were encountered by users, resulting in some scepticism as to the value of the new equipment. At first individual enterprises acquired single examples of NC machines and found that the problems outweighed the benefits, with the result that some users simply employed them as conventional machines or left them standing idle for long periods. Many examples of this are cited by Miroshnikov: thus in 1969 at enterprises of the administrations for the making of turbines, boilers and mining machinery, only 16 of 54 NC machine tools in stock were actually used as NC machines.[227] The problems encountered were also familiar in the West during the early years of NC: lack of skilled programmers and maintenance workers, unreliability of the control systems, low accuracy of some machines, inadequate production organisation; it proved difficult to devise a pattern of use of the machines which would enable the very high initial cost to be recouped during a reasonable period. The poor reliability of NC machines appears to have been a major source of their ineffective utilisation. The two worst problems were the chronic tendency of control systems to overheat when used continuously for a period of time during summer months, and the tendency of hydraulic systems to overheat because of the lack of a suitable industrial oil able to withstand high temperatures.[228] With experience, it was decided that use of NC machines was viable only if a number of machines was concentrated in one section and employed on parts of complex shape on a multi-shift basis.[229] Given these

---

[223] *Sots. Ind.*, 29 January 1974; for full details of an earlier projected version of this system see *Stanki i Instrument*, 1972, No. 4, pp. 7–8.
[224] *Mekhanizatsiya i Avtomatizatsiya Proizvodstva*, 1974, No. 7, p. 5.
[225] *Ibid.*, p. 10.
[226] *Mekhanizatsiya i Avtomatizatsiya Proizvodstva*, 1974, No. 7, p. 5.
[227] L. P. Miroshnikov, *Primenenie stankov s programmnym upravleniem*, Khar'kov (1972), p. 73, see also *Izvestiya*, 13 July 1972.
[228] L. P. Miroshnikov, *op. cit.*, pp. 40–2.
[229] *Stanki i Instrument*, 1966, No. 2, p. 8.

problems of the early years it is hardly surprising that an editorial in *Stanki i Instrument* in 1963 complained that 'the number of orders for programme controlled machines is still small, users underestimate the possibilities of PC machines'.[230] British experience was very similar; the journal *Engineering*, reporting that in 1964 NC was at last showing signs of gaining acceptance, observed that two years earlier it had still been a 'troublesome intruder'.[231]

The policy of concentrating NC machine tools in special sections devoted to the machining of particular classes of components now seems to be widely accepted in the Soviet Union. This path has been vigorously pursued in the machine tool industry itself; early in 1974 there were 72 such sections at enterprises of the Ministry of the Machine Tool and Tooling Industry, accounting for about 75 per cent of all NC machines installed in the ministry.[232] Three large leading enterprises under the ministry are to be set the aim of the maximum possible use of NC machine tools, raising the proportion of such machines in the total stock to 35–50 per cent.[233] To facilitate the introduction of NC in the machine tool industry a special technological centre has been created at Orgstankinprom.[234]

Some Soviet writers consider that the high prices of NC machines compared with the basic machines they replace hinder their diffusion.[235] According to a machine tool industry economist, Bitunov, high prices limit the use of NC machines at existing enterprises, which have insufficient resources for re-equipping, while at new enterprises designers still do not make sufficient provision for the use of NC. As most NC models have been introduced only in the last two or three years, a very large proportion are sold at high temporary prices. In 1973 only about 12 per cent of the total NC machine tool output of the machine tool ministry was sold at fixed, price-handbook prices; the remainder were sold either at 'one-off' prices (in the case of first prototypes of new models) or temporary prices (in the case of machines of new design assimilated for the first time in the USSR).[236] According to Bitunov, these factors limit the demand for NC machine tools and the consequent small volume of their production in turn limits the extent to which costs can be further reduced, and hinders measures for raising quality and the technical level.[237] In January 1973 the prices of machine tools with programme control were reduced on average by 25 per cent, compared with an average reduction for the machine tool industry as a whole of 9·8 per cent.[238]

Given their high initial price, the organisation of multi-shift work for NC machines is particularly important. In the USA shift work for NC machines is

[230] *Ibid.*, 1963, No. 3, p. 1.
[231] *Engineering*, 17 July 1964, p. 92.
[232] *Mekhanizatsiya i Avtomatizatsiya Proizvodstva*, 1974, No. 7, p. 5.
[233] *Ibid.*
[234] *Vestnik Mashinostroeniya*, 1970, No. 4, p. 7.
[235] L. P. Miroshnikov, *Voprosy Ekonomiki*, 1974, No. 6, p. 112; V. V. Bitunov, *Mekhanizatsiya i Avtomatizatsiya Proizvodstva*, 1974, No. 7, pp. 21–4.
[236] *Mekhanizatsiya i Avtomatizatsiya Proizvodstva*, 1974, No. 7, p. 24.
[237] *Ibid.*, p. 21.
[238] *Tseny i stimulirovanie nauchno-tekhnicheskogo progressa v sotsialisticheskikh stranakh* (1973), p. 171.

widespread; a survey in 1965 showed that half the user factories operated their machines for two shifts and about 30 per cent round-the-clock.[239] In the USSR desirability of multi-shift operation is acknowledged, but it is frequently not achieved; about 80 per cent of NC machines in the USSR are used only in a single shift.[240] A major reason for this poor utilisation is the lack of sufficient trained NC machine operators and specialists. According to Miroshnikov the provision of training for NC personnel is inadequate.[241]

THE PROBLEMS OF TECHNOLOGICAL LEVEL

This final section will attempt to draw some conclusions on the question of the comparative technical level of Soviet NC machine tools. Four distinct but closely related aspects of this problem are discussed: the path of technical development; innovation performance at different stages of the process in terms of both the speed of their completion and the quality of their fulfilment; the technical qualities of the NC machine tools themselves; and the utilisation of NC machines and its effectiveness.

THE PATH OF TECHNICAL DEVELOPMENT

In examining the development of NC machine tools in a number of countries, one is struck by the existence of distinct paths of technical development, shaped both by social and economic circumstances, and by more directly technical factors. Thus in the United States and Britain the initial development of NC was promoted by the need to satisfy certain specific requirements of production for military purposes. The nature of these requirements led to certain technical solutions in the form of, first, high precision, closed-loop, contouring type control systems of great complexity and cost, frequently fitted to specially designed machine tools, and, later, of very complex machining centres of limited application. In both these countries, however, a large share of output in unit terms took the form of relatively more simple drilling and boring machines with positioning type control. In Japan a very different path has been followed, with the building of NC machines for the general engineering user; the machines have a simpler, open-loop control with emphasis on general-purpose turning and milling types. To some extent this path has also been followed in Germany. Both Germany and Japan until recently were countries with relatively weak aerospace-military sectors. Japan has also played a prominent role in the development of new forms of programme control, notably DNC and CNC. The path of development of the French industry has been similar to that of Britain and the USA, again reflecting the existence of a large aerospace-military sector. The Soviet path of development appears to have combined elements of both the US-British and the Japanese approaches. An advanced sector has developed complex, high precision machines to meet the needs of the aviation industry and other high priority users, but at the same time a major emphasis has been placed on the batch production of

---

[239] *American Machinist*, 22 November 1965, p. 113.
[240] *Voprosy Ekonomiki*, 1974, No. 6, p. 114.
[241] *Ibid.*, pp. 114–16; L. P. Miroshnikov, *op. cit.*, pp. 88–93.

simpler, general-purpose models fitted with open-loop control systems for use in quantity in general engineering.

Many factors have shaped these strategies:
—the industrial structures of the economies;
—relative wage levels and availability of skilled machine tool operators;
—the level of development of the machine tool and electronics industries.

It is not possible to make meaningful assessments of the 'level' of different countries in terms of the path of development followed. The fact that the US machine tool industry builds complex, high precision NC machines while Japan and the Soviet Union tend to build simpler, cheaper, and somewhat less accurate machines does not in itself give grounds for stating that there is a technological lag between Japan and the USSR on the one hand, and the USA on the other. Technological level as such is involved only insofar as a particular path of development was shaped by supply considerations directly related to the technological capacity of a country. Thus, in the case of the USSR in the early years the bias towards relative technical simplicity may have been due to the limitations of the electronics industry.

INNOVATION PERFORMANCE

Information on the rate of completion of various stages of the innovation process is summarised in Table 4.44. The period of initial research in the Soviet

Table 4.44 *Rate of completion of stages of the innovation process (number of years)*

Stage of innovation process	USSR	USA	UK	Japan	FRG
(a) From start of research to first prototype	9	5	6	5	3
(b) From prototype to start of industrial production	7	5	7	8	6
(c) Diffusion—NC output as a propn. of total output:					
(i) From prototype to 0·1%	(10)	7	(7)	10	7
(ii) From 0·1% to 0·25%	1	1	(1)	1	2
(iii) Frpm 0·25% to 0·5%	1	2	2	1	4
(iv) From 0·5% to 1·0%	1	3	2	3	×
(v) From 1·0% to 1·5%	2	×	×	×	×
(d) Diffusion—Growth of NC machine tool stock					
(i) From prototype to 500 units	10	(7)	8	9	9
(ii) From 500 to 2000 units	2	(3)	3	2	4
(iii) From 2000 to 5000 units	2	3	(6)	4	(4)
(iv) From 5000 to 15000 units	2	4	×	×	×

× – not yet achieved; ( ) – estimate.

Source:
Compiled from tables and text above. In some cases additional data used from sources previously employed.

case was comparatively protracted, but it seems likely that serious research did not in fact take place until after the appearance of the first American prototype. Performance in the following stage was similar to that of the UK and Japan and, while diffusion was initially rather slow, subsequent rates have exceeded those of any other country. Similarly, the growth of the NC stock was initially rather slow (Britain, Japan and Germany presumably imported relatively more machines in the early years), but subsequent rates of growth have been high.

The changing position of the USSR in relation to other countries is indicated in Table 4.45, which represents an attempt to quantify the leads and lags for various stages of the innovation process and for the three major developments which have occurred since the first appearance of NC machine tools. By the prototype stage a six-year lag had already opened up in relation to the USA, but the situation was more favourable with respect to other countries. During the 1960s, however, the lag in relation to the USA increased to about nine years

Table 4.45 *Leads and lags: the innovation process and major developments*

Indicator	Year achd. by USSR	USSR in relation to:			
		USA	UK	Japan	FRG
(a) *Stages of the innovation process*					
—Start of research	1949	−2	−1	+4	+6
—First prototype	1958	−6	−2	=	=
—Start of industrial production	1965	−8	−2	+1	−1
—Diffusion: NC output as a proportion of total output:					
To 0.1%	1968	(−9)	(−5)	=	−4
To 0.25%	1969	−9	(−5)	−1	−3
To 0.5%	1970	−8	−4	=	+1
To 1.0%	1971	−6	−3	+2	++
To 1.5%	1973	++	++	++	++
—Diffusion: growth of NC stock:					
To 500 units	1969	(−10)	−5	−2	−2
To 2000 units	1970	−8	−2	−1	=
To 5000 units	1972	−8	(+2)	+1	(+2)
To 15000 units	1974	−5	++	++	++
(b) *Major developments*					
First machining centre	1971	−12	(−10)	−5	−10
First third generation control system	1973	−7	(−5)	(−5)	(−5)
First use of computer for control	1973	−6	(−4)	−5	(−4)

Number of years: USSR in advance, +; behind, −; estimate, ( ); Soviet lead, not yet achieved in other country, ++.

Source: As for Table 4.44.

and the situation in respect to other countries also deteriorated. The position with regard to the rate of diffusion was clearly serious by 1968–9 and the Soviet government's concern at that time is understandable. Since 1969 the diffusion lag has been largely overcome; progress has been extremely rapid, but the lag in relation to the American NC stock has not yet been eliminated. Since 1969 the Soviet industry has also been catching up on major technical developments introduced in other countries in the second half of the 1960s. Although approximate, the information presented also indicates that there has been a levelling of performance in Western Europe and Japan in relation to the USA.

The qualitative aspect of the fulfilment of the stages of the innovation process is much more difficult to assess without detailed study of the technical features of NC machines built at different periods. As noted above, the Soviet examples of 1958 were certainly well below the technological level achieved by that date in the USA, and while comparing quite favourably with the German, French and Japanese level of 1958, were inferior to those shown at the Paris exhibition of 1959. By 1965 when the Soviet industry first produced 50 units per year, the situation had probably deteriorated in relation to both the USA and other countries, to a large extent because of the backwardness of Soviet control system technology at that time. Technical characteristics are considered in more detail in the next section.

THE TECHNICAL QUALITIES OF NC MACHINE TOOLS

Given the paths of technical development followed, is it possible to identify areas of strength and weakness in the machines built by different countries, and can any quantitative expression be attributed to such 'leads' and 'lags'? Logically, the first element to be considered would be the basic machine tool, but as most Soviet NC machines are based on modifications of standard general-purpose machines, any assessment of technological level would involve an assessment of the level of Soviet machine tool technology in general, a task beyond the bounds of the present study. *A priori* one would expect that machine tools designed and built specially for NC to be superior to adapted machines and for this reason alone one would expect some inferiority of the average Soviet machine to its American, British or German equivalents which tend to be specially designed for NC. The latest Soviet NC models, such as the 16K20F3 of the Krasnyi Proletarii factory or the Gor'kii factory's 6R13F3 milling machine, are based on new general-purpose machines and the adaptation to NC presumably takes account of experience accumulated over a number of years. Thus one can expect that any lag in the basic machine tools is now being reduced, or even eliminated, as the new machines enter batch production. Available evidence does not permit any assessment of more specialised NC machines. It should be noted that until recently the accuracy of the basic machine may not have been a primary Soviet concern: the limitations of control systems put definite limits on the level of accuracy which could be achieved.

The second element is the control system and our evidence strongly suggests that until recently this has been the major weakness of Soviet NC technology. Independently of the influence of the overall path of development, several objective criteria enable a lag during the 1960s and early 1970s to be identified. First, the Soviet industry was slow compared with other countries in making the transition from valves to semi-conductors, and from the latter to integrated circuits. Second, in the Soviet press dissatisfaction was widespread with the range of functions controllable with the most generally used systems. Third, the fact that the Soviet industry has found it necessary to conclude a number of agreements with Western firms indicates the existence of a lag which the industry is anxious to overcome. Fourth, the well attested unsuitability of some systems for use under industrial conditions and the complaints of poor reliability point to design and construction weaknesses. Finally, the use of five channel punched tape as opposed to the more versatile eight channel tape employed in the West also indicates a lag. Many of these problems are now being overcome; third generation systems based on integrated circuits are in regular production, new systems provide for the control of a broader range of functions, and eight channel punched tape is being used. Cooperation with foreign firms has almost certainly improved the technology of linking systems to the machine tools. Thus the lag with respect to control systems must have been substantially reduced during the last three or four years, although it is not possible to quantify this improvement. The availability of new control systems of greater capacity has probably been a major factor in the rapid development of automatic tool changing features which has taken place in the last two years, and has also facilitated the development of machining centres. Despite this progress, the continued bias towards open-loop systems is evidently not simply a matter of preference, but is also influenced by the fact that the high quality DC motors of fine regulation, required for precision closed-loop systems, are not yet in production. The use of computers for machine tool control is now becoming of increasing importance generally. The Soviet industry was rather slow compared with the USA and Japan in developing third-generation computers for CNC and DNC. While such equipment is now in production, its cost is probably high and the volume of production inadequate. Nevertheless, the Soviet industry is undertaking interesting work in the creation of automated machining systems.

THE UTILISATION OF NC MACHINES

In assessing the overall technological level of Soviet NC, the utilisation of NC machines in the Soviet engineering and metalworking industry must also be considered. Many of the problems encountered have also existed in other countries, but, during the 1960s at least, Soviet industry was evidently not very successful in handling the organisational problems associated with NC machine tools, in particular those of securing skilled workers and specialists and ensuring multi-shift use of the machines. The fact that many programme controlled machine tools have been employed as ordinary universal machines clearly indicates a lag in NC technology. Today the machine tool industry itself appears to be playing a much more active role in promoting effective use of NC machines throughout industry and with the wider availability of NC equipment it is probable that utilisation is in fact gradually improving. The programming of NC machines also appears to have been backward. The reliance on manual programming methods indicates shortcomings in computer utilisation; these are now being overcome with the development of computing centres with time-sharing capability.

CONCLUSION

Although it is not possible to give quantitative expression to the relative technological level of Soviet NC machine tools, the general trend of development can be determined. In the initial period when control systems were based on an established radio-electronic technology, Soviet performance compared quite well with that of Germany and Japan, but lagged behind Britain and also, to quite a considerable degree, behind the USA, the pioneer of NC technology. During the 1960s the rate of diffusion and the technology lagged behind the achievements of the other main NC producing countries and the Soviet industry generally fell behind. This lag appears to have been associated with problems of developing suitable control systems, based at first on semi-conductor technology and later on integrated circuits. This widening gap gave rise to official concern and prompted government measures in 1968. Since 1968 intensive activity has substantially changed the situation. In terms of diffusion the Soviet industry has overtaken the other main producing countries, although this achievement was facilitated by a downturn in the rate of growth of output in capitalist countries. Since 1969 the technological level of Soviet NC machines and their control systems has greatly improved and the lag which appeared during the 1960s has now been considerably reduced. The economic situation in the main capitalist NC producing countries has promoted a search for new technical solutions leading to the development of computerised control of machine tools. The general state of the Soviet computer industry has imposed some constraints on progress in this field, but suitable computers have now been created and with intensified CMEA cooperation, in particular with the GDR, it seems likely that the Soviet machine tool industry will not only keep abreast of developments, but may move ahead in such areas as the creation of large automated production systems. This case study indicates that in the conditions of the Soviet economy technological lags can be very quickly narrowed and overcome once their existence has been acknowledged and priority granted to their elimination.

5. High voltage electric power transmission

Introduction
This chapter deals with the process of technological change in the field of high voltage electrical power transmission in the USSR as compared with Western Europe and the USA. It attempts to show that the USSR has moved from a position before the fifties of being a follower of technological trends in HV technology to a position in the early sixties of ranking roughly equal with that of the leading countries. It will be argued that the existence of such trends is understandable when set against geographical and economic conditions prevalent in the USSR as compared with some other countries. Since the later 1960s, however, the USSR has tended to be overtaken in certain respects by some Western countries.

Both high voltage alternating current (HVAC) and high voltage direct current (HVDC) technologies are examined. In both, the measure of technical progress will, in the main, be the voltage. This measure is useful because the level of voltage used in the power system reflects the standard of technology in design of switchgear, isolators, transformers, etc.; the problem of moving to a higher voltage is essentially one of improving the design of these basic equipments. Some consideration will also be given to the degree to which the USSR relies on imports or on its own efforts in know-how and equipment.

Power systems and their development in the USSR, 1917–mid-1950s
Essentially, a power system simply consists of a number of power stations connected one to another and to consumers by means of an electrical network. The system will typically have various types of power station running on different fuels and at different efficiencies. The electrical network will typically contain transmission lines of various voltage levels depending on distance of transmission and power transferred.

Transmission lines are either cables or overhead lines. The former are used mainly for low voltage distribution in towns and the latter for cross-country transmission at high voltages. This is basically because the capital cost of cable transmission is much higher and is only justified where overhead lines cannot be used; in any case the purely technical problems of cable transmission grow considerably as voltage increases.[1] In overhead transmission, the main items of equipment are the transmission towers and the isolators. The latter, which must both support the conductors and isolate them electrically from the tower structure, are usually made of porcelain, and are of a shape designed to maximise their insulating properties.

The modern power system will also contain substations, whose primary function is to transform the power from one voltage to another. The main items of substation equipment are as follows:

1. Step-up or step-down transformers, often with facilities for on-load voltage regulation.
2. Circuit breakers (electrical switches) whose function is to interrupt the power flow automatically and quickly should any fault occur on the transmission network.
3. Protection equipment to sense when fault conditions arise and automatically relay this to circuit breakers.
4. Devices for compensating for reactive power[2] flowing in the network.

Most power systems operate in the alternating current (AC) rather than the direct current (DC) mode, both for the reason that this is the most convenient method of generation, and because it is by far the easier way to transform current from one voltage to another. As will be seen, however, there are some applications where DC transmission has distinct advantages.

As a general rule, the more interconnected the power system—i.e., roughly speaking, the more dense the transmission network—then the more reliable and, up to a point, the cheaper will be the supply to any one consumer. Unfortunately, the measurement of the degree of interconnectedness, especially for the purposes of comparing different countries, is very difficult, and not amenable to accurate quantification. Several factors must be considered:

—the total length of transmission lines;
—the voltages used;
—the capacity of the system;
—the territory covered;
—the population supplied and how it is distributed.

In the Soviet Union power systems are classified into three types, roughly on the basis of size or area covered. Regional power systems (RPS—*Raionnaya*

---

[1] For cost comparison of overhead and underground lines see *CEGB Newsletter*, 1965, No. 60, p. 11.

[2] Reactive power is that power which must be supplied for inductive and capacitative elements in the network—i.e., power which does no real work but nevertheless involves an increase in current flowing in the network, and therefore an increase in power lost in resistive elements in the transmission lines.

Energeticheskaya Sistema) are the smallest, and in the area can cover several administrative regions (*oblasti*). In 1970 there were 93 such systems in the USSR.[3] They are the basic units for the interconnected power systems (IPS—*Ob"edinennaya Energeticheskaya Sistema*) of which 11 existed in 1970.[4] Eight of the IPSs were in turn connected together to form a united power system (UPS—*Edinaya Energeticheskaya Sistema*) for the European USSR. Ultimately the Soviet Union plans to form a UPS for the whole USSR.

The substantial development of power systems in what is now the USSR began in 1920 with the GOELRO plan. We are told[5] that power supply in pre-revolutionary Russia was predominantly characterised by a 'low' level of centralisation and power system interconnection and standardisation of voltages and frequencies. Maximum voltage in use in Russia from 1912 to 1920 was only 70 kV compared with 150 kV in the USA.[6] This is, however, only one measure of Russian backwardness, and a fuller characterisation of the state of pre-revolutionary Russian power systems remains to be made. By 1929 the provision of HV transmission lines was almost 19 times the 1913 level.[7] As a result of this rapid construction of power systems within 12 years of the revolution, pre-revolutionary capacity was contributing very little to the Soviet power industry.

At first power systems were constructed to serve only local needs: they were RPSs which were smaller in capacity and territory than present networks. During the GOELRO plan, and the first five year plan which overlapped it, six such systems were established, centring on large towns or industrial complexes: by 1935 over half the total installed capacity of the Soviet Union was in stations connected to these six systems. The Moscow regional system (Mosenergo) claimed to be the largest in Europe in terms of total capacity.[8] By this time the many advantages of interconnecting power stations by a general network had been recognised in the Soviet Union, and development along these lines was proceeding fast. This was also true of other countries; in Britain, the grid system was established in 1928–33, although in this case interconnection was designed to facilitate maintenance in each area rather than transmission between areas.[9]

The development of IPSs was first proposed in the second five year plan (1933–7) and was facilitated by the raising of the maximum voltage to 220 kV in the mid-thirties. The plan envisaged the formation of three IPSs—Southern, Urals and Central. It was not until 1940, however, that the first of these systems, the Southern IPS, began to be formed with the completion of a 220 kV line linking the Dnieper and Donbass RPSs.[10] The war delayed the construction

[3] (Ed.) P. S. Neporozhnii, *Elektrifikatsiya SSSR* (1970), p. 386.
[4] *Ekon. Gaz.*, 1972, No. 6, pp. 12, 13.
[5] (Ed.) B. I. Weitz, *Electric Power Development in the USSR* (1937), p. 8; for this book see p. 223, n. 104 below.
[6] B. I. Weitz, *op. cit.*, and W. T. Taylor, *Overhead Electric Power Transmission Engineering*, London (1927), p. 13.
[7] P. S. Neporozhnii, *op. cit.*, p. 323.
[8] V. Yu Steklov, *V. I. Lenin i elektrifikatsiya* (1970), pp. 316–17.
[9] *CEGB Newsletter*, 1965, No. 60, p. 2.
[10] P. S. Neporozhnii, *op. cit.*, p. 325.

of the Southern and Central IPSs and, with the shift of industry towards the east, by the mid-forties most progress had been made with the Urals' system. The shift to the east also brought about the more rapid development of RPSs in Siberia and central Asia. As yet, however, there was no attempt to interconnect these eastern RPSs since the great transmission distances necessary to do so tended to require higher voltages than were then available.

The years from the end of the war until the mid-fifties were in the main ones of reconstruction, completion and extension of the IPSs planned before the war. The radical advances in integrating power networks had to await the development of 400 and 500 kV transmission in the late fifties, which marked the beginnings of the creation of the UPS of European USSR and Siberia.

Basic features of HVAC and HVDC

As will be shown below, the operating voltages of HVAC transmission networks in USA, Western Europe and USSR have steadily increased. This trend stems basically from two major needs: the need to reduce costs and the need to cover increasing demand for electricity.

Costs can be reduced in two main directions. First, generating capacity can be concentrated into a smaller number of stations; from the point of view of the transmission network, this usually involves an increase in average transmission distances and average power transfers. Second, the degree of power system interconnection can be increased. This provides savings in three ways.

1. It increases reliability of supply to the consumer through making available alternative sources of supply in the event of breakdown of the normal source.
2. It reduces the amount of standby generating capacity needed.
3. It smoothes out the load variations of one power station or system and ensures that the increase in load at any point can be supplied by the lowest increment in generation plus transmission costs. This also generally involves increases in power transfers and transmission distances.[11]

These methods of cost savings are widely accepted and were embodied in both GOELRO and the subsequent five year plans.[12]

An increased operating voltage (AC or DC) makes it possible for transmission systems to cope with greater demands because resistive power losses in a conductor are reduced by reducing current: for the same power transfer, current can be reduced by increasing voltage. An increase in line length or power transfer results in a greater resistive power loss unless voltage is increased. The savings from lower power losses in the higher voltage variant must, however, be set against the greater costs of HV equipment and the

[11] This is shown for the USSR in J. P. Hardt, *Economics of the Soviet Electric Power Industry*, part of Ph.D Thesis, (1955) p. 122.
[12] See, for instance, L. Melent'ev, *Voprosy Ekonomiki*, 1967, No. 7, p. 4, for basic GOELRO principles.

greater losses due to corona discharge[13] on the line.

The basic constraints to be overcome in any increase in voltage are ensuring that the increased level of isolation can be achieved, and improving circuit breaker design. A substantial voltage increase generally involves a more than proportionate engineering effort in equipment design. For the British 400 kV grid in England, circuit breaker design effort was estimated to be about a hundred times that required for the 132 kV grid.[14]

The design of circuit breakers is perhaps the most important constraint, since such equipment is the lynch-pin of the system as far as reliability of supply is concerned.

> As operating voltage and size of networks increase, the consequences of switchgear failure are more serious. Thus there is not only a need for faster operating of breakers capable of handling higher fault currents, but the circuit breakers must be more reliable than their predecessors.[15]

As voltages move into the 1,000–2,000 kV range (ultra high voltage or UHV), however, other, mainly environmental, constraints become increasingly important.[16] Practical AC transmission at such voltages is technically feasible, but it involves the increase of transmission line tower heights, as well as problems of noise and high electrostatic fields around the lines.

Since the early fifties HVDC transmission has been given increasing attention as a replacement for HVAC in certain applications and as a complement to others. In purely technical terms the advantages of DC over AC are as follows.[17] First, with AC, reactive power losses in the transmission line increase as power transfers and line lengths increase; reactive power must also be supplied for capacitive and inductive elements in consumer loads. With DC this type of loss is completely eliminated. Second, when AC power systems work in parallel, stability, synchronisation and nominal frequency differences between systems become serious problems. Again, such problems do not arise with DC.[18] An HVDC single-circuit line consists of two (plus and minus) conductors (instead of three with HVAC), the earth or sea being used as a current return with consequent lower losses. In addition, since the DC voltage can be set at the peak rather than the mean voltage of the equivalent AC

---

[13] Corona discharge is the discharge associated with the breakdown of the air immediately around the high voltage conductor.

[14] *Electrical Review*, 29 January 1971, p. 145.

[15] *Ibid.*

[16] The terms 'Extra High Voltage' (EHV) and UHV are used fairly arbitrarily in the literature; here EHV is taken to mean voltages in the 400–1,000 kV range, UHV to mean any voltages above 1,000 kV.

[17] For a fuller account of the advantages of HVDC see P. G. Engström, *IEEE Conference on HVDC Transmission 19–23 September 1966*, Part 1, p. 84.

[18] The problem of maintaining a condition of stability in a power system essentially reduces to that of ensuring that mechanical power input is matched to electrical power output. This involves keeping generators in synchronisation and maintaining frequency standards. When two power systems are linked by an AC line the problems of keeping stability in both systems, as well as the newly formed system as a whole, are significantly increased. If the link is DC, these problems do not arise since the systems remain independent as far as synchronisation and frequency differences are concerned.

system, voltages can be higher. Finally, because current does not have to be supplied for reactive elements in the circuit, cables and overhead conductors can be smaller and therefore cheaper than for the equivalent AC voltage.

HVDC, however, also has several disadvantages. It cannot be transformed without prior conversion to AC, it requires costly AC/DC converter stations, and circuit breaker design is more difficult to the extent that no economically viable type has yet been designed.

Some Soviet authors have estimated the savings offered by DC in a general comparison of equivalent AC and DC overhead lines.[19] In respect of capital costs, a saving of roughly 30 per cent of the AC cost is forthcoming through a reduction in the use of conductor metal and of steel for line towers. As regards running costs, as reactive current need not be supplied, resistive power loss in a DC line is about half that in the equivalent AC line. Against these savings must be set the capital cost of AC/DC converter stations at the end of each line, and along its length, should intermediate substations be required. It follows that the longer the line, the greater the savings in metal and energy, and the more likely that these savings will outweigh the cost of converter stations.

There are four main applications of HVDC in power systems.

1. As explained above, in long-distance high-power applications where the emphasis is on uni-directional transfer of energy.
2. For submarine transmission where AC would have required intermediate reactive power compensating equipment, or where the cheaper cable costs of DC outweigh converter station costs (this, of course, applies equally to underground cable transmission).
3. As a link between AC systems, thus eliminating problems of stability and control of the equivalent AC link. In this context, the DC link also allows the connected power systems complete voltage and frequency independence. Thus, the systems are completely isolated except in respect of power transfer.
4. As a means of frequency conversion or of connecting power systems with different frequencies. In this case there is no need for the DC link to have any length, the converter station being the only element required.

HVAC DEVELOPMENTS UP TO 1960—THE INTRODUCTION OF 400–500 kV

Figure 5.1 shows the relative performance of the USA, USSR and Western Europe in the HVAC field from 1920 to the present day. By 1960 the USSR had introduced 500 kV, which was then the highest voltage in use in the world. The intention was to use this voltage as a basis for intersystem connection, and in particular, for the formation of the UPS of European USSR, and for a 2,000 km interconnection in Siberia.[20] The first 500 kV line joined the Central and Southern IPSs and this was followed by a line from the Lenin Volga Hydro power station to Sverdlovsk, thus joining the Southern and Urals IPSs.[21]

[19] L. P. Neiman, A. V. Pesse, and N. N. Shchedrin, *Elektricheskie Stantsii*, November 1962, p. 92.
[20] V. V. Burgsdorf, S. S. Rokotyan, and A. N. Sherentsis, *Elektrichestvo*, 1965, No. 1, p. 4.
[21] D. G. Zhimerin, *Istoriya elektrifikatsii* (1962), p. 372.

As early as 1940 the networks department of Teploenergoproekt (ODP TEP(Otdel Dal'nei Peredachi, Teploenergoproekt) had projected a 380 kV line to join the Kuibyshev hydro plant to Moscow.[22] The work was presumably interrupted by the war and it was not until 1950 that work on EHV resumed. This time the projected Kuibyshev–Moscow line was uprated to 400 kV, and was put into operation in 1956.[23] At this time, however, it seems there was doubt about the sufficiency of 400 kV to satisfy the demands of a 'fast-developing energy system' and in 1957 a decision was taken to adopt 500 kV for future long-distance AC transmission.[24] The justification for the decision lay in the relative cost estimates for 400 and 500 kV prepared by TEP. In these estimates a 400 kV line 850 km long was to cost about 155 million roubles whereas a 500 kV line 990 km long was to cost about 125 million roubles. Cost of transmission in the 400 kV variant was 0·29 kopeks per kWh as opposed to 0·17 kopeks per kWh for 500 kV.[25] 500 kV, it seems, was basically an extension of 400 kV, and involved no radically different techniques. After evaluating means of limiting internal overvoltages for 500 kV it was decided that allowable relative insulation levels could be reduced to accommodate the 100 kV increase in voltage.[26] In 1959 the first 500 kV line was commissioned and earlier 400 kV lines were uprated to the new level by 1963.[27]

These post-war developments in the Soviet Union were accompanied by similar trends in the West. In 1950, 220 kV was the highest voltage in use in Western Europe; at that time, the USA was leading the field with a 287 kV line. Sweden was the first country in the world to move into the EHV field when, in 1952, its first 400 kV line came into operation.[28] Interesting in this respect is the fact that the USSR and Sweden, the first countries to introduce EHV transmission, shared similar geographical problems, although on different scales. Both had a need to develop long-distance transmission; the USSR because of its size, Sweden because its cheapest and abundant hydro resources in the north were separated from the main population centres in the south.[29]

This situation can be contrasted with that in the UK, where transmission distances are comparatively short, the population more dense, and the need for development of high voltages much less urgent. In Britain, the density of population also leads environmental considerations to be of major importance,[30] a factor which is unlikely to have played much role in the case of the Soviet Union. By 1960, when the USSR had introduced 500 kV, the USA 345 kV, and Western Europe 400 kV, the CEGB was still relying on 275 kV (introduced in the early fifties) for bulk transmission, and had only just begun to consider an increase to 400 or 500 kV.

Elsewhere in the West, however, the late fifties saw increasing efforts put into

---

[22] S. S. Rokotyan, *Elektricheskie Stantsii*, 1967, No. 12, p. 3.
[23] *Ibid.*
[24] V. V. Burgsdorf, *et al.*, *op. cit.*, p. 8.
[25] D. G. Zhimerin, *op. cit.*, p. 371.
[26] V. V. Burgsdorf, *et al.*, *op. cit.*, p. 8, and S. S. Rokotyan, *op. cit.*, p. 3.
[27] S. S. Rokotyan, *op. cit.*, p. 4.
[28] G. Jancke, *Electrical Review*, 13 May 1960, p. 915.
[29] *Ibid.*
[30] *CEGB Newsletter*, 1965, No. 60, p. 9.

the development of EHV, and testing stations were set up by many Western electrical concerns.[31] The USA lost its lead by the maximum voltage measure by 1952 and did not recover parity until the late sixties. Nevertheless, by 1959, only the USSR, the USA, France and Sweden, were using HV transmission over 300 kV.

Design of insulators for EHV presented relatively few problems:

> ... the insulators for EHV systems are generally straightforward developments of the well-proven types used at lower voltages and no radically new designs have been necessary.[32]

The provision of circuit breakers for higher voltages was less straightforward. In 1955 a professional journal commented:

> The circuit breaker is still an element whose performance cannot be designed for, as is possible with transformers, rotating machinery etc. Performance has to be established by tests. Also it is well known that circuit breaker testing stations are so costly that even large switchgear firms cannot hope to install them on their own account.[33]

The transition to 500 kV was accompanied by a shift towards greater use of air-blast instead of the oil-filled circuit breakers, owing to technical problems in designing the oil-filled type for use with EHV.[34]

The fundamentals of the operation of air-blast circuit breakers were established in England,[35] but it was a French company, Delle Asthom, which, judging by its export orders for EHV circuit breakers, held the leading position. In 1953, the company supplied Sweden with small-oil-volume types for its 400 kV system,[36] and later, in 1957, when the USSR decided to adopt 500 kV, supplied five high power 525 kV air-blast types to the Soviet Union.[37] Thus in its development of 500 kV, the Soviet Union had to rely initially on foreign assistance in a relatively vital area. Since it seems that no more than five circuit breakers were imported, it must be presumed that further development of lines at this voltage was carried out with domestically produced circuit breakers, possibly modelled partly on the French design. And in many important equipment design areas the Soviet Union had made substantial advances, as a British Electricity Supply Delegation reported in 1956.[38]

For the development of 400 kV, the Elektroapparat factory, Sverdlovsk, had produced a new design of 400 kV air-blast circuit breakers, for which the Molotov Technical Institute had testing equipment.[39] The Zaporozh'e factory

---

[31] *Electrical Review*, 1 July 1955, p. 22.
[32] W. G. Robinson, *Electrical Review*, 1 January 1960, p. 2.
[33] *Electrical Review*, 2 September 1955, p. 441.
[34] *Electrical Review*, 4 June 1965, p. 859.
[35] *Electrical Review*, 28 October 1955, p. 858.
[36] *Electrical Review*, 4 June 1965, p. 859.
[37] Information obtained from private correspondence with Delle-Asthom.
[38] *Report on a Visit to the USSR 16th April–14th May 1956*, British Electricity Supply Delegation (issued by Central Electricity Authority), pp. 33–41.
[39] *Ibid.*, p. 38.

manufactured both transformers and isolating switches at 400 kV,[40] the Sovkabel' factory was manufacturing several lengths of cable for terminal connections in the 400 kV Kuibyshev–Moscow line, and reference was made to the use of Soviet-made 400 kV insulators.[41] While the extent to which the USSR depended on outside technical know-how for this equipment was not stated, one possible interpretation of these facts is that the Soviet Union had prepared quite extensively and of its own accord for the introduction of 400 kV. The decision in 1957 to uprate to 500 kV meant that 400 kV equipment designs could be used for, or easily extended to, 500 kV—except in the case of circuit breakers.

The relative diffusion of HVAC transmission lines in the six countries chosen is shown in the Appendix to this chapter, in Table 5A.1, and the accompanying Figure 5.1.

It emerges that by 1960, two per cent of the total Soviet HVAC line length was at 500 kV, which was not used in any other country at that time. But France and Sweden had the highest proportion of higher voltages: France had the greatest proportion of lines above 160 kV, and Sweden the greatest proportion at 300 kV and above. In absolute terms, however, the USSR, followed by the USA, had the highest total length of lines above 300 kV—partly reflecting the greater size of the two countries.

By 1960 lines of 300 kV and above per unit station-capacity were relatively weakly developed in the USA, compared to other countries such as the USSR, Sweden or France (see the Appendix, Table 5A.3). This subject will be taken up in more detail in the following section.

HVAC IN THE 1960s—THE MOVE TO 700 kV AND ABOVE

In 1961, the International Electrotechnical Commission accepted 500 kV as an international standard voltage[42] and, one year later, the same body chose both 700 kV and 750 kV as future standards.[43] The Soviet Union in the same year (1962) decided to take 750 kV as its standard,[44] and was projecting an experimental line of this voltage from the Konakov State Regional Power Station to Moscow, a distance of 90 km. Table 5.1 gives the main events in the development of 700 kV from then on. By 1971, when the Soviet Union was constructing a second 750 kV line from the Donbass to L'vov (1,200 km),[45] the USA had roughly 1,000 km of 765 kV already in operation (see the Appendix, Table 5A.1). The Donbass–L'vov line was still under construction in 1975, with completion planned for that year.[46] Thus in terms of the development of HVAC the position of the USSR had somewhat declined after its predominance in 1960.

[40] *Ibid.*, pp. 36, 38.
[41] *Ibid.*, pp. 34, 35.
[42] B. P. Lebedev, *Elektrichestvo*, 1963, No. 12, p. 34.
[43] *Ibid.*, p. 36.
[44] *Ibid.*, p. 34.
[45] *Energetika i Elektrifikatsiya*, 1971, No. 1, p. 3.
[46] *BBC Summary of World Broadcasts: USSR Weekly Economic Report*, 9 January 1975; 1 April 1975.

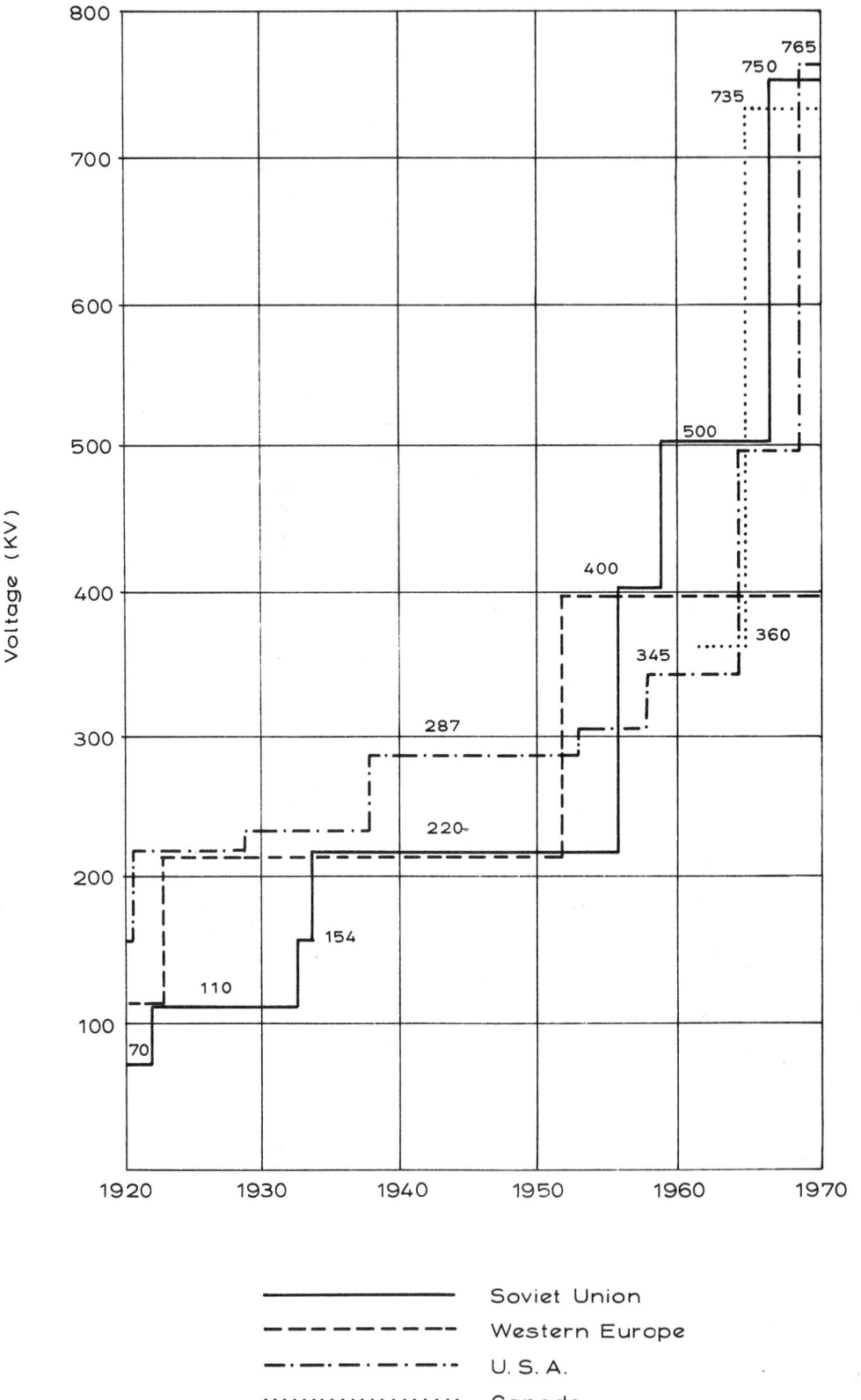

Sources:
USSR
1 First 110 kV line—1922: S. S. Rokotyan—in *Elektricheskie Stantsii*, 1967, No. 12, p. 2.
2 First 150 kV line—1932: *ibid.*
3 First 220 kV line—1933: *ibid.*
4 First 400 kV line—1956: D. G. Zhimerin, *Istoriya elektrifikatsiia SSSR*, (1962).
5 First 500 kV line—1959: Rokotyan, *op. cit.*, p. 4.
6 First 750 kV line—1967 (experimental): S. S. Rokotyan, *op. cit.*, p. 5.

USA
1 Up to 1960 information for the graph was taken from V. A. Venikov, *Electrical Review*, 29 January 1965, p. 158. Precise figures for some voltage levels were taken from the following: for 287 kV level, W. G. Robinson, *Electrical Review*, 1 January 1960, p. 3; for 345 kV level S. S. Rokotyan and S. A. Sovalov, *Elektricheskie Stantsii*, 1967, No. 11. p. 24.
2 First 500 kV line—1965: see V. V. Burgsdorf, S. S. Rokotyan, and A. N. Sherentsis in *Elektrichestvo*, 1965, No. 1, p. 7.
3 First 765 kV line—1969: see *Electrical and Electronics Abstracts*, January–June 1972, No. 15097, p. 793.

Western Europe
1 For developments up to 1950 see V. A. Venikov, *op. cit.*, p. 158.
2 First 400 kV line—1952 in Sweden: see Gunnar Jancke, *Electrical Review*, 13 May 1960, p. 195.

Canada
1 360 kV level: see B. R. Lebedev, *Elektrichestvo*, 1963, No. 12, p. 34.
2 First 735 kV line—1965: see *Electrical Review*, 1 January 1965, p. 2.

A Soviet engineer, Lebedev, writing in 1963, explained:

As a consequence of the increase in unit capacity of generators, the construction of a number of large thermal and hydro-electric power stations, and the growth in interconnection of grid-systems in the near future (1965–70), it became necessary to transmit power of the order of 2,000–3,000 mW single circuit over distances greater than 1,000 km. This problem is adequately solved from an economic point of view by the application of 750 kV AC.[47]

Another source confirmed these figures as the appropriate order of distance and power transfer for 750 kV.[48] In contrast, Lebedev gave the capability of 500 kV transmission as 700–1,000 mW over 1,000 km.[49]

Two factors, however, tended, in combination, to undermine the attraction of a full-scale introduction of 750 kV. First, assuming that the capabilities of 500 kV mentioned by Lebedev are somewhere near maximum, then relatively small demands were being put on the 500 kV level even by the late sixties, as the figures for the 500 kV lines existing in 1969 indicate (see Table 5.2). Moreover, 'for the majority of 500 kV lines, low loading ... was characteristic'.[50] One group of authors estimated in 1967 that the present 500 kV level would suffice

---

[47] B. P. Lebedev, *op. cit.*, p. 37.
[48] V. A. Shelest, *Ekonomika razmeshcheniya elektroenergetiki SSSR* (1965), p. 160.
[49] B. P. Lebedev, *op. cit.*, p. 37. See also for confirmation of order of magnitudes S. S. Rokotyan in *Elektricheskie Stantsii*, November 1960, Table 1, p. 93.
[50] I. S. Davydov, I. Ya. Mel'zak, and A. N. Sherentsis, *Elektrichestvo*, 1971, No. 2, p. 8.

for at least five to seven years ahead,[51] and another that its capabilities would be exhausted in 20–25 years.[52]

Second, and more important, much depends on the role that different voltages play in the grid system. HVDC which the Soviet Union began to develop in the early 1950s would be a strong competitor for 750 kV AC, as, indeed, would 500 kV AC; and for bulk transmission within systems, involving smaller capabilities and distances, 500 kV AC may well be more economic. But for bulk transmission in the Southern IPS, 750 kV had a strong and immediate claim. Here, there was heavy reliance in the past on 330 kV in contrast to the development of 500 kV elsewhere.[53] The transition from 330 to 750 kV was claimed to give an eight fold increase in capacity, so that with the growing demand on the system, the introduction of 750 kV was needed in the Southern IPS by as soon as 1970.[54] The existing 330 kV was to be used for primary distribution.

Table 5.1 *Main events in the development of 700 kV and above*

	1962	1963	1964	1965	1966	1967	1968	1969
USA		Prototype 750 kV line for AEP power system.[3]						AEP 765 kV line brought into operation.[7]
USSR		Projecting Experimental Konakov–Moscow line. (90 km 750 kV).[1]		Began construction of Konakov–Moscow line.[4]		Konakov–Moscow line commissioned.[6]		
Canada		Decision to uprate 500 kV project to 700 kV.[2]		735 kV line scheduled for commercial service.[5]				

Sources:
1 B. P. Lebedev, *Elektrichestvo*, 1963, No. 12, p. 36.
2 *Ibid.*, p. 34.
3 *Electrical Review*, 16 August 1963, p. 287.
4 V. V. Burgsdorf, S. S. Rokotyan and A. N. Sherentsis, *Elektrichestvo*, 1965, No. 1, p. 7.
5 *Electrical Review*, 1 January 1965, p. 2.
6 S. S. Rokotyan, *Elektricheskie Stantsii*, 1967, No. 12, p. 5.
7 *Electrical and Electronics Abstracts*, January–June 1972, No. 15097, *op. cit.*, p. 798.

[51] I. M. Vol'kenau, A. N. Zeiliger, A. I. Kolpakov, I. M. Markovich, P. E. Mironov, M. A. Sarkisov, *Elektricheskie Stantsii*, January 1967, p. 6.
[52] S. S. Rokotyan, S. A. Sovalov, *Elektricheskie Stantsii*, 1967, No. 11, p. 25.
[53] See map in (Ed.) P. S. Neporozhnii, *Elektrifikatsiya SSSR* (1970), p. 343.
[54] S. S. Rokotyan, S. A. Sovalov, *op. cit.*, p. 26. Also I. M. Vol'kenau *et al.*, *op. cit.*, p. 6.

Table 5.2 *Characteristics of existing 500 kV lines (1969)*

500 kV lines	European USSR	Urals	Siberia	USSR
Average length (km)	181	206	218	200
Maximum length (km)	297	414	340	414
Maximum capacities (mW)	300–920	—	200–710	—
Average maximum (mW)	640	—	480	—

Source:
I. S. Davydov, I. Ya. Mel'zak, and A. N. Sherentsis, *Elektrichestvo*, 1971, No. 2, pp. 7, 8.

I have found no comparative Soviet cost analysis of 330 or 750 kV schemes. But on the tenuous assumption that relative costs in the Soviet Union are similar to those in the USA (see Fig. 5.3 below) the case for 750 kV as against 330 kV for distances of the order of 1,200 km (the proposed length of the L'vov–Donbass line in the Southern IPS) seems overwhelming.

The extent to which 750 kV and other voltages are planned to grow in importance over the next decade in the Soviet Union is shown in Table 5.3. Over the decade 1970–80/82 only just over 4,000 km of 750 kV lines are to be introduced, in contrast to about 10,000 km of 500 kV (i.e., a rate of 500 kV construction roughly equalling that in the sixties (see the Appendix, Table 5A.1)). It seems, moreover, that only two 750 kV lines were to be constructed by 1975, one already under construction in the Southern IPS, and another linking the Central IPS with the North Western IPS.[55] In fact the 750 kV Donbass–West Ukrainian line was not yet complete by April 1975, though completion was still planned for the end of that year;[56] and by January 1975, construction of the second 750 kV line was apparently still at an early stage.[57] In comparison, six system-linking lines at 500 kV in European USSR were proposed, as Fig 5.2 shows.

Outside the Soviet Union, the only other countries to introduce voltages above 700 kV during the sixties were Canada and the USA. Most European countries envisaged moving to this new level, but only in the long term.[58] In the UK it was not until 1965 that existing 275 kV lines began to be uprated to 400 kV, and it was estimated that this level would suffice until the 1990s.[59]

In Canada, the initial role to be fulfilled by the 700–750 kV level was envisaged in 1962 as being to transmit power of about 4,000 mW from a new hydro station in the north to an industrial region in the south. Earlier, 500 kV had been proposed, but fresh analyses proved that 700 kV was cheaper.[60] Since Canada had been generally relying on 345–360 kV, the transition was to be

---

[55] (Eds.) A. S. Pavlenko, A. M. Nekrasov, *Energetika SSSR v 1971–75* (1972), see map, pp. 190–1.
[56] *BBC Summary of World Broadcasts: USSR Weekly Economic Report*, 1 April 1975.
[57] *Ibid.,* 28 January 1975.
[58] B. P. Lebedev, *op. cit.*, p. 34.
[59] S. Brown, *The Next 25 years in the Electricity Supply Industry*, CEGB, London, June 1971, p. 7.
[60] B. P. Lebedev, *op. cit.*, p. 34.

from the 300 kV to 700 kV ranges, as was the case in both the USSR and the USA. In the United States, one factor which led to the development of 765 kV was stated to be the lack of sufficient development of system-forming transmission lines, which in turn contributed to large system breakdowns[61] (see also data in the Appendix, Table 5A.3). In 1959, the USA had only 27 km of transmission lines above 300 kV per unit station capacity, compared with 83 km in the Soviet Union, 67 km in France, and 350 km in Sweden. Even allowing for differences in size between countries, the data still suggest a weak development of HVAC in the USA. It was not until the middle and late sixties that the situation began to be rectified with the introduction of 500 kV and later 765 kV (see Appendix Table 5A.3).

Table 5.3 *Lengths of lines at various voltages (in km)*

Voltage (kV)	1970	1975–6 (plan)	1980–2 (plan)
750	160	2000	4500
500	13140	17500	24000
330	13950	23000	33000
220	?	55000	58000

Source: P. S. Neporozhnii, *op. cit.*, p. 344.

The circumstances in which 765 kV was adopted in the USA were strikingly similar to those in the USSR. The AEP power system, like the Southern IPS, had been relying on 345 kV well into the late sixties and chose 765 kV.[62] By 1963 a 765 kV prototype line was in operation in the AEP system. In 1969, in an effort to keep pace with demand for electricity, the first fully operational segment was energised, and by 1972 about 1,100 km were in operation.[63]

Relative cost estimates of 345 and 765 kV for the AEP system are available and are illustrated in Fig 5.3. 765 kV was cheaper for all distances above 160 km at 2,000 mW, above 320 km at 1,500 mW, and above 480 km at 1,000 mW. Unfortunately comparable figures for 500 kV were not given. Thus for the AEP system, with an average transmission distance of 320 km, a required power transfer of 1,500 mW or more was necessary to make 765 kV cheaper.

No evidence has so far been found on the degree to which the Soviet Union depended on other countries for supply of 750 kV equipment and know-how. In 1963, Lebedev stated that 'we can for ourselves manufacture 750 kV equipment with higher requirements than those corresponding to international norms'.[64] At this time most countries, and in particular European ones, thought

[61] S. S. Rokotyan and S. A. Sovalov, *op. cit.*, p. 24.
[62] *Electrical Review*, 26 April 1968, p. 611.
[63] *Electrical and Electronics Abstracts*, January–June 1972, No. 15097, p. 798.
[64] B. P. Lebedev, *op. cit.*, p. 36.

Fig. 5·2 SYSTEM LINKING LINES FOR EUROPEAN U.S.S.R.

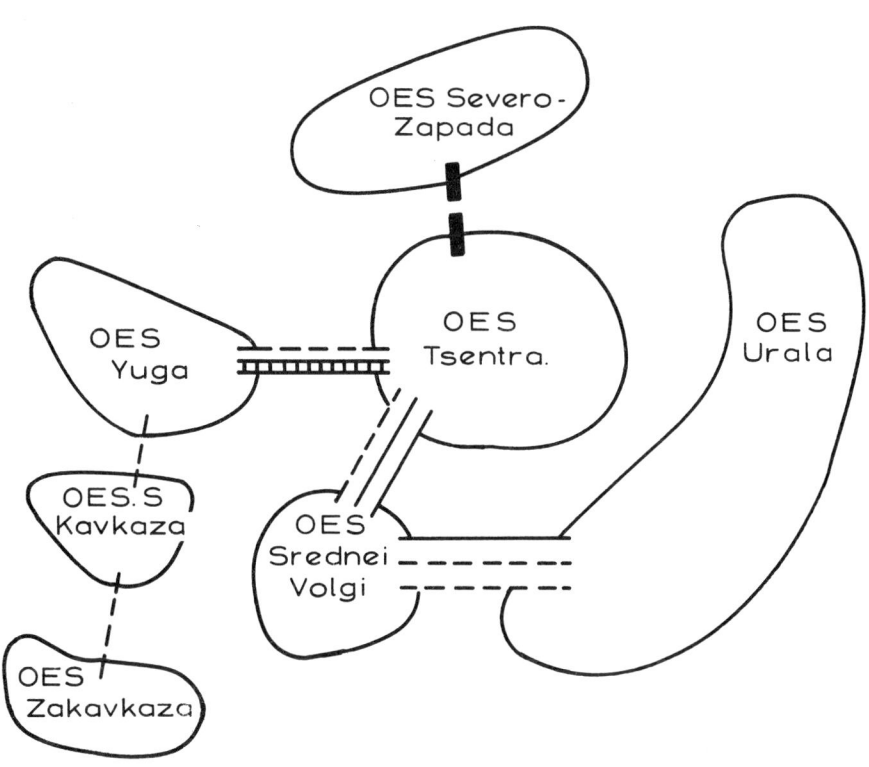

KEY

750 kv ▬ ▬

――――― 500 kv ― ― ― ― ―

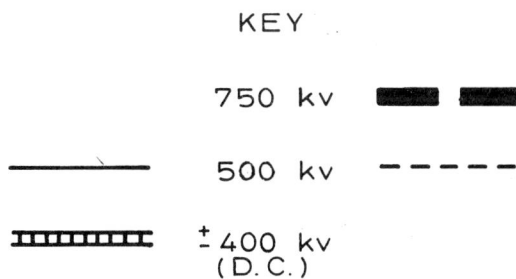   ±400 kv (D.C.)

Note: Only lines at 400 kv and above are shown here

Source: A.S. Pavlenko, A.M. Nekrasov, op. cit., p. 206

Fig. 5·3 RELATIVE COSTS OF 765 and 345 KV

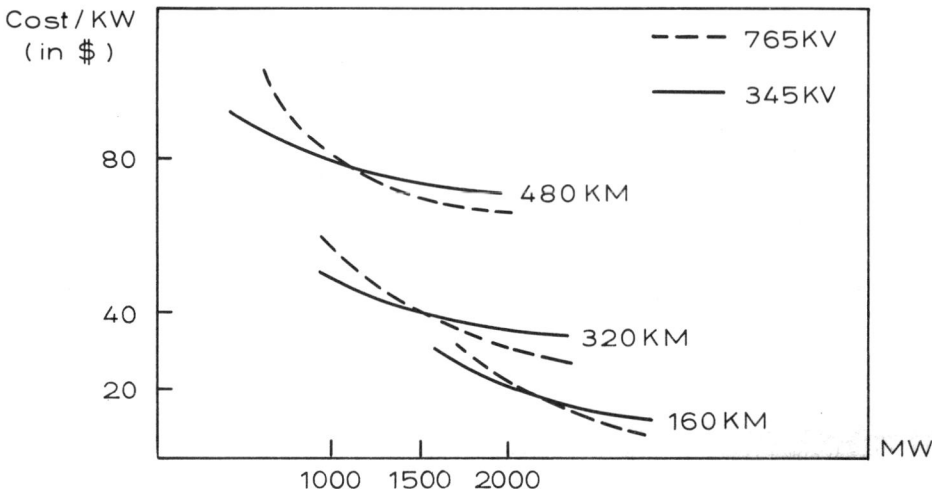

Source: Electrical Review 26 April 1968 p.611.
Note: curves are indicative only, being accurate only at break even points.

700 kV rather than 750 kV more suitable for them, and stated that they had no need to introduce 750 kV in the near future.[65] Only the USA, which at the time also preferred 700 kV to 750 kV, said that its factories were able and prepared to manufacture both 700 and 750 kV equipment. When the schemes were put into effect, however, it was the French company, Delle Asthom, which supplied circuit breakers to Canada (735 kV) and USA (765 kV),[66] and it seems likely that the USSR was able to produce the necessary 750 kV circuit breakers of its own accord.

During the 1960s the USSR continued to retain its leading position in the diffusion of HVAC as far as lines of 500 kV were concerned. In 1959, it was the only country to use 500 kV lines, which amounted to 2 per cent of the total length of line; by 1970 this proportion had increased to 5·1 per cent, a higher figure than the 4 per cent which the USA had then reached. For lines of 300 kV and above (including 500–750 kV lines), the position was markedly different. In 1959, 5 per cent of the length of Soviet lines came into this category; this was exceeded only by Sweden with 21 per cent and France with 8 percent. By 1970,

[65] *Ibid.*, p. 34.
[66] *Electrical Review*, 2 April 1968, p. 498.

the Soviet proportion had risen to 10·1 per cent; but this proportion was exceeded not only in Sweden (already 26 per cent in 1965) and France (13 per cent) but also in the UK (32 per cent), Germany, and the USA (both 11 per cent). There was thus a clear tendency in the 1960s for the Soviet Union to lose its leading position in the diffusion of the higher voltages in the HVAC range. The Soviet Union also tended to retain a remarkably high proportion of its line length at voltages of 160 kV and below: in 1959, it had a higher proportion of these lower voltage lines than Germany, Sweden and France; by 1970, the United States and the United Kingdom also held a smaller proportion of these lines than the USSR.

Two further interesting features emerge from Appendix Table 5A.3, which gives the lengths of HVAC transmission lines for every 1,000 mW of station capacity. First, except in the case of the UK, the total length of HVAC lines per unit capacity has remained roughly constant for each country during the sixties. This may reflect a certain 'appropriate' balance between network lengths and station capacities for each country for the period. Second, assuming this is the case, then Sweden, with roughly the same 'appropriate' balance as the USSR, relied to a much greater extent on voltage above 300 kV than did the USSR. This would be the case even assuming that the 1970 figures for Sweden are the same as those for 1965, and did not, as with other countries, show an increase in lengths of lines of 300 kV and above per unit station capacity. In this connection it must also be borne in mind that, for Sweden, the share of higher voltages is likely to be understated (see Appendix, note 3(d) to Table 5A.1).

HVDC

Although the first HVDC transmissions had been effected even before the Second World War, the 1954 Swedish installation was stated in a Swedish source to be 'definitely the first example of an installation of the type most likely to be adopted for other large-scale DC transmission'.[67] This appears, however, to be a patriotic exaggeration: by 1951, the Soviet Union had already set up an experimental installation from Kashira to Moscow longer than the 1954 Swedish line (see Table 5.4). By 1962, the USSR had commissioned the first part of a major intersystem DC link joining the Central IPS with the Southern IPS. Both installations were based on AC/DC-converter design philosophies different from those in the Swedish designs. The second Soviet line, in so far as it had no economic advantages and was the largest ever built anywhere, must, like the first line, be viewed more as an experience gathering stage in the development of HVDC. Although the first section was commissioned in 1962, the complete line was not brought to full projected power until 1965.[68]

Such great efforts were put into HVDC techniques in the Soviet Union owing to its serious imbalance of primary energy source distribution. Huge energy resources are located east of the Urals but population and capital are

[67] *Electrical Review*, 7 October 1955, p. 706.
[68] I. A. Syromyatnikov, Elektricheskie Stantsii, 1960, No. 11, p. 94.

concentrated in the west: eastern energy resources comprise about 90 per cent of total USSR reserves,[69] while the European regions have over 70 per cent of total population and produce about 75 per cent of total output.[70] In recent times there has been an increasing strain on European energy sources. 'The requirements of the European part for fuel can be sufficiently effectively ensured by using local high-calorie fuels ... only up to 1970'.[71] The predominant energy resources in the east are either hydro-power or low grade coal. In the former case energy has to be transported in the form of electricity, and in the latter case electricity is a more economical means of transport. Thus the problem is reduced to one of finding the most economical means of power transmission over distances of the order of thousands of kilometres. The cost estimates in Table 5.5 illustrate the advantages of HVDC power transmission in comparison with direct transport of fuel from various regions in the east and European USSR.

Thus, for both the Urals and the Centre, even allowing for a margin of error, the HVDC variants show definite savings as against transport of coal or gas. As regards the choice between DC and AC for long-distance transfer of electricity itself, many estimates of their relative costs have been made.[72] The most detailed estimates referring to the Soviet situation, prepared by ODP TEP, are presented in Table 5.6, for a line from Western Siberia to the Urals, capacity 4,800 mW ($28 \cdot 2 \times 10^9$ kWh) per year, and distance 2,400 km.

If these figures are correct, DC transmission is much more economical: the cheapest HVDC variant was calculated to be only about 60 per cent of the cost of the cheapest AC variant, both capital and current costs being lower.

The Soviet Union is conducting research and practical work on voltages of ± 750 kV DC, and, more ambitiously $\pm 1,100$ kV DC at the present time.[73] However, to date, only one long distance HVDC line is planned—that linking North Kazakhstan (Ekibastuz) to the Centre. The first stage of this line is due to be put on load in 1976,[74] and when in full operation in the 1980s, it will be the first step in forming the united grid system covering the whole USSR. Thus the period from the initial conception of the scheme to its implementation will have been at least 20 years, and it seems likely that the implementation of this highly ambitious project has received setbacks in the form of difficulties with mercury arc convertors and/or with the development of semiconductor convertor devices. Problems with mercury arc convertors have also figured prominently in Western experience (see p. 219 below). In fact these difficulties contributed to the lull in HVDC construction activity in the late sixties—a lull which is only now showing signs of disappearing. The major schemes at present under construction are shown in Table 5.4, the most important being the Coborra

---

[69] V. A. Shelest, *Ekonomika razmeshcheniya elektroenergetiki SSSR* (1965), p. 87.

[70] S. S. Rokotyan, *Elektricheskie Stantsii*, 1967, No. 12, p. 5.

[71] I. M. Vol'kenau, A. N. Zeiliger, A. I. Kolpakov, I. M. Markovich, P. E. Mironov, M. A. Sarkisov, *Elektricheskie Stantsii*, 1967, No. 1, p. 3.

[72] See for example, *The Electric Power Situation in Europe in 1955*, UN Economic Commission for Europe, Geneva (1957), p. 45.

[73] *Sots. Ind.*, 22 December 1973.

[74] *BBC Summary of World Broadcasts: USSR Weekly Economic Report*, 14 February 1975.

Table 5.4 *Details of first and some later HVDC transmissions brought into operation*

	Country	Date of Full Intro.	Voltage	Capacity	Distance	Places	Notes
1	France	1910	125 kV	20 mW	150 km	Montiers–Lyons	8 kV dynamos in series
2	Germany	Before 2nd WW	200 kV	15 mW	115 km	Elbe–Berlin	Mercury-Arc Convertors
3	Sweden	?	90 kV	6·3 mW	50 km	Trollhatten–Mallerad	MAC* Experimental
4	USSR	1951	200 kV	30 mW	112 km	Kashira–Moscow	MAC* Experimental cable
5	Sweden	1954	±100 kV	20 mW	90 km	Sweden–Gotland	MAC* Submarine
6	UK–France	1961	±100 kV	160 mW	64 km	Cross-Channel	MAC* Submarine
7	USSR	1962	±400 kV	750 mW	475 km	Volgograd–Donbass	MAC* Overhead line
8	New Zealand	1965	±250 kV	600 mW	40 km		
9	Italy	1965	200 kV	200 mW	104 km	Italy–Sardinia	
10	Japan	1965	±125 kV	300 mW			
11	Sweden–Denmark	1965	250 kV	250 mW	88 km		
12	Canada	1968	260 kV	312 mW		Vancouver	
13	USA	1970	±400 kV	1,440 mW		Celilo Sylmar	
14	Canada	(1975–76)	±450 kV	700 mW		Nelson River	MAC*
15	Mozambique	(1975)	±533 kV	1,500 mW		Caborra Bassa–Apollo	Thyristor Convertors

MAC – Mercury Arc Convertor.

Sources:
1 *Electric Power Situation in Europe in 1955*, UN Economic Commission for Europe, Geneva (1957), p. 46.
2 *Ibid.*, p. 47.
3 *Ibid.*, p. 47.
4 V. P. Pimenov and M. R. Sonin, *Elektrichestvo*, 1955, No. 7, pp. 93–9. Quoted in *Electrical Review*, 7 October 1955, p. 705.
5 *Situation and Prospects of Europe's Electric Power Supply Industry, 1962/63*, UN Economic Commission for Europe, New York (1964), pp. 71–2, Table 28.
6 *Ibid.*
7 S. S. Rokotyan, *Elektricheskie Stantsii*, 1967, No. 12, p. 5.
8 *Situation and Prospects of Europe's Electric Power Supply Industry 1962/63*, UN Economic Commission for Europe, New York (1964), p. 71–2, Table 28.
9 *Ibid.*
10 *Ibid.*
11 *Ibid.*
12 F. H. Last and R. M. Middleton in *CIGRE 24th Session 1972*, Vol. 1, paper 14–07, p. 2.
13 *Ibid.*
14 E. Jeffs in *Energy International*, October 1974, p. 17.
15 *Ibid.*

Bassa scheme in Mozambique, which, due to enter service in 1975, as well as having the highest voltage, will be the most powerful HVDC link in the world.

In the area of equipment design and manufacture several major technological difficulties had to be overcome. Although no economically viable HVDC circuit breakers have yet been developed, their application becomes important only in multi-terminal DC systems and, even here, alternative methods are

Table 5.5 *Relative costs of fuel transport and HVDC transmission*

	Variant	Incurred costs[a] (kopeks/kWh)
1.	HVDC 1,500 kV ($\pm$ 750 kV) from Ekibastuz (in Kazakhstan) to Centre, with GRES in Ekibastuz	0·48
2.	GRES in Centre	
	(a) using gas from Uzbekistan	0·53
	(b) using coal from Donbass	0·69
	(c) using gas from W. Siberia	0·66
	(d) using coal from Pechora (NW)	0·85
3.	HVDC 1,500 kV ($\pm$ 750 kV) Itat (W. Siberia) to Chelyabinsk (Urals), with GRES in Itat	0·46
4.	GRES in Urals	
	(a) using Kuznetsk coal	0·60
	(b) using gas from W. Siberia	0·58

Source:
I. M. Vol'kenau, A. N. Zeiliger, A. I. Kolpakov, I. M. Markovich, P. E. Mironov, and M. A. Sarkisov, *Elektricheskie Stantsii*, No. 1, 1967, p. 4.

Note:
a – incurred costs (*raschetnye zatraty*): $Z = C + E \cdot K$ where K are capital costs, C running costs, and E the reciprocal of the normative recoupment period, none of which were separately specified.

Table 5.6 *Relative costs of AC and DC: Urals to West Siberia*

Indicator	Transmission without intermediate substations		Transmission with two intermediate substations	
	AC 700 kV	DC $\pm$ 700 kV	AC 700 kV	DC $\pm$ 700kV
Capital cost (roubles/kW)	848	443	810	467
Energy loss (kWh $\times 10^6$)[a]	3·704	2·763	3·056	2·470
Cost of transmission (kop/kWh)	0·64	0·36	0·61	0·39

Source: Elektricheskie Stantsii, 1960, No. 11, p. 93.

Note: a – presumably this refers to kWh $\times 10^6$ per year.

possible.[75] In single point-to-point connections (the only application of DC to date), the switching function can be fulfilled by control of the convertor valves in stations at either end. Similarly, the design of HVDC transformers is still in its most primitive research stages—in the West at least.[76] In practice, therefore, DC has to be inverted to AC before voltages can be transformed.

While difficulties in the areas of DC circuit breaker and transformer design could be circumvented, the major obstacle to the development of HVDC lay in the design of convertor station equipment. Initially all such stations used mercury arc rectifiers and invertors. In the West, ASEA, the chief Swedish electrical manufacturer, carried out much of the initial development in this area and supplied convertor equipment for links from Sweden to Gotland,[77] from England to France,[78] from Italy to Sardinia,[79] and in Canada.[80] There is no evidence that the USSR relied on ASEA in any way for Soviet DC installations. Soviet development of HVDC convertor units was based on a policy of using single anode mercury arc valves,[81] while ASEA had pioneered the development of multi-anode types.[82] Since the USSR and Sweden were the first countries to construct HVDC installations, it must be presumed that the Soviet Union was by and large self-sufficient in the field of convertor development. In the late sixties in both the USSR and the West experience with mercury arc rectifiers tended to bring out serious problems. Zhimerin reports that in the Soviet Union these devices remained unreliable and expensive.[83] In the West it was admitted that many problems afflicted the early mercury arc schemes.[84] As a result increasing attention was paid to the replacement of mercury arc devices with thyristors, and by the late sixties the latter technology became a practical possibility. An English engineer, writing in 1969, gave an account of the advantages of thyristors over mercury arc valves. These included reduced maintenance, an ability to cope with large temperature variations, dispensing with expensive building structures and ventilation, space-saving, a simpler cooling system, and reduced protection requirements. On the other hand, thyristors involved higher losses.[85] By 1967 thyristors had virtually replaced mercury arc valves in industrial applications, but the only semiconductor device in an HVDC transmission application was a single experimental ASEA 50 kV-type installed in one leg of the mercury arc convertor station in the Sweden–Gotland link.[86] At this time 60 kV was seen as the upper voltage at which thyristors might compete. However, a year later in 1968, Siemens, a German company, successfully tested a 120 kV thyristor

---

[75] See A. K. David, *Proceedings of the IEE Power Record*, September 1972, p. 842.
[76] See *Electrical Review*, 22 January 1971, p. 129.
[77] *Electrical Review*, 18 August 1967, p. 234.
[78] *Ibid.*, 25 March 1960, p. 561.
[79] *Ibid.*, 18 August 1967, p. 234.
[80] *Ibid.*, 29 January 1965, p. 149.
[81] *Ibid.*, 18 August 1967, p. 234.
[82] *Ibid.*, 18 August 1967, p. 234.
[83] D. Zhimerin, *Sots. Ind.*, 21 June 1970.
[84] E. Jeffs, *Energy International*, October 1974, p. 17.
[85] *Electrical Review*, 19 December 1969, p. 908.
[86] *Ibid.*, 18 August 1967, p. 234.

convertor valve[87] and in 1971 a zero-length DC line, equipped throughout with 80 kV thyristor valves, was under construction in Canada.[88] This was the first installation to use thyristors completely for an HVDC transmission application.

Since 1968, all HVDC schemes ordered for application in the West, except one, were designed with solid-state convertors, although these were low voltage applications, the thyristor still not being able to compete with mercury arc devices at high unit voltages and currents. Nevertheless, at the present time, competition in the West between mercury arc and thyristor convertors is fierce. On the one hand, GEC, the main manufacturer of mercury arc convertors, claims that problems have largely been overcome and that performance has improved dramatically.[89] On the other hand, other convertor manufacturers such as ASEA, General Electric and Siemens-AEG-BBC have concentrated on thyristor development and produce mercury arc units only for replacement. Thus most manufacturers see the future in thyristor development, and, once semiconductors can be shown to work at similar ratings to mercury arc devices, then the eventual phasing out of the latter convertor types seems likely.

In the Soviet Union, following work done by the Krzhizhanovskii Energy Institute, two prototype thyristor units were produced. One was installed in a convertor station on the Volgograd–Donbass line and began operating in late 1973. In this application the thyristor bridge unit was rated at 100 kV, 900 amps, and operated in series with existing mercury arc valves.[90]

The evidence suggests, therefore, that developments in Soviet convertor technology lagged behind progress in the West, although a more accurate assessment requires further investigation of the relative success of the various schemes in operation. The gap between the first experimental Western and Soviet thyristor applications was up to six years. The diffusion of this technology was much larger in Western countries after ASEA, General Electric and Siemens had pioneered its development,[91] while in the Soviet Union no diffusion has yet taken place.

RECENT TRENDS IN HVAC

Since the beginning of the 1970s questions of raising voltages to new levels in the UHV range have increasingly been debated and investigated both in the West and the Soviet Union, and research facilities have been set up in the USSR and in several Western countries (including the USA, Canada, England, France and Italy).

In the Soviet Union, equipment operating at 1,150 kV was due to be constructed in the 1971–75 plan.[92] 1,150 kV is seen primarily as a replacement for 500 kV when this level becomes too low to handle increasing power

---

[87] *Ibid.*, 5 April 1968, p. 499.
[88] *Ibid.*, 22 January 1971, p. 119.
[89] E. Jeffs, *op. cit.*, p. 17.
[90] *Ibid.*, p. 20.
[91] *Ibid.*, p. 20.
[92] A. V. Shtern, *Energeticheskoe Stroitel'stvo*, 1971, Nos. 10–11, p. 138.

transfer. Thus, in one type of application, 1,150 kV could be used as a new intra-system maximum voltage level. The possibility is also recognised in the Soviet Union that UHVAC could be used also as a means of interconnecting one power system to another, so that it would compete with HVDC as a means of long-distance transmission between systems. The whole issue is hotly debated. At one extreme, doubt is expressed as to whether HVDC can compete at all with UHVAC.[93] The opposite and more frequently expressed view is that HVDC still retains its position particularly where large distances are involved.[94] In practice, the proposed line from Ekibastuz to Tambov at ± 750 kV DC is going ahead as planned and UHVAC does not seem to have been a challenge to its development. But a pilot line at 1,150 kV AC was due to be started towards the end of the ninth five year plan (1971–75),[95] and it was stated that intersystem connections are also to be made at 1,150 kV AC linking Central Siberia, North Kazakhstan and the Urals at some stage in the future.[96] It seems that these lines are to act mainly as a support to the HVDC scheme, and possibly provide a second line of attack should difficulties arise with HVDC.

Thus, the development of UHVAC in recent years has somewhat complicated the situation. UHVAC has been seen as both an intra-system transmission technique to replace 500 kV and eventually 750 kV, and a means of intersystem connection.

In the West, views differ on the appropriate voltage level to be employed. In 1972 delegates at the CIGRE meeting proposed that 1,200 kV should be adopted for international transmission standardisation. Italy, however, prefers 1,050 kV and the Canadians 1,100 kV.[97] Although such differences may seem inconsequential, in fact they conceal appreciable technological differences, and are also related to the needs of each country and voltage levels of existing systems. It is generally recognised that new transmission networks should be designed to operate at roughly 1·5 to 2·5 times the voltage of the existing network.[98] Thus, for instance, for the UK with its present 400 kV system, a jump to 1,200 kV would be too large a step, and an intermediate level of 750–800 kV might be more appropriate, although much depends on future power transfers required. This in turn depends on future levels of concentration of capacity, and the siting of large power stations.

In Italy, however, the requirements of transmitting large blocks of power between northern and southern regions have led to the decision to adopt 1,050 kV to overlay the existing 400 kV network, and plans are laid to have a 1,050 kV test line in operation by 1980.[99] In Sweden and France with 400 kV and 220

---

[93] G. M. Aleksandrov, *Elektrichestvo*, 1971, No. 1, p. 4.
[94] V. A. Ryl'skii, *Ekonomika mezhraionnykh elektroenergeticheskikh svyazei v SSSR* (1972), p. 74.
[95] *New Times*, 1974, No. 9, p. 14.
[96] S. S. Rokotyan, *Energeticheskoe Stroitel'stvo*, 1970, No. 4, p. 61.
[97] P. R. Howard, *Energy Policy*, September 1973, p. 157.
[98] *Ibid.*
[99] *Ibid.*

kV main transmission voltages, an 800 kV level is considered to be the most appropriate next step.[100] Generally, therefore, European countries would prefer to see international standards set below the 1,200 kV level.

On the American continent two 1,100 kV test lines are planned in the USA,[101] and in Canada it is planned to use 1,100 kV to 1,200 kV lines to connect generating complexes in the north to towns in the south.[102]

In the West some of the largest constraints to the development of UHV are seen as environmental ones. Since transmission by underground cable is extremely expensive, overhead lines must continue to be the main transmission medium. Apart from the familiar visual pollution effects of even larger transmission towers than those used for EHV, the move to UHV has given rise to two new problems. These are the electrostatic effects on people, vehicles, etc., passing under UHV lines, and also problems of noise from conductors when weather conditions are poor. If appropriate solutions to both these problems are found, then the major environmental concern seems likely to be that of the visual impact of increased tower sizes.

Thus, UHV is seen as an important new development in both the USSR and the West, though insufficient evidence is so far available on the comparative technological level of the different countries.

Summary and conclusions

In 1966, a comparative study of technological progress in USA and USSR was published.[103] It concluded that in the field of electric power transmission the rate of progress had been greater in USSR than in USA, a conclusion based on figures showing the relative diffusion of EHV AC transmission lines during the fifties and early sixties.

By 1959, the diffusion of higher voltages in Soviet power systems was measurably greater than in American systems. However, the USSR took second place to Sweden—both as regards the diffusion of higher voltages, and as regards the balance between network lengths at HV and station capacity.

In respect of innovation in HVAC, the Soviet Union after being a follower of HV trends had emerged at the turn of the sixties as a leader, since it was the first to introduce 500 kV. This achievement is reduced in proportion to the degree to which 500 kV was a simple extension of 400 kV, which had earlier been introduced by Sweden. Nevertheless, the little available evidence would seem to indicate that the USSR was largely self-sufficient in its development of the equipment for both 400 and 500 kV. Thus, given that the Soviet Union was

---

[100] *Ibid.*
[101] *Electrical World*, 1 July 1974, p. 40.
[102] P. R. Howard, *op. cit.*, p. 157.
[103] M. Boretsky, 'Comparative progress in Technology, Productivity and Economic Efficiency: USSR vs USA', *New Directions in the Soviet Economy*, USA Congressional Joint Economic Committee, p. 167.

considerably behind the USA in the HVAC field at the start of GOELRO,[104] Boretsky's conclusion is confirmed by this paper.

By the end of the 1960s, West Germany and the United States had joined the Soviet Union and France as leading countries in the diffusion of the higher voltages. Sweden however, remained well ahead of these four countries, and the UK proportion was now apparently the highest in the world.

Canada had been the first to introduce a voltage above 700 kV, closely followed by the USSR, and then the USA. In all three countries, 700 kV and above was adopted in systems formerly operating at a maximum of 300–360 kV. Soviet projections for the near future indicate that 750 kV is to have less significance than 500 kV had or will have. We have suggested that this is because the possibilities of 500 kV, with which the Soviet Union have had 10 years of experience, are not yet fully exploited both as an intra-system and intersystem voltage, and because of possible competition from HVDC in the sphere of intersystem connection.

No evidence was found that the USSR relied on foreign assistance for the development of 750 kV. Thus, in the absence of other information, it must be presumed that the USSR was able to manufacture the necessary equipment of its own accord. This conclusion is strengthened by the Soviet Union's own insistence that it had the facilities to do this.

As regards HVDC, Sweden and the USSR emerged as leaders in the field. In the Soviet Union its introduction stemmed from a need to gain experience with a view to its possible application in helping to solve energy and population distribution problems. For large east–west power transfers, HVDC was, according to Soviet estimates, about two thirds the cost of equivalent HVAC transmission and significantly cheaper than direct transport of primary energy resources.

In the area of convertor equipment development, essential to HVDC, what evidence there is indicates that the USSR was able to develop the necessary HVDC equipment of its own accord. The difficulties the USSR had with mercury arc rectifiers were similar to those experienced in the West. The

---

[104] That the USSR was considerably behind the USA during the 1920s is shown by B. I. Weitz in the following figures taken from 'Construction of the electric power system of USSR in the light of modern electrification', in (Ed.) B. I. Weitz, *Electric Power Development in the USSR* (1937), pp. 71 and 73 (translation of *Elektroenergetika SSSR*, published for the USSR Committee for International Scientific and Technical Conferences, at the Krizhizhanovsky Power Institute, Academy of Sciences of Soviet Union):

Lengths of lines at various voltage in mid-1920s (in km)

Voltage (kV)	USSR (1927)	USA (1926)
220	n.a.	1690
160	n.a.	n.a.
132	n.a.	5020
110	628	12670
66	n.a.	19560
60	75	14170
44	n.a.	12090
35	791	n.a.
33	n.a.	38360
22	550	16250
Totals	2024	119810

development of the alternative thyristor convertors seems to be lagging in the Soviet Union. Several commercial applications of this technology are already in existence in the West, the Swedish firm ASEA being particularly active in the field. The Soviet Union, it seems, has not yet progressed beyond the prototype stage.

The development of UHVAC is at a very early stage throughout the world, and it is not yet possible to assess the relative Soviet strength in the field; it seems at least to be keeping abreast with developments in the West.

Thus, the USSR moved by 1960 from a position in which it was a follower of technological trends in the HV field to one in which it ranks among the leaders, both in AC and DC fields. This is not surprising to the extent that HV technology is one area in which a country's performance as an innovator is to a large degree a function of its geographical and economic problems. Given the Soviet goal of, and effort put into, electrification, and given the country's size and energy distribution problems, the pressure was greater than in most countries to adopt higher HVAC voltages and to move to HVDC. In the 1960s, however, the Soviet Union has in several respects lost its leading position in the development and diffusion of HV technology.

Appendix

Table 5A.1 *Lengths of HVAC transmission lines, 1959–70 (in km)*

		100–160 kV	%	160–299 kV	%	300–400 kV	%	500 kV	%	735–765 kV	%	Total	%
USA[3]	1959	169697	84	28270	14	4970[4]	2	0	0	0	0	202937	100
	1965	210649	76	58398	21	8656	3	476	0·2	0	0	278179	100
	1970	249397	67	83081	22	25741	7	11433	3	950	1	370602	100
USSR[1]	1959[5]	83300	80	15400	15	3100	3	2400	2	0	0	104200	100
	1965	140000	74	35100	18	7110	4	8170	4	0	0	190380	100
	1970	190500	71	50800	19	13950	5	13140	5	160	0·1	268550	100
UK[2]	1959	12800	82	2750	18	0	0	0	0	0	0	15550	100
	1965	16500	70	6300	27	880	3	0	0	0	0	23680	100
	1970	17500	56	3750	12	9800	32	0	0	0	0	31050	100
FRG[3]	1959	20634	74	7376	26	0	0	0	0	0	0	28010	100
	1965	28849	70	10650	27	1296	3	0	0	0	0	40795	100
	1970	34383	65	12680	24	6232	11	0	0	0	0	53295	100
Sweden[3]	1959	7589	51	4286	28	3140	21	0	0	0	0	15015	100
	1965	9364	49	4878	25	4975	26	0	0	0	0	19217	100
	1970	n.a.		n.a.		n.a.		n.a.		n.a.		n.a.	100
France[3]	1959	9114	40	11694	52	1370	8	0	0	0	0	22695	100
	1965	9295	32	16221	57	3218	11	0	0	0	0	28734	100
	1970	9338	28	19662	59	4398	13	0	0	0	0	33398	100

Sources:
1. All USSR figures are taken from (Eds.) A. S. Pavlenko and A. M. Nekrasov, *Energetika SSSR v 1971–1975 godakh* (1972), p. 186.
2. UK figures for 1959 are for the end of the 1958–59 financial year and are taken from *Electricity Council Report and Accounts 1960–61*, pp. 172–3. The figures have been converted from miles to kilometres. UK figures for 1965 and 1970 are for the financial years ending in those years, and are taken from *Electricity Council Report and Accounts 1970–71*, pp. 146–7. Again, the figures have been converted to kilometres.
3. 1959 figures were taken from *Situation and Prospects of Europe's Electric Power Supply Industry in 1960/61*, UN Economic Commission for Europe, p. 74, Table 31. Since 1965 and 1970 data were not directly available, they were compiled as follows:
   Length of lines in existence in 1963 were taken from the *Electric Power Situation in Europe in 1964/65 and its Future Prospects*, New York (1966), UN Economic Commission for Europe, pp. 42–3. To these figures were added lengths of lines brought into service in 1964 at the various voltages. These figures were taken from the reference just quoted. Then lines brought into service during 1965 were added. The 1965 additions were obtained from *The 20th Survey of Electric Power Equipment*, OECD (1967), pp. 32–4.
   A similar procedure was followed to obtain the position in 1970. Yearly additions were taken from the following:
   The 21st Survey of Electric Power Equipment, OECD (1968), pp. 32–4 (for 1966).
   The 22nd Survey of Electric Power Equipment, OECD (1969), pp. 31–3 (for 1967–8 additions).
   The 24th Survey of Electric Power Equipment, OECD (1971), pp. 25–7 (for 1969–70 additions).
   The figures are subject to the following reservations:
   (a) All USA additions from 1967 onwards inclusive are OECD secretariat estimate.
   (b) 1970 additions for West Germany are OECD secretariat estimates.
   (c) 1964, 100–160 kV additions for West Germany were not available. Therefore, the 1963 situation was used also for 1964 position at this voltage.
   (d) The most important distortion in the figures derives from the fact that they do not take account of the lines withdrawn from service or uprated. Since these are likely to be at lower voltages, the proportion of low voltage lines is probably overstated for these countries. Thus from the point of view of seeing which countries have the highest proportions of the highest voltages, the USA, France, Sweden and W. Germany are cast in a more favourable light than had the correct figures been available.
4. The figure is for lines at 287–345 kV.
5. 1959 USSR data are for 1960.

Table 5A.2 *Total power station capacity[1] (in mW)*

	1960	1965	1970[3]
USA	185815	254520	316710
USSR[2]	66721	115033	166150
UK	34800	45700	65700
FRG	25900	37790	48950
Sweden	8955	11659	15480
France	20220	26195	37000

Sources:
1 Data for 1960 taken from *The Electricity Industry Survey—13th Enquiry*, OECD, Paris (1963), p. 37.
  Data for 1965 taken from *The Electricity Supply Industry—18th Enquiry*, OECD, Paris (1968), p. 41.
  Data for 1970 taken from *The Electricity Supply Industry—20th Enquiry*, OECD, Paris (1971), p. 46.
2 All USSR figures taken from *Narodnoe khozyaistvo SSSR v 1970 godu* (1971), p. 179.
3 All 1970 figures are estimates: provisional.

Table 5A.3 *Ratio of line lengths to installed capacity[1] (in km/kW × 10⁶)*

		100–160 kV	160–299 kV	300–400 kV	500 kV	735–765 kV	Total
USA	1959	910	150	27	0	0	1087
	1965	850	230	34	2	0	1114
	1970	790	260	81	35	3	1131
USSR	1960	1250	230	47	36	0	1563
	1965	1220	305	62	71	0	1638
	1970	1150	303	84	79	1	1616
UK	1959	368	79	0	0	0	447
	1965	361	138	19	0	0	518
	1970	266	56	149	0	0	417
FRG	1959	800	274	0	0	0	1074
	1965	765	282	34	0	0	1081
	1970	702	260	127	0	0	1089
Sweden	1959	848	477	350	0	0	1675
	1965	825	430	438	0	0	1693
	1970	n.a.	n.a.	n.a.	n.a.	n.a.	n.a.
France	1959	450	580	677	0	0	1097
	1965	353	619	123	0	0	1095
	1970	252	530	119	0	0	901

1 Table 5A.3 was compiled by dividing figures of Table 5A.1 by corresponding figures in Table 5A.2. Figures are rounded upwards.

6. The chemical industry: its level of modernity and technological sophistication

This case study is part of a broader survey of science, technology and innovation in the Soviet chemical industry. The aim of this first part is twofold. First, in order to establish a general perspective, it is intended to outline the distinctive features of the chemical industry and how they differ from other branches of the economy in the USSR and elsewhere. Second, a tentative attempt is made to evaluate how far the Soviet chemical industry 'leads' or 'lags behind' those of the most advanced Western countries in terms of its pattern of output and its scientific and technological achievements. For this purpose various interdependent criteria are used, which correspond closely to those outlined in chapter 1 above. It should be emphasised that this chapter is not meant to be more than a general survey. The information provided is not detailed enough to facilitate an assessment of sub-branches of the industry; to a large extent the data are also open to alternative interpretations.

Some distinctive features of the chemical industry in industrialised societies

At an abstract level, the chemical industry has been described as

> the archetypal matter-transforming industry ... [and] ... a principal exemplar of process technology. The function of a matter-transforming industry is to produce chemical or physical changes in the nature of substances; in effect it is a kind of *external* metabolism, whereby man converts the substance of his environment into technologically useful forms.[1]

The products of this transformation are extremely numerous and varied. One of the principal characteristics of the chemical industry is its heterogeneity. To

[1] D. W. F. Hardie and J. D. Pratt, *A History of the Modern British Chemical Industry*, Oxford (1966), p. 4.

a larger extent than many other major industries it does not restrict itself to a comparatively small number of products or services. Apart from the vast range of organic and inorganic chemicals, which constitute the 'intermediates' or 'building blocks' for the industry, there are the end products themselves—fertilisers, pharmaceuticals, synthetic materials (plastics, fibres, rubbers), paints, dyestuffs, industrial gases and so forth. The manufacture of each of these major end-products entails a distinctive process using highly specialised equipment.

For the most part the consumer does not come into direct contact with chemicals themselves but only with the end-products of the chemical industry and those of other industries for which chemicals have been used as raw material, reagent, etc., at some stage in their production. The chemical industry, in fact, is one of its own best customers. On the basis of West European data, the OECD estimated the following consumption pattern for chemicals:[2] chemical industry (25 per cent), agriculture (10 per cent), other industries (40 per cent) and the consumer (25 per cent). A broadly similar pattern prevails in the USSR, where approximately one third of total output is used within the chemical industry itself for further processing. Of the remaining two thirds, 80 per cent is used by other industries and agriculture, and 20 per cent is sold direct to the consumer.[3] Since the end of the Second World War there has been a tendency in Western countries for large chemical manufacturers to branch out into end-products.[4] The emergence of ICI as a major

Table 6.1 *Change in the chemical intensity (khimkoemkost') of 'gross social product' 1959–66*

Indicator	Expenditure on chemical products (in roubles) per 1,000 roubles of output in industry, construction and agriculture		Growth (at 1959 prices) (1959=100)
	1959	1966 (at 1959 prices)	
Gross social product	20·0	27·9	139
Industry	25·7	33·1	123
Construction	9·9	16·6	167
Agriculture	8·6	22·5	262

Source: See n. 5.

[2] D. W. F. Hardie and J. D. Pratt, *op. cit.*, p. 256.
[3] K. I. Klimenko and E. V. Petrova, *Ekonomicheskaya effektivnost' tekhnicheskogo progressa v tyazheloi promyshlennosti SSSR* (1971), p. 187. It should be pointed out that these proportions are a function of industrial organisation and accounting as well as of technology and final demand. If the industry was largely monopolised and intra-firm transactions were not reported, sales by the industry to itself would *appear* to be minimal; however, given the nature of the industry, such internal transactions would, in practice, be substantial.
[4] 'Chemical Industry', *Encyclopaedia Britannica*, Vol. 5 (1971), pp. 378–82.

producer of paints and synthetic fibres is an example of this. In the USSR, since the rapid planned development of the industry in the 1930s from a relatively low level, there has probably always been a relatively close link within the industry itself between the production of intermediate and end-products.

During the twentieth century, especially after the rapid post-war growth of the synthetic materials sector, the chemical industry has assumed a position of pivotal importance in advanced industrial economies. Indeed, it is part of the Soviet concept of 'the scientific and technological revolution' that the degree of penetration of chemical products into all branches of the economy indicates the extent to which an economy is a 'progressive' one. The Soviet interest in this question is reflected not only in numerous calculations of the increasing replacement of traditional materials by synthetic materials in specific areas of industrial production but also in overall measures of 'chemical intensity' (*khimkoemkost'*): Table 6.1 suggests that the Soviet economy (especially in agriculture, and construction) has become more chemically intensive in recent years, largely as a result of Khrushchev's chemicalisation drive of the early 1960s.[5]

The importance of the chemical industry within the economy as a whole can be gauged, to some extent, by its share of total capital investment. In 1968, the capital stock of the British chemical industry amounted to £2,200 million. In that year the chemical industry accounted for 15 per cent of gross fixed capital investment in manufacturing industry; 39 per cent of this was concentrated in capital intensive heavy organic chemicals and 15 per cent in plastics.[6] Due to the differing definitions of 'industry' in British and Soviet statistics it is difficult to obtain a comparable figure for the USSR. If, however, we use Bush's figures for cumulative gross fixed investment in industry at 1969 prices,[7] excluding investment in the extractive, fuel and energy branches, the share of the chemical and petrochemical industry in total investment for the period 1971–5 amounts to 13·3 per cent. This corresponds fairly closely to the British figure and suggests that the Soviet chemical industry occupies a position of roughly comparable importance within the economy.

In addition to capital outlays there is the important question of *capital intensiveness*, which is one of the characteristic features of the modern chemical industry. In 1968, capital investment per employee in the British chemical

---

[5] E. S. Savinskii, *Khimizatsiya narodnogo khozyaistva i proportsii razvitiya khimicheskoi promyshlennosti* (1972), p. 48. This source also discusses the proportion of chemical products in the costs of particular branches of industry (p. 45) and the use of traditional and polymer materials in the construction industry (p. 46).

[6] B. G. Reuben and M. L. Burstall, *The Chemical Economy*, London (1973), p. 109. According to S. Hays, *The Chemicals and Allied Industries*, London (1973), pp. 104–5, capital expenditure in the British chemical industry amounted to £222·5 million in 1968, almost one seventh of the total in manufacturing industry. In the United States between 1957–62, capital investment in the chemical industry was $22 milliard, about 13 per cent of all capital investment in industry, somewhat higher than the comparable figures for UK and FRG at this time, which varied between 10–12 per cent. See V. S. Sominskii, in *Zhurnal Prikladnoi Khimii*, Vol. 37, 1964, No. 1, p. iv.

[7] Keith Bush in *Soviet Economic Prospects for the Seventies*, Joint Economic Committee of US Congress, Washington DC (1973), p. 40.

industry was 2·5 times greater than that of manufacturing industry as a whole.[8] In the United States' chemical industry the value of fixed capital per employee in recent years has also been 2·5 times greater than the equivalent figure for industry in general.[9] The indications are that the Soviet chemical industry, in relation to other Soviet industries, is rather less capital intensive than this. In 1970, for example, the value of fixed capital (*osnovnye proizvodstvennye fondy*) per worker was 13 thousand roubles compared with a global average for industry of 7·2 thousand roubles (i.e., 80 per cent larger).[10] These indices are shown in Table 6.2. Two other interesting features should be noted. First, fixed capital per unit of output (*fondoemkost'*) is substantially higher for the chemical industry, as one would expect, and the extent to which it is higher than the *fondoemkost'* for industry as a whole (49 per cent) corresponds closely to a comparable figure of 50 per cent for the United States.[11] Second, although gross output per worker in the Soviet chemical industry exceeds that of industry as a whole, it is relatively less impressive than the other indices, and its excess over the average for all industry in fact declined somewhat between 1960 and 1970, especially between 1960 and 1965. Whereas in the United States the productivity of labour in the chemical industry exceeds that of all manufacturing industry by approximately 70 per cent, the equivalent Soviet figure is only 22 per cent.[12] The Soviet relative decline is associated with the heavy investment programme of the early 1960s; by the end of the decade the new capacity had begun to be assimilated, but labour productivity in the Soviet chemical industry, relative to that of Soviet industry generally, still falls far short of the American level.[13]

A further clue to the relatively low level of labour productivity in the Soviet chemical industry may be gleaned from an examination of the structure of capital investment. In the USSR, it appears that a considerable proportion of capital investment has typically been swallowed up by construction and assembly work.[14] Although the proportion of equipment (the 'active part'—*aktivnaya chast'*) in current investment increased during the early 1960s it was still, in the mid-1960s, substantially lower than the level prevailing in Western countries. This difference is at least partly due to different definitions of 'construction' and to higher building costs in the USSR relative to capital equipment. It is thus, to some extent, a pricing phenomenon.[15] In the United

---

[8] S. Hays, *op. cit.*, pp. 104–5.

[9] E. S. Savinskii, *op. cit.*, p. 30.

[10] *Ibid.*, pp. 26–7. (There are, of course, considerable variations in capital intensity *within* the chemical industry. Among the most capital intensive branches are: chemical mining, basic chemicals, dyes and fibres.) See *ibid.*, p. 32.

[11] E. S. Savinskii, *op. cit.*, p. 30.

[12] *Ibid.*, pp. 135–6.

[13] It is not clear, though, how this should be interpreted. It *could* mean that, given the relative capital:labour ratios, the United States non-chemical industry is relatively *inefficient*. I am grateful to Dr P. Hanson for bringing this point to my attention.

[14] V. S. Belyaev and N. V. Pen'kova, *Zhurnal Vsesoyuznogo Khimicheskogo Obshchestva im. D. I. Mendeleeva*, Vol. 12, 1967, No. 2, p. 146.

[15] If the difference could be entirely explained by these factors it follows that the widespread concern about the situation in the Soviet literature can be accounted for only by confusion or by a deliberate attempt to present a striking argument in order to attract more resources for investment in equipment.

Table 6.2 *Capital intensity and labour productivity in the Soviet chemical industry*

	1960			1965			1970		
	All Industry	Chemical Industry	As % of all Industry	All Industry	Chemical Industry	As % of all Industry	All Industry	Chemical Industry	As % of all Industry
Cost of fixed capital per unit of output (*fondoemkost'*) (rouble/rouble)	0·54	0·75	139	0·60	0·91	152	0·61	0·91	149·5
Fixed capital per worker (*fondovooruzhennost'*) (1000 roubles)	3·9	7·0	179	5·5	9·6	176	7·2	13·0	182
Output per worker (1000 roubles)	7·3	9·3	128	9·1	10·6	117	11·8	14·4	122

Source:
E. S. Savinskii *op. cit.*, p. 72.

States in 1965, for example, machinery and equipment accounted for 70 per cent of capital investment in the chemical industry and a similar pattern prevailed in other major Western countries.[16] The *capital stock (osnovnye fondy)* of the Soviet chemical industry in fact became *less* equipment-oriented during the course of the seven year plan (1959–65); equipment declined slightly from 38·7 per cent of capital stock to 37·8 per cent.[17] This was because the proportion of equipment in capital stock pre-1959 was higher than in subsequent allocations of capital investment, which included the massive building costs for construction of new enterprises, carried out during Khrushchev's chemicalisation drive.[18]

Table 6.3 *Structure of capital investment in the Soviet chemical industry (%)*

Year	Construction/ assembly work	Equipment	Other
1958	65·7	24·4	9·9
1960	62·4	30·8	6·8
1963	60·3	33·2	6·5
1964	59·4	34·7	5·9
1965	60·1	34·5	5·4

Source: See n. 14.

Another distinctive feature of the chemical industry, associated with high capital intensiveness, is its low labour intensiveness (*trudoemkost'*) relative to other manufacturing industries. The labour intensiveness of the Soviet chemical industry is said to be significantly lower than that of industry as a whole.[19] This is reflected in Table 6.4, which analyses the distribution of production expenditures in various industries.[20] For the chemical industry as a whole, expenditure on raw materials and energy forms a relatively high proportion of costs, while wages constitute a strikingly lower proportion than in other industries or in industry as a whole; the variation between sub-branches of the chemical industry is also marked.

The fact that the chemical industry is a high technology industry also influences the structure of the workforce. In April 1971, for example, over one third of the workforce in the British chemical industry was classified as administrative, technical or clerical staff;[21] this was substantially higher than

[16] A. G. Dedov and G. L. Kopen'kov, *Zhurnal ... Obshchestva im. Mendeleeva*, Vol. 12, 1967, No. 2, p. 213.
[17] V. S. Belyaev and N. V. Pen'kova, *op. cit.*, p. 147.
[18] This policy will be examined in detail in our second volume.
[19] E. S. Savinskii, *op. cit.*, pp. 135–6.
[20] S. Z. Pogostin, *Ekonomika i organizatsiya khimicheskogo proizvodstva* (1971), p. 143; see E. S. Savinskii, *op. cit.*, for a substantially similar account.
[21] S. Hays, *op. cit.*, p. 51.

Table 6.4 *Relative structure of expenditure on the manufacture of industrial products (%)*

Branches	a	b	c	d	e	f
Industry as a whole	66·8	6·0	20·9	3·3	3·0	100
Coal mining	17·6	3·9	64·1	6·3	8·1	100
Engineering	54·3	4·7	33·2	4·1	3·7	100
Chemicals:	69·0	10·0	14·3	3·5	3·2	100
Basic chemicals	60·0	8·2	21·8	5·0	5·0	100
Soda	21·5	30·0	33·5	8·0	7·0	100
Dyestuffs	65·0	7·6	17·5	3·9	6·0	100
Nitrogen	30·2	30·0	22·0	10·8	7·0	100
Paints	85·3	2·6	8·7	1·1	2·3	100
Tyres	88·7	2·4	6·5	1·2	1·2	100

Source: See n. 20.

Notes:
a – raw materials.
b – fuel and energy.
c – wages (including deductions for social insurance).
d – amortisation.
e – other expenditure.
f – total expenditure.

the overall proportion for British industry and is consistent with a comparable figure of between 35–40 per cent for other major Western countries.[22] It seems reasonably clear that workers (*rabochie*), including those concerned with repairs and transport, constitute a much larger proportion of total employees in the Soviet than in the British chemical industry, and that the proportion of administrative and technical personnel is correspondingly smaller. Hemy has produced some estimates for the USSR,[23] which are set out below as % of total employees:

	1958	1965	1970
Production personnel (rabochie)	77·0	74·0	72·0
Engineers and technicians (ITRs)	15·0 ⎫ 23·0	19·0 ⎫ 26·0	22·0 ⎫ 28·0
Administration and office staff	8·0 ⎭	7·0 ⎭	6·0 ⎭

[22] *The Chemical Industry: 1965*, OECD, Paris (1967) (quoted by G. Hemy, *The Soviet Chemical Industry*, London (1971), p. 151).
[23] G. Hemy, *The Soviet Chemical Industry*, London (1971), p. 136. Hemy's calculations are based on figures for 1955 (in N. N. Nekrasov, *Ekonomika khimicheskoi promyshlennosti* (1959), pp. 331–2) and for 1964–5 (in N. N. Kalmykov, *Ekonomika sovetskoi promyshlennosti* (1967), p. 248. Soviet statistical yearbooks do not give figures for *total* employment in the chemical industry or for the occupational composition of its personnel. The only source that comes near to this is *Trud v SSSR* (1968), pp. 84–5, which quotes figures of 991,000 (*rabochie*) and 153,000 (ITRs) for the chemical industry in 1966. Thus, in 1966, the proportion of ITRs in total employment in the Soviet chemical industry (excluding administrative and clerical staff) was 13·47 per cent, a figure that is lower than Hemy's estimate for 1965.

Table 6.5 *Percentage of all employees classified as administrative, technical or clerical in British industry in 1971.*

All manufacturing	26·3	Chemical and allied industries	37·2
All engineering	28·0	Mechanical engineering	28·0
Vehicles	26·0	Instrument engineering	35·9
Metal goods	20·4	Electrical engineering	41·7

Source: See n. 22.

These indicate that in 1970, just over one quarter of employees in the Soviet chemical industry were in white-collar jobs.[24]

Another major characteristic of the chemical industry, which in the Soviet view is intimately connected with the unfolding of the contemporary scientific and technological revolution, is the higher rate of growth of the chemical industry relative to that of industry as a whole. It is customary in Soviet writings to express this relationship as a 'ratio of relative growth', *koeffitsient operezheniya*—a term which is difficult to translate neatly into English. Table 6.6 shows that the Soviet ratio of relative growth is somewhat lower than that prevailing in Western countries with the exception of Japan. This, of course, is not unconnected with the fact that the overall levels of economic growth in the USSR and Japan were substantially higher than growth rates in the other countries and therefore is not necessarily to the discredit of the USSR; a low *koeffitsient operezheniya*, under these circumstances, is a burden that many countries would be willing to shoulder. However, the Soviet authorities are evidently concerned that this index should be improved, particularly since the ratio declined from 1·68 in the first half of the 1960s to 1·43 in the second half,[25] when overall growth accelerated slightly; this does not emerge in Table 6.6 due to the selection of time periods. As a consequence of this relatively rapid growth, the share of the chemical industry in the total volume of industrial production (at 1955 prices) increased from 5·8 per cent in 1965 to 7·1 per cent in 1970.[26] This share was scheduled to increase in the ninth five year plan (1971–5), which anticipated a ratio between 1·52 and 1·66.[27]

We now turn to that feature of the chemical industry which most concerns us in this study—the very prominent role played by research, development and innovation. The relatively high cost of chemical *research* can be partly explained by the extensive explorations and searches that must be carried out into complex chemical reactions and the properties of the substances that they produce. This is the breeding ground for the future commercial development of

---

[24] Even in the capital-intensive plastics sector, in the mid-1960s, 76·1 per cent of *enterprise* staff were *rabochie* and only 12·5 per cent were ITRs: *Ekonomicheskaya entsiklopediya: promyshlennost' i stroitel'stvo*, Vol. 3 (1965), p. 675.

[25] E. S. Savinskii, *op. cit.*, p. 67.

[26] K. I. Klimenko and E. V. Petrova, *op. cit.*, p. 185.

[27] S. Z. Pogostin, *Ekonomika i organizatsiya khimicheskogo proizvodstva* (1971), p. 18, stated that during this period the chemical industry would grow by 70 per cent compared with a growth of total industrial production of 42–6 per cent.

Table 6.6 *The ratio of relative growth of the chemical industry (koeffitsient operezheniya)*

	1951–65	1963–70
Great Britain	1·68[1]	2·1[3]
France	1·48[1]	1·75[3]
FRG	1·34[1]	1·9[3]
Italy	1·56[1]	2·0[3]
Japan	1·15[1]	1·0[3]
USSR	1·37[2]	1·49[4]
USA	1·70[2]	1·7[3]

Sources:
1 V. S. Sominskii, *Ekonomika khimicheskoi promyshlennosti* (1969), p. 15.
2 E. S. Savinskii, *op. cit.*, p. 158.
3 Calculated on the basis of figures quoted in S. Hays, *op. cit.*, p. 125.
4 K. I. Klimenko and E. V. Petrova, *op. cit.*, p. 184: this figure is calculated on the basis of the period 1960–9.

new products, but there is a large element of chance involved; the majority of chemical experiments do not yield anything of commercial value. The high cost of *development*, on the other hand, is due to the need for large-scale prototypes (compared with many other areas of technology where small-scale models can serve as prototypes)[28] to facilitate the transition of a chemical process from laboratory to full industrial scale. The scaling-up of chemical processes is more unpredictable than mechanical operations and requires a substantial amount of practical adjustment in addition to theoretical simulation. During the nineteenth century there were many failures and disasters resulting from indiscreet haste of manufacturers in setting up commercial installations without having first acquired the 'empirical accumulation of know-how'.[29] Christopher Freeman has assembled detailed case histories for 31 important innovations in the chemical industry. Two thirds of these cost £1 million from the first specification of the new process to commercial application (including the cost of the pilot plant but excluding the cost of the first industrial plant). Many innovations cost more than £5 million in R and D expenditures, particularly where a new product was involved as well as a new process, and over three quarters of these processes took more than four years to complete.[30] Jewkes also found that the cost of major innovations in the chemical industry (for example, the catalytic cracking of petroleum) was appreciably higher than that of most other industries, with the exception of the electronics and aerospace sectors (the development of the transistor in Bell Laboratories cost £28 million and the development of the Atlas missile cost over £3,000

[28] For example, in Germany and Britain, up to 1939, the early development of the jet engine cost only about $100,000 in each country; R. Schlaifer and S. D. Heron, *Development of Aircraft Engines and Fuels*, p. 90, quoted in J. Jewkes, D. Sawers and R. Stillerman, *The Sources of Invention*, London (1969), p. 161.
[29] J. Jewkes *et al., op. cit.*, pp. 156–7.
[30] C. Freeman, *National Institute Economic Review*, 1968, No. 45, p. 46.

million).[31] Thus, although small and medium-sized firms play a valuable innovatory role in the chemical industry and may be even more efficient in this respect,[32] the costs and risks of major innovations are borne by a relatively small group of corporations in each advanced country. In the United States, for example, the six largest firms account for 25–30 per cent of chemical R and D expenditures, while in West Germany, Bayer, BASF and Hoechst (the successor companies of the huge I. G. Farben combine) account for 50 per cent of the total.[33]

The chemical industry is, of course, a highly *research intensive* industry. Research intensiveness is conventionally measured in terms of R and D expenditure relative to turnover and the numbers of R and D personnel relative to total employment. The rankings of the British and American chemical industries, according to these criteria, are shown in Table 6.7. Three rather distinct categories can be identified: first, the aerospace and electronic sectors, which are by far the most research intensive; second, the instrumentation and chemical industries; and finally, the remainder of manufacturing industries, which lag well behind the four high technology areas. Another interesting feature of the chemical industry, which emerges from Table 6.7, is the relatively high proportion of R and D expenditure devoted to research in the chemical industry, referred to earlier.[34]

Absence of relevant data, inconsistencies and problems of statistical comparability make it difficult to obtain reliable Soviet figures. So far as the intensiveness of chemical R and D expenditure is concerned, Sominskii has referred tartly to 'an annoying lack of statistical data, which does not permit an accurate comparison to be made'. In his opinion, however, the proportion of resources devoted to R and D in the Soviet chemical industry is 'appreciably lower' than in Western countries.[35] To take another measure, Zaitsev and Lapin[36] state that, in 1968, 20 per cent of total allocations to science (*nauka*) were devoted to the fields of chemicals, petrochemicals and oil refining; this figure is large when we consider that in 1968, approximately 15 per cent of R and D expenditure in British manufacturing industry was devoted to chemicals,[37] but might be explained by the facts (a) that the *nauka* item as a whole probably excludes a substantial amount of costly expenditure on defence-related work and (b) that a large amount of expenditure under *nauka* goes to establishments outside the industry itself (for example, to institutes under the Academy of Sciences). When these factors are taken into account,

---

[31] J. Jewkes, *et al., op. cit.,* Appendix.

[32] Edwin Mansfield, in D. O. Edge and J. N. Wolfe (Eds.), *Meaning and Control*, London (1973), pp. 198–221.

[33] A. G. Dedov and G. L. Kopen'kov, in *Zhurnal Vsesoyuznogo Khimicheskogo Obshchestva im. Mendeleeva*, Vol. 12, 1967, No. 2, p. 214. Japan is an exception to this pattern; here R and D expenditure is distributed more evenly between chemical firms.

[34] The amount of medium- and long-term research is twice as high in the chemical industry as in industry in general, *Gaps in Technology: Plastics*, OECD, Paris (1969), p. 100.

[35] V. S. Sominskii, in *Zhurnal Prikladnoi Khimii*, Vol. 39, 1966, No. 6, p. 1223.

[36] B. F. Zaitsev and B. A. Lapin, *Organizatsiya i planirovanie nauchno-tekhnicheskogo progressa* (1970), p. 129.

[37] Based on S. Hays, *The Chemicals and Allied Industries*, London (1973), p. 99.

Table 6.7 Research intensiveness of British and US industries

	R and D expenditure as a proportion of turnover (%)		QSE engaged in R and D as a proportion of total employment (%)		Breakdown of British R and D expenditure (%)		
	UK/1958[1]	USA/1960[2]	UK/1968[3]	USA/1960[2]	Basic[4] Research	Applied[4] Research	Development[4]
Aerospace	21·3	17·7	1·68	7·3	0·8	37·8	61·4
Electronics, telecommunications and electrical engineering	6·1	10·5	2·21	3·7	3·2	14·0	82·8
Instruments	3·2	7·5	n.a.	3·8	n.a.	n.a.	n.a.
Chemicals	2·3	3·8	2·38	3·3	7·4	46·3	46·3
Mechanical engineering	1·5	3·6	0·29	1·8	0·9	18·8	80·3
Metal manufacture	0·5	1·2	n.a.	0·4	2·6	24·6	72·8
Total manufacturing	1·8	n.a.	0·45	n.a.	3·1	19·0	77·9

Sources:
1 *Report on the Census of Production 1968*, Part 156, Summary Tables, HMSO, London (1972), quoted in *Economic Trends*, No. 245, March 1974, p. xxxiv: variations in research intensiveness *within* the chemical industry are—pharmaceuticals (4·6 per cent of turnover), plastics (2·9 per cent), and general chemicals (2·1 per cent).
2 *Techniques in Sectional Economic Planning: the Chemical Industries*, United Nations Industrial Development Organisation (UNIDO), New York (1966), p. 46.
3 Calculated from figures in S. Hays, *op. cit*, p. 97.
4 *Ibid.*, p. 99.

the resulting figure may well produce a situation that is consistent with Sominskii's judgement.[38]

An estimate of research intensiveness in terms of manpower is also difficult to obtain since the Soviet statistical handbook does not quote a figure for the total number of employees in the chemical industry nor for the number of scientists (*nauchnye rabotniki*)[39] working in the industry. L. Kostandov, the Minister for the Chemical Industry, has stated recently that the number of *nauchnye rabotniki* in the branch exceeds 20,000.[40] Some estimates, which appeared recently in *Vestnik Statistiki* suggest that total employment in the Soviet chemical industry at the present time will be, very approximately, in the region of one and a half million.[41] On this basis, the proportion of professional R and D scientists in the total volume of employment in the Soviet chemical industry is about 1·3 per cent, much lower than a British figure of 2·38 per cent in 1968. However, these figures are not strictly comparable. The Soviet estimate would have to be adjusted upwards in the light of (a) a substantial amount of industrially-oriented chemical research carried on by *nauchnye rabotniki* outside the chemical industry, and (b) research and development work conducted at enterprise level within the chemical industry by staff, engineers especially, who are not classified as *nauchnye rabotniki*. It is quite obvious that there are considerable difficulties in obtaining comparable figures for both R and D expenditures and manpower, which arise out of the differing concepts of the 'chemical industry' from the viewpoint of the R and D process in the USSR and in Western countries.[42]

Arising directly out of intensive research and development activity, a final feature of the chemical industry should be mentioned—its rapid rate of product renewal.[43] This tendency is stimulated by the 'external dynamic' of new demands by other industries for chemical products (especially for products that can be substituted for costly traditional materials), but there is also a strong 'internal dynamic' peculiar to the chemical industry.[44] Modern chemical process routes are extremely complex and generally yield several subsidiary products, and so on. Thus, a spiral of product diversification is inherent in the economics of the industry. In some cases, the by-products have

---

[38] However, to offset the above adjustments a certain amount of R and D expenditure from industrial costs, credits and special funds within the discretion of the enterprise would have to be *added*.

[39] This category of workers is roughly equivalent to the British classification 'Qualified scientists and engineers' (QSE) working on R and D.

[40] L. Kostandov, in *Planovoe Khozyaistvo*, 1974, No. 3, p. 17.

[41] *Vestnik Statistiki*, 1974, No. 9, p. 94 quotes a figure of 1,667,000 for 1973. This figure only includes industrial personnel (*promyshlenno-proizvodstvennyi personal*), but it also includes the petrochemical industry. G. Hemy, *op. cit.* (1971), p. 136, calculated that total employment in the chemical industry in 1965 was 1·25 million.

[42] These complexities will be pursued in our second volume.

[43] An American study has shown that in terms of the ratio of sales of new products not in existence at an earlier date (1956–60) to total sales the chemical industry ranks third (16 per cent) after aircraft, ships and railroad equipment (35 per cent) and fabricated metals and instruments (17 per cent) (N. E. Terleckyj and H. J. Halper, *Research and Development. Its Growth and Composition*, New York, National Industrial Conference Board (1963), Table 19).

[44] *The Chemical Industry in the European Member Countries of the OECD*, OECD, Paris (1964), pp. 22–4.

become a point of departure for a whole new branch of industry, which may turn out to be more important than the manufacture of the original product. A more mundane explanation of rapid product renewal in the West can be found in the desire of firms to replace a product for which the patent has expired. Writing in 1967, Sominskii calculated that in West Germany 50 per cent of all chemical products supplied to the market by chemical firms were unknown 12 years earlier. Precise Soviet data are not available but Sominskii suggests that the rate of renewal in the Soviet chemical industry (an important indicator of successful innovation, in his view)[45] is appreciably lower than that prevailing in the USA, West Germany and Japan, and certainly not in excess of three to four per cent per year in terms of value.[46]

SUMMARY

In several respects the economic and organisational characteristics of the chemical industry are fundamentally different from those of manufacturing industry as a whole and constitute the hallmark of a high technology sector. The chemical industry is extremely heterogeneous in its range of products and technologies, it is capital intensive and wages form a relatively low proportion of running costs, white-collar workers are a relatively large component in total employment, the industry is research intensive (in terms of R and D manpower and expenditures), its product assortment is subject to rapid renewal and the rate of growth of the chemical industry tends to outstrip the overall rate of industrial growth. These dynamic factors push the chemical industry into a position of central importance in the economies of advanced countries and all branches of industry become increasingly dependent upon its products. The Soviet chemical industry follows this general pattern fairly closely, though less emphatically in terms of some indices than others.

From the point of view of research, development and innovation, the features described above make the chemical industry a particularly interesting and significant industry to study. It would be difficult to form any general impression about the technological level of a nation which purported to be highly developed, without taking the contribution of the chemical industry into consideration. This is emphasised by the fact that the chemical industry tends to be concentrated in the advanced industrial nations; six countries (USA, FRG, Japan, Britain, France and Italy) produce 75 per cent of the world's chemicals.[47] How Soviet chemical technology stands in relation to that of these other countries is an important element in the broader study of innovation.

---

[45] I am not so sure about the possibility of obtaining a meaningful measure of this. What is 'new' is not always easy to define, particularly in the USSR, where relatively well established products are often referred to as 'new' in order to justify higher prices.
[46] V. S. Sominskii, in *Zhurnal V.Kh.O. im. Mendeleeva*, Vol. 12, 1967, No. 2, p. 186.
[47] B. G. Reuben and M. L. Burstall, *op. cit.*, p. 126.

The diffusion of technology

MAIN PHASES IN THE DEVELOPMENT OF CHEMICAL TECHNOLOGY[48]

Elementary chemical techniques such as glazing, fermentation, soap making and the manufacture of paints and dyes have existed since ancient times. These techniques evolved through trial and error and were largely divorced from scientific explanation, although some of the work of the alchemists yielded useful information about the properties of particular chemical substances. The scientific basis of chemical reactions was established in the seventeenth and eighteenth centuries, notably through the contributions of Bacon (the definition of the element as the simplest form of matter), Lavoisier (the law of the conservation of matter), and Dalton (atomic theory). These represent the major premises of modern chemistry.[49]

PHASE ONE: THE INORGANIC CHEMICAL INDUSTRY

The modern chemical industry began to take root at the end of the eighteenth century and developed very rapidly during the nineteenth century. The 'industrial revolution' had increased the demand for chemical products, which were required in large quantities for the manufacture of soap, glass, cotton textiles and fertilisers. A number of timely innovations allowed the chemical industry to keep pace with this demand. The most important chemical products at this time were the alkalis, caustic soda and soda ash (sodium carbonate). Indeed, nineteenth-century writers often referred to the chemical industry as the 'alkali trade'[50] and in Britain it was the scene of a dramatic commercial battle between manufacturers using the Leblanc process and those using the more sophisticated Solvay process, pioneered by Ludwig Mond. The Leblanc manufacturers were eventually destroyed and were only able to hold on for so long because the hydrogen chloride gas emitted during the process could be used in the manufacture of bleaching powder; this advantage disappeared after 1894, when the Castner-Kellner electrolytic process enabled chlorine to be produced more cheaply. A second substance of vital importance during the nineteenth century was sulphuric acid, 'the pig-iron of the chemical industry', which was widely used in chemical processes (e.g., in the Leblanc process) and was also used in the manufacture of superphosphate fertilisers. The major innovation in the production of sulphuric acid, the lead chamber process, was developed by a Birmingham chemist, John Roebuck, at the end of the eighteenth century. This process was superseded during the last quarter of the nineteenth century, especially in Germany, by the catalytic contact process, which provided a very strong and pure acid (oleum) necessary for the growing

[48] This section, pp. 240–43, is based on Appendix Table 1, 'Landmarks in the evolution of the chemical industry to World War II', in R. Amann, 'The Soviet chemical industry: its level of modernity and technological sophistication', *CREES Discussion Paper* RC/C, No. 11, University of Birmingham, pp. 75–81. Unfortunately, for reasons of space, it was not possible to include this table in the present volume.

[49] See Appendix Table 2, 'Landmarks in modern theoretical chemistry prior to World War I', in *ibid.*, pp. 82–4.

[50] B. G. Reuben and M. L. Burstall, *op. cit.*, p. 13.

organic chemicals sector. Thus, by the end of the nineteenth century, most of the basic industrial processes for the manufacture of major inorganic chemicals had been established and the technology was widely understood.

PHASE TWO: THE EARLY ORGANIC CHEMICAL INDUSTRY

During the last quarter of the nineteenth century a new branch of the chemical industry began to emerge, which depended upon more sophisticated technology—the synthetic organic chemicals sector. Earlier products, such as dyes, fibres and drugs were organic in the sense that they were derived from natural living organisms. The new industry, on the other hand, was based on the element carbon, which could combine with other elements to form literally millions of compounds.

The birth of the organic chemicals industry can be traced back to W. H. Perkins' synthesis of the dye 'aniline purple' in 1856. But, despite pioneering work by individual British scientists, Germany became the dominant power in this and other branches of the chemical industry and by 1913 was producing 80 per cent of the world's output of dyestuffs.[51] One of the factors that lay behind this domination was the capacity to understand the theoretical complexities of organic chemistry and to apply this knowledge to industrial practice. Even in the manufacture of inorganic chemicals Ludwig Mond had often preferred to employ German chemists to work in his British plants[52] because of their more scientific approach. In the organic chemicals industry, knowledge of chemical theory and scrupulous experimentation were essential. In this respect, the training offered to aspiring British chemists was vastly inferior to that of Germany. Whereas Mond had received his first grounding in chemistry at a school where the great German chemist Bunsen had once taught, hardly any teaching of science existed in British schools. The first major scientific appointment in England was the election of Clerk-Maxwell to the chair of experimental physics at Cambridge in 1871.[53] Consequently, there were few science graduates in Britain and little or no training in the techniques of industrial research. British progress in organic chemicals was also impeded by the lack of pure concentrated sulphuric acid (the contact process had not displaced the lead chamber process in Britain at this time) and by the British patents laws, which allowed foreigners to register patents in London without attempting to exploit them.[54] In general, 'The British chemical industry remained in the nineteenth century while the German industry moved further and further into the twentieth.'[55]

Success by German firms such as Hoechst, Bayer and BASF in the development and manufacture of synthetic dyestuffs provided the foundation for further expansion into organic chemicals generally. Using their accumulated scientific knowledge and large financial resources these firms developed

---

[51] B. G. Reuben and M. L. Burstall, *op. cit.*, p. 16.
[52] J. M. Cohen, *The Life of Ludwig Mond*, London (1956), p. 147.
[53] *Ibid.*, p. 16.
[54] T. I. Williams, *The Chemical Industry* (1972), p. 68; B. G. Reuben and M. L. Burstall, *op. cit.*, p. 16.
[55] B. G. Reuben and M. L. Burstall, *op. cit.*, p. 17.

new pharmaceutical preparations (e.g., aspirin and salvarsan) and pioneered complex chemical processes such as the Haber nitrogen fixation process. This process, which handles gases at high temperature and pressure using highly specialised equipment, may be regarded as the first 'modern' chemical process; over £1 million was spent on its development before the first plant came onstream in 1913.[56] Moreover, Germany was the first country to make industrial research a national concern when it created the Kaiser Wilhelm Gesellschaft and its research institutes (now Max Planck institutes) in 1911.[57] Germany's scientific preeminence and the weaknesses which were exposed in Britain and France by war-time bottlenecks combined at last to provoke these governments into action. The research intensiveness of modern industry and that of the chemical industry in particular was now recognised; in 1915 the DSIR was created in Britain and this was paralleled in France by the formation of the Centre National de la Recherche Scientifique.[58]

PHASE THREE: THE SYNTHETIC MATERIALS INDUSTRY[59]

The synthetic materials industry has its origin in the nineteenth century when the properties of natural polymers, such as rubber and cellulose, began to be exploited. In 1830, for example, Goodyear developed the important vulcanisation process for making the characteristics of natural rubber more stable throughout changes in temperature. Most of the early plastics and manmade fibres (e.g., celluloid, rayon) were based on cellulose, usually in the form of wood-pulp. But the theoretical chemistry of polymer substances did not become understood until the inter-war period when the contributions of H. Staudinger (Germany) and W. H. Carothers (USA), for example, were particularly outstanding. This work paved the way for the manufacture of synthetic polymer substances, which found application in new plastics, fibres, adhesives, rubbers and paints.

Compared with the manufacture of inorganic chemicals and the early organic chemical industry, the 'polymer revolution' placed an even greater premium on research and development, and here again the German chemical industry occupied a leading place during the inter-war period. Christopher Freeman discounts raw material or cost advantages and attributes this German success almost entirely to technical superiority.[60] From 1925 to 1939 the huge I. G. Farben combine spent on average just over seven per cent of its turnover on research and development, more than it distributed in dividends, and employed over 1,000 qualified R and D personnel; it was without question one of the most research-intensive firms in the world and responsible for a large proportion of major innovations in the plastics industry.[61] The huge costs of R and D in synthetic materials (for example, it took eight years and cost $50

[56] *Ibid.*, p. 17.
[57] T. I. Williams, *op. cit.*, p. 82.
[58] *Ibid.*, p. 84.
[59] The dates of the first commercial production of the most important plastics and manmade fibres are set out schematically in Tables 6.19 and 6.21.
[60] C. Freeman, in *National Institute Economic Review*, 1963, No. 26, p. 22.
[61] C. Freeman, *op. cit.*, p. 33.

million to bring nylon to commercial production after the first industrial batch)[62] also led to the formation of large corporations in other countries. Firms such as Du Pont and ICI, formed in 1926, were in a better position to shoulder the substantial costs of innovation and were the source of a correspondingly large proportion of major developments. Extensive industrial research was needed to bring these ideas to commercial fruition and, although talented individuals could still make a contribution, patents statistics suggest that they increasingly did so as employees of large organisations.[63]

The Second World War acted as a further stimulus to the production of polymerised products, synthetic rubber, perspex and polythene being notable examples. But it was mainly in the post-war period with the mass production of synthetic materials using large-scale sophisticated plant that the 'polymer revolution' began to impinge on all branches of production. This is also associated with the growth of large petrochemical industries in Western countries which provided the organic intermediates for synthetic materials, previously based on ethyl alcohol (obtained through fermentation) or carbide-derived acetylene. The American petrochemical industry had begun in the 1930s but after the Second World War the West European countries began to create their own petrochemical industries. Petroleum-based chemicals caused a cost revolution in the synthetic materials industry[64] and rapidly became the main source of raw material; whereas in 1949 petroleum was the starting point for 11 per cent of British organic chemicals, in 1968 it was the basis of 87 per cent.[65]

'GENERATIONS' OF CHEMICALS?

This brief account of the major phases in the evolution of the chemical industry, suggests that groups of chemical products can be roughly distinguished by virtue of the technological complexity of the manufacturing process that produces them. Moreover, there has been a tendency over time for these processes to become more expensive to bring to fruition, due to the substantial costs of research, development and construction. If this perspective is accepted, it then becomes possible to assess, indirectly, the technological level of a particular nation's chemical industry by means of the composition of its overall output. Of course, this approach is rough and cannot be used inflexibly. Innovation in the inorganic chemicals sector, for example, did not cease absolutely at the end of the nineteenth century; during the twentieth century innovation has taken place in all branches of the chemical industry to a greater or lesser extent.[66] However, the extent to which a nation has a highly developed synthetic materials industry and petrochemical industry, in addition to a sound basis in traditional inorganic chemicals, can still be regarded as a

---

[62] B. G. Reuben and M. L. Burstall, *op. cit.*, p. 32.
[63] C. Freeman, *op. cit.*, p. 32 and J. Jewkes *et al.*, *op. cit.*, p. 88.
[64] In 1952, for example, ethylene based on ethyl alcohol cost £250 per ton to produce compared with £90 per ton for petroleum-based ethylene (B. G. Reuben and M. L. Burstall, *op. cit.*, p. 35).
[65] *Ibid.*, p. 33.
[66] Especially, for example, in pharmaceuticals (the most research-intensive branch of the chemical industry) and in the production of agricultural chemicals.

significant *general indicator* of its level of technological accomplishment in terms of diffusion.

The relevance of examining this general pattern of diffusion applies no less to the USSR, even though its pattern of development has been markedly different from that of Great Britain or Germany. Most historians agree that on the eve of the First World War the Russian chemical industry was still in its infancy. It is true that there were several notable developments during the nineteenth century. Russia was the first country in the world to have a large-scale industrial plant for the manufacture of sulphuric acid based on the catalytic contact process and the production of soda by the Solvay process was introduced as early as 1869.[67] But these advanced developments were anomalies when set against the overall structure and output of the industry at this time. The output of sulphuric acid in Russia in 1913 was eight per cent that of Germany and six per cent that of the United States; the output of dyestuffs (which were, in any case, based almost entirely on imported German intermediates) was three per cent of the equivalent German production; and the strategically important nitrogen industry barely existed.[68] The fact that a large segment of Russian industry was owned and managed by foreign businessmen gave rise to a number of negative features which inhibited indigenous development.[69] The chemical engineering industry in Russia was extremely feeble and most reactors, pumps and centrifuges were imported. Large quantities of minerals and basic chemical intermediates were also imported, while Russia's own rich deposits of coal and minerals remained relatively unexploited. It is remarkable, for example, that prior to the outbreak of the First World War not even one ton of benzol was obtained from the foreign-owned coking plants in the Donbas; apparently this was a deliberate policy on the part of German and Belgian managers to protect the market for exporting firms in their native countries.[70]

The undynamic attitude of the Tsarist bureaucracy towards the chemical industry received a rude jolt during the first few months of the war when it became clear that the industry was not equipped to supply the quantity of explosives that was required. New undertakings were brought into operation, sustained pressure was brought to bear on existing foreign firms and the output of benzene, toluene and xylene, for example, increased substantially up to 1917.[71] However, even these gains were arrested by the war damage and the further ravages of revolution and civil war. In 1921, much of Russia's chemical plant lay destroyed or idle; fewer than 15,000 men worked in an industry which produced about 40,000 tons of acids, 10,000 tons of alkalis and 25,000 tons of

---

[67] S. Z. Pogostin, *Ekonomika i organizatsiya khimicheskogo proizvodstva* (1971), p. 9.

[68] V. S. Sominskii, in *Zhurnal Prikladnoi Khimii*, 1967, No. 10, pp. 2109–15.

[69] In 1916–17, share capital (*aktsionnyi kapital*) in the Russian chemical industry has been estimated at 166·9 million roubles of which foreign capital amounted to 86·3 million roubles. N. P. Fedorenko and E. S. Savinskii, *Ocherki po ekonomike khimicheskoi promyshlennosti, SSSR* (1960), p. 19.

[70] V. S. Sominskii, *op. cit.* (1967): this story is also confirmed independently in V. N. Ipatieff, *The Life of a Chemist* (edited by X. J. Eudin, H. D. Fisher and H. H. Fisher), Stanford, Cal. (1946), p. 196.

[71] V. N. Ipatieff, *op. cit.*, p. 210.

salts. This state of affairs did not persist for long. During the New Economic Policy (NEP) in the 1920s, under Glavkhim,[72] the chemical industry in Russia began to be restored to its pre-war level. A prominent role in this was played by the concessions and technical agreements that were arranged between the new Soviet regime and foreign firms. By the late 1920s the task of restoration had been completed, and 77,000 workers were now employed in the industry compared with 50,000 in 1913.[73] During the decade of forced industrialisation that succeeded NEP, this base was developed very rapidly. In order to catch up with the leading Western countries, the creation and expansion of basic branches of the Soviet chemical industry were telescoped into a relatively short period; the development of an independent raw material base, the creation of substantial capacity for the production of basic inorganic and organic chemicals, fertilisers and even synthetic materials were all pursued simultaneously. Predictably, this policy gave rise to periodic bottlenecks due to discrepancies in the complex material balances underlying the industry. When we examine the pattern of output of the contemporary Soviet chemical industry what we are asking, in effect, is how far the USSR has been successful in assimilating processes of differing complexity within a relatively short timespan.

THE PATTERN OF OUTPUT OF THE SOVIET CHEMICAL INDUSTRY

Since the end of the 1930s, by which time the USSR had succeeded in laying a basic foundation in industrial chemicals, the Soviet chemical industry has continued to develop at a relatively rapid pace. The triumphant position of the German chemical industry in terms of its share of world output, came to an end with the Second World War, and the shares of the other major producing countries, with the exception of Japan and the United States have either fallen or remained constant. At the same time, the USSR has steadily increased its share of the world's output of chemicals.

In common with most Western countries, the rate of growth of the Soviet chemical industry exceeds that of other major branches of industry such as electric power, engineering, cement and steel (see Fig 6.1).[74] The exception is Japan, where the rapid growth of the automobile industry during the 1960s was an important factor contributing to the more rapid rate of growth of engineering and steel. In both the Soviet Union and Japan, the rate of growth of the chemical industry was appreciably higher than that of other major Western countries during the 1960s; despite the uncertainties of estimating in value terms there seems no good reason to dispute the Soviet official claim that it now possesses the largest chemical industry in Europe and the second largest in the world. A wide gulf still separates the two 'superpowers', however, and Soviet output per capita is evidently still far lower than in the other industrialised countries. In the next part of our study we show that the Soviet industry is also relatively backward in terms of its *pattern* of output.

[72] This was the department of Vesenkha responsible for the development of the chemical industry.
[73] *Trud v SSSR* (1968), p. 84. The figures refer to the territory contained within the USSR's 1939 frontiers.
[74] Based on figures in Appendix Table 6A.1.

Table 6.8 *Share of various countries in world output of chemicals (% in value terms)*

	1938	1949	1961	1964
USA	29	49·3	36·8	36·4
USSR	5·3	6·0	9·2	9·5
FRG	21·4[a]	4·9	7·9	7·3
Japan	5·6	1·9	6·9	7·0
UK	8·6	9·7	6·8	6·8
France	5·6	4·1	5·2	5·2

Source:
V. S. Sominskii, *Ekonomika khomicheskoi promyshlennosti* (1969), p. 18. There are some rather large discrepancies between these 1964 figures and those quoted in *Market Trends and Prospects of Chemical Products*, ECE, E69 IIE/Mim 6 (1965); in the latter, the Soviet proportion of world turnover is given as 11·8 per cent, that of Japan as 5·37 per cent and the combined figures for the USA and Canada as 34·8 per cent. These differences can probably be explained by the use of different exchange rates in making the estimates.

Note:
a – includes both East and West Germany.

TRADITIONAL INORGANIC CHEMICALS

This is a sector that performs well in comparison with Western countries. Since 1950, the total volume of soda ash (Na_2CO_3), a key alkali, produced in the USSR increased relatively rapidly and by 1972 the gap between the Soviet Union and the United States had been closed; whereas in 1950 the USSR was producing only 749 thousand tonnes of soda ash (roughly the same as the level of output in West Germany and France at this time) compared with 3,621 thousand tonnes in the United States, in the following 20 years the Soviet Union soared ahead of the West European countries and by 1972 had gained a slight lead over the Americans.[75] In per capita terms, a rather different picture emerges. Here, the Soviet Union has been catching up with the United States, West Germany and France (which is the leading per capita producer of soda ash) but it still lags behind these countries. Paradoxically, the only country that the USSR surpasses in terms of both total and per capita output is Japan; this is explained by the fact that the Japanese 'chemical miracle' has been heavily slanted towards the technologically sophisticated sectors of organic chemicals and synthetic materials where it has been aided by American know-how and stimulated by intense domestic competition.[76]

In the manufacture of sulphuric acid, a key acid, Soviet performance is respectable but relatively less impressive.[77] The Soviet rate of growth, between 1950–70, was the highest of all the major producing countries; in terms of total volume of output an ever widening gap has opened up between the USSR on the one hand and the West European countries and Japan on the other. But, in

[75] See Appendix Table 6A.2 for the full series of figures.
[76] B. G. Reuben and M. L. Burstall, *The Chemical Economy*, London (1973), p. 134. It should be noted, however, that the Japanese *rate of growth* of soda ash production was superior even to that of the USSR during the 1960s.
[77] This is based on Appendix Table 6A.3.

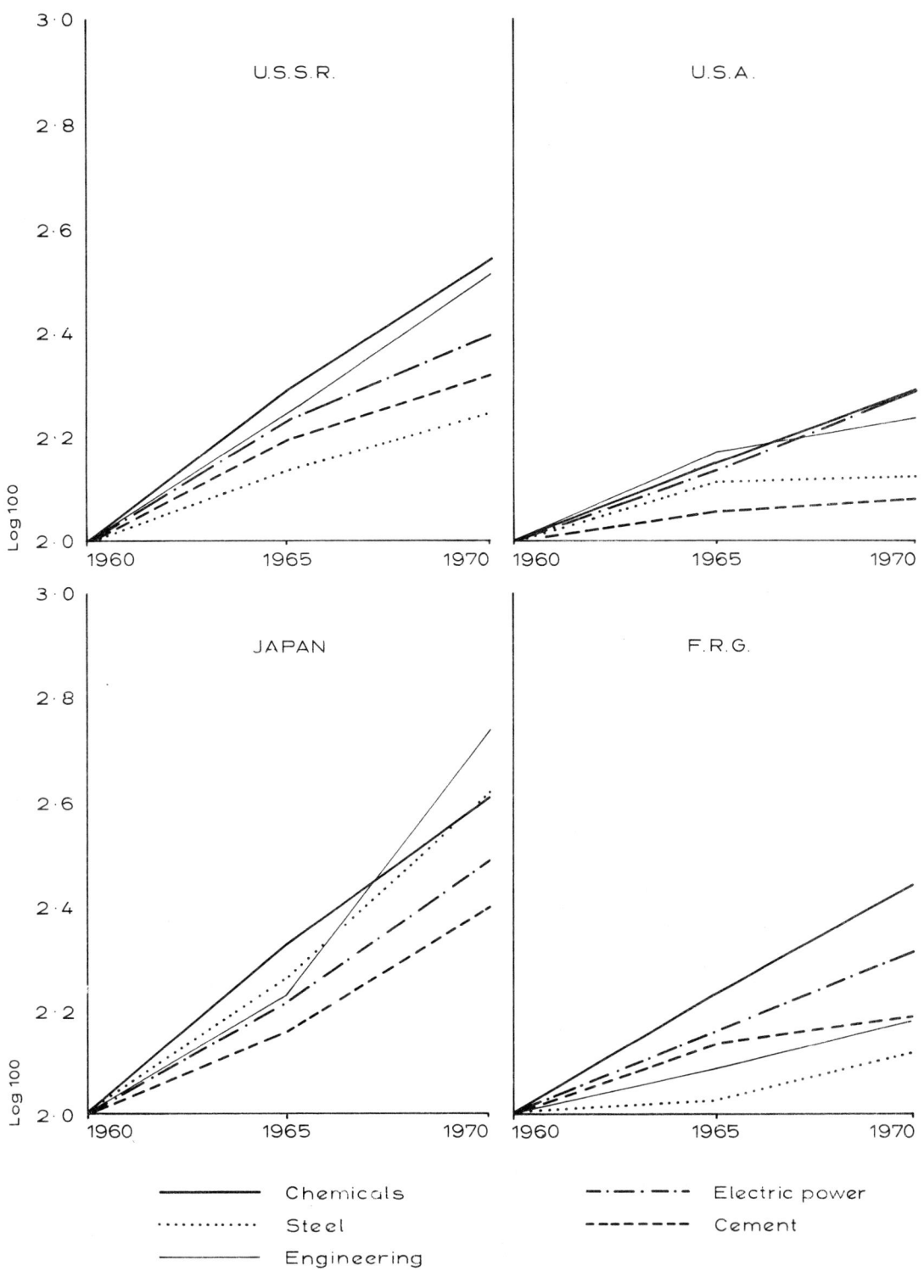
Fig. 6·1 COMPARATIVE GROWTH RATES OF KEY INDUSTRIAL PRODUCTS

Source: Appendix table 6A·1

per capita terms, the USSR still finds itself somewhat behind Japan and the West European countries, who are grouped together in an intermediate position, and substantially behind the United States. In 1972, US per capita output of sulphuric acid was more than double that of the USSR.

FERTILISERS

In the same way that soda ash and sulphuric acid can be taken to represent the traditional inorganic chemicals sector, superphosphate and nitrogenous fertilisers can be regarded as leading examples of traditional chemical end-products. The manufacture of mineral fertilisers has received considerable priority in the USSR:[78] the Soviet Union possesses and extremely rich raw material base and fertilisers are seen as an important means of improving the perennial problem of low agricultural yields. For nitrogenous fertiliser production, the USSR is able to draw upon its huge reserves of natural gas and petroleum; for phosphate fertilisers, it is able to exploit the large deposits of apatite and phosphorite (e.g., on the Kola peninsula) which collectively make the USSR the world's largest source of phosphate resources.[79]

Reliable comparative figures of fertiliser output are difficult to establish, particularly in the case of superphosphates.[80] Although most series inflate the Soviet figures in comparison with those of the United States, it nevertheless still seems probable that owing to the very rapid growth of superphosphate fertilisers in the USSR during the post-war period, the Soviet Union is now the largest world producer in both absolute and per capita terms.[81]

The manufacture of nitrogenous fertilisers requires rather more sophisticated technology. It depends in the first instance on the production of ammonia (usually via the Haber process or a refinement of it), which is an extremely soluble gas in water. The resulting solution of ammonium hydroxide combines with acids to form salts such as ammonium sulphate and ammonium nitrate. The Soviet rate of growth has again been extremely rapid relative to

[78] This aspect of chemicalisation will be discussed in our second volume.
[79] 'East Bloc Review', *European Chemical News* (Large Plants Supplement), 24 September 1971, p. 128.
[80] This can be seen clearly when the UN statistical series is compared with the much more detailed OECD series for Western countries. It emerges that Western countries, especially West Germany and Japan, produce large quantities of complex fertilisers (containing a phosphoric element) which are not classified as 'superphosphates'; Britain and France produce superphosphate fertilisers of different concentrations that do not consistently appear in the UN statistics. The most viable comparison is that between the USSR and the United States, but whereas the Soviet figure for superphosphates appears to include all types of phosphoric fertilisers, the American classification is much more narrowly defined. All these factors tend to inflate Soviet output relative to that of Western countries, so that only an approximate comparative measure can be obtained.
[81] See Appendix Table 6A.4. On the basis of percentage figures relating to the composition of Soviet fertilisers output, which were published in *Khimiya v Sel'skom Khozyaistve*, November 1974, it may be estimated that the output of single and triple superphosphates in the USSR in 1970 amounted to 2,118 thousand tonnes compared with the equivalent American figure of 1,921 thousand tonnes. However, in terms of production of *all phosphate fertilisers*, the respective figures are 3,940 (USSR) and 4,979 (USA). In order to obtain this latter comparison, the Soviet figure in *Narodnoe khozyaistvo SSSR* is assumed to be roughly equivalent to the total American output of phosphate fertilisers, recorded in *The Chemical Industry 1971–72*, OECD, Paris (1973).

that of major Western countries, particularly during the 1960s when a substantial amount of ammonia plant was imported from the West.[82] In 1950 the Soviet output of nitrogenous fertilisers was roughly in line with that of Japan and the West European countries, but by 1972 the USSR had shot ahead and was strongly challenging the United States. In terms of per capita output the USSR still lags behind the United States and France but has recently overtaken West Germany.[83]

HEAVY ORGANIC CHEMICALS

This is one of the key sectors of the modern chemical industry, providing the essential feedstocks for the manufacture of synthetic materials and other organic end-products. Many of the industrial processes involved here, especially those based on petroleum (e.g., catalytic 'cracking'), only became widely diffused after the end of the Second World War, stimulated by the rapid growth of plastics and synthetic fibre production. In terms of our earlier classification, the sophistication of petrochemical technology places it firmly in 'Phase 3' and the extent to which a country has assimilated these technologies is a sharp indication of its level of accomplishment. The paucity of Soviet statistics in this area is therefore a source of considerable frustration. Even Academician Fedorenko's excellent and detailed study of organic synthesis[84] fails to disclose these figures, though it surveys Western data with great thoroughness. An interesting exception to the traditionally reticent attitude of the Soviet statistical authorities in this matter occurred in 1973, when output figures for a number of organic chemicals were included in the Comecon Statistical Yearbook (see Appendix Table 6A.8). The figures refer almost entirely to basic aromatic hydrocarbons such as benzene, xylenes, phthalic anhydride, phenol and naphthalene. The general picture is that Soviet output in all cases represents less than half of equivalent American output; with respect to xylenes and phthalic anhydride, especially the former, it is substantially less than half.[85] While in the case of inorganic chemicals and fertilisers the USSR has succeeded in detaching itself from Japan and the leading West European countries and is challenging the USA in terms of total output, this is clearly not the case in the organic chemicals sector. Moreover, these figures can give only a very partial insight into Soviet performance in this area. At least two important elements are still missing. No output figures have been published as yet for key aliphatic hydrocarbons such as ethylene, acetylene, propylene and butylene; and a series crucial to the assessment of the level of process technology is missing: no firm indication is given of the structure of the raw material base (comprising oil, coal, natural gas and other sources) that is used in the manufacture of the chemicals for which statistics

---

[82] See pp. 262–5 below.
[83] See Appendix Table 6A.5 for detailed figures.
[84] N. P. Fedorenko, *Voprosy ekonomiki promyshlennosti organicheskogo sinteza* (1967).
[85] This is significant in two ways; first, because phthalic anhydride is an important feedstock for the manufacture of sophisticated polyesters and alkyd resins; second, because the figures suggest strongly that in the USSR phthalic anhydride is obtained primarily through the less technologically advanced naphthalene route.

have been published. The absence of satisfactory production figures suggests that the manufacture of heavy organic chemicals, and of petrochemicals in particular, is a relatively backward area in the USSR. But even without these figures it is possible to obtain a reasonably firm impression of the major characteristics of the industry. As early as 1948, 40 per cent of organic chemicals in the United States were derived from petroleum and natural gas. By the early 1960s the raw material base for the manufacture of organic intermediates had undergone rapid transformation in the other Western countries, despite their lack of native petroleum resources; in 1962, for example, 63 per cent of organic chemicals in Great Britain were derived from petroleum and in 1963, 85 per cent of French organic chemicals were obtained from this source. The early 1960s saw an especially rapid transformation of the Japanese raw material base; whereas in 1961, Japanese acetaldehyde was based almost exclusively on carbide-acetylene, by 1965, 71 per cent of this product was derived from petroleum.[86] In the Soviet Union, on the other hand, the main switch to petrochemicals began only in 1958–60[87] and at this time the industry was relatively underdeveloped, as Table 6.9 below shows. Only 7–8 per cent of the feedstocks used in the manufacture of synthetic fibres in 1960 were derived from petroleum, a remarkably low proportion, lower than the Soviet aggregate input figure for synthetic materials, which was itself relatively low. It is clear from Table 6.10 that even up to the present the Soviet Union has been much more dependent on coal and fermentation as sources of organic chemicals than is the case in Western countries; in this connection, it is interesting to note that the USSR is now the world's largest manufacturer of ethyl alcohol (the one basic organic chemical that regularly appears in *Narodnoe khozyaistvo SSSR*).[88] As Khrushchev amply demonstrated during the debate on chemicalisation, this pattern was not due to cost factors peculiar to the USSR;[89] the relative neglect of the petrochemical industry was partly due to technological conservatism. A British chemical engineer has suggested that 'Stalin, intent on building up heavy industry after the devastation of war and seeing coal miners as pillars of the proletariat, was either unwilling or unable to experiment with a petrochemical industry'.[90] It was also due in some measure to underdeveloped technology and perhaps also to the difficulty among planners of reconciling the priorities that should be attached to some petroleum products as sources of energy on the one hand or organic chemicals on the other.[91] Fedorenko echoes Khrushchev in claiming that 'The dominant pattern of technical progress in the modern chemical industry is the construction of highly economic (large scale) technological installations for the production of olefins and other intermediates necessary for the manufacture of diverse synthetic materials.'[92] In order to achieve this aim, the USSR

[86] N. P. Fedorenko, *op. cit.* (1967), pp. 344–6.
[87] *Ibid.*, p. 350.
[88] See Appendix Table 6A.6.
[89] The author intends to discuss this debate in his contribution to the second volume.
[90] B. G. Reuben and M. L. Burstall, *op. cit.*, p. 136.
[91] N. P. Fedorenko, *op. cit.* (1967), p. 352.
[92] *Ibid.*

imported very large items of plant and equipment during the 1960s (e.g., the ethylene and propylene crackers erected by Humphreys and Glasgow at Kazan and Polotsk)[93] and has been heavily dependent on the West in this area. Nevertheless, the rate of growth of 'basic organic synthesis' in the Soviet Union from 1960–70 was only marginally greater than that of the chemical industry as a whole. In comparison, the rates of growth of the organic chemical industry have been more rapid in major Western countries such as Japan and France than in the USSR.[94] With the exception of the United States (which already had a substantial base in 1960) the rate of growth of the petrochemical industry in Western countries has been particularly dramatic. Under these circumstances the Soviet Union continues to lag behind the West and given the size of its industrial economy remains a substantial importer of basic organic chemicals.

Table 6.9 *Proportion of petrochemicals used in the manufacture of synthetic materials (%)*

	1958	1965	1970 (plan)
USSR			
Aliphatic	25	56	75–80
Aromatic	12	20	60
Inorganic	16	50	60
	1958	1963	
USA			
Aliphatic	94	95	
Aromatic	70	86	
Inorganic	46	61	

Source: N. P. Fedorenko, *op. cit.* (1967), p. 28.

Table 6.10 *Structure of the raw material base for organic synthesis in 1970 (%)*

	USSR[2]	FRG[1]	Japan[1]	UK[1]	France[1]
Oil and natural gas	67·0	89·6	94·5	91·9	89·1
Coal	26·0	8·0	5·5	8·1	7·6
Other sources	7·0	n.a.	—	—	3·3

Sources:
1 *The Chemical Industry, 1971–72*, OECD, Paris (1973), p. 129.
2 G. F. Borisovich, Yu. T. Livshits and L. V. Koronnaya, *Khimicheskaya Promyshlennost'*, 1974, No. 3, p. 6.

[93] B. G. Reuben and M. L. Burstall, *op. cit.*, p. 136.
[94] See Appendix Table 6A.7.

SYNTHETIC MATERIALS

In this, the most technologically sophisticated branch of the modern chemical industry, the Soviet Union trails well behind the leading Western countries in terms of both total and per capita output (with the probable exception of synthetic rubber). Here again the pattern is different from that to be found in fertilisers and inorganic chemicals. In these more traditional areas the USSR had obtained fairly high levels of output by the beginning of the 1950s and in the subsequent period was either catching up with or overtaking the major Western producers, including in some cases the United States. On the other hand, the rates of growth of the synthetic materials branches have been extremely rapid in all countries, particularly during the 1960s. The USSR has not been able to improve its position substantially relative to the Western countries and in many cases its position has deteriorated. The manufacture of plastics and synthetic resins is a case in point. In 1957, the USSR produced 225·2 thousand tonnes of these products, considerably less than the United States (2,039 thousand tonnes) but in the same order of magnitude as the other Western countries. During the period up to 1972 the Soviet plastics industry grew more rapidly than the United States and British industries, but less rapidly than those of FRG, France and, particularly, Japan. As a result, the gap in both absolute and per capita terms between the Soviet Union and the latter countries has widened. In 1972, the per capita output of plastics and synthetic resins in the USSR was approximately one tenth of that in West Germany, one sixth that of Japan and one fifth that of the United States.

A further indication of the degree to which advanced technologies are diffused in the USSR is revealed in the composition of plastics output. A joint mission of the British Plastics Federation and the British National Export Council, which visited the USSR in March 1966, was particularly struck by the imbalance between the production of thermosets and the more sophisticated thermoplasts (e.g., PVC, polystyrene and the polyolefins) and judged that the

Table 6.11 *Output of polycondensation plastics per tonne of polymerised plastics*

	1950	1955	1960	1965	1970
USSR	3·05	3·7	5·0	3·4	1·9
USA	1·4	0·7	0·5	0·35	0·3
FRG	1·3	0·85	0·8	0·6	0·5
Japan	11·2	1·8	0·7	0·4	0·3
UK	2·8	1·1	0·7	0·5	0·4
Italy	5·1	0·7	0·6	0·4	0·3
France	3·8	0·7	0·6	0·4	0·3

Source: I. V. Rakhlin, *Naucho-tekhnicheskii progress i effektivnost' novykh materialov* (1973), p. 163.

output of thermoplasts was far below Soviet requirements.[95] For example, in 1959 the production of 'phenoplasts' (thermo-setting plastics) represented 36·86 per cent of plastics output in the USSR compared with 10·3 per cent in the United States;[96] in 1970, phenoplasts still accounted for 16 per cent of Soviet plastics output.[97] The distinctive general pattern of Soviet output is shown clearly in Table 6.11. Whereas the output of polycondensation plastics tends to be half that of polymerised plastics, or less, in the leading Western countries, these proportions are approximately reversed in the case of the Soviet Union.[98] Even in terms of making the transition from one group of products to the other, Soviet progress has been much less rapid than that of Japan, Italy or France. These general findings are confirmed by a previously unavailable series of production statistics, published recently in the 1973 Comecon Statistical Yearbook.[99] Here it emerges that the USSR is now the world's second largest manufacturer of the more traditional (and technologically less sophisticated) phenol-formaldehyde resins while at the same time it lags well behind the leading Western countries in the production of polystyrene, PVC and the polyolefins (polyethylene and polypropylene); in 1972, for example, Soviet output of polyolefins was approximately one tenth that of the United States, one seventh that of Japan and one quarter that of West Germany in *absolute terms*. In the case of all the more sophisticated polymerised plastics, Soviet output expanded from a very low level more rapidly than that of other countries over the whole period 1960–72. Some catching up has therefore occurred. The main peculiarity of the Soviet plastics industry, therefore, lies in the continued growth of traditional plastics and resins, which accounts for the pattern shown in Table 6.11. There seems to be no easily discernible rationale for this. During the last few years, NIITEKHIM and the Economics Division (*Otdelenie*) of the USSR Academy of Sciences have carried out a thorough investigation into the economic effect of introducing various plastics materials into different branches of the economy.[100] The main conclusion to emerge from the study was that polymerised plastics

---

[95] *British Plastics*, April 1966, pp. 223–4. The delegation also considered that the USSR was deficient in most kinds of plastics fabrication equipment (e.g., injection moulding machines, extruders, large capacity calenders, blow moulding machines, etc.). On the other hand, the journal subsequently ran a series of articles on Soviet plastics *research* which the editors presumably thought would be of interest to their readers; see *British Plastics*, June 1966, pp. 337–40; February 1967, pp. 86–9; October 1967, pp. 90–5; and March 1968, p. 86.

[96] See N. P. Fedorenko, *Ekonomika promyshlennosti sinteticheskikh materialov* (1961), pp. 537–612; this source gives a detailed breakdown of Soviet plastics output for the 1959 to 1965 (plan).

[97] I. V. Rakhlin, *op. cit.*, p. 163.

[98] The two most common criteria for classifying plastics are (a) the nature of the reaction used in the production of the plastic (addition polymerisation as against condensation polymerisation) and (b) the thermal properties of the plastic so produced (thermoplasts as against thermosetting plastics). Generally speaking, the more traditional materials such as phenoplasts and aminoplasts are thermosetting plastics produced through polycondensation; polyethylene, PVC and polystyrene, on the other hand, are all thermoplasts based on addition polymerisation. For further details and other criteria see G. I. Kutyanin, *Plasticheskie massy i khimicheskie tovary* (1971), pp. 42–7.

[99] See Appendix Table 6A.10.

[100] These detailed results are presented in *Plasticheskie Massy*, 1972, No. 1, pp. 3–11.

produce the greatest economic benefit and, correspondingly, that the most rapid growth in demand was for these products.

The pattern described above is also exemplified by the case of synthetic (i.e., non-cellulose) fibres, a sector of production which is openly acknowledged to be a problem one in the USSR. Russia began the 1960s with a level of output substantially below that of the United States and somewhat below that of the other Western countries. Output increased almost tenfold between 1960–70, faster than in any other Western country, and in this sense the USSR could be said to be catching up. From the late fifties onwards, the production of synthetic fibres also developed rapidly in the West, and by 1972 the USSR had not succeeded in overtaking any of the major Western countries in terms of total output. In per capita terms, the largest producer was the United States (11·6 kg per capita), followed by West Germany 10·7, Japan 10·3 and Great Britain 6·7; the USSR in 1972 produced only 0·96 kg of synthetic fibre per capita, a level achieved by all the major Western countries prior to 1960.[101] A further indication of technological level in this branch of production is the proportion of synthetic fibres in the total output of manmade fibres. In the West there has been a decisive and rapid switch during the last 20 years away from cellulose-based fibres towards those based on polymerised substances (terylene, orlon, etc.). The Soviet Union has also followed this pattern but is still relatively more dependent on non-polymerised raw materials. In spite of the fact that during the 1960s the production of Soviet synthetic fibres grew at a faster rate than artificial fibres, the output profile of the industry in 1972, like that of plastics, was relatively primitive compared with that of the leading Western manufacturing countries. Whereas in all the Western countries synthetic fibres accounted for well over half of total output, just under one third of Soviet chemical fibres were of the sophisticated polymerised kind, a pattern established in the USA by about 1957 and in other Western countries

Table 6.12 *The proportion of synthetics in total output of manmade fibres (%)*

	1950	1960	1965	1970
USSR	5·4	7·1	18·9	26·8
USA	8·9	39·7	53·8	72·3
FRG	0·5	18·5	39·2	68·8
Japan	0·4	21·4	43·2	66·3
UK	2·5	22·7	37·8	56·3
France	2·0	27·4	40·1	57·2

Source:
Figures obtained from Appendix Table 6A.11; the figures for France and the year 1965 come from the more comprehensive Appendix Table 11 in *CREES Discussion Paper* RC/C, No. 11, University of Birmingham.

[101] See Appendix Table 6A.11.

by the early 1960s. Moreover, as Table 6.13 shows, even within the sphere of synthetic fibres itself, the production of polyester and acrylic fibres was relatively underdeveloped. Thus, in the early 1970s the Soviet chemical fibres industry was still dominated by the viscose process, assimilated during the inter-war period. Again, as with plastics, this pattern is to some extent the consequence of a 'conservative'[102] policy of expanding the more traditional artificial (*isskustvennyi*) fibres sector at the same time as the synthetic (*sinteticheskii*) sector of the manmade fibres industry was undergoing rapid development. Thus, while Soviet output of synthetic fibres has proceeded at a rate which is roughly equivalent to that experienced by Western countries at a similar level of output, the output of artificial fibres has advanced more rapidly than that of most Western countries at a similar level. This trend explains the relatively slow changes in the composition of total output of chemical fibres, observed in Table 6.12.

Soviet sources do not give output figures for the production of synthetic rubber or even a growth rate that could be compared with Western statistics. One is, therefore, obliged to attempt rather crude estimates. These are based on output figures relating to the late 1930s, for which a few statistics are available.[103] In order to bring these figures up to date it is necessary to construct

Table 6.13 *The structure of Soviet manmade fibre production (%)*

	1960	1965	1970 (plan)
Total artificial fibres[1]	92.8	81.0	72.2
Viscose fibres[1]	85.1	72.4	64.2
Cuprammonia fibres[2]	5.1	4.3	n.a.
Acetate fibres[2]	1.5	4.3	n.a.
Total synthetic fibres[1]	7.2	19.0	27.5
Polyamide fibres[1]	6.6	16.3	18.7
Polyester fibres[1]	0.1	1.7	5.5
Acrylic fibres[1]	0.5	0.3	0.3

Sources:
1 L. I. Grameteeva, *Tekhniko-ekonomicheskie problemy razmeshcheniya vazhneishikh otraslei khimicheskoi promyshlennosti* (1970), p. 19.
2 E. P. Ivanova et al., *Ekonomika promyshlennosti khimicheskikh volokon* (1968), p. 23.

---

[102] Two alternative explanations for this phenomenon can be suggested. First, that the technological capability of the Soviet chemical fibre industry was such that artificial fibres had to continue to be developed rapidly because synthetics could not be substituted for them quickly enough. Alternatively, one could argue that, given the very low per capita output of *all* manmade fibres in the USSR relative to Western countries, there was a demand for artificial fibres which had yet to be met before the main switch to synthetics could take place. Both factors could be at work.
[103] According to V. Tikhomirov, *Oshibka Edisona* (1973), p. 103, output of synthetic rubber was 11,000 tonnes in 1934, 25,000 tonnes in 1935 and 40,000 tonnes in 1936; V. I. Kasyanenko, *Zavoevanie ekonomicheskoi nezavisimosti SSSR, 1917–1940* (1972), quotes a figure of 70,200 tonnes for 1937 (the figure which I have used in my estimates).

an appropriate growth rate. There is no straightforward way in which this can be done.[104] The only possibility is to construct a growth rate based on Soviet statistics of automobile tyre production on the grounds that this is the major outlet for synthetic rubber. This assumption is confirmed by Klimenko and Petrova, who show that approximately 70 per cent of Soviet synthetic rubber is used in the manufacture of automobile tyres, a proportion that remained relatively constant between 1958–70.[105] One important qualification should be added. Soviet tyre statistics are expressed in numbers of tyres produced and, since 1950, small tyres for light automobiles and motorcycles have become a larger proportion of the total.[106] On the other hand, from 1958–70 tyre production consumed a slightly diminishing share of synthetic rubber output;[107] but the net effect of using automobile tyres as a basis is probably to exaggerate the rate of growth of synthetic rubber output in the USSR.

The pattern of output and growth based on the above tentative estimates is almost the reverse of that to be seen in the field of plastics or synthetic fibres.[108] In the late 1920s the Soviet leadership decided that it wanted to free itself, as far as possible, from expensive imports of natural rubber, which would require large expenditures of precious foreign currency and would make the USSR dependent on foreign countries for supplies of a strategically important commodity. Thus, by the end of the decade, as a result of native research and development, the USSR had succeeded in creating a synthetic rubber industry that was relatively large for its time. Commercial production of synthetic rubber in the USA began in 1929, and by 1940 there were a dozen different rubbers being produced in the United States and Germany which were superior in their technical characteristics to the Soviet SKB type.[109] During the Second World War severance of natural rubber supplies led to the rapid increase in the output of synthetic rubber in Germany and the United States. America emerged from the war as the world's leading producer of synthetic rubber, a position that it has since maintained. Thus, in 1950, the United States and the Soviet Union were undisputedly the world's leading manufacturers in this field in terms of volume. The other Western countries did not begin to produce synthetic rubber in large commercial quantities until the late 1950s but, when they did begin, the rate of development was extremely rapid,

[104] Unfortunately the index numbers in *Narodnoe khozyaistvo SSSR* for the 'rubber-asbestos industry' cannot be used; this statistical category includes goods manufactured from natural rubber and asbestos products and according to the official definition the 'synthetic rubber industry' is not classified within this branch. In any case, these index numbers do not stretch back far enough in time. (N. P. Fedorenko, *Ekonomika promyshlennosti sinteticheskikh materialov* (1961), pp. 353–4).

[105] K. I. Klimenko and E. V. Petrova, *Ekonomicheskaya effektivnost' tekhnicheskogo progressa v tyazheloi promyshlennosti* (1971), p. 220.

[106] In 1950, tyres for heavy transport, tractors and agricultural machinery amounted to 86·4 per cent of the total whereas in 1969 this proportion had fallen to 66·9 per cent; the proportion of light automobile and motorcycle tyres correspondingly increased from 21·5 to 33·1 per cent. (Calculated from *Narodnoe khozyaistvo SSSR v 1969 godu* (1970), p. 216).

[107] K. I. Klimenko and E. V. Petrova, *op. cit.*

[108] See Appendix Table 6A.12. These estimates exceed those of the International Rubber Study Group in London, by, on average, 100 thousand tons per annum.

[109] A. C. Sutton, *Western Technology and Soviet Economic Development 1930–1945*, Stanford, Cal. (1971), p. 123.

especially in Japan. By the mid-1960s these countries had overtaken the USSR on a per capita basis and by 1972 Japan was threatening to overtake the USSR in terms of total volume of output. The rapidity of this advance can perhaps be contrasted with the relatively more sluggish performance of the Soviet Union in making up the leeway in the production of plastics and synthetic fibres. It suggests a greater flexibility and capability on the part of Western economies to generate and assimilate advanced technology in a relatively short space of time.

SOME GENERAL PERSPECTIVES

In the years following the Second World War the USSR has emerged as a major producer of chemicals and chemical products and it now possesses the second largest chemical industry in the world. As in most of the other leading industrial countries, the production of chemicals in the USSR has expanded at a faster rate than that of other key industrial commodities such as steel, engineering products, cement and electric power. As Fig 6.2 shows,[110] the growth profile of the Soviet chemical industry during the 1960s resembles that of chemical industries in Western countries; the most rapid rate of development has taken place in the synthetic materials sector, followed by fertilisers and inorganic chemicals. Moreover, apart from the phenomenal performance of Japan in the production of synthetic materials, the aggregate rate of growth of the USSR in each of these three sectors has exceeded that of Western countries. This picture is slightly deceptive because the 1960 Soviet production base of plastics and fibres in relation to Western countries was relatively much smaller than that of inorganic chemicals and fertilisers. This has meant that in absolute terms the Soviet Union has made rapid progress in the more traditional sectors of the chemical industry but its performance in the more sophisticated sectors has been less impressive. A rough impression of the overall level of sophistication in various countries can be obtained from Table 6.14, which lists a series of coefficients relating synthetic materials to inorganic chemicals.

No rash conclusions can be drawn from this about technological leadership, but it does suggest, in a striking way, how far the Japanese and West Germans have depended on the most advanced areas of chemical technology for the course of their development during the 1960s and the extent to which the Soviet chemical industry is still rooted in traditional technologies despite its large overall size.

[110] 'Synthetic materials' is an aggregate of plastics and synthetic fibres; 'fertilisers' is an aggregate of nitrogenous fertilisers and superphosphates; 'inorganic chemicals' is an aggregate of soda ash and sulphuric acid. In order to obtain this aggregate figure for the purpose of working out a joint rate of growth it was first necessary to calculate the output of each commodity in value terms. This was done by means of multiplying physical output by 1965 world prices (derived from *Vneshnyaya Torgovlya SSSR v 1965 godu*—average import/export prices for each commodity). In the case of synthetic materials, which are diverse, a few 'typical' products were chosen as a basis for the multiplier (synthetic staple fibre and PVC). Further sources of inaccuracy are: (a) the impact of price variations throughout the period (although 1965 prices were chosen to minimise this) and (b) the fact that Soviet foreign trade prices (much of this trade is done with Eastern Europe) are out of line with world prices. It is unlikely that any of these factors would substantially alter the picture presented above. The index numbers for organic chemicals come from Appendix Table 6A.7.

THE PATTERN OF SOVIET FOREIGN TRADE IN CHEMICALS AND CHEMICAL PLANT

A further insight into the diffusion of chemical technology in the USSR can be gleaned from its pattern of foreign trade. Here one can discover the extent to which the overall needs of the nation for chemical products are satisfied by domestic production and how far it is dependent upon imports from abroad. The pattern of this dependence is significant from a technological point of view. In this connection one should bear in mind that, in general, the advanced Western nations are net exporters of chemical plant and chemical products,

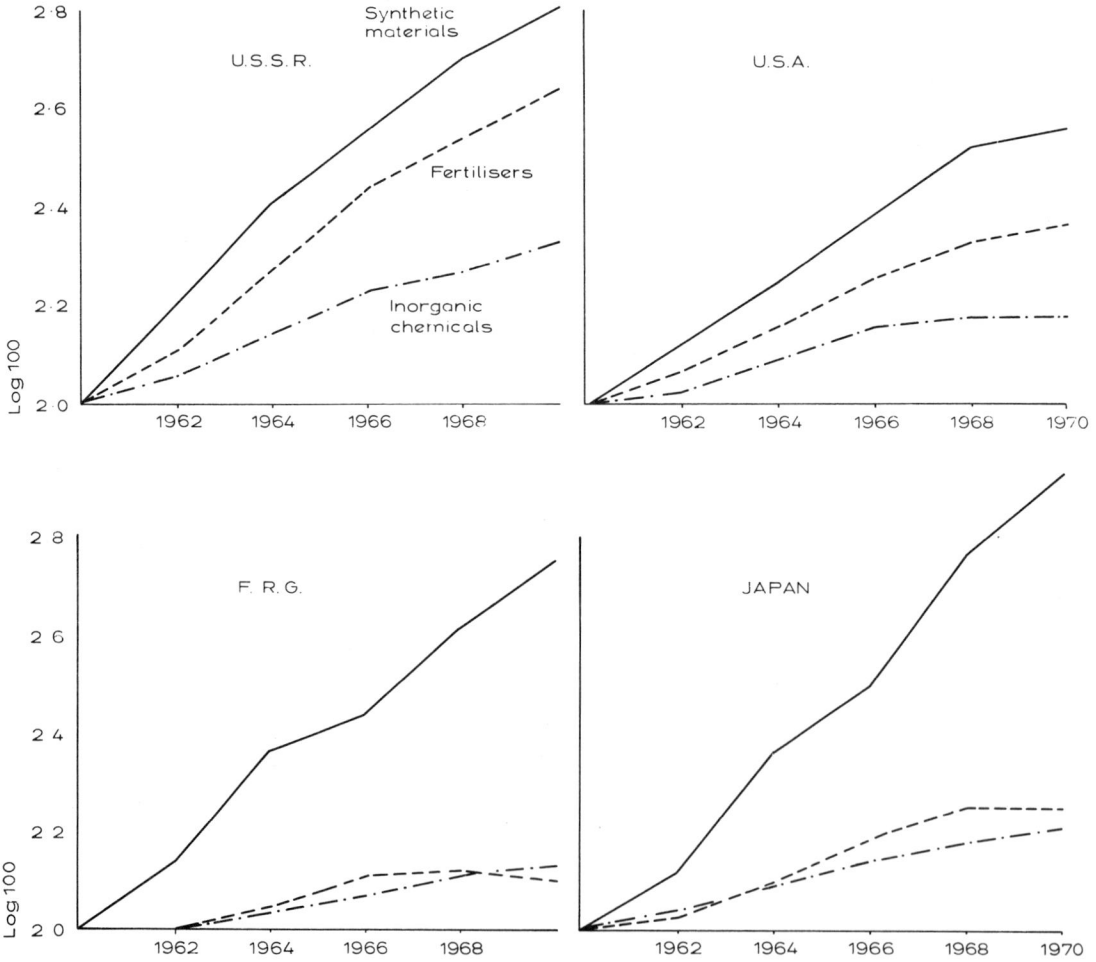

Fig. 6·2 COMPARATIVE GROWTH RATES OF SELECTED SECTORS OF THE CHEMICAL INDUSTRY 1960-1970

especially in sectors of production which require sophisticated technology.[111] As Table 6.15 shows, Great Britain is a net exporter in all sectors of the chemical industry except inorganic chemicals and fertilisers; this is almost the exact mirror image of the Soviet pattern, as will emerge below.

Table 6.14 *Weight of synthetic materials as a percentage of weight of inorganic chemicals*

	USSR	USA	Japan	FRG	France
1960	4·48	15·78	17·20	23·70	13·89
1961	5·19	15·87	17·61	27·57	14·20
1962	5·74	18·78	18·33	32·17	16·34
1963	6·47	19·99	23·21	34·71	18·08
1964	7·29	20·65	28·28	39·44	19·25
1965	7·73	21·09	31·02	43·92	19·87
1966	8·65	23·25	36·04	49·46	21·88
1967	9·53	23·78	45·61	58·00	23·63
1968	10·57	29·41	66·23	65·60	25·95
1969	11·29	32·58	75·33	75·62	30·37
1970	11·93	32·11	86·88	83·62	33·21

Source:
Calculated from Appendix Tables 4, 5, 10 and 11 in R. Amann, *CREES Discussion Paper* RC/C, No. 11, University of Birmingham.

TRADE IN CHEMICALS AND RELATED PRODUCTS

Unlike the leading chemical-producing nations of the West, the Soviet Union is a large net importer of chemicals and related products,[112] and the overall volume of its trade is less than that of those countries.[113] During the second half of the 1960s the Soviet deficit balance reduced slightly from 232·2 million roubles in 1965 to 163·9 million roubles in 1970 but by 1972 had risen once again beyond its 1965 level. To a large extent the size of this deficit is influenced by fluctuations in the export of fertilisers and in the import of synthetic rubber. The general picture that emerges from these statistics is that the production of

[111] See S. Hays, *The Chemicals and Allied Industries*, London (1973), pp. 128, 135; the overall trade balance in chemicals 1960–9 and in plastics materials is quoted in this source for France, West Germany, UK, USA, Japan and Italy. With the exception of Italy in 1969, all these countries were substantial net exporters (e.g. balance in 1969 was: France = £250 million; W. Germany = £2,040 million; UK = £690 million; USA = £2,268 million; and Japan = £45 million).

[112] See Appendix Table 6A.13. Soviet foreign trade returns exclude some categories which are sensitive (e.g., explosives) or negligible items of trade (e.g., fertiliser imports). This means that it is often difficult to reconcile the total figure for a given class of chemicals in *Vneshnyaya Torgovlya SSSR* with the aggregation of sub-totals. Each of the foreign trade tables in the Appendix is internally consistent and therefore these totals sometimes differ from those in *Vneshnyaya Torgovlya* and in other Appendix tables, relating to aggregated sub-groups of chemicals.

[113] In 1969, for example, the USSR exported 393·9 million roubles worth of chemicals and imported 586·8 million roubles worth. In the same year the equivalent figures for Western countries (in £ million) were: United States: (3,500/1,232), W. Germany (3,600/1,560), Japan (1,070/780), Great Britain (1,800/1,110), France (1,600/1,350) and Italy (820/975).

fertilisers is the only major branch of the chemical industry in which the USSR has a trading surplus; even in this sector the USSR tends to be an exporter of unprocessed minerals and fertilisers and an importer of more sophisticated agricultural chemicals. In paints, varnishes and photographic materials, Soviet imports have more than trebled since the mid-1960s, while exports have remained until recently at approximately the same level.

It is worth taking a more detailed look at the pattern of trade in relation to particular sub-sections of the Soviet chemical industry as they are classified in the foreign trade returns. An obvious example to take is that of 'chemical products' (*khimicheskie produkty*—classification number 30).[114] Here the pattern is typical. Due largely to its trade with Eastern Europe and the Third World, the USSR is a net exporter of basic organic and inorganic acids, coke chemicals (especially of benzol, exported to both Eastern and Western Europe) and wood chemicals; it has recently become a small net exporter of alcohols and aldehydes and its trade is roughly in balance with regard to intermediates for the aniline dye industry. But these exports of relatively simple products are not enough to offset the more substantial imports of alkalis, plastics and 'miscellaneous chemical products'. The large imports of alkalis are something of a puzzle in view of the high level of Soviet soda ash production but are

Table 6.15 *British trade in chemicals in 1971 (£M)*

	Exports	Imports	Balance
Inorganic chemicals	62·7	68·3	− 5·6
Organic chemicals	154·5	147·3	+ 7·1
Other chemicals	148·9	82·5	+ 66·4
Pharmaceuticals	161·1	42·2	+118·9
Toilet preparations	23·9	6·2	+ 17·7
Paints	21·6	5·9	+ 15·7
Soaps and detergents	23·8	9·4	+ 14·4
Synthetic resins	154·9	116·5	+ 38·4
Synthetic rubber	22·2	20·5	+ 1·7
Dyestuffs	58·7	35·7	+ 23·0
Fertilisers	3·6	25·9	− 22·3
Other chemical products (including pesticides and photographic chemicals)	121	34·9	+ 86·1
Total	956·8	595·2	+361·6

Source:
These figures are derived from the quarterly articles on the chemical industry in *Trade and Industry*; they are quoted by S. Hays, *op. cit.*, pp. 152–3.

[114] See Appendix Table 6A.14.

probably connected with larger requirements arising from the relative underdevelopment of synthetic detergent production in the USSR.

The Soviet Union obtains the largest proportion of its plastics imports from the West (notably from Italy, Japan and West Germany), while its trade with Eastern Europe in these products is roughly in balance.[115] As might be expected, Soviet exports to the West of plastics or plastics intermediates tend to be negligible; the typical pattern during the last half of the 1960s was for the USSR to export relatively simple intermediates, such as formaldehyde and urea, to Eastern Europe, and to import more sophisticated products, such as polyethylene and PVC additives, from the West. Trade in manmade fibres is perhaps the most striking example of one-sidedness.[116] The USSR does not export any manmade fibres. It has occasionally imported small quantities of cellulose-based fibre from the GDR, but the bulk of its imports, mainly of sophisticated synthetic fibres, come from the West; Great Britain has traditionally been the major supplier of these commodities. Even in the basic category 'miscellaneous chemical products' (classification 309) the USSR runs a substantial trade deficit[117] and this category includes such key chemicals as ethylene glycol, calcium carbide, acetone, textile auxiliaries, accelerators, reagents and catalysts.[118] Apart from exports of phosphorus to West Germany and Japan, Soviet exports to the West are negligible and those to Eastern Europe are very small. An interesting feature of trade in this category is that imports from the West have remained at more or less the same level since the mid-1960s whereas those from Eastern Europe have more than doubled. Eastern Europe now supplies the USSR with approximately three times the volume of these chemicals supplied by the West, and in 1972 imports from the GDR alone exceeded those from all the Western countries combined.

In view of the pattern of trade described above it is interesting to calculate how far the Soviet Union is dependent on imports of key chemical commodities, where figures for both trade in physical quantities and for domestic production are available. These calculations are set out in Table 6.16 and confirm what we would expect, namely that compared with Western countries the USSR is substantially dependent on imports of relatively sophisticated commodities such as synthetic fibres and dyestuffs (as opposed to crude intermediates); by the late 1960s self-sufficiency in synthetic rubber production had been achieved. On the other hand, apart from the curious exception of soda ash, the Soviet Union has established its independence in the manufacture of relatively simple products such as sulphuric acid and nitrogenous fertilisers.

In overall terms, however, the USSR remains dependent on imports of chemical products and especially of those requiring technologically advanced manufacturing processes, which come predominantly from the 'developed West'. There has been a steady upward trend in the proportion of chemical

---

[115] See Appendix Table 6A.15.
[116] See Appendix Table 6A.16.
[117] See Appendix Table 6A.17.
[118] G. Hemy, *The Soviet Chemical Industry*, London (1971), pp. 276–7.

Table 6.16 *Balance of trade as a proportion of domestic production (%)*

	1965	1966	1967	1968	1969	1970	1972
Synthetic (non-cellulose) fibres	−7.92	−8.54	−9.91	−9.23	−12.89	−13.47	−10.04
Sulphuric acid	+0.74	+0.98	+1.63	+1.80	+1.46	+1.78	+1.60
Soda ash[a]	−6.28	−11.77	−9.95	−13.05	−14.44	−13.70	−10.15
Synthetic rubbers[b]	−2.40	n.a.	+5.38	n.a.	+3.02	+4.99	+5.13
Synthetic dyes	−3.85	−3.72	−5.18	−9.33	−8.39	−8.23	−15.95
Nitrogenous fertilisers[c]	+2.56	+3.67	+4.50	+4.68	+4.36	+3.97	+3.33

Notes:
a – mainly imported from E. Europe.
b – calculated on basis of very rough estimates of domestic production (see Appendix Table 6A.12).
c – no imports are quoted in Soviet returns.

products in the total trade with the 'developed West'; this proportion (excluding manmade fibres) increased from 3.98 per cent in 1960 to 7.45 per cent in 1971, with a peak of 9.37 per cent in 1967.[119]

TRADE IN CHEMICAL PLANT AND EQUIPMENT

The USSR is also a substantial net importer of chemical plant and equipment, and therefore depends on foreign countries for a large part of the underlying technology of its chemical industry in addition to the manufactured goods that those processes produce.[120] The scale of this dependence is underlined by the fact that, since 1965, imports have on average accounted for just over one third of the chemical equipment introduced into the Soviet economy. Soviet exports of chemical equipment, which go predominantly to Eastern Europe, are a very small fraction of total imports. In 1969, for example, this proportion was approximately 3.5 per cent of *total* imports and only about 9.1 per cent of imports from Eastern Europe; substantial exports to Bulgaria during the early 1970s have improved the position considerably. During the last half of the 1960s it seemed that Eastern Europe had begun to supplant the West as the major source of chemical equipment, and in 1970 the Eastern bloc supplied the USSR with over 60 per cent of its imports. In retrospect, it appears that 1970 was a freak year. There were substantial imports of equipment from the West in 1971 and 1972 but the steady trend towards an increasing share of East European countries in the Soviet market persists.[121]

Despite the slight decline in the proportion of imports coming from the West, it is clear that many of these countries have played a key role in supplying

---

[119] *Soviet Economic Prospects for the Seventies*, Joint Economic Committee of the US Congress, Washington DC, June 1973, pp. 701–7. As the bulk of imports of chemicals and related products come from the 'developed West' the proportion of these goods in *total* imports from the world is correspondingly smaller; this proportion has also increased steadily from 2.66 per cent in 1960 to 5.62 per cent in 1971.
[120] See Appendix Table 6A.18.
[121] For analysis of this trend, see *East/West: Monthly Supplement*, Vol. I, No. 11, pp. 6–7.

the Soviet Union with process plant for the more technologically sophisticated sectors of its chemical industry. Until very recently the United States had not come into the picture as a major exporter of chemical plant to the USSR. Between 1960–66 American contracting organisations carried out 64 per cent of the world's contracts for chemical plant (including those on the US home market) but were not competing in socialist countries for political reasons. Great Britain, on the other hand, has always been among the major exporters in this field, and during the same period over 30 per cent of its export work was for socialist markets.[122] However, since 1969 the British share of the market has declined and Japan and Western Germany have been the main beneficiaries of the latest wave of Soviet plant imports.

The Soviet Union imported many large and expensive items of chemical plant from the West during the 1960s,[123] notably in the areas of petrochemical intermediates, plastics (particularly plant for the production of thermoplastics and plastic fabrication), agricultural chemicals (particularly pesticides and complex compound fertilisers) and manmade fibres. Among the most outstanding items were the £20 million contract signed with Simon-Carves in 1963 for high pressure polyethylene plant and the £31 million deal with the Polyspinners consortium, begun in 1964, for the construction of the huge polyester fibre plant at Mogilev;[124] in 1973, another large deal was completed with the West German contractors Salzgitter for a low density polyethylene plant to be erected at Kazan. A recent Soviet source has revealed that about one half of the country's total output of ammonia and acrylic fibres, over one third of its output of caprolactam and nearly all its output of polyethylene and polyester fibres is manufactured on the basis of imported equipment.[125] In contrast, most of the imports of chemical plant and equipment from Eastern Europe have been of a relatively simple and well established kind such as processes for the production of sulphuric acid, nitric acid and dyestuffs. At a more sophisticated level of technology there have been some imports of pharmaceutical plant from Hungary and of plastics fabrication equipment from the GDR, reflecting the specialities of these particular countries.

The substantial dependence of the USSR on the West for its advanced chemical technology naturally raises two intriguing questions: whether the Soviet Union will be in a permanently weak position in terms of its ability to catch up with the West in chemical technology, and how competitive its end-products are. On the first point, the general feeling among Western engineers who have worked on Soviet construction projects seems to be that imports of

---

[122] C. Freeman in *National Institute Economic Review*, 1968, No. 45, p. 35.

[123] The paragraph is based on a very full list of individual items of chemical plant that were imported from the West during the 1960s, which was brought to the attention of the author on an informal basis. The list is meticulously compiled from various press sources but even so is not absolutely complete, especially with regard to the paper and cellulose industry. It is, however, too substantial to be included here, though the compilers were in principle agreeable to this.

[124] It was planned that this plant would have a workforce of over 7,000 producing 70 different products with an output of more than 50,000 tons per annum, equivalent to total UK production before ICI's expansion in the late 1960s. This represented a great effort by Soviet planners to create a major outlet for large hydrocarbon resources and save annual imports by about £50 million (R. Milner, *Sunday Times*, 19 October 1969).

[125] *Vneshnekonomicheskie svyazi sotsialisticheskikh stran* (1974), p. 190.

Western plant are not in themselves sufficient to permit the USSR to close the technological gap with the West. This is especially true of some contractors (American firms, in particular, it seems) who take the view that '... everything down to the paper clips should be brought in from outside. This means that the plant actually works but that there is virtually no educational spin-off.'[126] At best, the Soviet Union may be able to *reproduce* Western technology (e.g., a polyester fibre plant like that at Mogilev) but by the time it is able to do so the level of technology in the West will have advanced to a new stage (e.g., future plant based on continuous polymerisation). The lesson here seems to be that the technology gap in chemical processes can only be narrowed by a native R and D effort culminating in successful innovation, which rapidly builds upon assimilated Western know-how. The prime example of a country that has pursued this strategy successfully is, of course, Japan. Almost all plastics materials were first produced in Japan under license by US subsidiaries and, subsequently, substantial amounts of know-how were purchased from abroad.[127] Whether the Soviet R and D system is flexible and fast moving enough to follow this path is a major theme and will be examined at length in the second volume of our study. On the second point, competitiveness in the manufacture of end-products, there seems little doubt that the innovating country enjoys considerable advantages in the short and medium term (i.e., 10–15 years). Even when its patents expire or it decides to break its monopoly by licensing the process abroad, accumulated expertise often allows the innovating country to maintain its superiority in terms of output range and quality.[128] These peculiarities of technological gap trade have been investigated by M. V. Posner,[129] who concludes that it is only in the long term that a country, even one with a relatively low wage structure, can equal or overcome this initial superiority; the classic example is that of nylon production in Great Britain where many years elapsed after the initial construction of capacity before imports from the United States finally ceased. Pursuing this theme, Hufbauer argues:

> The technological gap ... implies a 'pecking order' of trade. The country with the longest imitation lags must rely almost entirely on low-wage exports to pay for technological gap imports. The country which innovates can largely depend upon technological gap exports to pay for low wage imports. Between the two extremes, the nation's reliance upon low-wage exports and technological gap imports increases [with] its imitation lags.[130]

---

[126] B. G. Reuben and M. L. Burstall, *op. cit.*, p. 142.

[127] *European Chemical News*, 2 March 1962, quoted by C. Freeman, *National Institute Economic Review*, November 1963, p. 42.

[128] C. Freeman, *op. cit.* (1963), p. 22.

[129] M. V. Posner, 'International Trade and Technical Change', *Oxford Economic Papers*, October 1961.

[130] G. C. Hufbauer, *Synthetic Materials and the Theory of International Trade*, London (1966), p. 32. One might add that superior factor endowment is a point that should also be taken into consideration here (e.g., Soviet exports of fertilisers). In this case Western countries are able to obtain valuable natural resources in return for the products of technological expertise. The extent to which a technology is diffused after initial imitation is also an important element that will influence the pattern of trade.

Although this kind of analysis is not strictly meant to be applied to countries like the USSR where officially controlled trade (by physical controls and subsidised exports) produces distortions, it nevertheless fits in closely with the prevailing Soviet pattern.

Technological development

The approach adopted in the previous section was designed to assess how far advanced chemical technology has been diffused in the Soviet economy. This is *one* measure of 'technological level' but it is not sufficient to stand on its own since, as we have seen, a great deal of advanced technology has been imported into the USSR, which would considerably influence the overall output profile of the Soviet chemical industry and its pattern of foreign trade. We now consider the question of how far the Soviet Union is an independent inventor and how far it lags behind the advanced Western countries in bringing major developments to commercial fruition. The first part of this question can be approached through the examination of patents statistics. The second part can be approached by comparing the dates of the first commercial production of key chemical products (in this case, plastics and manmade fibres), and by examining the circumstances in which these developments took place.

PATENTS

PROBLEMS OF INTERPRETING PATENTS STATISTICS

'A patent may be defined as a statutory privilege granted by a government to inventors for a fixed period of years to exclude other persons from manufacturing, using or selling a patented product or from utilising a patented process or method.'[131] On the face of it, an analysis of patents statistics would seem to be an excellent means of assessing the level of inventive activity of a particular country. Unfortunately, there are several sticky problems that cannot be entirely eliminated, although they to some extent cancel each other out. In Western countries not all valuable inventions are patented. Some firms prefer to maintain secrecy rather than run the risk of competitors finding a technical or legal loophole in their patents protection;[132] moreover, national patent laws, while providing for temporary monopoly, usually contain powers to compel reluctant firms to license or manufacture.[133] This preference for secrecy leads to an *underestimate* of inventive activity. On the other hand, there are countervailing factors which lead to an *overestimate*. Innovating firms attempt to surround a valuable patent with several others, expressly to prevent potential competitors from exploiting the new product or process by an alternative method. Some large firms make a practice of buying up inventions

---

[131] *Gaps in Technology: Plastics*, OECD, Paris (1969), p. 106. Patents are granted as an exclusive privilege for between 15–20 years (varying between countries). Under the rules of the Paris Convention, foreigners enjoy equal rights with nationals. Applicants are allowed one year after applying for a patent in one member country before it becomes necessary to secure further protection by registering in the other countries.
[132] J. Jewkes, *et al.*, *op. cit.*, pp. 88–90.
[133] B. G. Reuben and M. L. Burstall, *op. cit.*, p. 130.

(issued in their own names in some countries such as the USA), not always to use them but to block competitors or bargain with them.[134] Finally, there is the problem of weighting. Many patents do not represent real advances in technology and have little commercial value, although here Freeman found that, for a large group of patents over a long period, distortions of this kind were relatively unimportant.[135] It is clear, however, that any conclusions based on patents statistics must be treated with caution, and where statistics of Soviet patents are brought into the picture, that caution must be extreme.

THE SOVIET PATENTS SYSTEM[136]

The registration of inventions in the USSR began as long ago as 1918 when a Committee for Inventions was set up. This was the predecessor of the present Committee for Inventions and Discoveries, established in 1955. However, these arrangements were largely of domestic significance. The participation of the Soviet Union in the international patents system only began in July 1965, when the USSR acceded to the Paris Convention. As Kosygin pointed out in his speech to the XXIII Party Congress, this move reflected a new stage in Soviet science and technology.[137] Due to the developed state of Soviet technical achievements it was felt that the USSR was in a position to earn foreign currency through the sale of licenses. At the same time, inspired by the Japanese use of licenses to promote advanced technology, the USSR had come to regard patents as an important source of information about technical progress.[138]

Soviet patents are two kinds: the author's certificates (*avtorskoe svidetel'stvo*) and the patent proper (*patent*). In the case of the former, the invention is bought by the state, and the inventor receives a payment based on economic return when the invention is used in the economy. In the latter case, the author retains exclusive rights for 15 years provided he pays the appropriate fees. As in many Western countries, the State can compel an inventor to sell the license, if he is unwilling to do so, and decide upon a suitable payment for him based upon its use.[139] In virtually all cases, the awarding of patents is confined to foreigners, and author's certificates are awarded almost exclusively to Soviet individuals or organisations. A particularly interesting feature of the Soviet patents system is that the Committee for Inventions and Discoveries actively seeks to get inventions incorporated into economic plans, allowing the inventor to receive his remuneration.[140] Thus in the USSR

---

[134] J. Jewkes *et al.*, *op. cit.*, pp. 88–90.

[135] With the help of a scientific consultant Freeman looked at all plastics patents taken out between 1790 and 1955 and identified 117 crucially important ones. In terms of distribution, the picture did not differ substantially from that emerging from a purely non-weighted approach. (C. Freeman, in *National Institute Economic Review*, November 1963, p. 38).

[136] I am indebted to my colleague M. J. Berry for his advice on this subject and for allowing me to read his unpublished paper, 'Inventions in the Soviet economy'.

[137] A. N. Kosygin in *XXIII S"ezd KPSS* (1966), Vol. 2, p. 62.

[138] M. J. Berry, 'Inventions in the Soviet economy' (unpublished manuscript).

[139] *Sbornik zakonodatel'nykh aktov i postanovlenii po izobretatel'stvu i ratsionalizatsii* (1965), pp. 20–1.

[140] M. J. Berry, *op. cit.*

pressure is exerted from the centre for the use of patents, whereas in Western countries the initiative comes from the individual firms who independently scan the patents literature.

Before examining the patents statistics it is worth pausing for a moment to consider a number of factors that might tend to depress the numbers of Soviet patents[141] (registered at home or abroad) and thus lead us to underestimate the level of Soviet inventive activity. M. J. Berry has suggested three main factors:

1. Inertia and ignorance of formalities.
2. Reluctance to spend foreign currency.
3. Inability to recognise what inventions might be commercially important in the West.

There is certainly considerable substance in each of these points, especially point 1, as a recent leading article in *Pravda* has emphasised.[142] It is worth mentioning a few other considerations, but it seems probable that these will, at best, only partially offset the factors described above. Given such a long established and well organised patents system it would seem unlikely that individuals and especially organisations would actually be unaware of the financial advantages to be gained from registering their inventions, though these financial incentives may not be sufficiently large to overcome inertia in many cases. So far as the registration of patents abroad is concerned, this is undertaken by the Committee itself, not by individual persons or organisations, so that ignorance of the formalities cannot be considered as a relevant factor. Moreover, in view of the Soviet fear that some of their best ideas are snapped up by opportunistic Western firms it seems unlikely that a professional organisation would be ignorant of what was of value (after all, they did award the patent in the first place!). Two further factors should be considered: first, secrecy (either commercial or defence), in which case the USSR is in a not dissimilar position to that of many Western countries or firms; second, triviality, in which case we are not underestimating the level of significant inventive activity.[143] In 1968, for example, the Paton Institute of Welding had 70 foreign patents to its credit.[144] One wonders whether this was due to the fact that the Paton Institute was 'keen' or 'patents conscious' or, alternatively, whether it had developed something worth patenting. In cases such as this, formalities have been mastered, inertia overcome and patents obtained. The answer to these conflicting possibilities is far from clear, but in the opinion of the author it seems that there are, on balance, serious obstacles facing Soviet organisations and individuals who wish to register their inventions abroad. This must be taken into consideration in the following analysis.

---

[141] Henceforth Soviet 'author's certificates' will be referred to as patents for the sake of simplicity.
[142] *Pravda*, 12 March 1974.
[143] According to *Izobretatel' i Ratsionalizator*, 1968, No. 1, p. 2, much of what passes as new technology from R and D organisations is only new to the Ministry or to the USSR.
[144] *Ibid.*, 1968, No. 11, p. 2.

SOVIET INVENTIVE ACTIVITY

The total numbers of patents granted to nationals in all fields of science and technology in the Soviet Union are fewer than in the United States but appreciably in excess of those granted by other major Western countries.[145] For patents awarded in the field of chemistry (see Table 6.17) a roughly similar picture emerges. Unfortunately, no American statistics are available, but the overall volume of inventive activity appears to be substantially higher in the USSR than in West Germany, Britain or France.[146] There are, however, two factors which seriously undermine the significance of this result. The first is an important technical point and relates particularly to the West German performance, which is surprisingly poor in the light of what we already know about the advanced state of German chemical technology from the previous section. The numbers of patents awarded by different countries vary according to the stringency of the procedures for checking out the originality of patents applications. In this respect, the procedures in Germany are particularly thorough, and many German firms prefer to gain immediate protection by first registering their invention abroad before they subsequently go through the lengthy 'search' in their own country.[147] The second point is methodological. These aggregate statistics provide no way of distinguishing between relatively important developments, which one ideally wants to get at, and mere trivia or hare-brained schemes.

One way of getting round this problem is to look at patents taken out in a foreign country because the expense of this operation is more likely to discourage frivolous applications. The London Patent Office was selected for this exercise partly because the detailed data are fairly accessible but also because London is an important centre in which to gain protection and to publicise inventions; it is the principal European base of American chemical contractors and the largest world centre for chemical design-engineering work.[148] Naturally, these statistics are biased in favour of British firms, whose comparative performance is thus to a large extent invalidated. The groups of patents that were selected for examination were 'Organic Chemistry' (Classification C2) and 'Macromolecular Compounds' (Classification C3),

---

[145] See Appendix Table 6A.19. In 1972, for example, the USA granted 51,515 patents to its nationals, USSR 38,523, Japan 29,101, France 10,767, UK 10,116 and W. Germany 9,642.

[146] The *Industrial Property* statistics for particular branches of technology do not give a breakdown between nationals and foreigners for the patents awarded by a country. In the USSR the distinction between author's certificates and patents is a very reliable guide to this distribution. The estimated distribution for the Western countries is based on a series of coefficients between patents *in all fields* granted to nationals and foreigners (in Appendix Table 6A.19). These estimates are, therefore, very rough.

[147] Belgium is favoured as a good country in which to file an initial patents claim because it grants patents without extensive search. A search is made only when the patent is challenged. Thus, the Belgian Patents Office avoids having to take trouble over patents that nobody is likely to pay attention to. As the total number of patents applications grows there is a trend towards this system in many countries. The British system is somewhere between that of West Germany and Belgium; an initial search is made for novelty but confined to a more limited period of time than is the case in West Germany. (This information was kindly supplied by the Birmingham Patents Office.)

[148] C. Freeman, *National Institute Economic Review*, 1968, No. 45, p. 37.

Table 6.17 *Grants of patents in chemistry, 1968–72 (including estimated breakdown between foreigners and nationals)*

	1968	1969	1970	1971	1972
USSR					
Total of patents granted	n.a.	n.a.	4192[a]	4724[a]	5398[a]
to nationals (estimate)	n.a.	n.a.	3664[3]	4063[4]	4498[5]
to foreigners (estimate)	415[1d]	269[2d]	528[3d]	661[4d]	910[5d]
FRG					
Total of patents granted	2947[1]	2960[2]	1354[3]	1803[4]	2929[5]
to nationals (estimate)	(1690)	(1626)	(671)	(824)	(1371)
to foreigners (estimate)	(1275)	(1334)	(683)	(979)	(1558)
UK					
Total of patents granted	7663[1c]	7565[2c]	7737[3e]	7741[4f]	8144[5]
to nationals (estimate)	n.a.	(1912)	(1952)	(1933)	(1925)
to foreigners (estimate)	n.a.	(5653)	(5785)	(5808)	(6219)
France					
Total of patents granted	8047[1b]	5471[2]	4650[3]	8015[4]	7220[5]
to nationals (estimate)	(2620)	(1758)	(1510)	(2134)	(1682)
to foreigners (estimate)	(5427)	(3713)	(3140)	(5881)	(5538)

Sources:
1 *Industrial Property*, No. 12, 1969, Statistical Annex, p. 11.
2 *Industrial Property*, No. 12, 1970, Statistical Annex, pp. 11–12.
3 *Industrial Property*, No. 12, 1971, Statistical Annex, p. 11.
4 *Industrial Property*, No. 12, 1972, Statistical Annex, pp. 14–15.
5 *Industrial Property*, No. 12, 1973, Statistical Annex, pp. 20–5.

Notes:
a – author's certificates plus patents.
b – includes patents of addition and special patents for medicaments.
c – figures relate to complete specifications accepted in 1968; this inflates the figure by no more than 850 (see original source for explanation).
d – these do not include 'author's certificates'.
e – refers to applications accepted, not patents granted.
f – includes patents of addition.

two of the most important sectors of the modern chemical industry.[149] On the basis of the source used (*Abridgements of Specifications*) it was not possible to identify the nationality of all the individuals listed; to do this accurately would require a very time-consuming search through the original patents specifications. Rather than compare the numbers of patents awarded on a global

[149] *Division C2* comprises: antibiotics, organic compounds containing boron, organic compounds, organo-metallic compounds, organic compounds containing phosphorus, cyclopentanophenanthrene compounds and vitamins. A large proportion of these patents are in the pharmaceutical industry where the Swiss-based firms CIBA and Hoffman-La Roche showed up particularly strongly.
Division C3 comprises: cellulose derivatives, epoxy resins, indiarubber, cellulose and compositions, graft polymers, proteins, plastic compositions, addition polymers, indiarubber compositions, condensation polymers, organic silicon compounds and polysaccharides. The statistics were compiled from the index in *Abridgements of Specifications*, The Patent Office, London, and cover the period 20 October 1971–13 December 1972.

national basis, it was decided instead to compare the numbers of patents awarded to organisations, which could be identified relatively easily from the source.

Of the total number of British patents granted to foreigners *in all fields* in 1972 the Soviet share was 1·69 per cent compared with 39·78 per cent for the United States and 19·81 per cent for West Germany.[150] In the fields of organic chemistry and macromolecular compounds a similar picture emerges. The total number of patents awarded to all Soviet organisations and institutes was less, and in some cases considerably less, than the total awarded to major individual Western chemical companies. This result needs to be offset against a number of other factors. First, as was mentioned earlier, the cumbersome nature of the German patents system forces German companies to seek protection abroad in the first instance. Moreover, the strong desire for legal protection is a characteristic of German chemical companies and in some cases their secretiveness is so great that they are reluctant to sell licenses even to other German firms.[151] These factors would tend to inflate the number of German patents. Second, the large number of patents awarded to ICI will to some extent be influenced by the fact that ICI is a British company. Third, to put the Soviet performance in true perspective, we must not forget that the Western chemical companies included in Table 6.18 are among the largest and most inventive in the world. In 1970, Du Pont was the largest chemical company in the world, with a workforce of 110,685 and annual R and D spending of $250 million; ICI, the world's second largest chemical company, employed slightly more workers but its annual R and D expenditure was just under half that of Du Pont.[152] However, even taking these factors into consideration, and the reservations about the Soviet patents system that were mentioned earlier, this is a very striking result indeed, and suggests that the level of significant inventive activity in the USSR falls far below that of Western countries in these advanced areas of chemical technology. It may be maintained that the Soviet Union only applies for foreign patents protection in the case of really outstanding developments. A very thorough and expert study would be needed to resolve this question, and if Soviet performance came within reasonable proximity to that of leading Western companies the general conclusion of Soviet backwardness suggested above would be open to more serious doubt. But when the total number of British patents granted to all Soviet organisations is less than the number awarded to the Sumitomo Chemical Company (the thirty-third largest chemical firm in the world, far removed from

[150] Calculated from *Industrial Property*, No. 12, December 1973, Statistical Appendix, pp. 14–15; the equivalent figures for other countries were: Japan 10·31 per cent and France 7·14 per cent.
[151] C. Freeman, 'Chemical process plant: innovation and the world market', *National Institute Economic Review*, 1968, No. 45, p. 46.
[152] The ranking of the world's 200 leading chemical firms can be found in *Chemical Age*; this source is quoted by B. G. Reuben and M. L. Burstall, *op. cit.*, pp. 128–31. According to V. S. Sominskii, in *Zhurnal Prikladnoi Khimii*, Vol. 39, 1966, No. 6, p. 1226, the three leading German firms (BASF, Bayer and Hoechst) are responsible for over 50 per cent of German chemical R and D expenditure. In the United States the top four chemical companies account for 60 per cent of R and D expenditure in industrial chemicals: (*Techniques of Sectoral Economic Planning: The Chemical Industries*, UNIDO, New York (1966), p. 46).

London, with a workforce of only 18,069 and an annual level of R and D spending of $16 million) it is difficult to doubt the general picture emerging from Table 6.18, although the exact proportions should naturally be treated with great caution.[153]

Table 6.18 *Patents granted to Soviet institutes and organisations by the London Patent Office and those granted to some major Western chemical companies, 1971–2*

Designation and world ranking of firms by size	Section C2 (organic chemistry)	Section C3 (macro-molecular compounds)	Total
USSR: total patents awarded to organisations	33	35	68
Du Pont (USA) 1	45	93	138
Monsanto (USA) 10	43	53	96
Farbenfabriken Bayer AG (FRG) 7	198	172	370
Badische Anilin und Soda Fabrik-BASF (FRG) 5	158	89	247
Imperial Chemical Industries (UK) 2	176	131	307
Rhone-Poulenc SA (France) 9	47	36	83
Sumitomo Chemical Company (Japan) 33	93	23	116

Distribution of patents awarded to the USSR, by different kinds of organisation:

Academy institutes: 29·5   (Of which: USSR Academy of Sciences including Academy of Medical Sciences, 23; Republican Academies, 6·5.)
Branch institutes: 30
Project/design and technological institutes: 4·5
VUZY (higher educational establishments): 2·5
Industrial enterprises: 1·5

Source: See n. 149.

THE FIRST COMMERCIAL PRODUCTION OF SYNTHETIC MATERIALS

The level of development of synthetic materials to a large extent defines how 'modern' the industry is in a technological sense. As J. D. Bernal has observed

> The solution of the mechanism of chain reactions and polymerisation by such chemists as Semenov and Melville represents one of the most important chemical achievements of the 20th century. Their application has given rise to a whole new branch of industry—artificial fibres and plastics. Many of these, especially nylon, are now commonplace, whereas even 30 years ago nobody had even dreamed of them.[154]

[153] An analysis of changes over time would also be useful for clinching this argument, since it was only in 1965 that the USSR acceded to the Paris Convention. Thus, if the numbers of Soviet patents registered abroad during the 1970s did not rise appreciably, it would be even more convincing evidence of the relatively low level of inventive activity in the USSR.

[154] Quoted from 'Science in History' (*Nauka v istorii obshchestva*, p. 438) by I. V. Rakhlin, *Nauchno-tekhnicheskii progress i effektivnost' novykh materialov* (1973), pp. 73–4.

Hufbauer refers to these related branches of industry as 'footloose';[155] though they depend on raw materials such as wood-pulp, petroleum and natural gas they are often to be found in countries with a high technical capability but lacking in the appropriate natural resources. The critical factor, therefore, is technological know-how and the economic benefits of acquiring this know-how are considerable. Sophisticated chemical processing adds highly to the value of the raw materials used; it has been calculated that the value per ton of carbon (from coal or oil) is increased 20 times or more between the input and the output stages in the case of synthetic materials.[156] The extent to which a nation initiates, or rapidly imitates, the manufacture of modern synthetic materials is therefore one of the sharpest and most accurate indicators of its level of chemical technology. It is, therefore, worth examining the evolution of Soviet synthetic materials production in some detail.

PLASTICS AND SYNTHETIC RESINS

While the bulk production of sophisticated polymerised plastics is a recent development, the origins of the plastics industry date back to the late nineteenth century. The first plastic, celluloid (a mixture of nitrocellulose and camphor), was first produced in the United States as long ago as 1872 when J. W. Hyatt sought a viable substitute for ivory in the manufacture of billiard balls.[157] This was followed after the turn of the century by the first commercial production of galalith (France, 1900) and of the important phenol-formaldehyde plastic-bakelite (United States, 1909).

In Tsarist Russia the plastics industry was virtually non-existent.[158] In 1916 at Orekhovo (Moscow region) a semi-handicraft factory situated in derelict premises began to produce bakelite (*karbolit*); it employed only 50 workers and produced at the rate of about 50 kilos a day.[159] This early commercial venture was apparently based on the laboratory synthesis of bakelite by the 'contact' method which had been developed by G. S. Petrov and his colleagues between 1909 and 1912.[160]

War and revolutionary turmoil interrupted the further development of the plastics industry in Russia and the threads were not picked up again until the latter years of NEP. At this time the Okhtinsk chemical combine in Leningrad was the focal point of activity. This was the oldest chemical factory in Russia and had been built in 1719, on the orders of Peter the Great, to manufacture gunpowder.[161] Due to its highly qualified staff, the Okhtinsk combine became a

---

[155] G. C. Hufbauer, *Synthetic Materials and the Theory of International Trade*, London (1966), p. 13.
[156] *The Chemical Industry in the European Member Countries of the OECD*, OECD, Paris (1964).
[157] *Encyclopedia Americana* (1970), Vol. 22, p. 221 ff.
[158] N. P. Fedorenko, *Ekonomika promyshlennosti sinteticheskikh materialov* (1961), pp. 537–612.
[159] V. S. Lel'chuk, *Sozdanie khimicheskoi promyshlennosti SSSR* (1964), p. 274; *Bol'shaya Sovetskaya Entsiklopediya* Vol. 33 (1955), p. 203.
[160] G. I. Kutyanin, *Plasticheskie massy i khimicheskie tovary* (1971), p. 93; *Stroitel'naya sel'skokhozyaistvennaya i meditsinskaya tekhnika, khimicheskaya tekhnologiya* (1971), pp. 178–83.
[161] M. I. Garbor in *Khimicheskaya promyshlennost SSSR* (1959), p. 80.

major centre of research into plastics in the late 1920s and was responsible for starting up the production of bakelite and celluloid.[162] At about the same time, the first industrial production of galalith (derived from the condensation of casein and formaldehyde) began in Moscow region. But these were only small beginnings. The substantial development of the Soviet plastics industry occurred during the industrialisation drive of the 1930s.[163]

The first two five year plans (1928–37) were largely concerned with consolidating the output of the three basic plastics begun under NEP (celluloid, galalith and bakelite) and with building up the R and D base of the industry. During the first five year plan, the capacity of the celluloid shop at Okhtinsk was enlarged to 1,000 tons per year and bakelite production was increased to 800 tons per year.[164] In 1929, the first University Department (*kafedra*) of Plastics in the world was created at the Leningrad Technical Institute under the direction of S. N. Ushakov and A. A. Vandsheidt, employing 180 scientific workers (*nauchnye rabotniki*).[165] A similar chair was created for G. S. Petrov at the Mendeleev Chemico-Technological Institute, Moscow, in 1932 and in the same year the first Soviet plastics research institute was set up in Leningrad, based on the Okhtinsk factory laboratory and specialising in polymerised plastics. However, several Soviet writers seem to agree that progress was disappointing during the first five year plan due to a lack of overall coordination and an insufficiently sound research base.[166] This situation began to be improved at the end of 1931 when the *obedinenie* 'Soyuzkhimplastmass' was created, not without some sniping from existing administrative departments, and the production of plastics became organised as a distinct branch of industry. Throughout the 1930s the rate of growth of the plastics industry was consistently higher than that of the chemical industry as a whole; while in 1931 the output of plastics and synthetic resins was 3,200 tons, by 1940 it had risen to 23,700 tons.[167]

During the third five year plan (1938–41) a significant diversification took place in plastics output. A start was made on the production of the 'aminoplasts' (urea or melamine formaldehyde), which were obtained by soaking wood pulp in a mixer with a solution of urea-formaldehyde condensate.[168] The production of these plastics thus began in the USSR approximately 10 years after production had started up in Great Britain, Germany, France and the United States. However, during the same period, preparatory work commenced on several of the more sophisticated polymer

---

[162] P. M. Luk'yanov and A. S. Solov'eva, *Istoriya khimicheskoi promyshlennosti SSSR* (1966), p. 145.
[163] P. M. Luk'yanov, *Kratkaya istoriya khimicheskoi promyshlennosti SSSR* (1959), p. 300.
[164] P. M. Luk'yanov, *op. cit.*, p. 336.
[165] V. S. Lel'chuk, *op. cit.* (1964), p. 289.
[166] See, for example, the comments of P. M. Luk'yanov, *op. cit.*, p. 337, and P. M. Luk'yanov and A. S. Solov'eva, *op. cit.*, p. 158.
[167] *Ekonomicheskaya entsiklopediya: promyshlennost' i stroitel'stvo* (1964), pp. 495–9; this figure is not consistent with the 10,000 tons for 1940 quoted by *Narodnoe khozyaistvo SSSR v 1969 godu* (1970), p. 214.
[168] P. M. Luk'yanov and A. S. Solov'eva, *op. cit.*, p. 180, and *Stroitel'naya... khimicheskaya tekhnologiya*, *op. cit.*

substances, which allowed the Soviet Union to substantially (but temporarily) reduce this imitation lag. Between 1936–41 a method of synthesising silicon compounds was developed by K. A. Andrianov 'long before this had been developed in other countries';[169] these compounds are very stable at high temperatures and applications in electronics, aviation and rocketry are particularly important. On the eve of Soviet involvement in the Second World War the first output of polyvinyl chloride (PVC) was obtained. This development had been carried out at the Mendeleev Chemico-Technological Institute in Moscow and the factory 'Kooperakhimiya' by P. I. Pavlovich and his colleagues.[170] Thus, by 1940, according to standard Soviet accounts of the evolution of the industry, 'practically all known plastics and articles derived from them had been assimilated'[171] and the plastics industry had become, 'for its time', a relatively developed branch of the economy.[172]

War once again interrupted the realisation of these embryonic developments, which were only consolidated during the post-war period of economic reconstruction. In 1947, for example, production of silicon polymers began on an industrial scale.[173] At the same time the manufacture of PVC was recommenced and one of the most important polyvinyl alcohols (polyvinyl butyral) began to be produced, based on preparatory work carried out by the Institute of Polymerised Plastics.[174]

During the 1950s the production of polyethylene and polystyrene began. Together with PVC, these are currently the most important plastic materials and account for well over half of the world's output of plastics.[175] However, the exact circumstances of their assimilation in the USSR on an industrial scale are shrouded in mystery in Soviet sources. In 1959, a Soviet source strongly implied that these vitally important polymerised materials were only produced on a semi-industrial scale or, at best, in very small quantities prior to Khrushchev's chemicalisation drive.[176]

The comparative dates for the first commercial production of important plastics materials in the USSR and the West are set out schematically in Table 6.19. Unfortunately, it has not proved feasible to calculate aggregate imitation lags based on Hufbauer's methodology, for three reasons.[177] First, the chronology of Soviet plastics production is not sufficiently precise at a number of crucial points. Second, information could not be obtained about the first production of a number of plastics included in Hufbauer's original table,

---

[169] M. I. Garbor, *op. cit.*, p. 93; the source does not refer to the lifelong researches of Professor Kipping of Nottingham University in this field, though this would not detract from the obvious achievement of Soviet scientists in developing these polymers in advance of Britain.

[170] *Ibid.*, p. 82.

[171] *Stroitel'naya ... khimicheskaya tekhnologiya, op. cit.* (1971).

[172] N. P. Fedorenko, *op. cit.* (1961), p. 537.

[173] M. I. Garbor, *op. cit.*, p. 93.

[174] *Ibid.*, p. 101.

[175] *Gaps in Technology: Plastics*, OECD, Paris (1969), pp. 27–8; the exact figures are: Polyolefins 22·3 per cent, PVC 23 per cent and polystyrene 10–11 per cent.

[176] M. I. Garbor, *op. cit.*, p. 84.

[177] G. C. Hufbauer, *op. cit* (1966), pp. 82–6. The lag in particular kinds of plastics is weighted according to the importance of the material concerned, defined in terms of its proportion in world trade in plastics.

though it is reasonably clear that many of these are produced in the USSR and to exclude them would distort the calculations. Finally, there is an ambiguity about the meaning of the term 'first *commercial* production' in a Soviet context. In the West this implies a sufficiently effective level of technology to produce profit and goods which are attractive to the consumer. In the USSR, on the other hand, it is conceivable that the products of a less satisfactory technological process could be heavily subsidised (e.g., in priority areas such as defence) or less attractive goods sold on a sellers' market; the achievement of 'commercial production' in the Western sense might, under these circumstances, take considerably longer than the attainment of 'industrial production' (*promyshlennyi masshtab*) in the USSR. Thus, in the light of the factors discussed above, any inaccuracies in Table 6.19 would simply be compounded by attaching a precise numerical weight to them. However, it is still possible to draw some useful conclusions from Table 6.19. In no case has the Soviet Union been the first country to produce a major plastic material on an industrial scale. For most plastics, with the notable exceptions of silicon polymers and PTFE (which have application in the high priority aerospace sector), Soviet production began much later than in the majority of leading Western countries, although in some individual cases Soviet production preceded that of Italy or Japan. This result seems to be consistent with the view of Freeman and Hufbauer that successful innovation at the level of the firm (or nation) makes further innovation more probable and cumulative expertise cuts down the time taken to assimilate foreign developments.[178]

A further question must be posed, however, that can shed a different light on Soviet performance. How far has the first commercial production of plastics in the USSR been due to 'indigenous imitation' and how far to the importation of foreign plant and equipment? If the latter has been influential it is obvious that this must be taken into account in assessing the level of Soviet chemical technology in this field.

The most significant contribution towards answering this question has been made by A. C. Sutton in his three-volume work on *Western Technology and Soviet Economic Development*,[179] which is based largely on Western materials such as government papers, technical journals, delegation reports and the reminiscences of engineers who have worked in the USSR. Sutton's provocative conclusion is that the Soviet chemical industry as a whole has acquired its technology by replicating or scaling up original Western designs and by seizing key items of industrial plant from the Eastern regions of Germany at the end of the Second World War. In order to develop the more sophisticated sectors of organic chemicals and synthetic materials, the USSR has been forced to import capacity on a considerable scale.[180] Significantly, these deals are usually arranged so that the Soviet side provides the buildings and power plant and the Western contractor provides the process, start-up and

---

[178] This view is developed by W. W. Leontief, in *Review of Economics and Statistics*, November 1956, quoted by G. C. Hufbauer, *op. cit.* (1966), p. 42.
[179] Vol. 1, *1917 to 1940*, Stanford, Cal. (1968); Vol. 2, *1930 to 1945*, Stanford, Cal. (1971); Vol. 3, *1945 to 1965*, Stanford, Cal. (1973).
[180] A. C. Sutton, Vol. 3, *1945 to 1965* (1973), pp. 144–6.

training programmes. Sutton comments that 'Such contracts are unusual in the West except perhaps in underdeveloped areas lacking elementary skills and facilities.'[181] In the particular case of plastics, Sutton claims that 'No indigenous large-scale plastics production has been traced, only pilot operations.'[182]

A systematic attempt was made to reconcile the evidence of Sutton and others with the dates of first commercial production set out in Table 6.19.[183] The result of this is presented in Table 6.20, which excludes data concerning the substantial Soviet purchases of process plant for organic intermediates and plastics fabrication. These results are not absolutely conclusive, but are probably sufficient to modify Sutton's sweeping claim quoted above. The first conclusion is that there is no evidence that Soviet plastics production during the inter-war or immediate post-war period was influenced by the import of Western plant and equipment. This does not mean that such imports did not take place but that it would be necessary for Sutton to demonstrate that they did before he could sustain his claim; his sparing use of Soviet materials has possibly led him to overstate his case. On the other hand, if Soviet sources acknowledged Western assistance and influence frankly instead of creating an impression that the USSR is a country pressing forward on all frontiers of technology in heroic isolation, it would make objective investigation a good deal easier. The second conclusion from Table 6.20 is that in all cases the huge quantities of plastics plant and equipment, imported during the late 1950s and 1960s, came after production had been established in the USSR on an industrial or semi-industrial scale. This might give rise to two possible interpretations: (a) that Soviet sources are incorrect in stating that *large-scale* production took place at these particular times and are fudging the distinction between full-scale industrial production and pilot-plant production (in which case Sutton is right) and (b) that if large-scale production did take place, the quality and cost factors were unfavourable. Unless some explanation of this kind is put forward it is difficult to see why the USSR could not build extensively on the basis of its own technology. These massive imports of foreign plastics plant must be borne in mind in relation to the data presented in Table 6.19 and, especially when the extensive training arrangements are taken into account, they indicate clearly the relative backwardness of Soviet technology in this field.

ARTIFICIAL AND SYNTHETIC FIBRES[184]

Like the plastics industry, the manmade fibres industry has its origins in the last quarter of the nineteenth century when the Count de Chardonnet began to manufacture nitrocellulose fibre in his factory at Bescançon. This fibre did not

---

[181] A. C. Sutton, Vol. 3, *1945 to 1965* (1973), p. 147.
[182] *Ibid.*, p. 165.
[183] The full list of Soviet purchases of chemical equipment purchased from the West, referred to in p. 263, was used in addition to Sutton; this is a more substantial source for the late 1950s and 1960s.
[184] A very useful list of comparative terms for manmade fibres (including Soviet terms) can be found in *Man-Made Textiles*, August 1963, p. 31; September 1964, p. 45; and September 1965, p. 44.

become widely diffused due to its poor quality, high cost and the danger of conflagration in the manufacturing process, and production finally petered out during the 1920s. The main breakthrough in the production of manmade fibres occurred in 1892 with the development of the viscose process by Cross, Bevan and Beadle (UK). This process, initially applied on an industrial scale in Britain by Courtaulds, formed the basis of the manmade fibres industry in all countries during the inter-war period.

In Tsarist Russia, the production of manmade fibres only existed on a semi-handicraft (*polukustarnyi*) scale. A small viscose silk plant, 'Viskoza', was started up in 1909 near Moscow, financed by Belgian and British capital.[185] This factory produced at the rate of 100 tons a year,[186] but was shut down in 1916 and preserved for future use after the war. Prior to 1923, a lease on the plant had been taken out by a group of private firms, but their attempt to start up production met with failure.[187] Production was successfully begun again in 1924, when 'Viskoza' became part of the Wool Trust (Vigon'trest), and it was due to the experience gained in carrying out this job that the first Soviet specialists began to emerge. In 1927, this small group of young engineers dismantled the old machines and replaced them with new ones of substantially greater capacity[188] and by 1929 Soviet output of viscose silk had reached its pre-war level.[189]

The importance of creating a large-scale manmade fibres industry was stressed by the central committee of the Soviet Communist Party in October 1927, and a special committee under V. V. Kuibyshev was set up to study the feasibility of this and to report its findings to the Council for Labour and Defence. Despite some conservatism within the textile industry itself,[190] the Council decided in March 1928 to press forward with the development of chemical fibres. During the first five year plan, three large new viscose plants were built at Leningrad, Klin and Mogilev, and by 1932 Soviet output of manmade fibres had risen to 4,000 tons compared with 221 tons in 1928/29.[191] These new plants were completed in less than two years, but all the machinery was imported from abroad, and technical assistance was needed from French and German firms to install the machinery and to put the plants into operation.[192] Referring to the level of Soviet fibre technology at this time, Academician P. P. Shorygin pointed out that 'The most elementary technological processes are not clear to us and we are not able to design factories without the help of foreign specialists.'[193] In order to improve this

---

[185] V. S. Lel'chuk, *Sozdanie khimicheskoi promyshlennosti SSSR* (1964), p. 274.
[186] N. P. Fedorenko, *Ekonomika promyshlennosti sinteticheskikh materialov* (1959), p. 415.
[187] P. M. Luk'yanov, *op. cit.*, pp. 300–1.
[188] G. E. Birger and A. A. Konkin, in *Khimicheskaya promyshlennost' SSSR* (1959), p. 115.
[189] P. M. Luk'yanov and A. S. Solov'eva, *op. cit.*, p. 145.
[190] Some administrators took the view that given Russia's abundance of land and labour but scarcity of capital, it would be more rational to extend the production of natural fibres (see *Izvestiya Tekstil'noi Promyshlennosti i Torgovli*, 1928, No. 1, pp. 4–6, quoted by V. S. Lel'chuk, *op. cit.* (1964), p. 275).
[191] P. M. Luk'yanov and A. S. Solov'eva, *op. cit.*, p. 161.
[192] G. E. Birger *et al.*, *op. cit.*, p. 115.
[193] *Khimiya i Sotsialisticheskoe Khozyaistvo*, 1932, No. 9, p. 73, quoted by V. S. Lel'chuk, *op. cit.*, p. 278.

Table 6.19 First commercial production of various plastic materials (in order of introduction in the USSR)

Major plastics	USSR	USA	Germany (FRG after 1945)	UK	France	Italy	Japan
Phenol-formaldehyde (bakelite: Soviet designation 'Karbolit')	1915 (on semi-handicraft scale: the original 'Karbolit'[1] factory was enlarged in 1925).	1909*	1910	1910	1916	1922	1923
Galalith	1925[3]	1919	1899*	1912	1900	1921	1927
Celluloid	1926[2]	1870*	1878	1877	1875	1924	1908
Cellophane	1936[9]	1924	1925	1930	1917*	1946	1929
Urea-formaldehyde	Soviet designation—'amino-plasts'. Production began during third five year plan (1938–41). But real development occurred during early post-war period 1946–50[10]						
Polyvinylchloride (PVC)	1940[4]	1929	1929	1928*	1930	1936	1935
Polyvinyl butyral	1947[8]	1933	1931*	1940	1940	1951	1939
Silicon polymers	1947[6]	1937*	1948		1942		1944
Polymethylmethacrylate (perspex)	Late 1940s[19] (wide diffusion of volume and range by 1959)[13]	1941*	1950	1952	1954	1955?	1951
High pressure polyethylene	Early 1950s[14]	1936	1930*	1933	1938	1937	1938
Polytetrafluorethylene (PTFE: Teflon)	1949[12]	1941	1944	1937*	1954	1952	1954
Polystyrene	Mid–Late 1950s[7]	1943*	1958	1945	1958	1955	1963
Epoxy resins (Switzerland—the innovating country)	Late 1950s[5]	1933	1930*	1950	1951	1942	1957
Linear polyethylene	1959	1947	1955	1955	n.d.	1958?	n.d.
Polycarbonate	(Produced, then, on a *semi-industrial scale*)[16] 1970	1956	1955	1959	1956	1954*	1958
Polyformaldehyde	(Produced on a *semi-industrial scale* (*polypromyshlennyi masshtab*))[17] 1970	1957*	1957*	n.d.	n.d.	n.d.	1959
	(Produced on a *semi-industrial scale* (*polypromyshlennyi masshtab*))[17]	1960*	n.a.	n.a.	n.a.	n.a.	n.a.
Polypropylene	(existing, then, only on an *industrial prototype*). (*promyshlennaya ustanoyka*)[17]	1957*	1957*	1959	1960	1957*	1961

Sources:
1. M. I. Garbor, 'Promyshlennost' plasticheskikh mass i sinteticheskikh smol', in *Khimicheskaya promyshlennost SSSR* (1959), p. 80.
2. *Ibid.*, pp. 80–91.
3. *Ibid.*, p. 81.
4. *Ibid.*, p. 82.
5. *Ibid.*, p. 91.
6. *Ibid.*, p. 93.
7. *Ibid.*, p. 96. (The source disclosed that polystyrene was being produced in the USSR at the time of writing (1959), implied that it was a recent development, but gave no exact indication of the first date of commercial production.)
8. *Ibid.*, p. 101.
9. G. E. Birger and A. A. Kontin, 'Promyshlennost' khimicheskikh volokon', in *Khimicheskaya promyshlennost SSSR* (1959), pp. 119–20.
10. P. M. Luk'yanov and A. S. Solov'eva, *Istoriya khimicheskoi promyshlennosti SSSR* (1966), p. 180.
11. G. E. Birger et al., *op. cit.*, p. 124.
12. M. I. Garbor, *op. cit.*, p. 103.
13. *Ibid.*, p. 102.
14. P. M. Luk'yanov, *Kratkaya istoriya khimicheskoi promyshlennosti SSSR* (1959), pp. 442–3.
15. P. M. Luk'yanov and A. S. Solov'eva, *op. cit.*, pp. 192.
16. M. I. Garbor, *op. cit.*, p. 97.
17. *Stroitel'naya, sel'skokhozyaistvennaya i meditsinskaya tekhnika, khimicheskaya tekhnologiya* (1971), p. 181.
18. *Encyclopedia Americana*, 1970, Vol. 22, p. 221 ff.
19. P. M. Luk'yanov, *op. cit.*, p. 420.

General notes:
The items in the table are in the chronological order in which they were introduced commercially in the USSR.
Not all the items in Hufbauer's tables are included because, for many, there is no information in Soviet sources; this may or may not mean that they were not produced.
Western dates can be found in G. C. Hufbauer, *Synthetic Materials and the Theory of International Trade*, London (1966), p. 131.
n.d. = not produced in 1966 but no data thereafter.
n.a. = not available.
* – innovating country.

Table 6.20 *Soviet purchases of plastics plant and equipment (excluding processes for intermediates and fabrication)*

Material	First Soviet large-scale production	Purchases
Cellophane	1936	1961—Plant built by Maurer (Switzerland) began production.
PVC	1940	1964—Plant installed by Japanese consortium: cost $14 million.* 1966—2 Plants built by Anger (Austria) began production: cost £4 million; total capacity 50,000 tons. 1968 onwards—installation of plant by Alkor Werke Lissman AG (FRG): cost £2·2 million. 1971—Delivery due of plant by Ishikawajiama–Harima (Japan): cost £600,000.
Polymethylmethacrylate (perspex)	Late 1940s	
High pressure polyethylene	Early 1950s	1965–66—Three plants installed by Plane Trade Ltd. (Switzerland). 1958—BASF license of polyethylene process sold to USSR.* 1962—Production began of plant built by Salzgitter (FRG): cost £5·6 million; capacity 24,000 tons. 1963—Contract for four plants signed by Simon-Carves (UK): cost £56 million; capacity equal to total British capacity in 1964.* 1966—two plants built by Salzgitter (FRG): cost £8·2 million; capacity 48,000 tons. 1971—Plant built by ENI (Italy).
Polystyrene	Mid late 1950s	1961—Plant built by Sterling Moulding Powders (UK): cost $12·1 million; capacity 10,000 tons.* 1962—Suspension process plant completed by Simon Carves (UK): cost £650,000; capacity 5,000 tons. 1965–66—Contract signed by Plane Trade Ltd. (Switzerland) for polystyrene plant plus associated dyeing and granulating facilities.
Low pressure polyethylene	1959 Semi-industrial scale	1962—Production began of plant built by Salzgitter (FRG): cost £4·7 million; capacity 24,000 tons. 1972—Contract signed by Salzgitter (FRG) for large plant: cost £15 million; capacity 120,000 tons.
Polypropylene	1970 industrial prototype	1967—Contract signed by Lurgi (FRG) for petrochemical plant for polypropylene.

Source:
Derived from Table 6.20 above; A. C. Sutton, *op. cit.*; and source described on p. 263 above.

Note:
* – these four items are referred to by A. C. Sutton and confirmed in other sources.

situation, a laboratory for artificial fibres had been organised at the Karpov Physico-Chemical Institute in 1929; this was the establishment at which the first Soviet specialists received their formal training. In 1931, the scientific base of the industry was further strengthened by the creation of the All Union Institute for Artificial Fibres (VNIIV) under the direction of Academician Shorygin. The aim of this institute, which was destined to play an important role in the development of Soviet manmade fibres, was to give technical assistance to the industry and, specifically, to release it from dependence on foreign specialists. By means of collaboration between the institute, factory laboratories and pilot plants (*opytnye ustanovki*), the intricacies of the viscose process were mastered, and Soviet capacity could now be extended without foreign assistance.[194]

The next major phase in the development of the Soviet chemical fibres industry began in 1935 when the government authorised the design and construction of large-scale enterprises to produce viscose staple fibre as well as continuous filament. The production of viscose staple had already been assimilated at the Leningrad plant between 1932–3, somewhat later than in the leading Western manufacturing countries, especially the United States and Germany. In order to design these plants, a new Project Institute (GIPROIV) was set up, and between 1935–7 this institute designed and supervised the construction of a large plant in Kiev with an annual capacity of 4,000 tons.[195] This factory introduced a semi-automatic method (machine-'OK') for trimming viscose filament within the centrifuge, devised by V. M. Aronovich and Z. F. Kipershlak, which represented 'a major landmark in the history of viscose fibre production'.[196] According to Soviet sources, all the process plant for the Kiev factory was designed by Soviet engineers and supplied by Soviet manufacturers, and the productivity of the machinery was said to be nearly double that of the earlier foreign models.[197] In 1935, a start was also made on the production of cuprammonium fibre at the Rostokinsk wool factory and subsequently three small factories were built using the same process; this fibre was not produced on a large scale, however, due to the high cost of the necessary raw materials (especially copper hydrate).[198] Thus, due to these rapid developments, the output of chemical fibres in the USSR between 1928–40 increased from 200 tons to about 11,000 tons,[199] of which 8,500 tons was viscose silk, 1,400 tons was viscose staple and the remaining 1,100 tons was cuprammonium fibre.[200] But high strength viscose cord was not produced until 1949, and acetate fibre and nylon, first produced in the USA and Italy in 1938, was also not produced in the USSR before the war. Hence, Soviet statements to

---

[194] G. E. Birger *et al.*, *op. cit.*, p. 118.
[195] G. E. Birger *et al.*, *op. cit.*, pp. 119–20.
[196] *Stroitel'naya, sel'skokhozyaistvennaya i meditsinskaya tekhnika, khimicheskaya tekhnologiya* (1971), pp. 173–8.
[197] G. E. Birger *et al.*, *op. cit.*, pp. 119–20.
[198] N. P. Fedorenko, *op. cit.* (1961), p. 422. Production of cuprammonium fibre in Great Britain ceased in the early 1950s for this reason and was supplanted by nylon (*Chambers's Encyclopedia* (1967), Vol. 5, pp. 616–19).
[199] N. P. Fedorenko *op. cit.* (1961), p. 416.
[200] G. E. Birger *et al.*, *op. cit.*, p. 120.

the effect that on the eve of the Second World War the chemical fibres industry was 'relatively developed for its time' may perhaps be an exaggeration.[201]

The war resulted in a serious setback for Soviet fibre production. Nearly all the fibre plants lay in the path of the German invasion and were either partly or fully destroyed. Output in 1945 was only nine per cent of that in 1940 and it was not until 1949 that the pre-war level of output was once again achieved.[202] From the perspective of world chemical fibre technology the war also represented a watershed between the cellulose-based fibres of the inter-war period and more sophisticated synthetic polymer fibres, such as nylon-6 and nylon-66, which were to be the main sources of development in Western countries during the subsequent period. The first Soviet production of nylon-6 ('kapron') began at Klin in 1948 and the rapid industrial assimilation was made possible by research on the synthesis of polyamides carried out during the war by I. L. Knunyants and his colleagues.[203] This initial development was not exploited further until after 1956 when GIPROIV produced a standard plant design (*tipovoi zavod*) and seven nylon-6 plants based on this design were built. In general, the late 1940s and early 1950s were a time of considerable diversification and technical progress in the Soviet manmade fibres industry. During this period the production of high strength viscose cord was begun at Mogilev on new centrifuge spinning machines;[204] the output of acetate filament began at Serpukhov (previously held back by the overall weakness of the organic chemicals sector, and by shortages of acetic acid and acetic anhydride, in particular); and 'kapron' staple was produced on a continuous line process which could be used for manufacturing staples from other polymers.[205] In 1947, a new method of continuous spinning, trimming and dyeing was devised by N. L. Livshits and after initial introduction at Kamensk, had, by the late 1950s, been introduced into all Soviet viscose factories.[206]

During the late 1950s the Soviet manmade fibres industry began to branch out into even more sophisticated polymer materials. The researches of Z. A. Rogovin and Z. A. Zazulina paved the way for the acrylic fibre 'Saniv' (a copolymer of acrylonitrile/40 per cent and vinyldenchloride/60 per cent) and the fluorine fibre 'Ftorlon' to be produced on an experimental scale.[207] A genuinely original and important Soviet innovation also occurred in the late 1950s; this was the creation of 'Enant' (a nylon-7 fibre derived from the polycondensation of amino-oenanthic acid) developed by the USSR Academy Institute of Organic Compounds, VNIIV and the State Institute for the Nitrogen Industry.[208] In July 1960, the first pilot plant for the production of the

[201] N. P. Fedorenko, *op. cit.* (1961), p. 416.
[202] *Ibid.*, p. 419.
[203] G. E. Birger *et al.*, *op. cit.*, p. 124.
[204] *Stroitel'naya ... khimicheskaya tekhnologiya*, *op. cit.* (1971); E. P. Ivanova *et al.*, *Ekonomika promyshlennosti khimicheskikh volokon* (1968), p. 20.
[205] N. P. Fedorenko, *op. cit.* (1961), p. 458.
[206] E. P. Ivanova *et al.*, *op. cit.*, p. 20.
[207] N. P. Fedorenko, *op. cit.* (1961), p. 491.
[208] The melting point of this fibre is higher than that of nylon-6 (220°C c/p 208°C), its moisture absorbance is less and its wash and wear properties are probably superior to those of either nylon-6 or nylon-66 (*Man-Made Fibres*, May 1966, p. 49). Unfortunately, there were some teething troubles in bringing this fibre to mass production (N. P. Fedorenko, *op. cit.* (1961), p. 460).

Soviet polyester 'Lavsan' (terylene) came on stream at Kursk and before the end of the year production began on an industrial scale. Research on this very important fibre started at the Academy Institute of Organic Chemistry in 1949, when a research team, led by Korshak, began its investigation into the complexities of polyester synthesis. In 1951, the project was handed over to VNIIV for further development. The term 'Lavsan' is, in fact, based on the initials of the laboratory in which the original preparatory work was done (i.e., *Laboratoriya Vysokopolimernykh Soedinenii Akademii Nauk*).[209]

By the beginning of the 1960s most of the major synthetic fibres (polyamide, polyester and acrylic) were produced in the USSR on either a semi-industrial or commercial scale. However, Soviet writers agree that, in terms of overall output and quality, Soviet synthetic fibre production at this time was only at an initial stage (*nachal'naya stadiya*) of development.[210] In order to meet growing Soviet demand, a *grandioznaya programma* was required, entailing the construction of 27 new large enterprises. During the seven year plan (1959–65) investment in manmade fibres was 8·5 times greater than that of the preceding seven years;[211] output of synthetic fibres grew at an average rate of 29·5 per cent per annum compared with an overall growth rate of 13·7 per cent for chemical fibres as a whole.[212] However, the production of high quality fibres in all spheres of the industry continued to be a problem. At the beginning of the 'chemicalisation drive', Fedorenko had referred to the comparatively inferior quality of Soviet viscose, acetate, nylon and cuprammonium fibres, placing the responsibility for this on shoddy workmanship, inadequate quality control and poor equipment.[213] It is very probable that the poor quality of intermediates was also an important factor.[214] These criticisms continued to be expressed in both Soviet and Western sources throughout the 1960s.[215]

The key phases in the development of the Soviet manmade fibres industry, and comparable events in Western countries, are set out in Table 6.21. It emerges clearly that in the case of all the most important manmade fibres, production began in the West before it began in the USSR;[216] in some cases, such as acetate fibre, the gap was as much as 17 years between the beginning of Soviet output and that of the last of the Western imitators. If we were to assess the Soviet achievement within the perspective of its own economic growing

---

[209] *Man-Made Textiles*, February 1963, pp. 44–6; the data for this article comes largely from *Khimicheskie Volokna*, 1960, No. 5, p. 3 and 1959, No. 2, p. 11.

[210] See, for example, N. P. Fedorenko, *op. cit.* (1961), p. 451.

[211] *Ibid.*, p. 419.

[212] The Soviet terms for making the conventional distinction between cellulose-based fibres and non-cellulose fibres are '*iskusstvennyi*' and '*sinteticheskii*' respectively. The figures in the text are quoted by E. P. Ivanova *et al., op. cit.*, p. 21.

[213] N. P. Fedorenko, *op. cit.* (1961), p. 505.

[214] M. I. Garbor, *op. cit.*, p. 132, argues that the inferior quality of Soviet fibres could be explained 'not so much by the backward level of our technology, but by the absence of high-quality intermediates'.

[215] See, for example, *Stroitel'naya . . . khimicheskaya tekhnologiya, op. cit.* (1971); *Man-Made Fibres*, May 1963, p. 60; *Ibid.*, October 1963, p. 72; *ibid.*, May 1964, p. 46; *ibid.*, January 1964, pp. 41–3; *ibid.*, September 1966, p. 60.

[216] The only exception to this is the semi-handicraft production of viscose silk in 1909, which preceded Japanese production by seven years; the first large scale production of viscose staple in the USSR and Japan occurred in the same year (1933).

pains we would naturally have to take into account such factors as initial backwardness, shortages of high quality intermediate products and extensive war damage.[217] Within a context of straightforward international comparison, however, it is clear that the Soviet Union is substantially backward in chemical fibre technology.

When the evidence of Sutton and others is reconciled with the data in Table 6.21 it also emerges that Western countries have played a decisive role in the development of the Soviet chemical fibres industry throughout its history (see Table 6.22). It is possible to divide these fibres into three groups according to the pervasiveness of foreign influence. First, there is the relatively small group of fibres (viscose staple and cuprammonium fibre) which, in the absence of evidence to the contrary, appear to be indigenous Soviet imitations (though how far the USSR was assisted by foreign firms through technical agreements which were still in force at the time that these fibres were introduced on a commercial scale, remains to be investigated). Second, there is another group of fibres (viscose silk, viscose cord and acetate fibre) for which substantial imports of foreign equipment followed fairly closely after first commercial production had begun in the USSR; this suggests strongly that the Soviet technology was not regarded by the Russians themselves as being of a sufficiently high standard in terms of cost or quality indices. In the case of the third group, which includes all the polymerised fibres, the acquisition of process plant from abroad actually *pre-dated* first commercial production in the USSR. This is the most extreme form of dependence on foreign technology. When one examines the cost and size of many of these plants it is difficult to disagree with the view expressed in one American journal that 'the USSR behaves as if it had no chemical industry at all';[218] the output of the Mogilev polyester fibre plant, alone, was scheduled to be the equivalent of the total British output of polyester fibres in 1964 and the physical size of the plant can be judged from the fact that it occupied an area roughly equal to that of Hyde Park.[219] The main weakness in the Soviet research-production cycle does not appear to be at the research end. Many foreign specialists have been impressed with the quality of Soviet research into manmade fibres, especially that undertaken by VNIIV.[220] Soviet technologists have also shown that they are capable of developing specialised fibres to be produced in small quantities. The most notable examples of such fibres are 'Enant' (nylon-7), 'Ftorlon' (a fluorine with copolymer) and 'Vinitron' (a combination of nitrocellulose and chlorinated PVC), all of which are original Soviet innovations. It is possible

[217] These first two factors are themselves a reflection of technical level in a broader sense; the third factor is not confined to the USSR.

[218] *Chemical Week*, 11 March 1961, quoted by A. C. Sutton, *Western Technology and Soviet Economic Development 1945 to 1965*, Stanford, Cal. (1973), p. 146.

[219] *Sunday Times*, 19 October 1969.

[220] During the 1960s, for example, the British journal *Man-Made Textiles* ran a series of articles informing their readers about the latest developments in Soviet research into manmade fibres. Due to the fact that British firms were doing a considerable amount of business with the USSR at this time, it is conceivable that there was an ulterior motive in this; though this interpretation is, perhaps, too machiavellian. American observers were also impressed with the level of Soviet research; see for example, M. Buras, *Chemical and Engineering News*, 31 July 1961, p. 134, quoted by A. C. Sutton, *op. cit.* (1973), p. 179.

that the development of these fibres has been a matter of high priority because, as Sutton suggests (following expert American assessment), their resistance to chemicals and photodegradation makes them valuable in military application (lightweight clothing, parachutes, etc.);[221] this interpretation appears to be consistent with the relatively early assimilation of PFTE and silicon polymers referred to in the previous section. But outside these specialised areas of high priority, the evidence strongly suggests that Soviet designers and engineers find great difficulty in making the transition from research or pilot plant production to full-scale mass production. As a result of this weakness the USSR lags behind the West in manmade fibre technology and requires large injections of foreign process plant to keep pace with the latest developments.

The level of research[222]

Having worked backwards through the 'research-production cycle' we now arrive at the first link in the chain—the ideas and scientific principles that, either directly or indirectly, form the basis of future technological developments. Again, it should be stressed that research is only one element in the overall assessment of technological level, and taken on its own it could be misleading. For example, Britain's contribution to applied chemical *science* during the nineteenth century was considerable, but its chemical *industry* lagged well behind that of Germany before the First World War.

There are two very difficult problems to be faced before an assessment of the level of Soviet chemical research can be attempted. The first problem is that of 'objectivity'. Though science is thought to be an international activity, characterised by dispassionate judgement, elements of national pride and ethnocentricity inevitably creep into the assessments of scientists and historians of science who write about the development of their discipline. Western and Soviet sources (particularly the latter) tend to stress their own contributions and, without expertise in the fields concerned, it is impossible to reconcile their different opinions. The only possible solution is to select the *best available* measures, which will minimise bias as far as possible. The second major problem lies in the *scope* of the assessment. Although there will usually be a fairly strong correlation between outstanding research contributions and the overall level of chemical science in a particular country, one must also bear in mind the possibility that a relatively underdeveloped country could produce a great scientist (for example, the Argentinian bio-chemist, L. Leloir), while a more developed country with abundant facilities and a sound overall performance may produce less than its share of outstanding scientific talent.

With the above problems in mind, the following approaches were applied in order to obtain a comparative measurement of the level of Soviet chemical research: (a) an assessment of outstanding contributions to chemical science,

---

[221] A. C. Sutton, *op. cit.* (1973), pp. 374–5.
[222] This section is only concerned with research *outputs*. The Soviet research *effort* in chemistry will be discussed in a future case study to be published in our second volume.

Table 6.21 *First commercial production of various chemical fibres (in order of introduction in the USSR)*

Type of fibre	USSR	USA[5]	Germany (FRG after 1945)[5]	UK[5]	France[5]	Italy[5]	Japan[5]
Viscose continuous filament	1909 (ceased production in 1913. Started up again in 1924)[1]	1905	1901	1900*	1903	1919	1916
Viscose staple fibre	1933[1]	1927	1916*	1925	1931	1931	1933
Cuprammonia staple fibre	1935[1]	1926	1897*	1904–55	1904–50	1925	1924
Nylon continuous filament (Soviet designation—'Kapron')	1948[1]	1938*	1941[6]	1941	1941	1938*	1942
Viscose cord (high strength filament)	1949[2]	1937	1935*	1936	1936	1939	1941
Nylon ('Kapron') staple fibre	Early–Mid 1950s (produced in 'significant' quantities by end of 1950s)[3]	1946	1950	1947	1941*	1942?	1950
Acetate continuous filament (staple not produced in USSR?)	1954[2]	1919	1907*	1921	1923	1930	1937
Polyester fibre (terylene: Soviet designation—'Lavsan')	Early 1960s (both produced on a semi-industrial scale in 1958 and were the priority areas for development in 7-year plan, 1959–65)[4]	1949*	1955	1950	1954	1954	1958
Acrylic fibre (Orlon: Soviet designation—'Nitron')		1944	1943*	1957	1955	1958	1957
PVC fibre	Not produced in USSR by 1970. But manufacture 'to be organised in the near future'[1]	1936*	n.d.	n.d.	n.d.	1952	1944
Polyvinyl alcohol fibre		1963?	n.d.	n.d.	n.d.	n.d.	1950*
Polypropylene fibre		1957*	n.d.	1961	n.d.	1957*	1962

Sources:
1 *Stroitel'naya, selskokhozyaistvennaya i meditsinskaya tekhnika, khimicheskaya tekhnologiya* (1971), pp. 173–8.
2 E. P. Ivanova et al., *Ekonomika promyshlennosti khimicheskikh volokon* (1968), p. 20.
3 N. P. Fedorenko, *Ekonomika promyshlennosti sinteticheskikh materialov* (1961), p. 500.
4 G. E. Birger and A. A. Konkin, in *Khimicheskaya promyshlennost' SSSR* (1959), p. 125.
5 G. C. Hufbauer, *Synthetic Materials and the Theory of International Trade*, London (1966), p. 132.
6 A. C. Sutton, *Western Technology and Soviet Economic Development 1945–65*, Vol. 3, Stanford, Cal. (1973), p. 181.

Notes:
* – original innovating country.
n.d. = not produced in 1966 but no data thereafter.

Table 6.22 *Soviet acquisition of manmade fibre plant from abroad (excluding equipment for intermediates, textile fabrication and finishing)*

Designation of fibre	Date of first large-scale production in USSR		Foreign acquisitions
Viscose silk	1909 (re-started in 1924)	1928	10-year agreement between USSR and Soieries de Strasbourg S-A for viscose technology. This firm built plants in Leningrad, Moscow and Mogilev (for the latter, equipment was supplied by Oskar Kohorn AG (Germany) which also had a technical assistance contract with the USSR).[1]
Nylon-6	1948	1944	German nylon-6 plant at Landsberg dismantled and shipped to USSR; probably rebuilt at Klin.[4a]
		1959	Contract with Zimmer (FRG) for nylon-6 plant; capacity 2,100 tons.[3]
		1960	Production started of two nylon-6 plants built by Snia Viscosa (Italy): total capacity 4,200 tons.[3]
Viscose cord fibre	1949	1959	Equipment supplied by Chatillon (Italy) for high tensile strength cord:[2] capacity 17,500 tons, cost £10 million.
		1963	Contract signed with Chatillon (Italy) for rayon tyre cord plant: cost £4·6 million; capacity 17,500 tons.[3]
		1964	Production began of viscose tyre cord plant built by Courtaulds (UK) Cost £9 million; capacity 17,700 tons.[3]
Acetate fibre	1954	1958	Supply of machinery and technical assistance by Courtaulds (UK) and Krupp (FRG): cost £3·8 million; capacity 5,400 tons.[3]
		1958	Cellulose acetate plant built by Courtaulds (UK): cost £3·8 million; capacity 6,800 tons.[3]
		1962	Contract signed with Courtaulds (UK) for cellulose triacetate plant; cost £6 million: capacity 7,000 tons.[3]
Polyester fibre	1960	1958–61	Polyester fibre plant ('Lavsan') built at Kursk by Krupp: cost $14 million;[5] capacity 17,500 tons.[3]
		1961	Output began of (part of) polyester plant at Kursk built by Von Kohorn (USA): capacity 2,450 tons.[3]
		1964	Contract signed with Polyspinners (UK) for polyester fibre plant at Mogilev: cost $140 million: capacity 50,000 tons.[3]
		1964	Contract signed with Krupp (FRG) for polyester fibre plant: cost £1·5 million; capacity 3,000 tons.[3]
Acrylic fibre	Semi-industrial scale 1958	1959	Purchase of Japanese Kanekalon and Acrylonitrile plant: cost $30 million.[6]
		1959	Contract signed with Courtaulds (UK) for complete acrylic fibre plant: cost £2·2 million; capacity 4,500 tons.[3]
		1967	Contract agreed with Courtaulds (UK) for acrylic fibre plant: to cost £9·4 million.[3]

Sources:
1 A. C. Sutton, *Western Technology and Soviet Economic Development, 1917 to 1930*, Stanford, Cal. (1960), p. 223.
2 *Ibid., 1945 to 1965*, Stanford, Cal. (1973), p. 170.
3 List of Soviet purchases of process plant based on press reports (informal source).
4 A. C. Sutton *op. cit.*, (1973), p. 181.
5 *Ibid.*, p. 182.
6 *Ibid.*, p. 184.

Note:
a – Sutton's view that the first production of nylon-6 in the USSR was based on the German Landsberg plant is consistent with the long gap in plant construction between 1948 and 1956, referred to earlier. Presumably, GIPROIV spent several years examining the operation of the original plant before it formulated its *tipovoi zavod*.

based largely on the awarding of Nobel Prizes and associated data,[223] (b) the extent to which the published results of Soviet research in chemistry are cited in foreign scientific journals and vice versa, and (c) a professional assessment of Soviet work in one particular key branch of modern chemistry (polymer science).[224] The possible sources of bias present in each of these approaches are reviewed below in the relevant sections.

OUTSTANDING CONTRIBUTIONS TO CHEMICAL SCIENCE

In Tsarist Russia, chemical research, as opposed to the chemical industry, was highly developed.[225] During the eighteenth and nineteenth centuries individual Russian chemists contributed significantly to the development of the discipline. In the latter part of the eighteenth century Lomonosov, 'the father of Russian chemistry', carried out important experimental work on the conservation of matter in chemical reactions and on other fundamental chemical principles (e.g., atomic theory and the kinetic theory of gases). During the nineteenth century the work of G. Hess on the laws of thermochemistry was important and the famous Kazan school of chemistry, founded by N. N. Zinin and A. M. Butlerov, produced a number of gifted scientists. The contribution of Butlerov himself on the structure of organic compounds was used to predict the existence of many unknown compounds and his researches laid the basis for the polymerisation of hydrocarbons of the ethylene series (of great importance in the twentieth-century plastics and synthetic fibres industry). During the last quarter of the nineteenth century, the greatest of all Russian scientists, D. I. Mendeleev, formulated his periodic table of chemical elements and confirmed it experimentally through the discovery of the new elements, Gallium, Scandium and Germanium; this was a landmark in the history of chemical science. Although Western and Soviet historians of science may differ on some matters of priority in scientific discovery, or on whom to include in their anthologies, there seems to be a general consensus that all the scientists mentioned above played a prominent role in the evolution of chemistry. Mendeleev, in particular, is generally regarded as one of the greatest chemists who ever lived and on a similar plane to other great chemists such as Boyle, Lavoisier and Dalton. Thus, before the revolution in 1917, a well established tradition in chemical research had been created in Russia.

It is more difficult to identify the major contributions to chemistry during the twentieth century by means of general histories of science and technology, partly because the subject became so vast and specialised, but also (in the case of the more recent work) because of uncertainties as to how these contributions would stand the test of time. Fortunately, at the turn of the century, a formidable administrative and scientific apparatus was created to make just this kind of objective assessment—the Nobel Prize Committee for Chemistry.

---

[223] The possibility of using a number of chemical engineering prizes was also considered, but this was discounted due to the relatively greater national partiality.

[224] See Appendix B, written by Dr A. Holt.

[225] This will be looked at in greater depth as part of a future case study of the history of Russian chemical science and the development of the chemical industry. The information in pp. 288−9 is based on Appendix 2 of the first draft of this chapter (*CREES Discussion Paper RC/C* No. 11).

The problem of how far the awarding of Nobel Prizes in chemistry is likely to be free from bias is discussed below, together with the results that emerge from a study of how these prizes have been awarded.

THE OBJECTIVITY OF NOBEL PRIZES AS A MEASURE OF OUTSTANDING CONTRIBUTIONS TO CHEMISTRY

The Nobel Prize for chemistry was first awarded in 1901.[226] Nobel laid down in his Will that the prize should be awarded for the most outstanding work in chemistry during the preceding year, although this rather stringent condition was subsequently interpreted to mean either (a) the most outstanding *recent* work, or (b) older work, the significance of which had recently become apparent. Unfortunately, this condition, even in its diluted form, was sufficient to deprive Mendeleev of the Nobel Prize in 1906 on the grounds that his work was outstanding but its significance had long been apparent.[227]

There is obviously a problem in deciding what work will be of enduring importance and the statutes lay down a number of strict conditions to assist the committee in making a decision that will stand the test of time; to be shown to have conferred immortality on a relatively mediocre chemist would be a source of embarrassment. First, the statutes insist that 'no consideration whatsoever shall be given to the nationality of the candidate'.[228] In order to ensure objective consideration of each candidate, Nobel donated a considerable sum of money in his Will to the creation of two institutes in the fields of physics and chemistry, to be administered by the Swedish Academy of Sciences. Part of the work of these institutes is devoted to the full-scale scientific investigation of each candidate's work and, in addition, the award-making body can appoint outside experts to help in the deliberations.[229] The nomination procedure is also designed to eliminate bias as far as possible. Apart from members of the Swedish Academy of Sciences and previous prize-winners, nominations are accepted from professors in at least six universities, 'selected by the Academy of Sciences with a view to ensuring the appropriate distribution of the commission over the different countries and their seats of learning'.[230] These institutions are chosen so that the main language regions of the world (including Russia, naturally) are represented.

It would be naïve to think that bias could be entirely eliminated from these kinds of judgement and there have certainly been occasions when the Nobel Committee has been severely criticised. However, the scientific and adminis-

---

[226] There were originally (before the prize for economics began to be awarded) five Nobel Prizes: the prizes for chemistry and physics, awarded by the Swedish Academy of Sciences; the prize for medicine and physiology awarded by the Caroline Institute; the literature prize, awarded by the Swedish Academy; and the Nobel Peace Prize, awarded by a special committee of the Norwegian Storting.

[227] H. Schuck *et al.*, *Nobel: The Man and his Prizes*, London (1962), p. 368. The prize went instead to the French chemist Moissan for his work on fluorine, and Medeleev died before his claim could be considered once again.

[228] *Ibid.*, p. 647; the statutes are included in full in the Appendix to this book.

[229] *Ibid.*, pp. 649–50.

[230] *Ibid.*, p. 349. About 450 people annually have the right to nominate candidates for the chemistry prize, of which just under one third are Swedes, p. 348.

OUTSTANDING CONTRIBUTIONS TO CHEMISTRY IN THE TWENTIETH CENTURY

In the opinion of the Nobel Committee on Chemistry, only one Russian chemist during the twentieth century has made an outstanding contribution to the subject that could surpass the claims of others. This event occurred in 1956, when the prize was awarded jointly to C. N. Hinshelwood (Britain) and N. N. Semenov (USSR) 'for their research on the mechanism of chemical reactions'. In terms of total numbers of prizes awarded between 1901–72 this places the USSR firmly in the 'third division' of scientific nations. However, this relatively simple way of looking at Nobel Prizes is not sufficiently exact. First, it does not show how the fortunes of particular countries have fluctuated over time. Second, it is also misleading because many of the prizewinners travelled about a good deal and spent long periods in different countries, thus making it difficult to decide which country should get the credit for the final achievement. By means of an extensive search through the career profile of each prizewinner it is possible to establish their place of work in each year between their first academic appointment and death or retirement at the age of 70,[231] and to calculate what proportion of total Nobel prizewinners (past or future) living in any given year were working in particular countries. This is a more exact measurement of the location of outstanding scientific talent than a simple 'league table', and the results of the survey are presented in Fig 6.3. Due to the fact that only recognised scientific talent is taken into account, the graph is at its least accurate at both extremes of the time scale. During the early period, scientists who might well have won a Nobel Prize at a later date, had they lived long enough, are not counted; during the recent period, the graph does not take into account future prizewinners who are at present engaged in mature research. Thus, the graph probably underestimates German dominance before the First World War and American ascendancy after the Second World War. However, on the basis of the available evidence, it is interesting to note the steep and steady decline of Germany as a world power in chemical science together with the dwindling significance of France; during the same period, Britain improved its position, and from the 1930s onwards the United States emerged as the main centre of outstanding research in chemistry. Soviet research (according to the criteria used in this exercise) has remained at a relatively low level throughout the whole period.[232] Confirmation of these 'centres of excellence' can be obtained by looking at the places where scientists worked, other than in their native country, prior to the award of the Nobel

---

[231] It was assumed that scientists continued to do useful work up to this age, although this obviously varies between individuals. The biographical information was obtained from E. Farber, *Nobel Prize Winners in Chemistry, 1901–61*, London (1962); *Science* (1962–72); *Who was Who in Science* and *International Who's Who*, 1967–8 and 1972–3; *American Men and Women of Science*, 12th edn, 1973. Only stays of one year or more are counted.

[232] The slight Soviet peak in 1961 is accounted for by the visit of R. B. Woodward (USA) to the USSR, although the length of his visit is not clear from the biographical sources used.

Prize.[233] The evidence suggests that up to the 1930s a few years spent at a famous German centre of chemical research such as Gottingen or Munich was considered to be almost an obligatory element in the training of a promising young scientist. This tendency was particularly widespread among continental Europeans, apart from the French. The post-war generation of scholars, who obtained fellowships and grants to work in leading British and American centres such as Berkeley, MIT, Harvard, London and Cambridge, began to achieve international stature during the last 10–15 years.

Table 6.23 *'League table' of Nobel Prizes in chemistry, 1901–72*

	Country	Points
Leading countries	Germany (GDR and FRG)	22
	Britain	15
	USA	15
Secondary countries	France	5
	Sweden	4
	Switzerland	3
Other countries	Argentina	1
	Canada	1
	Czechoslovakia	1
	Finland	1
	Holland	1
	Italy	1
	Norway	1
	USSR	1

Sources:
E. Farber, *Nobel Prize Winners in Chemistry, 1901–61*, London (1962); *Science* (1962–72): every year this American journal publishes feature articles on each of the prizewinners as the results are announced.

Rules for calculating the points:
1 In cases where the prize was *shared* for different contributions among scientists in the same country or different countries, each contribution was counted.
2 In cases where the prize was awarded *jointly* to scientists in the same country for the same common contribution, this was only counted once.
3 In cases where the prize was awarded *jointly* to scientists in different countries, a point was awarded to each country.

CITATION OF SOVIET CHEMICAL RESEARCH[234]

Another method of ascertaining the scientific standing of a nation is to see how often the results of its research are referred to in foreign scientific papers. An evaluation of what constitutes important work is thus undertaken by the scientific community at large rather than by a small committee of experts. This

---

[233] Foreign travel after the award of the prize was thought to be less significant than travel before the award, especially during the formative years of postgraduate and post-doctoral research. Scientists with an established international reputation would tend to receive generous invitations from countries anxious to *raise* the standard of their scientific work.

[234] See also Appendix 6B for an analysis of Soviet and Western research in polymer chemistry by means of citations.

Fig. 6·3 LOCATION OF KNOWN, OUTSTANDING SCIENTIFIC TALENT IN CHEMISTRY, 1901-1972

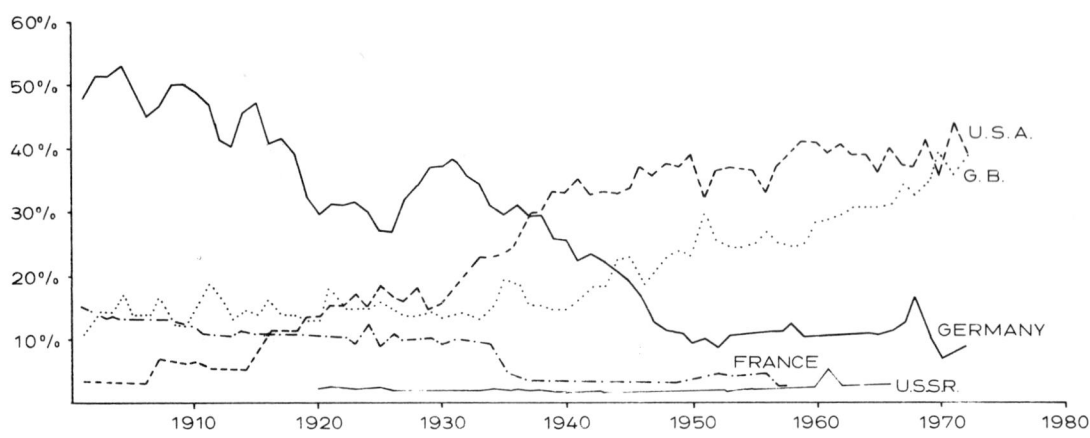

approach also has the advantage that it is not restricted to outstanding contributions but covers the whole spectrum of research activity in a particular field.

The main source of data for studies of this kind is the huge American publication *Science Citations Index*, which comes out four times a year and covers all the world's leading scientific journals. The *Index* is divided into three parts. The first part lists in alphabetical order the authors of all the articles that have appeared during a preceding period of three months, with full references. The second part classifies these articles in terms of key phrases or concepts. The third and largest section lists in alphabetical order all the papers which are cited by the original group of articles. This is a valuable source of information for scientists and cuts out a good deal of conventional bibliographical work. By picking out an initial selection of key articles and 'cycling' backwards through past issues of the *Index* it is possible to compile an exhaustive bibliography in a comparatively short time. It is also a valuable source of information for assessing the relative impacts on world science of individuals, institutions or nations.[235]

The objectivity of using citations as a measure of Soviet performance should not be overstated. It is possible that, on balance, a larger proportion of valuable (and, therefore, citable) work is kept secret in the USSR than in other countries. It is also possible that scientists are directed towards articles in their fields primarily through their membership of 'invisible colleges', and that

[235] For example, these data can be used for 'spotting' potential Nobel prizewinners or used by individual scientists to obtain a measure of their own significance in the field; now that a *Social Science Citations Index* exists, this pleasure is available to social scientists!

citations are solicited by means of the early circulation of offprints; the relative remoteness of Soviet scientists makes this kind of close association improbable and, therefore, they may not be cited as often as they deserve to be. The language barrier is also an impediment, but this should not be exaggerated. Several of the major Soviet chemical journals are translated into English in full.[236] Also, the huge publication *Chemical Abstracts*, like the *Science Citations Index*, covers all the world's leading chemical journals and is, in fact, even more exhaustive than the latter. The index of this publication is extremely detailed and is designed to refer the scientist to the relatively few articles which have been published recently within his precise field of research. In the case of a Russian article, he would read the abstract and thereafter it would be a relatively minor step either to read the Russian translated journal (if available) or to obtain a translation of the article from the original journal if it seemed to be sufficiently important.

In 1969 a very important study appeared in the USSR, which largely addressed itself to the measurement and interpretation of Soviet scientific performance by means of citations.[237] The study was remarkable both for the richness of its data and for the undoctrinaire manner in which the data were handled. With the help of a large panel of experts, the authors identified a number of key fields of contemporary science and proceeded to look systematically at the patterns of citations in the most important national journals relating to these fields. Only citations of major articles were considered; references to books, reviews, short communications and letters to editors were excluded from the sample. The authors demonstrated that the 'effectiveness' of Soviet science (i.e., the extent to which articles by Soviet scientists were cited in foreign journals) was substantially inferior to that of selected Western countries, despite the fact that the total output of Soviet work, referred to in *Physics Abstracts* and *Chemical Abstracts*, was second only to that of the United States.[238] Their evidence regarding chemical science is set out below in Table 6.24. Though these striking results would appear to cast doubt on the value of much of the chemical research undertaken in the USSR, Nalimov and Mul'chenko suggest that 'It is scarcely possible to accept the hypothesis that the level of our work is lower than the world scientific level'[239] (with the exception of some branches of analytical chemistry for which it is admitted that Soviet scientists are not adequately provided with modern equipment). Instead, the USSR is depicted as a country which, for various reasons, is cut off from the main information streams of world science and,

---

[236] For example, Russian Journal of Physical Chemistry (*Zhurnal Fizicheskoi Khimii*), Russian Journal of Inorganic Chemistry (*Zhurnal Neorganicheskoi Khimii*), Soviet Plastics (*Plasticheskie Massy*), Russian Chemical Review (*Uspekhi Khimii*) and Polymer Science USSR (*Vysokomolekulyarnye Soyedineniya*).

[237] V. V. Nalimov and Z. M. Mul'chenko, *Naukometriya* (1969). Much of the data for this volume first appeared in Z. B. Barinova *et al.*, *Nauchno-Tekhnicheskaya Informatsiya*, Series 2, 1967, No. 12, pp. 1–11.

[238] V. V. Nalimov and Z. M. Mul'chenko, *op. cit.*, p. 137. (These data on total output come from Derek Price, 'The Distribution of Scientific Papers by Country and Subject—a Science Policy Analysis'. Pre-print (1968)).

[239] *Ibid.*, p. 162.

consequently, suffers great losses in the circulation and channelling of ideas.[240] There is certainly some force in these arguments but it is doubtful whether impediments to the free circulation of information are sufficient in themselves to explain the Soviet position. The language barrier is an obvious obstacle to communication, but its impact should not be exaggerated. The fact that journals such as *Chemical Abstracts* cut down the information search to a few, well defined items has already been referred to.[241] Conversely, although the lack of direct contacts with foreign scientists is a problem, it is by no means certain that the *formal* information system in the USSR is markedly worse than equivalent arrangements in Western countries. The All Union Institute for Scientific and Technical Information (VINITI) has stimulated a good deal of interest in the West and it is not unknown, even, for Western scientists with a fluent knowledge of Russian to use *Referativnyi Zhurnal* as a guide to developments that are taking place in their field in other Western countries! Thus Soviet scientists are able to keep reasonably well up to date with developments abroad, and to incorporate them in their own work. To this extent, the judgement of Soviet research papers by Western standards is based not on how out of date they are, but on their intrinsic quality.

It is possible to draw conclusions from Nalimov and Mul'chenko's data that are more consistent with the view that the relatively low level of citation of Soviet chemical research in the West is due to more fundamental factors than the simple problems of language and communication. The work of Japanese chemists, for example, is considerably more 'effective' than that of their Soviet counterparts. Even when a Soviet journal in the sample is available in an English cover-to-cover translation (for example, the *Journal of Physical Chemistry*) it does not seem to stimulate a higher level of citation. On the other hand, it is significant that eminent Soviet scientists in the selected fields have a relatively high rate of citation in foreign journals,[242] which would tend to suggest that Western scientists exercise some rational discrimination in their selection of source materials; where important research results are presented by established Soviet scientists, the language barrier and cultural isolation do not appear to be insurmountable obstacles to communication. Finally, insofar as information problems are important, it might be argued that Nalimov and Mul'chenko have not pursued this matter to its logical conclusion. Poor information flows may, to some extent, reduce the level of citations, but they will also retard the *level* of scientific development itself. Presumably, it is on these grounds that the expensive improvements to Soviet information services, which the authors recommend, must be justified.

---

[240] Specifically, Nalimov and Mul'chenko point to the following factors: absence of direct, regular contacts with foreign scientists; delays in receiving foreign journals; insufficient knowledge of foreign languages by Russian scientists and vice versa; considerable delays in publishing articles in Soviet journals; badly organised libraries in the USSR; absence of specialised information centres in the USSR with computer retrieval systems.

[241] This might perhaps be contrasted with the diffuse nature of sources and research topics in the social sciences; here extensive research is needed, so independent linguistic skills on the part of the researcher are relatively more important.

[242] *Ibid.*, pp. 151–2.

Table 6.24 *The 'effectiveness' of Soviet scientific contributions in chemistry*

(a) *Distribution of citations of articles in physical chemistry by country of origin*

Country or Language	Percentage of articles in *Chemical Abstracts*	Citations in journals (%)			
		USSR	FRG	USA	France
USSR	28·0	44·0	3·6	2·7	4·8
GDR/FRG	5·0	11·0	29·0	5·6	7·6
English Language	38·0	38·0	58·0	88·0	64·0
Japan	3·9	1·5	1·2	0·8	4·1

Journals used: Zhurnal Fizicheskoi Khimii (this is translated into English); *Zeitschrift für physikalische Chemie* (FRG); *Journal of Physical Chemistry* (USA); *Journal de Chimie Physique* (France).

(b) *Distribution of citations of articles in molecular spectroscopy by country of origin*

Country or language	Percentage of articles in *'Chemical Abstracts'*	Citations in journals (%)			
		USSR	USA/Canada	FRG	France
USSR	18·5	51·0	4·2	2·6	4·7
English language	59·0	30·0	68·0	54·0	45·0
GDR/FRG	7·0	7·0	7·2	38·0	20·0
France	5·8	7·5	5·5	1·7	23·0
Japan	5·0	3·1	5·3	1·7	3·0

Journals used: Optika i Spektroskopiya (USSR), not available in English; *Berichte der Bunsengesellschaft für Physikalische Chemie* (FRG); *Journal of Molecular Spectroscopy* (USA); *Canadian Spectroscopy* (Canada); *Comptes Rendues Hebdomadaires de Séances de l'Academie des Sciences* (France); *Bulletin de la Societé Chimique de France* (France).

(c) *Distribution of citations of articles in analytical chemistry by country of origin*

Country	Percentage of total articles	Citations in journals (%)	
		USSR	USA
USSR	21·5	42·0	3·8
USA	20·5	18·0	61·4
GDR/FRG	10·0	10·0	5·9
Japan	6·8	2·5	2·7
UK	4·3	6·6	9·7
France	4·2	3·1	2·9

Journals used: Zhurnal Analiticheskoi Khimii (USSR), not available in English; *Analytical Chemistry* (USA); data on the total number of articles produced in each country comes from R. B. Fisher, in *Analytical Chemistry*, 1965, Vol. 37, No. 13, pp. 27a–37a.

Conclusions

In November and December 1967, the United Nations Industrial Development Organisation (UNIDO) organised an international symposium in Athens to discuss problems relating to the development of the chemical industry.[243] An attempt was made at the symposium to define the typical stages in the development of the industry, and the following typology emerged:

Phase 1
- (a) Production is restricted to simple chemical end-products (oils, paints, soaps, polishes, matches, cosmetics, simple pharmaceutical preparations, etc.).
- (b) Process technology is unsophisticated and skill requirements are low.
- (c) R and D in process engineering depends on sources outside the country. The domestic engineering industry can, at most, undertake maintenance and repair.

Phase 2
- (a) Sporadic backward integration to basic chemicals: ammonia, sulphuric acid and nitric acid.
- (b) Production of fertilisers (using native basic chemicals) and other agricultural chemicals; production of more sophisticated end-products (e.g., plastics, manmade fibres, detergents, dyes and explosives).
- (c) Some R and D (mainly confined to universities); capacity to design simple process equipment. The domestic chemical engineering industry is capable of producing simple items for plant construction (e.g., pipes, ducts, tanks, etc.).

Phase 3
- (a) A diversified base exists for heavy organic and inorganic chemicals (acids, alkalis, salts, industrial gases, petrochemicals); a broad range of organic intermediate products is produced enabling the synthesis of complex dyes and pharmaceuticals to be undertaken.
- (b) Substantial R and D capacity, although imported technology is still used to a considerable extent.
- (c) The domestic chemical engineering industry is capable of supplying 50–75 per cent of all process equipment required (e.g., blowers, compressors, pumps, mixers, pressure vessels, etc.).
- (d) Other industrial sectors are penetrated by chemical products; plastics, synthetic rubbers and fibres are adopted as structural components of industrial products.

[243] The proceedings of this symposium are published in *The Chemical Industry*, UNIDO Monograph, No. 8, New York (1969).

Phase 4

(a) Important new products and processes spring from domestic R and D specialising in particular lines, supplemented by international exchanges of advanced technology.

(b) Chemical engineering know-how is capable of translating laboratory innovations into efficient industrial production by devising new manufacturing processes and designing sophisticated equipment.

(c) The domestic chemical engineering industry can supply specialised and sophisticated equipment to the chemical industry, while not neglecting the advantages of international specialisation and exchange.[244]

According to the criteria used in earlier sections of this paper, it was suggested that the USSR lagged behind the main Western countries in its level of chemical technology. The UNIDO typology, on the other hand, suggests not only that the USSR is backward in terms of each of these particular indicators, but that its chemical industry is situated *at a wholly different stage of development*, corresponding most clearly to Phase 3. As we have seen, the output profile of the Soviet chemical industry as a whole is slanted towards relatively simple technologies and, even within the more advanced sectors, such as plastics and manmade fibres, it is the more traditional products that predominate. Large quantities of uncomplicated end-products, such as fertilisers, are produced in the USSR.[245] The Soviet Union also produces substantial quantities of basic organic and inorganic chemicals (particularly the latter), but the quality, range and quantity of these is not sufficient to prevent large imports of basic chemical reagents, in addition to more sophisticated products such as plastics and manmade fibres. The pattern of trade is quite unlike that of the most advanced Western countries, which are all net exporters of chemical products, especially of those requiring complex technological processes. In the key sectors of organic chemicals and macromolecular compounds, the Soviet level of significant inventive activity appears to be far lower than that of Western countries. Moreover, the USSR tends to lag behind these countries in bringing important synthetic materials to commercial production and, in some cases, Western assistance has been decisive in bringing this production about. However, irrespective of whether these achievements could be credited to the USSR or not, the typical pattern has been for initial production to be followed by substantial purchases of process plant from Western countries. Indeed, during the 1960s, the Soviet Union has supplied only about two thirds of its own process plant for the chemical industry as a whole. Compared with most Western countries, the

---

[244] *Ibid.*, pp. 44–6.
[245] The fertilisers produced using well established Soviet technology are of the relatively unsophisticated varieties; large imports of foreign plant and equipment have taken place during the 1960s for the manufacture of complex fertilisers.

Soviet research effort and total output of scientific papers are probably considerable, but the overall quality is such that it does not appear to have made a proportionate impact on world science. Also, the Soviet research effort does not seem to have generated any really important and original innovations, which could be successfully scaled up to mass production. Thus, there is a consistent pattern of backwardness highlighted by all the criteria adopted in this study.

Appendix 6A. Statistical Data

Table 6A.1 *Comparative growth rates of key industrial products in selected countries, 1960–70*

		1960	1965	1970
(a)	*Chemicals*			
	USSR (growth)	100	195[3]	349[4]
	USA (growth)[6]	100	142	195
	Japan (growth)[6]	100	214	408
	FRG (growth)[6]	100	169	276
		1960	1965	1970
(b)	*Electric power* (million kWh)			
	USSR[1]	292274	506672	740925
	growth	100	173	253
	USA[1]	891581	1226795	1739796
	growth	100	137	195
	Japan[1]	115500	192159	359549
	growth	100	166	311
	FRG[1]	119028	172340	242612
	growth	100	145	204
		1960	1965	1970
(c)	*Steel (1,000 tonnes)*			
	USSR[1]	65300	91000	115900
	growth	100	139	177
	USA[1]	90100	119000	119100
	growth	100	132	132
	Japan[1]	22100	41200	93300
	growth	100	186	421
	FRG[1]	34100	36800	45000
	growth	100	108	132

		1960	1965	1970
(d)	*Cement (1,000 tonnes)*			
	USSR	45520[5]	72388[2]	95248[2]
	growth	100	159	209
	USA	56063[5]	65078[2]	67753[2]
	growth	100	116	121
	Japan	22537[5]	32486[2]	57189[2]
	growth	100	144	254
	FRG	24905[5]	34133[2]	38325[2]
	growth	100	137	154
		1960	1965	1970
(e)	*Engineering*			
	USSR (growth)[8]	100	183	324
	USA (growth)[7]	100	147	173
	Japan (growth)[7]	100	173	546
	FRG (growth)[7]	100	122	151

Sources:
1 See Appendix Table 6A.1 above.
2 *UN Statistical Yearbook* (1971), pp. 285–6.
3 *Narodnoe khozyaistvo SSSR v 1969 godu* (1970), p. 210.
4 *Ibid., 1922–1972 godu* (1972), p. 169.
5 *UN Statistical Yearbook* (1969), pp. 276–7.
6 Calculated from *OECD Main Economic Indicators-Supplement 1966; Industrial Production*, pp. 24–5; ibid. (1972), pp. 24–5 (series 'chemicals and petroleum and coal products', ISIC 31–2).
7 Calculated from OECD, *op. cit.* (1966), p. 19 and (1972), p. 19 (series 'non-electrical machinery', ISIC 36/CITI 36).
8 *Narodnoe khozyaistvo SSSR 1922–1972 godu*, p. 176.

Table 6A.2 Production of soda ash in selected countries (1000 tonnes) (per capita figures in kg Na_2CO_3)

	USSR		USA		FRG		Japan		France	
	Absolute	Per capita	Absolute	Per capita	Absolute	Per capita	Absolute	Per capita	Absolute	Per capita
1950	749[1]	4·2	3621[7]	23·8	735[7]	15·4	165[7]	2·0	717[7]	17·2
1953	1194[5]	6·3	4426[8]	27·6	794[8]	16·1	275[8]	3·2	653[8]	15·3
1957	1618[5]	8·0	4184[6]	24·3	989[6]	19·2	392[6]	4·3	826[6]	18·6
1960	1887[1]	8·9	4093[6]	22·6	1117[3]	21·0	519[3]	5·6	840[3]	18·4
1963	2545[1]	11·4	4205[9]	22·2	1055[3]	19·0	641[9]	6·7	927[3]	19·4
1967	3149[1]	13·5	4180[9]	21·0	1158[3]	20·1	872[9]	8·8	1084[9]	21·9
1970	3668[2]	15·1	3859[9]	18·8	1334[4]	22·4	1230[4]	11·9	1419[4]	27·9
1972	3850[10]	15·5	3792[9]	18·1	1397[9]	23·4	1300[9]	12·0	1427[9]	28·0

Sources:
1 *Narodnoe khozyaistvo SSSR v 1969 godu* (1970), p. 211.
2 *Ibid., 1922–1972* (1972), p. 170.
3 *The Chemical Industry 1969–1970*, OECD, Paris (1971), Supplement, p. 10.
4 *UN Statistical Yearbook* (1971), p. 262.
5 *Promyshlennost' SSSR* (1966), p. 146.
6 *UN Statistical Yearbook* (1966), p. 280.
7 *Ibid.* (1959), p. 241.
8 *Ibid.* (1971), p. 262.
9 *Ibid.* (1973), p. 269.
10 *Narodnoe khozyaistvo SSSR v 1972 godu* (1973), p. 217.

Table 6A.3 Production of sulphuric acid in selected countries (1000 tonnes) (per capita in kg H_2SO_4)

	USSR Absolute	USSR Per capita	USA Absolute	USA Per capita	FRG Absolute	FRG Per capita	Japan Absolute	Japan Per capita	UK Absolute	UK Per capita	France Absolute	France Per capita
1950	2125[1]	11·8	11820[8]	77·6	1446[9]	30·2	2030[8]	24·5	1832[9]	36·4	1215[9]	29·1
1953	2969[6]	15·4	12703[8]	79·3	1897[5]	38·6	2685[8]	31·0	1905[5]	37·6	1180[5]	27·7
1957	4569[6]	22·7	14932[7]	86·8	2723[7]	52·9	3932[7]	43·3	2373[7]	46·1	1600[7]	36·1
1960	5398[1]	25·4	15915[4]	88·1	3170[4]	59·6	4452[4]	47·7	2745[4]	52·4	1983[4]	43·4
1963	6885[1]	30·8	18993[3]	100·2	3316[4]	59·8	4991[4]	52·0	2927[4]	54·6	2394[3]	48·7
1967	9734[1]	41·4	26141[3]	131·3	3778[4]	65·5	6284[4]	62·9	3234[4]	58·8	3227[3]	65·1
1970	12059[2]	49·6	26784[3]	130·4	4435[5]	74·5	6925[5]	66·9	3352[5]	60·2	3682[5]	72·5
1972	13685[3]	55·0	28165[3]	134·2	4735[3]	79·4	6714[3]	62·2	3449[3]	61·2	4114[3]	79·2

Sources:
1 *Narodnoe khozyaistvo SSSR v 1969 godu* (1970), p. 211.
2 *Ibid., 1922–1972* (1972), p. 170.
3 *UN Statistical Yearbook* (1973), pp. 264–5.
4 *The Chemical Industry 1969–1970*, OECD, Paris (1971), Supplement, p. 12.
5 *UN Statistical Yearbook* (1971), p. 258.
6 *Promyshlennost' SSSR* (1964), p. 143.
7 *UN Statistical Yearbook* (1966), p. 276.
8 *The Chemical Industry in the European Member Countries of OECD, 1953–1962*, Paris (1964), p. 49.
9 *UN Statistical Yearbook* (1956), p. 237.

Table 6A.4 Output of superphosphates (1000 tonnes, 100% P_2O_5) (per capita figures in kg)

	USSR[a]		USA[e]		Japan[d]		FRG[d]		UK[c]		France[c]	
	Absolute	Per capita	Absolute	Per capita	Absolute	Per capita	Absolute	Per capita	Absolute	Per capita	Absolute	Per capita
1950	538.8[6]	2.99	1668.1[3]	10.95	216.5[3]	2.61	63.1[3]	1.32	185[3]	3.68	147.9[3]	3.54
1953	677.6[1]	3.58	1938[1]	12.10	243[1]	2.80	72[1]	1.46	207[1]	4.09	167.9[2]	3.94
1957	1082.3[4]	5.33	2008.3[2]	11.68	324.3[2]	3.57	74.2[2]	1.44	172[2]	3.34	205.9[2]	4.65
1960	1207.4[1]	5.63	2046[1]	11.32	379[1]	4.06	68[1]	1.28	146[1]	2.79	214.5[1]	4.70
1963	1493.9[1]	6.64	2122[1]	11.20	303[1]	3.16	53[1]	0.96	105[1]	1.96	243.4[1]	5.09
1967	2806.1[1]	11.89	2665[1]	13.38	251[1]	2.51	73[1]	1.26	70[1]	1.27	254.4	5.13
1970	3636.6[7]	14.95	(1921)[9b]	(9.35)	(188)[9b]	(1.82)	(50)[9b]	(0.84)	(93)[9b]	(1.67)	(412)[9b]	(8.1)
1972	3940[10]	15.85	n.a.	n.a.	(167)[11b]	(1.55)	(57)[11b]	(0.96)	(93)[11b]	(1.66)	(502)[11b]	(9.67)

Sources:
1 *UN Statistical Yearbook* (1968), p. 285.
2 *Ibid.* (1960), p. 230 (figures have been re-calculated on the basis of 100% P_2O_5; concentrations given in the original figures were: USSR 19%, USA 18%, UK 19%, FRG 18%, Japan 17%).
3 *UN Statistical Yearbook* (1951), pp. 228–9 (figures re-calculated on the basis of 100% P_2O_5, concentrations given in original figures were: USA 18%, UK 18%, FRG 18%, Japan 16% and France 14%).
4 *UN Statistical Yearbook* (1966), p. 281.
5 *Ibid.* (1964), p. 273.
6 *Narodnoe khozyaistvo SSSR v 1969 godu* (1970), p. 212.
7 *Ibid., 1922–1972*, p. 171.
8 *The Chemical Industry 1969–1970*, OECD, Paris (1971) p. 92.
9 *Ibid., 1971–1972*, OECD, Paris (1973), p. 114.
10 *Narodnoe khozyaistvo v 1972 godu* (1973), p. 218.
11 *The Chemical Industry 1972–1973*, OECD, Paris (1974), p. 110.

Notes:
a – it is not clear from the Soviet figures (i.e., in *Narodnoe khozyaistvo SSSR*, which are aggregated to form the UN figure) whether all the phosphate fertilisers referred to are superphosphates. The Soviet figures are almost certainly overestimates and possibly gross overestimates in the light of notes c, d and e below.
b – these OECD statistics are not comparable with the preceding run of UN figures.
c – these figures do not include concentrated superphosphates before 1970.
d – both Japan and FRG produce a substantial quantity of *complex* fertilisers not included in these statistics.
e – apart from single and concentrated superphosphates, the USA produces an enormous quantity of other straight phosphate fertilisers (3,404,000 tonnes in 1971) which are *not* included in Table 6A.4; all the other Western countries also produce substantial quantities of these relative to their production of superphosphates.

Table 6A.5 Output of nitrogenous fertilisers (1000 tonnes, 100% nitrogen) (per capita figures in kg)

	USSR[a]		USA[b]		Japan[b]		FRG[b]		UK[b]		France[b]	
	Absolute	Per capita	Absolute	Per capita	Absolute	Per capita	Absolute	Per capita	Absolute	Per capita	Absolute	Per capita
1950	392[1]	2·81	996[2]	6·54	414·6[2]	5·0	464·7[2]	9·71	275[2]	5·46	259[2]	6·20
1953	n.a.	n.a.	1507[2]	9·41	592·2[2]	6·85	645·3[2]	13·13	315[2]	6·23	285·7[2]	6·70
1957	630[3]	3·10	2141[4]	12·45	882·8[4]	9·73	1047·4[4]	20·36	347[4]	6·75	510[4]	11·51
1960	1003[1]	4·68	2739[5]	15·16	1029·9[5]	11·05	1180·3[5]	22·18	448·9[5]	8·57	670·8[5]	14·68
1963	1759[9]	7·82	4012[5]	21·18	1298·7[5]	13·54	1269·4[5]	22·90	562·2[5]	10·48	912·5[5]	19·08
1967	3753[7]	15·90	6607[6]	33·18	2034·7[6]	20·36	1559·1[6]	27·02	855[6]	15·55	1233[6]	24·88
1970	5423[7]	22·33	8161[7]	39·45	2105[7]	20·33	1504·6[7]	25·27	747·9[7]	13·42	1351[7]	26·61
1972	6551[8]	26·36	8472[8]	40·36	2454[8]	22·75	1470·6[8]	24·67	816[8]	14·66	1472[8]	28·35

Sources:
1 *Narodnoe khozyaistvo SSSR v 1965 godu* (1966), p. 188.
2 *UN Statistical Yearbook* (1956), p. 243.
3 *Ibid.* (1965), p. 287.
4 *Ibid.* (1959), p. 227.
5 *Ibid.* (1966), p. 282.
6 *Ibid.* (1969), p. 259.
7 *Ibid.* (1971), pp. 263–4.
8 *Ibid.* (1973), pp. 270–1.
9 *Ibid,* (1963), p. 141.

Notes:
a – Soviet figures contain an unspecified amount of technical nitrogen.
b – output of fertilisers is measured from 30 June to 30 June. Following UN practice the year in which the first six months occur is the one that is designated in the table to be compared with the Soviet calendar year. This means that all Western figures for any given year in the table are exaggerated when compared with Soviet figures.

Table 6A.6 *Output of ethyl alcohol in selected countries (1000 hectolitres)*

	USSR		USA		FRG		Japan		UK		France
	Absolute	Per capita	Absolute	Per capita	Absolute	Per capita	Absolute	Per capita	Absolute	Per capita	Absolute
1962	20367[1]	9.26	23948[1]	12.8	n.a.	—	1662[1]	1.75	1547[1]	3.31	n.a.
1965	23604[1]	10.28	26320[1]	13.5	90.31[1a]		1930[1]	1.97	1765[1]	3.25	60[1a]
1968	26698[1]	11.25	25900[1]	12.9	1960[1]	3.38	2275[1]	2.25	1450[1]	2.62	61[1a]
1970	27964[1]	11.52	26185[1]	12.7	2218[2]	3.72	2069[1]	2.0	1747[2]	3.15	83[2a]
1972	27396[2]	11.02	21540[2]	10.26	2084[2]	3.49	2288[2]	2.12	1556[2]	2.79	n.a.

Sources:
1 *UN Statistical Yearbook* (1971), pp. 254–7.
2 *Ibid.* (1973), pp. 260–2.

Note:
a – In tonnes.

Table 6A.7 *Comparative aggregate growth rates of organic chemicals*

	USSR[b]		USA		Japan		FRG		UK		France	
	Total	Petro-chemicals	Total	Petro-chemicals	Total	Petro-chemicals	Total	Petro-chemicals	Total	Petro-chemicals	Total	Petro-chemicals
1960–61	100[1c]	n.a.	100[5]	100[5]	n.a.	n.a.	100[5]	100[5]	100[5]	100[5]	100[5]	100[5]
1963	n.a.	n.a.	152[5]	n.a.	100[5]	100[5]	113[5]	143[5]	133[5]	143[5]	122[5]	134[5]
1965	255[1]	n.a.	151[5]	137[5]	268[5]	254[5]	193[5]	260[5]	222[5]	265[5]	186[5]	214[5]
1967	303[2]	n.a.	176[5]	176[5]	316[5]	392[5]	219[5]	430[5]	296[5]	386[5]	242[5]	323[5]
1970	406[3]	n.a.	n.a.	n.a.	726[6a]	1114[5a]	331[6a]	676[6a]	377[6a]	n.a.	422[6a]	648[6a]
1972	453[4]	n.a.	n.a.	n.a.	882[7a]	1385[7a]	378[7a]	806[7a]	378[7a]	n.a.	479[7a]	772[7a]

Sources:

Soviet figures are based on the series 'Output of products of basic organic synthesis'; Western figures are based on the OECD series 'Production of aliphatic and aromatic hydrocarbons' (in total, and those derived from oil and natural gas). With the exception of Japan, the base year for all the Western index numbers is 1961.

1 *Narodnoe khozyaistvo SSSR v 1969 godu* (1970), p. 210.
2 *Ibid.*, 1968 (1969), p. 247.
3 *Ibid.*, 1922–1972 (1972), p. 169.
4 *Ibid.*, 1972 (1973), p. 216.
5 *Chemical Industry 1969–70*, OECD, Paris (1971), Supplement, pp. 6–7.
6 *Ibid.*, 1971–72, pp. 131 and 134.
7 *Ibid.*, 1972–73, pp. 136 and 140.

Notes:

a – these 1970 and 1972 figures include chemicals derived from natural gas as well as from petroleum.
b – no aggregate figures in absolute terms are quoted in Soviet sources. Morever, the growth rate for the 'petrochemical industry' quoted annually in *Narodnoe khozyaistvo* cannot be equated with the rate of growth of petrochemicals themselves; this refers to a broad administrative category.
c – 1960.

Table 6A.8 *Output of key organic chemicals*

	USSR	USA	FRG	Japan	UK	France
Methanol (1,000 tonnes CH_3OH)						
1965	482[1]	1306[2]	603[2]	470[2]	n.a.	123[2]
1970	1004[1]	2243[3]	863[3]	938[3]	n.a.	212[3]
1972	1130[1]	n.a.	1005[4]	1085[4]	n.a.	237[4]
Pure benzene (1,000 tonnes)						
1965	697[1]	2535[2]	478[2]	352[2]	n.a.	180[2]
1970	1036[1]	3873[3]	830[3]	1573[3]	463[3]	385[3]
1972	1148[1]	3665[4]	855[4]	1852[4]	624[4]	393[4]
Xylenes (1,000 tonnes)						
1965	94[1]	1000[2]	101[2]	91[2]	n.a.	96[2]
1970	233[1]	1475[3]	128[3]	871[3]	n.a.	91[3]
1972	293[1]	1742[4]	342[4]	929[4]	283[4]	163[4]
Phthalic anhydride (1,000 tonnes)						
1965	58.4[1]	n.a.	282[2]	269[2]	n.a.	70[2]
1970	93.1[1]	324[3]	187[3]	217[3]	n.a.	74[3]
1972	105[1]	n.a.	225[4]	240[4]	n.a.	76[4]
Phenol (1,000 tonnes C_6H_5OH)						
1965	247[1]	558[2]	156[2]	88[2]	49[2]	103[2]
1970	347[1]	797[3]	241[3]	185[3]	n.a.	123[3]
1972	381[1]	n.a.	228[4]	200[4]	n.a.	134[4]
Naphthalene (1,000 tonnes)						
1965	80[1]	344[2]	152[2]	70[2]	n.a.	35[2]
1970	112[1]	323[3]	149[3]	132[3]	n.a.	35[3]
1972	114[1]	326[4]	129[4]	105[4]	34[4]	31[4]

Sources:
1 *Statisticheskii ezhegodnik stran-chlenov SEV* (1973), pp. 105–10.
2 *The Chemical Industry 1965–66*, OECD, Paris (1967), pp. 173–7.
3 *Ibid., 1970–71*, pp. 141–5.
4 *ibid., 1972–73*, pp. 141–5.

Table 6A.9 *Output of plastics and synthetic resins (1000 tonnes) (per capita figures in kg)*

	USSR[a]		USA		FRG		Japan		UK		France	
	Absolute	Per capita	Absolute	Per capita	Absolute	Per capita	Absolute	Per capita	Absolute	Per capita	Absolute	Per capita
1950	67·1[1]	0·37	1034[3]	6·79	n.a.	n.a.	19[3]	0·23	n.a.	n.a.	n.a.	n.a.
1953	103·6[5]	0·55	1318[3]	8·23	n.a.	n.a.	54[3]	0·62	n.a.	n.a.	n.a.	n.a.
1957	225·2[5]	1·11	2039[7]	11·86	551[7]	10·71	356[7]	3·92	368[7]	7·15	182[7]	4·11
1960	311·6[1]	1·45	2850[7]	15·77	964[4]	18·11	737[7]	7·90	588[4]	11·23	347[4]	7·59
1963	567·2[1]	2·52	4068[7]	21·48	1434[8]	25·42	1069[6][b]	11·15	756[6]	14·09	508[6]	10·62
1967	1113·6[1]	4·72	6256[8]	31·11	1496[8]	43·3	2688[6]	26·90	1108[6]	20·15	890[6]	17·96
1970	1672·6[2]	6·89	8712[8]	42·52	4170[8]	70·02	5154[8]	49·78	1449[8]	26·0	1519[6]	26·92
1972	2042[8]	8·21	8894[8]	42·37	5289[8]	88·74	5573[8]	51·66	1608[8]	28·88	2100[8]	40·45

Sources:
1 *Narodnoe khozyaistvo SSSR v 1969 godu* (1970), p. 214.
2 *Ibid., 1922–1972* (1972), p. 172.
3 *The Chemical Industry in the European Member Countries of OECD*, Paris (1964), p. 51.
4 *The Chemical Industry 1969–70*, OECD, Paris (1971), Supplement, p. 29.
5 *Promyshlennost' SSSR* (1964), p. 146.
6 *UN Statistical Yearbook* (1971), p. 265.
7 *Ibid.* (1966), p. 283.
8 *Ibid.* (1973), p. 272.

Notes:
a – there is a discrepancy between Sources 5 and 1 where the series overlap; this is not sufficient to make a substantial difference.
b – there is a rather substantial difference between the UN figures for Japan in 1963, which appear in the 1966 and later handbooks; the later figures have been accepted as being more reliable.

Table 6A.10 *Output of key plastic materials*

	USSR	USA	FRG	Japan	UK	France
Phenol-formaldehyde resins (1,000 tonnes)						
1960	112[1]	215.7[2]	77.3[2]	n.a.	45.3[2]	23.4[2]
1965	177[1]	418[3]	100[3]	76[3]	52.2[3]	31[3]
1970	265[1]	500[4]	167[4]	219[4]	46.1[4]	54.4[4]
1972	285[1]	n.a.	173[5]	251[5]	41.8[5]	66.1[5]
Polyolefins (1,000 tonnes)						
1960	1.2[1]	542.5[2]	80.8[2]	n.a.	125.7[2]	37.1[2]
1965	57[1]	1552[3]	300.9[3]	454[3]	256.2[3]	116.2[3]
1970	267[1]	3160[4]	900[4]	1885[4]	496.8[4]	429.2[4]
1972	307[1]	n.a.	1176[5]	2098[5]	553.3[5]	646.3[5]
PVC and copolymers (1,000 tonnes)						
1960	24.8[1]	401.5[2]	172.7[2]	n.a.	107.8[2]	110.5[2]
1965	80.8[1]	n.a.	375.3[3]	483[3]	196.8[3]	213.4[3]
1970	160[1]	1425[4]	777[4]	1161[4]	314.5[4]	412.1[4]
1972	207[1]	n.a.	931[5]	1080[5]	344.3[5]	538.7[5]
Polystyrene and copolymers (1,000 tonnes)						
1960	7.5[1]	407.8[2]	n.a.	n.a.	49.6[2]	43[2]
1965	28.9[1]	n.a.	n.a.	125[3]	92.1[3]	83.5[3]
1970	82.2[1]	1320[4]	n.a.	668[4]	162[4]	191.9[4]
1972	98.5[1]	n.a.	n.a.	796[5]	184[5]	266.6[5]

Sources:
1 *Statisticheskii ezhegodnik stran-chlenov SEV* (1973), pp. 111–12.
2 Statistical supplement for 1953–64 data, *The Chemical Industry 1965–66*, OECD, Paris (1967), pp. 58–62. Figures for phenol-formaldehyde resins including cresylic plastics.
3 *The Chemical Industry 1965–66*, OECD, Paris (1967), p. 190.
4 *Ibid., 1970–71*, OECD, Paris (1972), p. 153.
5 *Ibid., 1972–73*, OECD, Paris (1974), p. 153.

Table 6A.11 *Output of chemical fibres (1000 tonnes) (per capita figures in kg)*

	1950	1953	1957	1960	1963	1967	1970	1972
USSR								
Total output:	24·2[1]	62·3[3]	148·7[3]	211[1]	308[1]	511[1]	623[2]	747[9]
Artificial	22·9[1]	57·9[3]	137[3]	196[1]	266[1]	395[1]	456[2]	508[9]
Synthetic	1·3[1]	4·4[3]	11·7[3]	15[1]	43[1]	116[1]	167[2]	239[9]
Synthetic per capita	0·01	0·02	0·06	0·07	0·19	0·49	0·69	0·96
Synthetic as % of total	5·4	7·1	7·9	7·1	14·0	22·7	26·8	32·0
USA								
Total output:	626	655	751	773	1189	1697	2251	3061
Artificial	571[8]	543[5]	517[6]	466[6]	612[5]	630[5]	623[5]	632[9]
Synthetic	55·5[8]	112[5]	234[6]	307[6]	577[5]	1067[5]	1028[5]	2429[9]
Synthetic per capita	0·39	0·70	1·36	1·70	3·05	5·36	7·93	11·57
Synthetic as % of total	8·9	17·1	31·2	39·7	48·5	62·9	72·3	79·3
FRG								
Total output:	162	175	258	281	373	495	723	801
Artificial	161[8]	169[5]	239[6]	229[6]	265[5]	243[5]	226[5]	160[9]
Synthetic	0·5[8]	6·5[5]	19[8]	52[6]	108[5]	252[5]	497[5]	641[9]
Synthetic per capita	0·02	0·13	0·37	0·98	1·95	4·38	8·35	10·75
Synthetic as % of total	0·5	3·6	7·4	18·5	28·9	50·9	68·7	80·0
Japan								
Total output:	115·5	242·5	481	551	701	1101	1550	1629
Artificial	115[8]	236[5]	439[6]	433[6]	462[5]	523[5]	522[5]	512[9]
Synthetic	0·5[8]	6·5[5]	42[6]	118[6]	239[5]	578[5]	1028[5]	1117[9]
Synthetic per capita	0·01	0·07	0·46	1·26	2·49	5·78	9·93	10·35
Synthetic as % of total	0·4	2·7	8·7	21·4	34·1	52·5	66·3	68·6
UK								
Total output:	172·4	189·8	225	269	326	433	599	627
Artificial	168[8]	181[5]	193[6]	208[6]	221[5]	239[5]	262[5]	253[9]
Synthetic	4·4[8]	8·8[5]	32[6]	61[6]	105[5]	194[5]	337[5]	374[9]
Synthetic per capita	0·09	0·17	0·62	1·16	1·96	3·53	6·05	6·72
Synthetic as % of total	2·5	4·6	14·2	22·7	32·2	44·8	56·3	59·6

Sources:
1 *Narodnoe khozyaistvo SSSR v 1969 godu* (1970), p. 214.
2 *Ibid.*, 1922–1972 (1972), p. 172.
3 *Promyshlennost' SSSR* (1964), pp. 148–9.
4 *Ibid.*, p. 112.
5 *UN Statistical Yearbook* (1971), pp. 240–1 and 243–4.
6 *Ibid.* (1966), pp. 263–4 and 266–7.
7 *Ibid.* (1948), pp. 202–4.
8 *Ibid.* (1959), pp. 207–9.
9 *Ibid.* (1973), pp. 246–50.

Table 6A.12 *Output of synthetic rubber (including crude estimates of Soviet output) (1000 tonnes; per capita figures in kg)*

	USSR[5]		USA		FRG		Japan		UK		France	
	Absolute	Per capita	Absolute	Per capita	Absolute	Per capita	Absolute	Per capita	Absolute	Per capita	Absolute	Per capita
1950	184·3	1·02	484[3]	3·19	n.a.	n.a.	n.a.	n.a.	n.a.	n.a.	n.a.	n.a.
1953	201·8	1·06	862·1[1]	5·29	6·3[1]	0·13	n.a.	n.a.	n.a.	n.a.	n.a.	n.a.
1957	324·7	1·60	1132[2]	6·60	12[2]	0·23	n.a.	n.a.	0·8[2]	0·01	n.a.	n.a.
1960	438·7	2·05	1459[2]	8·07	81[2]	1·52	19[2]	0·20	92[2]	1·76	17[2]	0·37
1963	570·4	2·53	1643[1]	8·63	112[1]	2·02	103[1]	1·07	127[1]	2·37	98[1]	2·05
1967	745·9	3·16	1943[1]	9·76	206[1]	3·57	281[1]	2·81	204[1]	3·71	189[1]	3·81
1970	877·5	3·61	2232[1]	10·87	302[1]	5·07	697[1]	6·73	306[1]	5·49	316[1]	6·22
1972	982·8	3·95	2455[4]	11·70	300[4]	5·03	819[4]	7·59	307[4]	5·51	368[4]	7·09

Sources:
1 *UN Statistical Yearbook* (1971), p. 252.
2 *Ibid.* (1966), p. 274.
3 *Ibid.* (1959), p. 228.
4 *Ibid.* (1973), p. 258.
5 See text pp. 255–6.

Table 6A.13 Soviet trade with the world in chemicals and related products (1000 roubles)

	Chemical Products (30)	Paints, dyes and tanning materials (31)	Photographic materials (33)	Fertilisers and Agricultural chemicals (34)	Rubber and rubber goods (35)	Artificial and synthetic fibres (51301/2)	Total
Exports							
1965	87891	11025	2537	93865	51119	—	246437
1966	93826	12499	2130	122329	60732	—	291516
1967	120717	13124	1919	137607	63875	—	337242
1968	140556	12562	1701	151025	67639	—	373483
1969	144902	14110	1472	156555	76902	—	393941
1970	165035	15427	1525	194454	86191	—	462632
1972	169486	21430	3792	121766[b]	76552	—	393026
Imports							
1965	173139	44691	14129	38618	179012	29151	478740
1966	169752	52660	11691	37402	185969	22181	478655
1967	188064	70309	14039	32022[a]	164481	27285	496200
1968	200116	87073	19005	38431[a]	153219[b]	23003	520847
1969	214803	102227	19953	45989[a]	173793[b]	30009	586774
1970	232131	114459	25942	43009[a]	172584[b]	38429	626554
1972	271733	146802	40008	44780[a]	119513[b]	30256	653092
Trade Balance							
1965	−85248	−33666	−11592	+55247	−127893	−29151	−232303
1966	−74926	−40161	−9561	+84927	−125237	−22818	−187139
1967	−67347	−57185	−12120	+105585	−100606	−27285	−158958
1968	−59560	−74511	−17304	+112594	−85580	−23003	−147364
1969	−69901	−88117	−18481	+110566	−96891	−30009	−192883
1970	−67096	−99032	−24417	+151445	−86393	−38429	−163992
1972	−102247	−125372	−36216	+76986	−42961	−30256	−260066

Source:
Vneshnyaya Torgovlya SSSR za (1965, 1966, 1967, 1968, 1969, 1970, 1973) god. For a discussion of this source see p. 259, note 112 above.

Notes:
a – includes *only* agricultural chemicals, the bulk of (34) in past years.
b – incomplete information.

Table 6A.14 *Soviet trade with the world in 'chemical products' (1000 roubles)*

Exports

	A	B	C	D	E	F	G	H	J
1965	5991	5520[a]	21118	16808	1811[a]	4705	3514	12867	72334
1966	5326	18683	15640	18770	5316	5165	3517	12391	84888
1967	5584	19934	23873	27498	8671	5051	3521	9270[a]	103402
1968	6713	22767	25208	32885	10721	4842	4733	12611[a]	120480
1969	6384	25891	20648	35160	11416	5232	5213	12555[a]	122499
1970	6755	30145	19398	36022	14823	5589	7377	16011[a]	136120
1972	14269	29742	21563	40726	11053	5227	6280	11352[a]	140212

Imports

	A	B	C	D	E	F	G	H	J
1965	25485	8139	—	55149	17307	1229[a]	4724	55194	167227
1966	30103	11007	—	45296	12577	695[a]	5050	56888	161616
1967	28478	8724	—	59043	14201	131[a]	5174	45498[a]	161249
1968	28927	7608	—	66320	12052	—	5217[a]	49965[a]	170089
1969	32002	8204	—	73413	10101	—	5629[a]	50333[a]	179682
1970	31583	9593	—	77966	8723	—	6654	61324[a]	195843
1972	36127	9825	—	96516	6378	—	6582	67538[a]	222966

Trade Balance

	A	B	C	D	E	F	G	H	J
1965	−19494	−2619	+21118	−38341	−15496	+3476	−1210	−42327	−94893
1966	−24777	+7676	+15640	−26526	−7261	+4470	−1453	−44497	−76728
1967	−22894	+11210	+23873	−31545	−5530	+4920	−1653	−36228	−57847
1968	−22214	+15159	+25208	−33435	−1304	+4842	−484	−37354	−49609
1969	−25618	+17687	+20648	−38253	+1315	+5232	−416	−37778	−57183
1970	−24828	+20552	+19398	−41944	+6100	+5589	+723	−45313	−59723
1972	−21858	+19917	+21563	−55790	+4675	+5227	−302	−56168	−82754

A – soda products and other alkalis (301).
B – salts of organic and inorganic acids (302).
C – coke chemical products (303).
D – plastics and materials for plastics productions (304).
E – alcohols and aldehydes (305).
F – wood chemical products (306).
G – intermediates for aniline dye industry (307).
H – miscellaneous chemical products (309).
J – total.

Source:
As in Table 6A.13.

Note:
a – incomplete information.

Table 6A.15 *Soviet trade in 'plastics and materials for production of plastics' with individual countries (1,000 roubles)*

	1965	1966	1967	1968	1969	1970	1972
Imports							
Developed West							
Austria	—	—	—	1282	1945	2650	3710
Belgium	473	5	146	2	200	143	6799
Denmark	—	—	—	—	—	—	—
France	565	—	973[a]	2161	3110	3797	4876
Italy	11296	10306	11744	13561	14216	13058	9291
Japan	13382	8112	5342	3782	14523	14670	13759
Netherlands	821	4526	8463	7093	2157	2086	2083
Sweden	—	—	—	—	—	—	—
UK	3880	4354	4488	3533	2792	6872	5333
USA	1477	953	627	1743	1688	1818	2492
FRG	5453	2672	11512	14666	10611	11240	19015
Total	37347	30928	43295	57823	51242	56334	67358
East Europe							
Czechoslovakia	4056	2575	1984	1878	2224	1939	1892
GDR	12260	8784	8561	8220	9098	8375	11695
Hungary	—	—	—	—	—	—	—
Poland	—	—	—	2091	3645	3865	2607
Romania	—	406[a]	1609[a]	443[a]	864[a]	1544[a]	5659
Total	16316	11765	12154	12632	15831	15723	21853
Grand total	53663	42693	55449	60455	67073	72057	89212
Exports							
Developed West							
Austria	54	55	74	11	—	—	—
Belgium	—	—	100[a]	—	—	—	—
Denmark	—	—	—	—	—	—	176
France	—	—	—	1382	990	—	—
Italy	—	—	—	—	—	—	—
Japan	—	—	—	—	—	—	—
Netherlands	625	292	313	431	376	449	610
Sweden	—	8[a]	216[a]	—	—	—	—
UK	—	—	—	—	202[a]	—	325
USA	—	—	—	—	—	—	—
FRG	—	—	—	—	—	—	—
Total	682	355	703	1824	1568	449	1111

East Europe							
Czechoslovakia	4998	2512	3262	3362	4177	5212	4828
GDR	4490	3152	2963	3563	4589	4646	5076
Hungary	2307	2656	3108	4395	6765	4402	5751
Poland	31	619	1478	1555	687	1289	2968
Romania	87	98	65	340	626	183	436
Total	11913	9037	10876	13215	16844	15732	19059
Grand total	12595	9392	11579	15039	18412	16181	20170
Trade balance							
West	−36665	−30573	−42592	−45999	−49674	−55885	−66247
East	−4403	−2728	−1278	+583	+1013	+9	−2794
East and West	−41068	−33301	−43870	−45416	−48661	−55876	−69041

Source:
As in Table 6A.13.

Note: a – incomplete information.

Table 6A.16 *Soviet trade in manmade fibres (artificial and synthetic) with selected countries (1,000 roubles)*

Imports

	1965	1966	1967	1968	1969	1970	1972
Developed West							
Austria	643	1466	1242	432	956	1917	376
Belgium	—	—	—	—	—	—	—
Denmark	—	—	—	—	—	—	—
France	2129	763	1365	988	1341	2412	1358
Italy	5344	1883	1933	1618	2024	2926	2107
Japan	6314	6826	2602	1944	2283	335	2598
Netherlands	—	—	181	—	—	—	—
Sweden	—	—	—	—	—	—	—
UK	8954	6727	9021	12801	12205	17242	12654
USA	1216	1093	6013	870	3094	3198	1456
FRG	—	133	788	846	530	1016	490
Total	24600	18891	23145	19499	22433	29046	21039
East Europe							
Czechoslovakia	—	—	—	—	—	—	—
GDR	2369	197	240	—	—	—	—
Hungary	—	—	—	—	—	—	—
Poland	—	—	—	—	—	—	—
Romania	—	—	—	—	—	—	—
Total	2369	197	240	—	—	—	—
Trade balance							
West	−24600	−18891	−23145	−19499	−22433	−29046	−21039
East	−2369	−197	−240	—	—	—	+3772
East and West	−26929	−19088	−23385	−19499	−22433	−29046	−17267

Source:
As in Table 6A.13.

Note:
There are no Soviet exports of these commodities apart from the export of 3,772 roubles' worth of artificial fibre to Hungary in 1972.

Table 6A.17 *Soviet trade in 'miscellaneous chemicals' with individual countries (1,000 roubles)*

Imports

	1965	1966	1967	1968	1969	1970	1972
Developed West							
Austria	—	—	—	—	—	—	—
Belgium	435	—	477	1132	1548	1585	4281
Denmark	—	—	—	—	—	—	—
France	367	325	1618	1272	2253	2493	1692
Italy	2636	3675	2739	4381	2071	2111	1032
Japan	—	—	—	—	—	—	—
Netherlands	3180	2390	4308	2508	3533	3056	1442
Sweden	—	—	—	—	—	—	966
UK	7453	6751	6508	9849	4371	4419	5235
USA	2031	2997	632	367	1838	1789	1612
FRG	3688	3283	2475	2183	1066	2576	2923
Total	19790	19421	18757	21692	16680	18029	19183
East Europe							
Czechoslovakia	4470	4859	6622	5779	7177	8357	10583
GDR	8855	10453	15426	15192	15414	18680	21857
Hungary	1213	—	—	—	—	—	—
Poland	1925	4045	2981	3757	3944	2915	4568
Romania	3505	2311	2392	2322	1965	1668	4173
Total	19968	21668	27421	26868	28500	31620	41181
Grand Total	39758	41809	46178	48560	45810	49649	60364

Exports

	1965	1966	1967	1968	1969	1970	1972
Developed West							
Austria	52	76	41	20	73	78	—
Belgium	—	—	—	—	—	—	—
Denmark	—	—	—	—	—	—	—
France	—	—	282	158	2	226	330
Italy	—	—	—	—	—	—	—
Japan	—	82	301	233	266	559	1837
Netherlands	—	—	—	—	—	—	—
Sweden	—	—	—	—	—	—	—
UK	189	326	13	9	110	127	—
USA	—	—	17	105	31	—	—
FRG	—	—	3409	4032	3098	1088	389
Total	241	484	4063	4557	3580	2078	2556

East Europe

Czechoslovakia	253	429	828	581	1074	1663	548
East Germany	332	26	527	1283	1241	1733	1018
Hungary	944	835	1170	—	—	—	—
Poland	251	163	267	104	466	678	462
Romania	323	716	35	102	43	42	670
Total	2133	2169	2827	2070	2824	4116	2698
Grand total	2374	2653	6890	6627	6404	6194	5254

Trade balance

West	−19549	−18937	−14694	−17135	−13100	−15951	−16627
East	−17835	−19499	−24594	−24798	−25676	−27504	−38483
East and West	−37384	−38436	−39288	−41933	−38776	−43455	−55110

Source:
As in Table 6A.13.

Table 6A.18 *Soviet imports and exports of chemical equipment (1,000 roubles)*

(a) *Summary of trade in chemical equipment*

	1965	1966	1967	1968	1969	1970	1972
Total Imports	187430	208031	262651	278596	230530	217955	374809
Imports from East Europe	87933	75379	103628	93621	90000	133496	148935
East Europe share in total imports	46.9%	36.2%	39.4%	33.6%	39.0%	61.2%	39.7%
Total Exports	4818	4711	5239	7006	8176	53158	61131

(b) *Imports of chemical equipment by origin*

Developed West	1965	1966	1967	1968	1969	1970	1972
Austria	5209	2540	2972	4094	6240	2488	971
Belgium	1747	106	9996	1829	5126	1997	1677
France	12131	17604	36357	42073	21564	10660	27613
Italy	15146	10190	10118	8742	32612	15678	6815
Japan	5949	18702	21356	21300	19316	16730	95258
Netherlands	1138	5384	26839	18559	1971	68	117
Sweden	173	553	2376	3544	3178	1266	2403
Britain	39315	52589	44930	59199	26819	13769	15903
West Germany	16512	22334	10895	22985	16955	13997	61616
USA	113	890	23	563	3130	1480	5319
Sub-total	97433	130892	165862	182888	136911	78133	217692
East Europe							
Czechoslovakia	43770	31411	46593	27917	34536	38394	34599
East Germany	16098	21183	28502	38899	32623	52705	67985
Hungary	8294	10095	9996	14984	10453	15407	11331
Poland	7718	10463	11241	11821	12388	19657	26556
Romania	12053	2227	7296	—	—	7333	8468
Bulgaria	—	—	—	—	—	—	—
Bloc sub-total	87933	75379	103628	93621	90000	133496	148935

(c) *Exports of chemical equipment to East European countries*

	1965	1966	1967	1968	1969	1970	1972
Czechoslovakia	1171	924	249	506	170	241	880
GDR	—	223	328	352	1138	7435	5959
Hungary	314	96	1371	1472	459	554	7951
Poland	2161	1349	619	765	418	980	5530
Romania	217	530	420	1725	2373	1564	8193
Bulgaria	663	1065	1425	1336	1936	21112	28196
Total	4526	4187	4412	6156	6494	31886	56709

Source:
As in Table 6A.13.

Table 6A.19 *Grants of patents to nationals and foreigners, 1968–72*

	1968	1969	1970	1971	1972
USSR					
Grants to nationals	24497[1a]	25859[2a]	30636[3a]	33534[4a]	38523[5a]
Grants to foreigners	956[1a]	767[2a]	1830[3a]	2098[4a]	2625[5a]
Total	25453	26626	32466	35632	41148
Proportion (%) of nationals in total	96·24	97·12	94·36	94·11	93·62
USA					
Grants to nationals	45782[1]	50395[2]	47073[3]	55988[4]	51515[5b]
Grants to foreigners	13320[1]	17162[2]	17354[3]	22328[4]	23293[5]
Total	59102[1]	67557[2]	64427[3]	78316[4]	74808[5]
Proportion (%) of nationals in total	77·46	74·60	73·06	71·49	68·56
Japan					
Grants to nationals	18576[1]	18787[2]	21403[3]	24795[4]	29101[5]
Grants to foreigners	9396[1]	8870[2]	9475[3]	11652[4]	12353[5]
Total	27972[1]	27657[2]	30818[3]	36447[4]	41454[5]
Proportion (%) of nationals in total	66·41	67·93	69·45	68·03	70·20
FRG					
Grants to nationals	12143[1]	12432[2]	6386[3]	8295[4]	9642[5]
Grants to foreigners	9026[1]	10191[2]	6501[3]	9854[4]	10958[5]
Total	21169[1]	22623[2]	12887[3]	18149[4]	10600[5]
Proportion (%) of nationals in total	57·36	54·95	49·55	45·70	46·81
UK					
Grants to nationals	n.a.	9807[2]	10343[3]	10376[4]	10116[5]
Grants to foreigners	n.a.	28983[2]	30652[3]	31178[4]	32678[5]
Total	43038[1]	38790[2]	40995[3]	41554[4]	41794[5]
Proportion (%) of nationals in total	n.a.	25·28	25·23	24·97	23·64
France					
Grants to nationals	15627[1]	10288[2]	8539[3]	13696[4]	10767[5]
Grants to foreigners	32363[1]	21732[2]	17758[3]	37760[4]	35450[5]
Total	47990[1]	32020[2]	26297[3]	51456[4]	46217[5]
Proportion (%) of nationals in total	32·56	32·13	32·47	26·62	23·30

Sources:
1 *Industrial Property*, No. 12, 1969, Statistical Annex, pp. 2–3.
2 *Ibid.*, No. 12, 1970, Statistical Annex, pp. 2–4.
3 *Ibid.*, No. 12, 1971, Statistical Annex, pp. 2–5.
4 *Ibid.*, No. 12, 1972, Statistical Annex, pp. 2–3.
5 *Ibid.*, No. 12, 1973, Statistical Annex, pp. 4–7.

Notes:
a – author's certificates plus patents (in the case of awards to Soviet citizens, the numbers of patents are either negligible or non-existent). For further discussion of this point see n. 146, p. 268 above.
b – not including re-issues.

Appendix 6B. An analysis of the level of Soviet polymer research by means of citations

Most chemists would agree that Soviet chemical research, both in general and in the specific area of polymer chemistry, contains work of a quality and reliability comparable with the best in the Western world. They will usually qualify this, however, with the statement that it is often difficult to assess the value of any specific piece of work and also that it will often be 'buried' in a mass of material of inferior quality. This is a very subjective judgement, of course, and one which is difficult to quantify.

The term 'level' has the initial connotation of quality, or degree of excellence. If this were the only relevant factor it might lead one to equate, say, Swiss chemical science with American chemical science, since judged on a subjective basis their intrinsic quality appears to be very nearly equal. However, the range and volume of American chemical science is many times that of Swiss chemical science and so the former far outweighs the latter in overall impact. Although in specific areas, of which one example is that of pharmaceutical and medicinal chemistry, Swiss chemical science is a potent force on the world scene, American chemical science is a potent force over virtually the whole field of pure and applied chemistry. It is necessary to find a basis which provides a result reflecting a combination of quality of performance and quantity of output; this should then give a measure of the comparative impact of the countries or areas that are being compared. Provided that these categories are comparable in size and overall impact, it might be permissible to try to arrive at some rating reflecting 'impact per unit size'. In the present study the results have been obtained in a manner which provides a comparison between Soviet research in polymer chemistry and that of Western industrialised countries as a whole.

The basis of the assessment in the present study is the analysis of bibliographic references contained in a sample of leading scholarly journals published in Eastern Europe, Western Europe and the United States. These references have been classified according to the country in which the journal containing them was published. Subject to a number of qualifications and safeguards, this gives a collective appraisal by a number of specialists of the quality of work emanating from these various countries. The citation of scholarly work arises in a number of different ways. A review article by its very nature aims to survey a particular field and the author will aim to quote all relevant work on the subject. In writing a research paper for publication an author will wish to quote the most important work in the field so as to set his own work in context, even though he may not aim to give such an exhaustive bibliography as is contained in a review. In both cases authors will be selective and will exercise a critical faculty in that they will aim to quote publications which they deem to contain work which is important, relevant and of appropriate quality, whilst omitting references to work of doubtful quality, validity or reproducibility. Omission of important references constitutes a reflection on an author's knowledge and judgement in his chosen field. In deciding what references to include, both from Eastern and Western sources, the individual scientist will often be guided by his knowledge and experience of work in progress in particular institutes and laboratories.

There are important problems which arise with this method.

1. The problem of scientific stock as compared with current achievement.
2. The problem of cross publication.
3. The problem of awareness and availability.
4. The problem of dual publication of the same piece of work.
5. The problem of separating primary publications from secondary publications.

These problems must be reviewed in turn.

SCIENTIFIC STOCK AND CURRENT ACHIEVEMENT

An examination of any paper or review article indicates, as is to be expected, that the great majority of references are to relatively recent work in the field. Nevertheless, some references may still be included to key publications of an earlier period, occasionally going back into the last century. There is little doubt that the position of Russian scientific research relative to that of the West has improved since Tsarist days and hence the question arises as to whether a simple count of all references might not, at least in part, reflect the greater past achievements in the West rather than the current position. To obtain a better assessment of the current position a count has been made of references dated 1960 or later, as well as the total number.

CROSS PUBLICATION

In assessing the relative amount of work emerging from various countries one should strictly examine all publications to find out where the work was carried out, rather than attributing it automatically to the country in which the journal containing it was published. Thus there are some Western publications (e.g., *European Polymer Journal*) which encourage contributions on a broad international basis and which do include papers from Soviet scientists and others in Eastern Europe. Even though the *European Polymer Journal* is published in Britain, it is obviously wrong to credit Western science with all references to this journal. In so far as this happens it will be misleading. The amount of such cross publication appears to be so limited, however, as not to significantly distort the results. The reverse phenomenon of Western scientists publishing in Soviet journals appears to be so rare as not to create any problem.

There is the further possibility that publications in Western or Soviet journals might originate from East European scientists or others from third world countries such as India. Again the overall proportion of such publications is negligible.

AWARENESS AND AVAILABILITY

It is axiomatic that authors will only make bibliographic references to work with which they are familiar, and it follows that an analysis of bibliographic citations will give a distorted picture if authors do not have equal access to all published work. The volume of work being published in polymer science, as in other branches of science, continues to increase at an alarming rate, so that individual scientists have increasing difficulty in keeping abreast of published work. This problem of 'current awareness' has two aspects: individuals must acquire knowledge of the existence of publications which interest them, and the journals which contain these publications must be available to them. No individual can read, or even scan, more than a small fraction of primary publications and therefore he has to rely more and more on secondary publications, such as abstracts and review articles. *Chemical Abstracts* continues to give extremely good coverage of world publications in pure and applied chemistry (including polymer chemistry) but it has now become so large itself that even scanning the abstracts has become onerous. Increasingly, the chemist and polymer chemist must rely on review articles, such as those in *Macromolecular Reviews*, and on computerised searching of abstracts. The final result seems to be that, while occasionally some single important publication may be missed, the overall coverage remains good, but there is a longer delay in becoming aware of many important publications than was the case in the past.

The language problem might appear to pose another barrier, particularly in limiting Western knowledge of Soviet publications, since few Western scientists have any knowledge of Russian. This problem appears to be adequately dealt with, however, by the translation facilities which are available. A number of more important Soviet journals, including the main polymer journal *Vysokomolekulyarnye Soedineniya*, are published as cover to cover English translations. For the remainder, the abstracts in *Chemical Abstracts* will usually be sufficient to indicate to the individual scientist whether any particular publication is sufficiently important to himself to warrant an individual translation. There is no difficulty in obtaining such translations, and lists of papers already translated are published regularly.[1] As regards the availability of journals they are now so numerous that small and medium-sized libraries are inevitably limited by cost as to the number they can subscribe to. In the West this problem seems to be dealt with adequately by such services as that offered by the British National Lending Library for Science. It is difficult to assess accurately the comparable position on awareness and availability on the Soviet side. *Chemical Abstracts* is certainly available in the USSR, but the abstracting services provided by the All Union Institute of Scientific and Technical Information (VINITI) are likely to be the major source of information about Western scientific work.

DUAL PUBLICATION

In recent years there has been an increasing tendency, particularly in the West, for results to be published briefly in a preliminary form, which is usually

---

[1] *Translations Register-Index*, National Translations Center, Chicago, USA.

followed by a more detailed publication at some later date.

This is even catered for by periodicals which publish nothing else but such preliminary publications, such as *Chemical Communications* and *Tetrahedron Letters*. No attempt has been made in this survey to distinguish between preliminary publications and substantial papers on the grounds that reference will be made to one or the other but not both.

PRIMARY AND SECONDARY PUBLICATIONS

The results of scientific research are nowadays almost always published in the form of papers in specialist journals rather than in book form. The work contained in these papers, the primary publications, is then mentioned again in abstracts, reviews and books—the secondary publications. The complexity of modern science is such that large numbers of secondary publications are essential for any country or group of researchers to function effectively. These secondary publications can simultaneously inflate and deflate the number of citations; reference to the secondary review *in addition to* the original article constitutes double-counting, and will inflate the total, whereas reference to the secondary review as a *substitute* for the numerous original papers to which it refers will tend to deflate the total. It is difficult to estimate the net effect of this but it should be pointed out that review articles are common in both the USSR and the West.

LENGTH OF PUBLICATIONS

In comparing bibliographic citations it is important to be sure that we are comparing like with like. Thus, if it were the general custom of Soviet scientists to publish in the form of large comprehensive papers work which in the West would be fragmented into two or more separate papers, it would be misleading simply to compare numbers of citations. In so far as the doctrine of 'publish or perish' appears to apply equally to individuals in the East and in the West, such a state of affairs is not to be expected, but some *quantitative* guideline is obviously desirable. A sample check was therefore made by comparing two general chemical journals (*Journal of the American Chemical Society*, 1972, and *Izvestiya Akademii Nauk, Seriya Khimicheskaya*, 1972) and two polymer chemistry journals (*Journal of Polymer Science*, Polymer Chemistry Edition, 1973, and *Vysokomolekulyarnye Soedineniya*, 1973) for average length of publication. In each case the average length of papers in the Western journals was higher than that of papers in the Soviet journals by over 50 per cent. *Journal of the American Chemical Society*, 1972, contained 585 publications in 2,557 pages whilst *Izvestiya Akademii Nauk, Seriya Khimicheskaya*, 1972, contained 846 publications in 2,746 pages (as translated by Consultants Bureau under the title *Bulletin of the Academy of Sciences USSR Division of Chemical Science*). While the page size of these two publications is very nearly equal, the American journal has more words per page and fewer pages which are part blank. *Journal of Polymer Science*, Polymer Chemistry Edition, 1973, contained 272 publications in 3,330 pages whilst *Vysokomolekulyarnye Soedineniya* (as translated by Pergamon Press under the title *Polymer Science*

USSR) contained 434 publications in 3,180 pages. For these two journals the number of words per page was very nearly equal. Although this is only a sample comparison, these four journals are among those which are most frequently quoted.

JOURNALS WHOSE CITATIONS WERE ANALYSED

For an initial comparison, two review journals and two polymer science journals were selected, one from the Soviet Union and one from the West in each case. An analysis of the citations in these should give some assessment of relative standing as seen from these two sides. However, within any given national group of scientists there will be, for a variety of reasons, a systematic bias in favour of quoting references to work appearing in their own national journals. Quite apart from national feeling or pride, it is much easier to keep abreast of work in one's own country than abroad. All libraries will stock the journals of their own country as a matter of course, personal contacts are much easier to develop and maintain, and attendance at conferences is much easier within one's own country. A count of bibliographic references from Western and Soviet publications, as detailed above, will give a picture distorted in two opposite directions. In an endeavour to assess the picture from some less partial standpoint a count has also been made of references in polymer articles in journals from Romania, Czechoslovakia, Hungary, East Germany and Poland. Scientists in these countries have historic ties with the West, which they try to maintain in spite of any current political differences. At the same time the current political ties between their countries and the Soviet Union will make for ease of accessibility to Soviet publications and closer ties with Soviet scientists in general. In their bibliographic citations authors from these countries refer extensively to Soviet and Western work and they would appear to come nearest to an unbiased appraisal of the relative value of work in the two large blocks.

COMMENTS ON THE RESULTS

The results of the analysis are summarised in Table 6B.1 where the bibliographic citations are listed according to the origin of the journals in which they are contained. It is immediately apparent that whilst Western authors only infrequently cite work published in Soviet journals the converse is not the case. Soviet authors include in their citations a high proportion of references to Western literature and, although this does not appear in the summarised data in Table 6B.1, they do so fairly consistently.

Even when allowance is made for the fact that the Western countries[2] have a population two and a half times that of the USSR, the discrepancy is still very striking. The East European journals which have been examined appear to show two types of citation pattern. The first four on the list show a consistent pattern of citations predominantly of Western work, even when the population factor is taken into account. The last two on the list (*Plaste und Kautschuk* and

[2] This category includes all the countries of Western Europe, Japan, Australia, Canada and the United States.

Polymery) are relatively more orientated towards Soviet work, but the balance is still tilted towards the West. There appear to be two reasons for the different pattern shown by these journals. The first is that the last two journals on the list deal more with applied science and technology while the first four deal predominantly with fundamental research. Even though the sections in *Plaste und Kautschuk* and *Polymery* which deal explicitly with such topics as 'polymer processing' have been omitted from this analysis, the remaining articles still have a bias towards applied topics. The other factor is that in *Plaste und Kautschuk* there is an abnormally high proportion of contributions from authors in Soviet laboratories. This problem of 'cross publication' has been mentioned above, and no general attempt has been made to correct for it since it is on a relatively small scale. In this case, however, of the 119 articles in *Plaste und Kautschuk*, 1973, 14 were from Soviet authors, and these contained approximately half of all the citations of Soviet work in this volume. The overall picture of the citations from East European journals is that in fundamental research they rate the West considerably more highly than the Soviet Union, while on applied research the West is still rated more highly, but the gap is not so great.

One additional result to emerge from this analysis is the change in comparative rating which has taken place in recent years. This is generally accepted as a fact in subjective judgements but it is interesting to have it confirmed in a quantitative manner. If the citations referring to work published before 1960 are compared with those relating to work dated 1960 or later, then it appears that, even when viewed through Western eyes, the relative ranking of Soviet polymer science is higher in the post 1960 period than before. An extended analysis on these lines might throw light on the long-term question of whether the Soviet Union can ever achieve real parity with the West.

Table 6B.1 Bibliographic citations[1] according to origin of publication cited

	Publications in which citations are analysed	Number of articles	USSR	All citations West[2]	Other[3]	Doubtful	USSR	Citations dated 1960 or later West	Other	Doubtful
Reviews	Reviews in *Macromolecular Chemistry*, Vol. 10, 1973	8	10	781	68	1	10	674	68	0
	Uspekhi Khimii Vol. 42, 1973[5]	11	6888	803	70	11	625	542	62	8
Polymer Journals	*Vysokomolekulyarnye Soedineniya*, Vol. 15, 1973	434	3055	2459	119	65	2812	1683	115	63
	Journal of Polymer Science, Polymer Chemistry Edition, Vol. 11, 1973	272	134	4005	241	4	121	3097	231	4
	Revue Roumaine de Chimie,[5] Vol. 18, 1973	8	22	97	14	3	17	79	13	2
	Collection Czech Chemical Communications,[5] Vol. 38, 1973	30	19	317	103	3	11	225	98	2
East European Journals	*Acta Chimica Academiae Scientarum Hungaricae*,[5] Vol. 75, 1972–3 Vols. 77, 78, 79, 1973	9	5	83	32	0	4	51	32	0
	Journal für praktische Chemie,[5] Vol. 315, 1973	4	0	42	12	0	0	28	12	0
	Plaste und Kautschuk,[6] Vol. 20, 1973 (GDR)	119	411	1157	364	11	360	844	359	11
	Polymery—Tworzywa Wielkoszasteczkowe, Vol. 18, 1973 (Poland)[7]	109	196	832	455	13	178	602	430	11
	Totals for all East European journals	279	653	2528	980	30	570	1829	944	26

Notes:
1 References to reviews, books, theses, patents, and papers in journals and magazines.
2 See Appendix 6B, note 2, on p. 324.
3 References to material published in countries other than the Soviet Union or the West, also references to proceedings of all conferences and symposia, personal communications, and unpublished work.
4 Those publications whose origin was not immediately apparent.
5 Polymer articles only.
6 Not including articles in the sections on 'Polymer processing and applications' and 'Paint materials'.
7 Articles in section 'Scientific research work' only.

Table 6B.2 *Results weighted for population of Soviet Union vs West (i.e. figures as in Table 6B.1 except that figures for Soviet publications are multiplied by 2½)*

	Publication in which citations are analysed	Number of articles	All citations					Citations dated 1960 or later			
			USSR	West	Other	Doubtful		USSR	West	Other	Doubtful
Reviews	Reviews in *Macromolecular Chemistry*, Vol. 10, 1973	8	25	781	68	1		25	674	68	0
	Uspekhi Khimii, Vol. 42, 1973	11	1720	803	70	11		1563	542	62	8
Polymer Journals	*Vysokomolekulyarnye Soedineniya*, Vol. 15, 1973	434	7638	2459	119	65		7030	1683	115	63
	Journal of Polymer Science, Polymer Chemistry Edition, Vol. 11, 1973	272	335	4005	241	4		303	3097	231	4
	Revue Roumaine de Chimie, Vol. 18, 1973	8	55	97	14	3		42	79	13	2
	Collection Czech Chemical Communications, Vol. 38, 1973	30	48	317	103	3		28	225	98	2
East European Journals	*Acta Chimica Academiae Scientarum Hungaricae*, Vol. 75, 1972–3 Vols. 77, 78, 79, 1973	9	13	83	32	0		10	51	32	0
	Journal für praktische Chemie, Vol. 315, 1973	4	0	42	12	0		0	28	12	0
	Plaste und Kautschuk, Vol. 20, 1973	119	1028	1157	364	11		900	844	359	11
	Polymery—Tworzywa Wielkoszasteczkowe, Vol. 18, 1973	109	490	832	455	13		445	602	430	11
	Totals for all East European journals	279	1633	2528	980	30		1425	1829	944	26

7. Industrial process control

Introduction

In purely numerical terms the control and instrumentation industry appears insignificant in the Soviet national economy: the instrumentation industry proper contributes only 0·7 per cent of the total industrial output, and computers (a closely related branch) a further 0·3 per cent.[1] But the significance of the industry is considerably greater than these figures suggest. As Brezhnev declared at the XXIV Party Congress,[2] instrumentation shares with electronics the distinction of forming the basis on which other industries can develop; he called these two industries 'catalysts of scientific and technical progress'. Control is also closely related to management and cybernetics which are playing a vital role in the current attempt to modernise Soviet economic organisation. The Ministry primarily responsible for the industry, Minpribor, has been an experimental ground for economic and organisational reforms since 1965. It was in Minpribor that the scheme for Automated Systems of Management and Control, ASU, to which the Soviet authorities have allocated a high priority, was first introduced at an industry or branch level known as OASU, and Minpribor also acts as a consultant and contractor for other OASU systems. As is appropriate for a leading industry, the control and instrumentation industry has always held a high, if not the highest, position in the main success indicators: growth of output, plan fulfilment and growth of labour productivity.

Only 60 per cent of instrument output is produced by Minpribor,[3] and the remainder by other ministries; the chemical industry in particular possesses substantial instrument manufacturing capacity. However, these 'user indus-

---

[1] In 1972, output of what the Soviet statistics treat as instruments amounted to 3,000 million roubles; of this the value of clocks and watches was 12 per cent. In addition, the output of computers (which almost certainly includes desk calculators) was 1,200 million roubles, all in round figures. (*Narodnoe khozyaistvo SSSR v 1972 godu* (1973).)

[2] *Pravda*, 31 March 1971.

[3] *Ekon. Gaz.*, 1973, No. 39, p. 12. Minpribor probably produces 60 per cent of the total Soviet output of instruments proper and computers, considered together.

tries' do not appear to contribute positively to control and instrumentation technology.

The present chapter is devoted to instruments intended for the measurement and control of industrial processes. This group is the largest of nine into which the Soviet instrument industry is traditionally divided in the statistics, and it represents 24 per cent of this industry's output.[4] Also it is a group more closely associated with the economy than any other group, and, consequently, it attracts a considerable amount of attention in the Soviet press.

THE ESSENTIALS OF CONTROL

The purpose of an automatic control system is to maintain the desired value of a quantity or condition by measuring its existing value, comparing it with the desired value, and employing the difference (deviation) to initiate action for reducing this difference. Thus the system operates as a closed loop, and is capable of functioning without human intervention.

There is a great variety of control apparatus. From the point of view of the form of auxiliary power employed, one can divide the whole spectrum into hydraulic, pneumatic and electronic (formerly electrical) instruments; from the point of view of the mode of operation one can distinguish between analogue and digital instruments (continuous and discrete in Russian). Instruments which form an integral part of the control loop are known as *on-line* instruments; others are known as *off-line*. As far as physical construction is concerned, modularisation is almost universally recognised today, i.e., systems are assembled from small single-function units or 'building bricks'. In such a system standardisation of signal used between the units is very important.

Process instruments are usually organised into 'systems', that is, ranges of mutually compatible units covering the many functions required in a process control scheme; throughout this chapter, the discussion is conducted very largely in terms of such systems.

Control systems vary in size from elementary on-off devices (e.g., room thermostats) to huge multi-computer hierarchical systems (such large systems, however, are still quite rare).

In order to explain in easy stages how a control system functions, let us consider a heat exchanger in which water is heated up by steam. Assume the exchanger is controlled manually by a human operator, as shown in Fig 7.1a. The operator feels the temperature of the hot water outlet pipe and adjusts the steam flow control valve so as to keep water at the desired temperature. His right arm performs the functions of measurement and conversion to the corresponding nerve signals, while his brain compares the measurement with the desired value and computes the correction required; his left arm executes the correction by means of the control valve.

[4] *Narodnoe khozyaistvo SSSR v 1973 godu* (1974), p. 281. The other eight groups and their share in the output of the industry are as follows: electrical 14 per cent, optical 13 per cent, clocks and watches 12 per cent, mechanisation and automation of engineering-technical work (this title covers, in fact, office equipment) 11 per cent, electronic 10 per cent, mechanical measurement 8 per cent, physical measurement 5 per cent, medical 3 per cent. Computers are not included in the instrument industry proper.

(a)

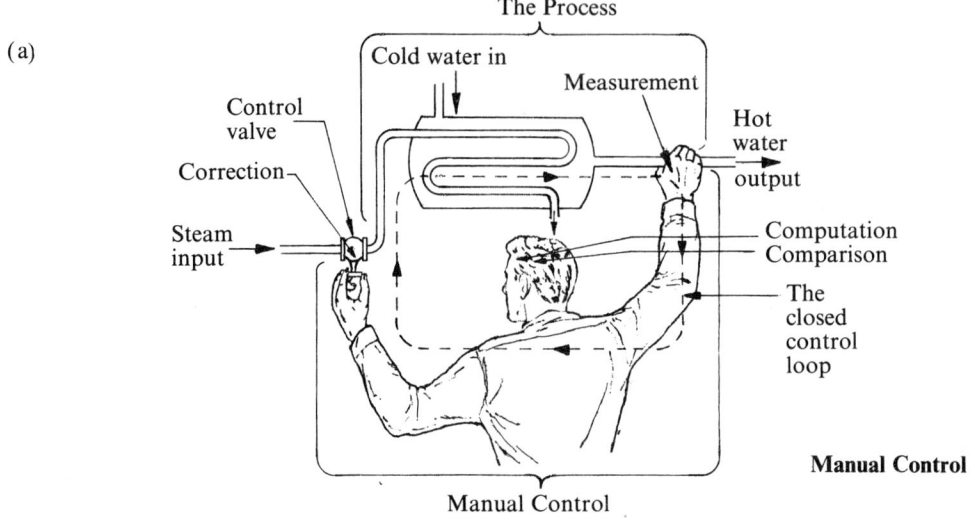

Manual Control

(b)

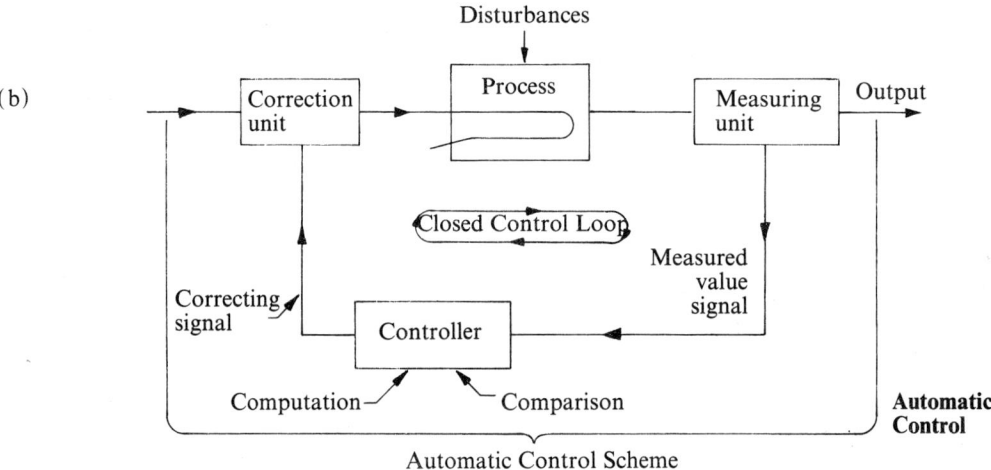

Automatic Control

Automatic Control Scheme

(c)

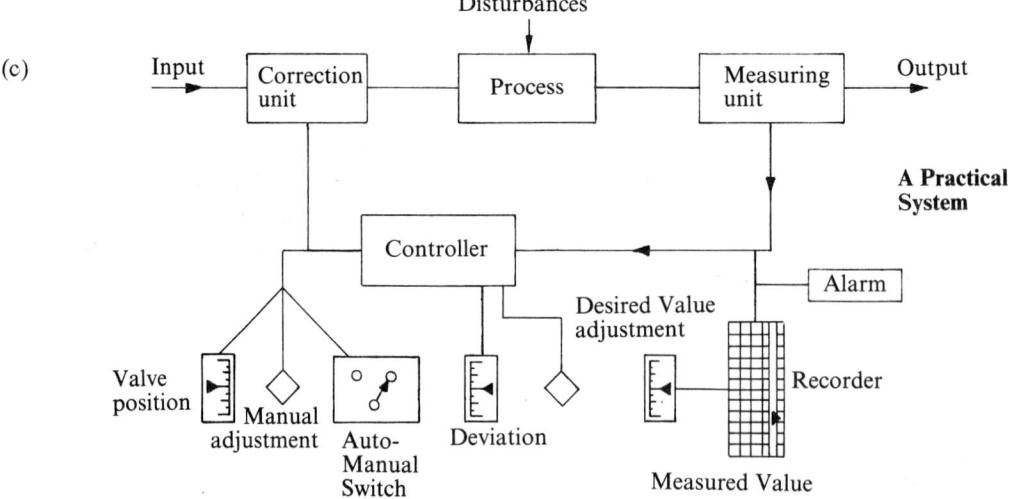

A Practical System

Fig. 7.1 Fundamentals of Process Control.
Source: D. M. Considine ed. *Process Instruments and Control Handbook*, McGraw Hill, 1957, pp. 10, 11.

These functions of measurement, comparison, computation and correction (with conversion of signal form, wherever appropriate) are inherent in most control schemes, manual or automatic. This is a closed-loop system consisting of the process (heat exchanger) and a 'control scheme' (the operator) in which there is a continuous unidirectional flow of information (clockwise on our diagram).

Two kinds of obstacles make the functioning of the operator as a process controller imperfect: disturbances and time lags. The main disturbances are load changes, i.e., changes in the rate at which the hot water is taken off. Other disturbances are introduced by changes in the temperature of the cold water supply and in the temperature and pressure of steam. In the present scheme, the operator has no knowledge of these disturbances until they start affecting his measured value. The situation can be improved by provision of more measurement points and control schemes for the disturbing parameters, but at an additional cost.

Time lags arise in this case because heat energy cannot be absorbed instantaneously by water, and therefore it is only after a certain time interval that a change in the steam flow produces a required change in water temperature. An additional complication is that the magnitude of time lag depends upon the load and upon the other disturbing parameters already mentioned. Due to lag there would be a general tendency for the operator to 'over-correct', but, in contrast to disturbances about which he had no prior knowledge, he can gain some knowledge about time lag by experience.

Dealing with disturbances and time lags is a typical broad task of most control systems.

These general features observed in manual control are also inherent in automatic control.

Let us now consider, for the same process, the replacement of the human operator by a simple automatic control loop consisting of only three units: a measuring unit, a controller and a correcting unit, as shown in Fig 7.1b (for simplicity the description is in terms of electronic instruments only). The measuring unit measures the controlled condition of the process (in this case the water temperature) and sends out to the controller a signal corresponding to the measured value. For instance, if the measuring unit range is 0 to 100°C and its output signal is 0 to 10 mA, when the water temperature is 92°C, the unit's output signal will be 9·2 mA. The measuring unit consists of two sections: a detecting element and a transmitter.[5] The detecting element (in our example probably a resistance thermometer, but one can also think in terms of the more familiar thermocouple) responds directly to the value of the

---

[5] Although the term 'transmitter' (*datchik*) is in daily use in control and instrumentation circles, evidently it has not been officially sanctioned in either East or West, possibly on the grounds that the distance over which the signal is transmitted in process control is normally very short by comparison with radio communication. Both British and Soviet official standards of control terminology (BS 1523:1967, Part 1, Definition 4005, and Sbornik Rekomenduemykh Terminov, Vypusk 77, *Elementy Tekhnicheskoi Kibernetiki*, 1968, p. 23) do not recognise this term, but recommend instead 'measuring element' and '*vosprinimayushchii blok*' respectively; the recommended alternatives are hardly ever used in practice. This problem was discussed in *Pribory i Sistemy Upravleniya*, 1972, No. 4, p. 16.

controlled condition (temperature) and converts it to a small electrical signal, which is then amplified and scaled in terms of the unified signal (0–10 mA) by the transmitter.

Physically, in our example, the detecting element would normally be housed in a metal pocket inserted into the hot water outlet pipe and the transmitter would be mounted nearby. The measured value signal travels by wires to the controller which may be some distance away in the control room, or equipment room.

The controller, like the human operator, performs comparison and computation. The incoming measured value signal is compared with the desired value (the specified value of the controlled condition which the automatic controller should maintain) and the difference between them, known as deviation, is obtained. Continuing with our example of a measured temperature of 92°C with an electrical signal of 9·2 mA, let us assume that the desired value is set to 90°C, represented by 9 mA, and therefore the deviation in terms of the signal is 0·2 mA, and, expressed as a percentage of the transmitter range, is +2 per cent.

Computation in the controller is performed upon the deviation signal, and the controller output signal is known as the correcting signal. The most widely used form of controller is the proportional-plus-integral form (commonly referred to as the two-term controller) which operates in accordance with the equation

$$V = k_1 \theta + k_2 \int \theta dt$$

where V is the correcting signal (i.e., controller output signal), and θ is the deviation (i.e., controller input signal).

In this way the output of the controller contains a factor proportional to the deviation and one depending upon the integral of deviation. The coefficients k_1 and k_2 (known as the proportional and the integral action factors respectively) are adjustable on each controller to take account of process characteristics (including the time lags, already mentioned) and of the objectives of a given control loop.

Integral action (the second term on the right-hand side of the equation) is the most useful tool of automatic control, since, as will be seen from the equation, it alters the correcting signal as long as deviation is present; and therefore is capable of dealing with disturbances much more effectively than the proportional action (the first term on the right-hand side) which for a given deviation produces only a fixed signal. The proportional-plus-integral form has been used throughout the history of process control, from the earliest days well into the computer age (process control computers are normally set to operate in accordance with the P+I algorithm). No better or even equal alternative has yet been invented.

A third term, the derivative action, is occasionally used in process controllers (then the controller equation becomes $V = k_1 \theta + k_2 \int \theta dt + k_3 d\theta/dt$) in order to provide a measure of disturbance anticipation, but it is not a powerful control tool.

The correcting signal is sent out to the correcting unit, and this then executes the command of the controller by adjusting the flow of steam in such a way as to eliminate the existing deviation in the temperature of water; in our example the steam flow will be reduced at a rate dictated by the controller. The correcting unit consists of an actuator and a control valve. While up till now we have been dealing in this example with devices working at very low energy levels, in order to operate the control valve a considerable force may be required from the actuator. In our example it may be found convenient to use an electro-pneumatic converter on the correcting signal, and a pneumatic actuator to operate the valve.

A control system of the simplicity shown in Fig 7.1b could not be used in practice because it is 'blind' (i.e., it has no indicators), and does not possess a control station.

These items are included in the diagram in Fig 7.1c, which shows indicators on the measured value, deviation (alternatively it could be the desired value) and valve position, and also a recorder and alarm unit on the measured value; the control station has facilities (a switch) for selecting either automatic or manual mode, for remote manual adjustment of the control valve, and for adjusting the desired value. With automatic control the closed loop is in operation, but with manual control the loop is broken and the control valve actuator is connected directly to the appropriate adjustment on the control station, so that the operator can select any required position of the control valve. Many process control loops are very much more complicated than described here, and they are interconnected to form an integrated system. In addition to the equipment already described, there is a considerable variety of auxiliary units which are mainly used in these more complex loops, such as signal summation and fan-out amplifiers, signal sources, analogue computing units (square-rooters, multipliers, dividers and squarers), memory units, limiters (which do not permit the signal to deviate beyond pre-set limits), auctioneering units (which select for transmission the highest or the lowest of the signals presented to it), delay units, etc. Numerically these units are often preponderant in a control system, although their function is less fundamental than, for instance, that of the controller.

Control systems of the individual processes vary appreciably in size, but a fairly typical process control system (for instance on a boiler) may contain about 80 transmitters and 30 actuators on the plant, nearly 400 modules in the equipment room cubicles (these modules contain mainly the auxiliary units and controllers), a control panel measuring perhaps 3×1 metres (carrying indicators, control stations, switches, pushbuttons, etc.) in the control room, as well as separate recorder and alarm annunciator panels.

Those familiar with communication engineering may note that throughout the present chapter little attention is being paid to the frequency pass-band of instruments under discussion. This is not an accidental omission, but a reflection of the fact that the significance of the signal pass-band is very much lower in process control than in telecommunication, mainly because the processes themselves are slow and their requirements regarding the frequency spectrum of control signals are very modest by telecommunication standards. The pass-band of an industrial process control instrument usually does not exceed 5 Hz, and in any event it should be kept below 50 Hz, in order to cut off the pick-up from electrical power lines. Both Soviet and Western specifications for process control equipment quote frequency pass-band only rarely.

To complete this introduction to process control technology, a number of crucial aspects of process control will now be described briefly under separate headings.

SIGNAL AND SUPPLY MEDIA: PNEUMATIC AND ELECTRONIC[6]

Pneumatic instruments which work on a supply of compressed air have the following *advantages*:
—low cost;
—their operation is easily understood and most faults can be detected visually;
—they can be safely used in explosion-hazardous atmostpheres.
Their drawbacks are:
—transmission distance is limited to only a few hundred feet—this is insufficient for a high percentage of modern sites;
—signal is not compatible with electronic computers;
—a considerable amount of maintenance is required.

Electronic instruments, by contrast, have the following *advantages*:
—high reliability, with a minimum of maintenance;
—practically unlimited transmission distance;
—direct compatibility with computers (provided a sensible signal form is used);
—utilisation of an easily available power supply.
Their drawbacks are:
—cost higher than for pneumatic instruments;
—safety in hazardous atmospheres possible only with careful design;
—electrical actuators have not yet been properly developed in Britain, although Germany and the USSR have been using them successfully for many years.

In conclusion it should be noted that today it is generally considered that, except for smaller plants, the net advantages of electronic instruments outweigh those of pneumatics; the overwhelming majority of advances in industrial process control are being achieved by means of electronics.

UNIFIED SIGNALS

In a system without a unified signal the output of, say, unit A can be connected, for instance, to the input of unit B only, and to no other unit of the system. On the other hand, in a unified-signal system the output of A can be connected to the input of B, or of C, or of D. Clearly, the flexibility of such a system is infinitely superior.

[6] Because of the very limited transmission distance, combined with the ability to deliver large forces, hydraulics is used almost exclusively in actuators, and there are no complete process control systems operating on this supply. For this reason hydraulics is not included in this discussion.

In the West, by about 1960, it was the common practice to use a unified current signal of 0 to 10 mA throughout the system. Refinements have gradually been introduced. First, a 'raised-zero' signal (4–20 mA) from the transmitter to the control room enabled this link to be made by two wires only, the same pair carrying both the signal and the transmitter supply; this has important safety consequences (see section on intrinsic safety). Second, a pulse signal from the control room to the actuator simplified the transfer procedure from automatic to manual control (it should be noted that the Germans and the Russians have always used such a signal to the actuators). Finally, with the advent of integrated circuits it has become advantageous to use a voltage signal (typically 0–5 V) inside the control room. Current thinking in the West accepts all these three refinements, although both national and international standardisation has been slow in incorporating this thinking. In the Soviet Union, no raised-zero transmitter signals are used, and there are no two-wire transmitters, even in the latest equipment (see pp. 355–9).

INTRINSIC SAFETY

Safe operation of electrical control equipment in explosion-hazardous areas, such as petrochemical works, can be achieved in one of three ways:
—contain the explosion within a suitably designed equipment box (explosion-proof housing, also known as fire-proof housing);
—isolate the equipment from the explosive mixture outside (by sealing, purging with inert gas, oil immersion or encapsulation);
—limit the energy which the electrical equipment can supply, below the level necessary to initiate an explosion, even under the conditions of every conceivable fault (and combination of such faults) arising. This is known as the intrinsically safe design, and in Russian as spark-proof (*iskrobezopasnyi*).

The third, intrinsically safe, design approach is clearly superior to the other two methods. First, it is very much cheaper in terms of both capital and running costs. Second, it is the only method which allows the control equipment to be opened up in the hazardous area for the purpose of maintenance or repair; the other two methods require elaborate procedures to be followed when disconnecting the equipment and removing it to a safe area for repair.

Intrinsic safety is a British concept, and it was developed in the mining industry in this country between the wars; in 1945 a British Standard was issued for it.[7] A valuable extension of the concept was provided by developing the 'safety barrier' (also a British contribution)[8] which allows wire connections to be made between intrinsically-safe equipment in the hazardous area and a higher-energy equipment in a safe area. A further extension, although less fundamental than the safety barrier, was the two-wire transmitter, referred to above, initiated in the USA, which makes it possible to mount the transmitter power supplies in a non-hazardous area.

[7] BS 1259, *Intrinsically Safe Apparatus and Circuits*.
[8] R. J. Redding, *Proceedings, Symposium on Flameproofing, etc.*, Institute of Electrical Engineers, London, 27 April 1962, p. 52.

Today the intrinsic safety concept offers a very effective means of achieving control equipment safety in hazardous atmospheres.

AMPLIFYING ELEMENTS
—thermionic valves;
—magnetic amplifiers;
—transistors;
—linear integrated circuits or operational amplifiers (abbreviated as linear ICs or op. amps.).

Valves were at first unique in providing a high input impedance, but are most inconvenient for industrial applications. For their electrodes, they require supplies of high voltage and a supply of very high current for their heaters. They dissipate a great deal of power, and this raises the temperature of other components in the vicinity. Their construction is delicate and their reliability low.

Magnetic amplifiers have the appearance of transformers wound with copper wire on iron cores, and weight is their inherent problem. In addition, they have poor input/output linearity and readily pick up mechanical vibrations (and transform them into electrical signals), but they suffer less from temperature effects than the valves do. 'Magamps' enjoyed their heyday in the post-war years, when the high degree of sophistication to which the Germans developed them for the V.1 and other weapons became known; but with the advent of the transistor their *raison d'être* largely vanished. Nevertheless, the Russians doggedly plan to use them well into the 1970s (see pp. 357–8).

The *transistor* is, of course, a semiconductor amplifying device, invented in 1947 by Shockley and others at Bell Telephone Laboratories. It represents a tremendous advantage by comparison with the two elements described above, because of its minute size and very modest power demands, but it requires a high degree of control of the manufacturing process (Soviet industry found this obstacle hard to clear) and its reliability in practice proved lower than expected.

An *integrated-circuit amplifier* is an assembly of transistors and other components of microscopic size, made from one semiconductor slice. By comparison with the transistor it represents a considerable simplification of the design and assembly, as well as cost economy; very high reliability is claimed for it relative to the number of equivalent 'discrete' components. Because of lack of standardisation among the manufacturers it has had a slow start even in the West, and the amount of practical experience with it is still limited. In Table 7.1 the advantages and disadvantages of these four elements are summarised.

The demands of semiconductor design and manufacture are formidable. In the United Kingdom the indigenous British-owned semiconductor industry has almost vanished, apart from two or three firms manufacturing on a limited scale, largely for in-house requirements. The vast majority of semiconductors used in this country are of American or Dutch origin in terms of design and finance, and are either imported or manufactured on British soil by foreign firms.

Table 7.1 *Amplifying elements for control systems*

Thermionic valves

Advantages: At present, none.
Disadvantages: High power dissipation.
Low reliability in industrial environment.
Temperature effect problems.

Magnetic amplifiers

Advantages: More reliable and less temperature-dependent than valves, but no advantages with respect to transistors.
Disadvantages: Weight.
Poor input-output linearity.
Readily influenced by vibration.

Transistors

Advantages: Small physical size.
Low power dissipation.
Reliability and suitability for industrial environment.
Disadvantages: Reliability not quite as high as expected.
Technology not easy to master.

Linear ICs (op. amps.)

Advantages: More compact, simpler to use and cheaper than transistors.
Expectation of very high reliability.
Disadvantage: Technology not easy to master.

THE MECHANISM OF INNOVATION IN PROCESS CONTROL

The technology we have described, in which individual elements are organised into compatible systems, imposes certain characteristic features upon the mechanism of innovation in the process control industry which distinguish it from other industries and, indeed, from other branches of instrumentation.

These features are found in both East and West and may be formulated as follows:

—Innovation happens in large steps or lumps ('systems'); the decision to innovate and the formulation of the specification of a new system involve much interdepartmental discussion and coordination.
—Longevity of the control equipment (site life) is high (20 years or more), and normally equal to the life of the main plant. However, sales (or production) life of any one type of equipment is fairly short in the West (3–7 years), but long in the USSR (over 20 years).
—Innovation (introduction of a new control system) takes place very largely on new sites only; on existing sites it may happen if the main process is altered (e.g., changing from solid to gas fuel).
—Diffusion of single new-design models on existing sites is usually not a practical possibility because of incompatibility with the existing associated equipment. Prototypes are occasionally taken on site trials, but on a purely

temporary basis, and the old equipment is restored on the termination of such a trial.

—Faulty components and worn-out moving parts are replaced continually through the site life of a control scheme. This does not affect the issue of innovation. But a high rate of replacement of components or whole units signifies low reliability, and is entirely a negative and undesirable parameter, in contrast to the positive use of the term 'high replacement rate' which is quite common in deliberations connected with the general subject of industrial innovation.

Analogue control up to 1970[9]

EARLY HISTORY (UNTIL c. 1950)

The development of indigenous industrial instrumentation and control in the Soviet Union had three main sources: measurements, precision mechanics and the power industry.

The famous chemist Mendeleev took a considerable interest in the problems of measurement, and it was due to his initiative that, on the basis of an earlier organisation, the Glavnaya Palata Mer i Vesov (Chief Board of Weights and Measures) was established in 1893 at St Petersburg (currently it is known as the VNII Metrologii); in 1901 it issued the first set of test specifications for electrical measuring instruments.[10]

The revolution changed little, except ownership, in the situation of the small instrument industry, and for the most part only simple instruments were manufactured, chiefly on the basis of foreign models. The effect of the first five year plan (1928–32) on the instrumentation industry was, on the whole, slight. During this period, however, precision mechanics entered the arena, and factories and laboratories were established in Leningrad and Moscow during 1927–9. In the early 1930s production started in Leningrad of the first equipment for automatically transmitting information over a distance ('*telemekhanika*' in Russian).

The second five year plan (1933–7) resulted in more substantial progress, in both engineering and organisation and the first modest control systems were developed. The power industry was the undisputed leader in this new development, and following the German practice in this field, the main emphasis was on the electrical control equipment (in contrast to the emphasis on pneumatic equipment, evident in the English-speaking countries). The first

---

[9] For information on process control equipment in current production, side-by-side with periodicals, visit reports, etc., two books have proved particularly useful: a handbook of industrial instruments and process control equipment, B. D. Kosharskii (ed.), *Avtomaticheskie pribory*, 2nd edition, Leningrad (1968), and a handbook of industrial analogue process controllers, Sh. E. Shteinberg *et al.*, *Promyshlennye avtomaticheskie regulyatory* (1973), both subsequently referred to here under the names of their principal authors as Kosharskii and Shteinberg respectively. A comparison of the content of these two handbooks shows, quite characteristically, that no new controllers have been put into production between the dates at which the handbooks were written (which can presumably be taken as 1967 and 1972 respectively).

[10] V. A. Rukhadze and A. T. Chervyakovskii, in *Pribory i Sistemy Upravleniya*, 1967, No. 10, p. 31. Unless otherwise stated, the material in the present section has been taken from this article.

automated hydro-electric power station was commissioned in 1932 at Erevan,[11] but the main impetus for developing automatic control was connected with innovation in boiler technology during 1938. Various more or less primitive controllers were at that time either already in experimental existence or were being developed,[12] but it was the VTI (All-Union Thermo-Technical Research Institute) which developed in this connection the electronic controller,[13] known as type ER, which proved most successful.

During the war the Soviet Union received a substantial amount of pneumatic control equipment from the West. The present author's employers, Bailey Meters & Controls Ltd, alone shipped to the USSR during 1942-5[14] instrumentation and control equipment for some 50 boilers at two dozen power stations, the boilers themselves also being supplied from Britain. The total steam generating capacity of these boilers was 8,000 tons/hr, which represents approximately 13 per cent of the total power industry steam generating capacity of the Soviet Union for 1945.[15]

Presumably other instrumentation and control firms in this country and in America also supplied their products to the USSR during the war.

Despite this temporary influx of foreign pneumatic influence, the Ministry of Power Stations reasserted after the war their determination to concentrate on electrical controls.[16] The VTI set about modifying their range of equipment and during 1949–51 produced a fundamentally new version of the ER controller,[17] in which the sensitive galvanometers were replaced by magnetic and valve electronic amplifiers.

The controller had been recognised as standard equipment in the power industry; it played a vital role in the country's economy and was produced in large numbers by the Energopribor and Komega (now MZTA) factories; its designers, V. D. Mironov, E. P. Stefani and N. I. Davydov received a State Prize for this achievement.[18] The controller possesses remarkable longevity, since it still remains in production in the Soviet Union (see pp. 349–52) and in satellite countries (in Poland, for instance, two factories manufacture it). Moreover, it provided a core for the MZTA system which constituted the standard control and instrumentation equipment for the Soviet power industry throughout the 1960s and well into the next decade (see pp. 342–4, 349).

The front of the VTI controller is shown in Fig 7.2 and, for comparison, the front of a contemporary British electronic controller (by Evershed and Vignoles)[19] is shown in Fig 7.3.

[11] *Avtomatika i Telemekhanika*, 1957, No. 11, p. 953.
[12] S. G. Gerasimov, E. G. Dudnikov, S. F. Chistyakov, *Avtomaticheskoe regulirovanie kotelnykh ustanovok* (1950), pp. 209–35.
[13] *Ibid.*, pp. 236–42.
[14] Thanks are due to Mr Ll. Young, a past director of the firm, for making available the relevant archives and for permission to utilise this material.
[15] *Narodnoe khozyaistvo SSSR v 1965 godu* (1966), p. 169; total USSR electrical power capacity in 1945 equalled 11,124 mW.
[16] E. P. Stefani, *Teploenergetika*, 1955, No. 9, p. 4.
[17] N. E. Davydov, *ibid.*, 1956, No. 3, p. 14; see also an instruction book: *Elektronnyi Reguliruyushchii Pribor tipa ER-111-59*, Moskovskii Zavod Teplovoi Avtomatiki (1960).
[18] V. D. Mironov et al., *Teploenergetika*, 1971, No. 7, p. 14.
[19] J. R. Boundy and S. A. Bergen, *ProcIEE*, Pt. II, Vol. 98, October 1951, p. 609.

340   TECHNOLOGICAL LEVEL OF SOVIET INDUSTRY

It will be seen that in the Soviet controller two thermionic valves are brought out through the front in the top corners, presumably in order to limit the temperature rise inside the controller, but they are probably unpleasantly hot to touch. The armatures of two relays, P2 and P4, also protrude through the

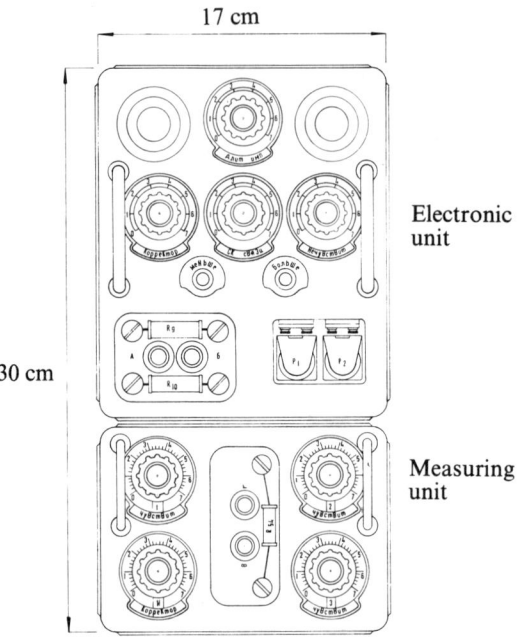

Fig. 7.2 Front of Soviet electronic proportional-plus-integral controller, launched in 1951.
Source: Instruction book published by Moscow City Sovnarkhoz, 1960.

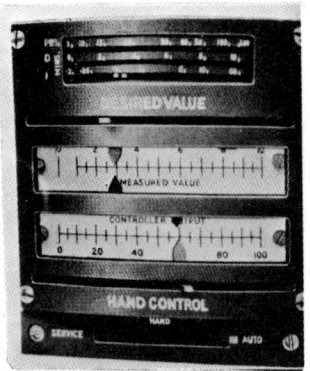

Fig. 7.3 Front of British electronic proportional-integral-derivative controller, with control station, launched by Evershed & Vignoles in 1949.
Dimensions: 18 × 16 cm.
Reproduced by permission of Evershed & Vignoles Ltd. and Kent Instruments Ltd.

front, so that the operator can see whether the relays work, but in this way they are exposed to dust and inadvertent handling (these relays proved highly unreliable and were later replaced). There are three selected-on-site resistors R9, R10 and R54, probably because suitable variable resistors were not available. Numerous other variable adjustments are mounted on the controller's front plate.

Although the front of the British controller (Fig. 7.3) occupies 40 per cent less area, it incorporates on it a complete control station (Hand/Auto switch, the indicators and Hand and Desired Value levers), in addition to the three term adjustments (P, I and D). Thus space utilisation is considerably better than in the VTI controller. Also in the British controller there are no hot valves protruding through the front, selected-on-site components attached to the front, or relay observation cut-outs. The controller, and the system of which it formed part, operated on a unified signal (0 to 15 mA) which the Soviet controller did not have. Probably the most important feature: the reliability of the British controller was very much higher. Its only disadvantage was that the control lever cut-outs facilitated the ingress of dust to a greater degree than one would like, although this was not crucial.

Circuit technology in the two controllers had certain similarities and, on the whole, the Soviet controller represented a fairly modest technological lag behind Britain, perhaps of the order of 2 or 3 years.

THE 'GREAT DEBATE' OF THE 1950s

A major review of the objectives of industrial control was carried out in the Soviet Union during the 1950s, and led to a fundamental review of policy. During this period automatic control was approached for the first time in a comprehensive manner, involving the whole process rather than its individual sections or 'loops'.

However intense internal discussions may have been, what was published resembled a series of monologues rather than a genuine discussion; as most published proposals were subsequently implemented, they evidently represented authoritative pronouncements on issues which had been decided beforehand, at least in principle.

The control and instrumentation debate was launched by a paper read in May 1950 by Trapeznikov at a general meeting of the Division of Technical Sciences of the USSR Academy of Sciences, which was later published in *Avtomatika i Telemekhanika*, with Lerner as co-author.[20] The authors advocated designing automatic control apparatus as a unified complex, which satisfied the requirements of various industries, and was constructed on the 'aggregate principle' (i.e., with standardised sub-assemblies), and use of the self-balancing principle.

With the advent of the Malenkov regime in 1953, as part of a general policy of improving the welfare of the people, a commission was appointed ostensibly to investigate how the control and instrumentation industry could contribute

[20] V. A. Trapeznikov and A. Ya. Lerner, *Avtomatika i Telemekhanika*, 1950, No. 4, pp. 226–50.

to raising the quantity and quality of production in the food and light industries.

The commission consisted of seven specialists (including Trapeznikov, Lerner and Ilin) and the representatives of five industrial ministries (most probably the traditional 'big five' users, namely: power, ferrous and non-ferrous metallurgy, chemicals and petroleum). Since the findings were published over Trapeznikov's name,[21] he probably chaired the commission. In practice, the commission looked at the whole of development and organisation of control and instrumentation, and made some momentous recommendations.

In technology it recommended encouragement of conveyor belt production methods, of all-electric control systems, and of '*kompleksnaya avtomatizatsiya*', that is, automatic control embracing the whole process. In the field of organisation it recommended that the production of the only general-purpose range of control apparatus at that time ('*teplovaya avtomatika*') should be transferred from the Ministry of the Electrical Industry to the Ministry of Machine Building and Instrument Construction, and that other industrial ministries should continue to produce control and instrumentation apparatus specific to their fields. These recommendations were fully implemented and have had a very long-lasting effect.

One notable omission from these recommendations, however, was any major move towards unification of the industry. In contrast, the need for unification found powerful expression four years later in 1958, in an article by M. E. Rakovskii and D. V. Svecharnik[22] demanding a coordinated technical policy in instrumentation, and proposing the creation of a unified state system of instruments and automation apparatus. This article also had a long-lasting effect and gave rise some years later to the GSP system, discussed later on in this chapter.

The article also discussed at length methods for achieving a contactless control apparatus without using transistors. Evidently at that time the chances of the instrument industry having transistors at its disposal must have been rated quite low; at that time in the West, development of the application of transistors to industrial instrumentation was already fairly advanced.

THE INTERMEDIATE STAGE (AUS, EAUS, MZTA)

As a first sign of modernisation, and as an embodiment of the 'aggregate construction' policy advocated by Trapeznikov, a pneumatic control system AUS (*Agregatnaya Unifitsirovannaya Sistema*) was developed during 1952–3 at NIITeplopribor.[23] Technically the system was comparable to the contemporary Western pneumatic systems; it was mainly intended for use in the chemical and petrochemical industries, where electrical apparatus is

---

[21] V. A. Trapeznikov, *ibid.*, 1954, No. 5, pp. 384–91.
[22] M. E. Rakovskii and D. V. Svecharnik, in *Priborostroenie*, 1958, No. 5, pp. 3–7. A further paper by V. V. Solodovnikov, *Priborostroenie*, 1958, No. 6, pp. 4–11, persuasively argued that control theory and cybernetics were opening up great prospects for process control.
[23] N. N. Mikhailov, *Priborostroenie*, 1957, No. 12, pp. 1–5.

undesirable because of explosion hazard. By comparison with the VTI controller, AUS was somewhat more advanced, as it operated on a unified signal (0·2–1·0 kg/cm²) and was more system-oriented, but as a pneumatic system it was inherently less versatile. AUS is still in production,[24] although a much more advanced pneumatic system (USEPPA-START) is also now available.

An electronic counterpart of AUS, designated EAUS, was announced in 1961[25] and operated on a unified transmission signal 0–5 mA. It was stated to be modularised; from the photographs this would, however, appear doubtful. The construction was generally unimpressive, and thermionic valves were used throughout (at that time in the UK fully transistorised systems had been operating for two years).

The EAUS was clearly intended as a stop gap; it was incomplete, and had no actuators and few transmitters. With time, however, it was substantially expanded, modernised, and acquired a permanent character; many valves were replaced by transistors and the physical design was tidied up appreciably. A special Design Bureau, SKB EAUS, renamed SKB SPA (*Sistem Promyshlennoi Avtomatiki*)[26] was set up at Cheboksary to work on improvements and manufacture took place at the ZEIM factory in the same town.[27] In view of the disastrously slow progress with the GSP system (see pp. 344–7), EAUS still remains the only electronic system with some pretence to comprehensiveness, working on a unified signal.

At the same time the power industry developed its own system around the VTI controller (see above); this became known as the MZTA system,[28] after the main manufacturer, the Moscow Factory for Thermal Automation (Moskovskii Zavod Teplovoi Avtomatiki). Reliability was improved,[29] but the main drawbacks still remained: the use of valves, the absence of a standardised signal, and non-modular construction. As a result of the latter, numerous versions of its component units have to be manufactured to cater for even slightly different requirements; the controller, for instance, is made in some 30 versions.[30]

Despite these fundamental drawbacks, the power industry has learned to live with the system; since its inception in the 1930s it has been playing a vital role in the national economy, and has been produced in very large numbers (at the MZTA factory alone 25,000 controllers were produced annually throughout the 1960s).[31] The system apparently provides tolerable, if

---

[24] Kosharskii, *op. cit.*, p. 417 ff.

[25] K. B. Arutyunov, *Priborostroenie*, 1961, No. 1, p. 23.

[26] *Teoriya, raschet i printsipy konstruirovaniya elektronnoi agregatnoi unifitsirovannoi sistemy*, Sbornik statei SKB SPA, vypusk III (1969), p. 2.

[27] Kosharskii, *op. cit.*, p. 633.

[28] Kosharskii, *op. cit.*, pp. 601, 608.

[29] V. D. Mironov (*Teploenergetika*, No. 6, p. 21) reported mean-time-between-failures, prior to the modifications, of only a few months; there is no information on the improved figure.

[30] Kosharskii lists 31 versions on pp. 606–8; a figure of 25 versions is suggested by the Joint Soviet-UK Working Group on the Problems of Electricity Supply and Transmission, Study Group No. 5, *Automation of Thermal Power Stations*, Meeting in USSR, July 1969, CEGB, London, 1970, p. 75.

[31] E. P. Stefani, *Teploenergetika*, 1965, No. 4, p. 2; also, Study Group 5, *op. cit.* p. 75.

unexciting mass automatic control. It is very cheap and allegedly costs less[32] than either one of the two pneumatic systems (AUS and USEPPA). (This is most unusual, since pneumatic systems are normally cheaper than electronic.) Nevertheless, the existence of a system of such antiquity is a tangible monument to the slowness of innovation in the Soviet industry. At the time of writing it is still in full production and constitutes standard equipment for new power stations. The 100 per cent equipment standardisation achieved in control and instrumentation by the Soviet power industry, while admirable in several ways, has evidently been allowed to continue far too long, and it has effectively blocked innovation.

Figure 7.4 shows the MZTA controller and a corresponding British controller currently supplied to the power industry (by Bailey Meters and Controls) and the attached table, 7.2, summarises their characteristics.

The table illustrates the very wide lag which now exists between the two countries in this area. This point can be reinforced by reference to the situation c. 1950 (see pp. 338–41) which shows that in the past 25 years the UK has made very substantial progress in this area, whereas the USSR has achieved only marginal progress.

Table 7.2 *Characteristics of electronic proportional-plus-integral controllers of 1960s*

Origin and type	Approx. year of development	Amplifying element	Signal	Construction	Weight kg	Power dissipation VA	Max. ambient temperature °C
Soviet VTI-MZTA	1950	Valves	Non-unified	Non-modular	12	22	40
British Bailetronic Mk. III. Bailey Meters & Controls.	1967	Transistors and ICs	Unified	Modular	5	12	70

THE STATE SYSTEM OF INSTRUMENTS (GSP)

The seeds of a unified control and instrumentation system sown by Rakovskii and Svecharnik in 1958 (see p. 342, above) took years to germinate; and evidently their powers of persuasion were unequal to those of Trapeznikov, who had normally managed to get rapid action on his ideas. From time to time various laudatory but vague references to the unified system appeared in the press, but it was not until 1964 that a fairly comprehensive 'blueprint' of the proposed system was published in an editorial of the journal of the control and instrumentation industry.[33] This stated:

[32] Private information.
[33] Editorial, *Priborostroenie*, 1964, No. 11, p. 1.

The State System of Instruments (GSP—Gosudarstvennaya Sistema Priborov) is a complete series of instruments and devices, embracing the whole field of Soviet instrument manufacturing, and it comprises the following: GSP for the measurement and control of industrial processes; GSP for the measurement of time; GSP for medicine, agriculture, etc.—i.e., GSP for all the vital branches of Soviet instrument manufacturing.

Each one of these systems ... can be divided into branches and sections ... for instance the GSP of measurement and control of industrial processes ... contains two sections:

(1) Instruments working on a unified signal throughout, intended for constructing ... control systems.

(2) Standardised series of instruments and devices for local applications.

The common tie between the two sections is provided by the modular principle of construction ... with maximum unification of parts and assemblies.

Apart from this, the first section is further divided into the following branches:

(a) Electrical (analogue and digital),
(b) Pneumatic,
(c) Hydraulic.

The article goes on to say how the modules are further classified in accordance with the functions they perform, and it outlines the development work already carried out.

Clearly this comprehensive scheme offered substantial rationalisation, modernisation and economy. It may indeed have been too ambitious and therefore potentially cumbersome, particularly as the non-industrial branches were also included.

As a first move, GOST standards were issued for the unified signals to be used between the system modules: GOST 9468–60 for pneumatic signals and GOST 9895-61 for the analogue electrical signals. The first of these fixed the pneumatic signal at 0·2 to 1·0 kg/cm^2; this was a mere confirmation of the status quo, since this signal range had been used throughout the continent, including the USSR, for decades. The electrical standard, drawn up by TsNIIKA, was, however, an ill-conceived document in that it included the signals in use at that time, and several others, so that the total number of 'standard' signals exceeded a dozen. Today only two of them, 0–5 and 0–20 mA, are considered as standard, but the GOST specification has not yet been amended.

A standard covering the general technical requirements for the industrial GSP was published only in 1967.[34] This standard falls far short of the requirements which British industrial instruments have to meet in relation to environmental conditions especially for maximum temperature, humidity, dust, and short high-amplitude pulses on the electrical mains supply. Thus the GSP instruments are not really suitable for more arduous environments, such

[34] *GOST 12997-67.*

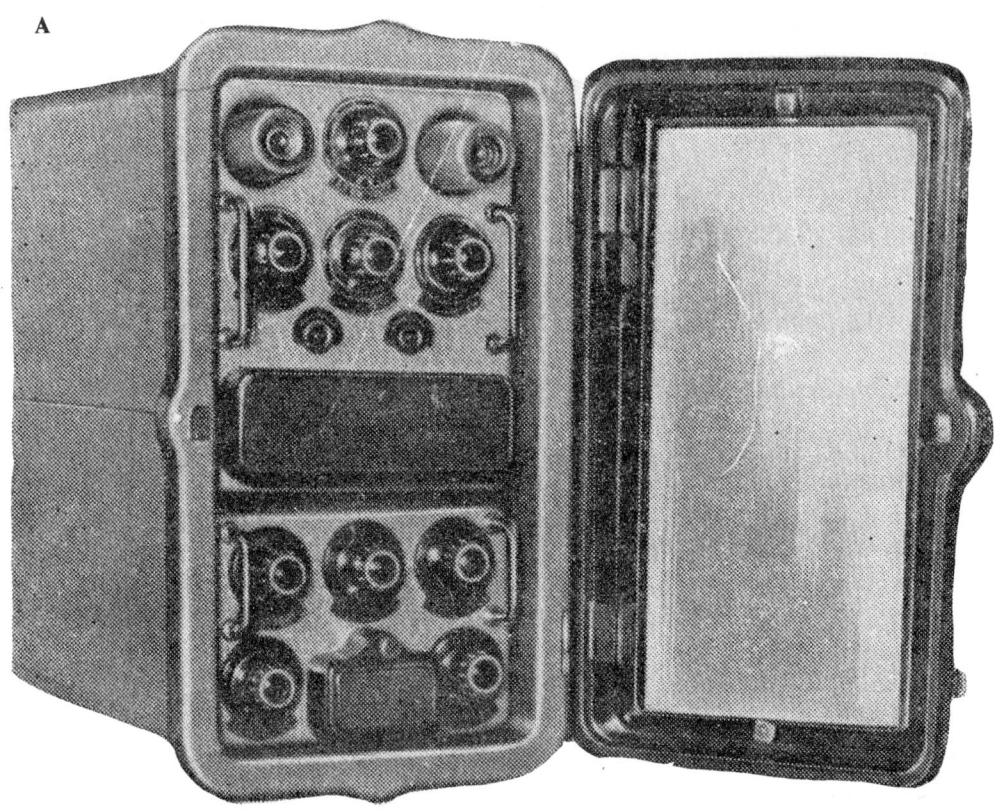

Fig. 7.4 Electronic proportional-plus-integral controllers of 1960s.
  A (above) Soviet MZTA controller. Front dimensions: 34 × 22 cm approx.
Source: Mashpriborintorg catalogue, 1969.
  B (below) British BAILETRONIC Mk. II controller module (pulled forward). Front dimensions: 18 × 19 cm approx.
Source: Catalogue of Bailey Meters & Controls, 1969.

as marine applications and sub-tropical locations, and if used in such locations would fail more frequently.

Details of the first range of GSP instruments produced were published in 1965;[35] this was a range of transmitters for pressure and associated parameters, developed at NIITeplopribor, in conjunction with the Manometer factory in Moscow. Although largely imitative in concept, the main source of inspiration being West Germany, the range appeared to augur well for the whole GSP system: the selection of transmitter types was very wide (over 800 versions), accuracy specification was quite impressive, and a degree of standardisation was reached in the design. Launching this range was hailed as a major achievement of the Soviet control and instrumentation industry. The first transmitters were produced in 1967[36] (thus the setting-up of their production took two years). Some years later, however, it was disclosed[37] that these transmitters proved unsuitable for more arduous industrial applications, and were rejected by the power and metallurgical industries due to their unreliability and sensitivity to vibration, dust and moisture; an alternative range is now being developed. Such a field failure of a whole range of instruments was probably due to either an inadequate specification or inadequate development, or both; quite possibly it had considerable repercussions.

The GSP electronic controller (which, by its nature, forms the backbone of a control system) also ran into difficulties, but at an earlier (probably batch production) stage. Despite announcements in 1967 and 1968 that the controller, type BFZA, was being produced at the MZTA factory,[38] the British delegation visiting the factory in 1969 found that production had not yet started.[39] The controller is not mentioned in a 1973 handbook of control equipment.[40]

Nothing more was heard about any other additions to the first stage of the GSP range. It will be seen below that a new approach to the GSP concept has again been attempted in the ninth five year plan (1971–5).

In conclusion it appears that GSP was in principle an excellent project, which suffered serious setbacks during its first stage. These setbacks were due, first, to the extremely slow progress made in the phases preceding production—policy formulation, drawing up of the specification, and development and design, and, second, to the failure of the transmitters, the first GSP range to reach the production stage.

DIVERSION TO PNEUMATICS (USEPPA)

During 1959–60, when the rest of the world had already turned away from pneumatics and was developing an intense interest in electronics, the Soviet

---

[35] A. L. Vainshtein et al., *Priborostroenie*, 1965, No. 10, p. 3; L. N. Shonin, *ibid.*, p. 5.
[36] *Pribory i Sistemy Upravleniya*, 1967, No. 9, p. 57, and Kosharskii, *op. cit.*, p. 71.
[37] I. Shendler, *Pribory i Sistemy Upravleniya*, 1972, No. 12, p. 31; V. D. Mironov et al., *ibid.*, 1974, No. 7, p. 27.
[38] *Pribory i Sistemy Upravleniya*, 1967, No. 10, pp. 4–5; *MPI Bulletin*, 1968, No. 2, p. 28.
[39] Study Group 5, *op. cit.*, p. 75.
[40] Shteinberg, *op. cit.*

Union embarked on a large project for a novel pneumatic control system which became known as USEPPA (*Universal'naya Sistema Promyshlennoi Pnevmoavtomatiki*). Initial work was carried out by Aizerman,[41] development at IAT lasted till 1964, and the system is being manufactured at the Tizpribor factory in Moscow.[42] After seeing the initial stages of USEPPA development, Professor Rosenbrock commented:

> The effort being devoted in the USSR to the theoretical and experimental investigation of pneumatic elements was surprising. It was taken by some to indicate that Russian solid-state electronic devices are not yet freely available (or perhaps not yet very reliable).[43]

Indeed, it is hard to find any other *raison d'être* for USEPPA, except that it was meant as a second-best alternative, enforced by the non-availability of semiconductors.

USEPPA has several characteristics normally associated with electronics: it is miniaturised and modularised, with plug-in units. The system is original, well engineered, and a credit to its designers. It is claimed[44] that it costs 30 per cent less than the conventional pneumatic equipment (AUS). This is surprising in view of its 'expensive' appearance (by Soviet standards). It suffers from drawbacks common to all pneumatics: limited transmission distance, transmission lags, and reduced computer compatibility. Its application on the home market is limited, since the largest user, the power industry, would not accept non-electrical controls; the same probably applies to the armed forces.

USEPPA represents the best achievement in Soviet process control so far, and, as far as the present writer is aware, until about 1970 no comparable system existed anywhere else in the world; the design team was awarded the Lenin prize.[45] The Schlumburger group (a large international concern with headquarters in France) took out a manufacturing licence for USEPPA,[46] but this does not appear to have been a commercial success, possibly due to the inherent limitations of pneumatic equipment. In the UK an agent has been appointed to market this system, but he does not attempt to sell it for installation outside the USSR, but only for re-export there.

USEPPA has a number of derivatives known under proprietary names. A whole control system assembled from these modules is known as START (*Sistema Avtomaticheskogo Regulirovaniya Zavoda Tizpribor*), a combined electronic-pneumatic scheme is known as PUSK-3,[47] and a range of logic

---

[41] J. H. Westcott, *Control*, 1960, No. 6, p. 93.
[42] V. A. Rukhadze and A. Ts. Chervyakovskii, *Pribory i Sistemy Upravleniya*, 1967, No. 10, p. 33.
[43] H. H. Rosenbrock, *Control*, 1960, No. 9, p. 95.
[44] *Nauchno-tekhnicheskie obshchestva SSSR* (1968), p. 392.
[45] Although the award has been frequently mentioned in Soviet publications, the names of the team are normally not given. However, in the handbook *Nauchno-tekhnicheskie obshchestva SSSR* (1968), p. 392, they appear as follows: M. A. Aizerman, T. K. Berende, T. K. Efremova, A. A. Tal, P. I. Atlas and P. P. Benediktov.
[46] K. N. Rudnev, *Vneshnyaya Torgovlya*, 1967, No. 6, p. 29, and a catalogue of the firm Prodest Elettronica S.p.A., Milano.
[47] G. T. Berezovets, *Priborostroenie*, 1966, No. 11, p. 1.

modules as AUSEDD (*Agregatnaya Unifitsirovannaya Sistema Elementov Diskretnogo Deistviya*).[48]

Although at the time of its launching USEPPA may have been a world-leader, by 1972 it had already been overtaken by Western development of miniature pneumatic instruments (see Fig 7.5).

MINOR SYSTEMS

In addition to the five major systems already described, several minor ones are worth noting; they are either incomplete, or fulfil special requirements.

1. The 'ferrodynamic' system, a crude range of stepping motor control apparatus produced for the metal-extraction industries at the KIP factory, Kharkov. (See Fig. 7.12a.)
2. The 'frequency' range of measurement apparatus, comprising transducers and special data-loggers. The transducers incorporate a vibrating wire (the frequency changes with applied pressure) and send out signals in the form of electrical oscillations. Superficially this may have the advantage of producing a quasi-digital signal for the benefit of the data-logger, but, in fact, handling a frequency signal in data processing is not much easier than handling a common analogue signal. This range is also produced at the KIP factory.[49]
3. USAKR—Unified System of Automatic Control and Regulation. This grand name covers a small antiquated electrical system with balancing motors, manufactured at the Zavod Elektropriborov, Kiev. There is no justification at all for the word 'unified' in its title.[50]
4. Another antiquated electrical system, the Teplopribor range, derived from the original VTI range (see p. 339 above), and manufactured at the Teplopribor factory in Chelyabinsk.[51]
5. The Kristall electro-hydraulic control system produced for small boilers at the MZTA factory, Moscow.[52]
6. A hydraulic and an electro-hydraulic system produced at the Teploavtomat factory, Kharkov.[53]

ANALOGUE EQUIPMENT IN CURRENT PRODUCTION: SUMMARY

Five major systems (two electronic, one mixed and two pneumatic) and at least six minor ones are in current production in the USSR. For a country with a planned economy this is a very large number of systems, and it reflects the well-known Soviet tendency to dissipate resources. If there was a market economy in the USSR, quite probably the number of control equipment

---

[48] Kosharskii, *op. cit.*, p. 383.

[49] The ferrodynamic and frequency systems are usually mentioned under a common heading, although they are quite distinct. Reference for both: Kosharskii, *op. cit.*, pp. 719–29; special frequency data loggers, *ibid.*, pp. 751–3. Shteinberg does not cover these ranges, since they are not of purely analogue nature.

[50] Kosharskii, *op. cit.*, p. 678; Shteinberg, *op. cit.*, p. 225.
[51] Kosharskii, *op. cit.*, p. 686; Shteinberg, *op. cit.*, p. 191.
[52] Kosharskii, *op. cit.*, p. 373; Shteinberg, *op. cit.*, p. 511.
[53] Kosharskii, *op. cit.*, pp. 355, 365; Shteinberg, *op. cit.*, pp. 430, 460.

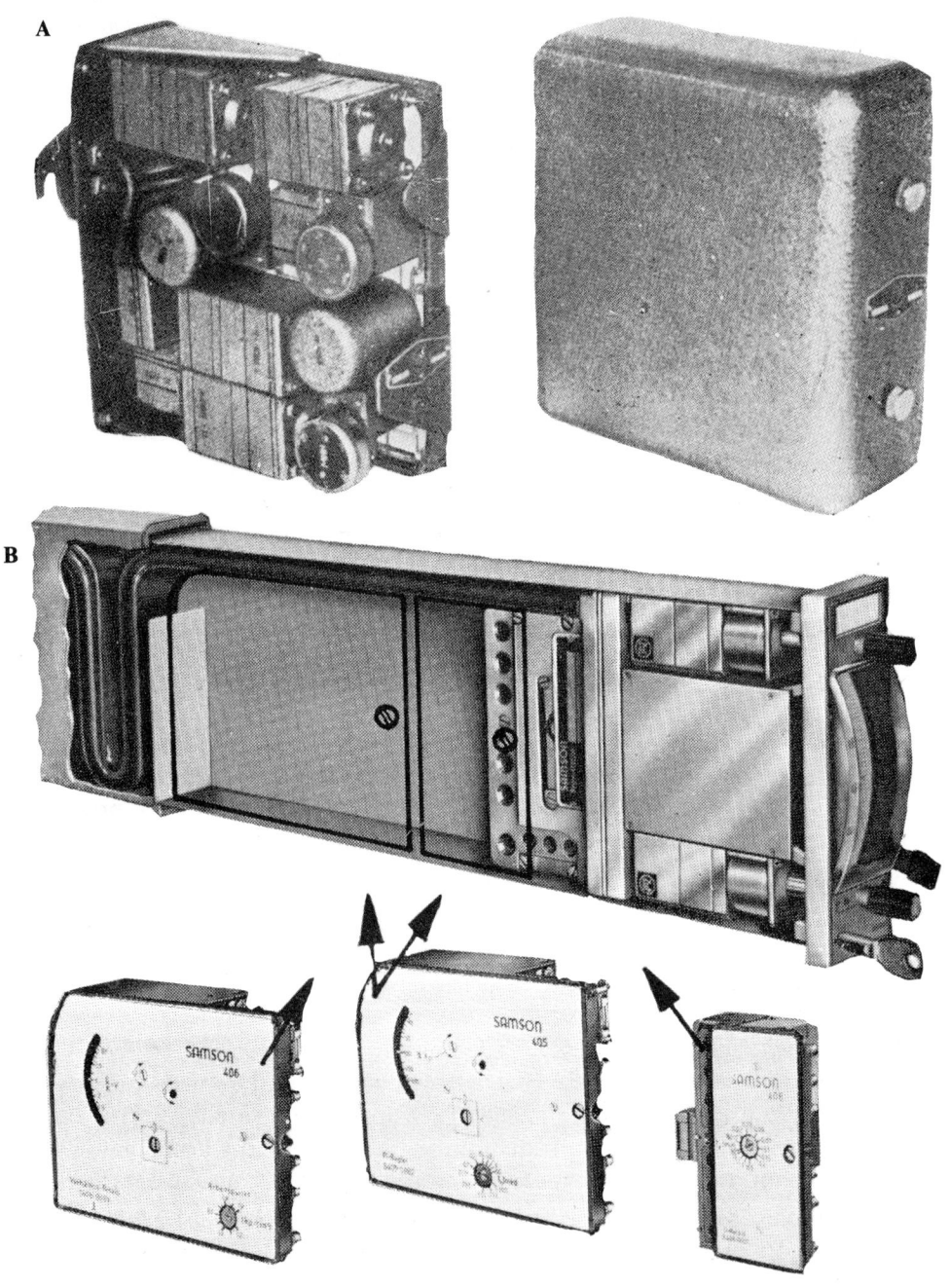

Fig. 7.5 Miniature pneumatic controllers

A (above) Soviet START controller built from USEPPA modules, 1964. The controller is 'blind', i.e. posesses no indicators. Dimensions: $20 \times 18 \times 8$ cm approx. Weight 7 kg.
Source: Mashpriborintorg catalogue.

B (below) Controller of the West German firm Samson, 1972. Lighter and smaller modules plug neatly into a control station. Dimensions: $50 \times 14 \times 5$ cm approx. Reproduced by permission of Samson Controls (London) Ltd.

systems would have been no higher. Moreover all the equipment in current production is of old design, and no new system has been launched since 1964. None of the electronic systems in current production is fully transistorised. Equipment used for the power industry (the largest user) is particularly antiquated.

The five major systems are as follows:

1. VTI-MZTA. Electronic with valves and magnetic amplifiers. Does not use a unified signal; construction is not modularised. Developed immediately after the war at VTI, modified somewhat at MZTA, and used in very large numbers by the power industry.
2. EAUS. Electronic, using magnetic amplifiers and some transistors. Employs a unified signal, construction is quasi-modular, although the modules are bulky. Developed *c.* 1960 at NIITeplopribor, but improved and expanded since then.
3. GSP1. Originally intended as a master system (first suggested in 1958) but due to very slow development never completed; further development has now been abandoned. Uses unified signals, electronic and pneumatic. The idea has been revived in the 1970s as GSP3 (see next section).
4. AUS. Pneumatic, developed 1952–3 by NIITeplopribor, still remains in production, despite the introduction of a modern pneumatic range (below).
5. USEPPA-START. Miniaturised pneumatic, developed 1959–64 by IAT and Tizpribor factory. An advanced system in its day, now overtaken by the West. Application limited due to the fact that it is not electronic.

A brief mention should be made of a Yugoslav-Soviet electronic process control system called SUPS, launched in 1973.[54] The Soviet Union provided market research and guidelines for the development of individual units, but the system was designed and is being manufactured in Yugoslavia. SUPS is modularised, uses largely discrete components (rather than integrated circuits) and unified signals (0–20 and 4–20 mA, but not 0–5 mA which is now the principal Soviet standard signal); it possesses two-wire transmitters for pressure and associated parameters, but not for temperature. Thus in technology it probably corresponds to the Western level *c.* 1969.

Since SUPS is not manufactured in the Soviet Union, and there has not even been any news of it being used there, it is not regarded by the present author as a Soviet system, and its existence is noted here merely to complete the record.

Thus, in summary, the situation by 1970 was a most unsatisfactory one from the Soviet point of view:

1. The control and instrumentation industry failed to adjust itself to the transistor age, and has been unable to reap the full benefit from this revolutionary technological breakthrough.
2. The State System of Instruments (GSP), which 10 years earlier carried high hopes of ending the technical backwardness, proved a failure.
3. No new system had been launched since 1964.

[54] S. V. Emelyanov *et al., Pribory i Sistemy Upravleniya,* 1973, No. 12, p. 8, and 1974, No. 1, p. 5.

4. The largest user—the power industry—remained firmly 'hooked' to an obsolete range of equipment, belonging to the 1950s.
5. Most control systems in production (pneumatic and electronic without a unified signal) were incompatible with computer management and therefore could not be incorporated into ASU schemes.

Equipment for the 1970s and 1980s (GSP and 'all-regime' RP2)

At the beginning of the 1970s, Soviet reviews of progress radiated satisfaction and confidence, while statistics presented a picture of plans over-fulfilled, outputs rising, and new instruments developed.[55] Nevertheless, the realisation of the true situation must have been there, because positive measures were taken to remedy it, by launching a far-reaching plan for the development of a new stage of the State System of Instruments, subsequently referred to in the present work as GSP3.[56] Quite probably the most compelling and final reason for not delaying further the long overdue modernisation of control and instrumentation equipment was the incompatibility of the present production ranges with the ASU concept[57]—a high priority undertaking. GSP3, despite continuing under what is now essentially a quite old title, represents for the Soviet control and instrumentation industry quite a revolutionary programme, full-scale implementation of which would have meant a spectacular narrowing of the lag behind the West.

Unfortunately it appears (see pp. 357–9) that this plan for radical innovation has been upset by the power industry opting out of it, to seek a slower road to modernisation. Such a move on the part of the largest user must have a marked effect on the level of innovation in the control and instrumentation industry as a whole.

In view of the potential importance of the innovation plan, it has been thought worthwhile to discuss it here in detail, even though information is incomplete and provisional, and even though we are dealing with plans which are still being executed, rather than actual accomplishments.

THE NEW GSP3 SYSTEM

The main areas of innovation incorporated in GSP3 appear to be:[58]

—large-scale use of integrated circuits as amplifying elements and of microconstruction techniques for other components;

[55] Perhaps the only exception to this rule was V. D. Mironov of the power industry (see p. 357 below) who did occasionally suggest that the situation left something to be desired, at least with reference to particular aspects of his industry.
[56] Soviet writers, for instance G. G. Iordan, *Pribory i Sistemy Upravleniya*, 1973, No. 9, p. 1, and *ibid.*, 1972, No. 12, p. 29, now distinguish two stages in the earlier development of GSP. This does not appear to be well founded, and is probably motivated by the desire to 'keep up' with the computer world, where the third generation had become available in 1973. In the opinion of the present writer, the 1970 GSP in reality constitutes the second generation, but for the sake of uniformity with the Soviet terminology, it is referred to as GSP3 in this work.
[57] This is evident in particular from V. D. Mironov, *Teploenergetika*, 1973, No. 4, pp. 2–5.
[58] G. I. Kavalerov, *Pribory i Sistemy Upravleniya*, 1972, No. 7, p. 1.

—use of standardised transmission signals;
—modular construction;
—unified system approach.

Such a programme identifies correctly the areas where innovation is of paramount importance (see p. 345 above). A speedy assimilation of all these principles in production, coupled with phasing-out the obsolete systems, could have wiped out the traditional lag of the Soviet control and instrumentation industry.

Evidently the broad outline for GSP3 must have been worked out by Minpribor some years earlier, before it was announced in 1972.[59]

The measure was foreshadowed at the XXIV Party Congress (March-April 1971)[60] and during 1970–1 IAT carried out broad market research for the system.[61]

Details of the overall system concept were published in 1972 by G. I. Kavalerov, head of the Technical Administration of Minpribor and editor of its journal.[62] As in the original GSP (see pp. 344–5 above), three branches were envisaged (electrical, pneumatic and hydraulic), and the scheme was intended to embrace industrial as well as non-industrial applications of control and instrumentation, and analogue as well as digital equipment. The same GOST standards remained in force, and therefore the criticisms raised in this respect earlier still apply to GSP3 (see pp. 345–7 above).

An important innovation in GSP was the provision of 'aggregated complexes' (AK), that is families of instruments, each capable of forming a system or loop for a specific purpose, each complex to be developed and manufactured by one organisation (typically a *Vsesoyuznoe ob"edinenie*, or department, of Minpribor).

There are altogether more than 20 AKs. Most are very narrow in concept (e.g., radiation pyrometers) or even sanction continued use of obsolete equipment (e.g., pre-third-generation computing devices), but Table 7.3 shows the six main complexes.[63]

The ASVT computer complex has been completed earlier than the other AKs, and has been in production since 1973. Of the others in the list the development of KTS LIUS appears to be most advanced, but apparently largely it comprises second generation technology (i.e., 'discrete' rather than integrated components).

The presence in the list of two process control complexes, ASKR and ASER, calls for comment as a clear case of duplication, contrary to the principles of GSP; this no doubt results from the existence in Minpribor of several

---

[59] A directive for developing one of the 'aggregated complexes' of the later GSP, KTS LIUS, was issued by Minpribor on 29 December 1967 (Order No. 389); see *Pribory i Sistemy Upravleniya*, 1972, No. 9, p. 1.

[60] *Direktivy XXIV s"ezda KPSS po pyatiletnemu planu razvitiya narodnogo khozyaistva SSSR za 1971–1975 gody* (1971), p. 27.

[61] G. I. Kavalerov, *op. cit.*, p. 1.

[62] *Ibid.*

[63] Except where otherwise stated, information in this list is from Kavalerov's article (see n. 58 above).

Table 7.3 *Soviet aggregated control complexes*

Abbreviation and the full Russian title of the aggregated complexes	Area of application of the aggregated complexes	Minpribor VO responsible for development and production
ASKR, Agregatirovannyi kompleks sredstv kontrolya i regulirovaniya.	General-purpose measurement and control apparatus for industrial processes.	Soyuzprompribor.[1]
ASET, Agreg. kompleks sredstv elektroizmeritel'noi tekhniki.	Electrical measurements.	Soyuzelektropribor.[2]
ASAT, Agreg. kompleks sredstv analiticheskoi tekhniki.	Chemical apparatus.	Soyuzanalitpribor.[3]
KTS LUIS, Kompleks tekhnicheskikh sredstv dlya lokal'nykh informatsionno-upravlyayushchikh sistem.	Data acquisition, processing and storage.	Soyuzpromavtomatika.[4]
ASER, Agreg. kompleks sredstv regulirovaniya.	Process control.	Soyuzpromavtomatika.[5]
ASVT or ASVT-M, Agreg. sistema sredstv vychislitel'noi tekhniki na osnove mikroelektroniki.	Computers, third generation.	Soyuzelektronschetmash.[6]

Sources:
1 Yu. I. Shendler, *Pribory i Sistemy Upravleniya*, 1972, No. 12, p. 29; L. A. Voronkov, *ibid.*, 1974, No. 8, p. 57.
2 M. S. Shkabardina, *ibid.*, 1973, No. 1, p. 14.
3 I. A. Dmitriev *et al.*, *ibid.*, 1972, No. 12, p. 31.
4 K. I. Didenko *et al.*, *ibid.*, 1972, No. 9, p. 1.
5 A. M. Khitrov, *ibid.*, 1974, No. 8, p. 54. There is some doubt as to whether the ASER range has full AK status, as it is not mentioned by Kavalerov in the quoted article.
6 A. I. Grishin, *ibid.*, 1973, No. 3, p. 16.

See also notes 58 and 63.

departments connected with process control. The status of the ASKR appears to be higher, as it is supposed to be a general-purpose process control system. But for a general-purpose system it is quite small, since only 30 units or modules are planned within it.[64] ASER lacks full recognition and is described as a mere 'experimental system'. But the few details so far published[65] indicate that it may be more advanced technically.

In 1972 the State Committee for Science and Technology considered[66] the GSP3 system and recommended that the experience and achievement of Minpribor in this area should be spread to other engineering ministries which produce a wide assortment of products, and that there should be a more intensive dissemination of information about this system. This represents a measure of recognition and success for GSP3. But GSP3 has also suffered a major setback, through the power industry's evident withdrawal from the

[64] Yu. I. Shendler, *Pribory i Sistemy Upravleniya*, 1972, No. 12, p. 30.
[65] A. M. Khitrov, *ibid.*, 1974, No. 8, p. 54.
[66] Postscript to the article by Kavalerov, in *Pribory i Sistemy Upravleniya*, 1972, No. 7, pp. 6–7.

scheme (already mentioned). In the recent writings of Professor V. D. Mironov, head of the department of thermal automation of VTI, the principal R and D establishment of the power industry, no mention is made of GSP either by name or implication,[67] and support is instead given for the alternative 'all-regime RP2', a slower road to modernisation, discussed below.

This defection of the largest customer for control and instrumentation production must obviously weaken the chances for the success of GSP3, and have a marked effect on the level of control and instrumentation in the USSR as a whole.

PHYSICAL CONSTRUCTION OF GSP3 (SYSTEM UTK)

Before turning to the RP2 system we must first consider in more detail the physical construction of GSP3. For housing the GSP3 a modular system known as UTK (*Kompleks Unifitsirovannykh Tipovykh Konstruktsii—*Complex of Unified Standard Designs) has been developed.[68] The information published so far on this system covers in detail only the equipment cabinets and their subdivision, down to the individual modules; no details have yet been published about the control panels, except for the broadest outline and overall dimensions.[69]

The UTK system was developed within Soyuzpromavtomatika of Minpribor, initially for the aggregated complex KTS LIUS; the design office SKB SAU at Kharkov acts as the 'head organisation' for UTK, responsible for the coordination of activities in this field. The development took place in accordance with Order No. 207 of Minpribor dated 3 September 1972,[70] but the OST standards for UTK had already been issued in 1971, so development must have started long before 1972.

The UTK system is very considerably influenced by Western standardisation of process control equipment, as is freely admitted by the Soviet sources cited here.[71] The key Soviet dimensions are, however, not identical with those accepted in the West: it is not known whether this is for engineering or political reasons.

Western influence can be seen at each of the three basic stages of assembly: basic modules, the crates holding the modules, and the cabinet holding the crates.

---

[67] For Mironov's writings, see p. 357, note 78, below.

[68] K. I. Didenko *et al.*, in *Pribory i Sistemy Upravleniya*, 1972, No. 8, p. 7, and in *ibid.*, 1972, No. 9, p. 6.

[69] In the housing of process control equipment, distinction should be made between the control panel, which constitutes an interface between the control system and the operator, and the 'blind' cabinets which house the bulk of control equipment, but are only occasionally attended by personnel. In addition there are transmitters and actuators, mounted on the process remote from the control room, but since no significant effort has been made in any country so far to standardise or modularise their construction, they are not discussed here.

[70] K. I. Didenko, *op. cit.*, 1972, No. 9, p. 6.

[71] West Germans pioneered the dimensional standardisation of control panel equipment, based on the multiples of 12 mm (DIN.43831, 1958), and their thinking and progress in this field have had a powerful influence throughout the world.

1. The basic modules are modelled on the CAMAC system,[72] developed by the Euratom organisation, but are not dimensionally identical. CAMAC is popular in the West European nuclear and scientific fields but not in process control. The International Electrotechnical Commission (IEC) includes CAMAC in its standards.
2. Crates, in common with all other known systems of standardisation, owe their origin to the British Post Office 19-inch 'rack', via CAMAC; but the basic width is 500 mm overall, while in CAMAC the corresponding dimension is 482·6 mm, equal to 19 inches exactly.[73]
3. There is no recognised Western standard with which to compare the UTK cabinet, nevertheless the common practice in the West, 10 crates per cabinet, is followed in the UTK.

From details published so far it would appear that UTK is very advanced as compared with other Soviet housing systems. If it is extended to cover the control panels, and adopted fully throughout the industry, it would wipe out the lag existing in this area with respect to the English-speaking countries; Germany would, however, remain in the lead.[74]

However, the UTK is not free from potential drawbacks.

1. It is probably too advanced for the Soviet Union; a compact CAMAC-type design largely depends for success on the availability of a wide range of highly miniaturised, reliable and generally well-made components, an area where the Soviet Union has always been particularly weak.
2. It appears that the status of UTK, in Soviet industry as well as within Minpribor, is not very firm. It is not covered by the State GOST standards, but only by the internal OST standards of Minpribor[75] (O stands for *otraslevoi*—branch). This could well raise difficulties with some users who may have different dimensional standards within their own ministries.
3. Soyuzprompribor of Minpribor itself, the department responsible for the general-purpose process control aggregated complex ASKR, is not among the organisations participating in the development of UTK.[76] This presumably means that it is not intended to extend this type of housing to the ASKR complex, a very severe restriction on the diffusion potential of the UTK system.

Thus in spite of its superficial superiority, the UTK housing system may not play such a significant role in the modernisation of the Soviet control and instrumentation industry.

---

[72] *CAMAC—A Modular Instrumentation System for Data Handling*, description and specification, Euratom, Switzerland, March 1969.

[73] International Electrotechnical Commission, Publication 297, 1969, p. 6.

[74] The 'raster' control panel technique launched in Germany about 1968 may represent the ultimate development in this field, given the general state of technology; as far as the present writer is aware, it has not yet been adopted or equalled by either UK or USA.

[75] In particular: basic dimensions OST 25-38-71, module boards OST 25-50-71.

[76] Soyuzprompribor does not appear on the list of organisations participating in UTK given by the head of SKB SAU, K. I. Didenko, in *Pribory i Sistemy Upravleniya*, 1972, No. 8, p. 7.

THE POWER INDUSTRY'S NEW SYSTEM (ALL-REGIME RP2)

It should be recalled here that the Soviet power industry, the largest user of process control equipment in the USSR, achieved complete standardisation in the range of control equipment known as VTI or MZTA, which is largely a product of the immediate post-war years, and therefore highly obsolete. There has been evident unwillingness to depart from this splendid position of complete standardisation. All innovation was effectively blocked, there was no experience within the whole industry of working with any other system, and when a possibility to innovate did arise (in the shape of GSP3), it predictably caused difficulties.

The alternative much slower road to modernisation being followed by the power industry is based on a control system which they call 'all-regime' (*vserezhimnaya*)[77] or RP2 (from the type designation of its proportional-integral controller) and which represents an updated version of the EAUS system developed by NIITeplopribor in 1960–1 (see p. 343 above); signals are unified, but it uses magnetic amplifiers on a large scale, housed in bulky units. From a formal point of view this system may constitute a part of the GSP3, but technically it differs drastically from the rest of it.

In his articles, which studiously avoid any mention of GSP3 (see p. 355 above), Mironov devotes a great deal of space to the all-regime system as the equipment of the future. Particularly, in a paper devoted to the implications of introducing the ASU-Energiya concept in the power industry,[78] he says:

> Considerable changes will have to be undergone by the automatic control apparatus. The accumulated experience of creating all-regime control systems on the basis of corresponding equipment* demonstrates the necessity for substantial review and expansion of its functional composition and improvement of its reliability.

Against the asterisk he adds the following footnote:

> We have in mind the development of the all-regime range of control apparatus based on the RP2 controller, carried out at the SKB SPA in Cheboksary.

Past experience shows that in control and instrumentation policy matters are not put in print in generally-available periodicals, unless a decision has already been made, at least in principle. Thus it can be taken that, although Mironov wants the system to be improved, approval has already been given for the power industry to use the RP2 system.

---

[77] This is only a convenient euphemism; in no sense is this system more 'all-regime' than any other system working on a unified signal.

[78] *Teploenergetika*, 1973, No. 4, pp. 2–5, especially p. 4. Mironov's other writings connected with this issue include: collections of articles by him and his collaborators, entirely devoted to RP2, *Teoriya, raschet i printsipy konstruirovaniya SKB SPA* (1969); on automatic control problems in large generating sets, *Teploenergetika*, 1970, No. 6, p. 2; a further description of RP2, *ibid.*, 1971, No. 5, p. 71; on the fiftieth anniversary of the VTI Institute, *ibid.*, 1971, No. 7, p. 14. ASU is a concept of hierarchical computer management and control (see p. 362 below).

Other evidence indicates that the RP2 system, with modifications requested by Mironov, may be produced at MZTA, a factory belonging to Minpribor, which works primarily for the power industry.[79]

No explanation has been provided in Soviet publications for the decision, or apparent decision, of the power industry, to opt out of GSP3 and adopt the 'all-regime' system, but the following factors may be speculatively suggested:

—Since the present VTI-MZTA equipment is not really capable of taking part in a computer controlled scheme, the planned introduction of ASU-Energiya in the power industry could have made it very urgent to modernise its automatic control equipment.
—The 'all-regime' system, even with the modifications requested by Mironov, may be available in production quantities earlier than the fully-fledged GSP3, so that adopting it will speed up the introduction of ASU.
—It could be argued that a change from VTI to GSP3 would be too drastic, and that personnel familiar only with instruments housed in cast-iron cases and using thermionic valves would not be able to operate and maintain integrated-circuit equipment in thin extruded-aluminium housings. Thus, a slower rate of progress is necessary.
—Mironov, who is a leading Soviet authority on control and instrumentation in the power industry, is strongly in favour of a signal level which is as low as

Table 7.4 *Technological characteristics of GSP3 and all-regime RP2*

Parameter	GSP3 for industries other than power	All-regime RP2 for power industry
Amplifying elements	Integrated circuits and hybrid (IC and discrete) assemblies. Quality of ICs and expertise in their use inferior to the West. Older aggregated complexes use transistors.	Magnetic amplifiers and transistors.
Transmission signals	Standardised, but no two-wire transmitters.	Reluctant use of such signals.
Modular construction	Advanced modular system for cabinets (UTK), no data on control panel	Large heavy units, hardly qualifying for the title of modules.
Unified system approach	Within aggregated complexes—yes, across complexes—no. The UTK housing system has no GOST status.	Insufficient information available.
Completion targets	KTS LIUS is being assimilated in production, but uses second generation technology; other complexes to be developed and partly assimilated by the end of 1975.	To be assimilated in production during 1973–5.

[79] The chief engineer of the factory described a new nameless system, with considerable resemblance to that described by Mironov, which it was to produce from 1973–5. (N. V. Shmelev *et al.*, *Pribory i Sistemy Upravleniya*, 1972, No. 12, p. 35).

possible and is prepared to accept 5 mA as the maximum level.[80] At the same time, he objects to the force-balance principle of operation in view of loss in reliability, due to increased complexity. These preferences of Mironov are met more fully in the 'all-regime' system than in the GSP3.

SUCCESS INDICATORS OF SOVIET SYSTEMS CURRENTLY UNDER DEVELOPMENT

Table 7.4 compares the technological characteristics of the two major systems.

The following illustrations (Figs 7.6, 7.7) show some of the equipment belonging to both GSP3 and RP2, as well as a British system also developed for the second half of 1970s. Soviet lags and leads in this area are listed on pp. 366–72 below.

Computer control

EARLIER ATTEMPTS

The subject of Soviet computers is dealt with in chapter 8. The present section deals only with process control and related applications of computers.

Despite the shortcomings of Soviet computers it is well known that they are being applied on an increasing scale to planning, management, data processing and other tasks in many branches of the Soviet economy.

Application of computers to the control, in a wider sense, of industrial processes has been passing through three distinct phases in the following chronological order:
1. Data processing and logging.
2. Direct digital control.
3. Hierarchical control and management (ASU).

The first digital computing devices entered industrial service in about 1960 as processors and loggers of data mainly for continuous processes, particularly in the chemical industry. This development was parallel to the West, but while in the West the data logger quickly disappeared as a separate 'breed' (except for a few small machines), and general-purpose computers are now used for data processing, in the Soviet Union it still survives. At least nine generic types of data loggers were in full production at the end of 1967, some being the original thermionic valve types, and one a pneumatic machine, a Soviet speciality, unknown in the West. While the ability to make a pneumatic computer is a tribute to mechanical manufacturing capacity, such a computer cannot really be regarded as a technological asset, since it is incompatible with nearly all other digital equipment; and, no doubt, it must be seen like the USEPPA system (see p. 347 above), as an enforced alternative to electronics.

The next stage of development was visualised as large-scale computerisation of closed-loop control (known in the West as DDC—direct

---

[80] See *Pribory i Sistemy Upravleniya*, 1967, No. 10, p. 38, and other articles by Mironov. He never justified his preference for low-level signals (which must reduce accuracy); a probable reason for his opinion is that Soviet semiconductors are unreliable at higher currents.

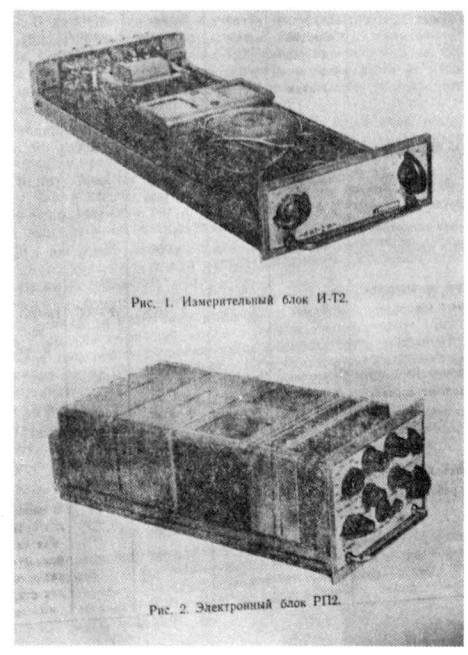

Fig. 7.6 Soviet control systems for the second half of 1970s.
  A (left) GSP3, with UTK housing. Two versions of the basic module (Spektr) and the crate taking 10 modules. This housing system is modelled on the Western CAMAC. The absence on module fronts of controls and, on the wider type, of test points, suggests that these are probably non-working 'space models'. The crate is empty. Dimensions: Module—$16 \times 16 \times 4$ cm approx; Crate—$50 \times 24 \times 18$ approx. Weight of module: c. 0·5 kg.
Source: Pribory i Sistemy Upravleniya 1972, No. 8, p. 9; 1972, No. 9, p. 4.
  B (right) Power industry's all-regime RP2. Large heavy units with magnetic amplifiers. Both units together form of the controller. Dimensions: Upper unit—$42 \times 16 \times 4$; Lower unit—$42 \times 16 \times 12$. Weight of the controller: about 15 kg.
In GSP3 and equivalent controller would probably occupy 3 or 4 modules.
Source: Teoriya, raschet (see n. 78 above), 1969 p. 6.

Fig. 7.7 A British general-purpose control system for the second half of 1970s. The declared objectives of the Soviet GSP3 are quite similar to this system's guidelines, but in contrast to GSP3 this system is equipped with 2-wire transmitters, it uses integrated circuits with a higher degree of expertise, and it has become available from production earlier (end of 1975). Each module: Front dimensions: 3.5×13 cm. Average weight: 0.3 kg.
Source: Catalogue of Bailey Meters & Controls Ltd.

digital control) of major industrial processes; this was the policy advocated, for instance, by Trapeznikov in 1960.[81]

The best-documented example of Soviet DDC is the system installed on the 200 mW boiler-generator set No. 6 at Zmievsk Power Station,[82] employing a 'Kompleks' type of system, designed by TsNIIKA, and based on a Minsk-2 computer. It was first put into use in June 1965, and finally commissioned in

---

[81] Automation and Mankind, *Proceedings, 1st Congress of IFAC, Moscow, 1960*, London (1961), Vol. 1, pp. 4–7. The DDC concept was basically of American origin, but its first full-scale application took place in this country in 1962 at an Imperial Chemical Industries factory at Fleetwood, Lancs.

[82] M. A. Duel, *Pribory i Sistemy Upravleniya*, 1967, No. 7, p. 20, and Study Group 5, *op. cit.*, p. 39.

December 1966. In addition to the auxiliary functions of data processing and sequence control of start-up and shut-down, the main part of the system provides continuous DDC of 32 boiler loops. A typical lack of confidence in the computer is, however, reflected in the provision of a 100 per cent standby with analogue controllers. In the West normal practice at that time was to provide $c.$ 25 per cent standby.

However, the quality of control obtained on these continuous processes with DDC was not superior to that obtained with ordinary analogue controllers,[83] and particularly in view of Soviet 100 per cent analogue standby provision, the DDC was clearly not an attractive proposition.

The case for DDC was stronger when optimisation could also be provided by the computer, but such instances were very rare; optimisation requires a very costly preliminary investigation, and it is often difficult to relate the optimisation concept to the actual process.

On the whole, therefore, there was disillusionment with the application of computers to the direct control of industrial processes, and cybernetics pointed the way to ASU.[84]

Before turning to ASU, a mention should be made of a Soviet practice which has never been employed in the West, the use of the computer on an industrial process purely as an 'operator's adviser'. The computer works off-line and produces the values of various settings, which are displayed, that the operator is recommended to adopt at a particular time. This practice started $c.$ 1960 as a means of popularising the computer, and was still in use in the mid-1970s. It would appear, however, that such schemes frequently give unacceptable advice and therefore produce a reaction opposite to the one intended.[85]

PROCESS CONTROL IN ASU (ASUTP)

Within the hierarchical computer Automated Management Systems ASU (Avtomatizirovannye Sistemy Upravleniya), control of the individual industrial processes is designated ASUTP (TP stands for Technological Processes); it forms the lowest level of the hierarchy and is subordinated to ASUP, that is the enterprise-level ASU, which, in turn, is subordinated to the industrial-branch level, the OASU.

ASUTP was introduced into the ASU scheme very late in the day. At first

---

[83] This information has been obtained from British delegations visiting USSR in the late 1960s, and is in keeping with the general tone of the Soviet press regarding DDC, and closely paralleled by the British experience.

[84] For instance, A. A. Liovin and V. V. Pavlov wrote in *Pribory i Sistemy Upravleniya*, 1972, No. 1, pp. 16–17: 'As an important result of the attempts made during the eighth five year plan it was established... that it is impossible to solve satisfactorily the problem of automatic control on the basis of a limited number of types of fixed-structure [i.e., second generation—ZAS] computers... At one time in the whole world and in our country it was considered worthwhile to connect a computer to the process, as this solved the problem of automatic control. But then it appeared that the introduction of computers is associated with many difficulties... Some attempted to overcome them, but others tended to seek areas of "simpler" computer applications, and management systems seemed simpler to introduce.'

[85] *Pribory i Sistemy Upravleniya*, 1967, No. 10, p. 53: *Trudy TsNIIKA*, 1969, No. 1, p. 5; *Pribory i Sistemy Upravleniya*, 1974, No. 5, p. 8, and private information from a British delegation of 1966.

disillusionment with direct digital control may have led the ASU 'founding fathers' to propose that the biggest economic effect from ASU could be obtained at the enterprise level, together with discontinuous processes, and in industrial management;[86] ASU was accordingly first introduced at the two higher levels. In Minpribor, for instance, 13 leading factories were transferred to ASUP during 1968–9 and the first stage of OASUpribor was formally accepted by the ministry in November 1970; but even the term ASUTP did not come into use until the second half of 1971,[87] and more concrete references did not appear until 1973.

At the same time a most unwelcome adulteration of the term has taken place, and now, 'according to some specialists and economists, all means of management directly interacting with technological equipment should be included in ASUTP'.[88]

In particular:

—Control schemes with computers, as well as those without them, are included under this heading; of the schemes which are being currently created, more than a half are without a computer.[89]
—In a majority of schemes with a computer, the computer is used only for data processing and logging.[90]
—Continuous as well as discontinuous processes are included.[91]

In summary, only 14 per cent of the currently developed ASUTP schemes 'correspond to the requirements of this class'.[92] This wide but undefined use of the term ASUTP makes it very difficult to draw any conclusions from the published information.

According to Zhimerin,[93] at the beginning of 1974 there were over 400 ASUTPs working in Soviet industry; from his text it is evident that some analogue process control systems are included in this figure, but the figure is clearly too small to include all such systems, and Zhimerin does not explain why some but not all analogue systems qualify to be included. Also the figure is too low to satisfy the definition quoted above by Maksarev, 'all means of management directly interacting with technological equipment'.

Several authors stress the high economic effectiveness of ASUTP. Significantly, however, it is not made clear in these pronouncements whether in the case of control, the comparison was being made with analogue control (as an obvious first alternative), or with manual control, which would be really quite an illogical and unfair basis of comparison.

Up to the time of writing a large percentage of computerised process control

---

[86] V. M. Glushkov, *Pravda*, 12 July 1964; V. M. Glushkov, A. Dorodnitsyn, N. Fedorenko, *Izvestiya*, 6 September 1964; V. Trapeznikov, *ibid.*, 25 September 1965.
[87] First mention in the present writer's records: N. Loskutov, *Sots. Ind.*, 20 September 1971.
[88] R. Maksarev, *Sots. Ind.*, 15 September 1973.
[89] *Ibid.*
[90] K. N. Rudnev, *Pribory i Sistemy Upravleniya*, 1974, No. 1, p. 1.
[91] R. Maksarev, *op. cit.*
[92] *Ibid.* He does not state his criteria for this statement.
[93] D. G. Zhimerin, *Ekon. Gaz.*, 1974, No. 2, p. 8.

schemes have been using second generation machines, particularly of the Minsk range.[94]

With the third generation computers, probably the best Soviet ASUTP scheme described so far is a data processing and control system on the 800 mW boiler-generator at Slavyansk Power Station;[95] it is claimed to comprise two M–6000 mini-computers, apparently without analogue back-up, and capable of improving the overall thermal efficiency of the generating set by 0·7 per cent, by means of continuous optimisation of combustion. Were these claims true, the Slavyansk system would have been in the same class as advanced Western computerised process control schemes. Unfortunately, the article in question does not appear to be a reliable source; it says very little about the control aspects (in contrast to data processing), does not clearly distinguish between plans and achievements, and appears to claim that the M–6000 computer was available already in 1971, which contradicts information from other sources. Moreover, it seems probable that the whole scheme remains only a paper exercise and has not been constructed or commissioned, because a British delegation which visited Slavyansk in 1975 did not see such a computer control system there and their hosts did not mention its existence or the intention of introducing it.

In conclusion, the unreliability of the sources makes it impossible to say to what degree process control is computerised within ASU, what is the technological level of these computer applications, and what degree of success these applications have achieved.

Critical review of the technology

DEVELOPMENT-PRODUCTION CYCLE

Case histories of the development and production of two Soviet systems and one British system are summarised in Table 7.5; each system embraces perhaps two dozen mutually compatible modules from which control schemes for industrial processes can be constructed by appropriate selection and combination. In all three cases the technology was almost completely new to the industry in question.

As Soviet examples the fastest and the slowest known cases of development are taken; the former is the miniature pneumatic system USEPPA, and the latter, the electronic branch of the GSP1 system.

As the British system the earliest transistorised electronic system, Bailetronic Mk.I, is taken; the present author took an active part in the development of this system at Bailey Meters and Controls Ltd.

Taking as the main comparison parameter the total cycle span, from formulation of the original idea to starting the development of an improved version, it will be seen that USEPPA took seven years, against Bailey's five; and if GSP1 completed its cycle (if it was not abandoned) and started the new one in 1972, its time would have been 14 years.

[94] K. N. Rudnev, *op. cit.*
[95] V. A. Dementev *et al.*, *Pribory i Sistemy Upravleniya*, 1972, No. 12, p. 18.

Generally, the rate of progress of USEPPA and Bailetronic were quite comparable, except that the development itself of the former took somewhat longer (six years, against Bailey's four).

This illustrates that on selected projects the Soviet Union can achieve a rate of development and assimilation which is comparable with the UK. However, these are exceptions, and the general rate is closer to that of the GSP, i.e., slower than the UK by a factor of approximately 2·5.

Attention should, moreover, be drawn to two further points. First, a system becomes really usable only when complete, i.e., when all its units are available. Second, in order to ensure a proper rate of innovation, after developing the original system the developers should be working on individual improvements or a complete new system. On both these counts the GSP fails badly, in contrast with the other two systems.

Table 7.5 *The rate of development of selected Soviet and British control systems*

	USSR		UK
	USEPPA	Electr. GSP1	Bailetronic
Idea formulated	About 1957	1958	1956
Development started	1958	About 1961	1957
Development finished	1964	Not finished	1961
Production started	1964	1967	1961
Development of improved version started	1964	About 1971/2 with GSP1 never completed	1961

It is not immediately clear why USEPPA was given such a high priority—and it must have had priority in the Soviet system in order to achieve such a rapid rate of development. As far as is known, USEPPA has no direct connection with armaments. It was probably supported both by the chemical industry (for which the system is chiefly intended), and as part of a drive to develop a pneumatic substitute for electronics in view of the slow progress with semiconductor devices (see pp. 347–8 above).

The highly disappointing history of the GSP system was related on p. 347 above.

The Bailetronic system was developed under normal commercial pressures, without special priorities or government support being involved, exclusively on the basis of the resources of a medium-size firm. The fact that progress with this system was faster than in the foremost Soviet case is characteristic of the differences between the two economic systems, at least in the case of control and instrumentation.

The manpower employed on the Bailetronic project did not exceed 20–25 men, including engineers, technicians, draughtsmen and model-makers; no comparable information is available for the Soviet projects, but, judging by general Soviet practice, their manpower could have been up to 10 times larger.

UK-GERMANY-USA COMPARISON

In the next paragraph technological levels of the Soviet and British control and instrumentation industries are compared, and in order to enable a broader assessment to be made, a similar comparison between Britain on the one hand and West Germany and the United States on the other hand is set out in Table 7.6. The British lead over Germany is mainly historical, and there does not appear to be a separate British lead over the USA apart from the common West-European lead, yet there are also positive leads of both these countries over the UK. One must therefore conclude as a net result that the UK lags behind West Germany and USA in the control and instrumentation industry, although the lag is not very pronounced. This means, however, that in general the Soviet lag with respect to these two countries is greater than with respect to the UK.

SOVIET LAGS WITH RESPECT TO THE UK

ANALOGUE CONTROL IN CURRENT PRODUCTION

Low level of automatic control. Despite continuous exhortations for at least two decades, the overall level of automation in process control in the USSR appears to be quite low, at least in the power industry.[96] In 1965 as much as 15 per cent of the total steam-generating capacity had no automatic combustion control. Moreover, between 1960 and 1964 this figure increased from 12 to 16 per cent, showing that some power stations without an automatic combustion control were built in the early 1960s. Also 60 per cent of boilers had no automatic control of steam temperature, and 25 per cent of pulverised-coal fired boilers had no automatic control of coal mills.[97] Of all the usual boiler controls it was only on feedwater that automatic control was provided in nearly 100 per cent of cases. The UK Central Electricity Generating Board does not publish comparable statistics, but are certain that no power station was built since the war without some form of automatic control on all the basic boiler control loops (combustion, steam temperature, feedwater and fuel feed). They estimate that by 1965 only a few per cent of the generating capacity had no automatic control on these loops.[98]

That the level of automation in general in the USSR is hardly rising at all, is shown by the fact that the percentage of workers engaged on non-automated manual operations decreases only very slowly, and the absolute numbers of such workers actually increase.[99]

[96] *Narodnoe khozyaistvo SSSR v 1965 godu* (1966), p. 173. The statistics have been discontinued in later issues.
[97] E. P. Stefani, *Teploenergetika*, 1965, No. 4, p. 2.
[98] Oral information.
[99] N. Rogovskii, *Trud*, 25 January 1974.

Table 7.6 Technological level of the British control and instrumentation industry compared with those of FRG and USA

	FRG	USA
British LEAD	— In intrinsic safety, although the degree of in-depth application is probably equal in both countries, the basic concept and its major extension (safety barriers) have been of British origin. — Contribution to the dimensional standardisation of equipment housing: the 19-inch rack (this is now of purely historical interest with little practical significance).	— Slow and reluctant introduction in the USA of the more recent European-originated valuable improvements and standardisation such as intrinsic safety, SI units of measurement, metrication.
British LAG	— World pioneering by Germany of the dimensional standardisation and miniaturisation of process control equipment. — Large-scale application in Germany of all-electric process control systems (in UK there is still no suitable electric actuator and this necessitates in conversion to either pneumatic or hydraulic actuators). The Soviet Union is in the same advanced position as Germany.	— Packaging and finish of instruments are in general very much superior in the USA. — USA positively leads in the area of electronic circuit components. — USA also leads in all aspects of computer engineering.

Slow innovation.[100] In control systems innovation is extremely slow, almost nonexistent. No new Soviet system has been launched since 1964, and there will probably be none until the next five year plan (1976–80). Not a single system has so far been taken out of production, although some systems were developed in the late 1940s. With the exception of the USEPPA-START system (which, however, is being at present rapidly out-distanced) and some peripheral equipment (see illustration on p. 373 below) current production in process control is characterised by a mass of antiquated equipment, virtually unusable by Western standards. Diffusion of significant modifications and improvements into production instruments is very slow, as can be seen from the handbooks of Kosharskii and Shteinberg.

It should be added here that in the Soviet 'non-system' instrumentation (which is outside the scope of the present chapter), such as laboratory instruments, optical and horological apparatus, the situation is significantly better, and, for instance, Soviet optical equipment and watches find a ready market in the West, albeit at the consumer level only.

In the British control and instrumentation industry a new system is produced on average every two or three years, and the average production life span of a system is about six years. In this way, rapid and steady technical progress is ensured, in marked contrast to the Soviet situation.

Standardisation. Apart from the power industry, the Soviet level of standardisation is extremely low, and far too many mutually incompatible, obsolescent and downright antiquated systems are in production.

Reliability. The current Soviet standard[101] calls for a mean time between failures (MTBF) of six years per 'instrument' (for earlier designs an MTBF of only a few months has been recorded).[102] If this figure is intended to apply to a single module, then it is not very impressive, because in British practice an MTBF of 50 years per module is quite normal.

Environmental conditions. Soviet instruments are designed to meet less stringent environmental requirements, particularly in the respects listed below.

— Maximum ambient temperature of 35 to 50°C, depending on class,[103] against the British requirement of 55 to 85°C.[104]

[100] In the New Year and other Minpribor reviews of progress and in articles by Rudnev statements can be found to the effect that innovation in the Soviet instrument industry is high; figures on the rate of innovation achieved are occasionally given. This may be seen as evidence invalidating the above paragraph. It should, however, be borne in mind that such statements concern the whole of Minpribor and not only process control instrumentation. Further it should be borne in mind that insignificant modifications are practiced on a considerable scale in the Soviet machine construction industry (including instrumentation) as a means of increasing prices. Quite possibly such modifications are also incorporated in the innovation statistics.
[101] *GOST 12997-67*, p. 6.
[102] V. D. Mironov, *Teploenergetika*, 1960, No. 6, p. 21.
[103] Soviet requirements are taken from the GSP standard *GOST 12997-67*; older ranges probably meet lower requirements.
[104] British requirements are taken from the Central Electricity Generating Board specification EES(1970), which is generally recognised by land-based users in other industries (different specifications apply in the marine field).

—Maximum relative humidity of 80 per cent, against the British figure of 100 per cent.
—Dust. Standard Soviet instruments are not dustproof (although there is a special class of dustproof instruments); in contrast, British instruments are normally dustproof.
—Mains supply transients. British electrical and electronic instruments are called upon to withstand short pulses of high amplitude (up to 1,000 per cent of nominal voltage) on their mains supply. No similar requirement appears in the Soviet specification. Since pulses of this nature are present in the mains of most industrial plants, British instruments should survive their repeated application without damage, but in Soviet instruments certain components could well be damaged at frequent intervals.

Physical construction. All the pre-1960 instrument designs still in production have an extremely heavy appearance, some actually use cast-iron fronts. Designs of the past decade are lighter, and some are even employing plastics, but they only imitate (like Soviet cars) American design features of some years earlier; there is no sign of any originality.

Also, modular construction is a rarity in Soviet systems, whilst in the UK it has been a well established rule for some years.

Signals. The vast majority of Soviet equipment operates without a unified signal; this prevents flexibility in control scheme design and virtually rules out computer participation. The Soviet Union does not produce any two-wire transmitters. In the UK unified signals have been in general use since about 1960, and every process control firm offers two-wire transmitters.

It is a measure of the backwardness of control and instrumentation in the Soviet power industry that Mironov, writing in 1973,[105] found it necessary to put considerable emphasis on advocating the use of a unified signal. In the UK similar discussions were taking place around 1956–8, i.e., about 15 years earlier.

Amplifying elements. In the Soviet equipment, thermionic valves and magnetic amplifiers are still in full use;[106] there are some transistors, but no fully transistorised system; as far as it is known, there are no integrated circuits in use in any of the standard production equipment.

In the UK the first fully transistorised control system was commissioned on site in 1959 and thermionic valves and magnetic amplifiers were phased out in the early 1960s, while integrated circuits came into use in the late 1960s, albeit on a limited scale.

Electrical equipment for hazardous atmospheres. The Soviet Union still relies mainly[107] on explosion-proof enclosures and purging; there are very few

[105] *Teploenergetika*, 1973, No. 4, p. 4.
[106] It is perhaps timely to recall that this section deals with current-production equipment. In particular, 'use' refers to the situation in production; due to the longevity of control equipment, the situation on sites is, in general, different, particularly in the UK.
[107] B. V. Nermenskii et al., *Apparatura i sredstva avtomatizatsii dlya vzryvoopasnykh sred* (1970), passim.

intrinsically-safe designs. In the UK, every control and instrumentation firm with system capacity offers intrinsically-safe equipment, while the two older methods are seldom used (as pointed out on p. 335 above, intrinsic safety has the advantages of enabling the instrument to be opened in a hazardous atmosphere, and also of being cheaper than the other two methods).

Summary. There is very little that the British control and instrumentation industry could learn from the Soviet side. In his regular reading of the Soviet control and instrumentation publications over 15 years, the present author can recall only two instances where he thought that the Soviet Union had a desirable instrument which was not available in the UK; one was a plating thickness gauge described *c.* 1960, and the other a portable gas-filled voltage indicator *c.* 1973—both non-system instruments.

ANALOGUE CONTROL CURRENTLY UNDERGOING DEVELOPMENT

1. *Power industry only (all-regime RP2)*.
—Large-scale use of *magnetic amplifiers* (rather than transistors and integrated circuits).
—Slow and reluctant introduction of *unified transmission signals*.
—Outdated *construction* in bulky units, not really qualifying as modules.

2. *Other industries (GSP3)*.
—Absence of *two-wire transmitters*.
—Lack of expertise in the application of *integrated circuits*.

COMPUTER CONTROL

1. *Second generation* computers still in full production (in the UK they were phased out about 1966) and use in process control.

2. Computers used off-line as *operators' advisers*, to little avail (a practice never adopted in this country).

SOVIET LEADS WITH RESPECT TO THE UK

ANALOGUE CONTROL EQUIPMENT IN CURRENT PRODUCTION

1. A miniature modular pneumatic system (USEPPA) was produced six years in advance of the West. However, as a pneumatic development, it is far less important than an electronic development would have been, and it has already been overtaken by the West.

2. Achievements by the power industry only:
—100 per cent standardisation, albeit on antiquated equipment.
—Commissioning time of control systems claimed to be very much shorter than in the UK.
—All-electric control systems used as standard for many years. The UK has not yet succeeded in developing a reliable actuator, and therefore even with

electronic control systems it is necessary to use pneumatic actuators. The Soviet Union in this respect followed the Germans who were the originators of all-electric systems.

ANALOGUE EQUIPMENT BEING DEVELOPED AT PRESENT

Although this is not yet a lead, it should be mentioned here that a full-scale introduction of the new system GSP3 would have meant a real advance, and a spectacular narrowing of the gap, perhaps to a lag of the order of only three years. However, in view of the power industry's choice of slower innovation, this can hardly materialise.

COMPUTER CONTROL

The ASU system is a sound concept, and its OASUpribor version represents, as far as the present author is aware, a unique attempt at the computer management of a complete industry comprising 300 enterprises and employing half-a-million men.

PRODUCTION

Conveyor-belt assembly lines are fairly common in Soviet instrument factories, but very rare in British factories.

SUMMARY OF LAGS AND LEADS

The most substantial Soviet lags are:
—Low level of automation of common processes.
—Slow or zero-rate innovation for long periods of time.
—Mass of antiquated equipment (analogue as well as computers) still in full production.
—Equipment at present developed for the power industry represents only a marginal improvement with respect to the equipment currently produced.
There are no substantial Soviet counterbalancing leads.

It must therefore be concluded that in this industry Soviet lags far exceed the leads.

The members of Study Group 5 were probably correct in 1970, when they assessed[108] that in analogue equipment the Soviet Union lagged behind the UK by some 10 years. Similarly a well informed article in the *Economist* assessed that Soviet computers were 10 to 15 years behind in quality and design.[109]

There is no indication that the gap has been diminishing since the early 1950s; on the contrary, it may well be increasing, as seen in the following comparison of progress with electronic controllers:

---

[108] According to their statement at a discussion meeting held on 5 March 1970 at the Institution of Mechanical Engineers in London. The statement does not appear in their published report.

[109] *Economist*, 12 December 1970.

	USSR	UK	USSR lag
First electronic controller	1951	1949	2 years
First transistorised controller	None up to 1974	1959	At least 15 years

Thus it would appear that in approximately two-and-a-half decades the Soviet lag increased by 13 years.

The illustrations in Figs 7.8–7.14 demonstrate some examples of retarded innovation in the Soviet control and instrumentation industry, and also some small-scale attempts to improve the situation.

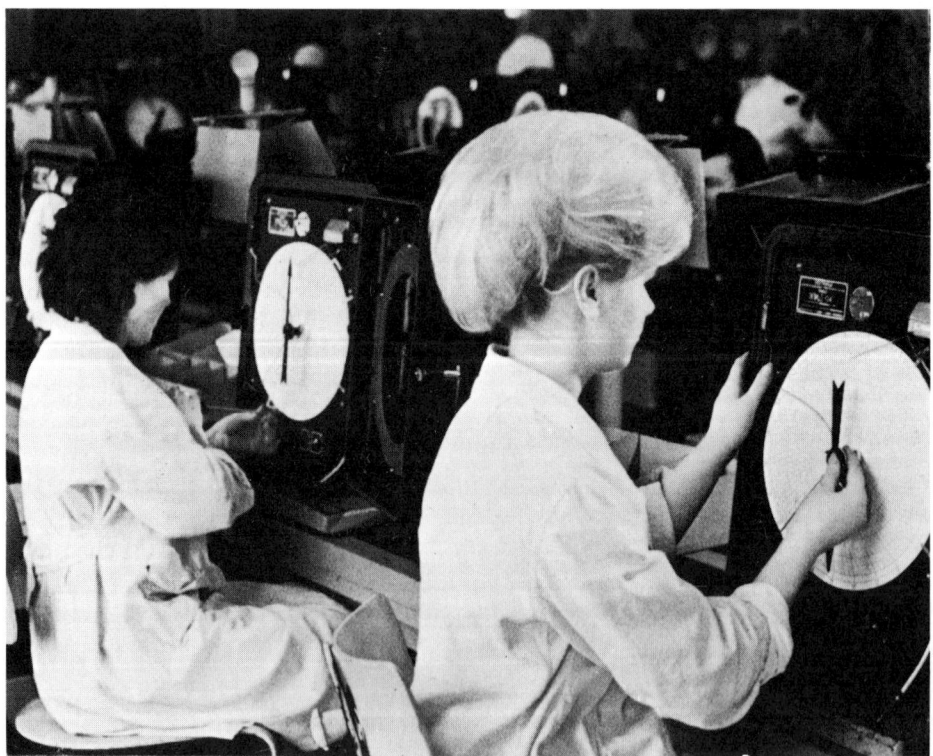

Fig. 7.8 A very old circular-chart recorder, originating from 1930s, still in full production in 1968 at the Manometer factory in Moscow, one of the most prominent and largest Soviet instrument factories.
Source: Mashpriborintorg Bulletin 1968, No. 2, p. 25.

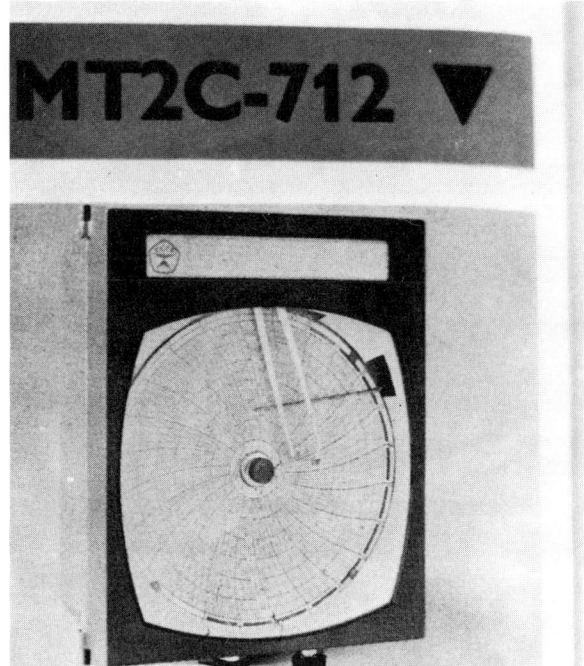

Fig. 7.9 A modern Soviet circular chart recorder.
In the top l.h. corner of the instrument is the quality-assurance emblem 'Znak Kachestva'
Source: Mashpriborintorg Bulletin, 1972, No. 1, p. 6.

Fig. 7.10 Antiquated heavy electrical Soviet control panel instruments. These were still in full production in 1970.
Source: Mashpriborintorg Bulletin, 1970, No. 8, p. 9.

Fig. 7.11 Instrument panel of pneumatic USEPPA-START system. By contrast, this system is provided with modern slim panel instruments (This particular illustration shows a signal indicator with high and low alarm settings).
Source: START catalogue, 1969.

A  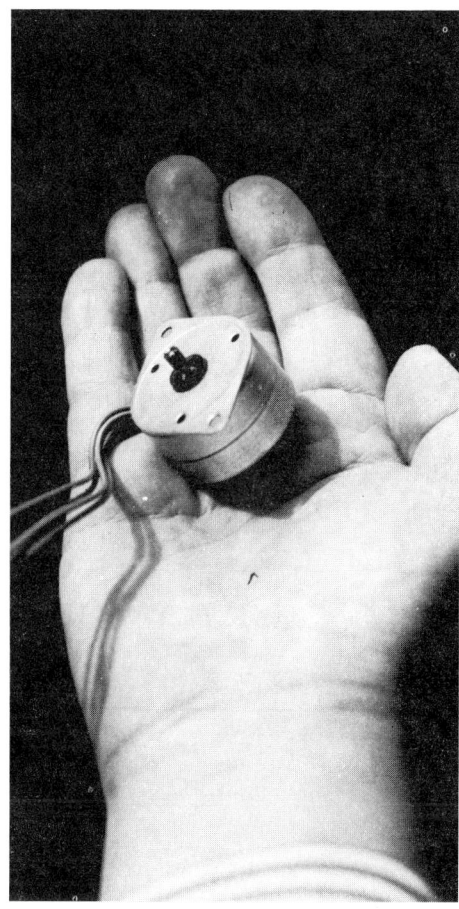 B

Fig. 7.12 Stepping Motors.
   A (left) Soviet
   B (right) British
The Soviet motor has an open construction (therefore requires careful protection against dust) and is excessively large. In the Soviet Ferrodynamic system such motors are used throughout the system. The British motor is miniature in size and it is fully enclosed. It is British practice only to use motors in the final system unit (actuator). As illustrated, the Soviet motor carries some additional fittings, whereas the British illustration shows the bare motor.
Sources: Mashpriborintorg Bulletin, 1970, No. 10, p. 7; Mullard motor photographed by the author.

Fig. 7.13 An extreme example of delayed innovation: thermionic valves used in a radio sonde, where a very strong case can be established for using transistors.
Source: Mashpriborintorg Bulletin, 1970, No. 8, p. 23.

Fig. 7.14 Adaption: the outer case of an American recorder on which a licence was taken out in 1930s, used to house another instrument over 30 years later.
Source: Mashpriborintorg catalogue, 1969.

8. Computer technology

Introduction

The impact of computer technology is very pervasive in advanced industrial societies. Possession of machines with powers of calculation and data storage thousands of times greater than those of human beings has been an essential prerequisite for most of the conspicuous scientific and technological achievements of the last decades. In the economic sphere, scarcely any aspect of production and management has been wholly untouched by the development of computer technology; an OECD report has asserted that 'the computer can be considered the key to the second industrial revolution'.[1] But as well as being carriers of innovation into other fields, computers have also undergone a series of discrete innovations in their own design and production. These two circumstances make computer technology an especially significant and interesting case study in evaluating the technology gap between different countries.

The modern electronic computer is descended from the calculating machines or engines of Babbage and other nineteenth-century inventors. The first electronic computer was completed in Germany, in 1941 by Konrad Zuse, and used for aircraft design in the Second World War. After the end of the war the ENIAC was completed independently in the United States in 1946, and this was followed by the EDSAC, developed in Manchester in 1949. In 1951 the first large-scale computer was put on the market in the United States. Work on the first Soviet digital computer began in 1948 in the newly founded Laboratory of Simulation and Computer Technology of the Ukrainian Academy of Sciences. By 1953 prototypes of several machines had been produced and by 1955 some of these were in serial production.[2] Of 19 countries

---

[1] OECD, *Gaps in Technology: Electronic Computers*, Paris (1969). In the USSR, computers are seen as an important element of the scientific and technological revolution.
[2] V. M. Glushkov, in *Soviet Automatic Control*, Vol. 15, No. 2, p. 1; G. Rudins, in *Soviet Cybernetics Review*, Vol. 4, No. 1, 1970, pp. 6–7.

considered in the OECD report on Electronic Computers, the Soviet Union was the fifth to have a working model.[3]

A modern computer system is made up of hardware and software. The hardware comprises the central processing unit (CPU) which carries out the arithmetical and logical operations, a small internal or operational memory containing data to which the processor has ready access, and a range of peripheral equipment. Peripherals include equipment for the input and output of data to and from the system; the external memory, which may be on tape, drum or disc; and a range of other equipment which can be included as required. The software is a set of programs which direct the computer to carry out operations of a particular kind. The software of a computer typically consists of an operating system which performs such basic functions as error detection and communication with peripherals; compilers, which translate instructions written in a high-level symbolic language, such as Algol, into the machine code understood by the computer, and thereby relieve the programmer of much time-consuming work; and applications programs which are geared to carry out the special operations required by the user.

Computer design has gone through three readily distinguishable phases, known as generations, and each generation is characterised by a different electronic technology used in the construction of computers. The first generation was based on cumbersome and unreliable vacuum tubes and valves. In the second generation these were replaced by individual semiconductors and transistors. In the third generation the separate components and wiring of the circuits were replaced by integrated circuits in which all the components and their interconnections are produced in miniature on a small ceramic plate.[4] Each successive generation has made possible faster speeds of operation and larger and faster internal stores. There have been parallel developments in peripherals and software. As a result of these developments there has been a startling reduction in the cost of computing. The same set of operations which cost $1·38 on a first generation American computer in 1954 would have cost $0·25 on a second generation machine in 1958 and only $0·035 on a third generation IBM 360 in 1965.[5]

In this chapter the development of Soviet computer systems and their diffusion throughout the economy are evaluated comparatively. This involves an examination of the technological level of the 'best' hardware and supporting systems available in the Soviet Union and the West over the last 20 years or so, and of the size and composition of the computer stock held in the Soviet Union and selected Western countries. In practice, the most advanced computer technology has, in recent years at least, emanated from the United States or from European subsidiaries of American firms.[6] Only the United Kingdom has

[3] OECD, *op. cit.*, p. 33.
[4] This account is largely taken from OECD, *op. cit.*, Annex II, pp. 179–85.
[5] OECD, *op. cit.*, p. 73.
[6] On the American lead in computer technology, see OECD, *op. cit.*, passim, and A. J. Harman, *The International Computer Industry*, Cambridge, Mass. (1971). On American penetration of the European computer industry and Europe's response, see Y. S. Hu, *The Impact of US Investment in Europe*, New York (1973), chs. 2 and 4, and N. Jequier, 'Computers' in (Ed.) R. Vernon, *Big Business and the State*, London (1974), pp. 195–228.

maintained an indigenous industry which, without American technical assistance, aspires to supply a full range of computers. For this reason, the comparison of technological levels is restricted to the USSR, United States and United Kingdom (for the period since 1960 only). The computer firms of other European countries and of Japan are not considered, although data on the diffusion of computers in these countries are presented. In the discussion of technological level I focus largely on the date of appearance of the prototype of a computer model, and ignore both the preceding stage of research and the process by which a computer is put into batch production. These omissions are due respectively to the author's lack of qualifications to assess levels of research and to the lack of adequate information on the gap between the appearance of a prototype and serial production.[7]

The reader of the present paper is referred to Richard Judy's invaluable earlier work on Soviet computer technology.[8] Since I have followed his method of comparison so closely, I have found it unnecessary to describe in detail the computer systems available in the USSR before 1968, the terminal date of his study; summary information on these systems is reproduced in the Appendix to this chapter and Judy's conclusions are outlined on pp. 397–9 below. The second section of this paper therefore discusses developments since 1968 in Soviet computers, dealing successively with CPUs, peripherals and software. The next section provides evidence on the diffusion of computer systems in the Soviet Union and other countries, while the last section compares Soviet and Western performance in developing and using computer systems.

PROBLEMS OF COMPARING TECHNOLOGICAL LEVEL

My comparisons concentrate largely on the performance of the central processing unit, and the particular index I use is the maximum number of operations per second of which each computer is capable. Unfortunately, this measure does not have the satisfying properties of objectivity or even of accuracy that it may appear to have. Several difficulties are involved.

First, there is the problem of comparability of data. The properties of Western computers have been evaluated and compared systematically over a number of years, and the definition of the measure can be made quite unambiguous. But the situation with Soviet computers is different. We have to rely on information published in the Soviet Union, some of it in the general press, and it is not always clear that the Soviet figures for operations per second are directly comparable with each other and with those used for Western computers. The published figures have been taken at face value, but it is important to bear in mind that they usually give the theoretical maximum rather than the maximum productivity actually achieved over an extended period.[9]

---

[7] Some scattered evidence on the latter point for the USSR is given on pp. 381–91 and in Appendix Table 8A.1, column 3 but it is not definite enough to warrant any firm conclusions.

[8] R. Judy, 'The Case of Computer Technology' in (Ed.) S. Wasowski, *East–West Trade and the Technology Gap*, New York (1970), pp. 43–71.

[9] This distinction can be very significant. The chief designer of the ES series of computers (see pp. 384–6 below) stated in an interview that the ES-1030 achieves higher actual speeds of operation than the Minsk 32, which is nominally rated as faster (*Nauka i Zhizn'*, 1973, No. 8, p. 6).

The second difficulty concerns the desirability of using the number of operations performed in the CPU of a computer as an index of the power of the system. The index gives the maximum number of operations per second and is therefore based on the simplest kind of operation, usually fixed-point addition. But most of the operations a computer is asked to perform are not of this type. What is needed is a synthetic index of performance based upon a weighted average of the time taken to perform the set of operations a computer will in fact be asked to perform. There is a substantial literature in the West on various ways of solving this problem, since information of this kind is necessary to assess the cost-effectiveness of alternative computer systems.[10] But in the Soviet case lack of data makes such a comparison impossible.

A broader but analogous objection can be made to the measure, on the grounds that it excludes such important attributes of a computer as the size of internal memory and access time to internal memory. Lack of data in the Soviet case again makes it impossible to construct an adequate synthetic index, but it should be noted that, according to Judy, the measure of operations per second tends to flatter Soviet achievements since they are relatively deficient in other areas.[11]

Subject to these caveats, in my view this measure does constitute a satisfactory measure of technological level. It is sometimes suggested that the USSR, the USA and Great Britain might not have the same requirements for large powerful computers, so that it would not be legitimate to take the productivity of the most powerful computers available in the three countries as an indicator of the extent to which technology was advanced. But in fact each country needs and uses computers for two main purposes, scientific and business. The former purposes require large processing capacities, the latter large data handling facilities. Historically, advances in computer technology have been associated with and usually have been necessary for the increase in the productivity of computers.[12] There is no reason to suppose, therefore, either that the USSR has special requirements for computers or that the productivity of computers is not, subject to the caveats mentioned above, a satisfactory measure of technological level. This does not, of course, imply that the Soviet *need* for computers has been equally strong as that of the United States. In 1968 Richard Judy supplied evidence that 'demand from Soviet industrial enterprises and other economic organisations has been rather weak', especially in relation to business-data processors, and that this low demand was in turn partly due to relative abundance of managerial labour.[13] A lower technological level in the USSR is not necessarily an indication of economic inefficiency.

[10] For example, W. Sharpe, *The Economics of Computers* (1969), pp. 297ff.
[11] R. Judy, *op. cit.*, p. 61.
[12] In the USSR great importance has recently been attached to the development of mini-machines capable of performing local calculations and passing larger problems on to centres equipped with powerful computers. This development tends to increase demand for powerful machines, at the expense of models of average productivity. Glushkov has recently stated that the USSR computer stock consists of too many middle-sized computers and too few large and small ones. (*Izvestiya*, 7 March 1974.)
[13] R. Judy, *op. cit.*, pp. 70–1.

The final point concerns the evaluation of the other elements of the computer system, notably the peripherals and software. A computer system should ideally be evaluated as a whole: its quality is usually determined by the weakest link in the chain of processor, peripherals and software. This makes it necessary to collect information on each link in the chain separately, combining it for an overall evaluation of the effectiveness of computer systems.[14] But at the same time, any relative successes in Soviet development of peripherals and software should not be given too much significance, as the needs of the system in these respects are largely determined by the capacity of the CPU. There is no need to attach powerful input-output devices to a slow computer; and processors have to reach a minimum level of capacity and internal storage to make it feasible to use high-level algorithmic languages. There is however a difficulty here, in as much as the importance of software in a computer system has increased continuously. In the West, technical developments, particularly integrated circuitry, have cut hardware costs immensely, and have made hardware a secondary consideration. It can be argued that the USSR and Western countries are in two different stages in the development of computer systems. In the USSR, hardware is still of primary importance, while in Western countries customers' software needs largely determine the shape of the hardware in a system.[15] On this argument, any comparison of technological level based either on hardware or on software may be misleading. In the comparisons that follow, greater prominence is given to hardware, the yardstick more appropriate for the level of development reached by the USSR, where comparisons are easier. At the same time it is important, even in the Soviet case, to consider software, to see if it is a restraining factor in the use of Soviet computers. Such analysis enables us to assess the ability of the USSR to exploit an innovation which requires simultaneous advance on a number of scientific fronts.

In summary, measures of technological level are used here which may rightly be criticised both in themselves and because the data are inadequate. The basic measures are backed up by further indices, the independent significance of which is open to some doubt. These factors should be borne in mind when interpreting the comparisons. But at the same time our measures have the great advantage of being readily quantifiable, and are, by the standards of technological comparisons in many fields, quite exemplary.

Soviet computer systems since 1968[16]

SECOND GENERATION COMPUTERS

The most important developments in this field have been the production of an improved member of the Minsk series, the Minsk 32, and the emergence of a family of computers in the ASVT series, the M-1010, M-2000 and M-3000.

The Minsk 32 has become probably the most widely used computer in the

---

[14] This approach was used by Judy in his earlier study (R. Judy, *op. cit.*).
[15] This point has been made by Mr R. Curnow of Sussex University.
[16] For details of all computers mentioned, see Appendix Table 8A.1.

data processing field and the most important second generation computer in the USSR. It emerged in 1967 and was in quantity production in 1968–9. An improved version is now capable of up to 65K operations/sec, which represents a twentyfold increase on the Minsk 23; internal storage capacity is up to 64K. The machine can be connected to as many as 136 input/output devices. The Minsk 32 is produced at the Ordzhonikidze Computer Factory, Minsk, by the Ministry of the Radio Industry, and has been awarded a quality mark (*znak kachestva*). At a conference on ASU (Automated Management Systems) in January 1972, plans were announced to equip the Minsk 32 with a disc unit in the first quarter of 1973,[17] but this has not happened.[18] It is said that 500–600 units are produced annually,[19] though by 1974 there were plans to discontinue production.[20]

The ASVT[21] series (*agregat sredstv vychislitel'noi tekhniki*) consists of four second generation models known as ASVT-D; the remainder, known as ASVT-M, are of the third generation. Originally the ASVT-D has only three models, the M-1000, M-2000 and M-3000, intended primarily for process control in ASUTP (automated systems for control of technical processes). We are told, however, that from the point of view of ASUTP they were not three models but one and a half, as the M-2000 and M-3000 were basically the same, and the M-1000 was not much good anyway. Apparently, not being a mini-machine, it could not replace such computers as the UM-1, the Dnepr-1 and the VNIIEM-3, and its capacity was not great enough to cope with larger-scale problems of a technological or economic nature.[22] Oddly, the M-1000 was not mathematically compatible with the M-2000. In view of these inadequacies the M-1010, a simpler and faster version of the M-1000, emerged soon after the appearance of the other three members in the range in 1969. The more powerful M-3000 model is used in the Sirena ticketing system, developed for Aeroflot.[23] The series was designed by the Severodonetsk Scientific Research Institute for Control Computers and is manufactured by the Ministry of Instrument Building, Means of Automation and Control Systems (Minpribor).

Two other second generation computers of interest are the Ruta-110 and the K-200. Manufacture of the former began in 1967 at Sigma, an *Ob"edinenie* of Minpribor in Lithuania. Production was largely by hand, slow and expensive. Demand for the machine ran at three times supply,[24] and the Ruta had no great impact. However, it was the first Soviet computer to incorporate a disc unit.

The K-200, produced by the Ministry of the Radio Industry (Minradprom) is a modest computer in quantity production in 1971. Wade Holland

---

[17] (Ed.) D. G. Zhimerin *et al.*, *Avtomatizirovannye sistemy upravleniya* (1972), pp. 36–7.
[18] *Mekhanizatsiya i Avtomatizatsiya Upravleniya*, 1975, No. 1, p. 4.
[19] *Soviet Cybernetics Review*, September 1973, p. 17.
[20] *Ekon. Gaz.*, 1974, No. 42, p. 8.
[21] Most of this information comes from *Pribory i Sistemy Upravleniya*, 1972, No. 1, pp. 16–19.
[22] Its deficiencies included loading of internal memory with trivial functions of centralised control, barely tolerable solution times and high cost.
[23] *Soviet Cybernetics Review*, January 1970, pp. 31–2.
[24] *Sots. Ind.*, 18 February 1972.

THIRD GENERATION COMPUTERS, OTHER THAN THE ES SERIES

The first Soviet computer based partly on integrated circuits was the Nairi-3, the design of which was completed in 1968–9 by the Erevan Scientific Research Institute of Mathematical Machines; a hybrid between the second and third generation, it still relied mostly on transistors. Intended for scientific, commercial and production control systems, the Nairi-3 is capable of 15–20K ops/sec and has an internal storage capacity of up to 32K. The machine is in manufacture at the Elektron factory in Erevan.

Apart from the Nairi-3, there are only two series of third generation computers in production. The Edinaya Sistema (ES) or Ryad series is discussed below. The other is a continuation of the M-series of computers of Minpribor. At first the series seemed to consist of three models; the M-4000, M-5000 and M-6000, known collectively as ASVT-M, but later others have been mentioned, notably the M-4030 and the M-400. In the ninth five year plan (1971–5) it is intended to install 400 M-6000s and 55 M-4000s in process control, but the M-4000 can also be used for economic purposes: a configuration is available for use in a ministry chief computer centre.[26] The M-6000 and the M-400 are described as small or mini-machines. The M-6000 would be used to control technological processes, for scientific work, to control operations in large-scale or serial production or as a data processing centre in a queuing system.[27] It is capable of 200K ops/sec. The M-400 is intended for use in scientific experiments and for control of technological processes which require high speeds in the collection and processing of data.[28] Its capacity is unknown. The function of the M-5000, technical details of which are also unknown, is to deal with book-keeping, supplies or other economic problems at comparatively small enterprises. In 1972 it was proposed to supply 20 enterprises of an industrial department (*ob"edinenie*) of Minpribor with M-5000 computers, whose task would be to solve local problems, and pass information on to the department.[29] The M-4030, the senior member of the range, is intended for use in automated management systems of various types—of technological processes, enterprises or ministries—and jointly with other members of the ASVT-M range, in hierarchical systems. Its maximum speed of operation is 330K ops/sec for addition with storage of operands in the register. The internal memory has a capacity of 128 Kbytes, capable of extension up to 512 Kbytes[30]: average cycle time is 2 microsecs. The M-4030 is compatible with respect to system of commands, presentation of data and the input/output interface with the ES series.[31]

---

[25] *Soviet Cybernetics Review*, September 1970, p. 32.
[26] (Ed.) D. G. Zhimerin *et al.*, *Avtomatizirovannye sistemy upravleniya* (1972), p. 388.
[27] *Ekonomika i Organizatsiya Promyshlennogo Proizvodstva* (Novosibirsk), 1973, No. 6, p. 36.
[28] *Pribory i Sistemy Upravleniya*, 1974, No. 11, p. 11.
[29] *Sots. Ind.*, 18 February 1972.
[30] A byte is a binary digit.
[31] *Pribory i Sistemy Upravleniya*, 1974, No. 11, pp. 18–19.

The first dates of serial production of the ASVT-M series are rather uncertain. In 1972, serial production of both the M-4000 and M-6000 was said to be beginning,[32] yet in 1973 Loskutov wrote that Minpribor was still preparing for production of the M-6000.[33] By 1974 the M-4030 and M-6000 were in serial production and the M-400 was to reach that stage by the end of the year.[34]

The ASVT-M range as a whole fills the gap between the ES series of computers, intended largely for economic data processing (see p. 385 below), and the other products of the instrumentation industry. ASVT is one of the elements in a coordinated range of industrial instruments and equipment known collectively as the State System of Instruments (*Gosudarstvennaya Sistema Priborov*—GSP).[35] Within this framework ASVT-M will offer a flexible choice of equipment for control at the level of the individual machine, the production line, the shop or the enterprise. The control system will be hierarchical, with the M-5000 and M-4030 standing at the top; the lower level will be made up of the M-6000 and M-400 computers and of other equipment for centralised monitoring and recording (M-40) and for collecting technical data (M-6010). All units will operate in real time.[36]

THE EDINAYA SISTEMA (ES EVM) SERIES OF THIRD GENERATION COMPUTERS

The most important series of third generation computers, known variously as the Edinaya Sistema (Unified System) or Ryad series, is the result of a programme of cooperation within CMEA (Council of Mutual Economic Assistance). At the XXV session of CMEA in December 1969,[37] an agreement for technical cooperation on the development of computers was signed by six countries: Bulgaria, Hungary, GDR, Poland, USSR and Czechoslovakia. Cuba joined the project later, in 1971. Three and a half years later the design stage of the project was substantially completed; a large exhibition of the ES range of equipment was held in Moscow in May 1973. The speed of completion is attributed to the combination of forces,[38] and Rakovskii, the President of the Intergovernmental Commission directing the project, has said:

> In truth, it is no exaggeration to say that in complexity, scale, orientation and concentration of forces, establishing the Unified System of computers is the biggest project in the history of fraternal cooperation of the socialist states.[39]

---

[32] *Ibid.*, 1974, No. 5, p. 1.
[33] *Planovoe Khozyaistvo*, 1973, No. 9, p. 35.
[34] *Pribory i Sistemy Upravleniya*, 1974, No. 11, p. 17.
[35] This is described in chapter 7, pp. 344–55 above.
[36] *Pribory i Sistemy Upravleniya*, 1974, No. 4, p. 4; 1973, No. 3, pp. 16–17. A real time computer system is one which processes data as soon as it is generated, and is therefore able to produce an immediate feedback.
[37] Work on the series in the USSR is said to have begun in 1967 (*Soviet Cybernetics Review*, January 1970, p. 32).
[38] *Ekon. Gaz.*, 1973, No. 27, p. 20.
[39] Quoted in *Nauka i Zhizn'*, 1973, No. 8, p. 4.

The ES range consists of a whole set of processors, peripherals and accompanying software, produced in all six countries.[40] Here I concentrate primarily on Soviet products, though it is worth noting that one of the chief effects of the project has been to equalise the technological level of production of computers and computer equipment in the Eastern European countries. Bulgaria, for example, used to import all her computer equipment, but in 1973 she was exporting such highly sophisticated equipment as disc units.[41] On the reverse side of the coin, there has probably been some levelling down of the more advanced countries, such as Czechoslovakia. In this section I consider only the central processors of ES machines; software and peripherals are dealt with on pp. 388–91 below. One of the objectives of the scheme has been to increase the level of specialisation of the member countries in computer production. It is clear from the published lists of a full set of equipment for each of the models (*komplektatsiya*) that while there is specialisation between models, there is little within the peripheral equipment of each model taken separately. The configuration of peripherals of the three Soviet-made processors has been published. All the items can be manufactured in the USSR and need not be imported. In the case of the Hungarian, Czech and German computers (respectively, the ES-1010, ES-1021, ES-1040) only a tiny proportion of the peripherals are imported to make up the standard configuration.[42]

Full details of the central processors of the ES range are given in the Appendix to this chapter. The smallest machine in the series, the Hungarian-made ES-1010, was tested by an international control commission in December 1972. It is also available in a mobile unit.[43]

The slightly larger ES-1020, made by the USSR and Bulgaria, was tested a year earlier and in 1972 went into serial production at the Ordzhonikidze Computer Factory, Minsk, and also at the Elektro-Mechanical Plant, Brest. It is intended for scientific and control purposes, and also for small automated management systems (ASU), and can work either alone or in conjunction with other machines. The Czech-made ES-1021, similar in capacity to the 1020, was tested in December 1972.

The medium-sized ES-1030, produced by Poland and the USSR, was tested in May 1972. Its size is such as to make it possible to use the more powerful operating system OS/ES (see p. 389 below). It is intended for scientific, economic and other uses, and the designers also propose to adapt it for use in a multi-machine complex and to link up two processors with a common core storage.

The ES-1040 is a German-built machine made at the Robotron factory 'on the basis of integrated circuits of their own manufacture'.

The more powerful Soviet-made ES-1050 is intended, like the even larger ES-1060, for large-scale economic planning uses at the level of ministries and

[40] See (Ed.) A. M. Larionov, *Edinaya Sistema EVM* (1974), passim.
[41] *Sots. Ind.*, 5 May 1973.
[42] *Pribory i Sistemy Upravleniya*, 1973, No. 10, pp. 2–5.
[43] This and the following descriptions are from *Pribory i Sistemy Upravleniya*, 1973, No. 10, pp. 1–5.

regions of the USSR.[44] The ES-1050, capable of 500K ops/sec, is in production, but the ES-1060 is still at the design stage. Wade Holland noted the reduction in projected speed of operation of this machine from 3 million ops/sec to the currently expected 1·5 million ops/sec.[45]

The Edinaya Sistema series is one of those cases, discussed in the last part of this chapter, where the range is very similar to an earlier American range, in this case the IBM 360 series, which first appeared in 1965.

It is difficult to gauge how many computers in the ES range have been produced. Of the Soviet models, the ES-1020 was in serial production in 1972,[46] and in 1973 'several dozens' of ES-1020 and 1030 machines were in use in the USSR and the ES-1050 had been transferred to quantity production.[47] It is, however, hard to believe that the alleged output target of 12,000–15,000 third generation machines could have been met by 1975, especially in view of the deficiencies in circuitry manufacture described by the Chief Designer, Larionov.[48]

The ES series has come in for some harsh criticism. Some of this concerns shortage of peripherals and the software lag, but Zhimerin, director of the All-Union Scientific Research Institute for Problems of Organisation and Management, noted early in 1974 that the core storage of the ES series computers was inadequate, and must at a minimum be doubled, or even trebled or quadrupled. He also observed, possibly in contradiction to the source quoted above, that serial production of the ES-1050 has been delayed, as had design of the ES-1060. Minradprom was held responsible for these failures.[49]

PERIPHERALS

Since 1968 the main development in this field has been the appearance of disc storage units, but there have also been improvements in other external memory devices and in input/output equipment.

Technical difficulties in preparing discs were overcome in 1968,[50] and a disc unit was first displayed in 1970,[51] seven years after the announcement of the IBM 1302 disc unit.[52] The first Soviet computer to be equipped with disc storage was the Ruta 110. The disc unit in question, the R-401, has a maximum capacity of 60 Mbytes with an average access time of 200 m secs, and channel data exchanges are not less than 30,000 bytes/sec.[53] This disc storage unit can also be used with the ASVT series computers.

Within the Edinaya Sistema range, there are eight separate disc packs of which two, the ES-5051 and 5060, are fixed, the remainder removable. The fixed disc units have capacities of 800 Kbytes, access times of 10 m secs, and

[44] *Ekon. Gaz.*, 1974, No. 2, p. 8.
[45] *Soviet Cybernetics Review*, September 1973, p. 8.
[46] *Pravda*, 9 June 1972.
[47] *Ibid.*, 8 May 1973.
[48] *Nauka i Zhizn'*, 1973, No. 8, p. 10.
[49] *Ekon. Gaz.*, 1974, No. 2, p. 8.
[50] R. Judy, *op. cit.*, p. 57.
[51] *Soviet Cybernetics Review*, September 1971, p. 26.
[52] R. Judy, *op. cit.*, p. 50.
[53] *Soviet Cybernetics Review*, July 1972, p. 22.

rates of data exchange of 100K and 150 Kybtes/sec respectively. The removable packs for which data are available (the ES-5050, 5052, 5056, 5058) have capacities of 7·25 Mbytes per unit.

Soviet input/output equipment has also undergone improvement in the last years. For example, line printers are now capable of speeds up to 900 lines per minute. Some comparisons with Western disc units and line printers are made in the final section.

TIME-SHARING

Time-sharing is an arrangement whereby a computer can be used for two or more concurrent operations. It fulfils one operation for a short time, and then moves on to fulfil another, the whole process being arranged in a strict sequence. The intention is to allow a more efficient use of the computer's time.

Time-sharing in the Soviet Union is the subject of a Rand Corporation report, prepared in 1971.[54] The author argues that time-sharing would be particularly advantageous in Soviet conditions, as it would permit large multi-machine computer centres to be established in place of 'small computer centres notorious for their primitive methods and inefficient use of machines'.[55] However, Doncov describes only five Soviet experiments in time-sharing, among them the automated system of management at the L'vov TV factory, which employs two Minsk 22 computers in a time-sharing mode, and the AIST[56] project at Novosibirsk Computer Centre. The latter comes in two variants, AIST-0 and AIST-1, of which the former relies on an M-220, the latter on a Ural 14 and a pair of BESM-6 control computers.[57]

Doncov identified three conditions for the development of general purpose time-sharing in the USSR:

1. The introduction of more modern computers with higher operation speeds and greater memory capacities (especially discs).
2. The development, dissemination and wider application of higher-level languages.
3. The development of appropriate input/output equipment.[58]

An additional requirement for reliable data transmission is receiving attention from the USSR Ministry of Communications.[59] All three conditions are satisfied actually or potentially, and the Soviet mastery of third generation computer technology is an important step forward.

Of second generation machines, the Ruta 110 can be time-shared,[60] and the Minsk 32 can fulfil three or four programs simultaneously. In the third generation the Nairi 3 has a time-sharing capability, and most significantly, the

---

[54] Boris Doncov, *Soviet Cybernetics Technology XII: Time-Sharing in the Soviet Union*, Rand, Santa Monica (1971).
[55] *Ibid.*, p. 8.
[56] AIST stands for *Avtomaticheskaya Informatsionnaya Stantsiya* (Automatic Information Station).
[57] V. I. Loskutov, *Avtomatizirovannye sistemy upravleniya* (1972), pp. 134–8.
[58] B. Doncov, *op. cit.*, p. 52.
[59] (Ed.) D. G. Zhimerin *et al.*, *Avtomatizirovannye sistemy upravleniya* (1972), pp. 42–5.
[60] *Soviet Cybernetics Review*, January 1970, p. 38.

ES range can be used in a time-sharing mode.[61] The special application program packages include facilities for time-sharing.[62]

In the United States, time-sharing made its first appearance in the early sixties and was disseminated in the following few years. We can say, inverting the argument of this chapter, that it is consistent with the lag between Soviet and American technology for a similar dissemination to take place in the USSR over the next few years.

SOFTWARE

Software has always been one of the major weaknesses of Soviet computer development. In 1972 Richard Judy endorsed his 1968 observation that the software lag of the USSR behind the United States was greater than the hardware lag.[63] In this section I assemble information on software for computers developed since 1968, concentrating on the third generation.

By October 1972, the operating system for the Minsk 32 had been developed to include 225,000 instructions and 175 books of instructions and description, but a compiler for Cobol was still undergoing tests in Minradprom.[64] (The Minsk 22 is said to be too small for effective use of Cobol,[65] hence program compatibility of the two models was of little value.) Minradprom was hoping to prepare an edition (*tirazhirovat'*) of the software for the Minsk 32, but as the *Pravda* commentator observed, to do that you have to have the software. The reason for the delay was transparent. The Institute of Mathematics of the Belorussian SSR, which was to develop software for the Minsk 32, did not receive a specimen of the computer until 18 months after it had gone into quantity production.[66] The lack of a compiler for Cobol is a particularly grave defect in a machine widely used in economic applications. In 1974 there were still complaints of inadequate software.[67]

The Ruta was equipped by 1971 with a unified system of software, including an assembly language compiler, a control program for simultaneous execution of up to three programs, a subroutine library and a compiler for Cobol.[68,69] Holland observes that 'the existence of such a complete software package is noteworthy since few Soviet computers are so equipped until several years after their introduction'.[70]

The Nairi 3 was designed to operate in languages both of the Minsk 22 and the earlier Nairi computers. Software for the latter is well developed.[71]

[61] *Pravda*, 8 May 1973.
[62] *Pribory i Sistemy Upravleniya*, 1973, No. 11, p. 6.
[63] R. Judy, *The Contribution of Computers and Mathematical Methods to Soviet Planning and Management* (1972), p. 6; unpublished paper, typescript kindly supplied by the author.
[64] *Pravda*, 9 October 1972.
[65] *Pribory i Sistemy Upravleniya*, 1972, No. 5, p. 6.
[66] *Pravda*, 9 October 1972.
[67] *Ekon. Gaz.*, 1974, No. 34, p. 7. A software handbook for the Minsk series in general was published in 1974.
[68] *Pribory i Sistemy Upravleniya*, 1971, No. 2, pp. 3–4.
[69] Apparently Cobol was chosen on the recommendation of a CMEA group on the use of algorithmic languages for economic information processing.
[70] *Soviet Cybernetics Review*, 1971, No. 5, p. 27.
[71] See M. Zharkov, *Programmirovanie dlya 'Nairi'* (1973).

The software situation with regard to the ASVT series was for some time unclear. In 1972 it was implied that the M-4000 lacked compilers for mnemo-codes, Cobol, Fortran and RPG,[72] yet Holland believed that at this time the M-6000 was provided to process control customers with a 'sufficiently developed system of software, including a compiler for mnemo-code, compilers for Fortran, Algol 60 and specialised languages, a set of programmes for control of information processing and input/output, interpreters of the machine languages of different ASVT models, etc.'[73] By 1974 the M-4030 had an operating system, the OS ASVT, which was an amended version of that for the M-3000 and M-4000. But this was intended only for users converting to the M-4030 from those models. The basic operating system for the M-4030 is a disc system, the DOS ASVT. This system includes compilers for Assembler language and for RPG, Algol, Fortran and Cobol.[74] The ASVT series as a whole is said to be compatible with the ES series,[75] to which we now turn.

As in other respects, the ES series was intended to avoid the mistakes and delays in software associated with earlier computers, yet already similar complaints are being voiced. The software system (*sistema matematicheskogo obespecheniya*—SMO) of the ES series consists of operating systems (*operatsionnye sistemy*—OS), servicing routines and batches of applications programs.[76,77]

There are four operating systems: OS 10/ES for the Hungarian ES-1010; MOS/ES (*malaya* OS) for the Czech model ES-1021; the disc system DOS/ES (*diskovaya* OS) for the ES-1020, 1030, 1040 and 1050 with low capacity memory; and the OS/ES for the same models with high capacity memory.

The OS 10/ES is used for single program data processing or scientific calculations. It contains control programs, service programs and compilers for Assembler (Autocode) and Fortran-IV. Like the MOS/ES for the ES-1021, the system was scheduled for completion in 1973.

The OS/ES is intended for the more powerful members of the series with core storage of more than 128 Kbytes. It is designed to make use of the bulk of application programs developed for the IBM 360 range—another indication of the kinship of the ES and IBM 360 series. Interestingly, divergencies of the peripheral equipment of the two ranges have imposed some limitations: for example, the ES-7030 line printer has 128 characters per line as against 132 in the IBM line printer.[78] A part of the OS/ES, embodying 700,000 instructions, has been tested on an experimental model of the ES-1050. Evidently this system is the key one in the software of the ES range, which it is described as transferring to a new higher plane:

---

[72] *Pribory i Sistemy Upravleniya*, 1972, No. 1, p. 5.
[73] *Ibid.*, p. 17.
[74] *Pribory i Sistemy Upravleniya*, 1974, No. 11, pp. 18–19.
[75] (Ed.) D. G. Zhimerin *et al.*, *Avtomatizirovannye sistemy upravleniya* (1972), p. 34.
[76] This section is based on information in *Pribory i Sistemy Upravleniya*, 1973, No. 11, pp. 4–6.
[77] All the machines in the series are program compatible upwards.
[78] *Pribory i Sistemy Upravleniya*, 1973, No. 11, p. 5.

this system makes it possible to satisfy the demands of a considerable part of the users of individual machines of the Unified System and of users of computer complexes built on the basis of them, and is a basis for the accelerated construction of equipment for teleprocessing, time-sharing and real-time operations, machine graphics and application programs.[79]

The disc system DOS/ES is useful for machines with operational memory in the range 64–128 Kbytes (i.e., the ES-1020 and 1030), and is intended for economic data processing. In 1973 it was credited with five compilers (for PL/1, Fortran, RPG, Assembler E and F). Under development were compilers for Fortran-IV and a compiler for Cobol which would also make it possible to use programs written in that language for the Minsk 32.

As far as special application programs are concerned, the Coordination Centre of the intergovernmental Commission which controls the ES project has established an information service providing data on programs used in automated systems of management (ASU) of the participating countries and other application programs. Software development for ASU has been spread throughout the countries in the project; for example, the Czechs have been assigned responsibility for the ASU subsystem for control of labour and wages; GDR for control of material-technical supply; the Hungarians for book-keeping.

In spite of the planning and coordination, the ES series has fallen victim to the same delays and shortcomings in software which have diminished the operational efficiency of virtually all other Soviet computers. In an article published in January 1974, Zhimerin described the software for the ES series as totally inadequate:

> At the moment, preparation of the operating system has not been finished, and there are no translators from three necessary algorithmic languages (Cobol, Algams, Fortran-Basic [*sic*]); this not only narrows the front of application of electronic computers, but also puts great difficulties in the way of composing external (*vneshnie*) programs.[80]

In order to remedy these defects, it was decided that Minpribor would set up in 1974 a science and production *ob"edinenie*, called Tsentroprogrammsistem, in Kalinin, which would develop algorithms for ASUs and supply customers with application programs on a contract basis.[81] According to Zhimerin, the *ob"edinenie* will be responsible for preparing programs for the Ryad series and for helping ASU personnel in using the programs.[82] A large number of organisations are involved in this effort. In April 1974 it was reported that Minradprom, Minpribor, Minvuz and the USSR Academy of Sciences were preparing a coordination plan to control work in the area, but by the end of

---

[79] *Ibid.*, p. 6.
[80] *Ekon. Gaz.*, 1974, No. 2, p. 8.
[81] *Ekon. Gaz.*, 1974, No. 16, p. 8.
[82] *Ekon. Gaz.*, 1974, No. 27, p. 8.

1974 there was still no basis established for concluding contracts with the *ob"edinenie*.[83]

Meanwhile there has been talk of the possibility of the USSR having recourse to foreign trade in this area. It was reported at one time that the USSR would like to buy from IBM the right to unlimited use of the 360 software in the USSR, in place of the conventional licensing arrangement. Any developments in this field, as in the case of trade in hardware, depend above all on political relations between the USSR and the USA, and it is unlikely that the Soviet Union is relying entirely on its ability to import technological knowledge from America.

The diffusion of computers

Information on the size of the Soviet computer stock is scanty, and in view of the great importance of computers in military as well as civilian applications this is scarcely surprising. Western writers have given various estimates of the computer stock, but they are subject to a number of uncertainties. There may be some ambiguity in the definition of a computer; for example, how large does a machine have to be to qualify as a computer? It is also important to distinguish between estimates of the total stock of computers and the stock in civilian uses only.[84]

The Soviet statistical handbooks and other official sources give a series for the value of output of computational equipment (*sredstva vychislitel'noi tekhniki*) for the years 1940, 1950, 1960, 1965–74. The figures include more than electronic computers; so much is clear from the small but positive entry in 1940. But the weighting of electronic computers must be very substantial in the later period. The plan figures for 1971–5 are also available.[85] These figures are supplemented by estimates presented to a Committee of the US Congress for the production of computers and data processing equipment for the period 1958–65. The actual increase in output of computers for the eighth five year plan (1966–70) was 480 per cent, the planned increase for the ninth five year plan (1971–5) is 260 per cent.[86] The available series are presented in Table 8.1; it will be seen that where they coincide, they are fairly consistent. The series give a good indication of the growth of Soviet production of computers: in value terms, output in 1973 was more than nine times 1965 output, at constant prices. In 1973, production of computers exceeded the targets of the five year plan for that year by more than 5 per cent, and was one of only 3 out of 50 categories to do so.[87] The editor of *Soviet Cybernetics Review* has estimated that the ninth five year plan calls for the production of 18,000 computers.[88]

[83] *Ekon. Gaz.*, 1974, No. 48, p. 8.
[84] The Ministry of the Radio Industry, one of the two computer producing ministries, is part of the military-industrial complex and no doubt supplies computers used in military applications and in space research.
[85] Alec Nove, *ABSEES*, July 1974, Special Section II, p. xx.
[86] *XXIV s"ezd KPSS: stenograficheskii otchet*, Vol. II (1971), p. 201.
[87] A. Nove, *op. cit.*, p. xxiv.
[88] *Soviet Cybernetics Review*, July 1971, p. 2.

Table 8.1 *Soviet production of computational equipment (million roubles)*

	Miller[1]	Nar. Khoz. SSSR[2]		Planned[3]
		1955 prices	1967 prices	
1940		0·3		
1950	47			
1958	60			
1960	73	79·9		
1961	93			
1962	127			
1963	160			
1964	187			
1965	267	245·3		
1966		288·0		
1967		376·8	260·8	
1968		519·0	359·7	
1969			485·0	
1970			709·7	
1971			879·4	
1972			1213·4	
1973			1634	1284
1974			2221	1602
1975				1999

Sources:
1 K. Miller in *New Directions in the Soviet Economy*, Joint Economic Committee of the US Congress, Washington DC (1966), p. 333. Coverage: production of computers and data processing equipment; recalculated from US dollars.
2 *Narodnoe khozyaistvo SSSR*, various years; *Ekon. Gaz.*, 1975, No. 5, p. 5. Coverage: computational equipment.
3 A. Nove, *ABSEES*, July 1974, Special Section II, p. xx. Coverage: computational equipment.

According to *Soviet Cybernetics Review*, the generally accepted Western estimate of the size of the Soviet computer stock in 1970 is 5,000–6,000.[89] Although this estimate has received some later confirmation,[90] it is not universally accepted. Judy suggests that, at most, 2,000 computers were in civilian use in the USSR in 1968.[91] Gerald Segal gives an estimate of 2,200 for total production in the 1960s.[92] The Diebold Institute estimates the total stock at the beginning of 1970 as 4,200.[93] Unfortunately, without knowing how unit costs have changed, it is impossible to check this figure by calculating annual production in units from the official value series and cumulating them for

[89] *Ibid.*
[90] J. Wilczynski, *Technology in Comecon*, London (1974), p. 114.
[91] R. Judy, in (Ed.) S. Wasowski, *East–West Trade and the Technology Gap*, New York (1970), p. 53.
[92] *The Times*, 24 May 1972.
[93] Quoted in J. Sláma *et al.*, *Ost Europa*, 1974, No. 2, p. 120.

annual estimates of the stock.[94] The absence of any such confirmation is especially regrettable, as, in the only case where Soviet figures for the size of the computer stock are given, for the Ukraine from 1958–72,[95] the figures seem to be too low for the 'generally accepted' estimate. The divergence is probably accountable in part to the absence in some estimates, including possibly the estimates for the Ukraine, of the stock in military use.

The information on stocks of computers held in other countries is subject to fewer uncertainties. In Table 8.2 the stock of computers in numbers of units is presented for several countries in various years. Figures in brackets give the number of computers per million population. The latter information is represented on the accompanying Fig 8.1. Alternate values of the Soviet lag behind selected countries with respect to computers per head in 1970 are given in Table 8.3, one based upon a Soviet stock in 1970 of 24·6 computers per million of population, the other based upon the lower estimate of 12·6, the value for the Ukraine.

These estimates are indirectly confirmed by a Soviet author who, in projecting demand in the USSR for automation equipment, including computers, bases his calculations on a Soviet lag behind the USA of 8–10 years.[96]

Table 8.2 *Stock of computers held in various countries*

	1960	1962	1965	1967	1970	1975 (plan)
France		285[1] (6·1)		2208[1] (44·3)	(90)[2]	
FRG		548[1] (10·0)		2963[1] (49·5)	(109)[2]	
UK		312[1] (5·8)		2252[1] (40·9)	(91)[2]	
US	5000 (27·7)[4]	7305[1] (39)		39516[1] (198·5)	70000 (344)[2]	
USSR	[120][4] (0·56)		[2000][3] (8·7)		[6000][3] (24·6)[2]	[22000][3] (84·6)
Ukraine	10[5] (0·24)	40[5] (0·94)	200[5] (4·4)		600[5] (12·6)	

(Figures in brackets give stocks per million population.)

Sources:
1 OECD, *Gaps in Technology: Electronic Computers*, Paris (1969), p. 126.
2 J. Sláma *et al., Ost Europa*, 1974, No. 2, p. 122.
3 Estimates, or estimated plans.
4 A. C. Sutton, *Western Technology and Soviet Economic Development, 1945 to 1965*, Stanford, Cal. (1973), p. 319 (for end of 1950s).
5 Mikulich *et al.*, in *Mekhanizatsiya i Avtomatizatsiya Upravleniya*, 1974, No. 4, p. 7.

---

[94] If we suppose unit costs have been constant and make certain other assumptions we can use the estimated target for total production in units in the 1971–5 plan, together with the actual and planned growth rates for the eighth and ninth five year plans and the Congress figures, to calculate output in each year. Cumulative output figures from 1958 are then as follows: 1960—270, 1965—1,530, 1970—7,216. If we allow for retirements from stock the 1970 figure is close to the 'generally accepted' estimate.
[95] *Mekhanizatsiya i Avtomatizatsiya Upravleniya*, 1974, No. 4, p. 7. The data are read from a graph and are subject to a margin of error.
[96] *Pribory i Sistemy Upravleniya*, 1974, No. 2, p. 2.

Table 8.3 *Soviet lag in computer stock*

Soviet lag in 1970 behind:	Higher estimate of Soviet stock	Lower estimate of Soviet stock
USA	9	11
United Kingdom	5–6	7
Federal Republic of Germany	5–6	7

Sources:
Based on Table 8.2 and Fig 8.1.

However, these figures tell only part of the story, for they count each computer as one unit and ignore the enormous disparity in computing power between different models of computers. What we would like ideally is an estimate of computing power in each country at any time, obtained by weighting each machine in the stock by its capability. One way of doing this is by taking the cost of the computer as an index of its capability, and using data on the value of stock. But this is subject to two drawbacks. First, the rate of technical progress in computer technology is very high, and this makes the relationship between computer power and the historic cost of the computer stock a complicated one. This problem makes it hard to create a quality-corrected quantity index from a sequence of value-of-stock figures.[97] Second, it would be impossible to establish a dollar-rouble exchange rate for computers and this prevents any cross-country comparison.

It is possible, however, to supplement the data on units of computers with very rough estimates of the proportion of the stock of computers of different generations. According to Sutton, in 1960 the American computer stock consisted almost entirely of second generation computers; indeed some early second generation computers had been removed from service by 1959.[98] In the USSR it seems certain that first generation machines were in service in the late 1960s and early 1970s. In 1970 a very high proportion of the American computer stock consisted of third generation machines; second generation computers were being retired at a rapid rate. In the USSR, in contrast, there was not a single third generation machine of domestic manufacture. In 1974 93 per cent of computers in the Ukraine were second generation models,[99] and of the remainder, the majority were probably first generation. In the USA a second generation computer had become a rarity. The Soviet lag was compounded by two factors. Output has grown at a slower pace in the USSR than in the USA, so that, even with the same lifespan of a computer, the Soviet Union would have a higher proportion of machines of earlier years. Second,

[97] Where data are adequate, a hedonic price index can be constructed and the desired quantity index calculated. This has been done by Chow for the US computer stock (*American Economic Review*, 1968, pp. 1117–30) and by Stoneman for the UK stock (P. Stoneman, *Technological Diffusion and the Computer Revolution*, Cambridge (1976)). But for the Soviet case the data are lacking and the Soviet price system would not in any case justify such an approach.
[98] A. C. Sutton, *op. cit.*, p. 318.
[99] *Mashinnaya Obrabotka Informatsii*, 1974, No. 18, p. 3.

the comparative scarcity of computers in the USSR has meant that they have been kept in service longer than elsewhere.

However, notwithstanding the relatively low endowment of the Soviet Union in computers, the rate of utilisation of such machines as there are is low. When asked by *Pravda* to explain the low rate of utilisation, Zhimerin mentioned several factors, including the inadequacy of software and peripherals and the lack of preparedness of enterprises receiving computers.[100]

Fig. 8·1 STOCK OF COMPUTERS PER MILLION INHABITANTS (LOGARITHMIC SCALE)

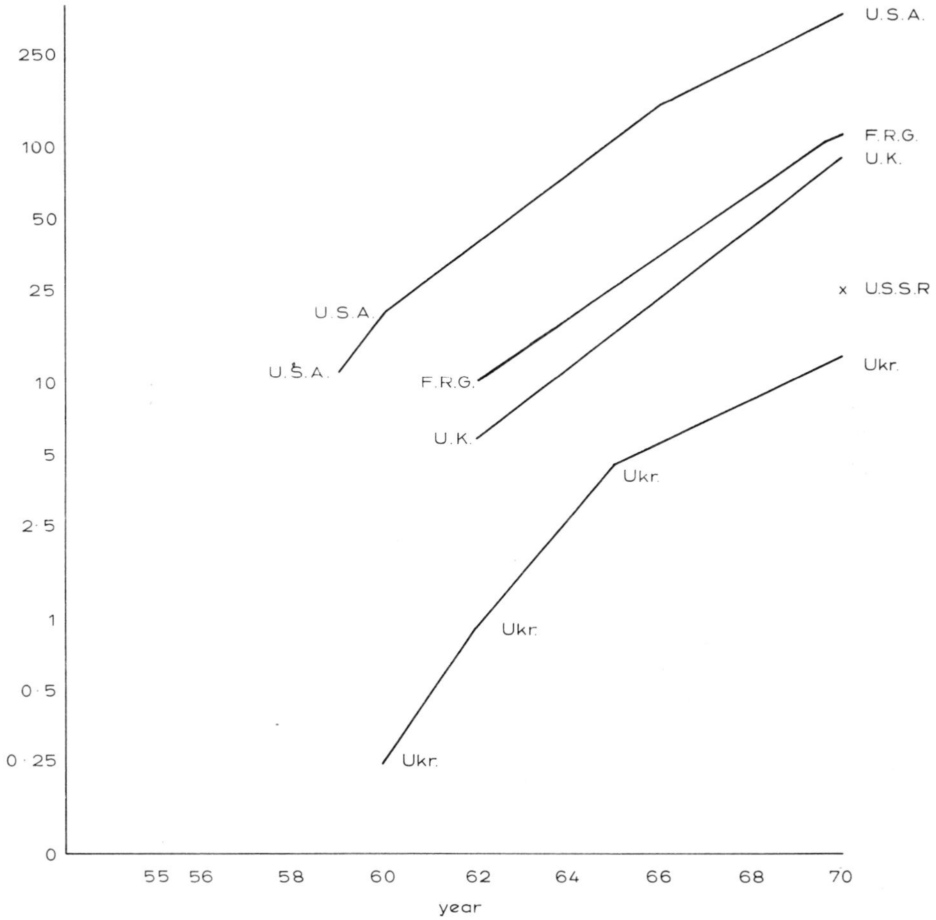

[100] *Pravda* reports that a Minsk 32 computer was left crated at a computer centre for six months before installation (*Pravda*, 9 October 1972).

Another factor to which Zhimerin has drawn attention is the lack of a centralised system for servicing and repair of computers.[101] The last point is borne out by an examination of the reasons for the down time or period of non-operation of computers used in management in the Urals regions: technical defects accounted for 52 per cent of the down time for the Minsk 22 and 80 per cent of the down time for the Minsk 32.[102]

Whatever the reasons, the rate of computer utilisation reported in 1972 was 10·3 hours per day (only 5·7 for the Minsk 32);[103] in 1973 the overall figure had increased to 10·7 hours, according to TsSU estimates.[104] These low rates undoubtedly exacerbate the computer shortage.[105]

THE AVAILABILITY OF PERIPHERALS

We have already considered the evidence on the technological level of Soviet peripheral equipment. This section considers the evidence on the supply of these peripherals to computer users. Although there is no definite statistical evidence it is clear that the inadequacy of the supply acts as a limiting factor in computer use. Two problems are perceived, the first relating to the production of peripherals, the second to the arrangements for their distribution.

The problem of production can be illustrated by two quotations from Soviet sources. An expert on automated management systems writes:

> There is a significant gap between the requirements of the national economy and the development and production of equipment for data collection, registration, storage and transmission and also of equipment for linking man and machine (displays, plotters, information tableaux, etc.). Even now the ministries are slowly mastering serial production of them. The peripheral equipment produced at the moment is unacceptably costly, which limits its application in the automated systems of management being established.[106]

A similar point is made by Zhimerin, a deputy chairman of the State Committee on Science and Technology:

> Unfortunately industry, primarily the instrument building factories, is still very slowly mastering the production of equipment and devices with technological parameters corresponding to the needs of automation. And although the M-6000 type computers produced by Minpribor satisfy the requirements of process automation, an inadequate list of external equipment limits the range of their application.[107]

[101] *Ekon. Gaz.*, 1974, No. 2, p. 8.
[102] *Planovoe Khozyaistvo*, 1972, No. 1, p. 65.
[103] *Ekon. Gaz.*, 1972, No. 4, p. 6.
[104] *Ekon. Gaz.*, 1974, No. 14, p. 8.
[105] This reference to low rates is in relation to the Soviet demand for computers. Some partial surveys of computer utilisation in Western countries suggest that they also have low rates of utilisation.
[106] V. I. Loskutov, in *Ekonomika i Organizatsiya Promyshlennogo Proizvodstva*, 1973, No. 6, pp. 36–7.
[107] *Ekon. Gaz.*, 1974, No. 2, p. 8.

The problems of development and production are compounded by the inadequacy of the distribution system for peripherals. It is a frequent refrain that computer centres are not supplied with an adequate range of peripherals: in Soviet terms there is no *komplektatsiya*, or making up of a full range of equipment. The lack of this essential condition for effective computing has been a subject of comment and criticism over an extended period. At the first All-Union Conference on Automated Systems of Management held in January 1972, the Minister of Heavy Machine Building for the USSR commented that the plan for supplying a computer centre covered only 30–40 per cent of the equipment needed, while the remainder had to be collected piece by piece: punch card and key-punching equipment from TsSU; other peripherals from Minpribor; and magnetic tape from the State Committee on Cinematography.[108] A highly critical article in *Pravda* took up the same theme.[109]

At the 1972 Conference, the Minister of the Radio Industry promised some improvement. From 1972 an industrial department of Minpribor, Soyuzpromavtomatika, through its trust Soyuzsistemkomplekt, has been able to supply its customers with a range of equipment stipulated in the project for an automated management system. The trust has made these so-called *kompleksnye postavki* (complete packages) according to a list agreed by Gosplan, Gossnab and TsSU USSR. However, this applies only to equipment produced within Minpribor. The director of the trust notes with regret that he has not yet obtained the rights of distributor (*fondoderzhatel'*) for the products of other ministries.[110]

Summary and conclusions

I noted earlier that this comparative study of computer technology follows roughly the same method as Judy's study which covered the period up to 1968. At that time, Judy concluded that

> Soviet computer technology started in the early fifties with a modest qualitative lag behind Western equipment. This lag lengthened into a serious gap by 1964, when Soviet technology was greatly inferior in all respects. Since 1965, with the announcement of the new Ural and Minsk systems, and the BESM-6, the gap has narrowed somewhat. Soviet computer technology remains quite inferior to the best in the West. Quantitatively the US appears to have about 50 times as many computers installed as does the Soviet Union which lags behind the United Kingdom, France, Germany and Japan as well as the United States. The gap separating contemporary Western computer software and that employed in the Soviet Union is enormous.[111]

[108] (Ed.) D. G. Zhimerin *et al.*, *Avtomatizirovannye sistemy upravleniya* (1972), p. 2.
[109] *Pravda*, 9 October 1972.
[110] *Sots. Ind.*, 14 August 1973.
[111] R. Judy, 'The case of computer technology' in (Ed.) S. Wasowski, *East–West Trade and the Technology Gap*, New York (1970), p. 62.

398 TECHNOLOGICAL LEVEL OF SOVIET INDUSTRY

Writing in 1972, Judy notes that 'the Soviet lag in software has been even greater than in hardware. In recent years the software situation has improved somewhat.'[112] In this section I review Judy's conclusions up to 1968 and continue the story to 1973.

Figure 8.2 shows the capabilities of the most powerful Soviet, American, and British computers generally available in the period since 1951 (since 1960 in the British case).[113] The drawbacks of this index are discussed above, but it is noticeable that the figure does confirm the relationship between the technological levels of the American and British industry which is generally accepted: rough equality in the early 1960s giving way to increasing American domination. Unfortunately the requirement of general availability is by no means unambiguous. We want to exclude machines of which only one or a very small number have been produced,[114] but it is not clear whether we should on this ground exclude the Soviet BESM-6 computer, of which only a small

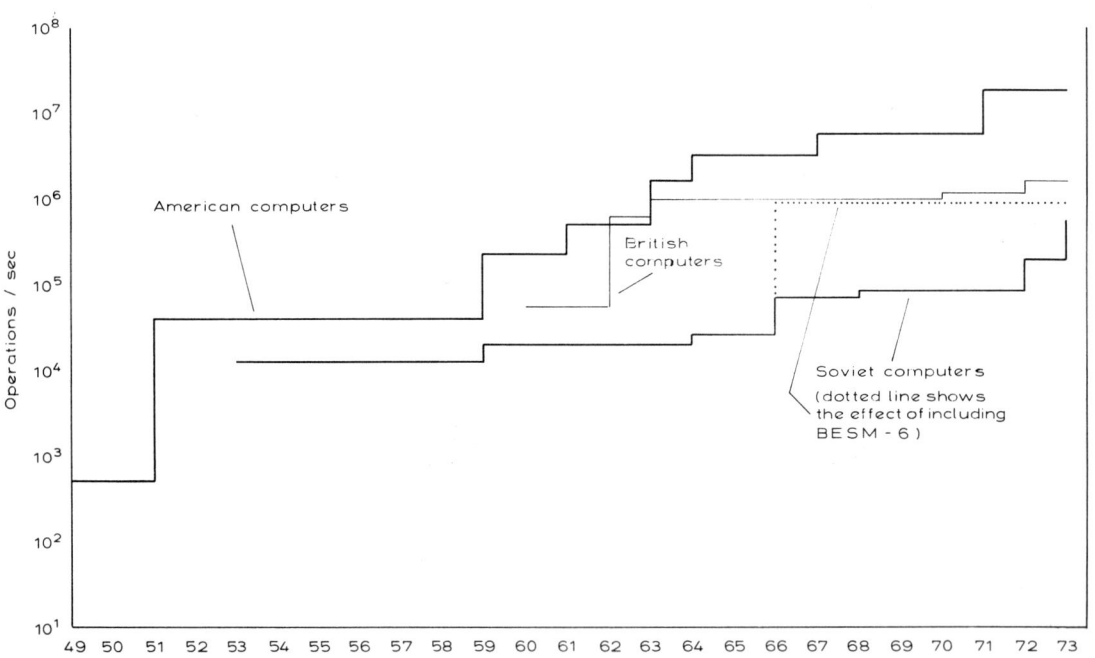

Fig. 8·2 COMPARISON OF MAXIMUM OPERATING SPEEDS OF BRITISH, AMERICAN AND SOVIET COMPUTERS 1949 - 1973

[112] R. Judy, *The Contribution of Computers and Mathematical Economics to Soviet Planning and Management,* unpublished paper (1972), p. 6.
[113] These data are taken from Appendix Tables 8A.1 and 2.
[114] Both these measures are introduced for selected years by Judy in his Table 7. R. Judy, *op. cit.* (1970), p. 61.

number have been available. Alternative measures of Soviet computer technology are shown in Fig 8.2, the dotted line illustrating the inclusion, the full line the exclusion, of the BESM-6. The significance of this decision is apparent from Figs 8.3 and 8.4. The top half of Fig 8.3 shows the vertical distance between our indices of American and Soviet technology as illustrated in Fig 8.2: it shows by how many times the best American computer is more powerful than the best contemporary Soviet computer. The bottom half of Fig 8.3 gives the more familiar measure of technological gap in terms of the lag in years of Soviet behind American technology. This is the horizontal difference between the Soviet and American indices in Fig 8.2. Figure 8.4 shows the same comparative information for Soviet and British computers.

Both parts of Fig 8.3 illustrate Judy's conclusion for the period up to 1968 that the gap between Soviet and American technology widened throughout the fifties and early sixties. In the second half of the sixties the lag was stabilised or even reduced, but since 1970 there is no sign of further reduction in the lag, which may even have increased. The comparison of Soviet and British performance in Fig 8.4 shows the same pattern, with the reduction in the gap slightly more pronounced in the sixties as British fell behind American technology.

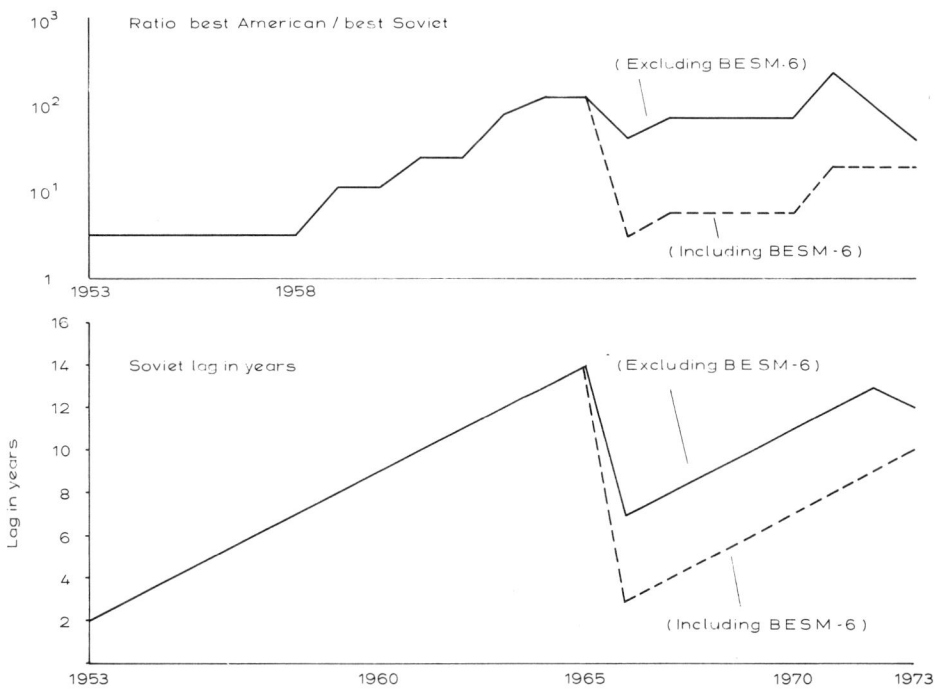

Fig. 8·3 COMPARATIVE PERFORMANCE OF BEST AMERICAN AND BEST SOVIET COMPUTERS

The method of comparison illustrated in Figs 8.3 and 8.4 suffers from the fact that a single computer tends to dominate whole sections of the curve covering a period of years. This makes the results sensitive to the inclusion of a particular computer. But we can check these results in cases where a Soviet machine has appeared with characteristics similar to those of an earlier American model. These comparisons are given in Table 8.4. They are based largely on similarities noted by Rudins.[115]

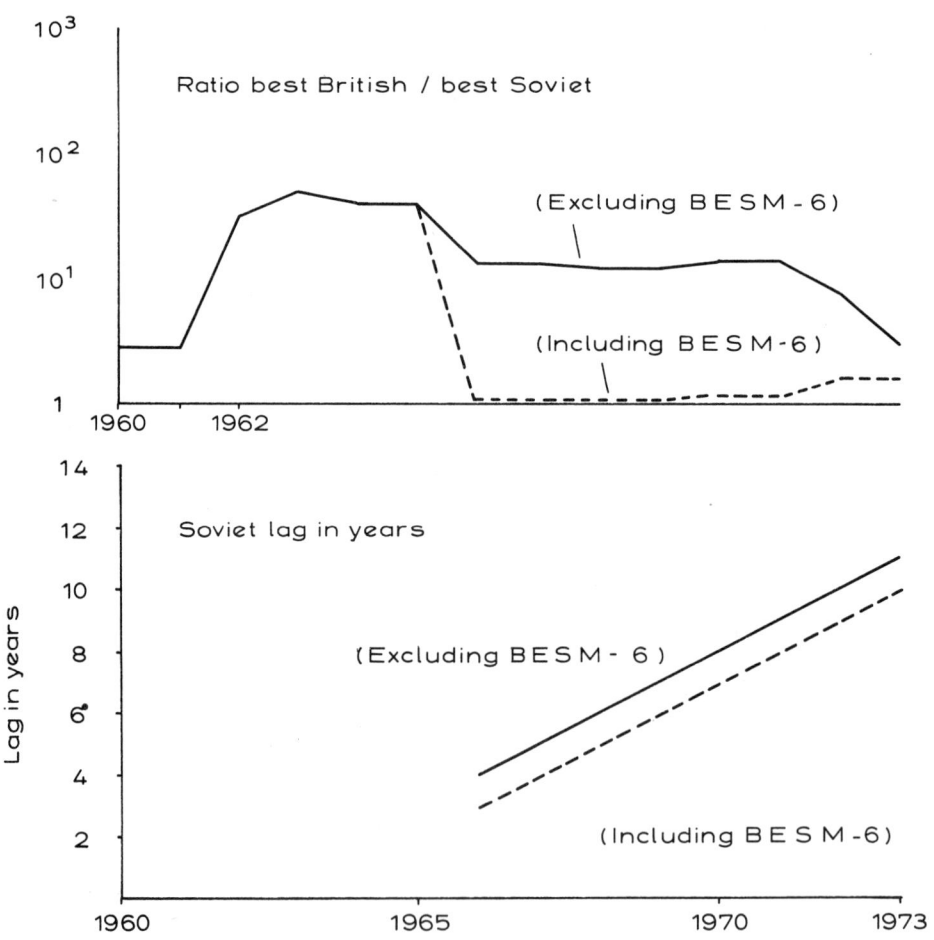

Fig. 8·4 COMPARATIVE PERFORMANCE OF BEST BRITISH AND BEST SOVIET COMPUTERS

[115] G. Rudins, in *Soviet Cybernetics Review*, January 1970.

Table 8.4 *Date of first production of comparable Soviet and American computers*[a]

American computer	Similar Soviet model	Date of appearance in USSR	Lag (in years)
IBM 650	Ural 1	1955	1
IBM 702	Ural 4	1962	7
IBM 1620	Nairi 1	1964	4
IBM 7094	BESM-6	1966	4
IBM 360 series	ES series	1972–3	6–8

Sources:
Data drawn from Tables 8.A1 and 8.A2.

Note:
a – comparison of dates of first American commercial installation and first Soviet industrial production.

The same general picture is corroborated by the lag of the USSR behind the United States in entering successive generations of digital computers.

Table 8.5 *Soviet lag in entering successive generations of computers*

	First generation	Second generation	Third generation
First Soviet computer	1952	1961	1972
First American computer	1946	1957	1965
Lag (in years)	6	4	7

Source:
See Table 8.4.

Of the lag between Soviet and American development in peripherals I consider only two examples. As we have seen the first Soviet disc unit was available in 1970, seven years after the first American unit. Characteristics of Soviet and American discs are shown in Table 8.6.

Table 8.6 *Characteristics of disc units*

Name	Country of origin	Date	Average access time microsecs	Data transfer rate Kbytes/sec
IBM 1302	USA	1963	165	184
IBM 3330	USA	1973	10–55	806
R 401	USSR	1970	200	30
ES 5056	USSR	1973	96	156

Sources:
R. Judy in (Ed.) S. Wasowski, *op. cit.* (1970), p. 50; *Computer Characteristics Review*, various issues; and sources cited in the text.

The other example is line printers, the technological level of which can be summarised in lines printed per minute (see Table 8.7).

Table 8.7 *Characteristics of line printers*

Name	Country	Date	Speed (lines per min)
IBM 1401	USA	1960	600
IBM 1403	USA	1965	1100
IBM 3211	USA	1973	2500
On Ural 4	USSR	1962	300/400
ATsPU 128 (ES 7030)	USSR	1967	400
On BESM-6	USSR	1967	700
ES-7032	USSR	1973	900
ES-7031	GDR	1973	1200

Sources:
As in Table 8.6.

These two examples suggest that the Soviet lag in development of peripherals in 1973 was about 8 or 10 years, at least as great in the lag in development of CPUs. There is moreover reason to believe that lack of some items of peripheral equipment is hampering the use of computer systems.[116]

There is further evidence to support Judy's view that software has recently improved. Compare, for example, the Soviet record in preparing software for the Minsk 32 with their record with the ES-1050. In the former case the computer was in serial production for 18 months before the software developer received it; in the latter case an initial batch of programs was available for testing on the prototype. Certainly the role of software grows more important with each successive generation of computer, and the achievement with the ES series is not remarkable by Western standards.[117] There are, moreover, delays in the preparation of application programs. Even so, the situation has greatly improved over the last five years as compilers for high-level languages have been prepared, and we can no longer with equal confidence identify the provision of software as a restraining factor in Soviet computer development.

In summary, the technological gap between the Soviet Union and the United States has continued since 1968. In respect of hardware and peripherals, there is no evidence that it has narrowed in the past five or six years; in respect of software, a substantial improvement has occurred in the USSR, from a previously low level. This rather bleak account of the situation so far might, however, be misleading about the future. Before 1968 Soviet computer technology operated under conditions of rather low priority and suffered many of the disadvantages of a competitive situation without reaping any of the compensating advantages. The degree of concentration of computer

[116] See above, p. 396.
[117] However, IBM had difficulty in preparing the operating system for its IBM 360 series.

production was low, even between ministries;[118] a large number of models were produced in small quantities, and in the field of R and D the fragmentation was even worse. Nor was the resultant lack of coordination compensated for by competitive innovation among rival organisations; the operation of the planning system tended to work in the opposite direction towards a conservative policy in innovation. But since 1968 a higher priority has been given to computer production and more effort has been put into coordination. A far smaller range of machines is produced in far greater numbers and there has been much greater emphasis on ensuring compatibility between different models and series. This effort has, at the least, prevented Soviet computer technology from slipping further behind American and created a more favourable outlook for the future.

A further question that must be discussed in assessing the lag is how far the Soviet Union has developed its computer technology independently and how far it has 'copied' from the West. In 1968 Judy observed that 'literally all significant technological innovations in computer technology have been made in the West'.[119] He also notes[120] that 'computer technology in the Soviet Union is virtually entirely imported from the West'.[121] Now it is virtually indisputable, as much now as in 1968, that all significant technological innovations have been made in the West, though the same is not true in the field of programming where the Soviet Union has made theoretical advances, even if they are not of great practical significance. But it is another question how far the Soviet Union has been able to copy particular Western developments rather than just follow their general orientation. The Co-com embargo on computer exports has only recently been relaxed, and for many years the only imports the USSR received were old computers scarcely more advanced than its own domestically produced machines. Of course, there may have been imports of more modern computers which circumvented the restrictions, but opinions differ as to how useful the possession of a single example of a third generation machine, for example, would be for copying purposes without the design drawings.[122] It is worth noting that the lag in computer development creates an additional reason for the Soviet Union to import Western equipment. An imported machine may be used for testing an application which will later rely on domestically produced computers. Without a more detailed comparison of Soviet computers and their alleged Western progenitors it is difficult to reach a firm conclusion on this issue.

[118] It is reported that in 1972 the ministry chiefly responsible for computer production accounted for only 41 per cent of total production. The situation was the same in the sixties. Compare the position in the West of IBM, which in 1962 accounted for nearly 70 per cent of the American stock of computers in value terms and 53 per cent of the rest of the world stock. (Yu. Subotskii, *Voprosy Ekonomiki*, 1972, No. 7, p. 16; Yu. Savinov and S. Medvedev, *Mirovaya Ekonomika i Mezhdunarodnye Otnosheniya*, 1973, No. 1, pp. 127, 130.)

[119] R Judy, *op. cit.* (1970), p. 64.

[120] *Ibid.*, p. 63.

[121] Sutton notes 'a substantial unity between his independently reached conclusions and those of Judy' (A. Sutton, *op. cit.*, p. 318).

[122] Dr L. Branscomb of IBM asserts that 'reverse engineering'—deducing manufacturing processes and design considerations from examination of the end product—is a generally over-rated concern, citing integrated circuits as an instance where it is not feasible (Lewis Branscomb, *Science, Technology and Detente*, mimeo (n.d.), p. 23).

Appendix

Table 8A.1 Characteristics of Soviet computers, 1953–73

Model	Date first produced (P) serially (S)	Generation	Internal storage	Access time to internal storage (microsecs)	Add time (microsecs)	Multiply time (microsecs)	Operations per second	Approx. cost (1000 roubles)	Primary application
1 BESM-1	1953 (P) 1955 (S)	1		10	80	270	7–8K		
2 Ural 1	1953 (P) 1955 (S)	1	1K drum	8000	1000	2000	100		
3 Strela	1953 (P) (S)	1	1K		500		2K		
4 BESM-2	1959 (P)	1	2K	6	70	230	8–10K	350	
5 M-20	1959 (P)	1	4K	6			20K		
6 Ural 2	1959 (P)	1	4K	12	180	470	5K	185	
7 Ural 4	1962 (P)	1	4K	12	130	470	5–6K	220	Data processing
8 Minsk 2	1962 (P) (S)	2	4K		156	272	5–6K	150	
9 Razdan 2	1961 (P) 1963 (S)	2	2K				5K	150	
10 Dnepr 1	1962 (P) 1964 (S)		½–2K				10K	100	Process control
11 BESM-4	1964 (P)	2	8K	10	47	95	20K	250	Scientific
12 M-220	1964 (P)	2	16K				20–28K	275	
13 Ural 11	1965 (P) (S)	2	16K				10K	100	Data processing
14 Ural 14	1966 (P) (S)	2	64K				10K	150	
15 Minsk 22	1963 (P) 1965 (S)	2	8K	24		200	6K	122	Production control
16 Minsk 23	1964 (P) 1967 (S)	2	40K	13			2–3K	165	Data processing
17 Razdan 3	1965 (P) 1967 (S)	2	16K	6			20K	540	Scientific
18 Nairi 1	1964 (P) 1965 (S)	2					2K	60	
19 Nairi 2	1966–7	2	2K	6		20	66K	64	Scientific Data Processing
20 Promin	1965 (P) (S)	2						500	
21 Mir 1	1966 (P) (S)	2	4K				250	53	Scientific
22 Vniiem 3	1966 (P)	2	32K	8			40–75K		Process control

Table 8A.1 (continued)

23 UM-1-NKh	1967 (S)	2	42K				1K	1100–1300	Process control
24 BESM-6	1966 (P) (S)	2	32–64K	0.8	1.1	1.9	1000K		Scientific
25 Minsk 32	1968 (P) 1969 (S)	2	16–64K	5M	15–40	15–130	65K	160	Data processing
26 Tbilisi-1	1968 (P)	2	4–28K	301	33	50	70–80K	370	
27 Ural 18	1965 (P) 1966 (S)	2	64K	20					Data processing
28 Ruta 110		2	8–10K		110	2200			
29 K-200	1971 (S)	2	24K	20	20	50	40K		
30 Nairi-3	1969	2/3	4–32K				15–20K		Scientific data processing
31 M-1000	1969 (P) (S)	2	4–32K	8	50	200	20K		Process control
32 M-1010	1969 (P) (S)	2	4–16K	10	20	150	50K		Process control
33 M-2000	1969 (P) (S)	2	48K	8	17	67			Process control
34 M-3000	1969 (P) (S)	2	96K	8	10	40	60K		Process control
35 M-4000	1972 (P) (S)	3					70K		Process control / Data processing
36 M-5000									Process control
37 M-6000	1972 (P) 1974 (S)	3	4–32K	2–5	5	50	200K		Process control
38 M-4030	1973 (P) 1974 (S)		128K	2	3	18	330K		Process control
39 M-400	1974	3							Process control
40 ES-1010	1972 (P) [in Hungary]	3	8–64K				10K		Data processing
41 ES-1020	1971 (P) 1972 (S)	3	64–256K	1	50–70	200–300	20K		Data processing
42 ES-1021	1972 (P) [in Czechoslovakia]	3	16–64K	1	25	175	40K		Data processing
43 ES-1030	1972 (P) 1973 (S)	3	122–512K	0·75	5–11	32–38	100K		Data processing
44 ES-1040	1973 (P) [in GDR]	3	256–1024K	0·45	1·4–2	7–8	300K		Data processing
45 ES-1050	1973 (P) (S)	3	128–1024K	0·4–0·6	0·65–2	2	500K		Data processing
46 ES-1060	Projected	3	256–2048K		0.5	1	1500K		Data processing

Sources:
R. Judy, in (Ed.) S. Wasowski, *op. cit.* (1970), p. 50; *Soviet Cybernetics Review*, January 1970; and sources cited in text.

Table 8A.2 *Some advanced American computers, 1946–73*

	Model	Date first produced	Generation	Internal storage K	Access time to internal storage*	Add time*	Multiply time*
A	Eniac	1949	1	1/50	1000	2000	2800
B	Whirlwind	1951	1	1		24	40
C	IBM 704	1955	1	4–32	12	24	12
D	IBM 705	1955	1	20–80	9	86	
E	IBM 7090	1959	2	32	2.2	4.4	
F	IBM 7030	1961	2	16–262	1	2	
G	IBM 7094 II	1964	2	32	1.4	2.8	
H	Philco 2000, Model 212	1963	2	32–64	1.5	0.6	
I	Control Data 6600	1964	2	32–131	1	0.3	
J	Univac 1108 II	1965	2	62–262	0.75	0.75	
K	IBM 360/90	1967	3	512–1024	0.75	0.18	
L	IBM 360/195	1971	3	1034–4096	0.054	0.054	

* – microsec.

Sources:
R. Judy, in (Ed.) S. Wasowski, *op. cit.* (1970), p. 50; *Computer Characteristics Review*, various issues.

Table 8A.3 *Some British computers, 1960–73*

	Model	Date first produced	Generation	Internal storage K	Access time to internal storage*	Add time*
a	AEI 1010	1960	2	4+	8.5	18
b	ICT/Atlas I	1962	2	16–262	2	1.6
c	English Electric Leo Marconi KDF 9	1963	2	4–32	6	1
d	GEC 90/300	1964	2	4–32	1.75	1.75
e	ICL System 4/70	1967	3	65–1048	0.7	1.1
f	ICL 1906A	1970	3	64–512	0.75	0.9
g	ICL 1906S	1972	3	131–524	0.3	0.6

* – microsec.

Source:
Computer Characteristics Review, various issues.

9. Military technology

Introduction

EARLIER STUDIES OF SOVIET MILITARY TECHNOLOGY

It is commonly argued that in the Soviet Union the level of military technology is higher than that of civilian technology. Forty years ago Trotsky wrote:

> where the purchasers are influential groups of the ruling bureaucracy the quality of the product rises above the average level . . . The most influential client is the war department. It is no surprise if the machinery of destruction is of better quality, not only than the objects of consumption, but also than the instruments of production.[1]

Foreign specialists have echoed this assessment of the relative levels of Soviet military and civilian technology. There may be disagreement about the exact relationship between military and civilian technology, or about the relative levels of Soviet and foreign military technology, but the general verdict is clear enough. Thus Sutton, at the end of his exhaustive study of Western technology and Soviet development, writes:

> Soviet innovation presents a paradox: an extraordinary lack of effective indigenous innovation in industrial sectors is offset—so far as can be determined within the limits of open information—by effective innovation in the weapons sectors.[2]

A similar assessment is to be found in much of the writing on the Soviet R and D effort.[3] Sometimes the assumption is implicit, sometimes explicit, but it

---

[1] L. Trotsky, *The Revolution Betrayed*, London (1937), p. 198.
[2] A. C. Sutton, *Western Technology and Soviet Economic Development 1945 to 1965*, Stanford, Cal. (1973), p. 361.
[3] For example, R. Perry, *Comparisons of Soviet and US Technology*, Rand R-827-PR, Santa Monica (1973), p. 35, writes: 'for a variety of reasons, the military R and D sector appears to be appreciably more efficient than its civil equivalent. At least for aircraft, the Soviet military

seems very rarely to be challenged. Given the importance of this assumption, surprisingly little effort has been made in the open literature to provide a systematic assessment of the relative levels of Soviet civilian and military technologies. The two case studies in this chapter—of tanks and intercontinental ballistic missiles (ICBMs)—are intended as a contribution to such an assessment. The case studies compare Soviet and foreign technologies; and it should be possible, by considering all the studies in this volume, to gain at least some indication of the relative levels of technology in the Soviet civilian and defence sectors.

Several comparative studies have been made of Soviet and foreign military technology. The United States Department of Defense has made public the results of some general investigations into the relative levels of Soviet and American weapons technology. According to Dr John Foster, the Director of US Defense Research and Engineering, estimates made in 1972 showed that the Soviet Union had technological superiority over the United States in 11 deployed weapons systems; that the Soviet Union had approximate parity with the United States in 4 systems; that the Soviet Union lagged behind in 17 systems. (The results are given in Table 9.1.) These results were based on a side-by-side comparison: for example, strategic air-defence interceptors were compared with similar aircraft on the other side but were not assessed in terms of their effectiveness in performing their assigned missions—for example, intercepting strategic bombers.

In 1973 Dr Foster declared that a recent study of the Soviet and American military-technological bases had shown that 'the technological superiority that the United States once possessed has been substantially reduced by the USSR'.[4] 167 areas were examined in the following broad categories:

1. Instruments, information and communication devices.
2. Weapons and weapon carriers.
3. Supporting science and technology.

The United States and the Soviet Union were estimated to be roughly equal in knowledge in 72 of these areas, and in application in 57. The United States was far ahead, or had some lead, in knowledge in 67 areas and in application in 73. The Soviet Union was far ahead, or had some lead, in knowledge in 28 areas, and in application in 37.[5] Among the conclusions drawn from this study were:

---

research and development system may actually be more effective than the US military research and development system.' See also N. Nimitz, *The Structure of Soviet Outlays on R & D in 1960 and 1968*, Rand R-1207-DDRE, Santa Monica (1974), pp. vi, viii; M. Checinski, *Osteuropa Wirtschaft*, 1975, No. 2, pp. 118–19; T. W. Wolfe, in *The Limitation of Strategic Arms. Hearings Before the Subcommittee on Strategic Arms Limitations Talks of the Committee on Armed Services*, US Senate 91st Congress, 2nd Session, 20 May 1970, Part 2, p. 60.

[4] *Department of Defense Program of RDT and E, FY 1974, Statement by DDR and E*, 93rd Congress, 1st Session, 1973, pp. 2–4.

[5] *Ibid.*, pp. 2–5. The areas considered were:

Weapons and weapon carriers	(66)	Land mobility	(7)
Aircraft	(16)	Conventional and nuclear weapons	
Missiles and space vehicles	(13)	(OTM)	(14)
Ships and submarines	(10)	High energy lasers	(1)

the altered relative status of the US and the USSR in national advancement evident in broad technological areas in 1950 and again in 1970 indicates that the Soviets have closed the 'technology gap' in several important respects.

Important US leads are in computer technology, integrated circuits, telecommunications, ship and submarine quieting techniques, and some designs of very strong fiber-reinforced composite materials.

Important Soviet leads are in chemical warfare defense techniques, high-performance integral rockets and ramjets, capability of land vehicles to cope with arctic conditions and difficult terrain, and aircraft maintainability.[6]

The studies on which these US Department of Defense conclusions are based have not been published. This is a serious drawback, for the categories of comparison are often so general as to be misleading if they are presented without further qualification.

The secrecy which surrounds military affairs in the West, and more especially in the Soviet Union, makes it difficult to obtain reliable data about weapons. Moreover, published information plays a role in the politics of deterrence and of defence budgets and is therefore liable if not to manipulation, then at least to biased interpretation. The danger of distortion is especially great where there is a strong political desire for clear and striking assessments of the military-technological threat, but the data on which the assessment is based are fragmentary and unreliable.[7]

Chemical warfare	(6)	communications devices	(49)
Supporting science and technology	(52)	Computers	(7)
Biomedical science	(11)	IR and acoustic sensors	(5)
Environmental science	(4)	ECM	(10)
Human resources science	(4)	Electron devices	(9)
Materials	(14)	Radar	(5)
Manufacturing processes	(13)	Telecommunications	(3)
Electrical power supplies	(6)	Lasers	(6)
Instruments, information and		Avionics	(4)

[6] *Ibid.*, pp. 2–6. In February 1976 the Director of US Defense Research and Engineering, Malcolm R. Currie, reported that the Soviet Union led the United States in: high-pressure physics, welding, titanium fabrication, high frequency radio-wave propagation, magneto-hydrodynamic power generation, anti-ship missiles, chemical warfare, artillery technology; the United States had a lead in: integrated circuit fabrication, computers, high-bypass-ratio turbofans, air-to-air missiles, numerically controlled machine tools, avionics, composite materials, inertial instrumentation, precision-guided weapons, satellite-borne sensor technology. Approximate parity existed in: high-yield nuclear weapons, aerodynamics, high-energy lasers. (*Aviation Week and Space Technology*, 6 February 1976, pp. 14–15.)

[7] Under Congressional pressure the Department of Defense de-classified papers which had been prepared in 1969 as a basis for an assessment of Soviet and American military technology. In spite of numerous deletions, these papers contain useful data on a wide range of weapons. They were nevertheless described in the *Congressional Record* as 'a rather mixed bag of assorted anecdotal accounts, containing useful and in some cases previously undisclosed data concerning comparative Soviet and American weapon systems, but not really providing a logical or intellectually sound basis for the kinds of explicit statements the DOD has been making about the Soviet technological threat and the size and rate of growth of Soviet military R & D' (*Congressional Record, Extensions of Remarks*, 4 August 1971, p. E8955). The papers are to be found on pp. E8955–63. An interesting illustration of the problems of assessing military technology is to be found in a report of the Brezhnev-Nixon summit meeting in July 1974: 'Mr Brezhnev led off with a Soviet analysis of the American nuclear arsenal. He emphasised the advanced American MIRV weapons, the overwhelming American superiority in bombers, and

Besides these general surveys of Soviet and United States military technology, some studies have attempted to assess in a systematic way the level of particular military technologies. A group at the Rand Corporation developed a methodology for quantifying technological trends, and applied this to a comparison of Soviet and American aircraft turbine engine technology between 1943 and 1971.[8] A set of parameters—performance measures and technical quality parameters—was selected to characterise the technology, and multiple regression analysis was used to estimate a trade-off surface of these parameters over time. On the basis of US data an equation was produced which provided a measure of the average technological trend, and the degree to which individual engines were ahead of—or behind—their time could be estimated by inserting the specific parameter values into the equation. Soviet data were then plotted against the American technological curve, and the results indicated that, apart from the earlier years, Soviet engines compared unfavourably with their American counterparts.

To provide a sound basis for comparison and avoid the 'index number problem', a Soviet technological curve was constructed and the American data plotted against it. This yielded similar results, showing that 'after 1967, all American engines are well in advance of the Soviet state-of-the-art trend in production hardware, sometimes by an indicated 4 to 6 years'.[9]

Even this result has to be treated with caution, however, because in the early years Soviet designers pursued objectives which were different from those of American designers or from those they themselves later adopted. The Soviet armed forces required high-performance, short-range interceptors rather than long-range fighters: until the late 1950s Soviet designers concentrated on high-thrust, large-volume engines which met their own design objectives but were 'appreciably more powerful and fuel hungry' than their American counterparts.[10] This point raises one of the most difficult methodological problems in comparing military technologies: it may be quite misleading to assume that designers of similar weapons are in fact aiming at the same goals. In this specific case, however, various other ways of comparing the data were used and the general conclusion remained that Soviet turbine engine technology lagged

---

Secretary Schlesinger's retargeting program—which, according to Brezhnev meant accepting the possibility of nuclear war. He expressed the view that the United States had a tremendous lead already, and was seeking to achieve a first-strike capability that would disarm the Soviet Union. He indicated that the Russians had to follow through with their program for putting multiple warheads on the newer missiles . . . The Americans were disheartened. One of those present said he thought that Brezhnev had been "sold a bill of goods" by the Soviet military. Subsequently, the Americans checked the Soviet assessment and discovered that it was technically justifiable, provided all the optimistic statements made by American admirals and generals about our new weapons were taken at face value and interpreted in what would be from the Russian point of view the worst possible fashion. On the next day, Sunday, Mr Nixon set out to reassure Mr Brezhnev. He went through the list of American strategic weapons, indicating weaknesses and strengths, the better to show that the United States had no intention—and, indeed, no rational interest—in starting any kind of nuclear war' (J. Kraft, *The New Yorker,* 29 July 1974, p. 70).

[8] See A. J. Alexander and J. R. Nelson, *Measuring Technological Change: Aircraft Turbine Engines*, Rand R-1017-ARPA/PR, Santa Monica (1972); R. Perry, *op. cit.*

[9] R. Perry, *op. cit.*, pp. 28, 32.

[10] *Ibid.*

Table 9.1 *Results of US Department of Defense comparison of Soviet and American military technologies in 1972*

The Soviet Union had technological superiority in the following deployed systems:
  anti-ballistic missile systems;
  fractional orbit ballistic missiles;
  (neither of these systems was deployed by the United States at the time)
  strategic air-defence interceptors;
  all aspects of civil and industrial strategic defence and recuperative planning;
  tactical anti-ship missiles;
  surface attack ships (excluding carriers);
  anti-aircraft and artillery systems;
  some armoured combat vehicles;
  medium and high altitude SAM (surface-to-air missile) air defences;
  surface-to-surface tactical missiles;
  heavy-lift helicopters.

Approximate technological parity was to be found in:
  tanks and anti-tank weapons;
  satellite tracking systems;
  satellite navigation systems;
  small arms.

The Soviet Union lagged in:
  ICBM guidance and penetration aids;
  strategic bombers;
  strategic submarines and SLBMs (submarine launched ballistic missiles);
  attack submarines;
  ASW (anti-submarine warfare) sensors and patrol aircraft;
  satellite communications systems;
  airborne surveillance sensors;
  defence-suppression weapons and systems;
  deep-strike tactical aircraft;
  aircraft carriers;
  guided ordnance;
  air-to-air superiority weapons;
  man-portable air-defence systems;
  close-support helicopters, aircraft and serial weapons;
  long-range logistic transports;
  artillery munitions.

Source:
Department of Defense Appropriations for Fiscal Year 1973. Hearings before a Sub-Committee of the Committee on Appropriations, US Senate, Part 1, US Government Printing Office, Washington DC, pp. 589–90.

behind American, and was, if anything, falling further behind.[11]

An analysis of Soviet and Western naval capabilities has been made at the Canadian Forces Maritime Warfare School.[12] This points out that two kinds of comparison are possible: side-by-side comparison of similar weapons systems, and the comparison of weapons systems in adversary situations. This is an important distinction:

> ... the infinite variety of adversary situations in modern warfare can ... make straightforward comparisons of comparative weapon systems capabilities either impossible, or alternately, misleading. To be sure, the anti-air (AA) weapons on one side may be compared to those on the other side—but the fact that such comparisons inherently disregard the adversary situation may lead to erroneous conclusions concerning the supremacy of one side over the other.
>
> In considering then, the example of a comparison of AA systems capabilities, it will be apparent that a meaningful comparison cannot be arrived at without considering the adversaries—in this case, the respective air threats. In such a comparison of apples with oranges and apples with bananas, the only realistic basis for comparison is the degree to which each side's appetite has been satisfied—in other words the degree to which each side's AA systems has [sic] met the challenge of the other side's airborne weapons and platforms.[13]

In spite of these methodological difficulties, the Canadian study reports the results of both adversary-situation and side-by-side comparisons.

The report of this study does not provide the data on which the comparisons were based, but the results can be set out briefly. It was found that 'in the aggregate area of hull form design, steel technology, and (possibly) overall systems engineering'[14] Western submarines were ahead, but the Soviet Union was closing the gap. In anti-submarine warfare too the Soviet Union lagged. The ability of Soviet ships, aircraft and submarines to detect and pinpoint a 'standard' submarine was found to be considerably less than that of Western forces. Their capability to attack, however, was only slightly less than that of Western forces: the Soviet Union had a superior capability to attack ASW surface forces, but fared less well against hunter-killer submarines. The Soviet Union was found to have a clear superiority in naval electronic warfare and in mine warfare. Soviet warships were found to have superior surface-to-surface capabilities, while Western surface-to-air capabilities were judged to be greater. In air-to-surface warfare no clear conclusion could be reached: Soviet aircraft had a considerably greater air-to-surface missile range than Western aircraft, but the latter could engage twice as many targets at shorter range. In

---

[11] *Ibid.*

[12] N. D. Brodeur, in (Eds.) N. MccGwire, K. Booth, J. McDonnell, *Soviet Naval Policy, Objectives and Constraints*, New York (1975), pp. 452–68. The term 'Western' here appears to mean 'NATO'.

[13] N. D. Brodeur, *loc. cit.*, p. 453.

[14] *Ibid.*, p. 455.

evaluating these differences account should be taken of the threat to be met: for example, it is presumably no accident that Western superiority in surface-to-air missiles is matched by the greater range of Soviet air-to-surface missiles.

The published report of the Canadian study is very brief, and provides too little information for its validity to be assessed; but, together with the Rand studies, it raises important methodological issues. The distinction between adversary-situation and side-by-side comparisons has already been noted. Another major distinction must now be drawn: that between military effectiveness and technological level. The Canadian study is concerned with military capabilities rather than levels of technology, and makes it clear that capabilities can be assessed properly only when the performance of a weapons system is examined within different tactical settings in order to see how well it performs the mission it is designed to perform.[15] It is important to note that there is no necessary relationship between level of technology and military effectiveness: a higher level of technology will not necessarily lead to a superior capability, while a better capability cannot be taken as evidence of a higher level of technology. Perry goes as far as to argue that the Soviet military R and D system is in some respects more effective than the American system, and that one consequence of this is the Soviet 'ability to create total systems with operational effectiveness not inferior to comparable US systems even though the Soviets are handicapped by inferior subsystems, such as engines and avionics'.[16]

Even if the substance of Perry's argument were wrong, the distinction implied here would remain valid. It is conceivable that, within certain limits, designers could create equipment not inferior to its foreign counterparts even though the level of technology embodied in the components parts was lower. In other words, design is the process whereby the whole is made greater than the sum of its parts; consequently it intervenes in the relationship between technological level and military effectiveness. Moreover, this argument can be extended from individual weapons systems to force structures: a more effective military force can be constructed from units that are individually less effective or embody a lower level of technology. This is an obvious point, since the effectiveness of a force is influenced by its size, deployment, military doctrine and so on. It is nevertheless an important point, although the relative role of quantitative and qualitative factors may be extremely difficult to determine.

The picture that emerges from these studies is a complex one. In general, the level of United States military technology is judged to be higher than the Soviet level, although the Soviet Union may lead in particular areas. In those areas where their categories of comparison coincide (anti-submarine warfare, submarine design and construction, surface-to-surface tactical naval missiles) the Canadian and Pentagon studies come to similar results. The picture can, of course, change quite rapidly in specific areas with the introduction of new

[15] The distinction between military effectiveness and level of technology is conceptually quite clear, but it might be misleading to hold to it too rigidly; while the operational mission may define what the appropriate technology is, the definition of the mission itself may be affected by the availability of particular technologies.

[16] R. Perry, *op. cit.*, pp. 37–8.

weapon systems. In the general surveys the impression is given that the Soviet Union is catching up and overtaking the United States, but in the one study where systematic comparison over time was made the Soviet Union was found to have lost an early lead to the United States.

What emerges more clearly, perhaps, from these studies is the need to take account of the distinctions between adversary-situation and side-by-side comparisons, and between military effectiveness and technological level. The methodological difficulties inherent in comparison cannot altogether be avoided, but their propensity to mislead can be minimised if these distinctions are borne in mind.

DESIGN PHILOSOPHY

Most studies of Soviet military technology suggest that the design and development of Soviet weapons are marked by several striking features, all of which indicate a coherent design philosophy. In the first place, there is wide agreement that Soviet designers concentrate on evolutionary progress, and continuous growth in small steps, while in the United States the tendency is to reach for larger, but less frequent, technological advances.[17] This means that Soviet progress comes as far as possible by modifying existing systems rather than by designing new weapons from scratch. Alexander points to the long series of modifications of the MiG-21, which incorporated new versions of engines, armaments and radar as they became available.[18] MccGwire finds a similar phenomenon in the design and development of naval weapons: the progressive introduction into service of a weapons system at different stages in its development, and long before it has reached the point where it would be deployed in the United States.[19]

There is wide agreement also that Soviet military design stresses commonality; Soviet designers try to use standardised parts and subsystems as much as possible.[20] For example, the Su-7 and Su-9 had common fuselages and tails, although the wings, armament and equipment were chosen for their different roles: the former was a ground attack fighter, the latter an all-weather interceptor.[21] A third and related feature of Soviet military equipment is that it is designed to 'minimum acceptable requirements'.[22] Designers concentrate on the simplest way of meeting operational requirements, and thus on the simplest appropriate technology. This is brought out strongly by two American naval architects in their comparative study of Soviet and United States destroyers.[23]

[17] See, for example, *DOD Program of RDT and E, FY 1974, Statement by DDR and E*, 93rd Congress, 1st Session, 1973, pp. 2–4; R. Perry, *op. cit.*, p. vi.
[18] A. J. Alexander, *R & D in Soviet Aviation*, Rand R-589-PR, Santa Monica (1970), p. 22.
[19] M. MccGwire, in (Ed.) M. MccGwire, *Soviet Naval Developments, Capability and Context*, New York (1973), p. 184.
[20] N. Nimitz, *op. cit.*, p. viii.
[21] A. J. Alexander, *op. cit.*, p. 22.
[22] A term used by J. Kehoe and H. A. Meier in a comparative study of Soviet and US destroyers, in *Soviet Naval Developments—III, Seminar 8–11 September 1974, Summary of Proceedings*, prepared by K. Booth, p. 57. A similar point is made by N. Nimitz, *op. cit.*, p. 57. See also J. W. Kehoe Jr, in *US Naval Institute Proceedings*, August 1975, pp. 57–65.
[23] J. W. Kehoe and H. A. Meier, in K. Booth, *op. cit.*, pp. 55–7.

It is to be found in other areas too: in aircraft design and radar technology, for example.[24] The emphasis on simplicity appears to be a matter of choice rather than of necessity, for where more sophisticated equipment is needed it can often be produced.[25]

One should beware of reading too much into these design practices. They must not be taken to imply that the defence sector produces few models of military equipment; in fact economising design practices seems to enable the defence sector to design and develop more models than would otherwise be possible.[26] Moreover, it should not be thought that technological progress comes through evolutionary change alone:

> the design philosophy of incremental change, marginal advance and design inheritance, if followed rigidly, would eventually lead to technological stagnation. Since these concepts are an established part of the system, discontinuous change must be sought through temporary deviations from the normal patterns in the form of crash programs, the establishment of problem-oriented *ad hoc* organisations and committees, temporary suspensions of the usual procedures, and high level political intervention.[27]

A similar picture emerges from MccGwire's reconstruction of Soviet naval shipbuilding programmes.[28] This suggests that decisions are handed down, and that inertia is the guiding principle until and unless the centre intervenes. Further, where existing programmes have had to be modified this has been done within clearly defined constraints. As a result, modification appears to encourage improvisation in the sense of fitting together elements that were not originally designed for each other: hence commonality and economy in the design and use of components.[29]

CONCLUSION

The two case studies in this chapter should provide some indication of the level of Soviet military technology, but they cannot give a comprehensive picture: the field is too large and complex to be assessed in this way. Table 9.1 above

---

[24] A. J. Alexander, *op. cit.*, pp. 23–4; *Congressional Record, loc. cit.*, p. E8962.

[25] This point is made about Soviet radar, *ibid.*

[26] Thus R. Perry, *op. cit.*, p. 13, writes that 'in recent years the aircraft development establishment in the Soviet Union has turned out more models of military aircraft than has the United States, has carried more of them to production, and has built them in larger numbers'. Details are given in *ibid.*, pp. 18, 19.

[27] A. J. Alexander, *op. cit.*, pp. 24–5.

[28] M. MccGwire, *loc. cit.*

[29] This chapter is concerned with the level of technology, rather than with explanations of why that level is what it is. Nevertheless, it is worth noting that some writers see these economising design practices as peculiar to the defence sector in the Soviet Union, and as evidence that the Ministry of Defence is a cost-conscious customer, able to impose its wishes on the design, development and production of weapons; and they see in this, rather than in an abundance of resources, the reason for the relatively higher level of technology in the defence sector. Thus N. Nimitz, *op. cit.*, p. viii: 'there is considerable evidence that the defense/space sector works for an economising customer. Defense hardware stresses design commonality, which makes development, production, and maintenance all cheaper; by contrast, the civilian sector is known for its propensity to reinvent not just the bicycle, but also the nuts and bolts that would hold it together.' See R. Perry to the same effect, *op. cit.*, p. 9.

gives some idea of the range of items to be found under the rubric of 'military technology'. Although reliable figures are not available for the weapons developed and produced by different countries, it is clear that the United States and the Soviet Union are in a class by themselves in terms of the range of military equipment they produce. The United Kingdom, France and the Federal Republic of Germany (and China too perhaps) are the next largest weapons producers.[30]

A distinction can be drawn between military and civilian technology in so far as the former is specifically designed for use in war. There is nevertheless a considerable overlap between the two, and the distinction becomes more blurred the closer one moves towards the research end of the research-development-production cycle.[31] Moreover, much of the development work on military technology may be indistinguishable from the civilian effort: this is true of computer technology, for example.[32] It seems to follow from this that the more advances in military technology come to depend on a large R and D effort, the greater will be the overlap between civilian and military. Any assessment of the relative levels of civilian and military technology will have to take account of distinctions within the large and complex field of military technology and bear in mind that military technology may depend on much the same research base and some of the same development effort as civilian technology.

Medium tanks

Tanks were made possible by the combination of armour, the machine-gun and the traction and internal combustion engines. They made their first appearance on the battlefield in September 1916 in the Battle of the Somme. Although their impact was at first slight, tanks showed during the First World War that they could overcome the immobility of trench warfare: they could break through barbed-wire defences; their armour plate neutralised the small arms bullet; they could cross open terrain and carry effective weapons into defended enemy positions. After the war several military theorists saw that tanks had made possible a new type of mobile warfare. These theories were adopted and put into practice in Germany and the Soviet Union in the 1930s. Elsewhere the development of tanks and tank forces was desultory, although much of the pioneering work on tank warfare had been done in Britain in the 1920s and early 1930s. Tank forces played a major role in the Second World War, and there has been little sign of their becoming obsolete in the post-war

[30] See the figures in *World Armaments and Disarmament, SIPRI Yearbook 1972*, Stockholm (1973), pp. 184–5.
[31] (Ed.) N. A. Lomov, *Scientific-Technical Progress and the Revolution in Military Affairs (A Soviet View)*, Washington DC (1974). (A translation of *Nauchno-tekhnicheskii progress i revolutsiya v voennom dele* (1973), p. 30.)
[32] F. A. Long, in (Eds.) B. T. Feld *et al., Impact of New Technologies on the Arms Race*, Cambridge, Mass. (1971), p. 278. This is not to say that military computers will in fact come from the same development programme; it is conceivable that military development work will be carried on in isolation from civilian work. See also on this general area, M. Leitenberg, *International Social Science Journal*, 1973, No. 3.

period, although the October War of 1973 showed the increased effectiveness of anti-tank weapons.[33]

The main combat features of the tank are its firepower, protection and mobility. The task of the tank designer is to find the 'optimal combination' of these properties, while maintaining certain important parameters, for example:

— the required mean specific pressure at the track to ground interface, otherwise muddy terrain or soft sand will create difficulties;
— the ratio between the bearing surface and the width of track, in order to provide optimal turnability;
— the dimensions of the vehicle must take account of the proposed mode of transport, for example, rail or air.[34]

These combat properties of the tank are not necessarily to be improved simultaneously: better armour protection may lead, for example, to lower mobility. In the interwar years different types of tank were developed in which the three combat properties were given different orders of priority. In Britain, infantry tanks were developed for use in conjunction with the infantry; mobility had low priority in their design. Cruiser tanks were developed with greater emphasis on speed and mobility. In the post-war period, however, this differentiation has given way to the idea of a main battle tank (MBT). Most countries now concentrate on a single type of battle tank, as well armed as possible (there are also light tanks which are used mainly for reconnaissance). Nevertheless, since tactical doctrine varies from one army to another, different emphases are still to be found in tank design.

SOVIET TANK DEVELOPMENT

No tanks were produced in Russia before 1917. During the Civil War the Red Army used Western-built tanks which they had captured from the Whites. The first Soviet tank was completed in August 1920. Experimental work was carried on during the 1920s, but few tanks were produced. In 1928 the Red Army had no tankettes and only 92 tanks, most of them light.[35] The military re-equipment programme which was launched at the end of the decade placed great emphasis on tanks and armoured forces. In July 1929 the post of *nachal'nik vooruzheniya RKKA* (Chief of Armament of the Red Army) was created, with responsibility for directing the process of re-equipment. At the same time I. A. Khalepskii, head of the Military-Technical Administration (*voenno-tekhnicheskoe upravlenie*) of the Red Army since its foundation in

---

[33] See R. M. Ogorkiewicz, *Armoured Forces*, London (1970); and *Design and Development of Fighting Vehicles*, London (1968); V. D. Mostovenko, *Tanki* (1958); P. A. Rotmistrov, *Vremya i tanki* (1972); K. Macksey and J. H. Batchelor, *Tank. A History of the Armoured Fighting Vehicle*, London (1970); H. Guderian, *Panzer Leader*, London (1974); *Voprosy strategii i operativnogo iskusstva v sovyetskikh voennykh trudakh (1917–40)* (1965), pp. 551–620; for a military analysis of the 1973 Middle East War see A. H. Farrar-Hockley in *The Arab-Israel War, October 1973 — Background and Events*, Adelphi Paper No. 111, International Institute for Strategic Studies, London (Winter 1974–75).
[34] A. Kh. Babadzhanyan, *Tanki i tankovye voiska* (1973), p. 39.
[35] *Istoriya vtoroi mirovoi voiny 1939–45*, Vol. 1 (1973), p. 270.

1924, became head of the newly created Administration of Mechanisation and Motorisation of the Red Army (*upravlenie mekhanizatsii i motorizatsii RKKA*) and chief of the armoured forces (*avtobronetankovye voiska*).[36]

At the end of 1929 Khalepskii set out on a tour of Europe and the United States, where he purchased several tank models that were to serve as the basis for Soviet designs in the 1930s.[37] Nevertheless, the Soviet tank production programme got off to a slow start because of the

> acute shortage of qualified cadres, the poor supply of high-quality steels, instruments and ignition apparatus to tank production, and the delay in specialisation and cooperation by the motor-tractor industry with tank-building.[38]

In 1930 and 1931, 910 tanks were produced. From 1932 to 1937, annual production averaged 3,255 tanks. In 1938 the number dropped to 2,271 but it increased again in 1939 and 1940 to almost 3,000 tanks a year. In September 1941, three months after the German invasion, a separate People's Commissariat of the Tank Industry was set up, with the tasks of evacuating tank factories to the Urals and speeding up tank production. About 1,780 tanks were produced in the first six months of 1941, and over 4,000 tanks in the second six months. The success of industrial mobilisation is shown by the production totals for subsequent years (see Table 9.2).[39]

In the 1930s Soviet tank production consisted primarily of the T-26 and BT light tanks, the T-28 medium tank, and the T-35 heavy tank. By the end of the second five year plan in 1937, 6,780 T-26s and 5,000 BTs had been produced; only several hundred T-28s, and several dozen T-35s were built.[40] (See Table 9.3 for the main characteristics of these tanks.) The turning-point in Soviet tank design came with the creation of the KV (Klimenti Voroshilov) heavy tank in 1939 and the medium T-34 in 1940. (The main characteristics are given in Table 9.3.) In August 1938 a meeting on tank construction in the Central Committee decided to procure new single-turreted medium and heavy tanks, with good armour protection, high manoeuvrability and powerful armament.[41] In 1939 an experimental model of the KV was produced. On their own initiative M. I. Koshkin and A. A. Morozov produced the T-32, a prototype of a pure tracked medium tank. In August 1939 the Main Military Council decided to approve the production of such a tank, and in 1940 the T-34 appeared. The initial production of these tanks was slow. In 1940 only 243 KVs and 115 T-34s were produced, although the plan was for 600 T-34s; in the first six months of 1941 393 KVs and 1,110 T-34s were turned out. The T-34 and

---

[36] *Ibid.*, p. 258; P. A. Rotmistrov, *op. cit.*, p. 45; M. V. Zakharov, *Voenno-istoricheskii Zhurnal*, 1971, No. 2, p. 4; *50 let vooruzhennykh sil SSSR* (1969), pp. 200–2.

[37] A. Sella, *Soviet Studies*, 1975, No. 2, p. 248.

[38] *Istoriya vtoroi mirovoi voiny 1939–45*, Vol. 1 (1973), p. 260.

[39] G. S. Kravchenko, *Ekonomika SSSR v gody velikoi otechestvennoi voiny (1941–1945 gg)* (1970), pp. 82, 172, 175, 285; V. A. Anfilov, *Bessmertnyi podvig* (1971), pp. 90, 95; P. A. Rotmistrov, *op. cit.*, pp. 88, 91; *Istoriya vtoroi mirovoi voiny*, Vol. 1 (1973), p. 214.

[40] P. A. Rotmistrov, *op. cit.*, p. 46; G. S. Kravchenko, op. cit., pp. 73–4.

[41] *Istoriya vtoroi mirovoi voiny*, Vol. 3 (1974), p. 384.

KV were modified in the course of the war; the T-34 was fitted with an 85 mm gun in 1943, thus becoming the T-34/85, and the KV became the IS (Iosef Stalin) tank and by 1944 was armed with a 122 mm gun.[42] (The main combat features of Soviet and German tanks in the years 1943–5 are given in Table 9.4.)

Soviet tanks in the 1930s were based on the models purchased abroad by Khalepskii. Mostovenko acknowledges this, but defends Soviet policy:

> in the foreign press the fact is often underlined that in 1931 we acquired abroad the 6-ton Vickers tank, the tracked-wheeled Christie tank and the Carden-Lloyd tankette, on the basis of which were created the T-26 and BT tanks and the T-27 tankette which became our army's equipment... There is no point in entering into polemics on this question, inasmuch as the practice of our tank construction, which created before the Second World War the best tanks in the world, is a more than convincing reply to such fantasies; let us merely note the following points:
>
> 1. The tanks acquired abroad were not those which received general recognition and entered the armoury, but models rejected by the foreign military leaders. The Vickers 6-ton tank and the Carden-Lloyd tankette did not become part of the British Army's equipment, while the Christie tanks were not approved by the US Army.
> 2. Similar tanks could have been built by us, but with great expenditure of time, since in our country the cadres of designers and production workers had not yet been trained. The international situation at that time forced our industry to set going in the shortest space of time the production of the necessary armoured equipment.
> 3. The design of all these tanks and especially of the Christie tank was seriously altered. The turrets were designed anew with the armament mounting, turning mechanisms and the bearing support, as was the main clutch, etc.[43]

It is generally agreed that with the development of the KV and T-34 tanks Soviet tank design outstripped the foreign experience on which it had hitherto relied.

The T-34 was recognised as the best medium tank in the world; its 'combination of mobility, protection and gunpower placed it well ahead of other tanks'.[44] The German tank commander General Guderian wrote of the tank battle north-east of Orel on 11 October 1941 that:

> numerous Russian T-34s went into action and inflicted heavy losses on the German tanks. Up to this time we had enjoyed tank superiority, but from now on the situation was reversed. The prospect of rapid decisive victories

---

[42] V. D. Mostovenko, *op. cit.*, p. 108; *Sots. Ind.*, 11 and 12 February 1975.
[43] V. D. Mostovenko, *op. cit.*, p. 95.
[44] R. M. Ogorkiewicz, *Design and Development of Fighting Vehicles*, London (1968), p. 34.

was fading in consequence. I made a report on this situation, which for us was a new one, and sent it to the Army Group; in this report I described in plain terms the marked superiority of the T-34 to our Panzer IV and drew the relevant conclusions as they must affect our future tank production.[45]

The T-34 was an offspring of the BT (Christie) series through several intermediate stages. It had a 76 mm gun with the high muzzle velocity for that time of 662 m/sec. It had a powerful diesel engine, the V-2, which could develop 500 bhp. This had been adapted from an aeroengine and its design completed in 1935. Experimental models were tested in one of the BT-5 tanks, and after further development the engine was installed in BT-7M tanks in 1938; the same engine was used in the KV tank. The diesel engine made it possible significantly to increase the range of the tank with the same fuel tank capacity, to simplify servicing, to reduce the risk of fire and to lessen interference with the radio. Other features of the T-34 were the large diameter track wheels, and the wide tracks which gave low ground pressure and thus good mobility. The T-34 had sloping armour which lessened the effect of anti-tank shells. Designs for welded hulls were worked out on the basis of electric arc welding techniques developed in 1939. The design was simple and thus suitable for mass production.[46]

The Germans responded at once to the appearance of the T-34. Officers at the front wanted a copy of the T-34 to be produced in Germany, but this was not feasible, because raw materials were in short supply and because essential elements—in particular the aluminium diesel engines—could not be mass-produced quickly. Instead, work was to continue on the Tiger heavy tank and to begin on the Panther.[47] Both these tanks appeared on the battlefield before the end of the war. It has been claimed that the Panther was greatly superior to other medium tanks when it appeared in 1943; it had greater speed and manoeuvrability than the T-34, and its 75 mm gun was 70 calibres long and had a muzzle velocity of almost 1,000 m/sec.[48] The rush to produce these tanks, however, made them often unreliable.[49] Soviet writers do not deny the excellence of the Panther, but they point, quite rightly, to the influence of Soviet tanks on its design. Besides, they assert the superiority of Soviet heavy tanks over both the Tiger and the Panther.[50]

---

[45] H. Guderian, *op. cit.*, pp. 237–8; V. A. Anfilov, *op. cit.*, p. 93, quotes another German general to the effect that the T-34's shells could pierce the armour of the German tanks at 1,500–2,000 m, while German tanks could strike Soviet tanks only at 500 m and could destroy the T-34 only if the shells struck the side or the rear.

[46] V. D. Mostovenko, *op. cit.*, pp. 110–12; the automatic flux-welding of tank armour was begun in 1942, before it was used in Germany or the USA; see E. Paton, *Reminiscences*, Moscow, p. 241.

[47] H. Guderian, *op. cit.*, p. 276.

[48] R. M. Ogorkiewicz, *op. cit.* (1968), p. 37.

[49] K. Macksey and J. H. Batchelor, *op. cit.*, pp. 118, 128–9.

[50] V. D. Mostovenko, *op. cit.*, pp. 147–8; G. S. Kravchenko, *op. cit.*, p. 47. Marshal Konev writes in a final assessment of the T-34 that 'from the beginning to the very end of the war there was not a better tank in any way . . . It was highly manoeuvrable, compact, small and low-built; these features made it less vulnerable and easier to conceal. To this we must add its high cross-country ability, its good engine and rather good armour. True, the early T-34 had an insufficiently powerful gun, but when this was replaced by a new and excellent 85 mm gun it destroyed all enemy tanks except the Royal Tiger' (*Year of Victory*, Moscow (1969), p. 111). In his *Western technology*

SOVIET AND WESTERN TANKS 1950–75

There have been no revolutionary changes in tank design since the end of the Second World War. In this period the Soviet Union has produced four medium tanks—the T-34/85, T-44, T-54/55 and T-62 (see Table 9.9)—and two heavy tanks—the IS-III and the T-10 (see Tables 9.5 and 9.6); the production of heavy tanks ceased in the 1950s. It is now reported that a new tank—called the T-72—is entering service, but it is not clear from the available evidence how far it differs from the T-62. All these tanks have used basically the same V-12 diesel engine with modifications to increase the horsepower when necessary; the T-54/55 and T-62 tanks have used the same torsion bar/flat track suspension system.[51]

In the West several main battle tanks (MBTs) have been produced since the war: the Centurion, Chieftain, Leopard, AMX-30, S-tank and the M46 to M60 series. These tanks can be divided into two generations: the Centurion, M46, M47 and M48 belong to the first post-war generation, while the other tanks were introduced in the 1960s (the main combat features of all these tanks are set out in Tables 9.5–9.7). Marshal Rotmistrov has pointed to the main design features of the Western tanks introduced during the 1960s:[52]

—classical layout with unchanged crew strength of four;
—main armament provided by 105–120 mm calibre, rifled guns, with new armour-piercing and high explosive shells;
—increase in the maximum and average rate of march, owing to an increase in specific power and use of better running gear, mechanical and hydro-mechanical transmission;
—lowering of weight without major weakening of armour protection, reduction in overall dimensions, mainly height;
—increase in basic ammunition load up to 50–63 rounds;
—protection for crew against the effects of nuclear weapons;
—increase in cruising range up to 400–500 km, as a result of greater fuel tank capacity and use of more economical diesel and multifuel engines;
—equipment for underwater driving, up to a depth of 4–5 m;
—infra-red sights and observation devices for night combat;
—greater reliability and simplified maintenance of main units and assemblies;
—greater first-round hit probability as a result of installing range finders, spotting guns and better sights.

---

and Soviet Economic Development 1930 to 1945, Stanford, Cal. (1971), pp. 240–2, 243, Sutton devotes a chapter to the military industries. His account of Soviet tank development stresses the Soviet debt to foreign designs, but makes little reference to the T-34 and KV tanks on which the Soviet claims to successful tank design rest. He writes that the T-32 was 'the basic Soviet tank of World War Two', although it was in fact only a prototype of the T-34. (This may, of course, be a misprint.) He has a table entitled *Soviet Tanks and their Western Origins 1930–45* which makes no mention of the T-34 or KV; the table is taken from *Oberkommando der Wehrmacht* records of March 1941—months before the German forces encountered the T-34.

[51] Defense Department's 1969 Comparison of United States and Soviet Weapon Systems and Technology, in *Congressional Record, Extensions of Remarks*, 4 August 1971, p. E8955.
[52] P. A. Rotmistrov, *op. cit.*, p. 227.

The latest Western tanks rely, as before, on steel armour, gun armament, piston engines, torsion suspension, mechanical and hydromechanical transmission. The two most innovative designs are the turretless Swedish 'S' tank, which has a crew of only three, and the US M60 A2, which is armed with the Shillelagh, a 152 mm tube capable of firing either high explosive rounds or a Shillelagh anti-tank missile.

No single criterion exists for the evaluation of tank design, but there is general agreement that the three major combat features of the tank must be taken into account: protection, firepower and mobility. These in turn can be broken down into the following categories:

Protection: armour;
NBC (nuclear, biological, chemical) protection.

Firepower: fire control;
calibre;
ammunition type;
ammunition load.

Mobility: speed;
range;
amphibious capability.

Soviet and Western tanks will be considered under these headings.[53] I have relied heavily on R. M. Ogorkiewicz's basic work, *Design and Development of Fighting Vehicles*, in this analysis.

Table 9.2 *Production of tanks during the Second World War*

Year	Germany	Britain	USA	USSR	Japan
1939	249[1]	969[1]	—[3]	2986[2]	462[1]
1940[1]	1460	1399	331	2794	1023
1941[1]	3256	4841	4052	6590	1024
1942[1]	4278	8611	24997	24668	1165
1943[1]	5966	7476	29497	20000	776
1944	9161[1]	n.a.	17565[1]	17000[1]	342[1]

Sources:
1 R. M. Ogorkiewicz, *Design and Development of Fighting Vehicles*, London (1968), p. 36.
2 *Voenno-istoricheskii Zhurnal*, 1960, No. 3, p. 22.
3 H. C. Thomson and L. Mayo, *The Technical Services. The Ordnance Department: Procurement and Supply*, Washington DC (1960), p. 225.

Note:
In the early part of the war a large proportion of Soviet output consisted of light tanks (T-50, T-60, T-70); in 1942 about one third of the tanks produced were light, but their production ceased in 1943. During the war as a whole 10·8 per cent of the tanks produced were heavy (KV and IS), 70·4 per cent were medium (T-34) and only 18·8 per cent were light. (G. S. Kravchenko, *op. cit.*, pp. 75, 286–7.)

[53] These headings are derived from the discussion in *Congressional Record* (see note 51).

Table 9.3 Soviet tanks of the 1930s

	T-26	BT-7M	BT-7	T-34	KV-1
Year of entry into service	1937	1939	1935	1940	1941
Weight (t)	10·5	14·6	13·8	26·5	47·5
Crew	3	3	3	4	5
Armament	1 × 45 mm 2 × 7·62 mm MG 1 × 7·62 mm AAMG	1 × 45 mm 2 × 7·62 mm MG 1 × 7·62 mm AAMG	1 × 45 mm 1 × 7·62 mm MG —	1 × 76 mm 2 × 7·62 mm MG —	1 × 76 mm 2 × 7·62 mm MG 1 × 7·62 mm AAMG
Armour thickness					
Hull, front and side	15–15 mm	20–13 mm	20–13 mm	45–45 mm	100–75 mm
Turret	15 mm	15 mm	15 mm	45 mm	95 mm
Engine (hp)	90	400	400	500	600
Max. speed (km/h)	30	60/86[a]	55/73[a]	51	35
Road range (km)	200	600–700	375–500	370	250

Source:
Istoriya vtoroi mirovoi voiny, Vol. 3 (1974), p. 420.

Notes:
a – second figure denotes speed on wheels.
MG – machine gun(s).
AAMG – antiaircraft machine gun(s).

Table 9.4 *Basic combat features of Soviet and German tanks 1943–45*

	Type	Weight (tons)	Armour (mm)	Armament	Max speed (km/h)
Soviet	T-34	28·5	45–50	85 mm gun and 2 MG	55
	KV	46	40–100	122 mm gun and 3 MG	35
	IS-3	46·5	60–120	122 mm gun, 3 MG, 1 AAMG	40
German	Pz Kpfw IV	24	10–50	75 mm gun and 2 MG	30
	Pz Kpfw V (Panther)	45	17–85	75 mm gun and 2 MG	50
	Pz Kpfw VI (Tiger)	56	28–100	88 mm gun and 2 MG	44

Source:
G. S. Kravchenko, *Ekonomika SSSR v gody Velikoi Otechestvennoi Voiny* (1970), p. 289.

Table 9.4a *T-50, T-60 and T-70 tanks*

	T-50	T-60	T-70
Weight (t)	13·5	5·75	9·05
Crew	4	2	2
Armament	45 mm	20 mm	45 mm
Armour thickness (max.)	37 mm	15 mm	60 mm
Engine (hp)	300 (diesel)	70 (petrol)	2 × 70 (petrol)
Max. speed (km/h)	60	43	51

Sources:
V. D. Mostovenko, *op. cit.*, pp. 113–141.
J. Milsom, *Russian Tanks 1900–1970*, London (1970), p. 165.

Note:
Units and assemblies already in use in certain trucks were selected for inclusion in the T-60 and T-70 (V. D. Mostovenko, *op. cit.*, p. 141).

PROTECTION

Armour. There exist various attitudes to armour; the Chieftain embodies the view that heavy armour is required for tank duels, while the AMX-30 reflects the view that thin armour is adequate since heavy armour will not increase the chances of survival on the battlefield. In general, however, the thickness of medium tank armour had not, until recently, increased greatly since the Panther, the frontal plate of which was 80 mm thick and inclined at 55 degrees to the vertical. More attention had been given to the distribution and sloping of the armour in order to increase ballistic protection. Soviet tank design had not moved to either of the extreme attitudes to armour (see Table 9.8). The most recent US and FRG battle tanks, however, are much heavier than their predecessors (see Table 9.7), and this reflects a change in the priority given to armour protection which has resulted from the development of composite

Table 9.5 *Combat features of battle tanks, c. 1950*

Parameters	T-34/85	IS-III	Centurion Mk.3[3]	M46[4]	M47
Entry into service	1944[1]	1945[6]	1948	1949	1951[7]
Combat weight (t)	32[1]	45·8[2]	50	44	46·2[5]
Crew	5[1]	4[2]	4	5	5[5]
Power-to-weight ratio (hp/t)	15·6[1]	12[2]	12·8	18·4	17·5[5]
Ground Pressure (kg/cm^2)	0·81[1]	0·82[2]	0·89	n.a.	0·97[5]
Armament:					
gun calibre (mm)	85[1]	122[2]	83·4 (20 pounder)	90	90[5]
MG (mm)	2 ×[1]	1 × 7·62[2]	1 × 7·62	n.a.	2 × 7·62[5]
AAMG (mm)		1 × 12·7[2]		n.a.	1 × 12·7[5]
Ammunition load (main round)	56[1]	28[2]	65	n.a.	71[5]
Stabilisation	No[1]	No[2]	Yes	No	No[5]
Engine (bhp)	500[1]	550[2]	640	810	810[5]
Max. speed (km/h)	53	40[2]	34	48	48[5]
Fording depth (m) prepared/unprepared	/1·30[8]	/1·30[2]	/1·40	n.a.	/1·22[5]
Night vision	No[1]	No[2]	n.a.	n.a.	IR (driver)[5]
Road range (km)	300/400[a]	190[2]	120	113	129[5]

Sources:
1 F. Wiener, *Die Armeen der Warschauer—Pakt—Staaten*, Munich (1971), p. 134.
2 F. M. von Senger und Etterlin, *Taschenbuch der Panzer*, Munich (1969), pp. 598–9.
3 F. M. von Senger und Etterlin, *The World's Armoured Fighting Vehicles*, London (1962), p. 282–3.
4 F. M. von Senger und Etterlin, *Taschenbuch der Panzer*, Munich (1969), pp. 456–7.
5 *Ibid.*, pp. 601–2.
6 F. Wiener, *op. cit.*, p. 136.
7 F. M. von Senger und Etterlin, *Taschenbuch der Panzer*, Munich (1969), pp. 456–7.
8 J. Milsom, *Russian Tanks, 1900–1970*, London (1970), p. 172.

Note:
a – the second figure refers to range with external fuel tanks.
IR – infra-red.

forms of armour; this change is not yet reflected in Soviet design. The material most commonly used for armour is nickel-chrome-molybdenum steel, although a manganese-silicon-chromium-nickel alloy has been used, especially in the Soviet Union. Cast steel armour is normal for turrets. For the hulls the steel is in the form of homogeneous, machinable quality rolled plate with an ultimate tensile strength of 140,000 to 160,000 lb/in^2 and a Brinell hardness number of around 300.[54] In this way the steel can best absorb the energy of heavy, high velocity projectiles without cracking, and at the same time resist their penetration. The Brinell hardness of Soviet armour was higher, but has now been reduced to 300, presumably to cope with new developments in anti-tank ammunition.[55]

[54] R. M. Ogorkiewicz, *Design and Development of Fighting Vehicles*, London (1968), p. 83.
[55] *Congressional Record, loc. cit.*, p. E8955.

Table 9.6 *Combat features of battle tanks, c. 1960*

Parameters	T-54/55 (T-55 features)	Centurion Mk. 9	M48 A2	M60 A1	T-10
Entry into service	1954[8] (T-55: 1957/8)[1]	1959[10]	1957[9]	1962[7]	1957[6]
Combat weight (t)	36[1]	52[2]	47.6[3]	48.1[4]	50[5]
Crew	4[1]	4[2]	4[3]	4[4]	4[5]
Power-to-weight ratio (hp/t)	16[1]	12.7[2]	18.2[3]	15.6[4]	11/14[5]
Ground pressure (kg/cm^2)	0.83[8]	0.94[2]	0.84[3]	0.78[4]	0.74[5]
Armament:					
gun calibre (mm)	100[1]	105[2]	90[3]	105[4]	122[5]
MG (mm)	2 × 7.62[1]	—[2]	1 × 7.62[3]	1 × 7.62[4]	1 × 12.7[5]
AAMG (mm)		1 × 7.62[11]	1 × 12.7[3]	1 × 12.7[4]	1 × 12.7[5]
Ammunition load (main round)	42[8]	64[2]	60[3]	63[4]	45[5]
Stabilisation	Yes[1]	Yes[11]	No[3]	No[4]	No[5]
Engine (bhp)	580[1]	650[2]	865[3]	750[4]	550/700[5]
Max. speed (km/h)	50[1]	34[2]	48.3[3]	48.3[4]	45[5]
Fording depth (m) prepared/unprepared	5.0/1.7[1]	amphib/1.07[11]	/1.22[3]	/1.22[4]	/1.30[5]
Night vision	IR[1]	IR[11]	IR[3]	IR[4]	IR[5]
Road range (km)	350[1]	190[2]	257[3]	498[4]	350[5]

Sources:
1 F. M. von Senger und Etterlin, *Taschenbuch der Panzer*, Munich (1969), pp. 286–7; 598–9.
2 *Jane's Weapons Systems 1973–74*, London (1973), p. 274.
3 F. M. von Senger und Etterlin, *op. cit.* (1969), pp. 601–2.
4 *Ibid.*
5 *Ibid.*, pp. 598–9.
6 F. Wiener, *op. cit.*, p. 137.
7 *Jane's Weapons Systems 1971–72*, London (1971), p. 247.
8 F. Wiener, *op. cit.*, p. 132.
9 R. M. Ogorkiewicz, *Design and Development of Fighting Vehicles*, London (1968), p. 88.
10 *Ibid.*, p. 43.
11 F. M. von Senger und Etterlin, *op. cit.* (1969), pp. 588–9, gives data for the Mk. 10, which apply also to the Mk. 9.

NBC protection. Soviet writing makes it clear that tank operations may have to be conducted in zones of radioactive and chemical poisoning. Exercises in preparation for operations on a nuclear battlefield began in 1954, and since then the Soviet armed forces have paid particular attention to training for nuclear warfare.[56] In 1955 US armoured units operated in the vicinity of nuclear explosions in Nevada, and their vehicles provided protection against blast and radioactivity.[57] But NBC protection is 'the area in which Soviet tanks excel'.[58] The Soviet Union has

[56] *50 let vooruzhennykh sil SSSR* (1968), p. 502; A. Kh. Babadzhanyan, *op. cit.*, chapter 6.
[57] R. M. Ogorkiewicz, *op. cit.* (1968), p. 41.
[58] Steven Canby, *The Alliance and Europe: Part IV Military Doctrine and Technology*, Adelphi Paper No. 109, London (Winter 1974–5), p. 4.

(1) designed tanks with the minimum cross section to reduce dynamic pressure damage (e.g., her T-54, T-55 and T-62 tanks are about a metre lower in silhouette than the US M-60); (2) provided radiation attenuating liners of lead and plastic for gaseous and neutron shielding (they also secondarily provide some spalling protection); (3) provided special ventilation systems; (4) provided tanks with automatic control units to seal the vehicle against blast effects and trigger the special ventilation system.[59]

In view of the continuity of design between the T-34 and the T-54/55 and T-62 tanks, it is likely that the silhouette has been kept low not in order to reduce dynamic pressure damage, but to reduce the size of the target for conventional projectiles. In any event it seems clear that the Soviet Union has done more to protect its armour from nuclear effects than the Western armies, which have not gone beyond providing their tanks with filter systems.

FIREPOWER

The choice of tank armament is dictated by the potential target. As armour grew in thickness, improved armour-piercing performance was needed. This could be obtained either by increasing the length of the gun barrel and raising the muzzle velocity of the armour piercing shot, or by increasing the calibre of the gun. Increases in gun calibre were more attractive than increases in muzzle velocity since 'large calibre, low velocity guns have an advantage over small calibre, high velocity guns because the performance of their projectiles falls off less rapidly with range'.[60] Clearly the most desirable objective is a combination of the largest possible calibre with the highest possible muzzle velocity. There are, however, limits to this kind of development. For example, the size of projectile limits the number that can be stowed in the tank, while the firing of large calibre, high velocity projectiles involves very large reaction forces on the vehicle. As a result of these limitations on gun development, the search for greater firepower has since the mid-1950s concentrated on the development of different types of ammunition and the improvement of fire control.[61]

Calibre. Ogorkiewicz shows that a high velocity (1,000 m/sec) large calibre (120 mm) gun would give an average reaction force of 100 tons. With the installation of muzzle brakes this could be reduced to 70 tons and thus made acceptable for a conventional tank of around 50 tons.[62] These data fit the Chieftain, which is the most heavily armed MBT, apart from the M60 A2. The 152 mm gun/launcher on the latter fires both the Shillelagh guided missile and conventional, medium velocity unguided projectiles.

[59] *Ibid.* The T-34/85 was 2·38 m high, while the T-54/55 and T-62 are only 2·25 m high. The Centurion Mk. 9 has a height of 2·97 m; the M60 A1 of 3·29 m; the AMX-30 of 2·29 m; the Leopard of 2·38 m and the S-Tank of 2·14 m. (See F. M. von Senger und Etterlin, *op. cit.* (1969), pp. 577–609 passim.) NBC protection is emphasised also in Soviet naval warship design. See J. W. Kehoe and H. A. Meier, *loc. cit.*
[60] R. M. Ogorkiewicz, *op. cit.* (1968), p. 57.
[61] R. M. Ogorkiewicz, *op. cit.* (1968), p. 55 ff.
[62] *Ibid.*, pp. 58–9.

Table 9.7 *Combat features of battle tanks, c. 1970*

Parameters	T-62	Chieftain	AMX-30	Leopard	S-tank	M60 A2	Leopard 2
Entry into service	1963–5[16]	1965[11]	1967[3]	1965[12]	1966/7[5]	1972/3[18]	1978/79[22]
Combat weight (t)	37[1]	52·2[2]	36[3]	39·6[4]	39[5]	58[6]	50·5[7]
Crew	4[1]	4[2]	4[3]	4[4]	3[5]	4[8]	4[7]
Power-to-weight ratio (hp/t)	15·6[1]	13·4[2]	20[3]	21[4]	18·7[5]	17[6]	30[7]
Ground pressure (kg/cm²)	0·80[1]	0·98[9]	0·77[3]	0·86[4]	0·94[13]	0·76[6]	0·83[7]
Armament: gun calibre (mm)	115[1]	120[2]	105[3]	105[4]	105[5]	152 gun/launcher[6]	105/120[7]
MG (mm)	1 × 7·62[1]	—	1 × 12·7[3]	1 × 7·62[4]	2 × 7·62[5]	1 × 7·62[6]	1 × 7·62[7]
AAMG (mm)		1 × 7·62[2]	1 × 7·5[3]	1 × 7·62[4]	1 × 7·62[5]	1 × 0·5[6]	1 × 7·62[7]
Ammunition load (main round)	45[1]	53[11]	50[3]	63[4]	50[5]	13 + 33[6]	
Stabilisation	Yes[1]	Yes[2]	No[15]	Added later[12]	No[15]	Yes[6]	Yes[15]
Engine (bhp)	580[1] (700[19])	700[17]	720[5]	830[4]	240 + 490[5]	750[6]	1500[7]
Max. speed (km/h)	50(55[19])	40[2]	65[3]	64[4]	50[5]	48[6]	68[7]
Fording depth (m) prepared/unprepared	3·96/1·40[19]	amphib/1·07[2]	4/2·2[3]	6/1·95[4]	amphib/1·5[5]	/1·22[6]	/1·1[7]
Night vision	IR	IR[2]	IR[3]	IR[4]	IR (driver)[21]	IR[20]	IR[7]
Road range (km)	350[10]/480[19]	402[2]	600[3]	600[4]	340[14]–400[15]	450[8]	640[15]

Sources:
1 F. M. von Senger und Etterlin, *op. cit.* (1969), pp. 598–9.
2 *Ibid.*, pp. 588–9.
3 *Ibid.*, pp. 582–3.
4 *Ibid.*, pp. 578–9.
5 *Jane's Weapons Systems 1973–74*, London (1973), p. 269.
6 *Jane's Weapons Systems 1974–75*, London (1974), p. 307.
7 *Ibid.*, p. 282.
8 *Tekhnika i Vooruzhenie*, 1974, No. 11, p. 44.
9 R. M. Ogorkiewicz, *op. cit.*, p. 196; *Jane's Weapons Systems 1973–74*, London (1973), p. 275 gives 0·92.
10 F. Wiener, *op. cit.*, p. 135.
11 *Jane's Weapons Systems 1973–74*, London (1973), p. 275.
12 *Ibid.*, p. 262.
13 This figure may be too high; R. M. Ogorkiewicz, *op. cit.*, p. 197, gives 12·8 lb/sq. in. (=0·90 kg/cm²).
14 R. M. Ogorkiewicz, *op. cit.*, p. 197 gives 320–400 km.
15 P. A. Rotmistrov, *Vremya i tanki* (1972), pp. 236–8.
16 See Table 9.9.
17 Mk. 5 Chieftains, which form the largest part of the British Army's tank force, have an 840 bhp engine (see *Jane's Weapons Systems 1973–74* p. 275).
18 *Jane's Weapons Systems 1973–74*, London (1973), p. 284.
19 *Jane's Weapons Systems 1976*, London (1975), p. 367. The latter figure denotes range with external fuel tanks.
20 *Ibid.*, p. 358.
21 *Ibid.*, p. 341.
22 *Ibid.*, p. 331.

Table 9.8 *Armour thickness on selected MBTs*

	Leopard	Centurion Mk. 10	T-54	M48
Driver's front mm/°	70/30	76/35	75/30	100/30
Side mm/°	n.a.	51/90	45/90	51–76/90
Turret front mm/°	60/	152/	105/	102/50

Source: F. M. von Senger und Etterlin, *op. cit.* (1969), pp. 577–609.

Since 1944 Soviet tanks have been armed with 85 mm, 100 mm, 115 mm and 122 mm guns. The first two and the last were adaptations of artillery guns, while the T-62's 115 mm gun was the first specially designed Soviet tank gun. The T-34/85 had a larger calibre gun than most contemporary medium tanks. The United States armed the M47 and M48 with a 90 mm gun in order to counter the T-34/85, while they were being introduced the T-54/T-55 was coming into service with a 100 mm gun. This gun was, however, inferior to the 105 mm gun, mounted on the Centurion Mk. 5 and subsequently on the M60 A1, Leopard and S Tank. The reason for the 105 mm gun's superiority lies not only in its calibre but also in its muzzle velocity, fire control and the type of ammunition used. It has been suggested moreover, that the T-62's smoothbore gun is not superior to the 105 mm gun (105/L7 A1), and thus certainly inferior to the Chieftain's 120 mm gun. According to one account of the October 1973 War in the Middle East 'the fresh shipments of Russia's latest T-62 tanks to Egypt and Syria need not have worried Israel. Trials of the two stolen T-62s have convinced Israel of their own Centurions' superiority,'[63] and later 'only the latest Russian T-62 tanks could match the Israeli armaments'.[64] In order to continue the comparison it is necessary to look at ammunition type and fire control.

Ammunition type. Until the Second World War the ammunition of main tank guns consisted of full-bore, monobloc, steel armour-piercing projectiles. During the war, however, new types of ammunition emerged which have since been further developed.[65] The two basic types are armour-piercing (AP) ammunition which relies on the kinetic energy of the projectile to pierce the

---

[63] *Insight on the Middle East War*, by the Insight Team of the *Sunday Times*, London (1974), p. 39.

[64] *Ibid.*, p. 143. Major-General Farrar-Hockley has assessed the performance of the T-62 in the Middle East War of 1973 as follows: 'the T-62 proved to be robust and manoeuvrable, but the smooth bore of its gun precluded accuracy at longer ranges. This may not be so important amongst the mists of Europe, but in the dry air of the Eastern Mediterranean, notwithstanding intermittent dust clouds, the spin-stabilised armour-piercing rounds of the Israeli tank guns were at the outset able to pick off the Syrian attackers at ranges in excess of 2,000 m. Then, as ranges lessened, the skill of the Israeli crews, combined with their tank-to-tank radios, which permitted warnings and target information to be exchanged, proved superior. Control within Syrian tank companies relied upon crude signals which were often misread. Accepting that the Syrian and Egyptian crews were less well trained than the Israelis—or than the tank corps of the Soviet Union—the T-62 did not manifest any particular advantage over the main battle tanks of Britain, the Federal Republic of Germany or the United States and was in many respects inferior, notably in having extremely vulnerable external fuel tanks' (*loc. cit.*, p. 33).

[65] R. M. Ogorkiewicz, *op. cit.* (1968), p. 59 ff.

Table 9.9 *Chronology of Soviet medium tank development since 1945*

1943	—	T-34/76 was fitted with an 85 mm gun and thus became the T-34/85.
1944	—	T-34/85 entered service.
1945/6	—	T-44/5 produced; it had a new turret on a lower and wider hull, a transversely-mounted engine and torsion-bar suspension. There seem to have been difficulties with the suspension system and production ceased in 1949. A few of the later T-44s were fitted with 100 mm guns.
1947	—	An improved version of the T-34/85 was introduced with changes in transmission, armour arrangement and fire-control and vision devices.
1952/3	—	Rotmistrov writes that the T-54 was received by the Army 'soon after the end of the war'. The T-54 began to appear in GSFG (Group Soviet Forces Germany) units (Wolfe). The T-54 was not used in the Korean War; the war may have slowed transition to the production of the new tank.
1953/4	—	T-54 appears in Soviet units (most sources). The first T-54 model had no fume extractor, stabilisation or snorkel equipment; only the driver had IF (infra-red) equipment. The T-54 had better protection and greater firepower than the T-34/85 but the same mobility and manoeuvrability. Maintenance was simplified and reliability improved.
1955/6	—	The T-54A appeared with vertical gun stabilisation and fume extractor.
1957/8	—	The T-54B appeared with gun stabilisation in both planes, IR for commander and gunner and snorkel equipment.
1958/9	—	The T-55 appeared with improved gun and more stowed ammunition, turret fitted with gyroscopic controls, built-in computer, factory-fitted IR and snorkel equipment, and a more powerful engine.
1963	—	The final T-55 production model began to enter service. It has a fairing around the commander's hatch, and the machine-gun has gone.
	—	T-62 first seen; enters service with GSFG (Wolfe). Similar to T-54/55 but has a 115 mm gun and a redesigned turret, and a more powerful engine.
1965	—	T-62 first delivered to Soviet units (most sources).
1971/3	—	M-1970 (T-72) in production. This appears to be an improved version of the T-62, with a new suspension system and the gun placed further back, but the same hull and the same gun calibre; it is also said to have improved mobility and armour protection. Some reports say it has a 122 mm gun; one report says the chassis is new and a departure from established Soviet practice in having six evenly spaced road wheels and four track support rollers; the glacis is similar in shape to that of the Chieftain, and the turret is similar to that of the T-62. The chassis is similar to the vehicle of the Ganef SA-4 surface-to-air missile. Latest reports indicate that the T-72 has a crew of three, weighs 40 tonnes, has an engine horse-power of nearly 1,000, and a road-speed of almost 80 km/h. (See *International Defense Review*, 1, 1976, pp. 24–6.)

Sources:
F. M. von Senger und Etterlin, *op. cit.* (1969), pp. 286–7, 298.
J. Milson, *Russian Tanks 1900–1970*, London (1970), pp. 108–16.
P. A. Rotmistrov, *Vremya i tanki* (1972), p. 220.
T. W. Wolfe, *Soviet Power and Europe, 1945–70*, London and Baltimore (1970), pp. 39, 40, 175.
Chairman JCS Gen. G. S. Brown, USAF, *US Military Posture for FY 1976*, p. 70.
A. Sidorenko, *Voenno-istoricheskii Zhurnal*, 1973, No. 3, p. 96.
F. Wiener, *op. cit.*, pp. 132–5.
Jane's Weapons Systems 1971–72, London (1971), pp. 245–6.
Jane's Weapons Systems 1973–74, London (1973), p. 317.

Note:
Information about Soviet tanks in the post-war period has to be taken largely from Western sources. For this reason it is extremely difficult to establish the stages of innovation since Western statements about innovation appear often to be projections back from the data on introduction into service on the basis of Western experience. Such estimates are useless for this study since it is precisely the comparability of Soviet and Western innovation processes that is at issue. There are discrepancies between the sources, and hence I have tried to indicate the range of evidence.

armour, and high explosive (HE) ammunition which relies on the focused blast energy of the projectile's explosive content to penetrate the armour. APCR (armour-piercing, composite, rigid) projectiles consist of a hard, high density core and a soft, low density jacket. These can achieve higher muzzle velocities than AP projectiles of the same calibre, because they are lighter, and can penetrate thicker armour because penetration is confined to the sub-calibre core. But because they have a lower ballistic coefficient their performance falls off more quickly with range than does that of AP projectiles. APCR ammunition was first used in 1941 with the 50 mm gun of the Pz Kpfw III. APDS (armour-piercing, discarding sabot) projectiles are similar to APCR, save for the fact that the outer jacket is discarded after the projectile leaves the muzzle. The armour-piercing performance of APDS is greatly superior to that of APCR at longer ranges, but the separation of the core from the sabot increases dispersion. APDS projectiles were first used in 1944 by the Churchill tank's 57 mm six-pounder gun. Very high muzzle velocities can be obtained with APDS projectiles; the 105 L/7 A1 gives a velocity of 1,450 m/sec.

Still higher velocities can be obtained with smooth-bore, fin-stabilised projectiles (as opposed to spin-stabilised projectiles from rifled barrels). The T-62's 115 mm gun fires APFSDS (armour-piercing, fin-stabilised, discarding sabot) projectiles. According to one authority

> fin-stabilised projectiles with a high supersonic velocity can be fired from smooth-bore barrels. Their muzzle velocity can then be in the region of 1,600 m/sec. Their hit probability is greater than that of the slow fin-stabilised projectiles. A special breechblock ring is not needed. But no other type of ammunition can be fired from the barrel. It is probable that the Soviets have taken this path with the T-62's 115 mm gun.[66]

Because of the dispersion of fin-stabilised projectiles, great increases in muzzle velocity do not necessarily lead to corresponding increases in the accuracy of fire. Moreover, the drag of the fins slows down the projectiles. This means that there is little to choose between APDS and APFSDS at long range. There is, however, the advantage that with a higher muzzle velocity the projectile can follow a flatter trajectory and thus improve hit probability.

HEAT (high explosive, anti-tank) projectiles penetrate armour by producing on detonation a high velocity jet of copper, the impact of which generates extremely high pressure. HEAT projectiles have to be of a relatively large calibre in order to be effective. They are, however, lighter in relation to their size than AP projectiles, and they need not be fired at high velocities since they are effective on impact. HEAT projectiles were first used by the Pz Kpfw IV in 1942. HESH (high explosive, squash-head) projectiles are used against armour and 'soft' (i.e., human) targets. They have a thin and soft casing which is squashed on impact so that the charge spreads out over the target and sticks to it. If people are the target the HESH projectile will have the normal blast effect; against armour it generates stress waves which fracture the inside of the armour. HESH ammunition is usually a secondary, not a primary, round.

[66] F. M. von Senger und Etterlin, *Taschenbuch der Panzer*, Munich (1969), p. 638.

Table 9.10 *Firepower and ammunition characteristics of Soviet and Western tanks*

Tank	Gun calibre[1] (mm)	Ammo. type	Muzzle velocity (m/sec)	Max. AT range (m)[a]	Rangefinder	Comments
T-34/85	85	AP[5]	792[5]		No	
M-47	90	HE[3] HVAP[2]	731[3] 914[3](1250)[2]		ORF[2]	
M48	90	HEAT[3] HVAP[3]	854[3]( ) 1235[3]	2000[3]	ORF[2]	Both of these tanks were armed with the M41 gun.
M48 A2	90	APC[6] APCR[2]	1250[2]	2000[3]	ORF[2]	
Centurion Mk. 9	105	APDS[2] HESH[3]	1450[2] 730[3]		RMG[2]	0·5 hit probability at 1800 m. (see comment on M60 A1, and R.M. Ogorkiewicz, *op. cit.*, p. 68).
T-54/T-55	100	APC[6]	900[2]		No[2]	Accuracy falls off rapidly above 1000m.;[3] 0·5 hit probability at 1000m.[4]
T-10	122	HEAT[7] HE[7] APHE[7]	900[7] 800[7] 905[7]	1200[7]	No	1200m is the maximum effective range.[7]
M60 A1	105	APDS[2] HEAT[3] HEP[3]	1450[2] 1150[3] 730[3]	2500[3]	ORF[2]	First round hit probability 85% at 1000 m; 40% at 2000 m; 25% at 2500 m.[3]
T-62	115	APFSDS[3] HEAT[3]	1600[3]	1500[3] 1000[3]	No.[2] Laser under development[3]	Low anti-armour range because of inaccuracy inherent in fin-stabilised rounds above 1500 m.[3]
Chieftain	120	APDS[2] HESH[3]	1370[3] 670[3]	3000[3] 8000[3]	RMG/laser[3]	
AMX-30	105	HEAT[2] HE[3]	1000[2] 700[3]	3000[3]	ORF[2]	Low muzzle velocity because of ammunition type.
Leopard	105	APDS[2] HEAT[3] HESH[3]	1450[2]	2500[3] 2000[3] 5500[3]	ORF[2]	As for M60 A1.
S-Tank	105	APDS[3] HE[3]	1500+[3]		RMG (discarded)/ laser (in development)[3]	
M60 A2	152	Shillelagh[3] HEAT[3]		3000[3]	laser[3]	Conventional round lacks accuracy above 1500m; tests have shown that conventional 105mm and 120mm high velocity guns have a faster response and are more accurate than the gun/launcher at all target ranges out to 2000 m.[3]

The abbreviations are for the following terms:

AP — armour piercing.
APC — armour-piercing, capped.
APCR — armour-piercing, composite, rigid.
APDS — armour-piercing, discarding sabot.
HVAP — hyper-velocity, armour-piercing (=APCR).
APFSDS — armour-piercing, fin-stabilised, discarding sabot.
HE — high-explosive.
HEAT — high-explosive, anti-tank.
HESH — high-explosive, squash-head.
HEP — high explosive, plastic (=HESH).
ORF — optical range-finder.
RMG — ranging machine-gun.

Sources:
1 Tables 9.5–9.7.
2 F. M. von Senger und Etterlin *op. cit.* (1969), pp. 577–609.
3 *Jane's Weapons Systems 1974–75*, London (1974), either under the entries for tanks or under the entries for tank guns.
4 S. Canby, *The Alliance and Europe: Part IV Military Doctrine and Technology*, Adelphi Paper No. 109, International Institute for Strategic Studies, London (Winter 1974–75), p. 17.
5 F. Wiener, *Die Armeen der Warschauer-Pakt-Staaten*, Munich (1971), pp. 131–7.
6 R. M. Ogorkiewicz, *op. cit.*, p. 197.
7 *Jane's Weapons Systems 1976*, London (1975), p. 369.

Note:
a – the figures given in the column for maximum anti-tank range may not be comparable. The figure for the T-62 may refer not to maximum range but to the range at which 0·5 hit probability is obtained (but see the assessment of the T-62 given in footnote 64). The M60 A1, Leopard and S-Tank will have the same hit probability as the Centurion Mk. 9 since they are armed with an almost identical gun. Ogorkiewicz (*op. cit.*, p. 68) argues that range is not a satisfactory criterion of tank performance, since tanks are most likely to acquire their targets at a range of 1,000–1,500 m. He shows (p. 138) that the 0·5 hit probability of a typical 105 mm gun can be increased from about 1,300 m to about 1,700 m with the aid of a rangefinder.

Most MBTs have some form of AP projectile as their main round: the French AMX-30 is the only exception. Until the T-62, Soviet tanks had AP or APC (armour-piercing, capped) rounds; the T-62 is equipped with fin-stabilised HEAT rounds as well as APFSDS projectiles. The Soviet Union was thus slow to introduce new types of ammunition into its tank design.

Fire control. Rangefinding is the main element in fire control. There are three basic types of rangefinder: optical, laser and ranging machine-gun. Laser rangefinders have been developed in recent years, and appear now to be fitted to the M60 A2, Chieftain, and S-Tank. The ranging machine-gun has been favoured by the British although its effectiveness stops short of 2,000 m: hence the move towards lasers for the Chieftain. Optical rangefinders come in various forms. A stereoscopic rangefinder was incorporated in the Panther in 1945, and in the M47 and M48 tanks. It proved to require a great deal of training, however, and was not much used in the field. Consequently a full-field coincidence (or super-position) rangefinder was installed on the M48 A3 and M60 tanks. The AMX-30 also had a coincidence rangefinder, and the Leopard a dual rangefinder which can be used either stereoscopically or coincidentally. The laser and optical rangefinders are linked to ballistic computers. Soviet tank design does not appear to have gone beyond a relatively simple stadiametric device which has been fitted into the commander's sight on the T54/55 and T62. This is reported to be 'little better than visual range estimation under battle conditions'.[67] Thus Soviet tank design appears to lag significantly in rangefinding equipment, although it has been reported that a laser rangefinder is being developed.[68]

A second important element in fire control is the *night vision equipment* which is fitted to most modern tanks. The first T-54 was fitted with infra-red (IR) equipment for the driver, while IR equipment was added for the gunner and commander on the T-54B. All the tanks in Tables 9.5–9.7 have IR equipment, with the exception of the T-34/84, M47 and the S-Tank. The equipment normally consists of IR sights and viewers and an IR searchlight.

An important element of tank design which is related to fire control is the *stabilisation of tank guns* so that tanks can fire accurately on the move. Guns are stabilised against angular motions of the vehicle by reference to gyroscopes mounted on the gun cradles. The T-54A had vertical gun stabilisation, and all Soviet tanks since then have had guns stabilised in both planes. British tanks since the Centurion Mk. 2 have had guns stabilised in both planes, and stabilisation controls have been fitted to the Leopard. US tanks have no gun stabilisation, although the Second World War M4 (Sherman) tank had stabilised power controls.

[67] M. C. Norman, in *Brassey's Annual 1968*, London, p. 195. See also Insight, *op. cit.*, Appendix: 'Apart from the characteristics of their different types of ammunition, the Russian tanks rely purely on visual judgement for rangefinding. The American tanks, by contrast, have a sophisticated device of optical prisms for estimating range, while the British Centurions have a neat and simple system of zeroing-in with tracer bullets from a range-finding machine-gun before the big round is fired (the system also corrects for side-winds). The next result is a gain of up to half a mile at long range, for comparable accuracy.'
[68] *Jane's Weapons Systems 1973–74*, London (1973), p. 274.

MOBILITY

Because of the demands placed upon them, special engines have had to be designed for tanks, and the development of a new engine is the most time-consuming part of tank development. The power-to-weight ratio sets a limit to the maximum speed of the tank and also governs its acceleration, climbing speed and average speed of travel in varied terrain. There are therefore good reasons for having as high a ratio as possible. In practice, however, there are limits to the amount of power tanks can use effectively, and until recently no more than about 20 gross bhp per ton has been aimed at (cf. Fig 9.1). Soviet tanks have used a diesel engine since 1939; in fact, they have used only one type of engine, the V-2, a water-cooled, 38·8 litre V-12 diesel. The British A-6 tank had used a diesel engine in 1928 but it was only in the mid-1950s that Western tank design turned to diesel engines, after recognising, rather late, the importance of fuel economy in tanks. The diesel gives the best combination of two basic indicators of engine performance: overall power, and specific effective fuel consumption. As Ogorkiewicz comments:

> The overall performance of the diesel engine is not easy to surpass, particularly from the point of view of fuel economy over a wide range of operating conditions, and its competitors are generally more complex as well as requiring a much more advanced level of technology to make them work.[69]

After the Second World War tank design in Britain, France and the United States concentrated on the application of fuel injection to spark ignition engines. The change to diesel engines in the 1950s gave a great increase in operating ranges (cf. Fig 9.2). Development has continued, chiefly on multifuel engines which can operate on gasoline as well as diesel oil, although these are still primarily diesel engines. Work has also been done on gas turbine engines, but the only tank to adopt a gas turbine engine is the S-Tank, which has a gas turbine and a comparatively small diesel combined in a two-engine power unit; the gas turbine is used only when additional power is needed.

A further indicator of mobility is provided by the nominal ground pressure exerted by the tank.

> Pressures range from about 7 lb/in² (0·49 kg/cm²) for light tanks and carriers to 14 lb/in² (0·98 kg/cm²), or more, for battle tanks. So far as soils are concerned, the nominal ground pressure should not exceed 10 lb/in² (0·70 kg/cm²) if vehicles are to perform adequately in muddy terrain or soft sand, and if it exceeds 13 lb/in² (0·92 kg/cm²) they are likely to experience difficulties under such conditions.[70]

The ground pressure depends on the design of the tank as a whole rather than on the engine, but is nevertheless an important indicator of mobility.

[69] R. M. Ogorkiewicz, *op. cit.* (1968), p. 94.
[70] *Ibid.*, p. 104.

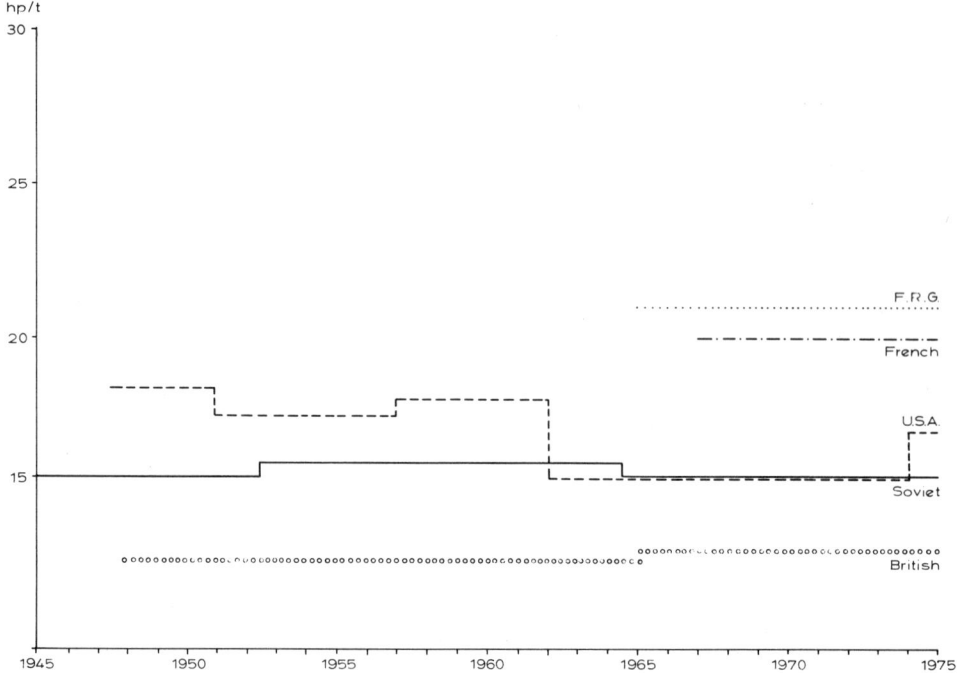

Fig. 9.1 DEVELOPMENT OF POWER TO WEIGHT RATIOS

The design of Soviet tank engines has not advanced greatly since 1945. Although clearly superior to Western tanks in speed and range in the early post-war years, Soviet tanks have been overtaken in both respects in the 1960s. This emerges clearly from Figs 9.2 and 9.3. Soviet tanks compare well with other MBTs in terms of nominal ground pressure (cf. Fig 9.4).

Amphibious capability. Most MBTs can ford, unprepared, in water between one and two metres deep. They cannot be made to float or operate submerged without the addition of special equipment. Work on the conversion of tanks for submerged operation began in Britain and Germany in 1939 and 1940. The Pz Kpfw III and IV were adapted for underwater operation in connection with Operation Sea Lion. The Soviet T-54B was the first tank to practise underwater crossing of rivers on a large scale. The T-54 is sealed for underwater operation and a small-diameter metal tube is erected on top of the turret to provide air intake. It can ford in up to 6 m of water. The tank is guided by radio from the bank. The M60 and AMX-30 can ford in up to 5 m of water with an extension tube fitted to the commander's turret. The commander can stand at the top and guide the tank.

Some tanks—the S-Tank, for example—carry collapsible fabric screens which can be erected so that the tank can float. These have the advantage of enabling tanks to cross water deeper than 6 m, but the disadvantages of making the tank very conspicuous and of taking time—say, 15 minutes—to erect.[71]

[71] *Ibid.*, pp. 105–7.

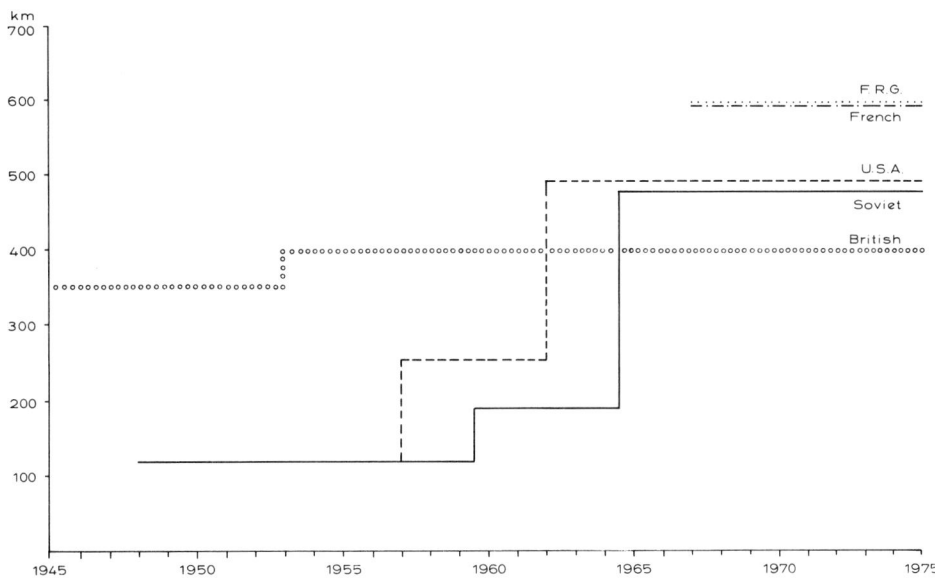

Fig. 9.2 TANK RANGE 1945 - 1975

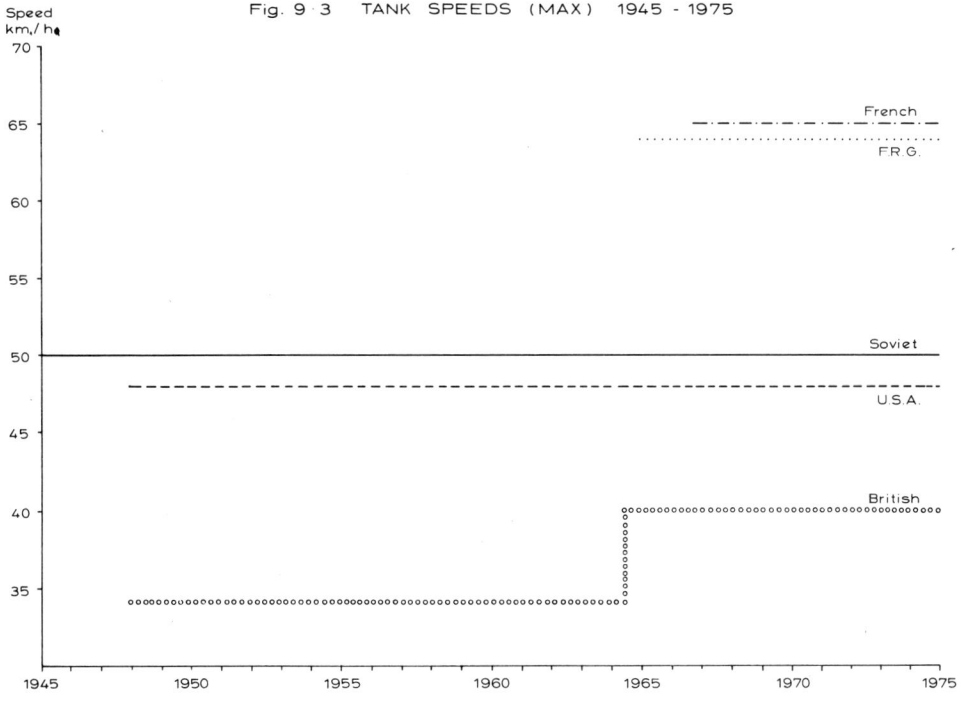

Fig. 9.3 TANK SPEEDS (MAX) 1945 - 1975

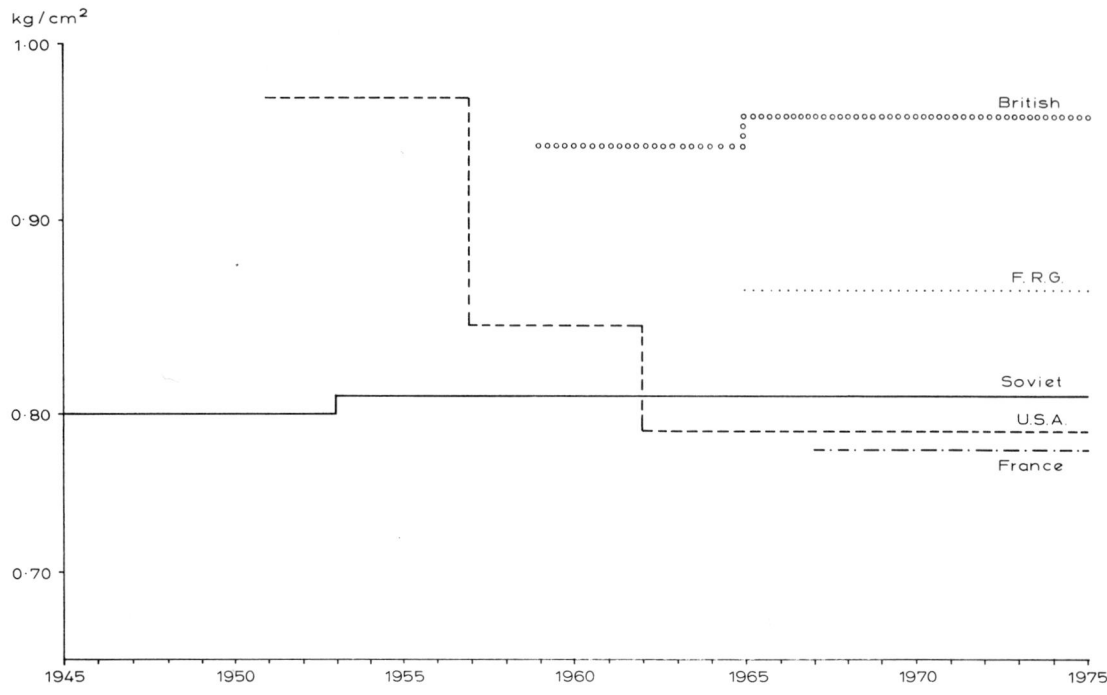

Fig. 9·4 NOMINAL GROUND PRESSURE 1945 - 1975

DESIGN PHILOSOPHY

The development of Soviet tanks fits very well into the general picture of Soviet design philosophy outlined in the introduction. Change has been evolutionary rather than revolutionary, and many features have been passed on from one generation to the next: the engine and transmission system are perhaps the clearest examples. Babadzhanyan indeed stresses that the designer should try to provide in his design for future development, and lay the basis for a family of vehicles.[72] Commonality is also apparent in Soviet tank design: thus only one gun was specifically designed for a tank—the others were borrowed from the artillery, and the basic tank engine was originally an aeroengine. The third feature of Soviet design noted in the introduction is also to be found in Soviet tanks: simplicity. A high priority is placed on rugged design, and reliability which is seen as more important than maintainability; consequently some components are used which are hard to service or replace. Ogorkiewicz writes that in the Soviet Union: 'the whole philosophy behind tank design has been that the ideal design is one which is just good enough and that anything better than that is a waste of time'.[73] Little attention seems to have been paid to human engineering:

[72] A. Kh. Babadzhanyan, *op. cit.*, p. 39.
[73] R. M. Ogorkiewicz, *Armoured Forces*, London (1970), p. 235.

the near-hemispherical shape of the turret on the T-54 and its successors is particularly worthy of note. None the less, while the hemisphere is the best shape for a turret, giving maximum protection for minimum weight, it is by no means compatible with the efficient working of a turret crew, especially where the turret ring diameter is limited. It seems likely that the turret crews in the T-10 or T-55, for example, are extremely cramped even in sitting positions ... crew efficiency must inevitably suffer.[74]

This assessment seems to be borne out by the comments of the Israeli soldiers who used T-54s and T-55s, which had been captured in 1967, in the 1973 Middle East War.[75]

The contrast between Soviet and Western design philosophies should not, perhaps, be overstressed. Western tanks have not seen revolutionary changes in the last 30 years. Moreover, there is some commonality to be found in Western tank designs; thus the British 105 mm gun has been adapted for use on the Leopard, the S-Tank and the M60. Nevertheless, a clear difference exists in design philosophies. It can be seen in the attitude to ruggedness and simplicity of design: Soviet tanks are, after all, rugged and simple *by comparison* with Western tanks. There has been a difference in the degree of design inheritance: innovation has been more marked in Western tank development. In some cases this has meant the adoption of technologies that Soviet tanks were already using, for example, diesel engines. In others it has involved the adoption of devices that Soviet tanks do not have, for example, optical rangefinders.

It must be remembered, however, that Soviet tanks are here being compared with the tanks of five foreign countries, and that the latter are likely to show more diversity and discontinuity in development than tanks from a single country. When Soviet tanks are compared with those of the USA or UK (countries which have produced more than one tank since the war) the difference in the degree of design inheritance diminishes: American tanks all have many features in common and show a strong evolutionary, rather than revolutionary, pattern of development; for example:

> the M47 through the M60 A1 tanks have used the same basic engine and transmission. In each model, and in many of the modifications between tank models, the engine and transmission have been improved; in the M60 and late model M48s a diesel engine has been used. All tanks ... have used torsion bars and rubber-tyred roadwheels; the medium and heavy tanks also have used a double-pin, rubber-bushed system.[76]

The same is true of British tanks: the Centurion went through 12 different marks. It is worth noting that the joint US-FRG MBT-70 project which was to have introduced many revolutionary elements into tank design had to be abandoned because it was proving too expensive.[77]

[74] M. C. Norman, *loc. cit.*, p. 196.
[75] Insight, *op. cit.*, appendix on the tank battle.
[76] *Congressional Record, loc. cit.*
[77] *Jane's Weapons Systems 1974–75*, London (1974), p. 308.

DIFFUSION

It is unfortunately extremely difficult to obtain reliable information about the production and diffusion of tanks, particularly for the years before 1965. Only rough estimates can be given here. The total Warsaw Pact tank 'park' has been estimated at 45,000 for 1968; it may have increased to 50,000 by 1975.[78] According to one source, 30,000 T-54/55s were produced up to 1963.[79] This would give an annual average rate of production of over 2,000 tanks, since the T-34/85 and the T-10 heavy tanks were also produced during this period. This suggests an expected life in peacetime of about 20 years for a tank; thus the oldest tanks now in service with the Warsaw Pact countries would have been produced in the early 1950s. In fact T-34/85s are still in service with Bulgaria, Czechoslovakia, Hungary, Rumania and the GDR; they may also be held in the Soviet Union for divisions of category III readiness.[80] If tanks held by non-Warsaw Pact states are taken into account the picture changes: in 1965 this number was about 1,500; and by 1975 it had risen to about 12,000.[81] Allowing for losses in the Middle East this suggests that about 12,000 tanks were exported by the Soviet Union to non-Warsaw Pact countries between 1965 and 1975. It is not clear whether production was increased to meet these exports; in any event it seems that in the post-war period Soviet tanks have been produced at an average rate of between 2,000 and 3,000 a year. This is not very far short of the level of production in the 1930s. It is not clear, however, how far the rate of production has fluctuated.

The total NATO tank force in 1975 was about 20,000.[82] The oldest tanks still in service were the M47 and the Centurion. The M47 was produced from 1951 to the mid-1950s. This suggests an expected peacetime life for a tank of something in the region of 20 years, which is similar to the figure for the Warsaw Pact. This would give an average annual production total of about 1,000 tanks, not allowing for exports or losses in war. In 1975 about 10,000 tanks built in NATO countries were in service with non-NATO forces.[83] Thus the world stock of these tanks would be about 30,000, which suggests an average annual production total of about 1,500 in the USA, UK, France and Federal Republic of Germany. This figure is in the same range as the figure of

---

[78] M. C. Norman, *loc. cit.*, p. 192; and *The Military Balance 1974–75*, London (1974), passim. The latter total is based on the number of tanks given for non-Soviet members of the Pact, and on the number of Soviet divisions (tank and motorised-rifle) multiplied by the number of tanks which each would have at full strength. This may well exaggerate the number. Moreover, Soviet tanks have been built in Poland and Czechoslovakia too. According to the Chairman of the US Joint Chiefs of Staff, the Soviet tank force consisted of 40,000 tanks in 1975; see *US Military Posture for FY 1976*, Washington DC, p. 70.

[79] *Jane's Weapons Systems 1971–72*, London (1971), p. 246.

[80] Soviet divisions have three degrees of combat readiness: Category I, between three quarters and full strength, with complete equipment; Category II, between half and three quarters strength, with complete fighting vehicles; and Category III, about one quarter to one third strength, possibly with complete fighting vehicles (*The Military Balance 1974–75*, London (1974), p. 9).

[81] *The Military Balance 1965–66* and *1975–76*, London (1965 and 1975), passim; these figures must be taken as very rough estimates indeed.

[82] *The Military balance 1975–76*, London (1975), passim. For the sake of convenience I am including French tanks in the NATO total.

[83] *The Military Balance 1975–76*, London (1975), passim.

1,300 which can be derived from data about the production of the most recent Western tanks.[84]

It seems to be characteristic of Soviet tank production that when a new model is introduced, production of the previous model does not cease. It is reported that production of the T-34/85 continued after the T-54 was introduced;[85] the T-55 is apparently still in production, although the T-62 was introduced in 1964/5.[86] Both of these tanks are being produced alongside the T-72. In the West, as a general rule, production of one model ceases when a new model is introduced. Within the Warsaw Pact the most modern tanks are sent, naturally enough, to the front-line forces in Central Europe. It is possible that the Soviet Union keeps at least two models in production so that it will not have to supply its most modern equipment to politically unreliable clients.

CONCLUSIONS

Ogorkiewicz has suggested that tank designs can be evaluated on the basis of their comparative effectiveness in performing the specific mission of destroying or 'killing' a hostile tank.[87] He has given the equation

$$P_{ss} = P_a \times P_s \times P_k$$

where P_{ss} is the probability of success, P_a the probability of being available, P_s the probability of survival and P_k the kill probability. These categories correspond roughly to, but are more precise than, the traditional concepts of mobility, protection and firepower. Lack of information makes it impossible to compare tanks on the basis of this equation, but several important points do emerge from Ogorkiewicz's analysis. The dominant factor in determining kill probability is the estimation of range—hence the importance of the fire control system, and in particular of the rangefinder; the performance of Soviet tanks must be seriously impaired by their lack of effective fire control equipment. The analysis shows also how important reliability is in attaining high availability and high hit probability; their rugged design may give Soviet tanks an advantage here.

Less precise methods of evaluation will have to be used here. Table 9.12 compares Soviet and foreign tanks in terms of their most important performance parameters for the years 1950, 1960 and 1970. This comparison is based very largely on the data in Tables 9.5–9.7, and is given as 'lag', 'lead' or 'equal', since more precise evaluation is not practicable. In most cases the data provide the basis for an objective assessment; in others—for example, armour protection—the evaluation is more subjective. The picture which emerged from this exercise is not especially sharp. But it suggests that by 1950 the

---

[84] *World Armaments and Disarmament. SIPRI Yearbook 1975*, Stockholm (1975), pp. 176–7. According to US intelligence estimates, the annual average rate of production of tanks between 1973 and 1975 was 2,600 for the Soviet Union and 450 for the United States (*Aviation Week and Space Technology*, 26 January 1976, p. 20).

[85] J. Milsom, *op. cit.*, p. 109.

[86] *US Military Posture for FY 1976*, p. 60; according to *Jane's Weapons Systems 1974–75*, London, p. 319, the T-55 is no longer in production.

[87] R. M. Ogorkiewicz, *Design and Development of Fighting Vehicles*, London (1968), Chapter 7.

T-34/85 had been overtaken by US tank design, although it was still superior to the best British tank; by 1960 the T-55 had given the Soviet Union a marked superiority over US and British tanks; by 1970, however, the T-62 had been overtaken by the M60 A2, the AMX-30 and the Leopard; it was superior to the Chieftain on all counts of mobility, but inferior in firepower. Thus the 1960s saw a relative decline—from a position of superiority—in the level of Soviet tank technology. (The parameters are all given equal weighting in this assessment, although military doctrine might well affect the priority given to different elements—firepower as against mobility, for example—in the design and evaluation of tanks as fighting vehicles.)

The change in the relative position of Soviet tanks in the 1960s can be assessed in terms of those features which Marshal Rotmistrov singled out as important in foreign tanks in the 1960s (see p. 421 above).

1. Soviet tanks, like all Western tanks except the S-Tank, have retained the classical layout.
2. Soviet gun calibre is greater than that of all foreign tanks except the Chieftain and the M60 A2 (see Fig 9.5), but the T-62's 115 mm gun is not thought to be more effective than the standard NATO 105 mm gun (see p. 429 above); Fig 9.5 suggests that the superiority of individual Soviet tanks in firepower has been lost since about 1960.
3. The Soviet Union lagged behind the Western tank-producing powers in introducing new types of AP and HE shells, although it now uses ammunition of both types.
4. Western tanks—in particular the Leopard and AMX-30—have overtaken Soviet tanks in speed and power-to-weight ratio (although the Chieftain lags behind). Soviet performance has changed little since 1945.
5. Since the T-34, Soviet medium tanks have been low and relatively light; of all the foreign MBTs only the S-Tank is lower and only the AMX-30 is lighter than their Soviet counterparts.
6. Figure 9.6 shows that Soviet tanks carry fewer rounds of ammunition than Western MBTs.
7. Soviet tanks are better protected against nuclear weapons effects.
8. Figure 9.2 shows that Soviet tanks have lost their lead in cruising range.
9. The Soviet Union was the first country to provide its tanks with underwater driving equipment; the M60, AMX-30 and Leopard followed suit later.
10. The Soviet Union was one of the first countries to provide its tanks with IR equipment for night combat.
11. The reliability and maintainability of tanks is difficult to judge.
12. The Soviet Union has lagged in the provision of rangefinding equipment, and this has affected first-round hit probabilities; Table 9.10 shows that while the T-55 has 0·5 hit probability at 1,000 metres, the Centurion Mk. 9 and M60 A1 have 0·85 hit probability at 1,000 metres and 0·5 hit probability at 1,800 metres.

By 1970 Soviet tanks were equal in four of these areas: night-vision equipment, layout, underwater driving and ammunition type. Soviet tanks have retained the classical layout although, as has been seen, conditions for the crew are more cramped than in Western tanks. The Soviet Union lagged in introducing new types of ammunition, but was the first to equip tanks for underwater driving, and among the first to provide IR equipment for night combat. In two areas—size and weight, and in protection against nuclear effects—the Soviet Union has enjoyed an overall lead. In five areas the Soviet Union has lagged behind the best Western tanks: cruising range, speed, gun calibre, rangefinding and ammunition load. In the last two of these areas the Soviet Union has always lagged. In gun calibre only the Chieftain 120 mm gun was superior to the T-62's 115 mm gun, but the 105 mm guns on the other tanks are not inferior in practice. In cruising range the Soviet Union was overtaken by the M60 A2, the AMX-30 and the Leopard; in speed too the AMX-30 and the Leopard have proved superior. It is difficult to compare the tanks in terms of reliability and maintainability.

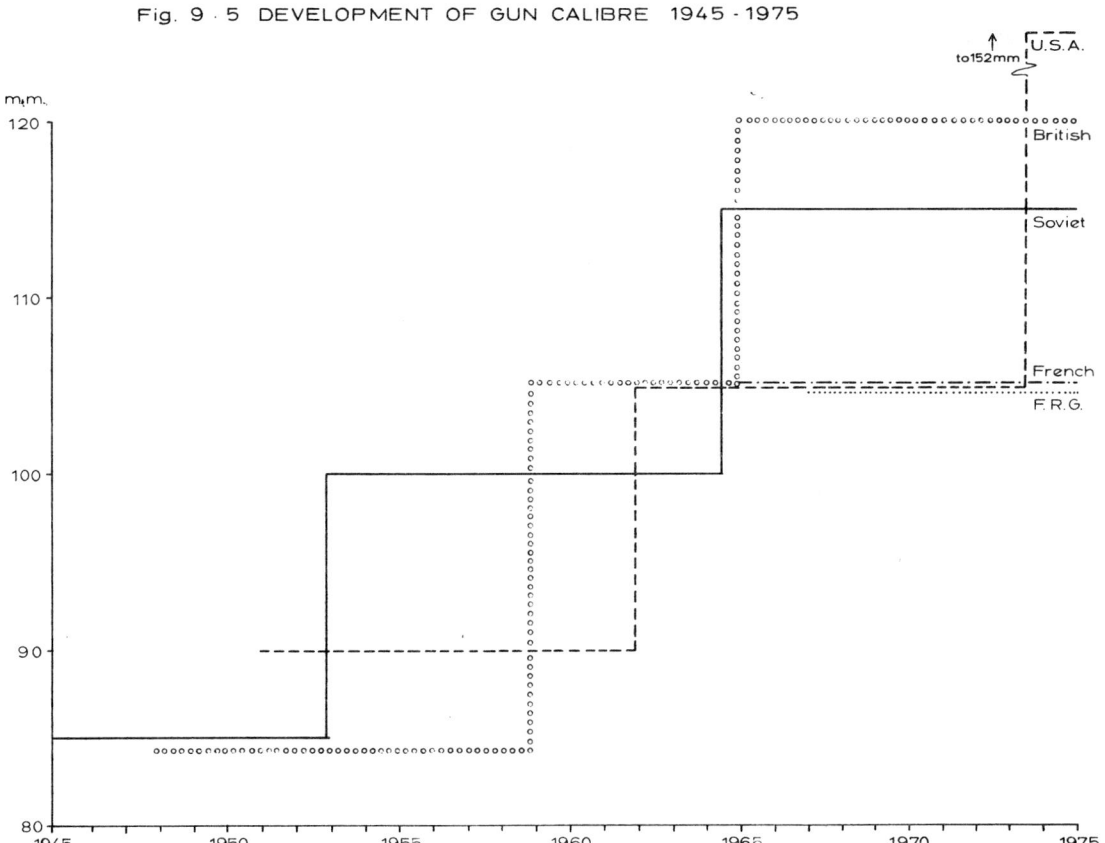

Fig. 9.5 DEVELOPMENT OF GUN CALIBRE 1945-1975

Table 9.11 Tank engines

Parameters	V-54 (Soviet; used in T-54)	Continental AVDS-1790-2 (USA; used in M60 A2)	Caterpillar LVMS-1050 (USA; experimental; development not completed)	Leyland L-60 (UK; used in Centurion and Chieftain)	Daimler-Benz MB 838 Ca-500 (FRG; used in Leopard)	Hispano-Suiza HS-110 (France; used in AMX-30)
Type	4-stroke diesel direct injection	4-stroke diesel direct injection	4-stroke diesel precombustion chamber	2-stroke diesel between the pistons	n.a.	4-stroke diesel
Combustion chamber					precombustion chamber	n.a.
Degree of compression	14.5	14.5	20.5	16.75	18	n.a.
Fuel	diesel	kerosene, diesel	multifuel	multifuel	multifuel	multifuel
Cooling	liquid	air	liquid	liquid	n.a.	n.a.
Supercharger	—	n.a.	n.a.	drive volume	drive centrifugal	turbocompressor
Degree of pressure increase in supercharger	—	1.9	3	1.4	n.a.	n.a.
Number and placement of cylinders	12-V-60°	12-V-90°	12-V-60°	6 in a series vertical	10-V-90°	12-180° horizontal
Max. power (hp)	520	750	1000	700	830	720
Revolutions at max. power	2000	2400	2800	2400	2200	2800
Litre power (hp/l)	13.4	25.6	58.2	36.8	22.2	n.a.
Overall volume (m³)	1.17	2.52[a]	0.92	1.41	1.86	n.a.
Overall power (hp/m³)	443	298[a]	1087	498	445	n.a.
Specific fuel consumption (g/hp-h)	180	159–172	n.a.	n.a.	180	n.a.
Coefficient of compactness (l/m³)	33.2	11.7[a]	18.7	13.5	20.2	n.a.
Engine weight (kg)	895	1710	1130	1400	1675	n.a.
Specific weight (kg/hp)	1.72	2.28	1.13	2	2.02	n.a.

Source:
Adapted from A. Kh. Babadzhanyan, *Tanki i tankovye voiska* (1970), pp. 80–1.

Note:
a – taking account of cooling system units.

Table 9.12 *Tanks 1950–70*

	USSR:USA			USSR:UK			USSR:FRG	USSR:Fr.
	1950	1960	1970[b]	1950	1960	1970[b]	1970[b]	1970[b]
Armour	−1	−1	?	−½	−½	−1	+1	+1
NBC protection			+1			+1	+1	+1
Calibre	−1	+1	0	+1	−1	−1	0	0
Rangefinding	−1	−1	−1	0	−1	−1	−1	−1
Stabilisation	0	+1	+1	−1	0	0	0	+1
IR	−1	0	0	0	+1	0	0	0
Speed	+1	+1	+1	+1	+1	+1	−1	−1
Cruising range	+1	+1	−1	+1	+1	+1	−1	−1
Power to weight ratio	−1	−1	−1	+1	+1	+1	−1	−1
Nom. ground pressure	+1	+1	−1	+1	+1	+1	+1	−1
	−2	+2	−1	+3½	+2½	+2	−1	−2

+1 – Soviet lead; −1 – Soviet lag; 0 – parity.

Sources:
Tables 9.5 to 9.7 above.
F. M. von Senger und Etterlin, *op. cit.* (1969), pp. 577–609.

Notes:
a – the tanks compared are: 1950: T-34/85, Centurion Mk. 3, M47.
                            1960: T-55, Centurion Mk. 9, M48 A2
                            1970: T-62, Chieftain, M60 A1, AMX-30, Leopard.
b – I have taken the T-62's 115 mm smoothbore gun as equivalent to the NATO 105 mm rifled gun.

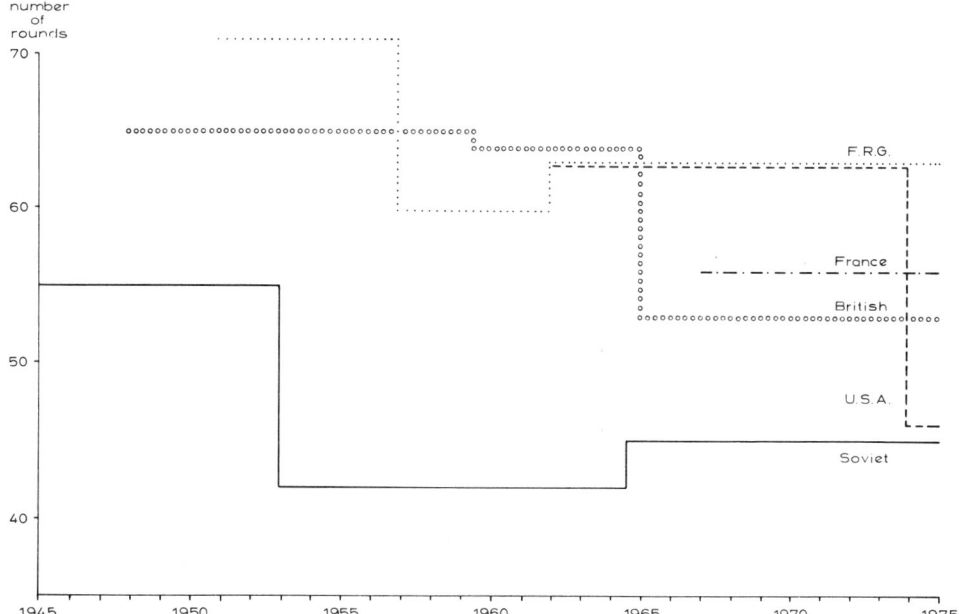

Fig. 9.6 ROUNDS OF AMMUNITION CARRIED BY TANKS 1945 - 1975

Table 9.13 *Summary of results in Table 9.12*

Parameter (USSR in relation to:)	USA	UK	France	FRG
Protection				
Armour	equal	lag	lead	lead or equal
NBC	lead	lead	lead	lead
Firepower				
Calibre	equal after lag	lag after lead	equal	equal
Firecontrol:				
—rangefinding	lag	lag	lag	lag
—stabilisation	lead	equal after lag	lead	equal
—IR	equal after lag	equal after lag	equal	equal
Mobility				
Speed	lead	lead	lag	lag
Cruising range	lag after lead	lead	lag	lag
Power-to-weight ratio	lag	lead	lag	lag
Ground pressure	lag after lead	lead	lag	lead

This study has attempted to assess the level of Soviet tank technology on the basis of a side-by-side comparison of individual tanks: military effectiveness has not been at issue. The data on which the study is based are sparse and unsatisfactory, and all conclusions must be treated with care. Nevertheless, a general picture does emerge from the fragmentary evidence: between 1950 and 1960 the Soviet Union strengthened its position *vis-à-vis* Western tanks, but between 1960 and 1970 lost its marked superiority to a new generation of Western tanks.

Intercontinental ballistic missiles

Intercontinental ballistic missiles (ICBMs) now form the major part of the strategic arsenals of both the United States and the Soviet Union (see Table 9.14).[88] The first ICBM was flight-tested successfully by the Soviet Union in August 1957 and by the end of 1958 the United States had followed suit. ICBMs were made possible by three main lines of technological progress: by the development of liquid-propellant rockets (LPRs) which could, by the end

---

[88] The term 'strategic arsenal' is not as clear as it might at first appear. The definition of strategic weapons has been a major issue at the strategic arms limitations talks (SALT). The term is usually taken to refer to ICBMs, SLBM (submarine-launched ballistic missiles), and strategic bombers (i.e., those with an intercontinental range). But the Soviet definition is different, in as much as IRBMs and MRBMs are counted as strategic weapons. Moreover, the Soviet Union wished to include in the SALT negotiations discussion of the so-called forward based systems (FBS), American weapons based outside the United States which could deliver nuclear strikes against the Soviet Union. US IRBMs and MRBMs were removed from service by 1965, while the Soviet IRBMs and MRBMs have been kept in service; their number has fluctuated between 600 and 800 in the 1960s and 1970s.

of the Second World War, deliver explosives far beyond the range of artillery and with a greater likelihood than bomber aircraft of reaching their target; by the creation of nuclear weapons during and after the War which made it possible to equip missiles with warheads of immense destructive power; and, especially after the invention of the transistor in 1948, by the development of guidance systems which made it possible to deliver the warheads to their targets with great accuracy. Since their introduction into service ICBMs have seen major innovations in guidance systems, engines, warheads and ground equipment.[89]

ICBMs are only one of many different kinds of military missile. In terms of the simple classification give by Fridenson (see Fig. 9.7) there are ballistic, guided, surface-to-surface missiles, usually with two or three stages and with various types of engine. Such missiles have to be classified also according to range: Table 9.15 gives Soviet and Western definitions. A further characteristic of ICBMs is that almost all the missiles discussed in this study—and all the ICBMs so far deployed—are launched from fixed sites. As ICBMs become more accurate they acquire the capacity to strike and destroy other ICBMs at their launch sites. This growing vulnerability may make fixed-site ICBMs obsolescent since submarine-launched ballistic missles (SLBMs) cannot be detected and are thus not vulnerable to surprise attack; fixed-site ICBMs may come to be replaced by land-mobile missiles.

In recent years a good deal has been published in the Soviet Union about the history of Soviet rocket research, and a detailed study could be made of the years 1921–45. But for the period that concerns us very little information has been made available in the Soviet Union. Although this paper is based exclusively on open sources, most of the data about Soviet missiles come directly or indirectly from Western intelligence sources and are subject to a margin of error. It is possible to draw a general picture of the development of Soviet ICBM technology by sketching in the main stages of the R and D programme to 1957 and the pattern of introduction of new weapons after that. The unreliability of the data should be borne in mind, however, since assessment of the technological level of Soviet ICBMs is very sensitive to the assumptions one makes about specific performance parameters.

There is broad agreement that ICBMs are designed to strike two kinds of target: area (or soft) targets such as large industrial and population centres, and point (or hard) targets such as missile silos.[90] In the United States the relative importance of these two types of target has been a central issue in debates about strategic doctrine. The policy of deterring an attack by threatening retaliation against soft targets calls for what is known as a 'countervalue targeting policy', and for missiles which can destroy area targets; a policy directed against hard targets, known as a 'counterforce targeting policy', requires missiles which can destroy the enemy's forces, and in particular his hardened missile silos. Soviet statements about targeting have

---

[89] For the history of rockets, see W. von. Braun and F. I. Ordway III, *History of Rocketry and Space Travel*, London (1967).

[90] Hard targets became important in the 1960s when the United States began to house missiles in underground silos which were designed to withstand the effects of a nuclear explosion.

consistently referred to both types of target. Strategic missiles are intended to deliver nuclear warheads against both hard and soft targets of strategic significance in the deep rear of the enemy.[91]

This section will attempt to assess the technological level of Soviet and United States ICBM forces in terms of the performance parameters of the missiles. It must be stressed that military effectiveness is not at issue, even though performance parameters will be used as a basis for assessing the level of technology.

SOVIET ROCKET RESEARCH TO 1945

The history of Soviet rocket research is treated in chapter 10, pp. 491–9 below, and only some of its main features will be mentioned here. Although Soviet rocket scientists and engineers seem to have been inspired by Tsiolkovsky's visionary writings about interplanetary travel it was the military who financed rocket research. The main centres of research—the Gas Dynamics Laboratory (GDL), the Group for the Study of Reactive Motion (GIRD), and the Reactive

Table 9.14 *Structure of strategic forces 1965–75*

	1965	1966	1967	1968	1969	1970	1971	1972	1973	1974
Strategic bombers:										
USA	(738)	(708)	697	646	581	517	565	525	496	496
USSR	155	155	155	150	140	140	140	140	140	140
ICBMs:										
USA	854	934	1054	1054	1054	1054	1054	1054	1054	1054
USSR	162	338	722	902	1198	1498	1527	1527	1547	1567
SLBMs:										
USA	464	592	656	656	656	656	656	656	656	656
USSR	—	—	—	32	128	224	336	444	564	636

Source:
SIPRI Yearbook 1974, Stockholm (1974), pp. 106–7. The IISS series is different, but the general picture is the same, except for the fact that the IISS series includes as SLBMs the SS-N-4 and SS-N-5 which are armed with nuclear warheads but have ranges of no more than 750 miles (see *The Military Balance 1975–76*, London (1975), pp. 71, 73).

---

[91] See the following sources: E. S. Fridenson, *Osnovy raketnoi tekhniki* (1973), p. 31: 'objects of strategic significance in the deep rear of the enemy (large military, industrial, political and administrative centres, the launching positions and bases of strategic missiles, control centres etc.)'; *Artilleriya i rakety* (1968), p. 224: 'strategic missiles are a means of delivering nuclear charges of great power and are capable of striking objects of strategic significance in the deep rear of the enemy (military, industrial, political and administrative centres)'; V. D. Sokolovskii, *Voennaya strategiya* (1968), p. 235: the Strategic Missile Forces 'can be used, if necessary, to solve the main strategic tasks of the war—to destroy the aggressor's means of nuclear attack—that basis of his military might, to rout the main groupings of his armed forces, and also to destroy all the enemy's vitally necessary objects'; (Eds.) N. Ya. Sushko and T. R. Kondratkov, *Metodologicheskie problemy voennoi teorii i praktiki* (1966), p. 147: 'the most important targets for nuclear strikes can be: the enemy's strategic nuclear weapons; the economic base of war; state and military administration; groupings of the enemy's forces'.

Table 9.15 *Classification of military ballistic missiles by range*

Western	
ICBM (intercontinental ballistic missile)	over 4000 miles
IRBM (intermediate range ballistic missile)	1500–4000 miles
MRBM (medium range ballistic missile)	500–1500 miles
SRBM (short range ballistic missile)	under 500 miles
Soviet	
Strategic missiles	over 1000 km
Medium-range (*rakety srednei dal'nosti*)	1000–5000 km
Long-range (*rakety dal'nego deistviya*)	over 5000 km
Intercontinental	5000–20000 km
Operational-tactical and tactical missiles	under 1000 km

Sources:
The Military Balance 1974–75, London (1974), p. 75.
E. S. Fridenson, *op. cit.*, p. 31.
Artilleriya i rakety (1968), pp. 224–30.

Research Institute (RNII)—received much of their support from the Red Army. Although the RNII was placed under the Scientific Research Sector of the People's Commissariat of Heavy Industry in April 1934 it retained its military orientation, working on powder rockets and their launchers, liquid-propellant rocket engines, winged (cruise) rockets and rockets for airplanes.[92]

There is a temptation to see all rocket research as no more than a step on the road to ICBMs and space launchers, and to assess Soviet research in the light of these goals alone. This is misleading, for the rocket research programme before 1945 pursued diverse goals. By 1939, indeed, the development of long-range liquid-propellant rockets had come to a virtual standstill: the GDL, which had maintained a distinct identity within the RNII, once more became a separate organisation and specialised in the development of liquid-propellant rocket boosters for aircraft, while the RNII now concentrated on powder rockets and in 1941 completed the development of the legendary 'Katyusha' rocket artillery.[93] None of this work was a direct contribution to the development of long-range ballistic missiles: the artillery rockets were free-flight rockets, used powder fuel and had a range of no more than 10 km. The GDL's work on rocket boosters for aircraft made an important contribution to the subsequent development of powerful liquid-propellant rocket engines, but it was not specifically directed towards that end, and the engines developed had a low thrust.[94]

[92] E. K. Moshkin, *Razvitie otechestvennogo raketnogo dvigatelestroeniya* (1973), passim.
[93] V. P. Glushko, *Development of Rocketry and Space Technology in the USSR*, Moscow (1973), pp. 18–19; E. K. Moshkin, *op. cit.*, p. 213; V. Tolubko, *Voenno-istoricheskii Zhurnal*, 1975, No. 4, p. 51.
[94] For example, in the development of turbopump propellant feed systems and gas generators—see E. K. Moshkin, *op. cit.*, pp. 212–31, and *The Soviet Encyclopedia of Space Flight*, Moscow (1969), p. 333.

Fig 9.7 *Classification of military rockets*

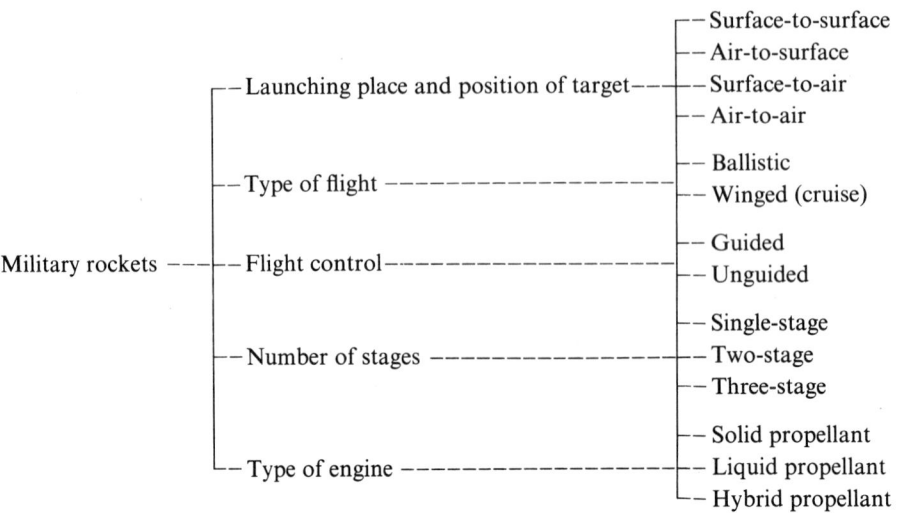

Source: E. S. Fridenson, *Osnoby raketnoi tekhniki* (1973), p. 25.

Germany, the United States and the Soviet Union each developed liquid-propellant rockets, rocket airplanes and rocket boosters for airplanes before 1945. Liquid-propellant engines were essential for long-range rockets, since at the time only liquid propellant could be stored in tanks on the rocket and fed into the combustion chamber in the required amounts during flight. Germany was much more successful than the other countries in this area: for all its many failings the A-4 (V-2) was far in advance of anything developed in the Soviet Union or the United States in terms of engine power, range, rocket design and guidance technology.[95] Only Germany used rocket aircraft at the front during the war; the number of missions was small and the role they played was not decisive. In May 1942 the first Soviet rocket plane, the BI-1, was flight-tested; it crashed on one of its first flights and little further work was done to develop rocket planes before 1945. Rocket boosters saw little operational use during the war, and none at all by the Soviet Union, although the RD-1 and RD-1X3 were given extensive ground and flight tests and put into serial production during the war.[96]

Soviet rocket research suffered many vicissitudes before 1945: the purges left their mark on the whole effort, and with the approach of war resources were shifted to work which promised quick results, in particular to rocket artillery

---

[95] E. K. Moshkin, *op. cit.*, pp. 183–8; W. von Braun and F. I. Ordway, *op. cit.*, p. 101; E. Klee and O. Merk, *The Birth of the Missile*, London (1965). The A-4 (V-2) weighed 13 tonnes, developed 30 tonnes of thrust in vacuum, was powered by LOX and ethyl alcohol, and could carry a payload of 1,000 kg over a range of almost 300 km; it had a miss distance of 3·5 miles over a typical flight path of 200 miles.

[96] W. von Braun and F. I. Ordway, *op. cit.*, passim: E. K. Moshkin, *op. cit.*, pp. 80, 213, 231–4; A. Yakovlev, *Tsel' zhizni* (1974), p. 418; F. I. Ordway III and R. G. Wakeford, *International Missile and Spacecraft Guide*, London (1960).

and rocket boosters for aircraft. Soviet achievements were modest when set beside the German programme, but there were intangible results which cannot be assessed easily. Of these the most important was the fact that a number of Soviet scientists and engineers had considerable experience in rocketry. There thus existed a cadre of experienced personnel who could provide the basis for the large-scale programme after the war.

THE DEVELOPMENT OF SOVIET NUCLEAR WEAPONS

Until 1940 nuclear physicists were the very model of an international scientific community: the rapid dissemination of experimental and theoretical results was not hindered by the claims of military security.[97] Soviet physicists were part of this community and contributed to the headlong progress of research: P. A. Cherenkov, I. M. Frank and I. E. Tamm were awarded the Nobel Prize in 1958 for their discovery in 1934 of the Cherenkov effect and its theoretical elucidation.[98] Consequently when Hahn and Strassmann discovered nuclear fission in 1938 the news of their discovery caused as great excitement in the Soviet Union as elsewhere.[99]

Nuclear fission had been achieved by bombarding uranium with neutrons, and its significance was increased when Joliot-Curie showed that neutrons were emitted in the fission reaction and thus indicated the possibility of a chain reaction. Research was now undertaken into nuclear fission in Leningrad, Khar'kov and Moscow, with the primary aim of attaining a chain reaction. Soviet physicists became increasingly aware of the military significance of nuclear fission, and their research was directed towards both slow and fast neutron chain reactions.[100] Before the German invasion Academician N. N. Semenov wrote to the People's Commissariat of Heavy Industry about the possibility of creating a uranium bomb, but the war put an end to any hope of such a project.[101]

The leading organiser of this direction of research was I. V. Kurchatov, head of the Laboratory of Neutron Physics at the Leningrad Physico-Technical Institute (LFTI). With the outbreak of war Kurchatov's laboratory was disbanded, and the equipment sent to Kazan with the rest of the LFTI. Kurchatov spent the first months of the war working on the degaussing of ships (demagnetising them to protect them against mines); early in 1942 he was made chief of the laboratory of tank armour.[102] Kurchatov did not consider it a mistake to stop research in nuclear physics; Flerov, who was one of his associates, thought differently. He was convinced that an explosive nuclear chain reaction was possible and in May 1942 he wrote to the State Defence

---

[97] See R. Jungck, *Brighter than a Thousand Suns* (Penguin, 1964).
[98] *Razvitie fiziki v SSSR*, Vol. 2 (1967), pp. 197–8.
[99] I. N. Golovin, *I. V. Kurchatov* (1967), pp. 38–9. This was also published in English in 1969 in Moscow. There are minor differences between the two editions. All page references are to the Russian edition.
[100] I. N. Golovin, *op. cit.*, 39–42; A. Kramish, *Atomic Energy in the Soviet Union*, London (1960), chapter 2; G. A. Modelski, *Atomic Energy in the Communist Bloc* (1959), pp. 26–30; P. T. Astashenkov, *Akademik I. V. Kurchatov* (1971), p. 138.
[101] I. N. Golovin, *op. cit.*, p. 42.
[102] I. N. Golovin, *op. cit.*, pp. 42–5; P. T. Astashenkov, *op. cit.*, p. 141 ff.

Committee that 'we must build the uranium bomb without delay'.[103] Kurchatov was not convinced that it would be right to devote resources to trying to build a uranium bomb, when the same resources could be used to achieve immediate results by less spectacular means—by improving armour for example.[104] In the meantime, however, the Soviet leaders had learnt of the US and German programmes to build the bomb, and at the end of 1942 the State Defence Committee put Kurchatov in charge of the development of a uranium bomb.[105] This work got under way in 1943. It seems that not all Soviet physicists were convinced of the urgency of the programme, and of those who were, not all were confident of its success.[106] Work began slowly, but expanded as new lines of enquiry opened up and the tide of war turned. The People's Commissariat of Nonferrous Metallurgy was given the task of supplying the programme with graphite and uranium.[107] The People's Commissariat of Munitions, headed by B. L. Vannikov, was brought in to help with the creation of a device which would shoot two parts of the uranium-235 isotope together to make them 'go supercritical'.[108] It was at this point that the R and D work was transformed into a large-scale engineering programme.

The Soviet programme received a major impetus with the completion and use of the United States' atomic bomb. On 24 July 1945, after the bomb had been tested but before it had been used, Truman 'casually mentioned' to Stalin at Potsdam that the United States had 'a new weapon of unusual destructive force'.[109] Stalin's reaction was to speed up work on the Soviet bomb.[110] By September a Scientific-Technical Council had been attached to the Council of People's Commissars to handle all questions involved in making nuclear weapons.[111] Vannikov was chairman of this Council and had as his deputies Kurchatov and M. G. Pervukhin, People's Commissar of the Chemical Industry. In 1946 this Council seems to have been supplemented by the First Main Administration of the Council of Ministers;[112] Vannikov was moved from the Ministry of Munitions and put in charge of this Administration. Beria was responsible for general supervision of the programme.[113]

Work now went ahead more rapidly. Kurchatov's plan was to accomplish a

---

[103] I. N. Golovin, *op. cit.*, pp. 46–8; P. T. Astashenkov, *op. cit.*, pp. 170–81. Flerov wrote to S. V. Kaftanov, who was responsible for science in the GKO (State Committee of Defence), and to Stalin. Kaftanov already knew something about nuclear fission. When he visited Berlin in 1939 he went to see Otto Hahn at his laboratory and talked to him about his work. (See R. Jungck, *op. cit.*, p. 236 ff.)

[104] I. N. Golovin, *op. cit.*, pp. 47–8.

[105] I. N. Golovin, *op. cit.*, p. 48 ff; P. T. Astashenkov, *op. cit.*, p. 170 ff.

[106] I. N. Golovin, *op. cit.*, pp. 51–2; P. T. Astashenkov, *op. cit.*, p. 172.

[107] I. N. Golovin, *op. cit.*, p. 56.

[108] *Ibid.*

[109] H. S. Truman, *1945: Year of Decisions*, New York (1965), p. 458.

[110] G. K. Zhukov, *Vospominaniya i razmyshleniya* (1969), p. 732.

[111] I. N. Golovin, *op. cit.*, pp. 58–9; P. T. Astashenkov, *op. cit.*, p. 229. V. Yemel'yanov, *Novyi Mir*, 1967, No. 2, p. 140.

[112] Vannikov remained People's Commissar of Munitions until 1946 (see A. Kramish, *op. cit.*, p. 117).

[113] A. Kramish, *op. cit.*, pp. 176–7; Beria had been given general supervision of the armaments and munitions industries during the war (see V. Kravchenko, *I Chose Freedom*, London (1947), p. 404). In July 1953, after Beria had been removed from power, the Ministry of Medium Machine Building was set up under Malyshev to take over the nuclear weapons programme.

chain reaction in a uranium-graphite system, set up plutonium production, test a model of the weapon under laboratory conditions (in so far as this was practicable) and check the calculations of a fast neutron chain reaction.[114] In December 1946 the first Soviet chain reaction was achieved, in spite of problems with the purity of the uranium. This first uranium pile—known as F-1—could not produce plutonium in sufficient quantities for a bomb; in January 1947 the foundations of the first production reactor were laid.[115] Research with the F-1 led to the development of a technology for the industrial extraction of plutonium from uranium; in spite of this, however, it proved extremely difficult to bring the first industrial reactor into operation. It went critical in 1948.[116] Enough plutonium was now produced for the design of the weapon to begin. Experiments were undertaken into the behaviour of a developing chain reaction, in order to ensure, as far as possible, the success of the first test. Two bombs were prepared. The first test took place on 29 August 1949 and was successful.[117]

The first Soviet explosion took place just over four years after the first American test; and the first Soviet chain reaction had taken place four years after the first American chain reaction. On this basis the Soviet Union and United States can be said to have taken about the same length of time to develop the atomic bomb. As evidence of political commitment to the development of the bomb, the setting up of the Vannikov Council in mid-1945 is comparable to the initiation of the Manhattan Project in August 1942.[118] If we take this as the basis for comparison, the Soviet Union took one year longer than the United States to develop the bomb. The British programme was slower (see Table 9.16).

The second stage in the development of nuclear weapons was the creation of the hydrogen, or fusion, bomb. Soviet work on this had begun before the first test of the atomic bomb. It is not clear whether a decision to develop the hydrogen bomb had in fact been taken before the end of 1949, but in any event the ground work had been done.[119] On 31 January 1950 President Truman announced that the United States would proceed with the development of the hydrogen bomb, and the first American thermonuclear device was exploded on 1 November 1952. The first Soviet thermonuclear explosion took place on 12 August 1953.

The Soviet Union claims that the explosion in August 1953 was the world's first test of a hydrogen bomb, since the US device consisted of equipment weighing 60 tons and was therefore not transportable—it could not be delivered to target as a bomb must be.[120] The Soviet device was exploded on top

[114] I. N. Golovin, *op. cit.*, p. 62.
[115] I. N. Golovin, *op. cit.*, p. 69.
[116] P. T. Astashenkov, *op. cit.*, p. 216.
[117] I. N. Golovin, *op. cit.*, pp. 75–6. Golovin gives 23 September as the date of the first test, but this was the date on which the United States announced that the Soviet Union had undertaken a test. The date of the test itself seems to have been 29 August. (See A. Kramish, *op. cit.*, pp. 122–3.)
[118] See R. Jungck, *op. cit.*, p. 110.
[119] I. N. Golovin, *op. cit.*, p. 76.
[120] I. N. Golovin, *op. cit.*, p. 77; P. T. Astashenkov, *op. cit.*, pp. 237–9; A. Kramish, *op. cit.*, p. 125.

of a tower and was deliverable;[121] the first deliverable American device was exploded on 1 March 1954. The Soviet bomb was not, however, as advanced as that exploded by the United States. It had a yield of 500 kilotons at most, while the yields of the first United States explosions were 10 and 15 megatons;[122] it was not until 1955 that Khrushchev spoke of megaton yields for Soviet weapons.[123] The first Soviet fusion bomb contained a small amount of thermonuclear material, which was ignited by a relatively large amount of fissionable material.[124] It was the first device to use lithium deuteride as a fuel, but it was not based on the Teller-Ulam idea and could not have been expanded into something bigger. According to Herbert York:

> it seems to have been a development step the US bypassed in its successful search for a configuration that would make it possible to produce an arbitrarily large explosion with a relatively small quantity of fissionable material.[125]

The Soviet Union made a slower start than the United States in nuclear testing (see Table 9.17).[126]

It is not clear what the Soviet Union learnt by espionage from the United

Table 9.16 *Steps in the developments of nuclear weapons*

	USA	USSR	UK	France
First chain reaction	Dec. 1942[4]	Dec. 1946[2]	Aug. 1947[1]	Dec. 1948[1]
First production reactor goes critical	1944[7]	1948[3]	1950[1]	1958–9[6]
First nuclear test explosion	16 July 1945[1]	29 Aug. 1949[4]	3 Oct. 1952[1]	13 Feb. 1960[5]
First thermonuclear test explosion	31 Oct. 1952[5]	12 Aug. 1953[2]	15 May 1957[5]	24 Sept. 1968[5]

Sources:
1 M. Gowing, *Independence and Deterrence, Britain and Atomic Energy*, Vol. 2, *Policy Execution*, London (1974), pp. 520–9.
2 I. N. Golovin, *I. V. Kurchatov*, Moscow (1967), pp. 66, 78.
3 P. T. Astashenkov, *Akademik I. V. Kurchatov* (1971), p. 216.
4 A. Kramish, *Atomic Energy in the Soviet Union*, London (1960), pp. 122–3.
5 *SIPRI Yearbook 1968–69*, Stockholm (1970), p. 243.
6 W. Mendl, *Deterrence and Persuasion*, London (1970), pp. 132, 149.
7 R. G. Hewitt and O. E. Anderson, Jr, *The New World 1939/46*, Vol. 1 of *A History of the US Atomic Energy Commission* (1962), p. 309.

[121] I. N. Golovin, *op. cit.*, p. 78.
[122] H. York, *Scientific American*, October 1975, pp. 110–11.
[123] A. Kramish, *op. cit.*, pp. 125–7.
[124] H. York, *loc. cit.*, p. 111.
[125] *Ibid.* For further discussion see H. York, *The Advisors. Oppenheimer, Teller and the Superbomb*, San Francisco (1976), pp. 87–93.
[126] A. Kramish, *op. cit.*, pp. 125–7.

States or British programmes, or how much time was thereby saved.[127] But it is clear that the existence of the US programme was an important factor in the evolution of the Soviet effort. The decision to embark on building the bomb was taken in 1942 only after reports had been received that the United States and Germany were already pursuing this goal. The dropping of the bombs on Hiroshima and Nagasaki demonstrated that the bomb could be built and that it was extremely powerful. Before 1945 many Soviet physicists had doubted that the bomb could be built; now the Soviet effort was intensified. Work on the fusion bomb was stepped up after the US explosion of November 1952.[128]

THE SOVIET MISSILE PROGRAMME 1945–57

In March 1945 the Red Army captured Peenemunde, the main German centre for rocket research and development. They found that almost all the leading scientists and engineers had fled to meet the US forces, taking most of the important documents with them. What was left was mostly wreckage.[129] The Soviet Union also inherited test and production facilities, although it is not clear how well equipped these were.[130] A team was sent from the Soviet Union in 1945 to examine what had been captured, and before the end of the year a decision had been taken to build versions of the A-4 (V-2).[131] In the Soviet Union

> already at the end of 1945 a special decision was taken to transfer several defence industries plants to a new footing (*na novyi profil'*)—the production of equipment and instruments necessary for setting up the serial production of rocket weapons. The direction and coordination of all this work was carried out by organs specially created under (*pri*) the Council of People's Commissars and in the Armed Forces, and headed by D. F. Ustinov, B. L. Vannikov and Chief Marshal of Artillery M. I. Nedelin.[132]

It appears that two groups worked on creating versions of the A-4 (V-2): one was drawn from the German engineers and technicians who had not fled to the West, the other was a Soviet group under S. P. Korolev.[133]

[127] In the late 1940s and early 1950s many wild statements were made about the benefits the Soviet Union gained from espionage. More recent writers suggest that the Soviet programme might have been speeded up by several months at most by the information gained from spies. (See G. A. Modelski, *op. cit.*, p. 33; Margaret Gowing, *Independence and Deterrence, Britain and Atomic Energy 1945–52*, Vol. 2, *Policy Execution*, London (1974), p. 150).

[128] I. N. Golovin, *op. cit.*, p. 77.

[129] W. von Braun and F. I. Ordway, *op. cit.*, p. 140; G. A. Tokaty-Tokaev, *Spaceflight*, October 1968, pp. 342–4.

[130] For differing accounts see A. C. Sutton, *Western Technology and Soviet Development 1945 to 1960*. Stanford, Cal. (1973), p. 271; and G. A. Tokaty, 'Soviet rocket technology', in Eugene M. Emme, *The History of Rocket Technology*, Detroit (1964), p. 279.

[131] *50 let vooruzhennykh sil SSSR*, (1968), p. 485; Irmgard Groettrup, *Rocket Wife* (1959), p. 12.

[132] Army General V. Tolubko, Commander-in-Chief of the Strategic Missile Forces, *Tekhnika i Vooruzheniye*, 1974, No. 11, p. 3. Nedelin was chief of the Main Artillery Administration at the time; in 1952 he was Deputy Minister of War for armament, presumably during the period when the R-2 was coming into service; and he was appointed first Commander-in-Chief of the Strategic Missile Forces when they were established as a separate branch of the armed forces in December 1959 (*Bol'shaya Sovetskaya Entsiklopedia*, Vol. 17 (1974), Col. 1198).

[133] P. T. Astashenkov, *Akademik S. P. Korolev* (1969), pp. 98, 102; Irmgard Groettrup, *op. cit.*, p. 12.

Table 9.17 *Nuclear explosions 1945–74 (announced and presumed)*

Year	USA A	USA U	USSR A	USSR U	UK A	UK U	France A	France U	China A	China U
1945	3	0								
1946	1	1(1)								
1947	0	0								
1948	3	0								
1949	0	0	1	0						
1950	0	0	0	0						
1951	15	1	2	0						
1952	10	0	0	0	1	0				
1953	11	0	2	0	2	0				
1954	6	0	2	0	0	0				
1955	13	2(1)	4	0	0	0				
1956	14	0	7	0	6	0				
1957	26	2	13	0	7	0				
1958	53	13(2)	26	0	5	0				
1945–1958	155	19(4)	57 +33[a]	0	21	0				
1959	0	0	0	0	0	0				
1960	0	0	0	0	0	0	3	0		
1961	0	9	30	2(1)	0	0	1	1		
1962	38	50(1)	41	1	0	2	0	1		
1963 (to 5 Aug.)	0	11	0	0	0	0	0	2		
1959– 5 Aug. 1963	38	70(1)	71	3(1)	0	2	4	4		
5 Aug.– Dec. 1963	0	14	0	0	0	0	0	1		
1964	0	28	0	6	0	1	0	3	1	0
1965	0	28	0	9	0	1	0	4	1	0
1966	0	40	0	14	0	0	5	1	3	0
1967	0	28	0	14	0	0	3	0	2	0
1968	0	37	0	12	0	0	5	0	1	0
1969	0	28	0	15	0	0	0	0	1	1
1970	0	30	0	13	0	0	8	0	1	0
1971	0	12	0	18	0	0	5	0	1	0
1972	0	7	0	22	0	0	3	0	2	0
1973	0	9	0	14	0	0	5	0	1	0
1974	0	6 +23[b] +18[c]	0	20	0	1	7	0	1	0
Totals	193	397(5)	161	160	21	5	45	13	15	1

A – atmospheric; U – underground and underwater (latter put in brackets).

Source: SIPRI Yearbook 1975, Stockholm (1975), pp. 510–11.

Notes:
a – Up to 1958. Dates unknown.
b – Conducted between 15 September 1961 and 20 August 1963.
c – Conducted from 1972–1973.

Note:
Some comments are necessary on these figures: the total numbers of tests are not publicly available; and more information is available about the US nuclear weapon testing programme than about the Soviet programme. The figures given are almost certainly too low, and, according to the *SIPRI Yearbook 1968–69*, Stockholm (1970), p. 245, the Soviet figure could be twice as large; the compiler of the original tables, M. Leitenberg, now thinks that the Soviet figure might be 30 per cent larger than that given in the table, but that it is unlikely to be twice as large (letter to the author, December 1975). This table does not distinguish between weapons tests and explosions for peaceful purposes; a breakdown may be found in *SIPRI Yearbook 1968–69*, p. 242, and *SIPRI Yearbook 1972*, p. 461 ff. The proportion of the latter probably does not exceed 15 per cent. It should be noted that in 1959 and 1960 there was a moratorium on testing, and on 5 August 1963 the Partial Test Ban Treaty was signed which prohibits tests in the atmosphere. France and China are not signatories. In 1974 a Threshold Test Ban Treaty was signed by the USA and the USSR which prohibits underground nuclear weapons tests with a yield of more than 150 kilotons, with effect from 31 March 1976.

The design work of these two groups was largely completed in 1946, and the German group—or at least part of it—was shipped to the Soviet Union in October of that year.[134] In October 1947 the rockets designed by the two groups were launched in the Soviet Union. The German-designed rocket flew about 300 km and landed in the target area.[135] The Soviet-designed rocket seems to have performed better: it was designated the R-1 (R: *raketa*, rocket).[136] This series of tests of the R-1 consisted of 11 launches,[137] and testing seems to have continued. The R-1 was introduced into military service, although it is not clear on what scale.

In 1946 further steps were taken towards the development of long-range missiles. First, the GDL-OKB, which had been working on rocket boosters for aircraft, now turned to the development of powerful liquid-propellant rocket engines for long-range missiles.[138] Second, S. P. Korolev, who had been V. M. Glushko's deputy for flight-testing at the GDL-OKB since 1942, was appointed head of his own OKB in August—this was probably a bureau for the design of rockets. Korolev had been a member of the team that went to Germany in 1945, and seems to have remained there until February 1947.[139] Third, in July 1946 the first missile unit in the Soviet Army was formed and officers from Guards Mortar Units, Artillery, the Air Force and Navy were sent to it in order to study how missiles might best be used in warfare.[140]

[134] *50 let vooruzhennykh sil SSSR*, p. 485; G. A. Tokaty-Tokaev, *Spaceflight, loc. cit.*, p. 344.
[135] I. Groettrup, *op. cit.*, p. 107.
[136] *50 let vooruzhennykh sil SSSR* (1968), p. 485.
[137] P. T. Astashenkov, *op. cit.*, p. 107.
[138] E. K. Moshkin, *op. cit.*, p. 231.
[139] P. T. Astashenkov, *op. cit.*, p. 99; A. Romanov, *Konstruktor kosmicheskikh korablei* (1971), p. 48.
[140] N. Krylov, *Voenno-istoricheskii Zhurnal*, 1967, No. 7, p. 21; Guards Mortar Units were the 'Katyusha' units.

In 1949 the Soviet Union began systematic launches of high-altitude rockets for geophysical and biological research. The engines for these rockets—among which were the V-2-A and V-5-V (V: *vysotnyi*, high-altitude) rockets—were designed by the GDL-OKB.[141] In the following year the R-2, an improved version of the R-1, was tested successfully. The improvements had been directed towards increasing the range and speed of flight of the rocket, as well as its firepower and accuracy. The R-2 is said to have attained 'higher tactical-technical indicators than the R-1', and to have entered the arsenal of the Soviet Army's rocket units.[142] Soviet sources do not say what these 'higher tactical-technical indicators' were, except that the R-2 had twice the range of the R-1, that is, over 600 km.[143] The next missile in this series appears to have been the SS-3 (Shyster). This was tested in 1953 or 1954, and entered service in 1955. According to one Soviet source it had a range of over 1,000 km;[144] and it is probably to this missile that Marshal Krylov referred when he wrote that 'by 1955 there were in the Soviet Army several missile units armed with medium-range missiles'.[145] Before the SS-3 was tested the decision seems to have been taken to develop an improved version, for in 1952 development work began on the RD-214 engine which powers the SS-4. Development was completed by 1957, and flight tests were begun in that year, although the SS-4 does not seem to have been deployed until 1959.[146]

The decision to build a missile which could serve as an ICBM and as a space launcher seems to have been taken in principle by 1947;[147] but the specific development decision was made in 1953, perhaps on the basis of the successful testing of the fusion bomb and of the SS-3.[148] In 1954 work began on the RD-107 and RD-108 engines, which were used to power the SS-6, the first ICBM and the launcher of Sputnik 1.[149] The SS-6 flew for the first time in August 1957. It had two stages: the first consisted of four RD-107 engines arranged around the missile, while the second consisted of the RD-108 in the main body. The SS-6 proved very successful as a space launcher, although it does not appear to have been designed primarily for this role.[150] It was less successful as a military missile. Technical difficulties are suggested by the sporadic testing of

[141] E. K. Moshkin, *op. cit.*, pp. 238–9.
[142] N. Krylov, *loc. cit.*, p. 19.
[143] P. T. Astashenkov, *op. cit.*, p. 109.
[144] P. T. Astashenkov, *op. cit.*, p. 109.
[145] N. Krylov, *loc. cit.*, p. 21: A. Lee and R. E. Stockwell, in (Ed.) A. Lee, *The Soviet Air and Rocket Forces*, London (1959), p. 150.
[146] *The Soviet Encyclopedia of Space Flight*, pp. 335–7; *Soviet Space Programs 1966–70*, Staff Report prepared for the use of the Committee on Aeronautical and Space Sciences, US Senate, US Government Printing Office, Washington DC (1971), pp. 139–40.
[147] In his various accounts of Soviet missile development G. A. Tokaty-Tokaev refers to a Governmental Commission for Long-Range Rockets (*Pravitel'stvennaya komissiya po raketam dal'nego deistviya*) which was set up in March 1947, and of which he himself was a member. This shows the Soviet interest in intercontinental missiles at the time, but the work of this commission did not provide the basis for the development of the Soviet ICBM. It was set up to examine and develop the Sanger/Bredt project for a glide bomber which would 'skip along the top of the atmosphere like a stone on a pond, reaching New York with a bomb load of six tons' (W. von Braun and F. I. Ordway, *op. cit.*, p. 119). (See G. A. Tokaty-Tokaev, *loc. cit.*, pp. 344–6).
[148] P. T. Astashenkov, *op. cit.*, p. 112.
[149] *The Soviet Encyclopedia of Space Flight*, pp. 333–5.
[150] P. T. Astashenkov, *op. cit.*, p. 119.

the missile during 1957 and 1958. The SS-6 had an unstable fuel which was difficult to store and therefore could not be launched quickly; its range was so short that it had to be based in the far north if it was to be able to strike the United States. It was deployed in very small numbers. It is unlikely to have become operational before 1960, and its operational state was probably far from satisfactory.[151]

SOVIET AND UNITED STATES ICBM PROGRAMMES 1957–75

After the launching of Sputnik, Wernher von Braun claimed that the United States could have developed an ICBM by the early 1950s if a full-scale programme had been embarked upon immediately after the war.[152] Until the mid-1950s, however, American interest in ballistic missiles was erratic and uncoordinated. The V-2 was tested and modified in the late 1940s, and in 1951 the US Army missile engineering team began design work on the Redstone tactical missile, which had a range of about 200 miles. A direct descendent of the V-2, Redstone was first test-fired in 1953, but did not enter service until 1958.[153] At the beginning of 1953 'no stated requirement for the ICBM existed, missile technology was primitive, a missile industry existed only in embryo form, and no service had committed itself to the acquisition of such a weapon'.[154] This state of affairs changed during the next three years. In 1954 the Atlas ICBM programme was initiated on the basis of R and D work which had been done since the war. In the following year two IRBM programmes were begun: the Joint Navy-Army Jupiter, and the US Air Force Thor.

This is quite different from the Soviet decision to give priority to missile development immediately after the war. The contrast should not be overstated, however. The Soviet programme was not just set up and left to proceed under its own momentum. It seems to have been expanded in 1947 after the successful tests of the R-1, and again after the 1953 decision to develop an ICBM.[155] The Soviet Union began to develop MRBMs about five years before the United States, but work on an ICBM started only one year earlier. The priority given to missile development after the war helps to account for the Soviet lead in flight-testing MRBMs and ICBMs. In some areas of missile technology the Soviet Union seems to have established a lead; for example in rocket engines (see Table 9.22 and 9.23).

Once the US programme gained momentum technical advances came very quickly in guidance systems, propellants, warheads and ground equipment. In 1957 the two IRBMs were flight-tested; in the following year the Atlas ICBM was launched. In 1957 the decision was taken to develop another ICBM, the Titan I, and this was launched in January 1959. By this time it had been decided

[151] T. W. Wolfe, *Soviet Power and Europe 1945–70*, Baltimore and London (1970), p. 182; *Soviet Space Program 1966–70*, p. 143; M. Armacost, *The Politics of Weapons Innovation: the Thor-Jupiter Controversy*, New York and London (1969), p. 213. N. S. Khrushchev, *Khrushchev Remembers. The Last Testament*, London (1974), pp. 46–8.
[152] M. Armacost, *op. cit.*, p. 24.
[153] F. I. Ordway and Ronald C. Wakeford, *International Missile and Spacecraft Guide*, London (1960), p. 13.
[154] M. Armacost, *op. cit.*, p. 27.
[155] N. Krylov, *loc. cit.*, p. 19.

Table 9.18 US land-based ballistic missiles

Name	Type	First tested	IOC	No. of stages	Engine type	Body dia. (m)	Length (m)	No. of warheads	Yield	Range (km)	CEP
Jupiter	IRBM	1957[5]	1958[5]	1[4]	Liquid propellant[4]	2·67	18·39	1[3]	n.a.	2400[3]	n.a.
Thor	IRBM	1957	1960[5]	1[4]	Liquid propellant[4]	2·44[3]	19·81[3]	1[3]	n.a.	2775[3]	n.a.
Atlas	ICBM	1958[5]	1959[3]	Core + Boosters[3]	Liquid propellant[3]	3·05[3]	ca 24[3]	1[3]	3 MT[3]	14500[3]	2mi[8] (for Atlas E)
— D			1959[5]								
— E Better guidance and more powerful engines		1960[5]	1961[3]								
— F Quick-firing model		1961[5]	1962[3]							10150[3]	
Titan I	ICBM	1959[5]	1962[2]	2[4]	Liquid propellant[4]	3·05[4]	29·87[4]	1[3]	n.a.	10140[3]	n.a.
Titan II	ICBM	1962[7]	1963[6]	2[2]	Liquid propellant[2]	3·05[2]	31·3[2]	1[2]	5–10 MT[1]	15000[2]	0·5[12]–0·75nm[9,1]
Minuteman I	ICBM	1961[2]	1962[2]	3[2]	Solid propellant[2]	1·8[2]	17[2]	1[2]	1 MT[2]	10000[2]	0·75nm[9]
Minuteman II	ICBM	1964[2]	1966[2]	3[2]	Solid propellant[4]	1·8[2]	18·2[2]	1[2]	1–2 MT[1]	11250+[2]	0·3[10]–0·35nm[9]
Minuteman III	ICBM	1968[2]	1970[2]	3[2]	Solid propellant[2]	1·85[2]	18·2[2]	3 MIRV[2]	×170 (200) KT[1]	13000[2]	0·2nm[10]–0·25[9] c. 100 m

IOC – Initial Operational Capability.
CEP – Circular Error Probable.

mi – miles.
nm – nautical miles.

Sources:
1 *The Military Balance 1975–76*, London (1975), p. 71.
2 *Jane's Weapons Systems 1974–75*, London (1974), passim.
3 *Jane's All the World's Aircraft 1962–63*, London (1962), passim.
4 V. I. Varfolomeev and M. I. Kopytov, *Proektirovanie i ispytanie ballisticheskikh raket* (1970), pp. 52–3.
5 W. von Braun and F. I. Ordway III, *History of Rocketry and Space Travel*, London (1967), pp. 129.
6 General G. S. Brown USAF, *US Military Posture for FY 1976*, Washington DC (1975), passim.
7 E. M. Emme, *The History of Rocket Technology*, Detroit (1964), pp. 151–3.
8 *Jane's All the World's Aircraft 1962–63*, p. 394.
9 L. E. Davis and W. R. Schilling, *Journal of Conflict Resolution*, 1973, No. 2, p. 233.
10 Statement by Congressman R. L. Leggatt, in *The Vladivostock Accord: Implications to US Security, Arms Control and World Peace, Hearings before the Subcommittee on Security and Scientific Affairs of the Committee on International Relations*, 24 and 25 June and 8 July 1975, Washington DC (1975), pp. 10, 11.
11 *The Defense Monitor*, 1974, No. 4, p. 7.
12 K. Tsipis, *Offensive Missiles*, Stockholm Paper No. 5, SIPRI (1974), p. 20.

to develop a second generation of ICBMs—Titan II and Minuteman I. Both Atlas and Titan I had to be fuelled with tons of LOX and kerosene minutes before firing, but the new ICBMs had storable fuel, liquid for Titan II and solid for Minuteman I. These missiles also had improved guidance systems and could be launched from underground silos. They entered service in 1962 and 1963, and by 1966 the first generation of missiles—Atlas, Titan I, Thor and Jupiter—had been phased out of service. Minuteman II, which had a longer range than Minuteman I and could deliver a heavier payload with greater accuracy, received its first flight test in 1964 and entered service in 1966. In the same year development of Minuteman III, which carries multiple warheads, was authorised; its first flight test took place in 1968, and it entered service in 1970.[156]

The pattern of deployment has been different in the Soviet Union. Generations of ICBMs have succeeded each other more slowly than in the United States; but in spite of the larger number of models deployed (ten as opposed to six in the US), a clear demarcation between generations can be discerned. The first Soviet ICBMs to be deployed in any quantity were the SS-7 and SS-8. Both are two-stage, liquid-propellant missiles with warhead yields of 5 mt; neither appears to be very accurate. They have been deployed both above ground and in silos. There have been three different models of the SS-7, marked by changes in the guidance system.[157] This was initially radio command, later inertial. Deployment of these missiles—sometimes referred to as the second generation of Soviet ICBMs—began in 1961 and ceased in 1964 or 1965.

Deployment of the third generation of Soviet ICBMs—the SS-9, SS-11 and SS-13—began in 1965. These missiles had improved accuracy, storable fuel and could be launched from underground silos. They were deployed in much larger numbers than the second generation of ICBMs, which were not, however, withdrawn from service. Of these missiles the SS-9 has excited most comment. It uses storable liquid propellant, is the most accurate missile of its generation, and is believed to have a warhead in the 20 to 25 mt range. Deployment of the SS-9 began in 1965 and continued until 1970. Tests of the warhead seem to have taken place during 1961 and 1962, and flight tests of the missile during 1963 and 1964.[158] Four models have been tested. Model 1 has a warhead yield of 18 to 20 mt, while Model 2 is thought to have a warhead yield of 25 mt. The deployed force consists primarily of Model 2. Model 3 is a fractional orbit bombardment system (FOBS), which could strike the United States by flying a partial orbit round the globe rather than the direct route. Model 4 carried three warheads, although they are not independently targetable; it has not been deployed. The test programme for this model suggests that the Soviet Union faced considerable difficulties with multiple

---

[156] See Tables 9.20 and 9.21.

[157] Chairman of the Joint Chiefs of Staff General George S. Brown USAF, *US Military Posture for FY 1976*, Washington DC (1975), p. 10; *Jane's Weapons Systems 1974–75*, London (1974), p. 12.

[158] The nuclear programme of 1961 included the detonation of very high-yield devices; it seems logical to see these as tests for the SS-9 or SS-10 warhead. One Soviet source refers to tests in 1963 and 1964 of a very accurate ICBM with a range of 12–13,000 km; this fits the SS-9 but not (on the available data) the SS-11. (*see 50 let vooruzhennykh sil SSSR* (1968), p. 505).

Table 9.19 Soviet land-based ballistic missiles

Name[a]	Type	First tested	IOC	No. of stages	Engine type	Body dia. (m)	Length (m)	No. of warheads	Yield	Range (km)	CEP
SS-3 Shyster	MRBM	1954[6]	1955[6]	1[2]	Liquid propellant[2]	n.a.	n.a.	1[2]	n.a.	800–1200[2]	n.a.
SS-4 Sandal	MRBM	1957[5]	1959[9]	1[2]	Liquid propellant[5]	1·6[2]	20·8[2]	1[2]	1 MT[2]	1800[2]	n.a.
SS-5 Skean	IRBM	n.a.	1964[2] (first seen)	1[2]	Liquid propellant[10]	2·4[2]	23·0[2]	1[2]	1 MT[2]	3000–3500[2]	n.a.
SS-6 Sapwood	ICBM	Aug. 1957[8]	1958–9[8]	2[7]	Liquid propellant[7]	10[7] (incl. boosters)	31·9[7]	1[7]	n.a.	n.a.	n.a.
SS-7 Saddler	ICBM	n.a.	1962[1]	2[2]	Liquid propellant[2]	3[2]	30–35[2]	1[1]	5 MT[9]	11000[2]	1·5–2·0nm[12]
SS-8 Sasin	ICBM	n.a.	1963[1]	2[2]	Liquid propellant[2]	3[2]	25[2]	1[1]	5 MT[9]	10–11000[2]	1·5nm[12]
SS-9 Scarp	ICBM	n.a.	n.a.	3[2]	Liquid propellant[2]	3[2]	35[2]			13000[2]	0·5–1·0nm[14] 0·7[12], 0·5[13]
Mod 1		{1963– 1964[4]	1967[1]					1[1]	18 MT[9]		
Mod 2			1966[1]					1[1]	25 MT[9]		
Mod 3	FOBS (not deployed)	1967[2]	1969[1]					1[1]			
Mod 4		1969[1]	1971[1]					3 MRV[1]	×4–5 MT[9]		
SS-10 Scrag	ICBM (not operational)	n.a.	1965[2] (first seen)	3[2]	Liquid propellant[2]	2·75[2]	37[2]	n.a.	n.a.	8000[2]	n.a.
SS-11 Sego	ICBM tested also as IRBM			2[2] (possibly 3)	Liquid propellant[2]	2[11]	20[10]			10000[2]	1·0nm[2]
Mod 1			1966[1]					1[1]	1–2 MT[9]		
Mod 2		1969[3]	1973[1]					1[1]	1–2 MT[9]		
Mod 3			1973[1]					3 MRV[1]	×200 (500) KT[9]		
SS-13 Savage	ICBM	1965[2] (first shown)	1969[1]	3[2]	Solid propellant[2]	1·7[2]	20[2]	1[2]	1 MT[9]	8/10500[2]	0·75[13]– 1·0nm,[14] 0·7nm[12]
SS-14 Scapegoat (Scamp with tracked erector/ launch vehicle)	IRBM	1965[2] (Scamp first shown) 1967[2] (Scapegoat first shown)	n.a.	2[2] (top two stages of SS-13)	Solid propellant[2]	1·4[2]	10·6[2]	1[2]	<1 MT[2]	4000[2]	n.a.

Missile	Type				Propellant			Warheads	Yield	Range	CEP
SS-15 Scrooge	IRBM	n.a.	n.a.	n.a.	n.a.	n.a.	n.a.	n.a.	n.a.	5-6000[11]	n.a.
SS-16	ICBM (land-mobile?)	1972[2]	1975[1]	3[10]	Solid propellant[1]	2[11]	20[10]	1[2] (MIRVs possible)	1 MT+[2]	8000+[2]	n.a.
SS-17	ICBM	1972[2]	1975[1]	2[2]	Liquid;[1] cold launch[1]	2.5[10]	25[10]	4 MIRVs[2]	×200 KT+[2]	9000+[2]	0.4nm[12, b]
SS-18 Mod 1	ICBM	1972[2]	1974[1]	2[10]	Liquid[10] cold launch[2]	3[2]	37[2]	1[9]	?20-50 MT[9,10]	12000[9]	0.25nm(?)[15]
SS-18 Mod 2			1975[1] (deployment started)					6-8 MIRVs[9] 1[9]	×1-2 MT[2,9]		0.50nm[16] 0.3nm[12]
SS-19	ICBM	1973[1]	1975[9] (deployment started)	2[2]	Liquid[2]	2.5[2]	20[2]	4-6 MIRVs[2]	200-400 KT[2]	9000+[2]	0.3nm[12]
SS-X-20 (Mobile?)	IRBM	n.a.	n.a.	2[10]	Solid[11]	2.5[10]	17[10]	MIRV?[10]	n.a.	4000[10]	n.a.

Sources:
1. General G. S. Brown, USAF, *US Military Posture for FY 1976*, Washington DC (1975), passim.
2. *Jane's Weapons Systems 1974–75*, London (1974), passim.
3. Admiral T. H. Moorer, *US Military Posture for FY 1975*, Washington DC (1974), p. 14.
4. *50 let sovetskikh vooruzhennykh sil SSSR* (1968), p. 505.
5. E. K. Moshkin, *Razvitie otechestvennogo raketnogo dvigatelestroeniya* (1973), p. 248.
6. N. Krylov, *Voenno-istoricheskii Zhurnal*, 1967, No. 7, p. 21; A. Lee and R. E. Stockwell, in (Ed.) A. Lee, *The Soviet Air and Rocket Forces*, London (1959), p. 150.
7. *Soviet Space Programs 1966–70*, Staff Report prepared for the committee on Aeronautical and Space Sciences, US Senate, Washington DC (1971), p. 135.
8. See p. 458 above.
9. *The Military Balance 1975–76*, London (1975), p. 71.
10. *Aviation Week and Space Technology*, 15 March 1976, p. 91.
11. *Air Force Magazine*, March 1976, p. 104.
12. Statement by Congressman R. L. Leggatt in *The Vladivostok Accord: Implications to US Security, Arms Control, and World Peace, Hearings before the Subcommittee on International Security and Scientific Affairs of the Committee on International Relations*, 24 and 25 July 1975, Washington DC, (1975), pp. 10, 11.
13. *The Defense Monitor*, 1974, No. 3, p. 7.
14. K. Tsipis, *Offensive Missiles*, Stockholm Paper No. 5, SIPRI (1974), p. 20.
15. See footnote 200.
16. *Aviation Week and Space Technology*, 24 September 1973.

Notes:
a – the Soviet names of these missiles are not known. The SS (surface-to-surface) numbers are attributed by the United States, while the codenames are given by NATO; the identification of codenames with numbers has been made by such sources as *Jane's* and *Aviation Week and Space Technology*.
b – this figure for the SS-17 is assigned on the grounds that the SS-17 is said to be less effective than the SS-19 as a countersilo weapon.

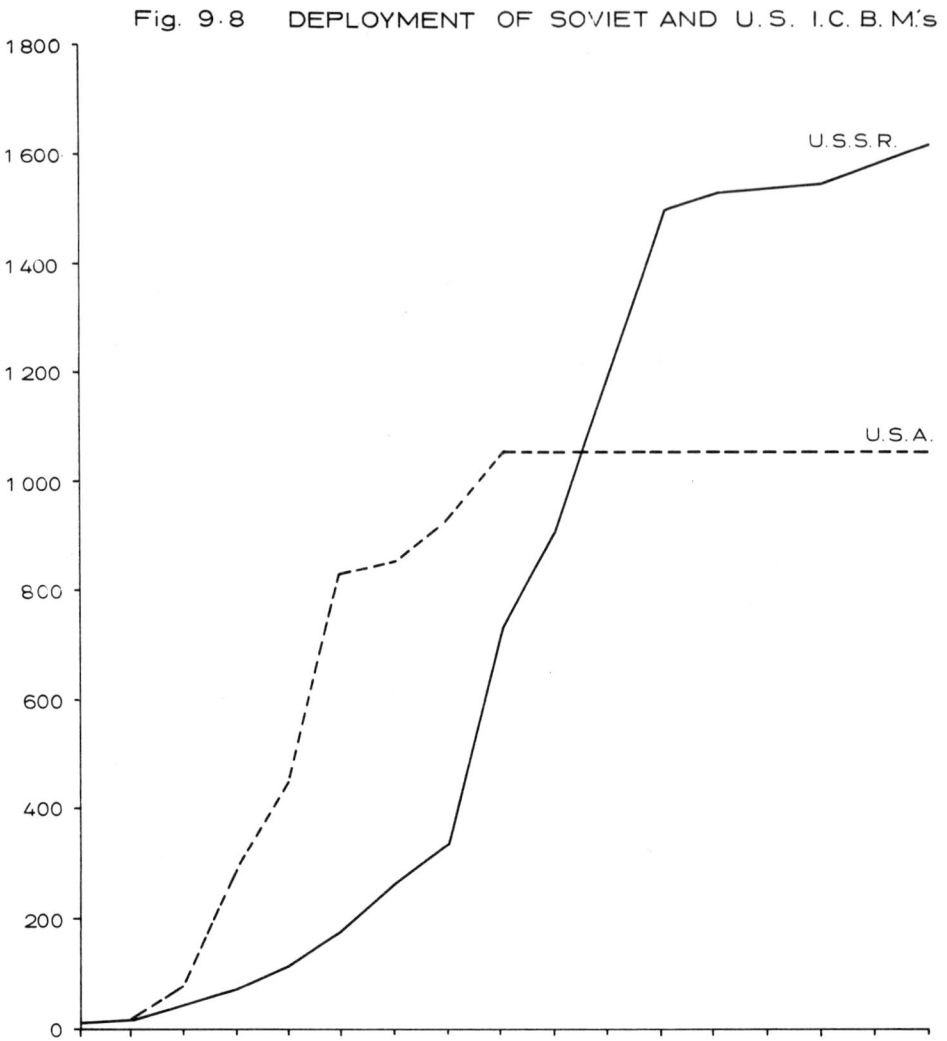

Fig. 9.8 DEPLOYMENT OF SOVIET AND U.S. I.C.B.M.'s

Source : Tables 9·20 & 9·21

warhead technology. Testing began in 1969 and stopped in 1970; it started again in January 1973 with re-entry vehicles of a much different design, but was not continued in 1974. This model has not been deployed, and the 1973 tests may have been component tests for the SS-18.[159] The SS-10, which has not been deployed, appears to have been a parallel and perhaps competitive development. It is reported to be slightly larger than the SS-9 and to be capable of carrying a larger warhead.[160]

[159] Chairman of the Joint Chiefs of Staff Admiral Thomas H. Moorer USN, *US Military Posture for FY 1975*, pp. 12–13; G. S. Brown, *op. cit.*, pp. 11–12.
[160] *Jane's Weapons Systems 1974–75*, p. 13; *The Military Balance 1969–70*, London (1969), p. 56, suggests a warhead with a 30 mt yield for the SS-10.

Table 9.20 Deployment of US ICBMs[b]

	1959	1960[1]	1961[1]	1962[2]	1963[2]	1964[2]	1965[3]	1966[3]	1967[3]	1968[3]	1969[3]	1970[3]	1971[3]	1972[3]	1973[3]	1974[3]	1975[4]
Atlas	?	18	(63)	(126)	(72)	(36)	—	—	—	—	—	—	—	—	—	—	—
Titan 1	—	—	—	(18)	(54)	(36)	—	—	—	—	—	—	—	—	—	—	—
Titan II	—	—	—	—	(18)	(54)	54	54	54	54	54	54	54	54	54	54	54
Minuteman I	—	—	—	(150)	(300)	(700)	800	800	700	600	500	490	390	290	(190)	(100)	—
II	—	—	—	—	—	—	—	80	300	400	500	500	500	500	500	(500)	(450)
III	—	—	—	—	—	—	—	—	—	—	—	10	110	210	(310)	(400)	(550)
Total:																	
Annual:[a]	?	18	45	231		436	100	80	220	100	100	10	100	100	100	90	150
					(−54)	(−54)	(−72)		(−100)	(−100)	(−100)	(−10)	(−100)	(−100)	(−100)	(−90)	(−150)[b]
Cumulative:	?	18	63	294	444	826	854	934	1054	1054	1054	1054	1054	1054	1054	1054	1054
IISS totals:	None	18	63	194	424	834	854	904	1054	1054	1054	1054	1054	1054	1054	1054	1054

Sources:
1 *The Military Balance 1969–70*, London, p. 55.
2 These figures are estimates, based on the fact that Minuteman I and Titan I became operational in 1962 and Titan II in 1963, and that Atlas and Titan I were phased out by 1965. See *Jane's All the World's Aircraft 1965–66*, London (1965), pp. 418, 427, which also gives the totals for Atlas and Titan I.
3 *SIPRI Yearbook 1974*, Stockholm (1974), pp. 106–7.
4 *The Military Balance 1975–76*, London (1975), p. 5.

Notes:
a – withdrawals are given below new deliveries.
b – brackets indicate estimates.

Table 9.21 Deployment of Soviet ICBMs[b]

	1959[1]	1960[1]	1961[1]	1962[1]	1963[1]	1964[1]	1965[2]	1966[2]	1967[2]	1968[2]	1969[2]	1970[2]	1971[2]	1972[2]	1973[2]	1974[2]	1975[3]
SS-6	(6)[b]	(12)	(12)	(12)	—	—	—	—	—	—	—	—	—	—	—	—	—
SS-7	—	—	(30)	(60)	(90)	(120)	150	150	150	150	150	150	139	139	139	139	139
SS-8	—	—	—	—	(30)	(60)	70	70	70	70	70	70	70	70	70	70	70
SS-9	—	—	—	—	—	—	42	108	162	192	228	228	228	228	228	228	228
SS-11 Mod 1	—	—	—	—	—	—	—	(10)	(330)	(470)	(720)	(950)	970	970	970	970	(278)
SS-13	—	—	—	—	—	—	—	—	(10)	(20)	(30)	(40)	60	60	60	60	(960)
SS-11 Mod 3	—	—	—	—	—	—	—	—	—	—	—	—	—	—	20	40	60
SS-16	—	—	—	—	—	—	—	—	—	—	—	—	—	—	—	—	40
SS-17	—	—	—	—	—	—	—	—	—	—	—	—	—	—	—	—	10
SS-18 Mod 1	—	—	—	—	—	—	—	—	—	—	—	—	—	—	—	—	10
Mod 2	—	—	—	—	—	—	—	—	—	—	—	—	—	—	—	—	—
SS-19	—	—	—	—	—	—	—	—	—	—	—	—	—	—	—	—	50
Total:																	
Annual:[a]	6	6	30	30	60 (−12)	70	72	76	384	180	296	300	29 −11	0	20	20	70 (−20)
Cumulative:	6	12	42	72	120	180	262	338	722	902	1198	1498	1527	1527	1547	1567	1617
IISS totals:	some	35	50	75	100	200	270	300	460	800	1050	1300	1510	1527	1527	1575	1618

Sources:
1 A more or less even pattern of deployment from the year of introduction into service up to the year 1965 has been assumed. No precise data are available about the number of SS-6 ICBMs deployed. I have assumed twelve, on the grounds that no more than 'a handful' are reported to have been deployed (see footnote 151). Not all series agree: the IISS data are given in the final line (see *The Military Balance 1969–70*, p. 55; *The Military Balance 1975–76*, p. 73); moreover, although the total number of SS-7s and SS-8s has been given as 209 for recent years, the composition of this total is sometimes given differently (see G. S. Brown, *op. cit.*, p. 10).
2 *SIPRI Yearbook 1974*, Stockholm (1974), pp. 106–7.
3 *The Military Balance 1975–76*, London, p. 8.

Notes:
a – withdrawals are given below new deliveries.
b – brackets indicate estimates.

Table 9.22 *US strategic missile engines*

Parameters	North American Rocketdyne MB-3	North American Rocketdyne S-3D	North American Rocketdyne LR 89-5	Aerojet-General LR 87-AJ-1
Thrust	68 t	68 t	75 t	136 t
Specific impulse	250 sec	245 sec	250 sec	150 sec
Oxidiser	LOX	LOX	LOX	LOX
Fuel	kerosene	kerosene	kerosene	kerosene
Developed	1955–59	–1958	1954–60	1957–59
Used	Thor	Jupiter	Atlas E first stage	Titan I

Sources:
W. von Braun and F. I. Ordway, *History of Rocketry and Space Travel*, London (1967), passim.
F. I. Ordway and R. G. Wakeford, *International Missile and Spacecraft Guide*, London (1960), passim.
Soviet Encyclopedia of Space Flight, Moscow (1969).
Jane's All the World's Aircraft, London, various years.

Table 9.23 *Soviet strategic missile engines, 1957–62*

Parameters	RD-107	RD-108	RD-214	RD-119
Controlled thrust in vacuum	102 t	96 t	74 t	11 t
Specific impulse	314 sec	315 sec	264 sec	352 sec
Oxidiser	LOX	LOX	nitric acid base	LOX
Fuel	kerosene	kerosene	kerosene products	unsymmetrical dimethyl hydrazine
Developed	1954–57	1954–57	1952–57	1958–62
Used	SS-6 1st stage	SS-6 2nd stage	SS-4	2nd stage of Kosmos space launch vehicle

Source:
The Soviet Encyclopedia of Space Flight, Moscow (1969), pp. 333–37.

The second missile of this generation is the SS-11, which is the most numerous in the Soviet strategic missile force. Deployment began in 1966 and ceased in 1970 or 1971. This is a liquid-propellant missile, with a warhead yield of 1 mt. It has been tested extensively, and has had more than 100 launches from operational silos.[161] Three models have been deployed. The first has been tested at intermediate as well as intercontinental range, and some missiles have been deployed in IRBM fields. Model 2 is the same as Model 1 but has been tested with penetration aids. Model 3 carries three warheads, which are not, however, independently targetable. In spite of having three warheads Model 3 is thought to be more accurate than Model 1.[162] The third missile of this generation is the SS-13, which is the first Soviet solid-propellant missile. It has been deployed in relatively small numbers. Table 9.20 shows how the

[161] T. H. Moorer, *op. cit.*, pp. 13–14; G. S. Brown, *op. cit.*, pp. 12–13.
[162] *Jane's Weapons Systems 1974–75*, p. 15.

deployment of third generation ICBMs stopped almost completely in 1971. Deployment of the second and third generations lasted about five years in each case.

The fourth generation of Soviet ICBMs—the SS-16, SS-17, SS-18 and SS-19—is now (1975) coming into service; flight testing began in 1972. This generation is marked by greater throwweight, improved accuracy and multiple warheads. The SS-16 is a solid-propellant missile and can be seen as a successor to the SS-13. It appears to have been tested as a mobile missile, and to have been launched from a silo.[163] It has a post-boost vehicle which could deliver MIRVs; by early 1975, however, it had been tested only with a single re-entry vehicle. The SS-17 is a liquid propellant missile which received intensive testing in 1973 and 1974. It can fit into the SS-11 silo, but is said to have four times the throwweight of the SS-11 because it can be cold-launched. It can carry four MIRVs, and has a computer-controlled re-entry vehicle arrangement.[164] The SS-18 has a throwweight 50 per cent greater than the SS-9, and has undergone extensive flight tests since 1972. Three models have been tested: models 1 and 3 carry a single warhead which some reports say has a yield equivalent to 50 mt. Model 3 has a warhead designed for high-speed re-entry into the atmosphere, and is thus more accurate than model 1. Model 2 has a MIRV bus containing from five to eight re-entry vehicles. By 1975 only model 1 had been deployed.[165] The SS-19 is a liquid-propellant missile which can deliver from four to six re-entry vehicles. Testing began in April 1973 and deployment in January 1975. It is said to have a throwweight about three to four times that of the SS-11.[166] If the data which have been published about these missiles are correct, this generation marks a major advance in Soviet ICBM technology: the SS-19 can be considered more or less equivalent to the Minuteman III, while the SS-18 has a combination of accuracy and yield which would make it far more lethal than any US ICBM.

ASSESSMENT OF THE LEVEL OF TECHNOLOGY

An assessment of the level of Soviet ICBM technologies requires a comparison with US technologies in terms of yield, accuracy, throwweight, multiple warheads, propellants, penetration aids, missile silos and countersilo lethality. It will become clear in the course of this examination that many different technologies have contributed to the development of ICBMs, and that consequently assessment of the level of technology is a complex matter.

YIELD

Nuclear explosions release enormous amounts of energy, and this energy is expressed in terms of TNT equivalent. The first fission bombs had a yield of 15–20 kilotons (kt); by the late 1960s warheads with a yield of up to 25

[163] G. S. Brown, *op. cit.*, pp. 13–15; *Jane's Weapons Systems 1974–75*, p. 16; but see our note 193.
[164] G. S. Brown, *op. cit.*, p. 15; *Jane's Weapons Systems 1974–75*, p. 16.
[165] G. S. Brown, *op. cit.*, p. 16; *Aviation Week and Space Technology*, 15 March 1976, p. 22. For the MIRV bus, see p. 475 below.
[166] G. S. Brown, *op. cit.*, p. 16; *Aviation Week and Space Technology*, 25 November 1974, pp. 18–19.

megatons (mt) had been deployed on the Soviet SS-9—more than a thousand times greater than the yield of the first nuclear weapons. Some reports suggest that the SS-18 has a warhead yield of 50 mt.[167] Most of the energy is released in the form of shock wave overpressures (pressures above normal atmospheric pressure), thermal radiation and nuclear radiation. A one-megaton weapon will create five pounds per square inch (psi) overpressure four kilometres from the point of explosion, and will destroy all houses in this area (50 square km). The thermal effects also spread out over a wide area: the heat released by a one-megaton explosion will ignite paper 14 km away.[168]

This means that nuclear warheads aimed against area (soft) targets do not have to be very accurate to be effective. It is different with such hard targets as missile silos, since these are designed to withstand very high overpressures and are immune to thermal effects. The overpressures created by an explosion are proportional to the cube root of the distance from the point of explosion. Consequently:

> the ability of a warhead to destroy a silo depends strongly on the accuracy with which it is delivered. As a matter of fact . . . the lethality of a warhead against a silo rises much more rapidly with improvements in accuracy than with increases in yield.[169]

High yield weapons are more suitable for area targets than for hard targets. Even here, however, high yield weapons are not always the most appropriate: four or five 1 mt weapons would destroy the same area as one 10 mt weapon. This makes it clear that total megatonnage is not a satisfactory basis on which to compare two missile forces.[170]

Table 9.24 shows that the Soviet ICBM force can deliver a greater megatonnage than the US force. At one time it appeared that Soviet ICBMs might be armed with even more powerful warheads. In 1961 Khrushchev declared more than once that the Soviet Union was developing bombs with yields of 20, 30, 50 and 100 mt.[171] He said that Soviet scientists favoured the 100 mt bomb as the most 'economical'.[172] In October an explosion equivalent to 58

---

[167] *Jane's Weapons Systems 1974–75*, p. 17; and H. S. Rowen, in *Aviation Week and Space Technology*, 15 September 1975, p. 54, implies a yield of 50 mt for the SS-18 by stating that there 'is nearly a factor of 1,000 between the yield of the smallest one [nuclear warhead] in the US strategic forces (Poseidon has RVs with a yield of 50 kt) and the largest one in the Soviet force'.

[168] J. P. Ruina, in (Eds.) Mason Willrich and John B. Rhinelander, *SALT The Moscow Agreement and Beyond* (1973), p. 45; the standard work on nuclear effects is (Ed.) Samuel Glasstone, *The Effects of Nuclear Weapons*, US Department of Defense and US Atomic Energy Commission, Washington DC (1962).

[169] K. Tsipis, *Offensive Missiles*, Stockholm Paper 5, Stockholm International Peace Research Institute (1974), p. 12.

[170] Sometimes a measure known as equivalent megatonnage (EMT) is used as a basis of comparison. This is derived by multiplying the number of warheads by the yield of the warheads to the power of two thirds. In this way account is taken of the fact that warhead yield to the power of two thirds is proportional to the area of shockwave destruction. This, however, is a cruder basis of comparison than the measure of countersilo lethality discussed below.

[171] R. M. Slusser, *The Berlin Crisis of 1961*, Baltimore and London (1973), p. 165.

[172] R. M. Slusser, *op. cit.*, p. 91; Khrushchev declared that scientists were pressing him to resume testing and thus end the three year moratorium (*ibid.*). Sakharov, however, tried to prevent the tests from taking place (see *Sakharov Speaks*, London (1975), pp. 32–3).

mt was created by a device for a 100 mt bomb (had the fusion materials been encased in uranium rather than lead, the yield would have been 100 mt or more).[173] Warheads with yields of this magnitude have not been deployed, perhaps because Soviet military planners came to realise that the development of US ICBMs in hardened underground silos made yield a less important criterion of effectiveness than before.[174]

Table 9.24 *Megatonnage deliverable by ICBM forces*

	USA	USSR
1960	54	n.a.
1961	189	150+
1962	618	300+
1963	966	600
1964	1528	900
1965	1340	1940
1966	1500	3280
1967	1840	5010
1968	1940	5900
1969	2040	7130
1970	2035	8800
1971	1985	8805
1972	1935	8805
1973	1885	8817
1974	1840	8829
1975	1715	8942

Source:
Calculated from Tables 9.18–9.21 with the following yields assumed: Titan I—5 mt; Titan II—10 mt; Minuteman II—2 mt; SS-9—20 mt; SS-11—2 mt; SS-18 mod.1—25 mt; SS-19 5 × 300 kt.

Note:
I have assumed Minuteman II and SS-11 to have 2 mt warheads, since they are said to be comparable missiles; the SS-11 warhead yield is, however, often given as 1 mt.

ACCURACY

Early ICBMs were set on course by radio command, but now all strategic missiles have inertial guidance systems which can sense, record and correct any deviation from the missile's flight path. Inertial guidance systems have, moreover, the advantage of immunity to electronic countermeasures. Hoag explains the principles of inertial guidance thus:

[173] R. M. Slusser, *op. cit.*, p. 389.
[174] This point is made by Michael MccGwire in 'Soviet Strategic Programmes', in (Eds.) Michael McGwire, Ken Booth and John McDonnell, *Soviet Naval Policy, Objectives and Constraints*, New York, Washington, London (1975), p. 496.

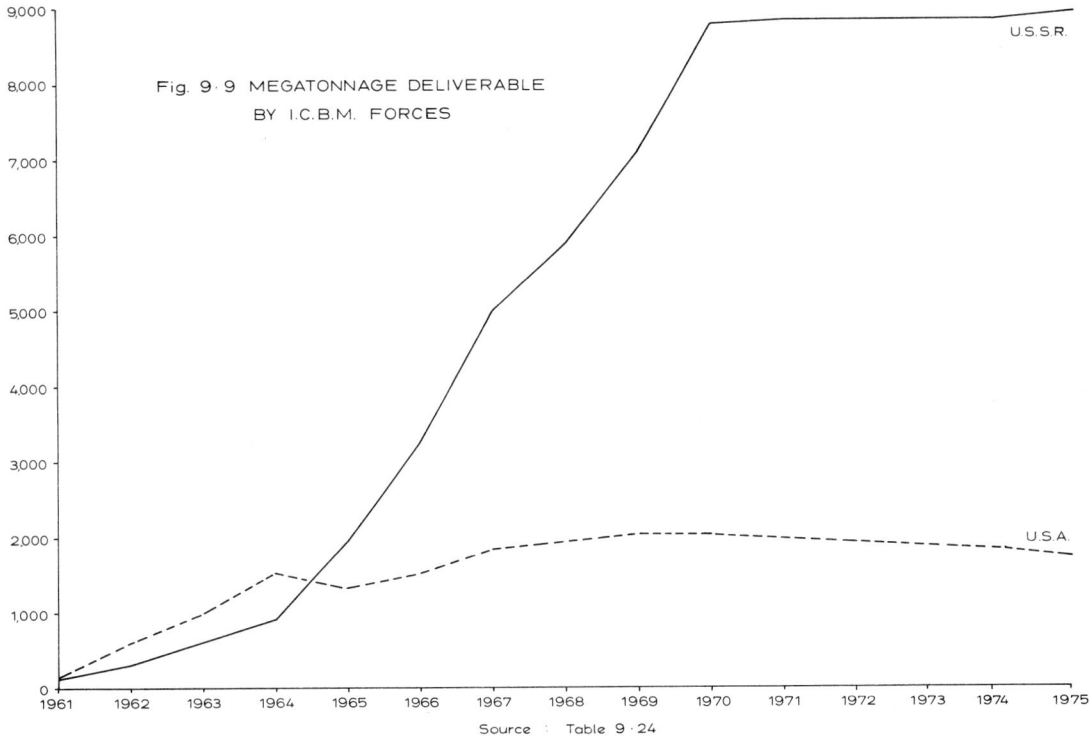

Fig. 9.9 MEGATONNAGE DELIVERABLE BY I.C.B.M. FORCES

Source: Table 9.24

a ballistic missile is propelled by a booster rocket to high velocity above the atmosphere where the warhead is released to coast freely in a ballistic path before it re-enters the atmosphere. Guidance of this weapon to hit a specific target occurs only during the rocket boost phase which extends over only the first few minutes of flight. Guidance measurements utilise inertial-sensing techniques. Inertial sensing is the application of Newton's laws as they operate in gyroscopes and accelerometers to measure all components of missile rotation motion changes and all components of missile translational motion changes due to all forces on the missile except gravity . . . the initial conditions of position (launch location) and velocity must be provided externally before launch. Guidance calculations determine the direction toward which to steer the rocket so that the velocity and position approach those conditions compatible with a free-fall coast to the target identified to the guidance system at launch. The rocket autopilot thrust-vector-control system responds to the guidance steering commands to control the direction of rocket thrust with respect to the missile in a stable fashion. When the proper conditions of position and velocity are achieved, the guidance system signals termination of rocket propulsion and the warhead is released.[175]

[175] D. C. Hoag, in (Eds.) B. T. Feld, T. Greenwood, G. W. Rathjens and S. Weinberg, *Impact of New Technologies and the Arms Race*, Cambridge, Mass., and London (1971), pp. 36–7.

In early missiles, accuracy was limited by guidance system errors; these have been eliminated to such a degree that further improvement in accuracy requires:[176]

—highly innovative design and fabrication of inertial sensors for extremely accurate, reliable and stable performance;
—accurate target location determination;
—survey of earth's gravitation field to measure anomalies, which guidance computation must then take account of;
—reduction of re-entry errors, either by reducing the drag on re-entry vehicles, or by providing them with terminal guidance systems.

With improvements of this kind CEPs of 30 m are thought to be possible. (The customary measure of accuracy is the Circular Error Probable (CEP): the radius of the circle, centred on the target, within which the warhead is expected to land with 50 per cent probability.)

Table 9.25 shows how rapid the rate of improvement in missile accuracy has been, and how the Soviet Union has lagged behind the United States. These figures suggest that the Soviet Union has faced considerable difficulty in developing highly accurate guidance systems. The scanty evidence available supports this view. The new Soviet ICBMs are the first to carry onboard computers, although US ICBMs have done so since 1962; and it is hard to see how high accuracies could be achieved without such computers. This indicates that the lag in electronics which can be observed in civilian technology is to be found in the military field too. It has been argued also that the Soviet Union lacks the very accurate metalcutting equipment essential in the construction of gimballed gyroscopes and accelerometers:

> missile accuracies of even one-half nautical mile require machining inertial guidance components to better than a few millionths of an inch and aligning them with precisions of a few micro-radians. The Soviet Union does not possess equipment capable of such precision, as evidenced by its recent efforts to acquire milling machines of that type in the United States.[177]

Another report suggested in 1973 that the Soviet Union was finding it difficult to master the gas-bearing inertial guidance technology which is essential for highly accurate MIRVed warheads.[178]

Accuracy has been a major concern of the Soviet missile programme since the immediate post-war years. When Korolev went to the Kremlin to report on the tests of the R-1 in 1947, Stalin asked him about the speed, range and height of the rocket's flight, and about its payload; but it was with particular interest that he enquired about the rocket's accuracy.[179] Greater accuracy is an evident goal of the latest generation of Soviet ICBMs. Since 1972 the Soviet Union has tested high-Beta warheads, warheads with high speed re-entry coefficients. The

[176] J. P. Ruina, *loc. cit.*, pp. 49–50.
[177] K. Tsipis, *op. cit.*, pp. 24–5.
[178] E. Ulsamer, *Air Force Magazine*, March 1973, p. 62.
[179] A. Romanov, *Konstruktor kosmicheskikh korablei* (1971), p. 49.

MILITARY TECHNOLOGY 473

United States had reached this point some years earlier. The warheads on the SS-9, SS-11 and SS-13 are thought to have terminal speeds of about Mach 1, whereas Minuteman III warheads have re-entry speeds in the region of Mach 10. This difference is important, for 'the slower the re-entry speed, the greater the effect of wind and other atmospheric factors on the vehicle's accuracy'.[180] The latest Soviet ICBMs are reported to have been tested with nosecones shaped for high speed atmospheric re-entry,[181] but it is not clear that these had been deployed by the end of 1975.

Table 9.25 *Accuracy of Soviet and US ICBMs*

USA

Missile	IOC	CEP (nm)	
Atlas	1959	2	⎤
Titan II	1963	0·75	⎦ 4 years
Minuteman I	1962	0·75	⎤
Minuteman II	1966	0·35	⎦ 4 years
Minuteman III	1970	0·20	] 4 years

USSR

Missile	IOC	CEP (nm)	
SS-7	1961–	(1·5–)	⎤
SS-8	1963	(2·0)	
SS-9	1966	0·5–1·0	⎬ 4 years
SS-11	1966	1·0	⎦
SS-13	1969	0·75–1·0	⎤ 9 years
SS-18	1975	0·25–0·5	⎦

Source: Tables 9.18 and 9.19.

[180] E. Ulsamer, *loc. cit.*, p. 63.
[181] *SIPRI Yearbook 1974*, p. 114. It is reported that three models of the SS-18 have been developed: two of these appear to have single warheads, one of which may be of high-Beta design (see *Aviation Week and Space Technology*, 9 February 1976, pp. 12–15). The Soviet Union is reported to have tested a manoeuvring re-entry vehicle (MARV) for its latest strategic weapons; this would be aimed at improving their accuracy (*ibid.*). MARV is under development in the United States too (*Jane's Weapons Systems 1974–75*, p. 6). A successful Soviet MARV would be evidence of considerable improvement in guidance and multiple warhead technologies.

Table 9.26 *Missile payload options*

	Example 1	Throwweight = 2000 lb
	For a single RV	Yield = 3 mt
	3 MIRVs	3 × 200 kt
	5 MIRVs	5 × 80 kt
	Example 2	Throwweight = 5000 lb
	For a single RV	Yield = 10 mt
	3 MIRVs	3 × 1 mt
	5 MIRVs	5 × 400 kt
	10 MIRVs	10 × 100 kt
	Example 3	Throwweight = 10000 lb
	For a single RV	Yield = 25 mt
	3 MIRVs	3 × 4 mt
	5 MIRVs	5 × 1·4 mt
	10 MIRVs	10 × 500 kt
	25 MIRVs	25 × 50 kt

Source:
J. P. Ruina, in (Eds.) M. Willrich and J. B. Rhinelander, *SALT The Moscow Agreements and Beyond* (1973), pp. 47–8. Ruina writes (p. 44) 'It should be noted that US assessments of Soviet strategic weapons systems generally infer bomb yields from what US technology can produce, rather than from direct observations of tests of bombs known to be installed in a particular Soviet system.' This inference can be drawn, however, only if one assumes that Soviet warhead technology is at the same level as that of the United States.

THROWWEIGHT

In recent years missile throwweight—the weight of that part of the missile above the last boost stage—has come to the fore in discussions of the American-Soviet strategic balance. Throwweight is influenced by various design features, including engine thrust, but is limited primarily by the total size or volume of the missile.[182] It can be doubled if a 'cold-launch' technique is used whereby the main booster is ignited only after a 'zero stage' has propelled the missile out of its underground silo.[183] Throwweight is an important parameter because of the payload options it creates: the greater the throwweight, the larger the scope for delivering multiple warheads and penetration aids. When multiple warheads are placed on a missile the total yield decreases, in part because more auxiliary equipment is needed in the re-entry vehicles in order to deliver the separate warheads to their targets; and the weight of this auxiliary equipment does not depend closely on the yield of the weapon. Table 9.26 shows the kinds of option available with different throwweights.

[182] J. P. Ruina, *loc. cit.*, p. 46.
[183] *Jane's Weapons Systems 1974–75*, p. 19. An important feature of the 'cold-launch' technique is that the silos can be used again: they can be reloaded with other missiles.

It is normally assumed that the yield-to-weight ratio (the ratio of explosive yield to throwweight) is the same for Soviet and United States missiles armed with single warheads. The yields given for Soviet missiles are not based upon monitoring of ICBM tests since, under the Partial Test-Ban Treaty of 1963, warheads cannot be tested in the atmosphere. The throwweight of Soviet missiles has to be calculated as a residual after other performance parameters and technical qualities of the missile have been assessed. Once this has been done the yield is calculated by assuming the same yield-to-weight ratio as for US missiles. This is a procedure which leaves room for error, and perhaps encourages estimates at the higher end of the range of possibilities. More important than this, however, is the possibility that the payload options listed in Table 9.26 are not in fact open to the Soviet Union. This is because the Soviet lag in electronics, and the consequent weight of auxiliary equipment on MIRVed missiles (arming and fusing mechanisms and guidance systems for the individual warheads, and engines and ejecting mechanisms for the 'bus' itself), may lead to a more rapid worsening of the yield-to-weight ratio than is to be found in US ICBMs.[184]

Notwithstanding these reservations, Soviet missiles have a much greater throwweight than US missiles. They are, in general, much larger, and some of the latest ICBMs can be 'cold launched', whereas no American ICBM uses this technique. The smaller size of the American ICBMs can be traced back to design decisions in the 1950s. In 1956 the US Atomic Energy Commission said that it could develop compact nuclear warheads with a high yield, and this influenced US missile design in the direction of smaller missiles.[185] Throwweights are not generally given for ICBMs, but some approximate figures can be worked out from open sources. Table 9.27 shows the Soviet lead quite clearly, and although the figures are very approximate they fit in with general statements to the effect that the Soviet missile force as a whole—SLBMs as well as ICBMs—has a throwweight two or three times greater than the US force.[186] (See Table 9.28.)

MULTIPLE WARHEADS

All the early ICBMs had single warheads, but in the 1960s MRV (multiple re-entry vehicle) and MIRV (multiple independently targetable re-entry vehicle) warheads were developed. This meant that a single launcher could deliver several warheads. MRVs fall in a fixed cluster, whereas MIRVs can be guided separately to their targets. York describes the operation of MIRV as follows:

> the multiple targeting of MIRV is accomplished by a device formally known as a Post Boost Control System (PBCS), but more commonly called a 'bus'.

[184] Barry R. Schneider and Stefan Leader, 'The United States–Soviet Arms Race, SALT and Nuclear Proliferation', *Congressional Record*, Vol. 121, No. 87, 5 June 1975, p. S9837.
[185] W. von Braun and F. I. Ordway, *op. cit.*, p. 129; Robert Gilpin, *American Scientists and Nuclear Weapons Policy*, Princeton, New Jersey (1962), p. 147.
[186] Rowens, *loc. cit.*, p. 54; *The Military Balance 1974–75*, London, p. 4, writes of a 'present' Soviet ICBM throwweight of 6–7 m pounds [2·7–3·2 m kg] and a US ICBM throwweight of 1–2 m pounds [0·45–0·9 m kg].

Initially, the main rocket booster puts the bus on a course that would cause it to impact somewhere near target number one. The bus contains several re-entry vehicles, a guidance and control system and some small rockets to modify its velocity so that it is aimed as precisely as possible along the orbit leading to target number one. When this has been accomplished, the bus gently ejects one of the re-entry vehicles. Then, while this first re-entry vehicle continues on its course to the first target, the bus guidance system instructs the propulsion units to modify its course so as to put it on an orbit leading to target number two. Another re-entry vehicle is then ejected and the process is repeated until each re-entry vehicle is *en route* to its prescribed target.[187]

The United States has enjoyed a clear lead in the development of multiple warhead technology. It first tested MRV in 1963, but has not deployed MRV warheads on an ICBM. (MRV warheads were deployed on the Polaris A-3

Table 9.27 *Throwweights of selected Soviet and United States ICBMs (kg)*

USA

Minuteman I	600[1]
Minuteman II	750[1]
Minuteman III	900[2,5]
Titan II	3600[1]

USSR

SS-11	750[3]
SS-9	4600[2]
SS-17	3000[3]
SS-18	7–7500[4,5]
SS-19	2800[5]

Sources:
1 V. I. Varfolomeev and M. I. Kopytov, *Proektirovanie i ispytanie ballisticheskikh raket* (1970), pp. 52–3.
2 J. P. Ruina, *loc. cit.*, pp. 47–8.
3 G. S. Brown, *op. cit.*, pp. 12–16.
4 *Aviation Week and Space Technology*, 25 November 1974, pp. 18–19.
5 J. Kraft, *The New Yorker*, 29 July 1974, p. 68.

Note:
The figures must be regarded as very approximate, since I have calculated them from the few hints and relationships given in the above sources.

[187] H. York, *The Origins of MIRV*, SIPRI Research Report No. 9, August 1973, Stockholm International Peace Research Institute, p. 7.

Table 9.28 *Total throwweight of Soviet and US ICBM forces (kg)*

	USA	USSR
1965	674,000	753,000
1970	870,000	2,700,000
1975	1,000,000	3,000,000

Source: Tables 9.20, 9.21 and 9.27.

Note: The figures are very approximate.

SLBM in 1964.) In 1968 testing of MIRV warheads began, and in 1970 the MIRVed Minuteman III became operational.[188] The Soviet Union has clearly encountered difficulties in the development of multiple warheads, as the test programme of the SS-9 Model 4 shows (see p. 464 above). Flight testing of MRV warheads began in 1968, but it was only in 1973 that they became operational on an ICBM—the SS-11 Model 3. Testing of MIRV warheads began in April 1973, and deployment of MIRVed ICBMs began in 1974, although it is unlikely that they became operational before the end of that year.[189] It has been suggested that the reasons for the Soviet lag in multiple warhead technology are the same as those which account for the lag in missile accuracy: backwardness in computers and electronics, and in precision grinding and cutting machinery.[190]

PROPELLANTS

With the exception of the SS-13 and SS-16, all Soviet ICBMs have liquid-propellant rocket (LPR) engines. All the Minuteman ICBMs have solid propellant rocket (SPR) engines. Of course, not all LPRs are like the RD-107 and RD-108: storable, stable liquid propellants have been developed in both in the Soviet Union and the United States. The United States appears to have made more advances in the development of SPRs. These have some advantages and some disadvantages, when compared with LPRs; it is not clear that one type of engine is decisively superior to the other.

In SPRs, all the propellant is stored in the missile's combustion chamber, so that no system is needed for supplying the propellant from the fuel tanks to the chamber. This makes the design of the missile simpler, and reduces the problems of maintenance and storage. It is possible to reduce the time required for launching by keeping the missiles charged with propellants. The disadvantages of the SPR are that it gives lower specific thrust than the LPR,

---

[188] H. York, *op. cit.*, p. 19. See Table 9.18.
[189] *Allocation of Resources in the Soviet Union and China—1975. Hearings before the Subcommittee on Priorities in Government of the Joint Economic Committee of the Congress of the US July 18 and 21 1975*, Part 1, p. 60.
[190] *Armed Forces Journal*, November 1973, p. 111; in 1972 the Soviet Union bought from the United States 164 precision grinding machines that can smooth miniature ball bearings to tolerances of 25 millionths of an inch. According to one reader these machines—of which the US had never owned more than 77—'are the key to our highly accurate, miniaturised ICBM guidance systems and the MIRVing of our warheads'.

Table 9.29 *Number of independently targetable warheads on ICBMs*

	USA	USSR
1965[1]	854	262
1966[1]	934	338
1967[1]	1054	722
1968[1]	1054	902
1969[1]	1054	1198
1970[1]	1074	1498
1971[1]	1274	1527
1972[1]	1474	1527
1973[1]	1674	1547
1974[1]	1854	1567
1975[2]	2154	1847

Sources:
1 *SIPRI Yearbook 1974*, Stockholm (1974), pp. 106–7.
2 Calculated from *The Military Balance 1975–76*, London (1975), pp. 5, 8.

Note:
MIRVs were first deployed by US in 1970, and by the USSR in 1975. The figures given above do not include MRVs.

and the engine's thrust value is more difficult to control. SPRs require constant monitoring in order to ensure that the missile's operational quality does not decline while the missile is in the silo.[191]

The United States had solved the major problems of SPRs by 1957.[192] Soviet ICBM deployment may suggest that the Soviet Union is not happy with its own SPR technology; or it may suggest that the sacrifice in specific impulse is not thought worthwhile for the benefits gained.

PENETRATION AIDS

Penetration aids are devices placed in warheads to help the missile penetrate anti-missile defences, or to reduce the warning time of an attack. Minuteman II was the first ICBM to be equipped with such aids; the first Soviet ICBM to be so equipped was the SS-11 mod 2, which became operational in 1973, seven years after Minuteman II. The most common aids are chaff, decoys and radar blackout equipment; these are designed to make it difficult for the enemy to locate or identify the significant elements of the attacking force. High speed re-entry technology can be seen also as a form of penetration aid. The fractional orbit bombardment system (FOBS) which the Soviet Union tested in the mid-1960s was designed to circumvent the United States' early-warning system by flying in partial orbit round the globe; by flying in such an orbit the missile could cut warning time substantially. This system is not thought to have been deployed (see 461 above).

[191] N. Zhemchuzhin *et al.*, *Meet Aerospace Vehicles*, Moscow (1974), pp. 219–25.
[192] Robert L. Perry, 'The Atlas, Thor, Titan and Minuteman', in (Ed.) M. Emme, *op. cit.*, pp. 155–7.

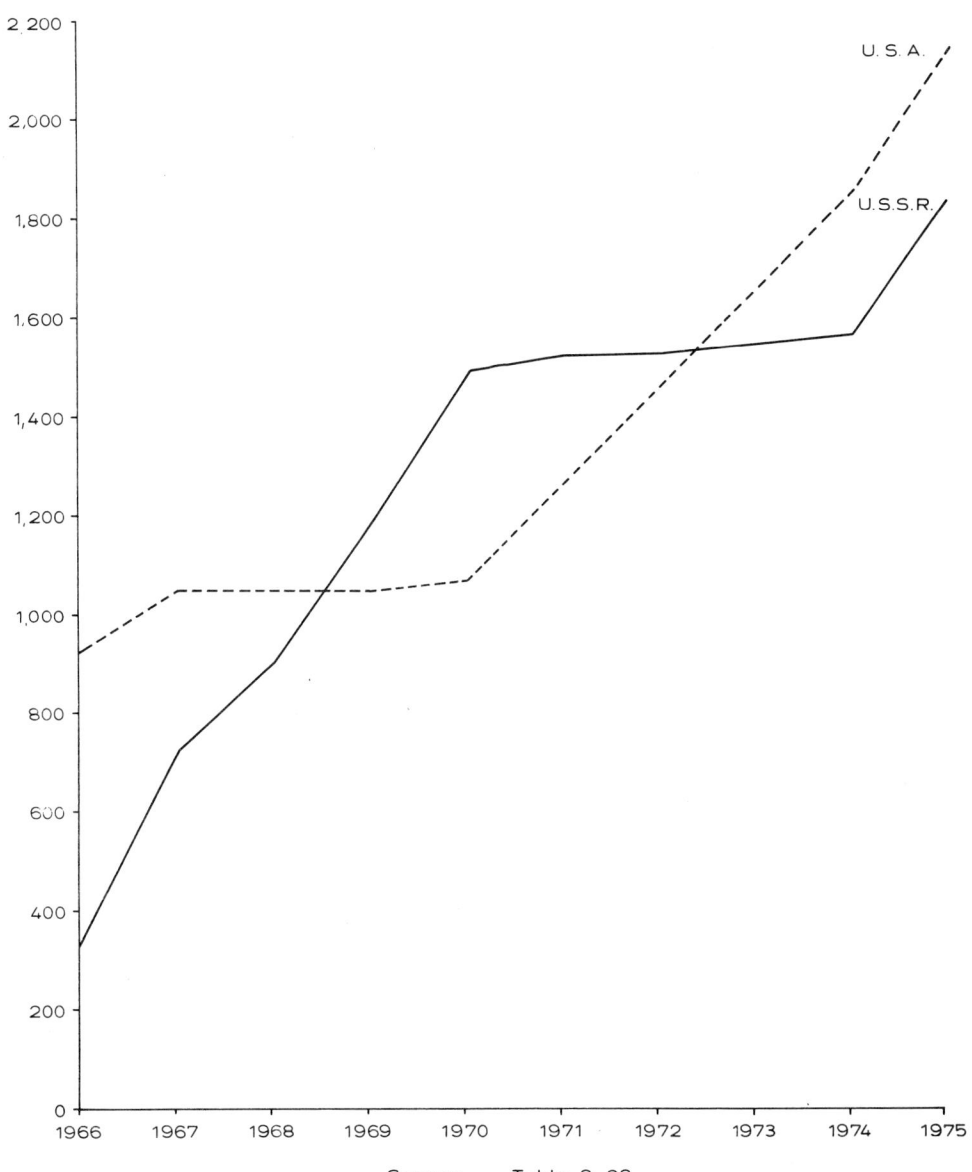

Fig. 9·10 NUMBER OF INDEPENDENTLY TARGETABLE WARHEADS ON U.S. AND SOVIET I.C.B.M.'s

Source   Table 9·29

MISSILE SILOS

The ability to retaliate in the event of attack is a central element in any deterrent strategy. Consequently both the Soviet Union and the United States have made efforts to ensure that their strategic forces—and each component of those forces—could survive an all-out attack. There are two main methods of enhancing the survivability of individual ICBMs: by placing them on mobile launchers, and by placing them in underground silos. The Soviet Union has apparently tested the SS-16 in a mobile mode, but no mobile ICBMs appear as yet to have been deployed.[193] Hardened underground silos have until now been a more important means of protection for ICBMs. They are constructed from concrete and steel and can withstand overpressures of up to 1,000 psi.[194]

The Soviet Union has lagged behind the United States in providing hardened silo protection for its ICBMs. All American ICBMs can be launched from underground silos: the Titan II and Minuteman II silos are hardened to 300 psi; Minuteman III silos are said to have a hardness of 1,000 psi.[195] The Soviet ICBM force, however, contains 140 SS-7 and SS-8 missiles at 'soft' sites, which may be said to have a hardness of about 5 psi. The remaining SS-7 and SS-8 missiles are thought to be in silos hardened to 100 psi. There is conflicting evidence about the hardness of the other silos: some, including the SS-9 silos, are said to be hardened to 100 or to 300 psi. The latest Soviet silos are said to have a hardness of 450 to 600 psi.[196]

COUNTERSILO LETHALITY

The comparison so far gives an unclear picture: it shows a Soviet lead in megatonnage and throwweight, and a US lead in missile accuracy and multiple warhead technology. The relative importance of these parameters can be assessed by examining the effectiveness of the different ICBMs as countersilo weapons. Various methods can be used to assess the destructiveness of a warhead to a missile silo.[197] Tsipis gives the following simple equation:

$$K = \frac{y^{\frac{2}{3}}}{(CEP)^2}$$

where K is the lethality of a re-entry vehicle (RV) to a silo, y is the yield of the RV expressed in megatons, and CEP is the accuracy of the RV expressed in nautical miles. This formula reflects the fact that countersilo lethality increases more rapidly with improvements in accuracy than with improvements in yield. It must be stressed that this is not an attempt to assess the effectiveness of the strategic arsenals since SLBMs are not included. Besides, a proper assessment of effectiveness would have to take account of other factors such as reliability, targeting and retargeting. The significance of K for this study is that it provides a crude but simple measure of countersilo lethality and thus a composite

[193] The mobile IRBM has, however, been deployed; in an interview in November 1974 General of the Army V. Tolubko, Commander-in-Chief of the Strategic Missile Forces, denied that a new mobile intercontinental system was being tested (*Nedelya*, No. 46, 11–17 November 1974, p. 4); it is now reported that the SS-16 has a mobile 'derivative', the SS-X-20, which consists of the top two stages of the SS-16, just as the SS-14 consists of the top two stages of the SS-13 (*Aviation Week and Space Technology*, 2 February 1976, p. 14).

[194] A good technical account may be found in V. Malikov, *Nazemnoe oborudovanie raket* (1971), ch. 2.

[195] *SIPRI Yearbook 1974*, Stockholm, p. 110.

[196] *SIPRI Yearbook 1974*, Stockholm, p. 111; Lynn E. Davis and Warner R. Schilling, *Journal of Conflict Resolution*, Vol. XVII, 1973, No. 2, p. 235; K. Tsipis, *op. cit.*, p. 21.

[197] L. E. Davis and W. R. Schilling, *loc. cit.*, gives a clear discussion of various methods, pp. 207–42. Tsipis' equation is given in *op. cit.*, p. 16.

indicator of ICBM performance. To destroy a silo hardened to 1,000 psi with 97 per cent probability, a K value of 108 is required; and with 90 per cent probability a K value of 71. To destroy a silo hardened to 300 psi with 97 per cent probability a K value of 45 is required; and with 90 per cent probability, a K value of 30.[198]

Table 9.31 gives the K values for US and Soviet ICBMs.[199] For most of the Soviet missiles I have used a range of estimates about yield since various figures are given in the literature; I have used accuracies towards the bottom end of the range of estimates: thus a CEP of 0·7 nm is assumed for the SS-9, and of 0·3 nm for the SS-18 and SS-19.[200] I have not calculated the CEP for missiles which had not been deployed by 1975. The results of these calculations are very sensitive to changes in the assumptions about accuracy. For example the SS-18 with a warhead yield of 25 mt has a K value of 95 for a CEP of 0·3 nm, and of 53 for a CEP of 0·4 nm; with a warhead yield of 50 mt the values are 150 for a CEP of 0·3 nm and 84 for a CEP of 0·4 nm. The assumptions I have made in Table 9.31 are unlikely to lead to an underestimation of Soviet K values.

Table 9.30 *Total K required to destroy each Soviet silo with probabilities of 97 and 90 per cent*

Silo hardness psi	K required per silo		No. of silos (S)	Total K.S	
	$P_k=0·97$	$P_k=0·90$		$P_k=0·97$	$P_k=0·90$
300	45	30	400	18000	12000
100	20	13	1100	22000	14300
			Total	40000	26300
300	45	30	1100	49500	33000
100	20	13	400	8000	5200
			Total	57500	38200

Source: K. Tsipis, *Offensive Missiles*, Stockholm Paper 5, SIPRI (1974), p. 23.

Note:
The lower half of the table has been adapted from Tsipis' data by changing the assumptions about the hardness of Soviet silos. The corresponding K values required to destroy each US silo with 97 and 90 per cent probability are 82,080 and 54,170, for 550 silos hardened to 1,000 psi, and 504 hardened to 300 psi.

---

[198] K. Tsipis, *op. cit.*, p. 22.
[199] I have not relied on Tsipis' calculations; in most cases my assumptions about yield and accuracy are different.
[200] Malcolm R. Currie, Director of US Defense Research and Engineering, is reported as saying of the SS-18 that 'we have information suggesting the possibility of a 0·25 mi circular error probability, although I personally feel that may be somewhat overstated' (*Aviation Week and Space Technology*, 25 November 1974, p. 19). It is interesting to note that the discussion which took place in the late 1960s about the threat posed by the SS-9 to the Minuteman force assumed a CEP of 0·25 nm for the SS-9; this has subsequently been seen as a mistakenly pessimistic (from the American point of view) assessment of the missile's performance.

Table 9.31 *K values of Soviet and US ICBMs*

Missile	Warhead yield (mt)	RV CEP (nm)	$K = \frac{y^{\frac{2}{3}}}{(CEP)^2}$	No. of RVs per missile (n)	K per missile (K.n)
Minuteman I	1	0·75	1·76	1	1·76
Minuteman II	2	0·35	12·9	1	12·9
Minuteman III	0·17	0·2	7·7	3	23·1
Titan II	10	0·5	18·5	1	18·5
SS-7/8	5	1·5	1·3	1	1·3
SS-9	18	0·7	14	1	14
	25	0·7	17·4	1	17·4
SS-11	2	1	1·6	1	1·6
	1	1	1	1	1
SS-13	1	1	1	1	1
	1	0·75	1·77	1	1·77
SS-16	1	0·4	6·25	1	6·25
	2	0·4	10	1	10
SS-17	0·2	0·4	2	4	8
	0·4	0·4	3·4	4	13·6
SS-18 mod. 1	18	0·3	76	1	76
	25	0·3	95	1	95
	50	0·3	150	1	150
SS-19	0·4	0·3	6	4/6	24/36

Several interesting points emerge from these calculations. Table 9.31 shows that before the present generation Soviet ICBMs, with the sole exception of the SS-9, lagged behind their US counterparts. Even the SS-9 does not have as great a K value as the Titan II or Minuteman III. With the introduction of the SS-18 and SS-19, however, the picture changes. The SS-19 appears to have a K value at least as great as—and perhaps greater than—that of the Minuteman III, while the SS-18 appears to be far more lethal than any US ICBM. Indeed, if the SS-18 has a K value of 150 it would be an effective countersilo weapon. Table 9.32 and 9.33 and Fig 9.11 give estimates of the total K value of the Soviet and United States ICBM forces between 1964 and 1975. These show a US lead from 1964 to 1970, and a rapid growth in that lead between 1970 and 1974 when Soviet ICBM deployment had come almost to a standstill and the Minuteman III was being introduced into the US strategic arsenal. The Soviet curve shows a sharp upturn in 1975 when the new generation of ICBMs started to come into service.

It must be borne in mind that these results are no more reliable than the assumptions on which they are based, but they do suggest that guidance and warhead technologies have a more immediate bearing than megatonnage and throwweight on the effectiveness of an ICBM force which includes missile silos among its targets. They indicate also that the latest generation of Soviet ICBMs mark a major step forward in Soviet ICBM technology.

Table 9.32 *K value of US ICBM force*

	1964	1966	1968	1970	1972	1974	1975
Atlas							
No. of missiles (n) ($K=0.5$)	36						
K.n	18						
Titan I							
No. of missiles (n) ($K=1.3$ (est.))	36						
K.n	47						
Titan II							
No. of missiles (n) ($K=18.5$)	54	54	54	54	54	54	54
K.n	999	999	999	999	999	999	999
Minuteman I							
No. of missiles (n) ($K=1.76$)	700	800	600	490	290	100	
K.n	1232	1408	1056	862	510	176	
Minuteman II							
No. of missiles (n) ($K=12.9$)		80	400	500	500	500	450
K.n		1032	5160	6450	6450	6450	5805
Minuteman III							
No. of missiles (n) ($K=23.1$)				10	210	400	550
K.n				231	4851	9240	12705
Total K	2296	3439	7215	8452	12810	16865	19509

CONCLUSION

The lack of accurate data makes it difficult to arrive at a definitive assessment of the level of Soviet ICBM technology. The difficulty is increased because some of the data to be found in open sources depend on the unproven assumption that the level of technology is the same in the Soviet Union as in the United States! This appears to be the way in which Soviet warhead yields and ICBM throwweights are estimated, for example. Such data cannot be used as reliable indicators of the level of Soviet technology since they already embody assumptions about that level. In other areas, such as missile accuracy, the range of evidence available makes it difficult to assess the relative Soviet and US levels.

A more fundamental difficulty also exists: uncertainty is inherent in assessing technologies which have not been put to the use for which they were designed. Even if the accurate data were available, it would not be possible to predict with certainty the outcome of a nuclear exchange. As Dr Fred Iklé, the Director of the US Arms Control and Disarmament Agency, has said 'all we know is that we do not know' what the effects of a large-scale war would be.[201] Iklé has pointed to several phenomena which were discovered only by accident.[202] One of these was the so-called 'fratricide' effect whereby warheads might be destroyed or diverted from their targets by other warheads which had already exploded. This makes it very difficult to calculate the effectiveness of MIRVs against hardened silos. Thus it is not only the lack of accurate data that should lead one to treat the calculations in this study with caution and scepticism.

Table 9.33 *K value of the Soviet ICBM force*

	1964	1966	1968	1970	1972	1974	1975
SS-7/8							
No. of missiles (n)	180	220	220	220	210	210	210
($K=1.3$) K.n	234	286	286	286	273	273	273
SS-9							
No. of missiles (n)		108	192	288	288	288	278
($K=17.4$) K.n		1879	3341	5011	5011	5011	4837
SS-11							
No. of missiles (n)		31	500	960	970	970	960
($K=1.6$) K.n		50	800	1536	1552	1552	1536
SS-13							
No. of missiles (n)				40	60	60	60
($K=1.8$) K.n				72	108	108	108
SS-17							
No. of missiles (n)							10
($K=13.6$) K.n							136
SS-18 mod. 1							
No. of missiles (n)							10
($K=95$) K.n							950
SS-19							
No. of missiles (n)							50
($K=30$) K.n							1500
Total K	234	2215	4427	6905	6944	6944	9340

[201] Quoted in *SIPRI Yearbook 1975*, Stockholm, p. 45.
[202] *Ibid.*

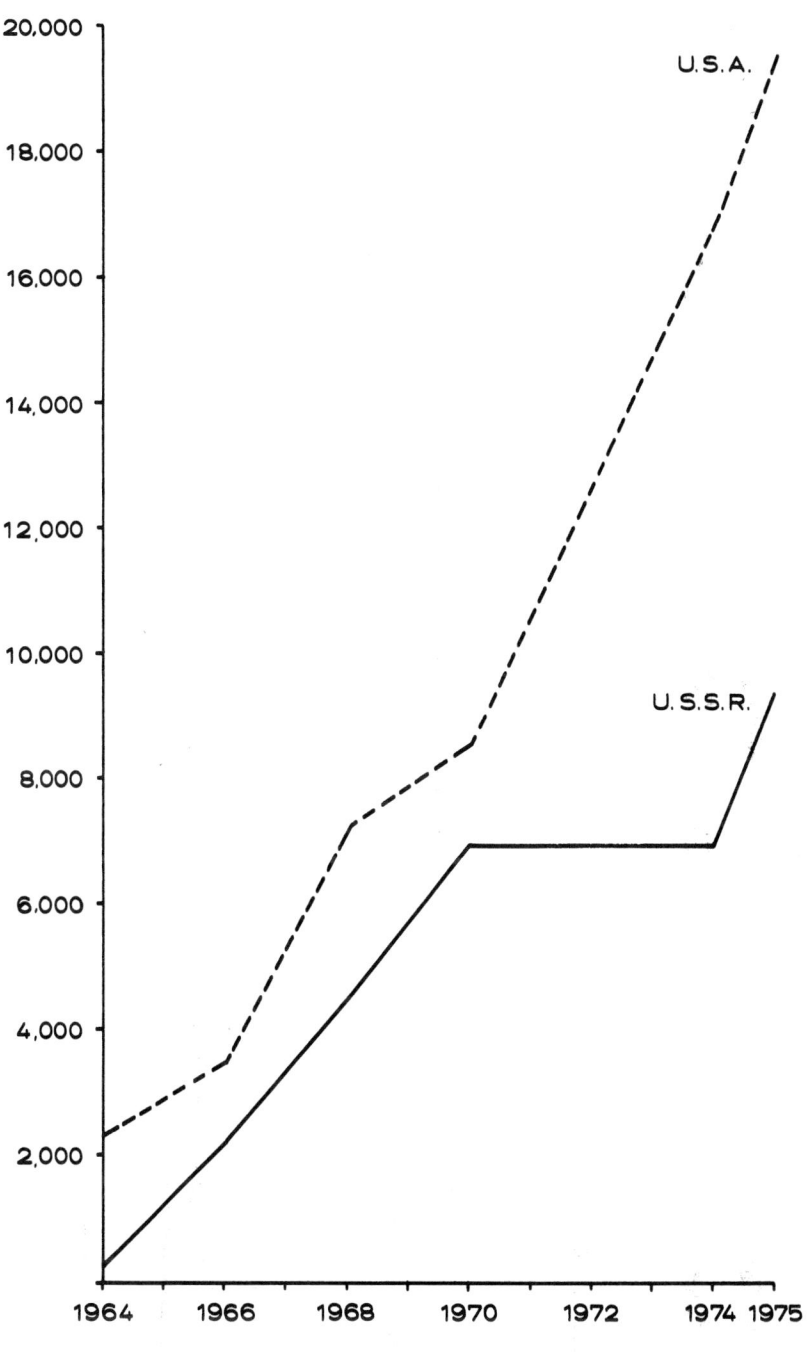

Fig. 9·11 K VALUES OF SOVIET AND U.S. I.C.B.M. FORCES

Source: Tables 9·32 & 9·33

It is nevertheless possible to draw some conclusions about the level of Soviet ICBM technologies. The first is that although the Soviet Union has had a number of spectacular firsts, almost all of these were in the 1950s. The selection of the landmarks in Table 9.34 is to some extent arbitrary and not all the landmarks chosen are of equal importance; but it is hard to see how the general picture might be changed to give a more favourable impression of Soviet technology. It is not clear, moreover, that the first ICBM launching or the first test of a deliverable thermonuclear device were in fact evidence of more advanced technology: the Soviet ICBM lead was quickly lost, while the Soviet thermonuclear test of 1953 represented a step in the development programme that the United States bypassed. Further, the MRBM which the Soviet Union tested in 1954 and deployed in 1955 had a range less than half that of the US IRBM tested in 1957 and deployed in 1958; the apparent Soviet lead of three years is illusory. It can be argued, however, that after the eclipse of the 1960s and early 1970s, the Soviet Union has caught up again: certainly the SS-18 seems (if the published data are correct) to have a far more lethal combination of yield and accuracy than any other ICBM. It is not the purpose of this study to examine what the wider significance of this apparent lead may be, but it does suggest a fluctuating relationship between Soviet and United States ICBM technologies. It is difficult to make a satisfactory assessment of the present state of affairs without more information.

Second, the available evidence suggests a rather slower rate of technological innovation in the Soviet Union than in the United States. For example, the rate of improvement in missile accuracy has not been much more than half the American rate, while the time between testing and deployment of MRVs was one year in the United States and five years in the Soviet Union; the time between testing and deployment of MIRVs seems to have been two years in both countries. In guidance system and multiple warhead technologies the United States has enjoyed a clear lead which the Soviet Union has been trying to eliminate. The Soviet Union has led in megatonnage and throwweight but this has been a result of American design decisions, and not due to the United States lack of the relevant technological capability. This distinction is important, for it shows how design decisions may compensate for a relatively lower technological base. The Soviet Union has attempted—with what success the lack of accurate data makes it difficult to say—to compensate for its backwardness in guidance and multiple warhead technologies by building very large missiles with a correspondingly large throwweight. It should be borne in mind that this is an area in which technological change is very rapid. The Soviet Union is reported to be developing a new generation of the ICBMs and to have begun testing manoeuvrable re-entry vehicles (MARVs) in order to enhance accuracy.[203] Nor has the US effort stood still. Work has continued on improving accuracy, and a new ICBM—the MX—is under development and could be deployed in the early 1980s. The final performance parameters of the MX are not clear, but it is likely to be cold-launched, to have a throwweight of

---

[203] *Aviation Week and Space Technology*, 29 March 1976, p. 15; 16 February 1976, p. 40.

over 3,500 kg and to be able to deliver more than a dozen warheads.[204]

Third, the design philosophy outlined in the introduction to this chapter emerges less clearly in Soviet ICBMs than in Soviet tanks. This may be because there is little information available about the component elements—engines, warheads, guidance systems—of the later ICBMs. Clear lines of descent can be seen in Soviet ICBM development; at the same time, however, the different generations are clearly marked off from one another, and each has seen improvements in the three main elements of the missile, as well as in ground equipment. This is not to suggest that the rate of deployment is tailored to stages in the R and D programme, but rather that when a new generation of missiles is required the most advanced developments are incorporated. This is rather different from the pattern of evolutionary designs and development to be found in Soviet tanks. A very important feature of Soviet ICBM history has been the development and production of competing (and redundant) models in each of the last three generations.

Fourth, there have been important differences between the Soviet and United States attitudes to the testing of missiles and their withdrawal from service. Between 1965 and 1975 the Soviet Union has conducted, each year, three or four times as many ICBM flight tests as the United States. This difference can be explained in part by the greater number of models developed and deployed by the Soviet Union during this period (seven as opposed to two

Table 9.34 *Landmarks in the development of ICBMs*

	USA	USSR	Soviet lag (−) or lead (+) in years
First nuclear chain reaction	Dec 1942	Dec 1946	−4
First production reactor in operation	1944	1948	−4
First atomic explosion	July 1945	Aug 1949	−4
First thermonuclear explosion	Nov 1952	Aug 1953	−1
First explosion of a deliverable thermonuclear device	Mar 1954	Aug 1953	+1
First M–IRBM test	May 1957	1953/4	+3
First M–IRBM operational	1958	1955	+3
First ICBM test	1958	Aug 1957	+1
First ICBM operational	1959	1960	−1
First MRV test (with SLBM in USA)	1963	1968	−5
First MIRV test	1968	1973	−5
First MIRVed ICBM operational	1970	1974/5	−4/5
First solid-propellant ICBM operational	1962	1968	−6
First cold-launched ICBM operational	—	1975	+

[204] *Aviation Week and Space Technology*, 19 January 1976, p. 13. And work is continuing to improve the performance of existing ICBMs.

by the United States); but the tests also include launches from operational silos for troop training.[205] The United States has been more willing to phase ICBMs out of service: Atlas, Titan I and Minuteman I have all been withdrawn. In the Soviet Union, only the SS-6, which was deployed in very small numbers, has been wholly taken out of service. By 1975 the United States had withdrawn about 1,000 ICBMs from service, while the Soviet Union had withdrawn no more than about 40 (see Tables 9.20 and 9.21). The Soviet practice may now be changing as the latest ICBMs, under the constraints of the SALT agreement, are placed in silos occupied by earlier models.

Fifth, the ICBM programmes have been organised and managed differently in the two countries. In the Soviet Union special bodies were set up in the Council of People's Commissars to ensure centralised control from the very beginning. On the military side the ICBM programme has been dominated by artillerymen, thus continuing the pre-war association of rockets and artillery. The Strategic Missile Forces were formed as a separate branch of the Armed Forces in 1959, although the SLBMs remain under Navy control. In the United States management and control have been more fragmented, and ICBMs and IRBMs came under the US Air Force during the 1950s only after considerable inter-service rivalry and an initial lack of enthusiasm for ballistic missiles on the part of the Air Force. In the United States, as in the USSR, missile-carrying submarines come under the Navy.

Final conclusions

In the introduction to this chapter the obvious point was made that two case studies could not provide a comprehensive guide to the level of Soviet military technology. It has also emerged in the course of our work that in each of the case studies methodological problems and the absence of reliable information make it impossible to arrive at definitive assessments of the level of technology. Two kinds of side-by-side comparison have been used. The section on tanks concentrated on a comparison of individual tanks whereas the study of ICBMs paid more attention to the characteristics of the force as a whole. This difference in approach has been dictated very largely by the kinds of data available, and also by the fact that comparison of tank forces would be methodologically more difficult than comparison of missile forces. (To reverse the approaches would lead to different results: taken as a whole the Warsaw Pact tank force would be superior—in terms of availability, survivability and kill probability—to the NATO force (other things being equal); if ICBMs are taken individually, the SS-18 appears to be the most lethal of all.) In spite of the difficulties some general conclusions about the level of technology can be drawn.

[205] In the late 1960s the Soviet Union conducted over 100 flight tests a year, compared with 20–30 by the United States (*Congressional Record*, 4 August 1971, p. E8958); in 1972 there were 61 Soviet ICBM launches, and 32 US launches; in 1973 130 to 30; in 1974 139 to 31 (*Aviation Week and Space Technology*, 19 January 1976, p. 15).

These studies do not show the Soviet Union gradually catching up or overtaking its foreign competitors: the relationship appears to fluctuate. In both cases the Soviet Union has first led the world and has then lost that lead. Soviet tank technology is generally recognised to have been supreme in the early 1940s but now Soviet medium tanks are inferior to the best foreign tanks. In the mid-1950s the Soviet Union appeared to have a clear lead in ICBM technologies, but this was quickly lost to the United States. On the other hand, the Soviet Union seems again to have closed the gap with its latest generation of ICBMs. This is, then, a shifting relationship, in turn influenced by arms race competition.

The evidence from Soviet ICBMs and tanks points up the features of design philosophy referred to in the introduction: commonality, design inheritance and simplicity. It would be wrong, however, to give them too much weight in explaining the patterns of Soviet weapons development: they are clearer for tanks than for ICBMs. The case studies also lend some support to the view that development was evolutionary, though punctuated by central intervention to set a programme going, or to change its direction. The pace of innovation may have been slower in the Soviet Union, but the inertia (or momentum) of the Soviet effort is reflected in gradual and steady progress. Diffusion has been less rapid in the Soviet Union, if only because of a greater Soviet willingness to keep equipment in service for longer than appears normal in the West (though this is truer of ICBMs than of tanks).

Although the focus of this chapter was relatively narrow, it emerges quite clearly that the defence sector is part and parcel of the Soviet system as a whole. Soviet history has left its mark on the Soviet military-technological effort: the development of the defence sector cannot be explained without reference to the industrialisation drive, the purges, the war and post-war stability. In a more specific way the strengths and weaknesses of Soviet technology as a whole are reflected in the defence sector: for example, strength in welding has affected tank technology: weakness in electronics and computer technology has influenced ICBM development. The defence sector cannot be seen as an isolated realm within the Soviet R and D system.

10. Rocketry: level of technology in launch vehicles and manned space capsules

Introduction

When the launching of the Soviet Sputnik was announced on 4 October 1957, the prestige of the USSR rapidly climbed to its highest point ever. The Soviet Union had established itself at this time as the pioneer in space exploration both in theory and practice.[1]

Since those days Soviet space exploration has grown into a vast, bold, sophisticated and expensive enterprise. In the present analysis we concentrate on a few important fields, particularly the development and performance of launch vehicles and piloted space capsules. Comparisons with the US space programme will be made where possible, with the aim of assessing the relative technological level of the two countries.

Our analysis is somewhat tentative owing to inadequate information. Although more material on certain parts of the Soviet space programme, such as rocket engine and spacecraft design, is being released by the Soviet authorities now than in the past, on some important aspects, including manufacture of rockets and spacecraft, space budget and manpower size, Soviet sources are silent. Such information as has been released is not always reliable. The reliability of the material published in the West is also often uncertain. The information is secondhand and often has traces of bias, and in one case information in a Western source is in complete contradiction with the Soviet evidence.[2]

[1] Only the occasional eccentric commentator such as Lloyd Mallan differed from this general view. In his *Russia and the Big Red Lie*, New York (1959), he simply dismissed Soviet space successes as a bluff.

[2] Thus L. Vladimirov in his *Russian Space Bluff*, London (1971), p. 43, claims that Yangel, an important German rocketry engineer, was captured by the Russians at the end of the Second World War and brought to the USSR. However, *Pravda* of 4 November 1973 published an article about Yangel, an outstanding Soviet space expert, Hero of Socialist Labour, Laureate of the Lenin Prize, a member of the CPSU, grandson of a Zaporozhe Cossack; Yangel was a student at the Moscow Aviation Institute in the 1930s and such data of his pre-war career are given as the theme of his degree work and the year he joined the CPSU. Clearly, only one of the two accounts can be right!

Comparative history of Soviet rocketry

Use of powder rockets for military purposes and fireworks has a very long history which started soon after the invention of gunpowder. For practical reasons the rockets used in space exploration are mostly liquid propelled and it is their development we are concerned with here. The end of the nineteenth and the beginning of this century witnessed important developments both in the West and Russia which facilitated the development of modern rocketry. They were, chiefly, rapid progress in metallurgy and engineering, extensive research into combustion engines and the liquification of gases.[3]

Table 10.1 *Pioneers in theory of reactive motion and first launchings of their LPRs (liquid-propelled rockets)*

	Started in the field	First works published	First LPR launched	LPR score
K. E. Tsiolkovsky (Russia)	1883	1903	—	
R. H. Goddard (USA)	1907	1914	16.3.1926	World's first LPR
R. Esnault-Pelterie (F)	1912	1913	—	
H. J. Oberth (G)	1922	1923	1929 (experimental)	
J. Winckler (G)	n.a.	n.a.	14.3.1931	Europe's first LPR
M. K. Tikhonravov (USSR)	1932	1935	17.8.1933	USSR's first LPR

Sources:
The Soviet Encyclopedia of Space Flight, Moscow (1969), pp. 129, 167–8, 284, 469.
(Eds.) S. A. Sokolova and T. M. Melkumov, *Pionery raketnoi tekhniki* (1972), pp. 11–12, 395, 569, 790.
A. Parry, *Russia's Rockets and Missiles*, London (1960), pp. 81, 96, 109, 116.

Prominent Soviet experts in rocketry hardly ever failed in their works to emphasise Tsiolkovsky's primary contribution to the theoretical roots of modern rocketry, and some Western specialists also give Tsiolkovsky the credit which he deserves: according to von Braun, 'Tsiolkovsky was the first to understand and develop the use of rockets in space travel'.[4] As early as 1883 Tsiolkovsky suggested the principle of reactive motion for interplanetary vehicles. Among his other pioneering contributions to the theory of rocketry is his formula describing the motion of a rocket and determining its characteristic velocity (the so-called Tsiolkovsky formula) published in 1903.[5]

[3] W. von Braun and F. I. Ordway, *History of Rocketry and Space Travel*, New York (1966), p. 40.
[4] *Ibid.*, p. 40.
[5] The most famous of Tsiolkovsky's published works are *Issledovanie mirovykh prostranstv reaktivnymi priborami* (Exploration of cosmic space by reactive devices) (1903), and *Kosmicheskie raketnye poezda* (Space rocket trains) (1929). His collected works were also published, in Russian, in four volumes (1951–64).

In the United States, Goddard started his independent research in the theory of reactive motion only in 1907 (i.e., almost a quarter of a century after Tsiolkovsky, of whose existence he was ignorant at the time), and launched his first LPR 19 years later. In 1926 he was followed by Oberth in Germany, whose experimental LPR was launched in 1929. Tsiolkovsky never reached that 'practical' stage in his own work. While the study of Tsiolkovsky's publications made Tsander's, Korolev's and other Russians' first steps in the theory of rocketry much easier, neither Goddard nor Oberth read Russian and they only learned about Tsiolkovsky at a later stage of their own independent research in rocketry (Oberth as late as 1925). The Goddard-Oberth professional relationship was almost completely non-existent. Oberth knew of Goddard's existence but did not know his work, and Goddard complained that he was unable to get acquainted with Oberth's work because it was secret.[6] The disputes about 'who-was-the-first' in particular inventions in rocketry[7] seem to support the notion that the isolation of the three men was more or less complete and that considerable overlaps in basic research had taken place. It seems reasonable to assume that the initial theoretical and practical achievements in rocketry were attained by the respective countries on their own initiative.

SOVIET ROCKETRY 1928–45

The Gas Dynamics Laboratory (GDL), the first Soviet R and D organisation in rocketry, was set up under the auspices of the Ministry of War in Leningrad in 1928, although its origins go back to 1921.[8] The chiefs of GDL between 1928 and 1933 were N. Tikhonravov (1928–30), B. Petropavlovsky (1930–1), N. Ilyin (1931–2) and I. Kleimenov (1932–3). One of its five departments (brigades) was working on solid-fuel rockets: the famous multi-barrelled 'Katyusha' rocket gun, used in the Second World War, was developed there. Another department led by V. P. Glushko was engaged on development of liquid propellant and electric rocket engines, and between 1929 and 1933 developed systems of chemical and pyrotechnical ignition as well as a method of internal cooling of combustion chamber walls. In this period the department developed and tested 52 LPRs (known as ORM-1 to ORM-52); some of these reached a thrust of 300 kg (see Table 10.2). In 1931, another organisation researching into rocketry was founded in Moscow. It was called the Group for Study of Reactive Motion, GIRD, and although it started as a semi-civilian or para-military establishment (it was subordinated to Osoaviakhim, Society for Assistance to Defence, Aviation and Chemistry), its semi-civilian status was soon brought to an end when, in 1932, due to a lack of funds, GIRD was transferred to the Soviet military, who were prepared to finance its activities.

[6] A. Parry, *op. cit.*, p. 117.
[7] E.g., the multi-stage rocket principle. Whereas Tsiolkovsky proposed it in a published work in 1929, Goddard had published it as early as 1914. However, Soviet sources claim that Tsiolkovsky had it worked out even earlier, but could not have it published because of 'unfavourable conditions of his time'. (A. Parry, *op. cit.*, p. 96.)
[8] V. Glushko, *Development of Rocketry and Space Technology in the USSR* (1973), p. 6; Soviet rocket research before 1945 is also discussed in chapter 9, pp. 448–51 above.

Tukhachevsky, then chief of armaments of the Red Army, took a great interest in the work of rocketry men and supervised and helped both GDL and GIRD.[9] The membership of the two organisations, together with branches of GIRD elsewhere in the Soviet Union, reached about 1,000 during the 1930s.[10]

The GIRD team which included Tsander, Tikhonravov and Pobedonostsev carried out both original research on reactive motion and the popularisation of rocketry. By the end of 1931 Tsander had developed the first Soviet prototype of an LPR engine, OR-1, and successfully tested it. Its thrust was only 0·145 kg and it was of a fairly elementary type, admittedly based on the principle of a blow-lamp. The OR-2, developed and built in 1931–2, was a more sophisticated version, with a thrust approaching 50 kg.[11] The first Soviet liquid-propelled engine actually used in a rocket was the '09' model, developed by Tikhonravov and flown in 1933.[12] The popularisation activities of GIRD included public lectures and courses on the theory of rocketry. While only three English language books on rocketry appeared in the 1930s in the West, several dozen titles came out in the Soviet Union in the same period.[13] A major German rocket expert, Helmut Groettrup, later commented: 'There is a much wider and more sensitive interest among the Russian intelligentsia for the rocket than among the German upper classes.'[14]

In September 1933, GDL and GIRD were officially merged into the Reactive Research Institute, RNII. Its military character was clearly set out in its statutory document, which stated its purpose as

> theoretical and practical work on problems of reactive motion with the aim of use of rockets in various fields of weaponry and the national economy. In particular, the Institute undertakes ... development and testing of rocket engines on solid, liquid and gas fuels.[15]

Although RNII was formally a single organisation, the GDL and GIRD teams preserved their separate identity and staff. The GDL group specialised in experimental rocket engines and between 1934 and 1938 built LPRs ORM-53 to ORM-102 with thrust power ranging from 100 to 600 kg (see Table 10.2); in 1939, it was separated from RNII (see p. 449 above). The GIRD group concentrated on designing rocket bodies and aircraft and on fitting the engines on the craft. Before the end of the Second World War RNII had designed the RD-103 rocket engine with a thrust of 900 kg, which was tested at altitudes up to 15,000 m.

---

[9] See a glossary of Soviet rocket research history in *Nauka i Zhizn'*, 1967, No. 10, p. 82; also P. Astashenkov, *Akademik S. P. Korolev* (1969), pp. 53–5; A. Romanov, *Konstruktor kosmicheskikh korablei* (1971), pp. 36–7.
[10] M. Stoiko, *Soviet Rocketry: The First Decade of Achievement*, New York (1972), p. 65.
[11] See *Nauka i Zhizn'*, 1967, No. 10, pp. 84–93.
[12] (Eds.) S. A. Sokolova and T. M. Melkumov, *op. cit.*, pp. 746–7.
[13] W. von Braun and F. I. Ordway, *op. cit.*, p. 61.
[14] I. Groettrup, *Rocket Wife*, London (1959), p. 161.
[15] *Aviatsiya i Kosmonavtika*, February 1974, No. 2, inside front cover. RNII was placed under the People's Commissariat of Heavy Industry in April 1934, but continued to retain its military character (see p. 449 above).

Table 10.2 *Development of major Soviet liquid propelled rocket engines (LPRs) 1929–45*

Developed by	Name type	F = fuel Ox = oxidant	Ignition	Thrust (kg)	Specific impulse (seconds)	Pressure in chamber	No. of chambers	Years of development	Used in
GIRD	OR-1		electric	0·145			1	1929–30	Prototype only[a]
	OR-2			50				1931–32	After 1935 in '07' and '216' rockets[b]
GDL (Gas Dynamic Laboratory)	ORM			6				1930–31	Stand tests only
	ORM-1	F = benzene Ox = oxygen	pyrotech	20	200			1930–31	Prototype only
	ORM-4 through -22	F = benzol, benzene: Ox = nitric acid, nitrogen tetroxide, liq. oxygen						1932	53 tests[c]
	ORM-23 through 49			100–250				1933	
	ORM-50	F = kerosene Ox = nitric acid	chemical	150	210			1930–33	Tikhonravov's '09' rocket[d]
	ORM-52	F = kerosene Ox = nitric acid		270–311	200	25 kg/cm²		1933	Rockets, torpedoes, planes
Team of GDL in RNII	ORM-53 through -70	F = kerosene Ox = nitric acid		100–500				1934–38	
	ORM-65	F = kerosene Ox = nitric acid	Manual + automatic	50–75	215	20–23 kg/cm²		1934–36	Korolev's rocket plane RP-318 and winged rocket 212
	ORM-65-1	F = kerosene Ox = nitric acid		122				1937	
	ORM-100 through -102	F = kerosene Ox = tetranitrom-ethane		80–100				1937–38	Experiments only
	ZRD			300				1936–40	Tikhonravov's rockets, rocket planes[f]
	RD-1	F = kerosene Ox = nitric acid	chemical	up to 1500			4	1941	Jets = PE-2, LA-7, YAK-3, SU-6[g]
	RD-1-X3 RD-2	F = kerosene Ox = nitric acid	chemical	300 200–700			2	1937–45 1945	Jets = LA-7P, SU-7, YAK-3, VK-105 PF[h]
	RD-3	F = kerosene Ox = nitric acid		900			3	1945	

Sources:
(Eds.) S. A. Sokolova and T. M. Melkumov, *Pionery raketnoi tekhniki* (1972), pp. 709–750.
V. P. Glushko, *Aviatsiya i Kosmonavtika* (1973), No. 6, pp. 32–43.
Nauka i Zhizn', 1967, No. 10, pp. 83–96.

Notes:
a – first Soviet LPR; air-breathing engine built on principle of blow-lamp by Tsander.
b – Tsander's design.
c – ORM-9 was of Glushko's design.
d – Glushko's design.
e – first LPR with controllable thrust of Glushko's design.
f – claimed world's first jet: tested 28.2.1940.
g – worked up to one hour—Glushko's design.
h – series production for jets—max. speed 795 km/h., max. altitude 7,800 m.

Despite the intensive Soviet R and D effort, no liquid-propelled Soviet missile was actually used in the Second World War. According to Vladimirov, after the execution of Tukhachevsky, the chief supporter of LPR in the leadership, nearly all the rocket experts were summarily executed in 1937–8; the few who survived, including Korolev and Voznesenskii, were required to work on aircraft design.[16] Another account published outside the Soviet Union, reports that Korolev joined Tupolev's '*Sharaga*' (a special aircraft design bureau, whose members were mostly prisoners) via a labour camp in Siberia, where he was involved in the pathetically inappropriate activity of gold-prospecting.[17] In view of this temporary downgrading of rocket research by Stalin, it is hardly surprising that the most advanced Soviet rocket actually used in the Second World War was the solid-fuel 'Katyusha', developed by RNII (see p. 449 above). Though highly effective, it was eventually dwarfed by the German long-distance V-1s and V-2s.

The development of American rocketry between the wars presents a sharp contrast. It was characterised by extreme individualism. The government had no interest whatsoever in rocketry. Goddard worked privately in secret: he did not join the American Interplanetary Society, founded in March 1930, or publicise his ideas, or help other rocket researchers.[18] His team consisted only of himself, his wife and three or four assistants. But both the scope of his activities and his success were amazing. Between 1914 and his death in 1945 he registered over 200 patents in rocketry, and in most of his inventions was ahead of the Soviet experts. Thus he developed a gyroscopic stabilization system for rockets in 1933–4 and successfully tested it in early 1935,[19] while the first Soviet gyroscopic stabilizer developed independently of Goddard, on which many experts must have been engaged, took three years (1935–7), and success was apparently fairly modest.[20]

The Soviet effort between the wars is perhaps more comparable to the German, which was also predominantly military in character, and supported by state (military) funds.[21]

At first, Hitler refused to finance the development of a long-range bombardment missile, but in 1936 he reversed his attitude, and in early 1937 the 25-year-old von Braun became the technical director of Peenemunde complex, a large new R and D centre. Their main task was to build a powerful LPR with a range of 120 miles; this was achieved in the early 1940s with the completion of the two V-1 and V-2 missiles. Comparing the RNII with Peenemunde, a Soviet military historian sourly commented:

---

[16] L. Vladimirov, *op. cit.*, p. 32.
[17] G. A. Ozerov, *Tupolevskaya sharaga*, Frankfurt (1971).
[18] W. von Braun and F. I. Ordway, *op. cit.*, p. 78; a detailed account of Goddard's work and times appears also in A. P. Dewey, *R. Goddard: Space Pioneer*, Boston (1962).
[19] W. von Braun and F. I. Ordway, *op. cit.*, p. 49.
[20] V. Glushko, *Aviatsiya i Kosmonavtika*, 1974, No. 3, p. 44–5.
[21] M. Stoiko, *op. cit.*, p. 68, quotes von Braun as saying: 'The Versailles Treaty had not placed restrictions on rockets, and the [German] Army was desperate to get back on its feet. We did not care much about that, . . . but we needed money and the Army seemed willing to give it to us.' See also, e.g., A. Parry, *op. cit.*, p. 114.

One of the reasons for the failure of rocket building in Hitlerite Germany may have been the organisation in Peenemunde, the gigantic centre of W. von Braun; the resources swallowed up by that centre were fantastic, but it did not live up to its expectations. Forty years ago our rocket men set up not an 'all-embracing' enterprise, but rather the one which was required: optimal.[22]

It takes a great deal of courage to associate Peenemunde, whose achievements in terms of both the speed of development and performance by far outstripped the parallel Soviet effort, with the 'failure' of German rocketry.

SOVIET ROCKETRY AFTER THE SECOND WORLD WAR

At the end of the Second World War German rocket facilities and personnel fell into American and Soviet hands. Numerous Western commentators and experts have subsequently tried to establish the scale of the German contribution to the development of rocketry in the two countries. The Americans openly admit the significance of von Braun and of his Peenemunde team in post-war American rocketry, but an assessment of the German contribution to the Soviet rocket effort is far more difficult, chiefly because it tends to be concealed and played down by Soviet sources. They have repeatedly insisted on the originality of their R and D and have persistently denied that Soviet rocketry benefited from German know-how, hardware and personnel. Thus, a biographer of Korolev, admitting that Korolev and other Soviet rocket experts 'were familiar with German war-time LP missiles', added that 'he did not, however, find in them any "discoveries"—they were all based on the ideas of K. E. Tsiolkovsky'.[23] At the same time this biographer and other Soviet commentators carefully conceal the fact that soon after the war Korolev was a member of a 15-man Soviet team of officers, rocket and aviation experts, including G. Tokaev-Tokaty, which spent almost two years in Germany studying rocketry and jet aviation.[24]

Another biographer describes the Korolev mission as 'a prolonged business visit abroad'.[25] In February 1947 Korolev began work as Chief Designer of OKB,[26] the Experimental Design Bureau which had been set up in 1944.[27] The first Soviet post-war long-range guided missile was built under Korolev's supervision immediately afterwards, and official tests came as soon as 7 November in the same year, the thirtieth anniversary of the October

---

[22] I. Chutko, in *Znamya*, 1973, No. 8, p. 184.

[23] P. Astashenkov, *op. cit.*, p. 99. By no means is a Goddardian view of the V-2 any less patriotic in the context of American rocketry: A. P. Dewey, *op. cit.*, p. 76, says, 'The V-2 missile... was almost identical with the Goddard rocket as described in this early publication' (reference is to Goddard's paper, 'A method of reaching extreme altitudes', published in 1919).

[24] See, for example, S. White's interview with Professor G. A. Tokaty in *The New Scientist*, 8 July 1971, p. 87; also G. A. Tokaev, *Comrade X*, London (1956), p. 287.

[25] A. Romanov, *op. cit.*, p. 48.

[26] A. Romanov, *loc. cit.* However, Korolev was appointed to this post earlier, 9 August 1946, presumably in his absence (see P. Astashenkov, *op. cit.*, p. 99).

[27] See V. Glushko in *Soviet Weekly*, 2 February 1974, and the same author in *Soviet News*, 23 October 1973. A. Isaev was OKB's first head.

Revolution, a mere 10 months after his return from Germany.[28] Some of the prototypes of Soviet rockets in that period are said to have resembled the V-2,[29] and Stoiko[30] suggests, without providing any reference, that in 1945–6 V-2 production was completely restored in the USSR, surpassing the level of German production in 1944 by 1946–7. Sutton claims that the post-war Soviet space industry has its roots in German rocket technology and concludes that 'Soviet rockets and missiles can be clearly traced to German V-2 technology and transferred production capabilities.'[31]

But he appears to be virtually ignorant of the development of Soviet rocketry between the wars and his account of the activities of German experts in the Soviet Union is, in some important respects, incorrect. He claims, for instance, that the return of the main group of the 6,000 German technicians brought to the Soviet Union in 1945 did not start until 1958.[32] But the wife of the most senior German rocket expert in the USSR, Helmut Groettrup, reports that by June 1952 all except the 'top 20' had been repatriated, Groettrup himself being back in Germany for Christmas 1953.[33]

Mrs Groettrup's diary clearly shows that the Germans in fact had hardly any connection with the main stream of Soviet developments in the field. In 1947, soon after the first tests of the V-2, not all of which were very successful, Groettrup deemed it necessary to lodge a formal complaint that the German management of his group was not being given any insight into the general policy and the future development of rocket work. This situation continued throughout. Groettrup's own project for an improved V-2, submitted in late 1947, was returned to him with Soviet comments and criticisms, while a parallel project, by a Russian, was approved and adopted for assimilation.[34] It can be inferred that even in 1946–7 the Russians had their own parallel rocket R and D in which the Germans did not participate, and which was at least as good in quality and speed as the Russian-supervised efforts by the German prisoners.[35] Nevertheless, the restricted nature of the information published about the immediate post-war years in the USSR may suggest that the benefit which they drew from German rocketry was not as insignificant as they claim.[36]

[28] Even to Romanov this must have seemed a somewhat hasty development of affairs because he deemed it necessary to add that 'without the pre-war experience accumulated by S. Korolev and designers of individual rocket parts, without the skill of other specialists, the task of constructing the rocket systems could hardly be solved so speedily. But, of course, the main basis of the success was the rapidly growing productive forces of the country' (*op. cit.*, p. 49).
[29] G. E. Wukelic, *Handbook of Soviet Space-Science Research*, New York (1968), pp. 7–8.
[30] M. Stoiko, *op. cit.*, p. 75.
[31] A. C. Sutton, *Western Technology and Soviet Economic Development, 1945 to 1965*, Stanford, Cal., pp. 254–79.
[32] A. C. Sutton, *op. cit.*, pp. 271–6. The first Soviet Sputnik was sent into orbit in October 1957. No references are given.
[33] I. Groettrup, *op. cit.*, pp. 152, 162, 163, 182. According to another account, by 1955 hardly any Germans worked in Soviet rocketry (see A. Parry, *op. cit.*, pp. 123–5).
[34] I. Groettrup, *op. cit.*, pp. 86, 107–8, 116.
[35] 'The Germans worked in very narrow fields, mutually isolated, always surrounded by 3:1 ratio of Russian fellow-workers. When sent back to Germany, they did not know much about the Soviet rocket programme, if anything' (A. Parry, *op. cit.*, p. 126).
[36] Sir Bernard Lovell's statement, also given without any references, that 'it appears that Korolev in 1947 had decided that the development of V-2 technology was not the way to create a Soviet long-range missile', seems distinctly premature (see his *Origins and International Economics of Space Exploration*, Edinburgh (1973), p. 18). The Soviet post-war missile programme is further discussed in chapter 9, pp. 455–9 above.

Soviet space policy

Before dealing with the technology side of the post-war Soviet space programme as such, a brief comment needs to be made about Soviet space policy, as indicated by the statements of persons closely connected with the Soviet space programme. The main problem which has aroused a lot of controversy in the West throughout the 1960s, and which has not yet been conclusively solved is whether the Soviets intended to undertake manned flights to the moon. As early as 1963, Sir Bernard Lovell stated, after a visit to the Soviet Union, that he was not convinced that the USSR thought it was scientifically or otherwise desirable to send a man to the moon; he considered that the manned mission to the moon was not the centrepiece of the Soviet space programme.[37] What is the centrepiece of the Soviet space programme, if such a thing exists at all, has always been something of a puzzle, and different Soviet experts and commentators offer different accounts of the order of priorities in the Soviet space programme.[38] In our view, although plans to build manned orbital stations were given high priority about 1965, manned flight to the moon never ceased to be seriously considered, however unclear its place among the Soviet objectives in space may be. Korolev, when submitting a long-term Soviet space programme on behalf of a group of scientists headed by himself and Keldysh in the early 1960s, argued that as rocket building advanced it would be possible to set up a permanent scientific station on the moon.[39] Zond 5 and 6, which undertook circumlunar flights between September and November 1968, were admittedly designed for manned flight.[40] In 1974 V. Ivanov and K. Kondrat'ev[41] described piloted stations of the Soyuz-Salyut type and automatic apparatuses as a step towards setting up a lunar observatory for the study of the earth (and other bodies). The evidence in fact appears to indicate that what has been delaying Soviet realisation of a manned flight to the moon is the lack of a powerful enough rocket carrier, as we shall see in the course of this chapter.

Soviet space technology

The technological side of a space programme can be roughly divided into the following categories:

[37] *Soviet Space Programmes, 1962–5*, US Congress Staff Report, GPO, Washington DC (1966), pp. 374–5.

[38] For some examples from the early 1960s, see *Soviet Space Programmes, 1962–5*; more recently, according to one account 'The total of the Soviet space programme is keyed to achieving an orbital station' (Moscow Radio, 3 May 1970, cited in D. L. Harvey and L. Ciccoritti, *US-Soviet Cooperation in Space*, Monographs in International Affairs, Center for Advanced International Studies, University of Miami (1974), p. 201); Academician Petrov a year later put automatic apparatuses in the first place, application satellites in the second and manned flights (including orbiting stations) in the third (B. Petrov in *Trud*, 9 April 1971); a few months later the President of the Academy of Sciences, M. Keldysh, talking about the Soviet space plans for the next few years, put manned orbital stations first and automatic apparatuses second—Moscow Radio, English Service for North America, 26 September 1971 (*BBC Monitoring Report*, 30 September 1971).

[39] A. Romanov, *op. cit.*, p. 62; for some other accounts of the Soviet plans for manned flights to the moon, see, for example, *Soviet Space Programmes, 1962–5, op. cit.*, pp. 367–8.

[40] M. Keldysh at a press conference, *Pravda*, 12 April 1969.

[41] V. Ivanov and K. Kondrat'ev, *Aviatsiya i Kosmonavtika*, 1974, No. 2, p. 5.

1. Launch vehicle building (i.e., construction of rocket engines and boosters).
2. Spacecraft (capsule) building, which usually takes into account the instruments and apparatuses for scientific research in space.
3. Ground and sea support systems (including launching, servicing and tracking facilities).

'Space spillovers'[42] and related matters might be included as a fourth category. Here we shall be primarily concerned only with the first two categories, though we shall not deal with instruments and apparatus for scientific research in space.

LAUNCH VEHICLE BUILDING

In the Soviet Union equal emphasis is being laid on both the scientific and technological aspects of rocket building—both are treated as an almost monolithic unity. Korolev himself is known to have been a theoretician of genius as well as an engineer, and always tended to see rocketry as an amalgam of complex problems which needed to be solved by a concerted effort of scientists and engineers, whom he considered equal partners.

In the USA, on the other hand, engineering has been considered the predominant activity in rocket building. Thus, von Braun stated: 'I believe an established missile programme, like Jupiter, has much more similarity with an industrial planning than with a scientific project ... I would say the Jupiter programme was 90 per cent engineering and 10 per cent scientific.'[43]

ROCKET THRUST AND DESIGN PHILOSOPHY

In this section the thrust of rocket engines is taken as the crucial indicator of performance. Thrust is basically the force developed by the rocket engines in the combustion chamber; in technical terms, it is equal to the reaction force minus the atmospheric pressure produced by the nozzle exhaust area. Both Soviet and American sources frequently report the thrust power of their rocket engines. Once the theory of reactive motion was worked out and the first prototypes of rocket engines on suitable fuels built by rocketry pioneers, the problem of increasing thrust became the chief concern of designers. The Soviet expert Glazunov emphasised in 1964 that the US did not possess rocket power comparable to that which thrust Voskhod into orbit, and concluded: 'thus the superiority of the Soviet Union is indisputable'.[44] The same type of reasoning has been used by the Americans since they successfully tested their Saturns IB and V. But thrust is not a mere device for intercontinental boasting; the more powerful the rocket, the more it can send into space.

The crucial developments are set out in Table 10.3. The Soviet lead in terms of thrust lasted over eight years, from 1957 to 1966: all Vanguard, Atlas, Thor and Titan 2 launch vehicles were less powerful than Vostok. But from 1967

[42] The term used by R. W. Campbell in his work *Space Spillovers in the Soviet Economy*, Manuscript, University of Indiana (1969).

[43] US Senate Committee on Armed Services Hearings: *Inquiry into Satellite and Missile Programs*, 85th Congress, 1st and 2nd sessions (1958), pp. 590–1.

[44] Quoted in *Soviet Space Programs 1962–5*, US Congress Report, Washington DC (1966), p. 82.

ROCKETRY: TECHNOLOGY IN LAUNCH VEHICLES AND MANNED CAPSULES

onwards, since the launching of Saturn V, the Americans have been consistently ahead.

What were the developments which gave the Soviet Union its initial advantage in thrust and hence space technology as a whole? As has been shown, the Soviet experts had accumulated extensive experience in rocket engine building since the early 1930s and their knowledge was further enriched by their familiarity with German rocket technology. Against this positive background the Soviet Union was driven to develop rocket technology by military necessity. According to several accounts the Soviet decision to concentrate on a high thrust for their launch vehicles had its origins in the initial Soviet handicap in nuclear technology. The first Soviet A-bomb was not tested until 1949, and by that time development of light nuclear warheads was already well under way in the United States. Rather than accept this lag and wait for lighter warheads to come, the Soviet authorities committed themselves to building a powerful guided intercontinental missile capable of carrying a heavy nuclear payload.[45]

The first Soviet ICBM was successfully tested in August 1957, two months before Sputnik I. The initial emphasis on high thrust was reflected in the Soviet design philosophy, which was quite different from the American. Sheldon[46] points out that the Americans, in order to make launch vehicles reach their potential, had to think very hard to minimise the weight of the structure wherever possible. Thus the Atlas tankers were so thin that they always had to be kept pressurized in order to prevent their collapse, while the walls of Vostok tankers were so thick that Soviet workmen could walk along their entire length without damaging them.

DEVELOPMENT OF LAUNCH VEHICLES

The chief launch vehicle upon which the Soviet space programme has been heavily dependent is referred to in the West as the Vostok system. It has undergone many changes since it was first conceived by Korolev and his Special Design Bureau in 1954. The Vostok power system is a combination of RD-107 and RD-108 liquid propellant rocket (LPR) engines. The early two-stage version (the power plant of the first Soviet ICBM and of the early Sputniks) consisted of a single central RD-108 (the first stage) surrounded by a cluster of four RD-107s (the second stage). This version developed a thrust of some 500 tons in 1957 and was later upgraded into a three- and finally a four-stage vehicle with thrust increasing to 650 tons (see Table 10.4). It can safely be said that it was this 'cluster' idea, the most daring and progressive of its time, which secured the Soviet Union its initial advantage in terms of rocket thrust.[47]

---

[45] For this plausible version of the history of the Soviet emphasis on high rocket engine thrust, see, for example, B. Lovell, *op. cit.*, p. 19. For Soviet nuclear developments, see chapter 9, pp. 451–5 above.

[46] C. Sheldon in *Soviet Space Programs, 1966–70*, US Congress, Washington DC (1971), p. 136.

[47] Vladimirov is rather ironic about the cluster principle and refers to it as a 'bunch of lilliputs', and a 'clumsy and not very reliable bunch system' (see L. Vladimirov, *op. cit.*, pp. 50, 51, 77–9). Yet the principle was good enough to be adopted some four years later by von Braun himself, when he conceived Saturn (see W. von Braun and F. I. Ordway, *op. cit.*, p. 167 and pp. 506–7 below).

Table 10.3 Development of thrust and payload capability of Soviet and US launch vehicles

Years	USSR				USA				Lead	
	Vehicle	Thrust (kg)	Payload capability (kg) E Orbit	Escape	Vehicle	Thrust (kg)	Payload capability (kg) E Orbit	Escape	Sov.	US
1957	2-stage Vostok	510000	1400						✓	
1958					Vanguard Atlas	17055 178616	1360			
1959	3-stage Vostok	600000	4750	1550	Thor-Delta	21500	>220		✓	
1960									✓	
1961									✓	
1962									✓	
1963									✓	
1964	4-stage Vostok	650000	7500	1650 (moon) 1180 (Venus)	Titan 2	240000	3175		✓	
1965	Proton	1800000	27000	7250	Saturn 1B Titan 3C	748230 1336300	16601 11340	6350	✓	
1966									✓	
1967					Saturn V	4048380	127008	45360 (moon)		✓
1968										✓
1969										✓
1970										✓
1971										✓
1972										✓
1973										✓

Sources:
Aviation Week and Space Technology, 11 March 1973, pp. 129–30.
V. P. Glushko, *op. cit.*, pp. 32–5, 39.
W. von Braun and F. I. Ordway, *op. cit.*, passim.
Soviet Encyclopedia of Space Flight, passim.

Another development before 1957 was the RD-214 first-stage LPR engine, which has been used only since 1962 in some launchings of the Kosmos programme with RD-119 as its second stage. Although not particularly powerful, RD-214 and RD-119 are nevertheless seen as technologically significant: RD-214 has the biggest thrust and specific impulse of all engines which burn hydrocarbon with nitric acid as oxydant, and RD-119 has the highest specific impulse of engines propelled by a high-boiling-point fuel.[48] The next rocket engine development did not occur until 1965, when the exclusively civilian Proton vehicle was launched. Having four times the payload capability and three times the thrust of the standard Vostok combination (designated group A in Table 10.4), Proton has been used relatively little, presumably for reasons of economy, mainly for heavy missions such as launching of space stations and in the Kosmos programme. The other two vehicles, group C in Table 10.4 (Skean or SS-5), and group E (Scarp or SS-9), were developed as military missiles modified for civilian space use by 1964 and 1966 respectively, making a total of five launch vehicle types used in the Soviet space programme so far.

In order to be able to meaningfully compare the development of, say, the Vostok launch vehicle (on which relatively more information is being released by the Soviet authorities) with the American Saturn, we have to consider the period of development of the respective vehicle types. The Soviet sources report that the period of development of the first ICBM engines including construction and testing was 1954–7. Assuming that the boosters took the same period to build we can infer that the earlier period, between 1947 (the year of the decision to build a long-range LP missile) and 1953 was spent on solving scientific and technological problems, working out alternatives and selecting the most satisfactory prototype for the final stage. Stoiko[49] offers the following estimates of the average duration of the individual stages of Soviet launch vehicle development: engine—five years; vehicle—three years; systems tests— one year; then follow flight tests in the Pacific and product improvement, including the addition of upper stages. He suggests that, taking into account overlaps between individual stages, it takes the Soviet Union some seven years from the authorisation of engine development to systems test completion. However, the upgrading of Vostok by adding the upper stages took at least seven years (1957–64) and that was a continuous process in which

---

[48] V. Glushko, *op. cit.*, p. 35. Specific impulse (SI) or specific thrust is an indicator quite different from thrust (T), mentioned above, and it is defined in the *Soviet Encyclopedia of Space Flight*, Moscow (1969), p. 431, as 'the impulse developed by a rocket engine per 1 kg of earth weight (at sea level) of consumable propellant'. It is measured in seconds and the basis formula for its calculation is:

$$SI = \frac{T}{mg} \quad \text{where} \quad \begin{array}{l} T \text{ is absolute thrust;} \\ m \text{ is mass exhausted from nozzle each second;} \\ g \text{ is gravitational acceleration.} \end{array}$$

The specific impulse (SI) is the most important measure of rocket performance because it shows the amount of propellant required to give a rocket lifting capability. In practice SI is influenced by such parameters as geometry of the nozzle, outside pressure, and the type of propellant used, as well as by various features of the engine (design, cooling, etc.). The values of SI of modern rockets vary from 270 to 400 seconds.

[49] M. Stoiko, *op. cit.*, p. 101.

Table 10.4 *Some characteristics of major Soviet launch vehicles*

Vehicle name, type	Years of development	Designer	Engine(s)	Stage	Propellant	Thrust (kg)	Diam. (m)	Length (m)	Payload capability (kg) Earth orbit	Payload capability (kg) Escape	Year first flown
A, SS-6 Vostok (Sapwood)	1954-7	GDL/OKB	4 × RD-107	1	LOX/kerosene	408000	3·0	19·0	1400		1957[a]
	1954-7	GDL/OKB	1 × RD-108	2	LOX/kerosene	102000	2·95	28·0			
				Shroud			2·60				
			5	Overall		510000					
A-1 SS-6 Vostok (Sapwood)	1954-7	,,	4 × RD-107	1	LOX/kerosene	408000	3·0	19·0	4750	1550	1959[b]
	1954-7	,,	1 × RD-108	2	LOX/kerosene	102000	2·95	28·0			
	1957-9	,,	1	3	LOX/kerosene	90000	2·60	3·4			
				Shroud				6·6			
			6	Overall		600000					
A-2 SS-6 Vostok (Sapwood)	1954-7	,,	4 × RD-107	1	,,	408000	3·0	19·0	7500	1650 (moon)	1964[c]
	1954-7	,,	1 × RD-108	2	,,	102000	2·95	28·0		1180 (Venus)	
	1957-64	,,	1	3	,,	140000		10·0			
				Overall		650000		13·6			
B-1	1952-7	,,	4 × RD-214	1	Nitric acid + hydrocarbon	74000	1·65	20·3			1962[d]
SS-4 (Sandal)	1958-62	,,	1 × RD-119	2	LOX/dimethyl-hydrazine	11000	1·65	8·5	135-450		
				Shroud			1·65	3·3			
				Overall		85000					
C-1 SS-5 (Skean)		,,		1			2·48	19·1	450-900		1964[e]
		,,		2			2·48	8·5			
				Shroud			2·48	4·6			
				Overall			2·48				
D Proton	-1965	n.a.		1	LOX/UDMH	1530000	4·0	40·0	12200-27000	5440-7250	1965[f]
		n.a.		2		270000	4·1	12·0			
		n.a.		3		n.a.		14·0			
				Shroud							
				Overall		1800000					
F SS-9 (Scarp)	-1966	n.a.		1			3·0	20·7	4500		1966[g]
		n.a.		2			3·0	11·7			
		n.a.		3			3·0	8·9			
				Shroud			3·0	6·1			
				Overall			3·0				
G	Currently under development	n.a.		1		4500000-6350000		21·0	136000	31750-40800	[h]
		n.a.		2		1530000	4·0	40·0			
		n.a.		3		270000	4·1	12·0			
				4				13·6			
				Shroud							
				Overall							

LOX – Liquid oxygen; UDMH – Unsymmetrical dimethyl hydrazine.

Sources:
C. S. Sheldon in *Soviet Space Programs, 1966–70*, Staff Report for Use of Committee of Aeronautical Space Sciences, Washington (1971).
V. Glushko, *op. cit.*, pp. 32–5, 39.
Aviation Week and Space Technology, 11 March 1973, pp. 129–30.

Notes:
a – used in Sputniks 1–3.
b – used in Luna 1–3; Vostok manned flights, Launch weight = 327,000 kg.
c – used in Voskhod–Soyuz; Kosmos; Luna 4–14; Mars*, Venera.*
d – used in Kosmos (Science).
e – used in Kosmos (Navigation).
f – used in Zond; Luna.
g – used in Kosmos (FOBS: Fractional Orbit Bombardment Satellites).
h – not successfully tested yet; US reported test failure in 1969. Intended for lunar and interplanetary flights.

*More advanced four-stage modification; number of engine nozzles increased by one.

'overlapping' is not easy to define. Because process improvement beyond the standard vehicle stage followed somewhat different patterns in the Soviet Union and the United States, we shall distinguish between (a) the period of development of the basic (standard) vehicle type and (b) the period for upgrading the vehicle, including the addition of stages. In the case of Vostok these were four years and seven years respectively.[50]

Let us now turn to the American Saturn series of vehicles (see Table 10.5). The initial idea of Saturn as a cluster of a number of new Jupiter engines around Redstone and Jupiter propellant tanks was conceived by W. von Braun in 1957, and approved by the US Department of Defense a year later. There were three basic vehicles in the Saturn series: Saturn 1, Saturn 1B and Saturn V, the ultimate purpose being their use for launching Apollo spaceships for the manned exploration of the moon, manned orbital stations and heavy satellites. Saturn 1 and Saturn 1B were two-stage vehicles.

Saturn 1's first stage has eight Jupiter engines, each developing a thrust of 85–90 tons. The second stage has six engines with a thrust of 6·8 tons each. The first static test of Saturn 1 took place in May 1961, and the first flight test in October of the same year, but the first payload was not sent into orbit until January 1964. Hence period (a) for Saturn 1 was four years, the same as for Vostok.

Saturn 1B underwent its first tests in April 1965 and the maiden flight followed in February 1966, so that period (a) for Saturn 1B was some seven to eight years. Its J-2 first stage engine develops a thrust of 90·7 tons and was used in the second and third stages of Saturn V. Both Saturn 1 and Saturn 1B were used for launching Pegasus military satellites and dummy Apollo spaceships. The first stage of the Saturn V moon vehicle is a cluster of five powerful F-1 engines which produce a thrust of 680 tons each. The development of this epoch-making engine started in 1959 and was completed by late 1967, when the first flight of the composite three-stage vehicle took place. That period for Saturn V was therefore some eight years. The second stage had five J-2 engines taken from Saturn 1B and the third stage had only one J-2.

Despite a different accommodation of engines in the individual vehicle stages of Vostok and Saturn V, the main similarity between the two is the cluster idea. W. von Braun must have come to the conclusion that it was the only technically feasible way to solve the problem of deriving maximum thrust from the first vehicle stage. Apart from this, the differences rather than resemblances between Vostok and Saturn V are more striking. Whereas Vostok's powerful two-stage 'basis' (built between 1954 and 1957) was subsequently upgraded, with later addition of the upper stages (1957–64) which improved rocket thrust and existing technology only marginally, the

[50] There are indications that the Soviets are in the process of developing their 'super vehicle' with a thrust of some six thousand tons (see Table 10.4 and reports on test failures of this vehicle in *Aviation Week and Space Technology*, 17 November 1969, p. 26; 9 April, p. 21, and 14 January 1974, p. 12). This can easily be seen as a logical extension of the Soviet efforts in rocketry. The period of development of this vehicle is undoubtedly considerably longer than that of Vostok or any other known Soviet launch vehicle. However, due to the total Soviet secrecy about this vehicle (as well as of others in their process of development), its whereabouts and specifications can only be a matter of guessing and reliance on the Western intelligence reports.

reverse was true of Saturn development: the most powerful first stage of Saturn V was the latest and longest development (1959–67), while the upper stages had been built and flown earlier as stage two of Saturn 1B (dummy Apollo and Pegasus missions). In other words, whereas the Soviet rocket engine designers have adopted a comparatively conservative approach of building upon a well-tested standard piece of technology with modest results, the Americans have followed the path of longer, but more sophisticated, innovation which facilitated carrying out a stupendously ambitious and publicly declared space goal, namely, landing on the moon. And here we come to another major difference between vehicle development in the two respective countries. Whereas any interested person could follow the American successes and failures in the process of Saturn V development, the circumstances of development of Vostok were kept entirely secret until success was assured. This feature of Soviet space vehicle development still appears to apply at present, considering, for example, the virtual absence of any public knowledge of Proton launch vehicle characteristics (see Table 10.4), as well as the total secrecy surrounding the development of the Soviet 'super vehicle'. (See note 50, p. 506.)

EXPLOITATION OF LAUNCH VEHICLES

From Table 10.6, which shows Soviet and US launchings by type of launch vehicle used, some significant inferences can readily be made. First, the Soviet Union has made very extensive use of Vostok. Whereas the US had a different type of launch vehicle for each manned flight programme (with the exception of Apollo and Skylab, both of which used Saturn), the Vostok-type launch vehicle was used for all three Soviet manned programmes. In 1961, the year of Gagarin's flight, the Vostok-type vehicle was the only one used by the Soviet Union in their entire civilian space programme, apart from much less important meteorological and geophysical rockets. By that time the US had already flown five types of launch vehicles and were leading in number of space launchings. The USA lost its lead in the annual number of space launchings in 1967 and the total number of launchings fell below that of the USSR in 1971. By the end of 1973, the total number of launchings was over 100 less than in the case of the Soviet Union. Yet the Soviet Union had used only five as compared to the US six types of launch vehicles (this excludes the US Redstone and Jupiter which flew only a few times). The average intensity of use of carriers is thus considerably higher in the Soviet Union than in the USA.[51] Vostok was particularly important. As is shown in Table 10.6, except in 1971 Vostok was always responsible for over 50 per cent of the annual number of Soviet launchings, and between 1957 and 1973 Vostok sent into space more than three fifths of all Soviet payloads. Thor, the most frequently used carrier in the United States, sent up less than 50 per cent of US payloads.

[51] Using Table 10.6 (and excluding Redstone and Jupiter launchings) we can calculate the average launch ratios per vehicle in the two countries and weigh them against each other. The average intensity of use of Soviet launch vehicles is over 40 per cent higher than that of US vehicles. Thus: $\frac{707}{5} \div \frac{593}{6} = 1\cdot 43$.

Table 10.5 *Some characteristics of major US launch vehicles*

Vehicle name	Years developed	Designer	Stage	Engines	Propellant	Thrust (kg)	Diam. (m)	Length (m)	Payload capability Earth orbit	Escape	Year first flown
Vanguard	1955–7	Martin	1		LOX/kerosene	12247	1·14	21·95			1958[a]
			2		Nitric acid/UDMH	3402					
			3		Solid	1406					
			Overall			17055					
Atlas SLV-3(C)	1958	GDCA	1	2 × LR-89-7	LOX/kerosene	152409	3·05	20·12	1360		1958[b]
		GDCA	2	1 × LR-105-7	LOX/kerosene	26308					
						178616					
Thor-Delta	1959		1		LOX/kerosene	78000		25·28	220+		1959[c]
			2		Nitric acid/UDMH	3500					
			3		Solid						
Titan-2	1962–4		1	2 × LR-87	Nitrogen tetroxide +kerosene-50	195048	9·14	31·39	3175		1964[d]
			2	2 × LR-91-AJ-9	Nitrogen tetroxide +kerosene-50	45360					
			Overall			240408					
Saturn 1B	1959–65	McDouglas	1	8 × H-1	LOX/kerosene	743904	6·52	24·48	16601	6350	1965[e]
		Chrysler	2	1 × J-2	LOX/kerosene	104328	6·61	17·8			
			Overall			848232					
Saturn V	1959–67	Boeing	1	5 × F-1	LOX/kerosene	3433752	10·06	42·06	127008	45360	1967[f]
		NAR	2	5 × J-2	LOX/liq. hydr.	510300	10·06	24·84			
		McDouglas	3	1 × J-2	LOX/liq. hydr.	104328	6·61	17·80			
			Overall			4048380					
Titan-3C	1962–71	Martin-Marietta	1	2 × UA1205	Solid	1088640	9·14	37·8	11340		1965[g]
			2	2 × LR-87	Nitrogen tetroxide +kerosene-50	195048					
			3	2 × LR-91-AJ-9	"	45360					
			4	2 × AJ10-138		7258					
			Overall			1336306					
Agena (upper stage)		Lockheed		1 × Bell 8096	Nitric acid/UDMH	7260	1·51	7·09			[h]
Centaur (upper stage)		GDCA		2 × RWRL 10A-3	LOX/liq. hydr.	13610	3·05	9·14			[i]

LOX – Liquid oxygen; UDMH – Unsymmetrical dimethyl hydrazine; Liq. Hydr. – Liquid hydrogen.

Sources:
Aviation Week and Space Technology, 11 March 1973, pp. 129–30.
Soviet Encyclopedia of Space Flight, Moscow (1969), passim.
W. von Braun and F. I. Ordway, *op. cit.*, passim.
Soviet Space Programs, 1962–5, US Senate Staff Report, GPO, Washington (1966), passim.

Notes:
a – used in Vanguard.
b – used in Midas, Samos, Ranger, Mariner, Mercury.
c – used in Explorer, Tiros, Discoverer, Telstar, Early Bird, ERTS, Nimbus, HEDS.
d – used in Gemini.
e – used in dummy Apollo, Skylab 2–4.
f – used in Apollo and Skylab 2–4.
g – used in DSCS, IMEWS.
h – used with Atlas, Thor, etc.
i – used with Atlas, Titan, etc.

Can one launching of a vehicle into space by one country be equated with one launching by another country? It would be absurd to suggest that the launching of the Apollo 11 mission to the moon on 16 July 1969 is equivalent to any Soviet Luna or Zond mission aimed at the moon, simply because of the absence of a Soviet attempt at manned lunar landing. Similarly incomparable were the Soviet Lunokhod remote-controlled vehicle missions, because the Americans, presumably due to their emphasis on the manned exploration of the moon, had not developed a vehicle of the Lunokhod type. Thus we can see that the different objectives of the respective countries' space programmes make it extremely difficult, if not impossible, to compare the individual Soviet and American launchings meaningfully. There are, of course, important exceptions, such as space programmes which run parallel to each other. Here, a table of 'space firsts' can reveal some interesting results (see Table 10.8) and 'lags' in terms of years could be calculated on the basis of a comparison of the data. For instance, in orbital stations, the Soviet Union is two years ahead: the Soviet Salyut 1 was launched on 19 April 1971, the American Skylab on 14 May 1973. But total number of launchings by the respective countries over a period may mean very little. Thus between 1969 and 1973, the years of historical advances by the US Apollo missions, numerically the US had launched into space a mere two fifths of what the Soviets did (see Fig 10.1A). In terms of the total weight launched into space (Fig 10.1B) the Soviets have always been ahead, despite the Apollo (1969–73) and Skylab (1973) 'heavy missions'.

In short, while both the number of launchings and the weight sent into space do indicate the degree of space activity, they do not necessarily suggest very much about the quality of technological advances in space.[52] Launching statistics should therefore be treated carefully, and no sweeping conclusions should be drawn from Fig 10.1.

SPACECRAFT BUILDING

The fleet of Soviet spacecraft from Sputnik to Polet and Soyuz is very impressive both in total number and in the performance of the individual craft. Development has been rapid. Sputnik was a fairly simple sphere with a few instruments on board; Soyuz and the latest types of interplanetary probe are sophisticated heavy apparatuses.

In spacecraft building as in the development of launch vehicles we can observe what observers like Stoiko have called the 'add-on' philosophy, which is technologically an important feature of the Soviet programme. Instead of building a new type of spacecraft for each mission, the design, materials, apparatuses and other equipment are basically the same.[53] According to requirements of various missions, new parts are incorporated into and

---

[52] A Czech commentator asked 'Why do we build such a quantity of artificial satellites and boost them up into orbit as cosmic litter?' ('K čemu slouzi kosmonautika?' ('What does space science serve?') in *Tvorba*, 1975, No. 3, p. 19). (He may have meant the Soviet Kosmos programme, although, for obvious reasons, he did not say this in so many words.)

[53] Spacecraft shell/body, service systems, on-board equipment, central circuitry, power supply systems—all of these are roughly the same (see M. Stoiko, *op. cit.*, p. 103).

Table 10.6 *Soviet and US launches by types of launch vehicles (1957–73)*

Type of launch vehicle	1957		1958		1959		1960		1961		1962		1963		1964		1965		1966		1967		1968		1969		1970		1971		1972		1973		Total per vehicle
	1	2	1	2	1	2	1	2	1	2	1	2	1	2	1	2	1	2	1	2	1	2	1	2	1	2	1	2	1	2	1	2	1	2	
USSR																																			
Vostok (A-SS-6)	2	100	1	100	3	100	3	100	6	100	13	65	13	76.5	22	73.3	33	68.7	34	77.2	36	54.5	41	55.4	43	61.4	43	53.1	40	48.2	45	60.8	53	62.4	431
Sandal											7	35	4	23.5	7	23.3	7	14.6	7	15.9	13	19.7	6	8.1	6	8.6	18	22.2	12	14.5	12	16.2	10	11.8	127
Skean															1	3.4	6	12.5			4	6.0	6	8.1	6	8.6	10	12.3	17	20.5	13	17.6	14	16.5	77
Scarp																			2	4.5	11	16.7	7	9.5	3	4.3	6	7.5	7	8.4	2	2.7	1	1.2	39
Proton																	2	4.2	1	3.4	2	3.1	4	5.4	4	5.7	4	4.9	7	8.5	2	2.7	7	8.1	33
Total launches	2		1		3		3		6		20		17		30		48		44		66		74		70		81		83		74		85		
Cumulative total	2		3		6		9		15		35		52		82		130		174		240		314		384		465		548		622		707		
USA																																			
Redstone	3	60.0																																	4
Jupiter	1	20.0	2	20.0																															5
Atlas	1	20.0					1	6.2	8	27.6	14	26.9	9	23.7	16	28.2	14	22.2	32	43.8	14	23.6	7	15.7	5	12.5	2	6.9	4	12.5	5	16.1	4	17.4	138
Thor			6	60.0	14	87.6	18	62.1	35	67.3	25	65.8	29	50.2	31	49.2	22	30.1	28	47.5	19	42.2	21	52.5	15	51.8	13	40.6	12	38.7	6	26.1	294
Scout							1	6.2	2	6.9	3	5.8	4	10.5	7	12.6	5	7.9	8	11.0	6	10.2	5	11.1	2	5.0	3	10.3	5	15.6	5	16.1	1	4.3	56
Titan															2	3.6	10	15.9	10	13.7	9	15.3	10	22.2	8	20.0	8	27.6	8	25.0	7	22.6	8	34.8	80
Saturn 1 (& 1B)																	3	5.4	3	4.8	1	1.4			2	4.4							3	13.1	12
Saturn V																					1	1.7	2	4.4	4	10.0	1	3.4	2	6.3	2	6.5	1	4.3	13
Total launches	5		10		16		29		52		38		57		63		73		59		45		40		29		32		31		23		
Cumulative total	5		15		31		60		112		150		207		270		343		402		447		487		516		548		579		602		

1 – number of launches by the vehicle; 2 – % of launches in the given year.

Sources:
C. S. Sheldon in *Soviet Space Programs, 1966–70*, US Senate Staff Report, GPO, Washington DC (1971), p. 131.
Flight International, 22 July 1971; 27 January, 20 July 1972; 15 February, 19 July 1973; 27 March 1974.

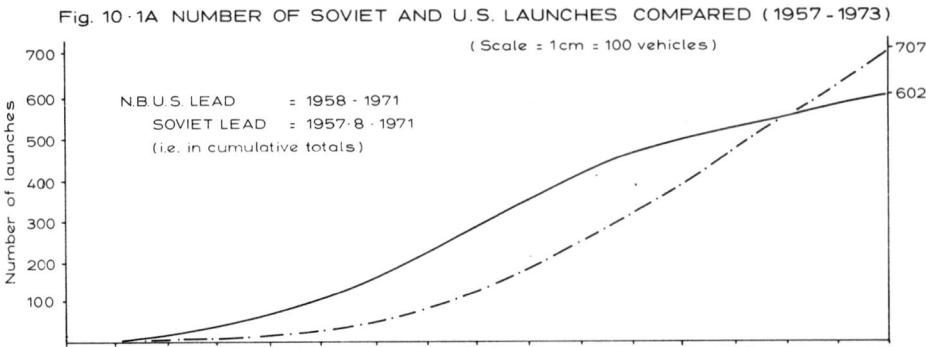

Fig. 10·1A NUMBER OF SOVIET AND U.S. LAUNCHES COMPARED (1957-1973)

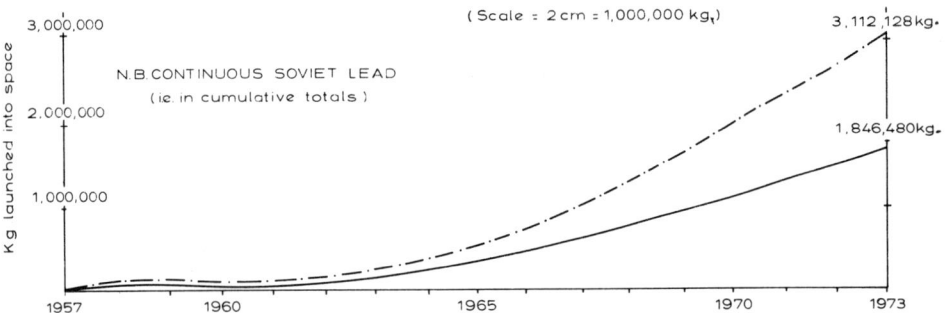

Fig. 10·1B WEIGHT LAUNCHED INTO SPACE BY SOVIET UNION AND U.S. (1957-1973)

Sources: All Royal Aircraft Establishment tables by J.A.Pilkington, P.G.King-Hele, and H.Hiller; up to 1973 Table of Earth Satellites, Vol.1 1957-68; Vol. 2 part 1 1969; Vol. 2+3 1970+1971. Vol. 2 part 4 1972; Space Vehicles Launched During 1973

unnecessary parts dropped from the standard spacecraft type. Stoiko divides all Soviet spacecraft up to 1971 into five categories (genotypes) according to the way they were assembled and the character of the missions: (a) early Sputniks (1–10), (b) lunar genotype (some Lunas, Kosmoses with scientific and navigational tasks, Elektrons); (c) planetary genotype (Luna, Molniya Comsat, Kosmos Metsat, Zond, Venus, Mars), (d) manned genotype (Vostok, Voskhod, Soyuz, Polet, Kosmos reconnaissance), (e) Proton.

Due to the publicity given to manned space flights, more is usually said and written about the genotype (d), manned spaceships. We shall also concentrate on this category because, technologically, it represents one of the most advanced parts of both the Soviet and American space programmes.

DESIGN PHILOSOPHY

The initial Soviet advantage in high thrust of their launch vehicles also had an impact on the design of their spacecraft. The Soviet designers were not as limited as were their American counterparts with respect to maximum weight of spacecraft.⁵⁴ They built the Vostok capsule as well as Voskhod and Soyuz as a heavy solid shell which potentially ensured a higher degree of safety during spaceflight and re-entry, and also enabled the adoption of an earth-like atmosphere both in composition and pressure. This possibility was not open to the Americans who could not afford to build a heavy capsule which would withstand high pressures.⁵⁵ Low-pressure oxygen was used in all the US spacecraft, including the first prototype of Apollo, until the disaster in January 1967. The death of the three US astronauts can be seen as a sad penalty paid for the initial lack of option in US spacecraft design.

Another important aspect of Soviet design philosophy was the role of the cosmonaut on board a spaceship. Again it was the concern of Korolev and others with the safety of cosmonauts during the flight which led to the emphasis on automatic rather than manual control of the spacecraft. Accordingly, the Vostok capsule was designed to be guided automatically through signals from the earth and the cosmonaut was supposed to use the manual controls only in case of an emergency.

Mark Gallai,⁵⁶ a prominent Soviet test pilot who was associated with the training of cosmonauts for the Vostok programme, revealingly describes the 'debate' on automatic vehicle manual control and its outcome. Prior to Gagarin's flight, there was some uncertainty among Soviet scientists as to the effects of weightlessness on man's work capacity. A German scientist, Troebst, caught their attention by his description of the possible effects of 'cosmic awe', which, he argued, could cause irrational behaviour of a cosmonaut during flight which might even lead to self-destruction. Korolev and his colleagues decided not to take any chances. Manual control had to be included in the spaceship system in case of emergency, but switching over from automatic to manual control was deliberately made more complex. One suggestion was that the cosmonaut would first have to dial a three-figure number on a six-figure dial; the number was to be transmitted to him by radio from ground control only in the case of emergency. Korolev, however, vetoed the radio part of the idea apparently because of the risk of a radio transmission failure and, instead, a 'sealed envelope' idea was adopted: the three-figure number, typed on a piece of paper, was to be placed in a sealed envelope and stuck on the wall of Gagarin's cabin just over his head.⁵⁷

The Soviet emphasis on automatic control prevailed even after Gagarin's and other Vostok flights. Professor Ivanchenko argued that a spacecraft controlled from the earth should 'pioneer' space for man, on his behalf, up to the point of absolute safety, and only then should man take over. He saw the main deficiency of the US Apollo programme in the fact that the Apollo spacecraft could not be controlled from the earth without assistance from the

---

⁵⁴ The weight of the Vostok capsule was 3·5 times higher than that of Mercury, and similarly Gemini was considerably lighter than Voskhod (see Table 10.7).

⁵⁵ J. Mansfield, *Man on the Moon*, London (1970), p. 85.

⁵⁶ Mark Gallai, in *Soviet Literature*, 1974, No. 1, pp. 139–48.

⁵⁷ This system, a result of the 'Troebst syndrome', was actually installed in the Vostok spaceship. Only there was foul play: Gallai admits that both he and a Vostok designer named Ivanov whispered the secret number, '125', into Gagarin's ear shortly before take-off.

crew.[58] Thus a Soviet cosmonaut is not so much a pilot of his spacecraft as a well-trained passenger, who keeps an eye on the automatic systems and, broadly, is expected to take over the controls manually only in the case of an emergency. A US astronaut, on the other hand, is deliberately made a more integral part of the spacecraft control and guidance systems.

ENGINEERING

US spacecraft designers drew heavily on the experience with US rocket planes which flew in the 1950s.[59] In the stage of spacecraft design there was also some collaboration on a consultancy level between US designers and future astronauts.[60] Neither element seems to have played any significant part in the design of Soviet spacecraft.[61]

The simplicity of design of their spacecraft has always been stressed in the Soviet Union. This approach was initially seen in the West as one of the main advantages of Soviet spacecraft engineering.[62] The Soviet cosmonaut Feoktistov, who had an opportunity to see the Apollo command module cabin and found it far more complex than that of Soyuz, commented:

> it seems to me that US specialists are not very rational in transferring from aviation the entire pattern of arrangement of instrumentation of the cockpit, without consideration for the specific features of space flight. The control scheme of Soyuz seems to me to be more simple, logical and therefore more efficient.[63]

Vostok, the first manned spacecraft designed for 10-day operation, consisted of two modules (sections): the command module (CM) and an instrument section with a retro-rocket engine. Only the former was a re-entry body. The cosmonaut, wearing a space suit, sat in an ejector seat in the CM. The CM, weighing 2·4 tons, had three hatches: parachute, technological and circular, the latter serving as an exit. There were about 240 electric bulbs, 56 electric motors, over 6,000 transistors and 800 relays and switches on board the spacecraft.[64] After re-entry the cosmonaut was ejected and landed by parachute, i.e., separately from the CM.

Voskhod, a multi-seater spacecraft which flew three men on the first flight and two on the second, was somewhat more sophisticated than Vostok. The

[58] V. Ivanchenko, *Izvestiya*, 28 May 1969.
[59] W. von Braun and F. I. Ordway, *op. cit.*, p. 204.
[60] W. Shelton, *Soviet Space Exploration*, New York (1968), pp. 126–9.
[61] When Korolev showed a group of Soviet cosmonauts the finished Vostok capsule for the first time in 1960, their reported reactions were those of astonishment (see A. Romanov, *op. cit.*, p. 68). On the grounds of Titov's and some other cosmonauts' suggestions some minor alterations are believed to have been made later (see W. Shelton, *op. cit.*, pp. 127–9).
[62] 'The tendency to overengineer a system and the consequent increase in its unreliability have plagued and embarrassed American rocket and missile experts again and again. The Soviets on the other hand, by stressing fundamental principles and simplicity in design, have apparently developed the happy faculty of making proper executive decisions in the manufacture or use of computer-based control systems and determining which of several automatic controls is best.' (F. J. Krieger, *Missiles and Rockets*, 24 April 1961).
[63] *Aviation Week and Space Technology*, 8 December 1969, p. 23.
[64] *The Soviet Encyclopedia of Space Flight*, pp. 494–6; M. Stoiko, *op. cit.*, pp. 174–5.

differences between the two have been given very different interpretations. A Soviet writer, Romanov, reports that Voskhod was roomier and more comfortable and that the 'short-sleeve' atmosphere and the soft landing of the crew within the capsule were important improvements.[65] But the emigré writer Vladimirov argues that Voskhod was basically the same size as Vostok[66] and that the Voskhod crews did not wear space suits because this was the only way to squeeze as many as three men in. The crew were brought down in the capsule because they could not be ejected at an altitude of some 7,000 metres without their space suits.[67] Owing to the inadequate official Soviet information, we are unable to determine which of the two interpretations is closer to the truth. The entry on Voskhod in the *Soviet Encyclopedia of Space Flight* is scantier in technical data than that on Vostok, containing nothing beyond a short list of improvements.[68] The same brief account is invariably given in all those Soviet sources consulted which deal with Voskhod, including the daily press.[69] The first spacewalk in history, undertaken by Leonov from Voskhod 2, which was provided with an air lock for that purpose, was certainly a spectacular achievement. Nevertheless, one cannot escape the tentative conclusion that, unless a substantial piece of information about Voskhod has not been released, the differences in engineering between it and Vostok were negligible compared with those between the Gemini and Mercury spacecraft in the USA. Gemini, unlike both Voskhod and Mercury, was provided with equipment and rocket engines to perform manoeuvring in orbit, including approaching other spacecraft.[70] Gemini also had an onboard computer for complex calculations of rendezvous guidance, an installation which represented a significant step towards the advanced Apollo features. Numerous other improvements have been discussed elsewhere.[71]

Only the Soyuz spacecraft was considerably different from the Vostok and Voskhod capsules in sophistication of design and engineering. The weight of Soyuz was a closely guarded secret until around 1970;[72] it was released in the Soviet Union only after some Western experts had made fairly precise estimates of it. The spacecraft consists of three modules as against two in Vostok/Voskhod: orbital (OM), command (CM) and Service (SM).[73] Like

[65] A. Romanov, *op. cit.*, p. 122. He also mentions a new instrument without giving details, an additional system for the ship's orientation with an ion sensor, and improved radio and TV apparatus.

[66] The difference in weight between it and Vostok is small (see Table 10.7). The exact size of Voskhod has never been officially announced in the Soviet Union.

[67] L. Vladimirov, *op. cit.*, pp. 128–9. Space suits were re-introduced after the death of three Soyuz 2 cosmonauts due to a leak in the capsule shortly before re-entry.

[68] *The Soviet Encyclopedia of Space Flight*, *op. cit.*, pp. 493–4.

[69] In contrast to this poor level of information on spacecraft in the Soviet press in general, one often finds technically irrelevant information, such as the moves of a chess match between ground control and the space craft crew.

[70] This level of engineering was reached by the Soviets only in Soyuz.

[71] See, e.g., W. von Braun, *op. cit.*, pp. 210–11; J. Mansfield, *op. cit.*, pp. 98–9, and various NASA Technical Reports.

[72] *The Soviet Encyclopedia of Space Flight*, published in 1969, does not give it, although the weights of all other Soviet and US spacecraft are shown there.

[73] For some comparisons between Soyuz and earlier Soviet spacecraft, see, for example, V. Seleznev, *Aviatsiya i Kosmonavtika*, 1968, No. 12, pp. 13–14.

Gemini and Apollo, Soyuz has considerable manoeuvring capabilities; these were performed by Soyuz missions 4 and 5, 6–8, and 10–11. Manoeuvring is, however, limited to an altitude of 1,300 km, owing to the capacity of its power sources, so that Soyuz would not be suitable for travel to the moon or for interplanetary flights. The only means of visual orientation for Soyuz crews are manually referenced periscopes or sensors. This again shows that the Soviet designers did not transfer aircraft cockpit apparatus to spacecraft.[74] The crew of Soyuz, unlike that of Apollo, does not have any control over the timing of launch abort operations; even during flight they have a fairly limited command, control and trouble-shooting capability. This is chiefly because crew communication with the spacecraft systems is not very extensive. There is no equivalent in Soyuz of the Apollo digital computer which permits the crew to communicate directly with all systems on board. The drum-type preprogrammed sequencers in Soyuz do not enable the crew to input any other commands than those already present in the drum.[75] All this to some extent reflects the initial Soviet design philosophy of automatic safety for the crew. But it should be borne in mind that no Soviet or any other automatic apparatus is yet sophisticated enough (if it ever will be) to ensure absolute safety. No automatic apparatus can be absolutely reliable,[76] nor can it foresee every possible hazard in the course of flight. Professor Ivanchenko's argument (see text and footnote 58, p. 514) does not consider these points. It could be argued that the safety of a Soviet cosmonaut may not be ensured to the same extent as that of an American who can actually over-rule the computer and is free to do so whenever he deems it necessary.[77]

The launching of Salyut, the Soviet and the world's first orbital station, has been seen as a practical step towards the realisation of Tsiolkovsky's old dream of 'settlements in space'.[78] The programme of manned orbital stations was probably adopted in the Soviet Union in the mid-1960s, when several articles on the subject appeared in the Soviet press after the launching of Polet I. At that time, however, the ultimate purpose of such stations was by no means made clear. One writer[79] called them 'orbital cosmodromes' and saw their sole function as launch bases for interplanetary flights. Others emphasised the use of orbital platforms for scientific research.[80]

Salyut I was launched on 19 April 1971. Its weight of nearly 19 tons required the Proton launch vehicle to place it in orbit.[81] It had its own rocket engines

[74] See *Aviation Week and Space Technology*, 28 January 1974, pp. 36–41.

[75] *Aviation Week and Space Technology*, 21 January 1974, pp. 38–42.

[76] Two examples from Soviet space flight experience illustrate this point: the automatic retro-system failed before re-entry of Voskhod 2, and Belyaev had to use manual control; a failure of an automatic apparatus caused Komarov's death when, after re-entry, one of the Soyuz 1 parachutes did not open.

[77] Thus Aldrin of Apollo 11 kept an eye on the onboard computer with his slide-rule throughout the whole flight and actually stepped in when the computer got overheated and failed to respond to crew commands before landing on the moon.

[78] A. Blagonravov, *Pravda*, 26 April 1971.

[79] G. Pokrovskii, *Grazhdanskaya Aviatsiya*, 1965, No. 3, pp. 12–13.

[80] B. Lyapunov, *Kryl'ya Rodiny*, 1964, No. 8, pp. 23–4; K. P. Feoktistov, *Tekhnika Molodezhi*, 1965, No. 1, p. 3.

[81] The station consisted of the following sections: (a) scientific, equipped with apparatus and systems for execution of scientific and technological experiments; (b) command module (CM),

which, on command from the earth, could boost the station to higher altitudes to prevent premature decay. This happened twice before Salyut was left to decay in the autumn of the same year. In the course of the Salyut spaceflight, Soyuz 10 and 11 docked with it for 5½ hours and 22 days respectively and the crews performed valuable experiments which contributed to several fields of space science and technology.[82] With Salyut I and Soyuz 10 and 11 the Soviets had scored another important space 'first' which was matched by the US with Skylab only two years later. Skylab was seen in the Soviet Union as a 'programme analogous to Salyut both in spacecraft design and mission tasks'.[83] Hardly anyone in the Soviet Union seemed to fully appreciate that Skylab in many respects, including weight and the sophistication of the equipment, was superior to Salyut.[84]

Conclusions

The Soviet space programme from 1957 onwards has been built on the firm basis of pre-war Soviet experience with liquid-propelled rockets, enriched by the familiarity of Soviet experts with German rocketry. Already at the early stages Soviet research and development in rocketry was sponsored by the military authorities and, like von Braun,[85] Korolev and other Soviet rocket experts had to justify their work in terms of the military significance of their end products.[86] This pattern did not change after the war—the launch vehicle which sent the Sputniks and Vostoks into earth orbit or beyond was a Soviet ICBM modified for those purposes and Soviet commentators do not conceal this fact.[87] Korolev received his first Hero of Labour medal from Khrushchev in 1956 'for service in development of defence technology',[88] obviously for his work on the Soviet ICBM. At that time the Soviet public as well as the world at large were ignorant of his existence. Only his death in early 1966 enabled his move out of oblivion. The name of the present holder of the position of Chief Designer is secret once again.[89]

---

used for flight control of the Soyuz-Salyut system; (c) service module (SM), containing systems ensuring normal conditions for the work and rest of the crew. The capsule of the Soyuz spacecraft was also used for work and rest. (See *Ekon. Gaz.*, 1971, No. 24, p. 2. *Pravda*, 8 June 1971).

[82] An interesting and useful account of Salyut 1, Soyuz 10 and 11 missions is given by P. Smolders in his *Soviets in Space*, London (1973).

[83] See interview with Shatalov, the man in charge of training of Soviet cosmonauts, in *Pravda*, 12 April 1973.

[84] See Table 10.7 for some comparisons.

[85] W. von Braun, *op. cit.*, p. 108. He claimed to have been arrested by the Gestapo on the charge that he had undertaken research into space rather than into rockets for military use.

[86] After the launching of the very first Soviet LPR in August 1933, Korolev wrote in GIRD's wall newspaper, *Raketa*: 'the collective of GIRD must mobilise its efforts in order to get out of the rocket its full capacity so that it can be passed over to the Workers' and Peasants' Red Army... it is essential that other types of rockets be created... Soviet rockets must win over the environment' (A. Romanov, *op. cit.*, p. 34).

[87] *Soviet Weekly*, 23 October 1973, p. 447; V. Glushko, *op. cit.*, p. 20.

[88] See A. Romanov, *op. cit.*, p. 102.

[89] In the first edition of *Konstruktor kosmicheskikh korablei*, published in 1969, Korolev's deputy is referred to as '*zamestitel*' (deputy), (p. 139). The second edition (*op. cit.*, 1971, p. 140), goes as far as 'Vasilii Mikhailovich'. We can reasonably speculate that, should poor Vasilii Mikhailovich die before the third edition is sent to print, we shall at last know his whole name. ('Vasilii Mikhailovich' fits none of those who have been tipped in the West as Korolev's deputies: Yangel was Mikhail Kuzmich, Isaev Alexei and Glushko is Valentin Petrovich.)

Table 10.7 *Soviet and US manned spaceflight data (1961–73)*

Spacecraft	Date of first flight	Number of flights	Average weight[a] (kg)	Total weight (kg)	Total man hours hrs:min	Total number of men/women in spaceflight[b]	Total number of orbits	Fatalities[c]
USSR								
Vostok	12. 4.1961	6	4723	28339	382.21	6	243	—
Voskhod	12.10.1964	2	5501	11002	124.55	5	31	—
Soyuz	23. 4.1967	12	6567	78808	4563.16	25	1201	4
Salyut	19. 4.1971	1	18600	18600	[j]	(3)	[j]	—
Total	—	21	—	136749	5070.32	36		4
USA								
Mercury[d]	20. 2.1962	4	1363	5453	3.24	4	34	—
Gemini	23. 3.1965	10	3652	36515	1929.41	20	604	—
Apollo	11.10.1968	11	49145[e]	512027	7506.03	33	330 (Earth) 365 (Moon)	3
Skylab	14. 5.1973	4	74783[f]	117828[g]	12347.57[h]	9	2477[i]	—
Total	—	29		671823	21847.05	66	3810	3

Sources:
C. S. Sheldon, *United States and Soviet Progress in Space: Summary Data Through 1973 and a Forward Look*, Congressional Research Service, Library of Congress (1974), pp. 27–30.
H. Hiller, J. A. Pilkington, *Table of Space Vehicles Launched During the Years 1958–72*, Technical Report 73006, Royal Aircraft Establishment (1973), passim.
J. A. Pilkington, *Space Vehicles Launched during 1973*, Royal Aircraft Establishment (1974), passim.

Notes:
a – except for Apollo and Skylab spacecraft, average weight has been calculated as total weight of all spacecraft flown divided by number of flights.
b – cumulative 'repetitive' total – number of crew on each flight considered, irrespective of how many times the same individual may have flown.
c – figures refer to number of deaths suffered in spacecraft. Break-down is as follows: Soyuz I (Komarov), Soyuz II (Dobrovolsky, Volkov, Patsaev); Apollo 6 (Grissom, White, Chaffee). Apollo 6 crew not included in total number of men in spaceflight as they died in spacecraft on pad during tests.
d – two suborbital flights of the Mercury programme (5.5 and 21.7.1961) disregarded.
e – weight in flight of Apollo 7 (Command/Service Module only, 20,581 kg.) excluded for calculation of average weight, but included in total weight.
f – figure refers to weight of Skylab body (Skylab 1) which was placed into orbit and occupied by successive crews of Skylab 2, 3 and 4.
g – figure includes weight of Skylab body (Skylab 1), and Skylab 2, 3 and 4 spacecraft.
h – total manhours in flight by crews of Skylab 2, 3 and 4.
i – total number of orbits by crews of Skylab, 2, 3 and 4.
j – included in calculations for Soyuz (Soyuz 10 and 11 missions).

The military character of Soviet rocket research and development before and after the Second World War resulted in certain special features which are largely lacking in the US space programme. There is no clear-cut division between the civilian and military parts of the Soviet space effort: all rockets have always been for both science and defence and the Soviet authorities have not sufficiently distinguished between the two themselves.[90] Military matters have always had a high priority in Soviet and Russian history and we believe that this partly explains both the high level of technology and the secrecy of the Soviet space programme. The talents and zeal of such distinguished pioneers as Korolev, Glushko, Isaev and Tikhonravov would have been wasted had not the Soviet military given them what they needed for pursuing their hobby, which eventually developed into one of the biggest assets of Soviet technology and prestige. Those men became chiefs of the Soviet space effort, giving it their supreme expertise, and a sense of long-term and purposeful dedication and continuity, features lacking in the US space programme. Paradoxical as it may seem, we believe that there is a lot more of a shock-type of effort in the US than in the Soviet space programme. The US self-imposed challenge, in the shape of the ambitious Apollo programme and sudden cuts in the NASA budget and manpower once the moon was reached, is perhaps the most typical illustration of this point.[91]

However, in terms of innovating technological advances in rocketry, the United States challenge-driven approach, supported with excellent engineering skills, has proved to work to the American advantage.

The Soviet reliance on well-tested hardware, illustrated above by the example of the Vostok launch vehicle, and the slow development of new, more powerful rocket carriers (since Proton in 1965 there has been no successful test of any vehicle with a higher thrust) cost the Soviets their leading position in rocket thrust. On the other hand, the slow, heavy-going path the Soviet Union has followed in the development of their new launch vehicles has undoubtedly some interesting implications. If a vehicle type is developed which proves to be highly successful and then an intensive use is made of it for many years, as has been the case with the Vostok carrier, vehicle standardisation, its manufacture on a production-line basis and thus economies of scale are the concomitant factors. This type of approach is, however, highly conservative. While the United States was catching up with the USSR in terms of rocket thrust

---

[90] This point is illustrated by the Soviet Kosmos programme, which is claimed to be scientific in its entirety, although a large part of it are evidently military satellites with reconnaissance mission tasks, Fractional Orbit Bombardment System (FOBS), etc. (For a detailed discussion of this aspect, see *Soviet Space Programs, 1966–70*, pp. 323–50; F. J. Krieger, *Soviet Astronautics, 1957–62*, Rand 3595, Santa Monica (1963), p. 13.) Soviet military chiefs are on record as testifying that the Soviet Strategic (Rocket) Forces, which differ from those in the US by being a separate service within the Soviet Army, have been responsible for launching all Soviet payloads ever sent into space (see M. Stoiko, *op. cit.*, p. 191).

[91] For details, see *Aviation Week and Space Technology*, 12 and 19 January 1970. Throughout the discussion in this paper we have deliberately avoided 'space race' and 'moon race' type of issues which, in our view, were conceived in the USA and do not therefore have a direct bearing on the Soviet space programme.

between 1957–66[92] by developing its Atlases, Thors, Titans and Saturns, the Soviet Union was improving its Vostok launch vehicle and modifying a few other military missiles for civilian space use. Only when the thrust of the Vostok was stretched to its limits by continuous upgrading, and when it had become clear that it could not successfully carry out the ambitious Soviet plans in space, was high priority given to the development of a new, more powerful vehicle, Proton. Even Proton, however, did not bridge the gap which by 1967 had divided the USSR from the USA in terms of thrust. A new launch vehicle was therefore conceived and we are still waiting to hear about its successful tests. Once the details of the development are known the period (a) (see p. 506 above) will prove to have been extremely long.

Clearly, Soviet reliance on traditional hardware and efforts to standardise work in the long run to their disadvantage: the absence of a powerful Soviet launch vehicle of the Saturn V type restricts the choice of alternatives in realising the most prominent of present declared Soviet intentions in space, placing large manned stations into earth's orbit.[93] There are basically two ways of orbiting a large space station. Either its entire body is launched on a carrier rocket and placed into orbit, or the station is assembled in space from parts brought from the earth. Soviet writers, obviously aware that the Soviet Union does not possess a powerful enough rocket for the first alternative, argue for the second, on the grounds that 'it is possible to use existing rocket carriers'.[94] This type of thinking about long-term Soviet plans can hardly be said to reflect any serious innovating intentions in rocket technology. Having the same 'existing rocket carriers' in mind, Korolev in 1957, with respect to future Soviet manned flights and other Soviet space plans, could perhaps have said and felt the same. It should, however, be borne in mind that in the Soviet Union, unlike the United States, there has always been reluctance to talk about concrete steps in innovation in rocket technology.[95]

Soviet design philosophy is also quite different from American. Its origins go back to the post-war years when military considerations were of primary

[92] In the early 1960s, the US was well aware of a considerable Soviet lead in thrust power of space vehicles. Thus, McNamara, US Defense Secretary, admitted in August 1962 that: 'We are behind in certain space developments, particularly those associated with large booster capabilities', while D. B. Holmes, then Director of the NASA Office of Manned Space Flight, said the Soviet Union had a 5:1 advantage over the US in payload capability (see US Congress, *Soviet Space Programs, 1962–5*, Washington DC (1966), pp. 87, 95).
[93] In spite of Soviet emphasis on automatic unmanned space vehicles for basic scientific research and discoveries, research into near-the-earth environment through manned flight remains a high priority and placing large stations into earth orbit is clearly No. 1 of the Soviet manned flight intentions (see A. Leonov in *Pravda*, 13 April 1971; B. Petrov in *Trud*, 9 April 1971; G. Pokrovskii, *Grazhdanskaya Aviatsiya*, 1965, No. 3, pp. 12–15).
[94] I. Belyakov, Y. Borisov, A. Tkachev, in *Sots. Ind.*, 15 January 1974.
[95] Shatalov's ability to comment on the Skylab even before its launching could have no parallels in the US with respect to any single Soviet programme in preparation, with the exception of the very recent joint US-Soviet Union space programme. Thus the development of the new powerful rocket mentioned above (p. 506) is never discussed in published Soviet material and all information about it released in the West is based on intelligence reports and/or on experts' speculations (see *Aviation Week and Space Technology*, 17 November 1969 and 14 January 1974). It is similar with the space shuttle: US shuttle engineering has been given favourable treatment in the Soviet press, but there have been very few hints published about Soviet efforts in that direction (see, e.g., *Sots. Ind.*, 18 November 1973; B. Petrov, *Soviet News*, 17 April 1973).

Table 10.8 *Major space 'firsts' (1957–73)*

Event	USSR Mission	USSR (Launch) date	Event	USA Mission	USA (Launch) date
First AES	Sputnik 1	4 Oct. 1957	Van Allen radiation belts	Explorer 1	1 Feb. 1958
Orbiting Geophysical lab	Sputnik 3	15 May 1958	Earth shape measured	Vanguard 1	17 Mar. 1958
Farside Lunar picture	Luna 3	4 Oct. 1959	Orbiting Solar observatory	OSO 1	7 Mar. 1962
Venus probe	Venera 1	12 Feb. 1961	Data from Venus	Mariner 2	27 Aug. 1962
Man in space	Vostok 1	12 Apr. 1961	Geodetic satellite	Anna 1B	31 Oct. 1962
Group space flight	Vostok 3, 4	11,12 Mar. 1962	Lunar close-up pictures	Ranger 7	28 Jul. 1964
Mars Probe	Mars 1	1 Nov. 1962	Mars pictures	Mariner 4	28 Nov. 1964
Woman in space	Vostok 6	16 Jun. 1963	Micrometeorite satellite	Pegasus 1	16 Feb. 1965
3-seat spaceship launch	Voskhod 1	12 Nov. 1964	Lunar orbit pictures	Orbiter 1	10 Aug. 1966
Spacewalk (EVA)	Voskhod 2	18 Mar. 1965	Lunar trenching	Surveyor 3	17 Apr. 1967
Cosmic ray measurements	Proton 1	16 Jul. 1965	Coloured picture of full earthface	Dodge	1 Jul. 1967
Lunar surface picture; soft landing on Moon	Lunar 9	31 Jan. 1966	Point stabilized orbiting astro-observatory	OAO 2	7 Dec. 1968
Venus landing	Venera 3	1 Mar. 1966	Manned Lunar orbit, Live Lunar TV broadcast	Apollo 8	21 Dec. 1968
Venus atmospheric probe	Venera 4	12 Jun. 1967	Man on the moon, return of Lunar rocks	Apollo 11	16 Jul. 1969
Automated return of Lunar sample	Luna 16	12 Sep. 1970	Long-life Lunar surface sensors	Apollo 12	14 Nov. 1969
Automated Lunar roving lab	Luna 17	10 Nov. 1970	Coloured pictures of Jupiter	Pioneer 10	3 Mar. 1972
Orbital station	Salyut 1	19 Apr. 1971	Radiotelescope, farside Moon	Explorer 49	10 Jun. 1973
Venus soil analysis	Venera 8	27 Mar. 1972			

Sources:
Pravda, 4 October 1967.
C. S. Sheldon, *United States and Soviet Progress in Space: Summary Data Through 1973 and a Forward Look*, Library of Congress (1974), p. 13.
G. A. Tokaty, *Quest*, No. 23, 1973, pp. 34–7.

importance, and it also stems from Soviet concern with maximum safety for manned flight programmes and maximum economy in other programmes,[96] aspects which led to Soviet emphasis on control from the earth. As far as maximum safety is concerned, the weaknesses of that line of thought have been discussed above. The simplicity of Soviet space apparatus was thought to be a good thing by F. Krieger as early as 1961. Since then much has changed, of course, yet Soviet design simplicity is still being referred to as an advantage, although increasingly less so in the West. Simplicity is one of the conditions of standardisation in Soviet space technology, which is seen as very important, especially in connection with their plans to build large orbiting stations.[97]

In the course of our discussion of Soviet manned capsules we have shown that the emphasis on simplicity has led to a lower degree of computerisation on Soviet spaceships as compared with American. The Soviet cosmonaut is more a passenger than a pilot of his ship. In terms of new technology, the systems of Soyuz are not as remote from Voskhod and Vostok as the Apollo systems are from Mercury and Gemini. Owing to the lack of a suitable carrier rocket the Soviet Union has not yet tested a capsule for manned flight to the moon, although from numerous Soviet sources it is obvious that such a manned trip is indeed the Soviet intention in the long run.

In the course of the 1960s, at least in the fields of rocket technology and manned spacecraft, the bold and pioneering Soviet space programme gradually ceased to maintain its high initial tempo *vis-à-vis* the dynamic US Gemini and Apollo programmes. The main reason for the loss of this tempo, and thus of the lead in these fields, is the peculiar Soviet pattern of innovation and design philosophy, both of which are very different from the American counterparts. The solid basis is still there, but the lead is lost—however temporary that may be. We believe that the Soviets would need an equivalent of the US Saturn V or space shuttle in order to be able to execute their stated space plans as effectively as the Americans have been capable of executing theirs. From the 'Old Guard' of Soviet rocket pioneers only Glushko survives. It remains to be seen what the 'New Guard' is like.

---

[96] See B. Petrov, *New Scientist*, 6 May 1971, pp. 308–10.
[97] See, for example, I. Belyakov *et al.*, *op. cit.*, p. 3.

11. Technological level and quality: machine tools and passenger cars*

Introduction

This chapter is devoted to assessments of the technological level of Soviet engineering products which focus on quality and reliability, with particular reference to general-purpose machine tools, in view of their importance in the Soviet industrial economy, and to passenger cars, because of the recent large quantity of resources allocated to this industry by Soviet industrial policy-makers.

A review of Western and Soviet published comments on the comparative technological level and quality of machine tools is presented first. These comments by specialists provide a general indication of the strengths and weaknesses of Soviet machine tool technology. In the second section a different approach is adopted: Soviet general-purpose machine tools are examined in detail from the point of view of their standards of accuracy and also their reliability in use. The third and final section is devoted to a brief examination of the quality and reliability of Soviet passenger cars with particular reference to the Moskvich range built by the Lenin Komsomol factory in Moscow.

Western and Soviet assessments of the technological level of Soviet machine tools with special reference to quality

In chapter 4, we were largely concerned with the Soviet stock and output of machine tools in comparison with American stock and output, using mainly quantitative criteria. Here we collect together some of the views expressed on the technological level of Soviet machine tools, with special reference to quality, and assess the evidence available. Clearly, more detailed work of the kind carried out by M. R. Hill in the section of this chapter on the quality and reliability of Soviet general-purpose machine tools is needed before firm general conclusions can be reached. In both Soviet and Western writings there

* Pp. 523–30 were written by M. J. Berry, and pp. 530–63 by M. R. Hill.

are comments on the quality and technological level of Soviet machine tools; although some of these writings tend to have built-in biases, they give some indication of comparative levels.

Improvement of quality has been treated in the Soviet press as one of the most important problems confronting the industry ever since 1955.[1] Increasingly since then stress has been placed on improved quality and higher levels of accuracy, as well as on increased productivity through automation and other methods.

WESTERN COMMENTS

Western comments on Soviet machine tools can be roughly divided into three main groups: first, comments published in the Soviet journal *Stankoimport Review*; second, comments made by journalists and others visiting the USSR; and, last, other assessments made by Western writers.

Table 11.1 *Some foreign comments on Soviet machine tools from recent issues of* Stankoimport Review

Country of speaker	Issue No.	Comments
Finland	36	In general Soviet machine tools competitive (p. 32); 1A616—convenient in use, retains original accuracy for long time (p. 32)
	40	Technical level corresponds to world level. We are satisfied with productivity, accuracy, durability of Soviet machine tools (p. 16)
France	40	KU-107—distinguished by reliability in use and modern design (p. 32)
FRG	39	2N55.6T75—specialists spoke highly of high production capacity, reliability and ease of operation (p. 30)
Japan	38	Soviet vertical turning and boring mills—bought after comprehensive comparative investigations. Can successfully compete with models produced by leading machine tool manufacturers (pp. 8–10)
		Great many advantages—excellent steel and other materials, heat treating techniques on a very high level. Excellent combination of handsome appearance, ample rigidity and ideal design. Accuracy is splendid—on a level with the very best machine tools made in Japan (pp. 8–10)
		They are above comparison with similar heavy duty models of other manufacturers (pp. 4–7)
Italy	39	Milling machines—very impressive. Special purpose and heavy duty machines—have no equals in Italy (pp. 18–21)
GDR	39	Precision automatics from USSR meet all requirements of best world practice (p. 15)
Norway	38	Good opinion of operational features (p. 23)
Bulgaria	40	5K32—very accurate; 1713 unusually high accuracy; 1P16, 1B10P, 1A12P—very accurate and highly productive, easy to control, quick to reset. No repair needed (pp. 9–13)

Note: this journal comes out quarterly; issue No. 40 was published Spring 1975.

---

[1] See, e.g., editorial in *Pravda*, 29 June 1955.

Recent issues of the journal *Stankoimport Review* included a number of comments made by foreign specialists and factory personnel—buyers and users of Soviet machine tools—and these have been collected in Table 11.1. No unfavourable comments have been traced, which is not unexpected since this journal serves as a publicity and advertising medium for the Soviet machine tool industry. In general there is agreement on the competitiveness of Soviet machine tools; they are considered to be reliable, accurate and easy to operate.

In the years since 1955 a number of visits have been made to the Soviet Union by Western technical journalists, machine tool producers, etc. As far as can be judged from a small sample of English language publications, their reports tend to be mainly descriptive, with few direct comparisons. Such comparisons as are included are both favourable and unfavourable, with some emphasis on the latter. Some of these Western views are given in Table 11.2.

Table 11.2 *Some Western views on Soviet machine tools*

Soviet Union backward or criticised	Soviet Union equal, ahead or praised	Source
Jig boring machine		*Metalworking Production*, 8 November 1956, p. 1814
Single spindle autos	Multispindle autos	*Ibid.*, p. 1816
Ultrasonic machines	Electro-erosion machines	*Ibid.*, 16 November 1956, p. 1859
	Horizontal boring machines	*Machinery and Production Engineering*, 10 February 1971, p. 204
'Large machines'		*Ibid.*, 10 March 1971, p. 364
'Operator aids'		*Ibid.*, p. 366

It is interesting to compare the general impressions of two visitors separated by an interval of some 13 years. In 1958 a writer in *Machinery*[2] concluded that new Soviet machines exhibited at fairs such as Leipzig were 'of modern design and of good finish and so far as can be ascertained from a visual inspection are of a high standard of quality'. In 1971 a writer in the same journal[3] concluded:

> in some instances the more basic standard types of machines designed to suit Russian industry were not always appropriate to the UK user and do not always incorporate the operator aids (for example for setting) which are increasingly being demanded in this country.

Although by and large the impression of Western correspondents visiting the USSR can be summed up as mildly critical, there are several more severe

---

[2] *Machinery*, 2 July 1958, p. 5.
[3] *Ibid.*, 10 March 1971, p. 366.

assessments. A chapter on machine tools by Sutton in his work on Soviet technology[4] quotes an article by Joseph Gwyer published in 1958:

> the bulk of current models turned out by the Soviet industry approach in make-up, speeds, rate of feed the US models made during the late 1930s and during World War II. Since then the United States has made considerable advances in machine tool technology.[5]

Sutton concludes that:

> by importing prototypes of advanced machines from the West the Soviets can, with little effort, keep abreast of world developments in this field. Thus, although the Soviets may lag by a few years at any one time, the effect over the long run is to keep Soviet machine tools more or less on an equivalent basis to current world technology.[6]

Sutton ascribes Soviet achievements in this field to lend-lease, Soviet acquisitions in Germany, and the import and duplication of advanced Western machines.

Another major criticism of the level of Soviet machine tool technology appeared in *American Machinist* in 1962, when the editor, Burnham Finney, attacked an article by Vladzievskii, the then director of ENIMS, who had claimed in an article that 'Russia is superior in certain types of machines' and that 'Western European firms willingly buy our machine tools and that American users would purchase Soviet machines except that our State Department restrains them'.[7] Finney claimed that the average US machine tool was 'far more sophisticated than the Russian' and was 'bigger, heavier, more powerful, more automatic and able to produce more'. In a subsequent article[8] Finney dealt with a number of specific points in Vladzievskii's article, rejecting his claims for Russian priority in developing ultrasonic machines, and that Russian balancing machines 'have no equals in the world', and he pointed out that Soviet gear cutting machines are copies of Gleason machines. He rejected the claim that for programme control and copying arrangements Russia is 'on the same level with the best machine building firms of the West' as untrue where numerical control is concerned. He asked what the Soviet Union was doing in a wide range of advanced machinery techniques and why, if their own industry was superior, the Russians were keenly interested in buying sophisticated American tools. *American Machinist* subsequently published a translation of a reply by Dr Vladzievskii,[9] but Finney's introduction showed him still unable to accept Vladzievskii's claims.

[4] A. C. Sutton, *Western Technology and Soviet Economic Development, 1945 to 1965*, Stanford, Cal. (1973), pp. 303–17.
[5] A. C. Sutton, *op. cit.*, p. 303.
[6] *Ibid.*, p. 304.
[7] *American Machinist*, 2 April 1962, p. 67. Original article not traced.
[8] *Ibid.*, 16 April 1962, p. 101.
[9] *American Machinist*, 7 January 1963, p. 53 and pp. 55–7.

The criticisms by both Sutton and Finney, however true they may be, were insufficiently documented and substantiated.

There is some indication that Western attitudes to the Soviet industry have changed over time. The period up to 1955 was largely one of ignorance. From 1956 to the early 1960s was the period of discovery and concern; this was the time of the post-Sputnik reaction to Soviet technology, of the Melman report[10] and frightening stories that the Soviet Union was about to swamp world machine tool markets.[11] Melman was mainly concerned with the quantity of Soviet machine tools produced, but also claimed that 'limited information suggests that the quality has been found acceptable by West European machine tool users'.[12] As time went on, the Soviet Union showed itself able to easily absorb the large number of machine tools it was producing, and the 'threat' was largely discounted. At the same time the value of Soviet machine tool imports began to climb steeply and remained far ahead of exports. As has been shown above (chapter 4, pp. 146–60), Soviet trade has continued to be largely with Eastern Europe and exports to the advanced capitalist countries remain low.

SOVIET COMMENTS

Soviet assessments of their own machine tools range from superficial generalisations which tend to concentrate on Soviet successes to detailed comparisons of technical specifications of Western machines and detailed criticism of their own machine tools.

In the late 1950s, in contrast to Gwyer's remarks quoted above, most Soviet machine tools were claimed to be up to date.[13] This claim was supported by two specific surveys of general-purpose machine tools. In one survey it was found that of 338 general-purpose machine tools produced in 1958, 303 models in their basic specifications—power, speed, level of automation and weight—were on a par with foreign machines.[14] Shortly afterwards an ENIMS survey was said to show that of the 270 most common machine tools 20 models were superior to foreign models, 40 were inferior and the remaining 210 were about the same.[15] More recently it was claimed that 86 per cent of machine tools were of modern standards.[16] It seems clear that such comparisons depend on specifications rather than performance and tell us little about the quality of the machines produced. Some interesting data published in another source give a rather different impression. Three different aspects of the machine tool are compared—design, manufacture and accessories (*prinadlezhnosti*). It was found, according to this study, that in design 76 per cent of Soviet machine

---

[10] S. Melman, *Report of the Productivity of Operations in the Machine Tool Industry in Western Europe*, OEEC, Paris (1959).

[11] See, for example, *Iron Age*, 10 March 1960, pp. 121–3.

[12] S. Melman, *op. cit.*, p. 16.

[13] See, for example, N. S. Acherkyan *et al.*, *Razvitie stankostroitel'noi i instrumental'noi promyshlennosti v SSSR* (1958), p. 18.

[14] P. M. Pen'kov, *Razvitie tipazha i struktura vypuska metallorezhushchikh stankov i komplektuyushchikh prinadlezhnostei k nim v 1958 i 1959–65 gody* (1958), p. 16.

[15] *Vestnik Statistiki*, 1960, No. 6, p. 26.

[16] *Vestnik Mashinostroeniya*, 1968, No. 1, p. 87.

tools were at world level, for manufacture 72 per cent, while for accessories only 8·3 per cent.[17] Figures for different types of machine tools are given in Table 11.3.

Other Soviet sources tend to support the view that although design may be satisfactory, the manufacture of machine tools leaves much to be desired and is probably worse than indicated in Table 11.3, and that for the accessories the situation is also far from satisfactory. Two authoritative sources are particularly critical of Soviet performance in this field.

Table 11.3 *Percentage of output on level of best foreign machine tools*

	Design	Manufacture	Accessories
Lathes	78·6	86·6	1·3
Drilling and boring machines	77·7	53·3	11·1
Grinding machines	70·7	48·7	24·4
Gear forming machines	80·7	94·7	7·0
Milling, planing, broaching machines	76·3	79·5	1·1

Source: (Ed.) A. M. Vilenskii, *op. cit.*, p. 90.

At the All-Union seminar for machine tool designers and technologists in 1965 it was claimed that although design of machine tools equalled the best foreign standards the quality of production in a number of cases was not so good.[18] As a result, most Soviet machine tools have too short a service life to first major overhaul—often half that of foreign machine tools. Although the norms for accuracy were much the same as in foreign standards, Soviet machine tools used the full range of tolerances in manufacture, which affected the length of time over which the original accuracy was maintained. There was insufficient strengthening of surfaces likely to wear, and not enough protection from dirt and damage. Many components after a short period of work changed their geometrical form because of residual internal stresses. The external finish and general appearance (*arkhitektonika*) of machine tools were unsatisfactory. Investigations had shown that the insufficiently high quality of machine tools could be explained: by design defects because of unsystematic work at machine tool factories in eliminating the defects which emerged in using the machine tools; by low quality of components (electrical and hydraulic equipment, spindle bearings) which reduces the reliability, durability and accuracy of machine tools; and by defects in the process of manufacture of machine tools and the organisation of their production. Strengthening technology was not used nearly enough and, when used, was sometimes used incorrectly.

[17] (Ed.) M. A. Vilenskii, *Ekonomicheskie problemy povysheniya kachestva promyshlennoi produktsii* (1969), p. 89.
[18] See (Ed.) I. M. Shakhrai, *Materialy vsesoyuznogo seminara konstruktorov i tekhnologov stankostroeniya* (1965), p. 3, from which the following is taken.

Hardening of slideways was very rarely used. Processes for manufacturing gears were also criticised. Little attention was paid to the protection of open rubbing surfaces from getting dirty with swarf and abrasives.

The second major source criticising the quality of machine tools was based on a 'technological level' study carried out by the Gosplan USSR Economics Research Institute on a wide range of products.[19] Machine tools were one of the areas considered to be backward. The quality of manufacture, materials and items used was found to be poor. Almost all universal machine tools were said to be not up to foreign standards—they were backward in smoothness of movement, noise, finish and service life to first major overhaul. Their ability to retain initial accuracy was only one third to one fifth that of foreign machine tools.

Several of these fairly severe criticisms are supported in other sources. Two cases may be singled out: first, electrical and hydraulic components and, second, the accuracy of machine tools. In the late 1950s the quality of the electrical and hydraulic components was extremely low,[20] and they came in for particularly severe criticism at the 1958 machine tool industry conference.[21] One of the participants was led to comment that he was surprised that 'there haven't been more complaints about electrical equipment'. Some of the equipment in production 'belongs to the 1930s and ruins our machine tools'.[22]

Accuracy is also a constant theme;[23] there seems to be general agreement that there is considerable room for improvement.

These two problems would appear to have differing causes. In the case of components the factory has either to make its own, often in unsatisfactory conditions, or to obtain them from other factories, some of which supply a wide range of industries and have little interest in satisfying the particular needs of the machine tool industry.

In the case of the accuracy of machine tools, it would seem to be a problem of improving the quality of the manufacturing process both by capital investment in, for example, temperature-controlled shops, and also by using other manufacturing techniques. These latter tend to be labour-intensive and because of existing success criteria are often unattractive to Soviet management. The impact of improved quality on labour productivity is indicated by the fact that machine tools for export are made to a higher standard than machines for the home market. The Minister for the Machine Tool and Tooling Industry, Kostousov, has estimated that if 'Krasnyi Proletarii', the largest machine tool

---

[19] See contribution by K. I. Klimenko in *Materialy vsesoyuznoi konferentsii po ekonomicheskim problemam nauchno-tekhnicheskogo progressa, Vypusk pervyi* (1970), p. 118.

[20] See, for example, A. E. Prokopovich, *Sovremennye konstruktsii metallorezhushchikh stankov* (1957), p. 25; A. E. Prokopovich, *Sostoyanie i perspektivy razvitiya stankostroeniya* (1956), p. 23 and p. 26.

[21] See V. T. Zusman, *Trebovaniya k elektrotekhnicheskoi promyshlennosti po obespecheniyu razvitiya stankostroeniya na period 1959–65 gody* (1958), p. 11, and *Otraslevoe soveshchanie po stankostroeniyu, Minsk, iyul' 1958: Vystupleniya uchastnikov soveshchaniya (sokrashchennaya stenogramma)* (1958), pp. 9, 11 and 136.

[22] *Ibid.*, p. 136.

[23] A. E. Prokopovich, *op. cit.* (1957), p. 7; *Otraslevoe soveshchanie ...*, p. 125; A. E. Prokopovich, *Stankostroenie v 1959–65 gody* (1959), p. 38; *Vestnik Statistiki*, 1963, No. 12, p. 69; L. S. Blyakhman, *Shagi reformy*, Leningrad (1969), p. 128.

producer in the Soviet Union, with an output of some 15,000 centre lathes and other machine tools, were to produce all its lathes to export standard, their production would be almost halved.[24]

Soviet sources also contain assessments of particular types of machine tools, and some of these have been collected together in Table 11.4. The types of machine tools listed tend to bear out much of what has been written above, namely that the Soviet Union lags behind in certain areas of precision machine tools and in particular in grinding and gear generating machines.

CONCLUSIONS

This section has examined the views of Western and Soviet writers on the quality and technological level of Soviet machine tools. The views of Western writers are rather mixed. Some at least consider Soviet machine tools to be clearly behind the West, although their evidence does not always appear to be adequate. A similar wide range of views is found among Soviet writers, but Soviet specialists clearly recognise that Soviet machines are behind their Western competitors in many respects, although in design they may be their equals. The frequent absence of influence from the end user on the manufacturers often means that the finished product lacks some of the refinements essential on Western machines. The manufacture of advanced machine tools of high quality has proved particularly difficult.

The quality and reliability of Soviet general-purpose machine tools

The quality, reliability and service life of engineering products depend to a very great extent upon the dimensional and shape tolerances of the component parts from which they are assembled. The economic production of these engineering components to the requisite dimensional parameters depends, in its turn, on the performance characteristics of available metalcutting machinery. General-purpose machine tools are widely used in almost every industrial sector in the USSR, as in the Western industrialised economies, to produce engineering components in unit, small batch and medium batch production conditions.

This section, consequently, surveys the technical characteristics of selected Soviet general-purpose machine tools from the following standpoints: dimensional capacity, accuracy, reliability, and other performance characteristics. Information for this survey has been obtained from relevant Soviet state standards, and other published Soviet material. This has been supplemented by impressions from a user's viewpoint of the quality, reliability and service life of selected Soviet general-purpose machine tools.

[24] *Opyt organizatsii raboty predpriyatii Ministerstva Stankostroitel'noi i Instrumental'noi Promyshlennosti v novykh usloviyakh planirovaniya i ekonomicheskogo stimulirovaniya i izuchenie perspektivnykh voprosov razvitiya otrasli* (1968), p. 15.

Table 11.4 *Soviet views of technological level of Soviet machine tools*

Year	Soviet Union backward Type of machine tool, etc.	Soviet Union ahead or equal Type of machine tool, etc.
1948	Gear shaving, diamond boring, honing machines, gear grinders with abrasive worm, centreless grinders with large working wheels[1]	
1958	Jig boring, grinding, gear grinding machines.[2] Automated lines with PC.[3] Machine tools for car industry.[4] Multispindle automatic honing machines with automatic cycle; automatic loading and removal,[5] gear forming machines,[6] gear hobbing; internal grinders, external broachers[7]	Automation of control[10]
1959		Gear cutting machine, Gleason type[11]
1963	Precision machine tools, grinders, automatic lines using batch produced machines[8]	
1966		1K62 centre lathe[12]
1968		Electrochemical and electrophysical machining,[13] 1K62, 1A616 centre lathes[14]
1970	Ultrasonic and spark erosion machines; programme control with computer[9]	Heavy machine tools[15]
1971		Measuring machines[16]

Sources:
1. Quoted in *The Engineer's Digest*, August 1952, p. 254.
2. P. M. Pen'kov, *Razvitie tipazha i struktura vypuska metallorezhushchikh stankov i komplektuyushchikh prinadlezhnostei k nim v 1958 i 1959–65 gody* (1958), p. 21.
3. O. V. Spasskaya, *Osnovnye zadachi tekhnologii stankostroeniya i smezhnoi promyshlennosti na 1959–65 gody* (1958), pp. 42–3.
4. *Otraslevoe soveshchanie po stankostroeniyu, Minsk, iyul' 1958: Vystupleniya uchastnikov soveshchaniya (sokrashchennaya stenogramma)* (1958), p. 47.
5. *Ibid.*, p. 113.
6. *Ibid.*, p. 120.
7. M. E. Mardanyan, *Osnovnye napravleniya razvitiya konstruktsii metallorezhushchikh stankov* (1958), p. 16.
8. *Planovoe Khozyaistvo*, 1963, No. 12, p. 69.
9. *Materialy vsesoyuznoi nauchnoi konferentsii po ekonomicheskim problemam nauchno-tekhnicheskogo progressa, Vypusk pervyi* (1970), p. 114.
10. P. M. Pen'kov, *op. cit.*, pp. 25–6.
11. Yu. P. Konyushaya, *Tekhnicheskii progress i sozdanie material'no-proizvodstvennoi bazy kommunizma* (1959), p. 79.
12. D. S. L'vov, *Osnovy ekonomicheskogo proektirovaniya mashin* (1966), p. 18.
13. *Opyt organizatsii predpriyatii Ministerstva Stankostroitel'noi i Instrumental'noi Promyshlennosti v novykh usloviyakh planirovaniya i izuchenie perspektivnykh voprosov razvitiya otrasli* (1968), p. 14.
14. Yu. D. Matevosov, *Ekonomicheskie problemy razvitiya stankostroitel'noi promyshlennosti v Zakavkaz'e*, Erevan (1968), p. 154.
15. *Vestnik Mashinostroeniya*, 1970, No. 3, p. 12; *Stanki i Instrument*, 1970, No. 4, p. 25.
16. I. I. Knyazitskii and G. V. Ust-Shomushskii, *Tochnye, nadezhnye, ekonomichnye*, Odessa (1971), pp. 21, 39.

THE ROLE OF SOVIET STATE STANDARDS FOR MACHINE TOOLS
THE LEGAL STATUS OF SOVIET STATE STANDARDS

Engineering standard specifications are documents, approved by a recognised authority at the relevant level, in which are specified rationalised dimensional parameters and defined quality characteristics of industrial articles. Standardisation is used in all industrial countries as a means of promoting production specialisation through variety reduction and ensuring that the quality of manufactured articles is to an acceptable level.[25] 'State standards' (*gosudarstvennye standarty* (GOST)), used in the USSR, differ from national standards in most Western countries in terms of their legal status. They are used as a means of ensuring that the state is provided with supplies of products of adequate quality, and they automatically form a legal framework for Soviet purchasing orders and contracts,[26] as well as carrying penalties for non-observance.[27]

SOVIET ORGANISATION AND PLANNING OF STATE STANDARDS

State standards are approved by a specialised body established for this purpose—the State Committee of Standards of the Council of Ministers of the USSR (Gosstandart SSSR). The State Committee is divided into administrations or departments, which carry out overall coordination of the drafting of standards for particular sectors of industry. The State Committee has also established a number of scientific-research institutes to assist its relevant administrations and departments by executing research work into the general problems of standardisation, and checking the technical content of draft standards prior to approval. The practical work of drafting standards is delegated by the State Committee to research and development organisations already established within the appropriate industrial ministries and responsible to these for product improvement. The standardisation activities of these organisations, which are likely to form a comparatively small part of their overall research and development effort, are coordinated through a chapter of the Annual Plan for National Economic Development, covering standardisation tasks, the fulfilment of which is a compulsory requirement for delegated organisations. Once a state standard has been drafted, approved and published, its introduction into industrial practice is checked by inspection organisations of the State Committee of Standards, which periodically check

---

[25] See M. R. Hill, *The Quality Engineer*, Vol. 37, No. 2, pp. 39–43, for an account of standardisation procedures in the centrally planned economy.

[26] Each Soviet state standard (GOST) specification contains the statement 'non-observance of these requirements is against the law'. The legal status of state standards is also stressed by the present Chairman of the State Committee of Standards of the USSR, V. V. Boitsov in (Ed.) V. P. Mysnichenko, *Povyshenie kachestva izdelii v mashinostroenii* (1968), pp. 14 and 15.

[27] In the RSFSR, for example, the continued manufacture of articles having a lower quality than that specified in the relevant state standard can lead to the dismissal of the enterprise's director, chief engineer and quality controller, or their sentence to one year's directed labour, or up to three years' loss of freedom (see B. G. Andreev, *Ekonomicheskoe znachenie povysheniya kachestva produktsii*, Leningrad (1968), p. 180).

the degree of adherence to standards by production organisations within their own locality.[28]

A recently established innovation in Soviet state standardisation practice is the State Quality Certification system. Samples of individual models of products are submitted by the industrial ministries responsible for their manufacture to the Quality Certification Department of the State Committee of Standards. These samples are then checked for conformity with technical requirements of similar items produced internationally. Furthermore, the factories responsible for their manufacture, together with their major suppliers, are inspected by a commission appointed by the Quality Certification Department, for adherence to high quality manufacturing methods and inspection procedures. Should the submission be successful, the product is awarded a 'Mark of Quality' (*Znak kachestva*) for a fixed time period, and the relevant enterprise receives extra bonuses for producing certified items. This procedure was established in 1965, and lists of those articles having 'Mark of Quality' approval began to appear in the Soviet technical press from 1967 onwards. It can be seen as a tendency towards attention to the design and manufacturing details of individual product models, as a means of product quality improvement, rather than broad technical requirements for a range of different models all belonging to the same product type range,[29] as defined by conventional state standards.

SOVIET STATE STANDARDS FOR MACHINE TOOLS

A study was carried out by the author in 1969[30] to determine the range of products of the Soviet machine tool industry for which state standards had been published. This survey revealed that, according to one source, GOST standards had been published for approximately 95 per cent of total Soviet machine tool output.[31] It was consequently considered important to evaluate the technical requirements specified in these Soviet documents, since they would have an important influence on the quality of machine tools supplied to the Soviet engineering industry.

It was initially assumed that these requirements would be similar to those adhered to in contemporary international industrial practice, in view of the drafting of these documents being delegated to the Experimental Scientific-Research Institute of Metalcutting Machine Tools (ENIMS) in Moscow, an internationally acclaimed institute in the field of machine tool research and product development.[32]

[28] See V. V. Tkachenko, *Metodika i praktika standartizatsii* (1967), p. 190. A description of procedures relevant to the drafting, publication and observation of machine tool standards is given in M. R. Hill, *The Quality Engineer*, Vol. 37, No. 2, pp. 39–43.

[29] For a description of the introduction of the 'Mark of Quality' system see V. V. Tkachenko, *op. cit.*, p. 257; *Standarty i Kachestvo*, 1968, No. 6, pp. 44 and 45; *Ibid.*, 1971, No. 6.

[30] M. R. Hill, *Standardisation Policy and Practice in the Soviet Machine Tool Industry*, unpublished Ph.D. Thesis, University of Birmingham (1970), pp. 82–92. An updated survey of Soviet state standards for machine tools was published by the present author in *Production Technology: Abstracts and Reports from Eastern Europe*, 1973, No. 45, pp. 3–9.

[31] (Ed.) V. V. Boitsov, *Standartizatsiya v narodnom khozyaistve* (1967), p. 173.

[32] R. Amann, M. J. Berry, R. W. Davies, in *Science Policy in the USSR*, OECD, Paris (1969), p. 422.

Time did not permit a complete study of standards relating to every product type, and hence attention was concentrated on widely used types of centre lathes and horizontal knee and column milling machines, i.e., basic machinery required for the processing of the bulk of engineering components of almost every shape.

DIMENSIONAL CAPACITY

Soviet state standards for machine tool basic dimensions specify those parameters which determine the size range of workpieces that the machine is capable of processing, and the envelope of surface dimensions which it is capable of generating. State standards specifying basic dimensions have been published for every type of machine tool produced in the USSR, relevant parameters being selected from a series of standardised preferred numbers extending over the required dimensional range.

The major dimensional parameters of individual machine models, consequently, correspond to those specified in the appropriate state standard relating to dimensional requirements for that product type. The 1K62 centre lathe produced at the Krasnyi Proletarii factory in Moscow, for example, conformed to GOST 440-57 in terms of its maximum workpiece diameter of 400 mm. It was produced in various bed lengths to conform to the standardised requirement of a minimum distance between centres of 710 mm, and a maximum distance between centres of 1,400 mm, for 400 mm 'swing' centre lathes.[33]

Such standardisation has provided a technical basis for product variety control, which, when combined with the Soviet policy of centrally administered production, resource allocation and the establishment of large machine tool factories, enabled large batch production technology to be used for the manufacture of those models of general-purpose machines having a high demand.[34] The economic benefits arising from such a standardisation policy could easily have been lost through increased utilisation costs, however, if the appropriate standardised parameters had limited the dimensional capacity of the machine tools in relation to consumer requirements.

A study[35] of a sample population of workpieces processed on centre lathes in Soviet factories in a broad range of engineering industries suggested that this was far from the case, however. Data was collected by ENIMS on a sample of 2,200 typical components processed on centre lathes having maximum workpiece diameters varying between 320 and 630 mm, located in 29 different engineering factories. The data was then classified in terms of proportion of time occupied by each group of centre lathes in the processing of each size

[33] M. R. Hill, *op. cit.* (1970), pp. 96–100.

[34] The labour force of the average Soviet machine tool factory about 1970 was 2,000 men (see *Planovoe Khozyaistvo*, 1971, No. 8, p. 21). Furthermore, 32 Soviet machine tool factories in 1965 produced machines in quantities greater than 1,000 per year, including 8 factories producing more than 3,000 machines per year (see V. P. Chernykh, *Vliyanie spetsializatsii na uroven' proizvoditel'nosti truda* (1965), p. 32). In the UK, on the other hand, the average labour force in the British machine tool industry varied between 150–200 (see C. F. Pratten, *Economies of Scale in Manufacturing Industry*, Cambridge (1971), p. 162).

[35] *Stanki i Instrument*, 1964, No. 3, pp. 27–33.

group of components. The results of this survey indicated that 80 per cent of machine processing time was accounted for by the production of components having a maximum diameter equal to 16 to 21 per cent of the maximum machine capability for turning work between centres, and 50 to 62 per cent of the maximum machine capability for chucking work.

The results of this survey suggested, therefore, that the Soviet standardised requirements for machine tool capacity were more than adequate to meet user demands. It may be considered, on the other hand, that such data also suggested that the Soviet machine tool industry was tending to produce machine tools having a dimensional capacity far in excess of user requirements. Such a phenomenon, if shown to exist, would not only have been a direct result of the requirements of the existing state standards, however, but also a consequence of the relative quantities of output of each machine type and size within the overall range.

This phenomenon was found not to be peculiar to Soviet industry: a British Workpiece Statistics Survey[36] obtained similar results for the utilisation of general-purpose machine tools, the dimensions of which are not specified by national standards. This British survey found that 80 per cent of turned workpieces were less than 12 inches in diameter, although 80 per cent of centre lathes had a maximum workpiece capacity greater than 12 inches in diameter. Both surveys also showed similar variations for centre lathes between processed workpiece lengths, and distance between centres.

These results, although not directly comparable because of the different bases used for grouping of data, do indicate that the apparent tendency to produce over-capacity machines in the USSR is not necessarily an inherent fault of either Soviet standardisation policy or Soviet aggregate planning methods, but probably caused by lack of communication between machine tool manufacturers and users, preventing a closer definition of user requirements, or conservatism by both makers and users alike. Similar factors probably account for the analogous results obtained from British surveys, and it is still an open question whether machine dimensional parameters closer to the majority, instead of the near totality, of user requirements would be acceptable to either Western or Soviet buyers, who would anticipate full utilisation from their more recent purchases.

It is interesting to note, however, that several Western European machine tool firms now offer a short bed chucking lathe in their product range, and a similar variant is also now produced (16P20; 500 mm centre distance) by the Krasnyi Proletarii factory.[37] The response of both Western European and Soviet users to such machines is not yet known.

MACHINE ACCURACY

The basic requirement of any general-purpose machine tool, assuming that it has adequate overall capacity, is its capability to produce components of required accuracy. This capability depends, in turn, on the precision with

---

[36] G. C. Tinker, *The Utilisation of Machine Tools*, MTIRA, Macclesfield (1968).
[37] *Stankostroenie SSSR, Vypusk 1; Stanki tokarnoi gruppy* (1970), pp. 24, 25.

which major elements can be moved and positioned in relation to one another. Soviet state standards relating to machine tool accuracy specify, therefore, relevant tests and acceptable tolerances for errors in the movement and positioning of machine elements. Similar tests are also carried out on their finished products by Western machine tool enterprises, usually based on testing procedures and permissible errors recommended by two internationally acknowledged experts in this field: Salmon[38] and Schlesinger.[39] Some Western countries have also followed a similar procedure to the USSR by incorporating these recommendations into their national standards.

ACCURACY OF SOVIET MACHINE TOOLS PRODUCED BEFORE 1970

Soviet machine tool standards can be technically assessed by comparing the technical requirements embodied in these documents with those of similar documents used in Western countries. A short exercise of this type was consequently carried out by the author in 1969,[40] paying particular attention to the accuracy requirements for horizontal knee and column milling machines and centre lathes specified by GOST 13-54 and GOST 42-56 respectively, because of the comparatively high proportion of Soviet machine tool output accounted for by these machines.[41] In view of the lack of relevant British national standards specifying accuracy requirements for horizontal milling machines and centre lathes at that time, it was decided to assess the technical requirements of the Soviet documents by a comparison of accuracy requirements specified by GOST 13-54 and GOST 42-56 for widely used models of machines, namely a 400 mm swing centre lathe and a 1,250 mm table size milling machine,[42] with those adhered to by British manufacturers of similar machine models (see Table 11.5).

The results of this comparison are shown in Table 11.6, columns 1, 2 and 7, and Table 11.7, columns 1, 2 and 7, for the milling machine and centre lathe respectively; information on the Salmon and Schlesinger tests is provided for comparative purposes (see Table 11.6, columns 5 and 6; Table 11.7, column 6).

[38] P. Salmon, *Machines-Outils; Reception, Verification*, Paris (1954).
[39] G. Schlesinger, *Testing Machine Tools*, London (1966).
[40] M. R. Hill, *op. cit.* (1970), pp. 212–47.
[41] In 1965, for example, turning machines, including turret and capstan lathes, accounted for almost 30 per cent of total Soviet machine tool output for that year, while milling machines accounted for 12 per cent of total Soviet machine tool output (see (Eds.) N. M. Oznobin *et al.*, *Sovershenstvovanie struktury promyshlennogo proizvodstva* (1968), p. 136, quoted in M. J. Berry, *Research, Development and Innovation in the Soviet Machine Tool Industry*, unpublished research report, Centre for Russian and East European Studies, University of Birmingham (1974), pp. B7–B9).
[42] The 400 mm swing 1K62 centre lathe and its variants are produced in quantities of 13,000 per year (i.e., more than 50 per cent of the total turning machine output, and hence some 12–15 per cent of the total Soviet output in 1965, using Oznobin's previously cited proportions combined with a total 1965 output figure of 186,130) (*Narodnoe khozyaistvo SSSR v 1968 godu* (1969), p. 257). The output of the 6M82 range is more difficult to estimate, however. Production planning data quoted in V. A. Anufriev *et al.*, *Krupnoseriinoe proizvodstvo frezernykh stankov* (1965), suggest that a total of 10 machines of the 6M82 and 6M83 (1,600 × 400 mm table sized machines) are produced daily (i.e., 3,000 machines annually).

Table 11.5 *Condensed specifications for the selected Soviet and British machines*

1. *Milling machines*

	6M82G	British No. 2 machine[a]
Table length (mm)	1250	1500
Table width (mm)	320	355
Table travel: longitudinally (mm)	700	700
transversely (mm)	260	300
vertically (mm)	380	500
Range of spindle speeds (r.p.m.)	31·5–1600	18–1800
Number of spindle speeds	18	24
Range of feeds (mm/min)	25–1250	12–1500
Number of feeds	18	16
Main motor power (kW)	7	7·46
Feed motor power (kW)	1·7	2·25

2. *Centre lathes*

	1K62	British 8-in machine
Max. workpiece diameter (mm)	400	n.a.
Swing over bed (mm)	430	432
Distance between centres (mm)	710, 1000, 1400	1370–1980
Swing over carriage (mm)	220	304
Spindle bore (mm)	45	76
Range of spindle speeds (r.p.m.)	12·5–2000	37–600
Number of spindle speeds	24	8
Number of feeds	42	45
Range of feeds (mm per rev.)	0·07–4·1	0·04–1·25
Main motor power (kW)	10	5·65

Source: Manufacturers' specifications.

Note:
a – the metric specification for the British machine was obtained by a conversion from the inch specification.

A fuller discussion of this comparative assessment is given in Appendix A (pp. 556–8), from which it can be tentatively concluded that although there were many tests for which similar tolerances were specified by both the appropriate Soviet state standard and the British manufacturers' testing standards, there were also several important tests for which closer accuracies were required from the British-built machines. Further study of tolerances of certain important component design elements used in the construction of the milling machine model, selected as a typical model of Soviet general-purpose machine tools, revealed certain inaccuracies which could be traced to the technology of their manufacture. A fuller discussion of this aspect of the study is given in Appendix B (pp. 558–61). Many of the general conclusions relating to the accuracy of the Soviet-built milling machines were reinforced by the results of a detailed study of the performance of these machines reported in the Soviet technical press, and certain Soviet industrial economists and engineers made general comments pertaining to the general accuracy levels of Soviet-built machine tools during this period.[43]

It is important to note, however, that a British user of Soviet milling machines visited by the author expressed general satisfaction with the alignment accuracies of these machines (see Appendix C for a more detailed discussion), and a machine shop foreman of the author's acquaintance expressed satisfaction with the accuracy in use of a 1K62 centre lathe purchased in 1968. This suggested that the differences in precision noted at the machine acceptance stage between the selected British and Soviet machine tools, as defined by alignment test requirements, did not necessarily affect the ability of the Soviet machine tools to meet accuracy requirements of typical users.

Furthermore, the previously cited workpiece statistics surveys carried out on centre lathes in the USSR[44] and the UK[45] tended to indicate that the accuracy requirements for turned parts were not significantly different between the British and Soviet engineering industries, and hence that Soviet lathes were as capable as their British counterparts of meeting the majority of users' accuracy requirements, in spite of differences in alignment test demands. As explained above, the procedures used to collect data for these surveys, and the subsequent treatment of data, were not directly compatible, but useful comparisons were nevertheless obtained from a study of the data collected. The data obtained from these surveys are tabulated in Table 11.8.

Since the classes of accuracy were not directly compatible, the actual tolerances required for appropriate Soviet 'classes of accuracy' were obtained from a Soviet engineering handbook. As shown in Table 11.9 below, the actual dimensional tolerance required for a turned shaft of a particular class of accuracy depended upon the diameter of that shaft.

Since the Soviet survey suggested that 80 per cent of processing time on centre lathes of 400 mm workpiece diameter was accounted for by workpieces of less than 300 mm diameter, while the British survey indicated that 80 per cent of workpieces processed on centre lathes were less than 12 inches in diameter, the following information was estimated from Tables 11.8 and 11.9 for the comparative accuracy demands of the bulk of workpieces processed on 400 mm swing centre lathes in the British and Soviet engineering industries (see Table 11.10).

The information contained in Table 11.10 suggests that differences in alignment test tolerances between the selected British and Soviet centre lathes,

---

[43] See N. V. Ol'khovskii, *Machines and Tooling*, 1968, No. 8, pp. 24–7, for a detailed account of performance tests carried out on a 6M82 milling machine. See also K. I. Shelkovyi and A. Z. Lavrenev, in (Ed.) V. P. Mysnichenko, *op. cit.*, p. 99, and B. G. Andreev, *op. cit.*, pp. 68 and 69, for general remarks on the accuracy of Soviet-built machine tools.

[44] *Stanki i Instrument*, 1964, No. 3, pp. 27–33.

[45] G. C. Tinker, *op. cit.*

recorded above, were not apparently reflected in lower accuracy demands for turned parts processed on centre lathes in the Soviet engineering industry. This tentative conclusion was to be expected, however, since centre lathes are frequently used as semi-finishing machines for components to be subsequently hardened, and ground to high precision and surface finish. The basic requirements of centre lathes in such working conditions are high rates of metal removal, and the ability to produce component dimensions to sufficient accuracy for subsequent finishing on grinding machines.

ACCURACY OF SOVIET MACHINE TOOLS PRODUCED SINCE 1970

New standards for accuracy of milling machines and centre lathes were published in 1972 for introduction into industrial practice at the beginning of 1974. These standards include tests similar to those previously specified, and tolerances similar to those specified in the pre-existing standards for 'machines of normal precision', although some tolerances are more demanding than those previously specified in GOST 13-54 and GOST 42-56 respectively. The new standards also contain provision for the requirements of machines conforming to 'machines of improved precision' (*stanki povyshennoi tochnosti*) in which the allowed tolerances for each test are approximately 0·6[46] of those allowed for the normal precision machines. In the case of centre lathes, tolerances are also included for 'high precision machines' (*stanki vysokoi tochnosti*), which are even more demanding than those of the 'improved precision' machines (see Table 11.6, columns 3 and 4, and Table 11.7, columns 3, 4 and 5).

It is difficult to estimate the proportion of each accuracy class of machines currently being manufactured in the USSR, and hence the degree to which the introduction of these standards will lead to an overall improvement in quality level of centre lathes and milling machines produced by the industry.

The central planning system should be capable, theoretically, of enforcing the production of 'improved precision' and 'high precision' machine tools, but this is partly dependent on adequate resources having been made available to factories to improve their technology of production, and an adequate incentive being provided through the pricing system. There is evidence to suggest that such resources and incentives are being made available, since one Soviet publication[47] states that the basic model of centre lathe now being large-batch produced in place of the 1K62 'normal precision' centre lathe, is the 'improved precision' (P class) 16B20P machine. 'Ultra-high-precision' (*stanki osobo vysokoi tochnosti*) and 'normal precision' (N class) variants are also included in the Krasnyi Proletarii factory's product range, although the latter model is not provided with mechanised cross-slide feed as a standard feature.

[46] See M. O. Yakobson, *Tekhnologiya stankostroeniya* (1966), p. 8, and I. M. Kucher, *Metallorezhushchie stanki* (1966), p. 157.
[47] *Stankostroenie SSSR, Vypusk 1, Stanki tokarnoi gruppy* (1970), pp. 24, 25.

Table 11.6 Geometric alignment accuracy tests and tolerances for knee and column milling machines having overall dimensional specifications listed in Table 11.5 above

Limits of permissible error

(Column 1)	(Column 2)	(Column 3)	(Column 4)	(Column 5)	(Column 6)	(Column 7)
Alignment Test to GOST 13-54[1]	Tolerances to GOST 13-54[1]	Tolerances to GOST 17734-72[2] 'Normal' Precision Machines (N class)	'Improved' Precision Machines (P class)	Tolerances to Salmon Tests for Precision Knee and Column Milling Machines[3]	Tolerances to Schlesinger Test for Horizontal Knee and Column Milling M/cs[4]	British Manufacturer's Tolerances for the No. 2 Machines[5]
Test 1. Table flatness test	0·030 mm per metre in any direction (Table concave only)	0·030 mm per metre (Table concave only)	0·020 mm per metre (Table concave only)	0·030 mm per metre in any direction (Test 4)	No test	No test
Test 2. Test discontinued						
Test 3. Relative perpendicularity of longitudinal and lateral table movement, measured in the horizontal plane	0·020 mm in 300 mm length	0·020 mm per 300 mm length (Test 2)	0·012 mm per 300 mm length (Test 2)	0·020 mm per 300 mm length (Test 7)	No test	No test for horizontal machines
Test 4. Table parallelism when moved longitudinally	0·030 mm for the complete table travel (of 700 mm for the 6M82G machine)	0·030 mm for the complete table travel (Test 3)	0·020 mm for the complete table travel (Test 3)	0·020 mm per 500 mm length of travel (Test 5a)	0·020 mm per 500 mm of travel, 0·010 mm per additional 500 mm (Test 4)	0·024 mm (Test 8-1)
Test 5. Parallelism of the table surface when moved in a transverse direction	0·020 mm (table rising)	0·025 mm table rising (Test 4)	0·020 mm table rising (Test 4)	0·020 mm table rising (for table traverse up to 300 mm) (Test 5b)	No test	0·020 mm (Test 8-5)
Test 6. Parallelism of the side faces of the Tee-slot and longitudinal table movements	0·035 mm on 700 mm length of table	0·030 mm for complete table travel (Test 5)	0·020 mm for complete table travel (Test 5)	0·020 mm per 500 mm (Test 8)	Total error of 0·030 mm, 0·020 mm per 300 mm (Test 8)	0·030 mm (Test 1-8)
Test 7. Axial spindle float (for the spindle cone)	0·020 mm	0·010 mm (Test 6)	0·006 mm (Test 6)	No test	No test	No test
Test 8. Axial float of the spindle nose face	0·025 mm	0·020 mm (Test 7)	0·012 mm (Test 7)	0·010 mm (Test 2b)	0·020 mm (Test 2b)	0·020 mm (Test 7-3)

Test	Source 1 (GOST 13-54)	Source 2 (GOST 17734-72)	Source 3 (Schlesinger)	Source 4 (Salmon)	Source 5 (British company)
Test 9. Radial runout of the spindle cone	0·010 mm at the spindle nose, 0·020 mm at a distance of 300 mm from the spindle nose	0·010 mm at the spindle nose, 0·020 mm at a point 300 mm from the spindle nose (Test 8)	0·006 mm at the spindle nose; 0·012 mm at a distance 300 mm from the spindle nose (Test 1)	0·010 mm at the spindle nose, 0·020 mm at a distance 300 mm from the spindle nose (Test 3)	0·010 mm at the spindle nose, 0·020 mm at a distance 300 mm from the spindle nose (Test 7-1)
Test 10. Radial runout of the spindle external locating surface	0·015 mm	0·010 mm (Test 9).	0·006 mm (Test 9)	0·010 mm (Test 2a)	0·010 mm (Test 7-2)
Test 11. Squareness of table centre Tee-slot to the spindle axis	0·020 mm on a length of 300 mm	0·020 mm on a length of 300 mm (Test 10)	0·012 mm on a length of 300 mm (Test 10)	0·020 mm on a length of 300 mm (Test 7)	0·020 mm on a length of 300 mm (Test 8-8)
Test 12. Parallelism of the table surface to the spindle axis	0·030 mm on a 300 mm length, table rising (for table widths greater than 160 mm)	0·025 mm on a 300 mm length (table rising) (Test 11)	0·016 mm on a 300 mm length (table rising) (Test 11)	No test	0·020 mm on a length of 300 mm (table rising)
Test 13 & 14. For vertical machines only					
Test 15. Perpendicularity of the table to the column slideways	0·020 mm in 300 mm along the table axis. 0·030 mm in 300 mm for transverse table movement	0·025 mm on a 300 mm length in both longitudinal and transverse direction (Test 13)	0·016 mm in 300 mm for both longitudinal and transverse table movement (Test 13)	0·020 mm in 300 mm for both longitudinal and transverse table movement (Test 9 and 10)	0·020 mm in 300 mm for both longitudinal and transverse table movement (Tests 8-12 and 8-13)
Test 15. Parallelism of the overarm to the spindle	0·025 mm in 300 mm (for table widths greater than 160 mm)	0·020 mm over a 300 mm length in both horizontal and vertical directions (Test 15)	0·012 mm in 300 mm (Test 15)	0·020 mm over a 300 mm length in both the horizontal and vertical directions (Test 11)	0·020 mm in 300 mm (Tests 8-18 and 8-19)
Test 17. Coaxiality of the bracket arm with the spindle	0·030 mm on a 300 mm length (for table widths greater than 160 mm)	0·030 mm on a 300 mm length (Test 16)	0·020 mm on a length of 300 mm (Test 16)	0·020 mm on a 300 mm length (Test 12)	0·020 mm on a length of 300 mm (Test 8-23)

Sources:
1 GOST 13-54.
2 GOST 17734-72.
3 G. Schlesinger, *Testing Machine Tools*, London (1966), pp. 43 and 46.
4 P. Salmon, *Machines-Outils, Reception, Verification*, Paris (1959), pp. 80–3.
5 Testing procedure of selected British company.

Table 11.7 *Geometric accuracy alignment tests and tolerances for centre lathes*

Limits of permissible error

(Column 1) Alignment Test to GOST 42-56[1]	(Column 2) Tolerances to GOST 42-56 (maximum workpiece diameter up to 400 mm)[1]	(Column 3) Machines of 'normal' precision (N class)	(Column 4) Machines of 'improved' precision (P class)	(Column 5) Machines of 'high' precision (V class)	(Column 6) Tolerances to Schlesinger Tests for Finish Turning Lathes up to 400 mm Height of Centres.[3]	(Column 7) Tolerances to British Manufacturer's Acceptance Tests for 216 mm Height of Centre Machine[4]
		Tolerances to GOST 18097-72[2]				
Test 1. Bed flatness along its length	0·020 mm per metre 0·040 mm over the complete length	0·040 mm over complete length (Test 1.2)	0·025 mm over complete length (Test 1.2)	0·016 mm over complete length (Test 1.2)	0·020 mm per metre (Tests 1a and 1b)	No test
Test 2. Bed level in transverse direction	0·020 mm per metre 0·030 mm per metre over complete length of saddle movement. (No twist permitted).				±0·020 mm per metre. No twist permitted (Test 1c)	No test
Test 3. Straightness of saddle movement in horizontal plane	0·020 mm per metre. 0·030 mm over complete length of saddle movement	0·025 mm over complete length of saddle movement (Test 1.1)	0·016 mm over complete length of saddle movement (Test 1.1)	0·010 mm over complete length of saddle movement (Test 1.1)	0·020 mm per metre (Test 2)	No test
Test 4. Parallelism of tailstock slideways with saddle movement	0·020 mm per metre, 0·025 mm over the complete length, considering horizontal and inclined slideways only	0·025 mm over the complete length in horizontal plane 0·040 mm over the complete length in the vertical plane (Test 1.4)	0·016 mm over the complete length in the horizontal plane 0·025 mm over the complete length in the vertical plane (Test 1.4)	0·010 mm over the complete length in the horizontal plane 0·016 mm over the complete length in the vertical plane (Test 1.4)	0·020 mm per metre (test 3)	0·0125 mm per metre (Test 1)
Test 5. Radial runout of the spindle nose sleeve	0·010 mm	0·010 mm (Test 1.5)	0·007 mm (Test 1.5)	0·005 mm (Test 1.5)	0·010 mm (Test 5)	0·0075 mm (Test 3)
Test 6. Radial runout of the spindle nose bore	0·010 mm at spindle nose. 0·020 mm at 300 mm from spindle nose	0·010 mm at spindle nose. 0·016 mm at 200 mm from spindle nose (Test 1.8)	0·007 mm at spindle nose. 0·010 mm at 200 mm from spindle nose (Test 1.8)	0·005 mm at spindle nose. 0·007 mm at 200 mm from spindle nose (Test 1.8)	0·010 mm at spindle nose. 0·030 mm at 300 mm from spindle nose (Test 7)	0·0075 mm at spindle nose. 0·013 mm at 300 mm from spindle nose (Test 5)
Test 7. Axial spindle slip	0·010 mm	0·008 mm (Test 1.6)	0·005 mm (Test 1.6)	0·003 mm (Test 1.6)	No test	No test
Test 8. Cam action of the spindle flange	0·020 mm	0·016 mm (Test 1.7)	0·010 mm (Test 1.7)	0·007 mm (Test 1.7)	0·010 mm (Test 6)	0·0075 mm (Test 4)

Test	Col 1	Col 2	Col 3	Col 4	Col 5	Col 6
(Test ?. Spindle axis parallel to the direction of saddle movement)	0·030 mm per 300 mm in vertical plane: 0·012 mm per 300 mm in horizontal plane	0·030 mm in vertical plane: 0·008 mm per 200 mm in horizontal plane (Test 1.9)	0·020 mm in vertical plane: 0·005 mm per 200 mm in horizontal plane (Test 1.9)	0·008 mm in vertical plane: 0·003 mm per 200 mm in horizontal plane (Test 1.9)	0·030 mm per 150 mm (Test 9)	0·0125 mm per 300 mm in vertical plane: 0·015 mm per 300 mm in horizontal plane
Test 10. Spindle axis parallel to top slide	0·030 mm per 100 mm length: 0·04 mm per 300 mm length	0·020 mm per 100 mm length; 0·035 mm per 300 mm length (Test 1.10)	0·012 mm per 100 mm length; 0·020 mm per 300 mm length (Test 1.10)	0·010 mm per 100 mm length (Test 1.10)	0·030 mm per 150 mm (Test 9)	No test
Test 11.	Not applicable to medium sized centre lathes					
Test 12. Parallelism of the tailstock sleeve taper with the bed	0·030 mm per 300 mm in both the horizontal and vertical planes	0·020 mm per 200 mm in both horizontal and vertical planes (Test 1.13)	0·016 mm per 200 mm in both horizontal and vertical planes (Test 1.13)	0·010 mm per 200 mm in both horizontal and vertical planes (Test 1.13)	0·030 mm per 300 mm in the vertical plane; 0·020 mm per 300 mm in the horizontal plane (Tests 11a and 11b)	0·0125 mm per 300 mm in the vertical plane; 0·02 mm per 300 mm in the horizontal plane (Tests 10 and 11)
Test 13. Parallelism of the tailstock sleeve to saddle movement	0·030 mm per 100 mm in the vertical plane. 0·010 mm per 100 mm in the horizontal plane	0·010 mm per 50 mm length in vertical plane. 0·008 mm per 50 mm length in horizontal plane (Test 1.12)	0·010 mm per 50 mm length in vertical plane. 0·006 mm per 50 mm length in horizontal plane (Test 1.12)	0·010 mm per 50 mm length in vertical plane. 0·004 mm per 50 mm length in horizontal plane (Test 1.12)	0·02 mm per 100 mm in vertical plane. 0·01 mm per 100 mm in the horizontal plane (Tests 10a and 10b)	0·0125 mm in vertical plane. 0·01 mm in horizontal plane (Tests 8 and 9)
Test 14. Axis of centres parallel with bed in a vertical plane	0·060 mm	0·030 mm (Test 1.3)	0·020 mm (Test 1.3)	0·012 mm (Test 1.3)	0·020 mm (Test 12)	0·020 mm (Test 12)
Test 15.	Not applicable to medium sized centre lathes					
Test 16. Axial movement of leadscrew	0·010 mm	0·008 mm	0·005 mm	0·003 mm	0·010 mm (Test 13)	0·0075 mm (Test 15)
Test 17. Screwcutting accuracy of the machine (Pitch error of thread cut on the machine)	0·025 in 100 mm length, 0·050 mm in 300 mm length	0·035 mm in 300 mm length (Test 1.14)	0·020 mm in 300 mm length (Test 1.14)	0·012 mm in 300 mm length (Test 1.14)	±0·020 mm per 50 mm length	No test
Test 18. Machine turns round and parallel N.B. work held in chuck	Ovality 0·010 mm Parallelism 0·010 mm in 100 mm length	0·008 mm out of round; 0·020 mm per 200 mm length out of parallel	0·005 mm out of round; 0·012 mm per 200 mm length out of parallel	0·003 mm out of round; 0·008 mm per 200 mm length out of parallel	0·010 mm (deviation from roundness) Parallelism 0·020 mm per 200 mm	0·0075 mm (deviation from roundness) Parallelism 0·020 mm per 300 mm (Test 16)
Test 19. Facing Test	0·020 mm on a 300 mm diameter face	0·015 mm on a 200 mm diameter face (Test 2.2)	0·010 mm on a 200 mm diameter face (Test 2.2)	0·005 mm on a 200 mm diameter face (Test 2.2)	0·020 mm on a 300 mm diameter face (Test 15)	0·0125 mm on a 300 mm diameter face (Test 14)

Sources:
1 *GOST 42-56*; 2 *GOST 18097-72*; 3 G. Schlesinger, *op. cit.*, pp. 54–55; 4 Testing procedure of selected British company.

Table 11.8 *Accuracy requirements of samples of components processed on centre lathes by the Soviet and British engineering industries*

UK data		USSR data	
Accuracy requirement	% of sample requiring this accuracy	Accuracy requirement	% of sample requiring this accuracy
<0·001 inch	15	2nd class of accuracy	21
0·001–0·003 inch	26	3rd class of accuracy	46
0·003–0·005 inch	16	4th class of accuracy	15
0·005–0·010 inch	22	5th class of accuracy	15
>0·010 inch	18	7th class of accuracy	3

Sources:
See notes 44 and 45.

Table 11.9 *Tolerance requirements for Soviet turned shafts*

Class of accuracy	Shaft external diameter range in mm						
	18–30	30–50	50–80	80–120	120–80	180–260	260–360
	Tolerance in mkm						
2nd class of accuracy	14	17	20	23	27	30	35
3rd class of accuracy	45	50	60	70	80	90	100
4th class of accuracy	140	170	200	230	260	300	340
5th class of accuracy	280	340	400	460	530	600	680

Source:
Compiled from O. P. Mamet, *Kratkii spravochnik konstruktora-stankostroitelya* (1964), pp. 302–5.

Table 11.10 *Estimated comparative accuracy requirements of components processed on 400 mm swing centre lathes in the Soviet and British engineering industries*

UK data		USSR data	
Accuracy requirement	% of sample requiring this accuracy	Accuracy requirement	% of sample requiring this accuracy
<0·001 inch (25 mkm)	15	<35 mkm	21
0·001–0·003 inch (25–75 mkm)	26	35–100 mkm	46
0·003–0·005 inch (75–125 mkm)	16		
0·005–0·010 inch (125–250 mkm)	22	100–340 mkm	15
>0·010 inch (250 mkm)	18		
		>340 mkm	18

A comparison of the age of machine tool state standards operating for 1969 and 1973, respectively, indicated that significant changes had occurred in the age structure of standards over this period, with a smaller proportion of standards greater than 10 years old in 1973 than in 1969 (see Table 11.11 below). Since newer standards appear to demand high quality, it is likely that the overall quality level of Soviet-built machines has consequently improved significantly in recent years.

Table 11.11 *Comparative age structure of Soviet machine tool state standards*

Age of standards	No. in each category 1969	1973
Less than 5 years	50	65
5–10 years	26	43
10–15 years	22	10
Greater than 15 years	13	4

Sources:
Ukazatel' gosudarstvennykh standartov SSSR v 1969 godu and *Ukazatel' gosudarstvennykh standartov SSSR v 1973 godu*. These directories are published by the State Committee of Standards of the Council of Ministers of the USSR.

OTHER ASPECTS OF MACHINE TOOL SPECIFICATIONS

Although adequate dimensional capacity is an obvious prerequisite for satisfactory machine tool utilisation, and geometric alignment accuracy has been a useful measure of machine tool quality for many years, these parameters are only one of many that affect satisfactory machine tool performance. Other factors include working speed and feed range, noise levels, vibration characteristics, performance of control systems, and machine reliability and service life.[48] A brief discussion of these factors is given below.

SPEED AND FEED RANGE

Machine tool speeds and feeds are also standardised in the USSR, using geometric series. The standard[49] specifies a number of geometric factors from which an appropriate value can be selected, and allowable variations between actual and theoretical spindle speeds caused by the practical constraints encountered in gearbox design. Maximum spindle speeds for particular machine tool types and sizes are also sometimes specified by the appropriate

[48] See (Eds.) F. Koenigsberger and J. Tlusty, *Specifications and Tests of Metalcutting Machine Tools*, Manchester (1970), for a full discussion of important parameters to be included in comprehensive machine tool specifications.

[49] The appropriate document is *Machine Tool Industrial Branch Standard (Otraslevaya Normal') N11-1*, published by the Ministry of the Machine Tool and Tooling Industry for use by subordinate design organisations and factories.

standards for basic dimensions of the appropriate machinery (see pp. 534–5 above). The chief advantage of this standardisation policy would appear to arise from savings in process planning costs by users, as a result of a likely rationalisation of machine speeds across different machine models within the same broad product type range. The author has found no evidence to suggest that the advantages arising from this policy have been lost through failure of standardised requirements to conform to user requirements. A study of the utilisation of the 1K62 centre lathe reported by L'vov suggests that the standardised cutting speeds and feeds available from that machine were more than adequate for the majority of user requirements. His study found that only 50–60 per cent of the machines' possible working speeds, and 17–20 per cent of the machines' available longitudinal feeds, were used over the sample of 100,000 components included in his survey.[50] Similar information was also obtained by Tinker[51] in his study of the utilisation of centre lathes in the British engineering industry, thus reinforcing further a general conclusion of similarities in parameters and utilisation characteristics for British and Soviet centre lathes.

NOISE LEVELS

In recent years, noise has become an increasingly important factor in machine tool acceptance and utilisation, as the effect of industrial noise on employee health has become more widely recognised and understood on an international scale. In 1967, the Soviet Ministry of the Machine Tool and Tooling Industry, published an 'industrial branch standard'[52] drafted by ENIMS. This document was based on a Soviet state standard specifying the methods to be used for the determination of machinery noise levels[53] which took into account the recommendations of Technical Committee No. 43 of the International Standards Organisation. Hence it would appear that Soviet practice in the field of acceptance tests, and limits, for machine tool noise levels are at an advanced stage in relation to contemporary international practice.

VIBRATION

A further factor influencing machine tool performance which has received increasing attention in recent years is machine tool vibration, and the limitations placed on economic metal removal rates by unacceptable levels of vibration. This topic is clearly a complex one, and the large quantity of papers in the Soviet technical press[54] suggest as high a level of research activity in this field as in the West. To the author's knowledge, however, no state standard for

[50] L'vov reports these data in *Ekonomicheskie problemy povysheniya kachestva promyshlennoi produktsii* (1969), p. 74, referring to his earlier study of the 1K62 centre lathe (see D. S. L'vov, *Osnovy ekonomicheskogo proektirovaniya mashinostroeniya* (1966).
[51] G. C. Tinker, *op. cit.*, p. 13.
[52] See *Machine Tool Branch Standard (Normal' Stankostroeniya) N89-40*, published by ENIMS in 1967.
[53] *GOST 11870-66*.
[54] A large proportion of papers in *Stanki i Instrument* and *Vestnik Mashinostroeniya*, both leading technical journals published by the Ministry of the Machine Tool and Tooling Industry, relate to the general theme of machine tool vibration.

machine tool vibration testing has yet been published in the USSR, although there is evidence to suggest that tests for unacceptable vibration levels are carried out by Soviet machine tool factories.[55] Standardised requirements for static stiffness, however, is a long-established Soviet practice. It is difficult, therefore, to assess current Soviet industrial practice in this field except by a detailed study of the relevant literature and performance testing of individual machine tool models, which were clearly beyond the resources of the current research study.

CONTROL SYSTEMS

The performance of contemporary machine tools is also influenced by the performance of the control systems fitted to them, particularly as the design of an appropriate control system may reduce certain errors inherent in machine construction. This section, however, has been concerned with a study of manually operated machine tools only, in view of the long-established Soviet experience in the design and production of these machine types. Soviet numerical control systems are discussed by J. M. Cooper in chapter 4, but the whole area of Soviet machine tool control systems still remains an area for further study.

MACHINE RELIABILITY AND PERFORMANCE IN SERVICE

In order to obtain an impression of the reliability and service life of Soviet-built machine tools, and supplement the previous study on machine accuracy, an initial survey was carried out of a small sample of Soviet-built machines in operation in a British factory.[56] These machines had been purchased during 1971 and could therefore be considered as typical general-purpose machine tools produced during 1970. The machines selected were widely used 'standard' machines, which the Soviet machine tool industry could be expected to produce satisfactorily in view of its long experience and the availability of these machines for export. The information was obtained by discussion with production and maintenance management personnel, and machine tool operators within the factory. The machines surveyed included a 2N55 radial drilling machine, the 6M82 range of milling machines similar to the type previously compared for alignment accuracy on pp. 536–9, and a 3G71 surface grinding machine. An abbreviated report of this survey is given in Appendix C below.

The overall impression gained from this initial study of a user's experience of Soviet built machine tools was that of design shortcomings and insufficient quality control during their manufacture. These design and production shortcomings in their turn produced problems in the productive capabilities of the machines when operating, and a low degree of reliability.

[55] Anufriev *et al.* mention vibration testing of completed machines at the Gor'kii Milling Machine Factory, under defined conditions of machine speed, feed, and depth of cut (see V. A. Anufriev *et al., op. cit.*, pp. 142–7).
[56] M. R. Hill, *Experience in the Use of Soviet General-purpose Machine Tools*, unpublished report, Centre for Russian and East European Studies, University of Birmingham (1973).

Even though the price of the machines studied was estimated to be only one half to two thirds of that of a British-built machine of similar overall specification, it was considered that the initial price benefits were lost through poor productivity and reliability. The maintenance reports indicated that a set of five Soviet-built machines required 200 hours of maintenance fitters' time (i.e., an average of 40 hrs per year per machine), whereas the remaining 1,000 machines had 18,500 hours available for their maintenance. These figures can only be considered as approximate, however.

In addition to the electrical safety problems referred to in Appendix C the electrical contactors were considered to be incorrectly designed in relation to their intended tasks, and problems were consequently encountered. The problem of quality of supply of electrical parts for Soviet machine tools has been mentioned elsewhere by the assistant chief designer of the Krasnyi Proletarii factory[57] and also by a Leningrad author complaining of faulty electromagnetic clutches used in the 1B124 and 1B136 single spindle autos.[58]

Doubts were also raised as to the service life of these machines in relation to their ability to maintain accuracy. Similar remarks have also been expressed by Soviet authors, but in relation to the wear resistance of materials used for important machine tool components.[59] It is important to stress, however, that the information contained in this section of the paper has been obtained from discussions in one firm only, selected because of its convenient location and well documented maintenance records. Hence, the machines surveyed represent an extremely small sample of Soviet machine tools of this type. Other buyers of Soviet machine tools are reported to be extremely satisfied with their purchases. One buyer of a Soviet-built boring machine, which was reported to have worked trouble-free for 24 hours daily over the previous 18 months since purchase, is on record as stating that 'if a Russian machine tool fits your specification, there are no grounds for worrying about quality and reliability'.[60] It is apparent, therefore, that a larger study is required, using controlled experiment conditions if possible, before any valid conclusions can be reached concerning the reliability characteristics of the types of Soviet machine tools described in this section of the paper. There is evidence to suggest that many of the shortcomings in quality noted in the products studied have now been removed, as a consequence of improvements in the requirements of state standards, general product development, and the use of the 'Mark of Quality' procedure.

The Gor'kii Milling Machine Factory, for example, has recently replaced the 'M' range (of which the 6M82 was the basic model), produced since 1959,[61] with an 'R' range, having the following features:[62]

—a safety clutch in the traverse feed mechanism;

[57] *Standarty i Kachestvo*, 1967, No. 10, p. 87.
[58] B. G. Andreev, *op. cit.* see note 27 above, pp. 140 and 141.
[59] K. I. Shelkovyi and A. Z. Lavrenev, in V. P. Mysnichenko, *op. cit.*, p. 99.
[60] *Metalworking Production*, November 1971, pp. 93–5.
[61] V. A. Anufriev et al., *Krupnoseriinoe proizvodstvo frezernykh stankov* (1965), p. 5.
[62] *Stankostroenie SSSR, Vypusk 6, Frezernye, strogal'nye, protyazhnye stanki* (1970), pp. 4, 5, 10.

—periodic automatic lubrication of vertical slideways;
—hermetic sealing of the knee to prevent the entry of coolant;
—roller bearing support for the leadscrew;
—sleeve clamping by means of a split ring;
—other features to customer's order.

A large proportion of the components from which this new range is built, however, are common with those previously used in the 'M' range. For example, 83·3 per cent of the components in the 6R82 machine are common to the 6M82.

The factory is also introducing a 'T' range of milling machines into its product range, which have the following features:

—a more compact main drive transmission system;
—wider speed and feed range;
—higher stiffness of drive transmission;
—hydraulic clamping of cutter and workpiece;
—automatic cycling for table movement.

It is interesting to note that the first machine tool enterprise to receive 'Mark of Quality' approval was the Krasnyi Borets Factory, Orsha, which manufactured the 3G71 surface grinding machine, discussed above. The 'Mark of Quality' approval was granted to the Model 3711 ultra-high-precision grinding machine in 1967, and the article[63] describing the approval procedure noted that modifications were necessary to those components influencing its accuracy and stiffness (presumably of the spindle head), and the hydraulic system was also improved. The 'Mark of Quality' procedure also appears to have been a significant factor in the improvement of product quality in the machine tool industry as a whole, 39 models of machine tool having been granted 'Mark of Quality' approval by 1973.[64]

A further major factor which has influenced the quality and working characteristics of finished machine tools has been the quality of components purchased from outside suppliers. Companies operating in market and mixed economies are able, to a certain extent, to select their sources of supply. Soviet enterprises, on the other hand, have not usually possessed the right to critically select their sources of supply, since transfer of supplies between enterprises has been usually coordinated by appropriate central supply organisations at industrial ministry level for intra-ministry transfers, and state level for inter-ministry transfers. The consequent likelihood of receiving supplies of inadequate quality and inconsistent delivery has been traditionally countered in the USSR by Soviet enterprises manufacturing more of their own component and semi-fabricate requirements than is usual in many Western countries.

The State Committee of Standards has recently strengthened the role played by its 'local inspection organisations' to consequently improve the quality of

[63] *Standarty i Kachestvo*, 1968, No. 6, pp. 44 and 45.
[64] *Ukazatel' gosudarstvennykh standartov SSSR v 1972 godu.*

supplies of components and sub-assemblies required for important end-products, by carrying out integrated quality assurance programmes. This work includes inspection of specialised enterprises producing important components used in a particular finished article, in addition to those enterprises responsible for its final stages of manufacture.

An example of such a programme was that developed for the machine tool industry in 1968, in which more than 30 suppliers of hydraulic, pneumatic and lubricating units, roller bearings, electrical equipment and raw materials, were inspected, in addition to 68 specialised factories engaged in the final manufacture of machine tools.[65] This integrated quality assurance programme, and the implementation of recommendations resulting from it, is almost certain to have resulted in an improvement in quality of current machine tool output.

Passenger cars

PRODUCT MIX SINCE 1970

A list of the basic models of Soviet passenger cars currently in production, together with a short account of their technical characteristics, is shown in Table 11.12 below. This table shows that the ZAZ 966, Moskvich 412 and GAZ 24 had been assimilated into production during the early years of the 1971–5 plan, and that the new Zhiguli range had been added to the industry's product range.

The product range of the Soviet motor industry can be considered narrow in terms of the number of basic models within the range, compared with contemporary Western industry practice. This feature is partly to be expected, however, in view of the usual Soviet engineering design and manufacture policy of concentrating the production of a rationalised range of products into a small number of factories. It is useful to note, however, that in spite of the apparent narrowness in range, the 'small car' (Soviet classification) is well catered for by the Moskvich 412 and Zhiguli cars of contemporary design, while the 'medium car class' is filled by the Volga.

During the years of the 1971–5 five year plan the product mix structure of the Soviet motor industry has been continually changing as a consequence of new capacity being introduced during the early years of the plan. A reliable target figure[66] for 1975 would appear to be 1·3 million passenger cars distributed in the following way:

Zaporozhets range	—150,000 per year
Moskvich range	—420,000 per year (produced at the Lenin Komsomol Automobile Factory, Moscow, at the rate of 200,000 per year, and at the Izhevsk Automobile Factory, at the rate of 220,000 per year)

[65] *Standarty i Kachestvo*, 1968, No. 1, p. 6.
[66] B. V. Vlasov et al., *Ekonomicheskie problemy proizvodstva avtomobilei* (1971), p. 46; production in fact reached 1·2 millions in 1975.

Zhiguli range —600,000 per year
Volga and Chaika ranges — 75,000 per year

TECHNICAL CHARACTERISTICS OF SOVIET PASSENGER CARS

In their short discussion on the technical characteristics of Soviet passenger cars, Vlasov et al.[67] note that

> when assessing the technical level [of automobiles], it is necessary to realise that each national manufacturer designs automobiles primarily for its own country's road and climatic conditions and demands in use.

With special regard to Soviet passenger cars, they then note that

> nationally [i.e., Soviet] produced automobiles designed for our [i.e., Soviet] demands in use, have, as a rule, a higher intrinsic weight and road clearance, and a lower maximum speed and brake horse power per litre of engine capacity; they are also less demanding with regard to the quality of fuels and lubricating oils.

In addition to the need to adapt Soviet vehicle specifications to meet the requirements of Soviet environmental conditions, Vlasov et al. mention a general tendency towards contemporariness of Soviet-designed vehicles with their Western counterparts, stating that

> the latest models of [Soviet] passenger cars correspond, in terms of technical characteristics, to the best [international] models, as shown [in Table 11.13 below] in which basic parameters of the Moskvich 412 are compared with equivalent foreign models.[68]

Furthermore it is also claimed that the Soviet motor industry, when designing its more recent models of passenger vehicle, has paid a certain amount of attention to the demands of export markets for improved internal and external finish, and other technical characteristics.[69]

As explained above, the USSR has devoted the majority of passenger car manufacturing resources towards commencement of the manufacture of the Zhiguli range of vehicles, and expansion of output of a modified and updated Moskvich range. In view of the Soviet decision[70] to purchase design and production process planning expertise for the Zhiguli range from Fiat, it is likely that the specifications of this vehicle are contemporary to their Western counterparts in view of its design being based on the successful Fiat 124, and production plant selection for the Tol'yatti factory being based on Fiat

[67] B. V. Vlasov, op. cit., p. 44.
[68] Ibid.
[69] Ibid.
[70] D. M. Bloch, *The Soviet Automobile Industry*, unpublished extended essay, CREES, University of Birmingham (1971), p. 3. (Bloch quotes *Pravda*, 2 July 1965 and *Trud*, 16 August 1966 as his information sources.)

Table 11.12 Technical characteristics of Soviet passenger car product range (1971)

	Model	Dimensions (mm) Length	Width	Maximum hp	Maximum speed (km/h)	Number of places	Fuel consumption (litres/100 km)
Zaporozh'e Automobile Factory	ZAZ 966	3730	1535	45/40	120	4	8·2
Lenin Komsomol Factory, Moscow	Moskvich 408	4090	1550	55/50	120	4–5	9–10
	Moskvich 412	4090	1550	80/75	140	4–5	9·5–10·5
Izhevsk Automobile Factory	IMZ (same as Moskvich 412)	4090	1550	80/75	140	4–5	9·5–10·5
Volga Factory, Tol'yatti	Zhiguli	4030	1625	80/60	140	4–5	9–10
Gor'kii Automobile Factory	GAZ-21 (Volga)	4810	1800	80/75	130	5	14
	GAZ-24 (Volga)	4735	1800	110/98	145	5	13–14
	GAZ-14 (Chaika)	5700	2000	240/200	170	7	10–23
Likhachev Factory, Moscow	ZIL-114	6284	2074	330/300	190	7	n.a.

Source:
B. V. Vlasov, *Ekonomicheskie problemy proizvodstva avtomobilei* (1971), p. 47.

experience. The Moskvich range, on the other hand, are Soviet-designed vehicles,[71] although manufacturing expertise was purchased from Renault.[72] It is useful therefore, to consider qualified assessment of this vehicle as a means of initially estimating the capabilities of the Soviet passenger car industry in terms of design and manufacture to specification. This model was given exhaustive tests in the UK by *Motor* magazine and the Consumers' Association in 1973.

Table 11.13 *Comparative technical data relating to the Moskvich 412 and passenger cars produced by Western European firms*

	Moskvich 412	Vauxhall Victor 1600	Opel Rekord	Fiat 125
Engine capacity (cc)	1478	1599	1432	1608
Maximum power (hp)	75	72	58	90
Maximum speed (km/hr)	145	150	133	160
Time to accelerate from rest to 100 km/hr (sec)	18·8	n.a.	20·5	14·8
Wheel base (mm)	2400	2590	2670	2506
Minimum road clearance (mm)	178	150	130	120
Number of lubrication points	14	4	none	none
Fuel consumption (litres/100 km)	9·1	10·0	9·9	11·2
Horse power per litre	50·9	45	38·8	56
Specific weight (ratio of the total weight of the car to the maximum power) (kg/hp)	17·9	19·4	23·9	14·9
Service time to first major overhaul (km)	125	n.a.	n.a.	100

Source: B. V. Vlasov, *op. cit.*, p. 45.

The well documented *Motor*[73] road test consisted of an examination of the car's overall features and its various performance and handling characteristics. The report was favourable in terms of the car's acceleration, tractability and top speed, compared with Western-built cars of similar engine capacity, and particularly with cars sold for the same price as the Moskvich 412, since the sales price of this car (£717) was roughly equivalent to that charged for 'economy' cars such as the Renault 4, Hillman Imp and BLMC Mini, and some £300 cheaper than Western-manufactured cars of similar engine capacity. A further advantageous feature of the car was its passenger accommodation and luggage space when compared with similar cars in the same price range, but this was to be expected in view of the car's engine capacity. The road test pointed to poor braking, however, and poor road-holding and imprecise

[71] D. M. Bloch, *op. cit.*, p. 13. (Bloch quotes *Avtoexport*, 1968, No. 12, p. 17, as his information source.)
[72] A. C. Sutton, *op. cit.*, pp. 197, 198.
[73] *Motor*, 28 July 1973, pp. 5–9.

steering, particularly at higher speeds. The fittings and finish were considered to be average, but this combined assessment was obtained as a compromise between 'well equipped for such a cheap car . . . surely the most comprehensive standard tool kit in the world', and, on the other hand, 'the paint-work of our L-registered test car was already bubbling and badly pitted . . . Inside we can only describe the decor as cheap and rather nasty.' A final, rather surprising point, was the poor ventilation and heating in the car, in view of the severe Russian climate.

Hence an overall assessment of the car from *Motor*'s road test points towards a 1½ litre car capable of adequate performance, with large passenger space, well equipped for driver maintenance, but lacking in good finish. These characteristics would point well towards Soviet requirements for a car intended for public transport and some private use by drivers trained in car maintenance, although poor ventilation and bodywork finish may lead to shortcomings in use in the Russian climate. Furthermore, in view of the large Soviet demand for passenger cars, it is natural for Soviet policy to concentrate on the immediate expansion of production of an adequate model for Soviet conditions, particularly in terms of ruggedness,[74] rather than delay production to remove design imperfections viewed from the Western marketing standpoint, particularly as its low selling price in the West would compensate for many of these.

Tests on the Moskvich range were also carried out by the Consumers' Association during the autumn of 1973. A sample of 11 cars were selected ranging from one car as purchased, to one car having 30,000 miles recorded on the odometer. With nine of these cars tested by Consumers' Association inspectors, the unpublished results of this survey, approached mainly from the viewpoint of driver safety, are summarised in Table 11.14 below. The general picture that emerged from this study was that of lack of attention to safety features in both design and manufacture of this passenger car, caused, in many cases, by lack of attention to detail during assembly. The main areas of concern were steering and braking, most of the cars tested having serious brake defects requiring immediate attention, usually as a result of leaks in the hydraulic system or poor fitting of cylinders. Hence, although one Soviet source considered the Moskvich 412 to be equivalent to the Vauxhall Victor 1600 in terms of several technical characteristics (see Table 11.13 above), CA tests suggested that it was far below this 1962 British-designed model in terms of general safety.[75] The author is informed that the British distributors of this

---

[74] The ruggedness of the Moskvich 412 is demonstrated by its successes in two rallies, namely the London–Sydney rally in 1968, run over a distance of some 16 thousand kilometres in difficult terrain, and the London–Mexico rally in 1970, run over a distance of some 26 thousand kilometres through 25 countries of Europe, South America and Central America. All four of the Moskvich cars completed the London–Sydney course, and the Soviet team was one of four to complete the course, from 12 contestants. In the London–Mexico rally, three Moskvich 412s were amongst the 22 competitors completing the course from 96 entrants. The three Soviet cars were placed second, third and fourth in their class. (See A. M. Tarasov, *Avtomobil'naya promyshlennost'—narodnomu khozyaistvu* (1971), p. 53.)

[75] See *Motoring Which*, October 1965, pp. 106–27, for a detailed account of a road test on the 1599 cc Vauxhall Victor 101.

vehicle have attempted to correct these faults by the establishment of a well equipped centre for checking and testing of vehicles prior to their release to local distributors.

It is important to note, however, that in spite of the design and manufacturing faults detected in the sample of Moskvich 412s tested by the Consumer's Association in 1973, the safety record of the 20,000 cars of this model sold in the UK was defended publicly by the Managing Director of the importing company,[76] while a London correspondent of *Izvestiya* recorded reports of satisfactory service from other British Moskvich owners.[77] This model of passenger vehicle is also reported to have been awarded a certificate of safety in Paris in 1970,[78] and several of the improvements to the braking system of this model recorded in the Soviet technical press in 1974 may have removed some of the previous design-based causes of the model's shortcomings.[79] Furthermore, the safety test technique at the Lenin Komsomol Factory

Table 11.14 *Frequency of detected faults in nine models of Moskvich 412*

Location of defect	Type of defect Potential defect requiring remedial action	Defect to be put right at owner's convenience	Extreme defect requiring urgent attention	Extreme defect requiring immediate attention
Engine and cooling systems	2	8	0	2
Fuel and exhaust	1	1	2	0
Transmission	2	2	2	2
Braking system	0	1	3	15
Steering mechanism	0	15	3	11
Suspension and wheels	3	11	3	4
Body	14	4	1	0
Miscellaneous	6	3	0	1
Total	28	45	14	35

i.e. 35 extreme defects were observed in the 9 cars tested, 2 in the engine and cooling system, 0 in the fuel and exhaust systems, etc.

Source: Tests conducted by Consumers' Association, 1973 (unpublished).

---

[76] *The Guardian*, 17 October 1973.
[77] *Izvestiya*, 4 October 1973.
[78] Tarasov (the Minister for the Automobile Industry of the USSR) notes that 'as a result of extensive tests carried out at the French state vehicle testing station in 1970, the Moskvich 412 was awarded an international safety certificate' (A. M. Tarasov, *op. cit.*, p. 52).
[79] *Avtomobil'naya promyshlennost'*, 1974, No. 11, p. 45, records that:

Following the introduction of an amplifier and pressure regulator in the rear brake hydraulic system, and the installation of separate brake activation and fault signalling, the Moskvich fully met international market demands, including those operating in Sweden. For further improvement of effectiveness and reliability, disc brakes with four piston pot calipers on the

(Moscow) has been substantially improved since 1974 with special attention being paid to improvement of body impact resistance and interior safety, in order that the Moskvich meet West European market requirements.[80] This general increase in attention to model safety in the Moskvich 412 is also reflected in other Soviet designs of passenger cars which are reported to incorporate up-to-date safety features.[81] The GAZ 24 (Volga), for example, is reported to have improved safety features, with brakes, suspension, road-holding and handling brought up to the standards of Western vehicles.[82]

Appendix 11A. Assessments of the accuracy of Soviet-made horizontal knee and column milling machines and centre lathes

HORIZONTAL KNEE AND COLUMN MILLING MACHINES

Table 11.6 above lists the allowable tolerances for specified tests for the Soviet-built general-purpose knee and column milling machine as specified in Table 11.5 above, tested to GOST 13–54, and the British-built machine as also specified in Table 11.5 above. Tolerances for similar tests specified by the internationally accepted Schlesinger and Salmon tests are also quoted for comparative purposes. For the majority of the specified tests, the British manufacturer specified an identical tolerance to that of Schlesinger, while some of the tolerances specified for the Soviet machine were sometimes wider than those specified by Schlesinger.

The finished accuracy of the spindle assembly was the same in both machines (see test 9) although the precision of finish grinding of the outer surface of the spindle was lower for the 6M82 (the tolerances for test 8 and test 10 are wider for the Soviet machine than its British counterpart).

The permissible deviations from slideway flatness were also similar for both machines (see test 4 and test 5). The Soviet machine was less accurate than its British counterpart, however, in terms of parallelism between the table surface and the spindle axis in the vertical plane (see test 12); perpendicularity of table movement to column slideways in longitudinal and transverse directions (see test 15); and parallelism of the bracket bore to the over-arm (see tests 16 and 17).

---

front wheels and separate hydraulic brake activation, guaranteeing not less than 60 per cent of full effectiveness from one of the hydraulic circuits only, have been introduced.

Further reports indicate that a vacuum amplifier, separate circuits, front disc brakes, and force regulated rear brakes were introduced into the new model of Moskvich 2140, a modernised version of the Moskvich 412, from the end of 1975 (see *Avtomobil'naya promyshlennost'*, 1976, No. 1, p. 40).

[80] *Avtomobil'naya promyshlennost'*, 1974, No. 11, pp. 41–3.

[81] *Ibid.*, particularly p. 42, where the Soviet author refers to energy absorption tests for steering columns for the GAZ 24 (Volga) and Zaporozhets ranges.

[82] D. M. Bloch, *op. cit.*, p. 15. Bloch quotes M. A. Yushanov, *Avtomobil'naya Promyshlennost'*, 1968, No. 8, pp. 13–17, as his information source. A. M. Tarasov (*op. cit.*, p. 53) notes similar improvements.

The parallelism of the table traverse to the spindle axis requires further discussion, however, in view of differences in testing procedure between the two manufacturers. For the Soviet machine, the parallelism of the table cross-traverse to the spindle axis, measured in a horizontal direction, could be calculated by working through the following series of accuracy tests.

Test 11 specified that the error in perpendicularity of the centre Tee-slot to the spindle axis would not exceed 0·02 mm per 300 mm table length, the British manufacturer also specifying a tolerance of 0·02 mm per 300 mm for this parameter. The parallelism of the Tee-slot to the longitudinal table movement was within 0·035 mm for the complete table travel of 700 mm for the Soviet machine (see test 6) while the British manufacturer specified a tolerance of 0·030 mm. Hence, for the Soviet machine, the longitudinal table movement could be a total of 0·02 mm in 300 mm, and 0·35 mm in 700 mm, out of perpendicular with the spindle axis. Test 3 specified that the mutual perpendicularity of table movement must not exceed 0·02 mm in 300 mm. Hence, since there was no direct check on the parallelism of the table cross-traverse in the Soviet testing procedure, this parameter could amount to the equivalent of 0·050 mm in 300 mm in those circumstances where the errors in tests 3, 6 and 11 were all additive. The British manufacturer, however, checked the cross-traverse of the table to the spindle axis directly, specifying a permissible error of 0·02 mm in 300 mm (this test, and its associated tolerances conformed to test 6 of the Schlesinger testing procedure), but there was no evidence of a similar test in the GOST 13–54 procedure.

CENTRE LATHES

Table 11.7 above lists the tolerances for accuracy tests specified by GOST 42-56, which included testing procedure and tolerances for centre lathes having maximum workpiece diameters up to 400 mm, the 1K62 centre lathe falling into this category (see Table 11.7). These are compared with internationally used testing procedures specified by Schlesinger, and by a British manufacturer producing the lathe specified in Table 11.5 above.

The testing procedure adopted by both Soviet and British manufacturers corresponded to those recommended by Schlesinger, except for test 7 of GOST 42-56 which was not specified by Schlesinger for medium sized centre lathes, and tests 1 and 2 of GOST 42-56, for which Schlesinger specified that the spirit levels should be mounted on the bed.

In several cases, the allowance specified by GOST 42-56 was equal to that specified by Schlesinger, except for test 6, where GOST 42-56 specified a tighter tolerance for the radial run-out of the spindle nose bore; test 8, where Schlesinger specified a tighter tolerance for the cam action of the spindle flange; test 9 where GOST 42-56 specified a tighter tolerance for parallelism, in the horizontal plane, between the spindle axis and the saddle movement; test 10 in which Schlesinger specified a tighter tolerance for parallelism, measured in the vertical plane, between the spindle axis and the top slide; and test 12 in which Schlesinger specified a tighter tolerance for parallelism, in the horizontal plane, between the tailstock sleeve taper and the bed. Tighter accuracies in the vertical plane were also specified by Schlesinger for tests 13 and 14.

In general, the tolerances specified by the British manufacturer were tighter than those specified by both Schlesinger and GOST 42-56, particularly for parallelism, in a vertical direction, between the spindle axis, tailstock cone and axis of centres respectively, and the machine bed slideways. In test 9, however, the parallelism of the spindle axis to the saddle movement, measured in the horizontal plane, was more accurate in the Soviet standard than the British manufacturer's testing specifications. For test 13, the machines had equal accuracy for parallelism of tailstock sleeve to saddle movement, in the horizontal direction.

Appendix 11B. Assessment of the accuracy of major design elements of the 6M82 milling machines

Appendix 11A above described a comparative assessment of the alignment accuracies of selected Soviet-built machine types with their Western counterparts. This appendix investigates the accuracy of those specified component design elements which directly influenced the final alignment accuracy of the assembled milling machine, namely slideway flatness, parallelism and perpendicularity; parallelism of spindle cone and spigot to bearing diameters; and spindle bearing race run-out. The precision of these elements is influenced to a certain extent by available production technology, and attention has been paid to this wherever considered relevant. The 6M82 horizontal milling machine was selected for this more detailed study in view of the existence of a publication[1] which described the technology of manufacture at the Gor'kii Milling Machine Factory, and the finished accuracy of important components (e.g., spindles, tables, columns) prior to assembly. Furthermore, manufacturing standards were also made available by the British producer of the No. 2 milling machine used as a basis for comparison with the 6M82.

SLIDEWAYS

The alignment tests for the assembled machines indicated that although the allowable errors in longitudinal table movement were approximately the same for both machines, there was a difference in tolerances for transverse table movement relative to the spindle axis and also the perpendicularity of table movement to column slideways. The tolerances for these alignment tests were considered to be influenced by three important factors: the perpendicularity of the column slideways relative to the spindle axis; the relative perpendicularity of the vertical and transverse knee slideways; and the parallelism of the table surface to the horizontal slideways of the saddle and knee.

A comparison of these elements for both the Soviet and British machines are shown in Appendix Table 11B.1 below. This table shows that the differences between the cross-traversing and climbing accuracies of the machines could be

[1] V. A. Anufriev et al., *Krupnoseriinoe proizvodstvo frezernykh stankov* (1965).

traced to the tolerances to which the column, knee and saddle slideways were finished; the tolerances of perpendicularity of the column slideways to the spindle; and the tolerances of perpendicularity of the column to the knee cross-traverse slideways.

A study of the manufacturing standards of the producer of the British-built No. 2 milling machine, and information on the manufacturing technology used in the Gor'kii Milling Machine Factory, revealed that the same sequence of finishing and assembly was used for both the Soviet and British machines, namely: finishing of the column relative to the spindle axis; fitting of the knee to the column; building up of the table and saddle assembly, and the fitting of this unit to the knee.[2]

In the British machine, however, the column knee and saddle slideways were scraped using the spindle axis as datum, while in the Soviet machine the column was finish ground, and the vertical slideways of the knee scraped to bring its ground saddle bearing surface parallel to the spindle axis.[3] It appeared that the Soviet technique of column grinding was inherently less accurate than the scraping procedure carried out by the British company, although the grinding of the knee horizontal slideways and the table surface parallel to the saddle slideways appeared to be sufficiently accurate (see Appendix Table 11B.1 below and test 5, Table 11.6 above). The experience of other British machine tool firms has shown, however, that the majority of slideway scraping operations can be replaced by less labour-intensive and more productive grinding processes, without any loss of accuracy. The lower level of accuracy of

Appendix Table 11B.1 *Comparison of accuracy of slideways*

Design element	Accuracy of Soviet manufacturer[1]	Accuracy of British manufacturer[2]
Perpendicularity of column slideways to spindle axis	0·015 mm in 300 mm	0·007 mm in 914 mm
Relative perpendicularity of the knee slideways	0·020 mm in 300 mm	0·005 mm in 300 mm
Parallelism of table surface to saddle slideways	0·020 mm in 300 mm	0·0127 mm over table width (greater than 300 mm)

Sources:
1 V. A. Anufriev *et al.*, *op. cit.*, p. 127.
2 British Company's Manufacturing Standards.

Note:
Since the longitudinal accuracy demands were similar for the Soviet and British machines, it was decided to omit accuracy for the table slideways from this comparison.

---

[2] British Company's Manufacturing Standards and V. A. Anufriev *et al., op. cit.*, pp. 91, 126.
[3] V. A. Anufriev *et al., op. cit.*, pp. 19, 131, 202.

slideway grinding in the Soviet factory appeared therefore to reflect on the accuracy of slideway grinding machines in that factory, and not on slideway grinding as a production process.

SPINDLE ASSEMBLY

The alignment tests relating to spindle rotation for this machine indicated that in certain respects, the spindle assembly of the Soviet machine was likely to be less accurate than that of its British counterpart.

The allowable radial run-out of the cylindrical surface of the spindle nose was 0·015 mm for the Soviet machine compared with 0·010 mm for the British machine, while the axial movements of the spindle nose face were 0·025 mm and 0·020 mm respectively for the Soviet and British machines. The radial run-out

Appendix Table 11B.2 *Comparison of bearing, spindle and housing tolerances*

	Front bearing			
	Soviet Class A double-row roller bearing[1]		Skefko Class SP NN30K double-row roller bearing[2]	
O.D.	170 mm		170 mm	
Bore	110 mm		110 mm	
Width	45 mm		45 mm	
Bore limits	+ 0	−20 mkm	+ 0	−10 mkm
Shaft limits	+ 0	− 6 mkm	+25 mkm	+ 3 mkm
Radial run-out for inner race	7 mkm		5 mkm	
O.D. limits	+ 0	−25 mkm	+ 0 mkm	−13 mkm
Housing limits	+ 4 mkm	−10 mkm	− 4 mkm	−21 mkm

	Rear bearing			
	Soviet Class B angular contact bearing		Skefko bearing No. 7215 B	
O.D.	130 mm		130 mm	
Bore	75 mm		75 mm	
Bore limits	+ 4 mkm	−19 mkm	+ 0 mkm	−12 mkm
Shaft limits	+10 mkm	0 mkm	+12 mkm	− 7 mkm
Radial run-out for inner race	12 mkm		10 mkm	
O.D. limits	+ 0 mkm	−13 mkm	+ 0 mkm	−15 mkm
Housing limits	+14 mkm	−10 mkm	+ 4 mkm	−21 mkm

Sources:
1 V. D. Myagkov, *Dopuski i posadki*, (1966), pp. 417–19.
2 Bearing Manufacturer's Catalogue.

of the spindle nose cone was equal for both machines, however, which suggested that inaccuracies in the Soviet machine were more likely to be caused by inaccuracies in the finish machining of the spindle spigot than from radial movement of the inner bearing race relative to the spindle housing in the machine column.

Appendix Table 11B.2 illustrates the differences in accuracy for those features which influence the final accuracy of the spindle assembly. The Soviet manufacturer used a different bearing arrangement[4] to that designed for the British machine, but from a comparison of the double-row roller bearings used to support the front end of the spindle, it appeared that the radial run-out tolerance of the inner race would be only slightly higher than that of a British bearing of the same type and size which would be selected by the British manufacturer for use in an identical bearing arrangement. The selection, by the Soviet designer, of a Class S bearing in place of a Class A variant as currently selected, would result in a radial run-out variance identical to that of the British bearing quoted in Appendix Table 11B.2.

The selection of bearings also influenced the accuracy to which the spindle journals were finished. The dimensions shown in Appendix Table 11B.2 illustrate that although the bearing inner and outer races, respectively, were finished to a higher degree of accuracy for the British bearing, the spindle journal and bearing housing tolerances in the Soviet machine compensated for this, thus making the tolerance bands for the fitted surfaces tend to become equal.

Appendix 11C. Soviet-built machine tools in use in Britain

This appendix describes the experience in use of Soviet-built radial drilling machines, knee-type milling machines and surface grinding machines, in a British machine tool factory. The information was obtained by discussion with production management personnel, maintenance management personnel and machine tool operators within the factory. The information obtained relevant to each type of machine is described below.

RADIAL DRILLING MACHINES

1.1 The radial drilling machine used by this factory was a 2N55 type, produced, as far as the author is aware, at the Odessa Radial Drilling Machine Factory, although a factory nameplate could not be found on the machine.

1.2 The machine was considered to be adequate for the tasks it was being used to carry out, namely the drilling of machine base components.

1.3 The machine had been found to be reliable in service, although it was not being operated under particularly intensive conditions.

[4] V. A. Anufriev *et al., op. cit.,* p. 16.

1.4 The machine, as delivered, did not comply with the relevant alignment requirements, but the majority of these faults could be traced to a basic error in the alignment of the spindle relative to the column. Once this fault had been corrected the machine was found to correspond with the remainder of the alignment requirements, which were considered to be adequate.

1.5 As delivered, the machine was provided with rubber hose to serve as electrical conduit. This was considered to be inadequate for safety reasons.

KNEE-TYPE MILLING MACHINES

2.1 The milling machines used by this factory were horizontal, vertical and universal knee-type milling machines (model numbers 6M82, 6M12P and 6M82 Sh), all produced by the Gor'kii Milling Machine Factory.

2.2 As delivered, these machines were considered to be unsafe for several reasons. As in the case of the radial drilling machine previously referred to (see section 1.5 above), the machine was only provided with rubber hose in the place of adequate electrical conduit. Furthermore, the main motor guards did not comply with the relevant Factory Act and the main motors were not correctly earthed according to the relevant British Standard. The spindle stop electrical system was also considered to be unsafe, since the emergency stop did not appear to be automatically retained, and if not retained manually for a sufficient time for the transmission inertia to be removed, the spindle could run-on when manual pressure was removed from the stop button.

2.3 The alignment accuracies of these machines were considered to be adequate and the machines were found to conform to these accuracies, but the design of the machine dials prevented the machine from being accurately set repetitively, particularly in a vertical direction, and hence prevented the machine from being used to its best advantage.

2.4 Some machines were found to vibrate excessively at spindle speeds within the range of 1,000 r.p.m.

2.5 The machines were considered to be underpowered, thus preventing high metal removal rates from being obtained when these were required, with consequent detrimental effects on man and machine productivity.

2.6 Of the six milling machines purchased, only three were used under normal production conditions. Of these three, one was unusable at the time of the author's visit as the table had been removed for remachining of the slides because of scoring, possibly caused initially by casting porosity. On the remaining two machines, approximately 30 hours maintenance work had been required since the end of the six-month guarantee (approximately 12 months), the bulk of this time being devoted to problems of traverse which could be caused by problems in the traverse mechanisms themselves, or table lubrication, or the casting porosity problems previously referred to.

SURFACE GRINDING MACHINES

3.1 The Soviet-built surface grinding machines used by this factory were model number 3G71 Universal High Precision Surface Grinding Machines produced by the Krasnyi Borets Machine Tool Factory, Orsha.

3.2 One of the main sources of difficulty in the use of the machines was the possibility of losing spindle head alignment, a source of inaccuracy which could not be compensated for by the skill of the operator as in the case of the milling machine dials previously referred to (see section 2.3 above). The initial alignments of the machines were considered to be adequate as were the accuracy of components used in their construction, but the spindle bearings themselves were considered to be weak in construction, thus causing distortion, which consequently caused the temperature of the bearing arrangement to increase and spindle inaccuracy to be further increased through thermal distortion. The problem was further increased by difficulties encountered in adjustment of the rear thrust bearing to exact squareness, thus further aggravating the thermal distortion problem previously referred to.

3.3 The hydraulic unit for this machine was considered to be underpowered for the performance required from it, thus preventing maximum table feed rates from being obtained, consequently reducing the output rates obtained from these machines.

3.4 The device used for table way lubrication was not considered to be adequate, thus causing problems in service, including table way pick-up which may also have been aggravated by the casting of the components used in its construction.

3.5 The downtime for these machines since the six month guarantee period (approximately 12 months) had been extremely high, varying between 54–70 hours per machine.

Index of Names

Alexander, A. J., 414
Andrianov, K. A., 274
Aronovich, V. M., 281
Arrow, K., 6
Babadzhanyan, A. Kh., 438
Babbage, C., 377
Bacon, F., 240
Bek, A., 62
Bergson, A., 10–12, 14, Table 1.1
Beria, L. P., 452
Berliner, J. S., 21, 47, Table 1.1
Bernal, J. D., 271
Berry, M. J., 27, 267
Bitunov, V. V., 192
Boitsov, V. V., 164
Bolassa, B., Table 1.1
Boretsky, M., 15–21, 47–8, 121, 127, 129, 222
Boyle, R., 288
Braun, W. von, 459, 491, 496, 500, 506, 517
Brezhnev, L. I., 328
Brooks, H., 24
Brubaker, E. R., Table 1.1
Bulganin, N. A., 152
Burks, R. V., 21
Bush, K., 229
Butlerov, A. M., 288
Carothers, W. H., 242
Cherenkov, P. A., 451
Cohn, S. H., Table 1.1
Cooper, J. M., 547
Dalton, J., 240, 288
Daukas, A., 121, 126, 127–8
Davies, R. W., 27, 30
Davydov, N. I., 339
De Barr, A. E., 188
De Chardonnet, Count, 276
Denison, E., 5
Desai, P., 14, Table 1.1
Doncov, B., 387
Fedorenko, N. P., 249, 250, 283
Feoktistov, K. P., 514
Finney, B., 526
Flerov, G. N., 451
Foster, J., 408
Frank, I. M., 451
Freeman, C., 235, 242, 275
Fridenson, E. S., 447

Gagarin, Yu. A., 507, 513
Gallai, M., 513
Gleizer, L. A., 162
Glushko, V. P., 457, 492, 519, 522
Glyazer, L., 10
Goddard, R., 492, 496
Gomulka, S., 22
Griliches, Z., 5
Groettrup, H., 493, 498
Guderian, H., 419
Gwyer, J., 157, 526, 527
Hahn, 451
Hanson, P., 11–12, 29
Hemy, G., 233
Hess, G., 288
Hill, M. R., 523
Hinshelwood, C. N., 290
Hitler, A., 496
Hoog, D. C., 470
Hocke, J., Table 1.1
Holland, W., 382, 386, 388
Hufbauer, G. C., 25, 264, 272, 275
Hyatt, J. W., 272
Iklé, F., 483
Isaev, A. M., 519
Ivanchenko, V., 513, 516
Ivanov, V., 499
Ilyin, N., 492
Jewkes, J., 235
Joliot-Curie, F., 451
Jorgensen, D., 5
Judy, R., 43, 65, 379, 380, 388, 392, 397, 399, 402, 403
Kaldor, N., 6
Kapitsa, P. L., 38
Kavalerov, G. I., 353
Keldysh, M. V., 499
Khalepskii, I. A., 417, 418, 419
Khrushchev, N. S., 40, 152, 229, 232, 250, 273, 454, 469
Kipershlak, Z. F., 281
Kleimenov, I. T., 492
Klimenko, K. I., 256
Khunyants, I. L., 282
Komzin, B. I., 3, 8–9
Kondrat'ev, K., 499
Korolev, S. P., 455, 457, 472, 492, 496, 497, 499, 500, 501, 513, 517, 519, 520

INDEX OF NAMES

Korshak, V. V., 283
Kosharskii, B. D., 368
Koshkin, M. I., 418
Kostandov, L. A., 238
Kostousov, A. I., 529
Kosygin, A. N., 266
Kotkovskii, I., 9
Krieger, F., 522
Krylov, N., 458
Kuibyshev, V. V., 277
Kurchatov, I. V., 451, 452
Kyn, O., Table 1.1
Lapin, B. A., 237
Larionov, A. M., 386
Lavoisier, A. L., 240, 288
Lebedev, B. P., 209, 213
Leloir, L., 287
Leonov, A. A., 515
Lerner, A. Ya., 341, 342
Lewis, R. A., 27
Livshits, N. L., 282
Lomonosov, N. L., 288
Loskutov, V. I., 384
Lovell, B., 499
L'vov, D. S., 546
MccGwire, 414, 415
Maksarev, R., 363
Malenkov, G. M., 341
Mansfield, E., 7
Maxwell, D. C., 241
Melville, H., 271
Mendeleev, D. I., 288, 289, 338
Mets, A. F., 117
Mikhalevskii, B. N., Table 1.1
Minasian, J., 7
Mironov, V. D., 339, 355, 357, 358, 359, 369
Miroshnikov, L. P., 177, 191, 193
Mond, L., 240
Moorsteen, R. H., 13
Morozov, A. A., 418
Mostovenko, V. D., 419
Mul'chenko, Z. M., 293-4
Nabseth, L., 25, 31, 53
Nalimov, V. V., 293-4
Nelson, R., 3
Nobel, A., 289
Norris, K., 5
Nutter, W., 48
Oberth, H., 492
Ogorkiewicz, R. H., 422, 427, 435, 438, 441
Palterovich, D. M., 126, 127, 132, 138, 144, 145
Pavlovich, P. I., 274
Perkins, W. H., 241
Perry, R., 413
Pervukhin, M. G., 452
Petropavlovsky, B. P., 492
Petrov, B., 181
Petrov, G. S., 272, 273
Petrova, E. V., 256
Pobedonostsev, Yu. A., 493
Posner, M. V., 25, 264

Powell, R. P., 13
Rakovskii, M. E., 342, 344, 384
Ray, G. F., 25, 31, 53
Roebuck, J., 240
Rogovin, Z. A., 282
Romanov, A., 515
Rosenbrock, H. H., 348
Rotmistrov, P. A., 421, 442
Rudins, G., 400
Sakharov, A. D., 21-2, 23
Salmon, P., 536, 556
Sargent, J., 7
Schlesinger, G., 536, 556, 557, 558
Segal, G., 392
Semenov, N. N., 271, 451
Servan-Schreiber, J. J., 23
Seton, F., Table 1.1
Sheldon, C., 501
Shorygin, P. P., 277
Shteinberg, Sh. E., 368
Sláma, J., 36, 53, 157
Solov'ev, Yu. P., Table 1.1
Solow, R., 3-5
Sominskii, V. S., 237, 238, 239
Stalin, J. V., 9, 13, 42, 452, 472, 496
Staudinger, H., 242
Stefani, E. P., 339
Stoiko, M., 498, 503, 510
Sutton, A. C., 27-9, 31, 275-6, 285, 394, 407, 498, 526
Svecharnik, D. V., 342, 344
Tikhonsavov, M. K., 492, 493, 519
Tinker, G. C., 546
Tokaev-Tokaty, 497
Tolkachev, A., Table 1.1
Trapeznikov, V. A., 9, 341, 342, 344, 361
Trotsky, L. D., 407
Truman, President H., 452, 453
Tsander, F. A., 492, 493
Tsiolkovskii, K. E., 448, 491, 497, 516
Tsipis, K., 480
Tukhachevsky, M. N., 492, 496
Tupolev, A. N., 496
Ushakov, S. N., 273
Ustinov, D. F., 455
Vaizey, J., 5
Vandsheidt, A. A., 273
Vannikov, B. L., 452
Vladimirov, L., 496, 515
Vladzievskii, A. P., 164, 526
Vlasov, B. V., 551
Vogel, H., 36, 53, 158
Wagener, H. J., Table 1.1
Weitzman, M. L., 13-15, Table 1.1
York, H., 454
Zaitsev, B. F., 237
Zazulina, Z. A., 282
Zhimerin, D. G., 219, 363, 386, 390, 395, 396
Zinin, N. N., 288
Zuse, K., 377

Subject Index

Academy of Sciences, USSR, 238, 390; Economics Division, 253; Institute of Organic Chemistry, 283; Institute of Organic Compounds, 282; Technical Sciences Division, 341
Acetate fibre, 281, 282
Acrylic fibres, 255, 282
Aeroflot, Sirena computerised ticket system, 382
Aerospace industry, research intensiveness, 236; *see also* NC machine tools, development
Agglomerates, *see* Iron Ore
Alcatel, 163, 184
Algol, 389
Alloy and quality steels, 104–8; heat treatment, 104; output, 104; special steels, 105; stainless steel, comparative output, 104; *see also* Synthetic slag
American Interplanetary Society, 496
American Machinist, 526
Aminoplasts, 273
Ammonia, Haber process, 242, 248
Apatite, *see* Phosphates
Apollo space programme, 506, 507, 510, 513, 515, 519
Arms Control and Disarmament Agency, US, 483
Artificial fibres, *see* Manmade fibres
Artificial Fibres, All-Union Institute for (VNIIV), 281, 282, 284
ASEA, 219, 220
Assembler, 389, 390
Astrakhan, 181
ASU, *see* Automated Systems of Management and Control
ASU-Energiya, *see* MZTA; RP2
ASUTP, *see* Automated Process Control Systems
ASVT-M series of computers, second generation (M-1000, M-3000), 382, 389; third generation (M-4000, M-6000), 190, 191, 353, 364, 383–4, 389
Atomic bomb, Manhattan project, 453; Soviet development of, 451–3, 455; and launch vehicles, 501; policy and organisation, 451–3
AUS control system, 342–3, 351
Austria, 96
Author's certificates (*Avtorskie svidetel'stva*), *see* Patents, Soviet system of
Automated Process Control Systems (ASUTP), history and definition, 359–64; number of systems, 363; at Slavyansk power station, 364
Automated Systems of Management and Control (ASU), 1972 All-Union Conference on, 397; Branch systems (OASU), and Minpribor, 328, 363; and control instruments, see ASU-Energiya; ASUTP; software for, 390
Automation, in iron and steel industry, 118; *see also* Blast furnaces; Open hearth steelmaking; Rolling mills; in power stations, 367; Soviet level of, 367; *see also* ASUTP
Aviation industry, Ministry of, 167–8; *see also* Machine tools, stocks; NC machine tools, output
Avionics, 47, 62
Avtomatika i telemekhanika, 341
Azovstal', 83, 116

Bailetronic Mk 1 control system, 364–5
Bailey Meters and Controls Ltd., 339, 344, 364
Bakelite, 272–3
Balancing machines, 526
Ballbearings, machine tools for production of, 156
Ballistic missiles, American: Atlas, 235, 459, 461, 488, 500, 501, 519; Jupiter, 459, 507; Minuteman, 461, 468, 477, 478, 480, 482, 488; MX, 486; Polaris, 477; Redstone, 459, 507; Thor, 459, 500, 507, 519; Titan, 461, 480, 482, 488, 500, 519; and backwardness of electronics industry, 472, 475, 477, 489; deployment, 463–8; design philosophy, 487; FOBs, 463, 478; general, 446–7; mobile, 480; organisation of programmes, 488; priorities, 459; propellants, 477–8; retirement and withdrawal, 488; Soviet: SS-3, 458; SS-4, 458; SS-5, 503; SS-6, 458–9, 488; SS-7–8, 461, 480; SS-9, 461–7, 469, 480, 481, 482, 503; SS-11, 461–7, 477, 478; SS-16–19, 468, 469, 473, 480, 481, 482, 486, 488; submarine-launched (SLBMs), 475, 476, 489; targeting, 447–8; technological level, 46–7; technological parameters: accuracy, 470–3, 480–1, 486, countersilo lethality, 480–2, multiple warheads (MIRVs, MARVs), 461, 468, 475–7, 486, penetration aids, 478, throwweight, 474–5, 486–7, yield, 453–4, 468–70, 486; *see also* Computers; Missile silos; Nuclear weapons; Rocket engines; Rockets
BASF, 236, 242
Bayer, 236, 241
Belgium, 100
Bell Telephone Laboratories, 235, 336

SUBJECT INDEX

Bench and floor grinding machines, *see* Grinding machines
Bending and forming machines, *see* Machine tools, metalforming
Bendix, 182
Benzene, 244, 249
BESM-6 computer, 387, 398, 399
Bezhitsk works, 101
Blast furnaces, 86–91; age structure, 89; automation of, 118; blast enrichment, 90; blast temperature, 89; coke consumption, 90–1; computer control, 91, 118; design philosophy, 62; efficiency of operation, 89; evaporative cooling, 91; level of mechanisation, 91; size of, 40, 86–9; use of natural gas, 90; use of raised pressure, 89–90; withdrawal of, 89
Boche, 159
Boring machines, average weight, 157; jigboring, imports, 153; production capacity, 153; NC, *see* Machine tools, stocks; NC machine tools, output; proportion of machine tool stocks, 174
Braun, W. von, 459, 491, 496, 500, 506, 517
Brest Electro-Mechanical Plant, 385
Brezhnev, L. I., 328
British National Lending Library for Science, 322
Broaching machines, 154
Bulganin, N. A., 152
Butlerov, A. M., 288

CAMAC, *see* Control instruments
Canadian Forces Maritime Warfare School, 412
Capital retirement, rate of, 58
Cars, output, 550; product mix, 550; technical characteristics, 551–6, and environmental conditions, 551, and export markets, 551; *see also* Moskvich; Motor industry; Volga; Zhiguli
Cast iron, *see* Ironmaking
Celluloid, 272, 273
Central Statistical Administration (TsSU), 123, 397
Centre National de la Recherche Scientifique, 242
CES, *see* Production functions
Chaika motorcar, 550
Charentsavan Boring Machine Factory, 168
Chelyabinsk Metallurgical Factory, 105
Chemical industry, distinctive features of, 229–39: capital and investment, 229–31, consumption pattern, 228, labour, productivity, 230–2, structure, 232–3; R and D, role and cost of, 234–5; rate of product renewal, 238–9; relative rate of growth, 234, 245; research intensiveness, 236–8; role in industrial economics, 229–31, history, 240–3; and industrialisation, 245; innovation in, and competitiveness, 263–4, costs and risks, 235; output, role of imported plant, 263, technological slant of, 297; pre-Revolutionary, 244; priority, 42–3; technological level, 42–3; indices of, *see* Petrochemical industry; synthetic materials; *see also* Chemical research; Chemical technology; Inorganic chemicals; Inventions; Organic chemicals; Petrochemical industry; Plastics and resins; Synthetic materials; UNIDO
Chemical intensity, of Soviet economy, 229
Chemical research, centres of, 290–1; level of, 287–94, citations as index of, 291–4, 320–5, problems of assessment, 287–8; Nobel Prizes and, 39, 288–91; outstanding contributions to, 288–90
Chemical technology, diffusion, 240–65; history, 240–3; Soviet dependence on imports, 262–3, 266, 297, *see also* Imports
Cherepovetsk factory, 95, 116, 117
CIGRE, 221
Cincinnati Milacron, 182
Cincinnati Milling Machines, 182
Cinematography, State Committee for, 397
Circuit breakers, design effort relative to voltage increase, 203; for EHV, 206–7, 700–765 kV, 213–14, Soviet development of, 206; function of, 200; for HVDC, 204, 218–9; and system reliability, 203
Citations, *see* Chemical research
CMEA (Council for Mutual Economic Assistance), *see* Comecon
Cobb-Douglas, *see* Production functions
Cobol, 388, 389, 390
Coborra-Bassa scheme, *see* HVDC
Co-com Strategic Embargo, *see* Computers; NC machine tools, development
Comecon (CMEA; SEV), comprehensive programme of, 170; and ES-series computer project, 384; Statistical Yearbook, 249, 253; structure of machine tool production, 146; *see also* Czechoslovakia; GDR; Machine tools, metalcutting; NC Machine tools, programming
Commerce, US Department of, 156
Commercial production, problems of definition, 275
Communications, Ministry of, 387
Communist Party, Central Committee, 23, 27, 118, 277, 418; XXIII Congress (1966), 266; XXIV Congress (1970), 328, 353
Computers, in ballistic missiles and spacecraft, 472, 477, 489, 515, 516, 522; and Co-com Strategic Embargo, 403; demand for, 380; design history, 378; diffusion, 397; impact of, 377; innovation in, 403; lag in production, 60–1; organisation of production, 403; output of, and stocks, 391–5; peripherals, 386–7, disc units, 387, 401, line printers, 387, 389, 402; production and supply problems, 396–7; priority, 43–5, 402; problems of comparison, 379–81; and process control, *see* ASUTP; second generation, *see* ASVT-M; K-200; Minsk; Ruta; service facilities, 396; Soviet dependence on foreign technology, 65; technological lag, 43–5, 397–400, 401–2; third generation, *see* ASVT-M; ES; Nairi-3; time-sharing, 387–8; *see also* Software
Conservatism, technological, 58; and launch vehicles, 519; and petrochemicals, 250
Consumer goods, and steel production 108–10
Consumers' Association, 553; *see also* Moskvich
Continuous casting of steel, 101–4; comparative growth of output, 102; diffusion, 119; general, and history, 40, 60, 102, 103; Soviet development of, 101–2
Control Computers, Severodonetsk Research Institute for, 382
Control and instrumentation industry, economic importance, 328; innovation in, 337–8, 344; organisation of, 342; output, proportion of total industrial, 328; number of systems produced, 350; Soviet, development of, 338–9; and transistor technology, 351; *see also* Control instruments; Minpribor
Control instruments, 336, 369, 370; amplifying elements: integrated circuits, 336, magnetic, 336, 369, transistors, 336, 342, 343, 351, 369, valves, 336, 343, 369; analogue, pre-1970, 338–352; development, 338–9, 364–5, *see also* MZTA; RP2; State Systems of Instruments; USEPPA; VTI; influence of electric power industry, 338–9, diffusion, of, 337–8; electronic: advantages and drawbacks, 334, level of transistorisation, 351; SUPS system, 351, *see also* Bailetronic Mk I; EAUs; MZTA; RP2; State Systems of Instruments; VTI; environmental conditions, 368–9,

370; innovation in, 368; intrinsic safety, 334, 335, 370; lags vs UK, 369–70, 371–2; leads vs UK, 370–2; minor systems, 349; physical construction, 369, 370; pneumatic, advantages and drawbacks, 334, *see also* AUS, USEPPA; policy review, 341–2; rate of replacement 338; reliability, 368; signals, 334, 343, 345, 358–9, 369, 370; standardisation, 368; Soviet technical backwardness, 44–5
Converters, AC/DC, development of, 219–20; and HVDC, 204, 215, 217; mercury-arc, 217–20, 223–4; thyristor: advantages and limitations, 219–20, diffusion of, 220, Soviet lag in, 220, 223–4
Cooperation agreements, 170
Council for Labour and Defence, 277
Courtaulds, 277
CPSU, *see* Communist Party
Cuprammonium fibre, 281
Cutter grinders, *see* Grinding machines
Cutting-off machines, *see* Other machines
Czechoslovakia, 10, 117, 170

Defense, US Department of, 156, 408, 409, 506
Delle Asthom, 64, 206, 214
Demag, 61, 104
Detergents, output, 261
Diebold Institute, 392
Diffusion of technology, comparative studies of, 28; as index of technological development, 25; Soviet pattern of, 58; *see also* Chemical technology; Computers; Control instruments; Converters, AC/DC; EHVAC; Group Technology; HVAC; Launch vehicles; Manmade fibres; NC machine tools; Oxygen steelmaking; Petrochemical industry
Direct digital control, *see* Automated Process Control Systems; NC Machine tools, control systems
Direct reduction of iron ore, *see* Ironmaking
Dnepr-1 computer, 382
Donbass-West Ukrainian 750 kV line, 211
Drilling machines, NC, 174; proportion of machine tool stocks, 174; reliability and accuracy, 561–2
DSIR, 242
Du Pont, 243, 270
Dyestuffs, German preeminence in, 241; Soviet dependence on imports, 261
Dynamo steel, *see* Rolled steel, electrical

East Germany, *see* German Democratic Republic
EAUS control system, 343, 351
Economic efficiency, 10–13, Table 1.1; obstacles to, 12
Economic growth, rates of, 9–10, 13, 52, Table 1.1; sources of, 9–10, 52; *see also* R and D
EDSAC, 377
EHVAC (400–800 kV), 206–15, 222–3; general: diffusion, 207, 211–12, 214–15, as index of technical progress, 222, rates of, 222, 223; insulators for, 206; Soviet innovation performance, 222, 224; Soviet technical self-sufficiency, 213–14, 222, 223; 400 kV, introduction of, 205; Soviet development of, 206–7; 500 kV: capabilities and demands on, 209–10, 223; diffusion, 214–15; as international standard, 208; introduction of, 204, 205; relative costs, and adoption, 205, 207; system-linking lines, 204, 211; 700, 750 kV: Canadian introduction of, 211–12; capabilities and relative advantages, 209–10; as Soviet standard, 208; Soviet transmission lines, 208; 765 kV, in USA and relative costs, 212
Ekibastuz-Tambov DC line, *see* HVDC

Electrical Industry, Ministry of, *see* Electrotechnical Industry, Ministry of
Electrical power systems, general, 199; level of interconnectedness, 201; pre-Revolutionary, 201; Soviet, *see* GOELRO; Interconnected Power Systems; Regional Power Systems, United Power System
Electric arc steelmaking, 100–1; economics of, 100–1; share of total steel output, 100; size of furnaces, 100, 103, 119
Electric Drive Units, Novosibirsk Scientific Research, Project-Design and Technological Institute of, (NIIKE), 181
Electric power industry, and control instruments, *see* Control instruments; MZTA; RP2; State Systems of Instruments; criteria of technological level, 41–2, 199; *see also* HVAC; HVDC
Electric Power, Ministry of (formerly Ministry of Power Stations), 339
Electronics industry, backwardness of, 45, *see also* Ballistic missiles; research intensiveness, 236; *see also* NC machine tools, development
Electro-slag remelting, *see* Remelting of steel
Electrotechnical Industry, Ministry of (formerly Ministry of Electrical Industry), 342
Elektroapparat Factory, 206
Elektron Factory, 383
Elektron satellite, 512
Enant, *see* Nylon
Energopribor Factory, 339
Engineering, 192
ENIAC, 377
ENIMS, *see* Metalcutting Machine Tools, Experimental Scientific Institute for
ES (Ryad) series of computers, 187, 384–6, 388, 402; organisation of development and production, 384–5; technical problems, 386; *see also* Software
Evershed and Vignoles, 339
Excello, 182
Exports (from USSR), of chemicals, 260–1; of chemical technology, 262; of metalcutting machine tools, 157–9; *see also* Foreign trade
Extra-High Voltage Alternating Current, *see* EHVAC

Factor endowment, 11
Factor productivity, growth of, Table 1.1; as index of technological level, 3
Ferranti, 163
Ferro-alloys, *see* Ironmaking
Ferrous Metallurgy, Ministry of, 83, 104, 118
Fertilisers, economic priority of, 248; output, nitrogenous, 248–9, superphosphate, 248; raw material base, *see* Phosphates
Fiat, 551
Fighter aircraft, MiG-21, 414; Su-7, Su-9, 414
Finsider, 112
Five-year plans, 1st (1928–32), 83, 201, 273, 338; 2nd (1933–7), 201, 273, 338, 418; 3rd (1938–41), 273; 8th (1966–70), 116, 391; 9th (1970–5), 132, 142, 168, 170, 174, 220, 234, 391, 550
Fluorine fibres (Ftorlon), 282, 284
Fondootdacha, *see* Recoupment
Foreign technology, Soviet dependence on, general, 29–30, 63–6; *see also* Chemical industry; Chemical technology; HVAC; Imports; Iron and Steel industry; Machine tool industry; Rolling mills; Technology transfer
Foreign trade, in chemicals: balance and breakdown, 259–60, as index of diffusion of chemical technology, 258–9, manmade fibres, 261, miscellaneous chemicals 261, plastics, 261; in chemi-

cal technology, as index of diffusion of chemical technology, 258; composition, 28; as index of technological level, 146–59; in metalcutting machine tools, 146–7, 152–9, 527; *see also* Exports; Imports
Forgings, 84
Fortran, 389, 390
Foundry equipment, import of, 153
Fujitsu-Fanuc, 163, 181

Galalith, 272, 273
Galvanised steel, coating technology, 116
Gas Dynamics Laboratory (GDL), 448, 449, 457, 458, 492, 493
GDL, *see* Gas Dynamics Laboratory
GDR, *see* German Democratic Republic
Gear Cutting machines, 526, 530; demand for, 154
Gemini spacecraft, 515, 516, 522
General Electric, 163, 182, 220
German Democratic Republic (GDR), 10, 117; and NC machine tools, 170–1, 190; trade in chemicals, 261
Gidding and Lewis, 182
Gidroavtomatika Experimental Factory, 186
Gidroprivod Factory, 186
GIPROIV, 281, 282
Gipromez, 98
GIRD, *see* Group for the Study of Reactive Motion
Glavnaya Palata Mer i Vesov (Chief Board of Weights and Measures), *see* Metrology, All-Union Research Institute for
Glavtyazhstankoprom, *see* Heavy Machine Tools, Chief Administration for
Goddard, R., 492, 496
GOELRO plan, 201
Gor'kii Milling Machine Factory, 162, 169, 170, 183, 196, 548, 558, 559, 562
Gosplan, *see* State Planning Commission
Gossnab, 397
Gosstandart, *see* Standards, State committee of
GOST, *see* State Standards
Grinding machines, 530, 547, 563; accuracy, 563; Bryant Centalign B and LZIO, 156–7; and polishing machines, proportion of Soviet–UK trade, 152; precision, 549; proportion of machine tool stocks, incl. bench and floor grinders, polishing machines, sharpening machines, tool and cutter grinders, 133; reliability and performance, 563; tool and cutter grinders, output, 154; *see also* Machine tools, demand for
Group for the Study of Reactive Motion (GIRD), 448, 492, 493
Group Technology, rate of diffusion, 54
Growth, economic, *see* Economic growth

Hammers, *see* Machine tools, metalforming
Heavy Industry, People's Commissariat of, 451; Scientific Research Sector of, 449
Heavy Machine Tools, Chief Administration for (*Glavtyazhstankoprom*), 158; trade in, 158–9
Hicks, *see* Production functions
Higher Education, as index of technological development, 24; Ministry of (Minvuz), 390
High-technology industries, concept of, 45; development of, and innovation in, 47; *see also* Science-based industries
High Voltage Alternating Current, *see* HVAC
High Voltage Direct Current, *see* HVDC
Hitler, A., and German rocket programme, 496
Hoechst, 236, 241
Humphreys and Glasgow, 251
Hungarian Machine Tool Research Institute, 171

HVAC, general: development of, 42, diffusion, 60, role of Western technology, 64; 400–800 kV, *see* EHVAC; 800 kV and above, *see* UHVAC
HVDC, costs of vs HVAC, *see* Reactive power; Resistive power losses; development of, 42, 215, and location of Soviet energy resources, 215–6, 223; disadvantages vs HVAC, 204, *see also* Circuit breakers; Converters; main applications, 204; Soviet research in, 216; Soviet and Swedish lead in, 223; switching of, 219; transmission lines: 203–4, Coborra-Bassa scheme, 218, Ekibastuz-Tambov 750 kV, 216, 221, intersystem, 215; transformers, 219; *see also* United Power system
Hydrogen bomb, 458; development, 453–5; yield, *see* Ballistic missiles

IAT, 348, 353
IBM, 386, 389, 391
ICBMs, *see* Ballistic missiles
ICI, 228, 243, 270
I.G. Farben, 236, 243
Il'ich Works, 115
Imports (to USSR), of chemicals, 260–1; of chemical technology, 249, 262–3, 266; of metalcutting machine tools, 146–52; *see also* Foreign trade
Industrial production, problems of definition, 61
Industry, sectors of, *see* e.g. Iron and Steel industry
Innovation, as index of technological development, 25; *see also* Ballistic missiles; Chemical industry; Computers; Control instruments; High technology industries; Launch vehicles; Military technology; NC machine tools, development; Traditional industries
Inorganic chemicals, *see* Fertilisers, Soda ash; Sulphuric acid
Instrument Building, Means of Automation and Control Systems, Ministry of (Minpribor, formerly Ministry of Machine and Instrument Building), 168, 181, 328, 342, 353, 354, 356, 382, 383, 384, 390, 396, 397; economic performance, 328; proportion of instrument output, 328; Soyuzpromavtomatika, 355, 397; Soyuzsistemkomplekt, 397; *see also* ASU
Insulators, *see* EHVAC
Integrated circuits, 336, 378, 381
Integrated complex machining systems, 190–1
Interconnected Power Systems, early development of, 201; numbers, 201; post-war construction, 202; and Second Five Year plan, 201–2
Intermediate industries, concept of, 42
International Electrotechnical Commission, 208
International Standards Organisation, 546
Inventions, in chemical industry, 297; and use of patents, 265–6
Inventions and Discoveries, State Committee for, 266
Ironmaking, 84–92; cast iron, 84; direct reduction of iron ore, 91–2; ferro-alloys, 84; KShS process, 92; pig iron, 84; *see also* Blast furnaces; Iron ore
Iron ore, 85–6; agglomerates, 85; direct reduction of, *see* Ironmaking; iron content, 85; output of, 85; pellet production, 86; sinter, 85, 118; *see also* Blast furnaces, Ironmaking
Iron and steel industry, administration, *see* Ferrous Metallurgy, Ministry of; development, 83; general, 39–41; organisation of research, 62; output, 83, 84, 92–3; priority, 39–40; role of Western technology in, 64; size of, 83; *see also* Ironmaking, Steelmaking
Isolation, and voltage increase, 203
Izhevsk Automobile Factory, 550
Izvestiya, 555

Jigboring machines, *see* Boring machines

K-200 computer, 382–3
Kaiser Wilhelm Gesellschaft, *see* Max Planck Institutes
Kapron, *see* Nylon
Karaganda Combine, 116
Karpov Physico-Chemical Institute, 281
Katyusha rocket artillery, 449, 492, 496
Kazan, 251
Kendrick, *see* Production functions
Key innovations, definition of, and technical progress, 15–21
Khartsyzsk Tube Works, 112
Khimkoemkost', *see* Chemical intensity
Khrushchev, N. S., 40, 152, 454, 469; chemicalisation drive, 229, 232, 250, 273
Kiev Institute of Automatics, 181
Komega Factory, *see* Moscow Factory for Thermal Automation
Kooperakhimiya Factory, 274
Korolev, S. P., 455, 459, 472, 492, 496, 497, 499, 500, 501, 513, 517, 519, 520
Kosior Factory, 164
Kosmos satellites, 503, 512
Krasnoe Sormovo Works, 102
Krasnyi Borets Factory, 549, 563
Krasnyi Oktyabr' Works, 101
Krasnyi Proletarii Machine Tool Factory, 168, 169, 170, 171, 174, 183, 196, 529, 534, 535, 539, 548
Krivoi Rog, 83
Krupp, 92
Krzhizhanovskii Energy Institute, 220
KShS process, *see* Ironmaking
Kuibyshev, Hydroelectric plant, 205; –Moscow transmission line, 205, 207
Kuibyshev, V. V., 277
Kurchatov, I. V., 451, 452
Kursk, 92

Labour productivity, as index of technological level, 3; and technical progress, 20, 21; and unemployment, 11; *see also* Chemical industry, distinctive features; Rolling mills
Lada, *see* Zhiguli
Lathes, 534, 535; accuracy of, 536, 538–9, 557–8; automatic, *see* Machine tools, demand for; centre, proportion of exports, 157; NC, 174; proportion of machine tool stocks, 133; relieving, 153; turning mills, average weight, 157
Launch vehicles, American, *see* Ballistic missiles; criteria of technological level, 45; development, 501, 519–20; diffusion, 55–7; exploitation, 507–10, 519–20; innovation in, 519–20, 522; problems of comparison, 510; thrust, and design philosophy, 501, Soviet leads and lags, 500–1, 519–20; *see also* Ballistic missiles; Proton; Rocket engines; Rocketry; Vostok
Lavsan, *see* Terylene
L-D process, *see* Oxygen steelmaking, 92
Lebedinsk Mining and Concentration Works, 92
LEMZ, *see* Leningrad Electro-Mechanical Factory
Leningrad Electro-Mechanical Factory (LEMZ), 181, 183
Leningrad Physico-Technical Institute (LFTI), Laboratory of Neutron Physics, 451
Leningrad Polytechnical Institute, 162
Leningrad *sovnarkhoz*, 179
Leningrad Special Design Bureau of Machine Tool Building, 181
Leningrad Technical Institute, 273
Lenin Komsomol Automobile Factory, 550, 555
Lenin Prize, 348
Lenin Volga Hydroelectric Power Station, 204
Leontief paradox, 25
LFTI, *see* Leningrad Physico-Technical Institute
Licenses, Soviet trade in, 266; for USEPPA control system 348; *see also* Patents
London Patent Office, 268
Luna space probe, 512
Lunokhod, *see* Space programmes
L'vov Milling Machine Factory, 168
L'vov TV Factory, 387

Machine and Instrument Building, Ministry of, *see* Instrument Building . . . Ministry of
Machinery, 525
Machine Science, Institute of, 162
Machine tool industry, history, 121; organisation of production, 140–1, 168–70, 181–2; Soviet, and role of foreign technology, 64–5; *see also* Machine tool and Tooling Industry, Ministry of
Machine tools, accessories, 527–8
Machine tools, demand for: effect on technological level, 145; and foreign trade, 153; grinding and polishing, 154; lathes, 144–5, 154; non-progressive and progressive, 143–4; official, and Gosplan, 143, 144; precision, 144; and Seven Year Plan, 153; and technological conservatism, 144
Machine tools, metalcutting, NC, proportion of output, 173; output of, 122; quality, 155–7; trade in: balance and breakdown, 152, 157–9, with capitalist countries, 146–7, 152–9, 527, with Comecon, 147, 157, 527, with Yugoslavia, 157, *see also* Exports; Imports; *see also* Boring machines; Broaching machines; Drilling machines; Gearcutting machines; Grinding machines; Lathes; Machining centres; Milling machines; *Other machines*
Machine tools, metalforming, imports, 153; proportion of machine tool stocks: bending and forming machines, hammers, presses, riveting machines, 137–8, as index of technological level, 129, and Ninth Five Year Plan, 132
Machine tools, output, technological structure of, 144–5; by value, 127–9; by weight, 126–7
Machine tools, quality of, 58–9, 523–50, accuracy, 528, 529, 536–45, components, 528, 529, 548, and design, 528–9, 547, influence of Western technology, 526, reliability and durability, 528, 529, 547–50, Soviet comments on, 527–30, speeds and feeds, 545–6, standard of manufacture, 528–9, 529–30, 547, vibration, 546–7, Western comments on, 524–7, 530; *see also* Drilling machines; Grinding machines; Lathes; Milling machines
Machine tools, stocks, age structure, 125; breakdown: by industry, 122–3, methodological problems, 139, studies of, 132, by type, 132, *see* Boring machines; Grinding machines; Lathes; Machine tools, metalforming; *Other machines*; by weight and value, 126–9; differences between, 139–46; and foreign trade, 154–5; of NC machine tools: in aviation industry, 176–7, in machine tool industry, 174–5, 177–8, proportion of machining centres, 178; pattern of use, 122, 123; size of, 122
Machine tools, unit construction, and NC, 165
Machine Tool and Tooling Industry, Ministry of, 137, 167, 546
Machining centres, attitudes to, 181; development, 179–81; GSP machine, 162; Milwaukee Matic II, 179; *see also* Machine tools, stocks; NC machine tools, output
Magnitogorsk, 40, 95, 115

SUBJECT INDEX

Malenkov, G. M., 341
Manhattan project, *see* Atomic bomb
Manmade fibres, artificial: first commercial production, 281–2, quality, 283, *see also* Acetate fibres; Cuprammonium fibre; Nitrocellulose fibres; Viscose silk; diffusion of, 53, 60; first industrial production of, 276–85, priority of, 277, research, quality of, 284, Soviet technological backwardness in, 283–4; synthetic: first commercial production, 281–2, growth and structure of output, 254–5, 283, petroleum-derived feedstocks, 249–50, quality, 283, Soviet dependence on imports, 261, *see also* Acrylic fibres; Fluorine fibres; Nylon; Polyester fibres; Terylene; Vinitron; Western technology and Soviet production, 284–5
Mannesmann, 61
Manometer factory, 347
Manpower, scientific, as index of technological development, 24; trends in development of, 27
Mark of Quality (*Znak kachestva*), *see* Standards, State Committee of
MARVs, *see* Ballistic missiles
Massachusetts Institute of Technology (MIT), 161
Mathematical Machines, Erevan Scientific Research Institute for, 383
Mathematics, Belorussian Institute of, 388
Max Planck Institutes, 242
Mechanisation, *see* Blast furnaces
Melman Report, 527
Mendeleev Chemico-Technological Institute, 273, 274
Mercury-arc converters, *see* Converters, AC/DC
Mercury spacecraft, 515, 522
Metalcutting Machine Tools, Experimental Scientific Institute for (ENIMS), 526, 533, 546; machine tool surveys, 527, 534; and NC machine tool development, 162, 164, 170, 181, 187, 190
Metrology, All-Union Research Institute for (VNII Metrologii, formerly Chief Board of Weights and Measures), 338
Middle East War (1973), 429, 439
Military technology, design philosophy, 414–5, 489, *see also* Tanks; Ballistic missiles; diffusion, 489; innovation in, 489; level of, 45–7; and military effectiveness, 413, Western assessments of, 407–16; overlap with civilian technology, 416; R and D, effectiveness, 413, effort, 416; and Soviet system, 489; *see also* Ballistic missiles; Naval technology; Tanks
Milling machines, 547, 548–9; accuracy of, 536, 537–8, 539, 556–7, 562; imports of, 153; NC, 174; proportion of exports, 157; proportion of NC stock of machine tool industry, 178; reliability and performance, 562; safety, 562; *see also* NC machine tools, output
Ministries, *see under relevant sector of industry*
Minpribor, *see* Instrument Making, Means of Automation and Control Systems, Ministry of
Minradprom, *see* Radio Industry, Ministry of
Minsk computer, 118, 187, 361, 364, 382, 387, 388, 396, 397, 402
Minvuz, *see* Higher Education, Ministry of
Mir computer, 118
MIRVs, *see* Ballistic missiles
Missile silos, 459, 479–80
MIT, *see* Massachusetts Institute of Technology
Mogilev Polyester Fibre Plant, 284
Molins, 190
Molniya satellites, 512
Molotov Technical Institute, 206
Moscow Factory for Thermal Automation (MZTA, formerly Komega Factory), 339, 343, 347, 349, 351, 357–8
Moscow Machine Tool and Tooling Institute (Mosstankin), 162
Moskvich motorcar, 550; output, 550; Renault technical assistance, 553; safety, and record, 554–5, improvements to, 555; technical characteristics, 551, *Motor* magazine test, 553–4, Consumers' Association test, 554–5
Mosstankin, *see* Moscow Machine Tool and Tooling Institute
Motor industry, machine tools for, 156; and steel production, 108
Motor magazine, *see* Moskvich
Munitions, People's Commissariat of, 452
MZTA control system, 339, 351; and ASU-Energiya, 358; development by electric power industry, 343; production of, 343; technical characteristics, 343
MZTA Factory, *see* Moscow Factory for Thermal Automation

Nairi 3 computer, 118, 383, 387, 388
Naphthalene, 249
NASA, 519
National Grid (UK), 201
Natural gas, in iron and steel production, *see* Blast furnaces; Open-hearth steelmaking
NC machine tools, control systems: 181–91, 547; closed loop, 185–6; computer numerical control, 189–91; development, and cooperation agreements, 170, lags in production and design, 182, relationship between electronics and engineering sectors, 163–4, 182, research and, 181, semiconductors and ICs, 182–3, 197, 198; direct numerical control, 189; foreign, Soviet use of, 184, 197; positioning and contouring, 174, 175, 184–5; production of, 181; recent developments in, 188–91; technical policy, 185–6; *see also* Integrated complex machining systems
NC machine tools, development: and aerospace industry, 161–2, 165–7, 168; and Co-com Strategic Embargo, 167; and Comecon, 170–1, 198; first Soviet prototypes, 162; general, 161–71; innovation, 194–6; organisation of, 164, 183; paths of, 193–4; *see also* ENIMS; Machining centres
NC machine tools, diffusion, 54, 196, 198
NC machine tools, output, by aviation industry, 175; product range, 168–70, 174; by volume: growth, 163, 168, 170, 171; proportion of boring machines, 175; proportion of lathes, 178; proportion of machining centres, 180–1; proportion of milling machines, 175
NC machine tools, programming, codes and media, 183; Comecon and development of, 188; computer-aided, 187–8, 196–7; problems of 186–7
NC machine tools, stocks, *see* Machine tools, stocks
NC machine tools, usage, concentration of stock, 192; organisational problems, 197–8; and prices, 192; problems of, 191–2; shift operation, 192
NEP, *see* New Economic Policy
New Economic Policy (NEP), 245, 272, 273
NIAT, *see* Technology and Organisation of Production, Scientific Research Institute for
NIIKE, *see* Electric Drive Units, Novosibirsk Scientific Research, Project Design and Technological Institute of
NIITEKHIM, 253
NIITeplopribor, 342, 347, 351, 357
Nitrocellulose fibres, 276

Nitrogen Industry, State Institute for, 282
Nitrogenous fertilisers, *see* Fertilisers
Nobel Prizes, history and administration, 289–90; *see also* Chemical research; Swedish Academy of Sciences
Nonferrous Metallurgy, People's Commissariat of, 452
Novoe naznachenie (*The New Appointment*, novel by A. Bek), 62
Novo-Lipetsk, 98–9, 102, 103, 115, 116
Novo-Tula Factory, 83, 102
Nuclear power, Soviet lead and lag vs USA, 60
Nuclear weapons, *see* Atomic bomb; Hydrogen bomb
Numerically controlled machine tools, *see* NC machine tools
Nylon, 281; -6 (Kapron), 282; -7 (Enant), 282

OASU, *see* Automated Systems of Management and Control
Obsolescence, Soviet rate of, 58
Odessa Radial Drilling Machine Factory, 561
OECD, 34, 163, 377, 378; studies of technological level, 24, 26
Okhtinsk Chemical Combine, 272, 273
Olivetti, 159
Open-hearth steelmaking, 93–6; furnaces, computer control, 118, design philosophy, 62, efficiency, 96, 99, oxygen enrichment, 95, size, 40, 94–5, two-bath, 95, use of natural gas, 95; production trends, 93
Ordzhonikidze Computer Factory, 382, 385
Ordzhonikidze Machine Tool Factory, 171, 180
Organic chemicals, feedstocks, 249–50; history of, 241–2, output, *see* Benzene, Naphthalene; Phenol; Phthalic anhydride; Xylene; problems of data, 249; Soviet lag in, 250
Organisation and Management, All-Union Scientific Research Institute for Problems of, 386
Orgstankinprom, 192
Orlon, *see* Acrylic Fibres
Osoaviakhim, *see* Society for Assistance to Defence, Aviation and Chemistry
Osteuropa Institut, 28
OST standards, *see* State Systems of Instruments
Other machines (including Cutting-off machines, threading and tapping machines, unit head machines), proportion of machine tool stocks, 136–7
Oxygen steelmaking, 96–100; converters, automation, 97–8, economies of scale, 98, L-D process, 96, 99, Q-BOP process, 99, SIP process, 100, size of, 97–8; diffusion, 52, 60, 96–7, 99, 119; general, 40; *see also* Scrap

Partial Test-Ban Treaty, 475
Patents, 265–70, 496; British, numbers granted to Soviet organisations, 268–70; definition of, 265; German system of, 268, 270; as index of national inventive activity, 265; Soviet, numbers of, 267–8, system of, 266–7
Paton Welding Institute, 106, 107; number of foreign patents, 267
Peenamunde, 455, 496–7
Pegasus space programme, 506, 507
Pelletisation of iron ore, *see* Iron ore
People's Commissars, Council of, 452, 455, 488
Perspex, 243
Petrochemical industry, as index of technological level, 243–4; and polymer revolution, 243; Soviet, development of, 250–1; technology, and diffusion of, 249, 250–1; *see also* Conservatism, technological

Pheno-formaldehyde resins, 253
Phenol, 249
Phosphates, Soviet resources, 248
Phosphorite, *see* Phosphates
Phthalic anhydride, 249
Physics, Institute of, 162
Pig iron, *see* Ironmaking
Pipes, *see* Tube and pipe making
PL/1, 390
Planing machines, average weight, 157
Plastics and resins, development of Soviet R and D, 273–4; first commercial production of, problems of assessment, 275, 275–6, Soviet performance in, 275, *see also* Aminoplasts; Polyethylene; Polystyrene; Polyvinyl butyral; PTFE; PVC; Silicon polymers; and five year plans, 1928–37, 273; growth of output, 252–3; lag in production, 253; structure of production, 252–3; traditional, continued output growth, 253; *see also* Synthetic materials
Plessey, 163, 182, 184
Polet spacecraft, 510, 512, 516
Polishing machines, *see also* Grinding machines
Polotsk, 251
Polyester fibres, 254–5
Polyethylene, 243, 274; import of plant, 263
Polymerised plastics, Institute of, 274
Polyolefins, 252
Polyspinners Consortium, 263
Polystyrene, 252, 274
Polyvinyl butyral, 274
Power Stations, Ministry of, *see* Electric Power, Ministry of
Pravda, 267, 524
Precision machine tools, 530, 539; output, 144
Presses, *see* Machine tools, metalforming
Pressings, 84
Process control, basic principles, 329–36
Process control instruments, *see* Control instruments
Production functions, and analysis of economic growth, 13; methodological problems, 14–15; and R and D, 6–7; and technological level, 3; types of, CES, 14, Table 1.1, Cobb-Douglas, 9, Table 1.1, Hicks, Table 1.1, Kendrick, Table 1.1
Programming, *see* Software
Proton launch vehicle, 507, 516, 519, 520; thrust and payload capability, 503
PTFE, 275
Purges, *see* Rocketry
PVC, 252, 274

Q-BOP process, see Oxygen Steelmaking
Quality Certification, State System of, *see* Standards, State Committee of
Quality of Soviet production, see Cars, technical characteristics; Machine tools; Standards, State Committee of; State Standards
Quality steel, see, Alloy and quality steel

Radio Industry, Ministry of (Minradprom), 382, 386, 388, 390
Rand Corporation, 36, 47, 62, 387, 410
R and D, economic growth, 7, 33–4; efficiency, 33–4; effort and Frascati definition of, 26; expenditure, 7–8, 24; as index of technological level, 7–8, 24; organisation, 62; return on, 10; *see also*, Chemical Industry; Military technology; Plastics; Production Functions
Reactive power, 203, 204
Reactive Research Institute (RNII), 448, 449, 493
Recoupment (*fondootdacha*), as measure of Soviet technical progress, 8–9

Red Army, 455; first missile units, 457; Mechanisation and Motorisation, Administration of, 417; Military-Technical Administration, 418; *see also* Rocketry
Regional Economic Councils (*Sovnarkhozy*), abolition of, 54
Regional Power Systems (RPS), development of, and GOELRO, 201; Dnieper RPS, 210; Donbass RPS, 201; Moscow RPS, 201; numbers of, 201
Remelting of steel, Electron beam, 107; Electroslag, 106–7; Plasma, 107; Vacuum arc, 107; Vacuum induction, 107
Renault, 553
Research, performance, as index of technological development, 24–5; quality of, 38–9; *see also* Chemical research
Research and Development, *see* R and D
Research intensive industries, 236
Research-production cycle, 36
Resistive power losses, 204
Riveting machines, *see* Machine tools, metalforming
Robotron Factory, 385
Rocket engines, 457–9; '09' model, 493; F-1, 506; Jupiter, 506; ORM series, 492–3; RD series, 493, 502–3
Rocketry, German influence on, 497–8, 501; history of Soviet, 1928–45, 492–6; military orientation of, 448–9, 493, 517–19; pioneers of, 491–2, *see also* Goddard, Tsiolkovskii; postwar, 497–510; purges and, 450, 496; R and D, 492–3, *see also* GDL, GIRD, RNII; Red Army and, 449, 492–3; *see also* Ballistic missiles; Launch vehicles; Rockets; Rocket engines; Spacecraft; Space programmes; Space stations
Rockets, development, 448–51, 455–9; early Soviet, 457, 458, 459, 472; liquid propelled, 446, 449, 450, 459; rocket aircraft, 449, 450; *see also* Ballistic missiles; Launch vehicles; Rocket engines; Rocketry
Rolled steel, 108–17; cold reduced, 108; cold rolled, 108, 115–16; electrical, 117; product range, 114; quality, 114; sectional, 108, 111; sheet, 110–11, 114, 115–16, 117, 119; structure of output, 108–11; *see also* Rolling Mills; Tinplate; Tube and pipe making
Rolling mills, age of 112–13, 115; automation, 115, 116, 118; cold rolling, 115–16; computer control, 118; foreign technology, 117; labour productivity, 112, 115; roughing, 113–14; strip, 115; structure of stock, 112; wire, 114; *see also* Tinplate; Galvanised steel
Rostokinsk Wool Factory, 281
RNII, *see* Reactive Research Institute
RP2 control system, 355; adoption by electric power industry, 357–9; and ASU-Energiya, 357, 358; comparison with GSP3 system, 359
RPG, 389, 390
Ruta computer, 382, 387, 388
Ryad series of computers, *see* ES
Ryazan Machine Tool Factory, 170, 187

Saab, 185
SALT, *see* Strategic Arms Limitation Talks
Salyut space station, 499, 510, 516–17, 520
Salzgitter, 263
Saniv, *see* Acrylic fibres
Saturn 1/1B launch vehicle, 500, 506
Saturn V launch vehicle, 500, 501, 506, 519, 520
Schlumburger group, 348
Science, bias in history and concepts of, 287
Science-based industries, 59; *see also* High-technology industries

Scientific manpower, *see* Manpower, scientific
Science and Technology, State Committee for, 168, 354
Scientific publications, Soviet output of, 38–9; types of, abstracts, 322, preliminary publications, 323, primary and secondary, 323, research papers, 320, review articles, 320
Scientific and Technical Information, All-Union Institute for (VINITI), 294, 322
Scientific and Technological Revolution, and chemical industry, 229, 234; Soviet concept of, 229
Scrap, cost of, 99; demand for, 100; in oxygen steelmaking, 99, 100; preparation of, 40
Second World War, and chemical industry, 228, 243, 245, 249, 256, 274, 275, 282; and continuous casting of steel, 101; and control equipment, 339; and electric arc steel, 100; and rocket research, 493, 496; and Soviet reconstruction, 48; and tanks, 419–20
Seven year plan (1959–65), and manmade fibres, 283; *see also* Machine tools, demand for
Sharpening machines, *see* Grinding machines
Sibelektrotyazhmash, 181
Siemens, 116; and NC control systems, 163, 184; and thyristor converters, 219, 220
Sigma *ob"edinenie*, 382
Silicon polymers, 274, 275
Simon-Carves, 263
Simulation and Computer Technology, Laboratory of, 377
Sinter, *see* Iron ore
Sinumeric, 159
SIP process, *see* Oxygen steelmaking
SKB process, *see* Synthetic rubber
SKB SAU, 355
SKB SPA (formerly SKB EAUS), 343, 357
Skylab, 510, 517
SLBMs, *see also* Ballistic missiles
Society for Assistance to Defence, Aviation and Chemistry (Osoaviakhim), 492
Soda ash, Leblanc process, 240; output, 246; Solvay process, 240, 244
Sodium carbonate, *see* Soda ash
Software, 388–91; for ES (Ryad) series, 389–91; imports of, 391; lag in, 44, 402; organisations of provision, 390–1; *see also* Algol, Assembler; Cobol; Fortran; PL/1; RPG
Soviet Cybernetics Review, 391, 392
Sovkabel' Factory, 207
Sovnarkhozy, *see* Regional Economic Councils
Soyuzkhimplastmass, 273
Soyuzpromavtomatika, *see* Minpribor
Soyuzprompribor, *see* Minpribor
Soyuzsistemkomplekt, *see* Minpribor
Soyuz spacecraft, 499, 506, 513, 515–16, 517, 522
Spacecraft, 510–17; design philosophy, 62–3, 510–14, 520–2; engineering, 514–17, Soviet simplicity of, 514, 520–2; role of cosmonaut, 513–14, 515, 516, 522; weight and internal atmosphere, 513; *see also* Computers; Gemini; Launch vehicles; Mercury; Polet; Soyuz; Space stations; Space programmes; Voskhod; Vostok
Space programmes, Lunokhod, 510; military influence on, 45–6; number and weight of launchings, 54; Soviet policy, 499, 522; tempo of, 522; *see also* Apollo; Launch vehicles; Pegasus; Spacecraft; Space stations
Space stations, *see* Salyut, Skylab
Sputnik, 458–9, 490, 501, 510, 512, 517
Stainless Steel, *see* Alloy and Quality steels
Stalin, J. V., 9, 13, 42; and development of atomic

bomb, 452, 472; political heritage and economic efficiency, 12; and rocket research, 496
Standards, *see* State Standards
Standards, State Committee of (Gosstandart), and Mark of Quality, 533, 548, 549; and organisation of State Standards, 532–3; quality inspection organisations, 549–50; *see also* State Standards
Stanki i instrument, 192
Stankoimport Review, 158, 524, 525
Stankokonstruktsiya Factory, 190
State Committee for Cinematography, *see* Cinematography, State Committee for
State Committee for Inventions and Discoveries, *see* Inventions and Discoveries, State Committee for
State Committee for Science and Technology, *see* Science and Technology, State Committee for
State Committee of Standards, *see* Standards, State Committee of
State Defence Committee, 452
State Planning Commission (Gosplan), 168, 397; Economic Research Institute of, 529
State Standards (GOST), 532–5; legal status, 532; for machine tools, 59, accuracy, 536, 539, 557–8, 558–61, age structure of, 545, dimensional parameters, 533–6, 538; organisation of, *see* Standards, State Committee of; for State Systems of Instruments, 345, 353
State Systems of Instruments, GSP-1, 342, 343, 344–7, 384, rate of development, 364–5, setbacks, 347, 351, specifications and GOST standards for, 345; GSP-3, 352–6, GOST and OST standards for, 355, 356, innovation in, 352–3, rejection by electric power industry, 352, 354–5, 358, Western influence on, 355–6
Steelmaking, *see* Alloy and quality steels; Continuous casting; Electric arc steelmaking; Open hearth steelmaking; Oxygen steelmaking; Remelting of steel; Rolled steel; Rolling mills; Tinplate; Tube and pipe making
Sterlitamak Drilling Machine Factory, 168
Strategic Arms Limitation Talks (SALT), 488
Submarines, *see* Naval technology
Sulphuric acid, catalytic contact process, 240–1, 244; lead chamber process, 240, 241; output, 248
Sumitomo Chemical Company, 270
Superphosphate fertilisers, *see* Fertilisers
Sverdlov Factory, 162, 164, 179, 180, 181, 184
Sverdlovsk, 204
Swedish Academy of Sciences, 289
Synthetic fibres, *see* Manmade fibres
Synthetic materials, 252–8; growth of output, 252; history of, 242–3; as index of technological level, 243, 272; Soviet lag in, 252; R and D costs, 242–3; *see also* Plastics and resins; Synthetic rubber
Synthetic rubber, output and development of, 255, 256–7; SKB process, 62; Soviet self-sufficiency in, 261
Synthetic slag, 105

Tank Industry, People's Commissariat of, 418
Tanks, ammunition, 429–34; design constraints, 417; design philosophy, 438–9; diffusion, 57, 440–1; engines, 435; exports of, 440; fire control, 434, 441, 442, 443; firepower, 427–34; layout, 442, 443; mobility, 435–6, 441, 442, 443; production, 418, 440–1; protection, 424–7, armour, 424–5, 441, 451, NBC, 426–7, 442; service life, 440, Soviet development of, 417–21; Soviet models: BT, 418, 419, 420, KV (IS), 418, 419, 420, T-10, 421, 439, T-26-32, 418, 419, T-34, 418, 419, 427, 442, German wartime assessment of, 419–20, technical and combat characteristics of, 420, T-34/85, 421, 429, 440, 441, 442, T-35, 418, T-54/55, 421, 426, 427, 429, 434, 436, 439, 440, 441, 442, T-62, 421, 426, 429, 431, 434, 441, 443, T-72, 421, 441, technological level of, 46, 57; Western models: AMX-30, 421, 424, 434, 436, 442, 443, Centurion, 421, 429, 434, 439, 440, Chieftain, 421, 424, 427, 429, 434, 442, 443, Leopard, 421, 429, 434, 439, 442, 443, M46-60, 421, 422, 426, 427, 429, 434, 436, 439, 440, 442, 443, Panther, 420, 424, 434, S-tank, 421, 422, 429, 434, 435, 436, 439, 442, Tiger, 420
Technical progress, assessments of, aggregated, 15–21, disaggregated, 23–6; and capital investment, 5–6; concepts of, 3–7, 10, 15; and GNP, 6; in heavy industry, Table 1.1; and key technologies, 15; rates of, 15–20, 21, Table 1.1; *see also*, Key innovations; Labour productivity
Technological agreements, 65
Technological conservatism, *see* Conservatism, technological
Technological level, assessments of, 26–9; and economic effectiveness, 37–8; methods of assessment, 2–21, 23–6, 30–1; problems of comparison, 31–3; and quality of production, 37; selection of criteria, 37; *see also* Ballistic missiles; Chemical industry; Factor productivity; Foreign trade; Labour productivity; Production functions; R and D; Synthetic materials, Tanks
Technology, advanced, definition of, 31–3, and economic effectiveness, 37–8; concepts of, 2
Technology gap, 21–3, 66
Technology and Organisation of Production, Scientific Research Institute for (NIAT), 181
Technology transfer, 29–30
Telephones, number installed in economy, 49
Teploenergoproekt, 205, 216
Terylene (Lavsan), 282–3
Thermo-Technical Research Institute, All-Union (VTI), 339, 355
Threading and tapping machines, *see* Other machines
Thyristor converters, *see* Converters, AC/DC
Tinplate, 116, 119
Tizpribor Factory, 348, 351
Toluene, 244
Tol'yatti, 551
Tomsk Mathematical Machines Factory, 181
Tooling, priority of, and specialisation, 141
Traditional industries, concept of, 39–40; development of, and innovation in, 47; growth, 52; priority, 59
Transformer steel, *see* Rolled steel, electrical
Transistors, 447; development costs, 235; *see also* Control instruments; Control and Instrumentation industry
Transmission costs, 202
Transmission lines, cable and overhead, general, 200; overhead, and UHVAC, 222; submarine, 204
Truman, President H., 452, 453
Tsiolkovskii, K. E., 448, 491, 497, 516
TsNIChermet, 102, 105
TsNIIKA, 345, 361
TsSU, *see* Central Statistical Administration
Tube and pipe making, 112
Tukhachevsky, M. N., and Soviet rocketry, 492, 496
Tupolev *sharaga*, 496
Turbine engines, 47
Turning mills, *see* Lathes

UHVAC (800 kV and above), constraints on usage, 203, 222; research in, 220; Soviet performance in,

224; Western debates on voltage levels, 221–2; 1,150 kV, role of, 220–1
Ultra-High Voltage Alternating Current, see UHVAC
Ultrasonic machines, 526
UM-1 computer, 382
Unemployment, see Labour productivity
UNIDO, see United Nations Industrial Development Organisation
United Nations Industrial Development Organisation (UNIDO), development typology of chemical industry, 296–7
United Power System, 201, 202; and HVDC, 217
Unit-head machines, see Other machines
Ural computer, 118, 387, 397
Ural-Kuznetsk Combine, 83
US Air Force, 488
US Arms Control and Disarmament Agency, see Arms Control and Disarmament Agency, US
US Department of Commerce, see Commerce, US Department of
US Department of Defense, see Defense, US Department of
USEPPA pneumatic control system, 44, 364–5, 370; derivatives, 348–9; development and characteristics, 348
US Navy, 488
USSR Academy of Sciences, see Academy of Sciences, USSR

V-1 rocket, 496
V-2 rocket, 496, 498; American model, 459
Vacuum arc remelting, see Remelting of steel
Vacuum degassing of steel, 106
Vacuum induction remelting, see Remelting of steel
Verkh-Isetsk Metallurgical Factory, 117
Vestnik statistiki, 238
Viscose silk, 277, 280, 282
Viskoza, 277
Vigon'trest, see Wool Trust

VINITI, see Scientific and Technical Information, All-Union Institute for
Vinitron, 284
VNIIEM-3 computer, 382
VNII Metrologii, see Metrology, All-Union Research Institute for
VNIIV, see Artificial Fibres, All-Union Institute for
Volga (GAZ-24) motorcar, 550; improvements to safety, 555–6; output, 551
Voskhod spacecraft, 500, 512, 513, 515, 522
Vostok launch vehicle, 500, 501–3, 519–20; exploitation, 507; stages of development, 503; thrust and engine configuration, 501–3, 506–7
Vostok spacecraft, 512, 513, 514, 515, 522
VTI, see Thermo-Technical Research Institute, All-Union
VTI (ER) control system, development, 339–41, 343; and electric power industry, 339; technical assessment of, 340–1

Warner and Swasey, 182
Warsaw Pact, 440, 441, 488
Western Technology and Soviet Economic Development, 275–6
Wool Trust (*Vigon'trest*), 277
World War II, see Second World War

Xylene, 244, 249

Yugoslavia, 157

Zaporozh'e Factory, 206
Zaporozhets motorcar, 550
ZAZ 966 motorcar, 550
ZEIM Factory, 343
Zhiguli motorcar, 550; and Fiat technical assistance, 551–3; output, 551
ZIL Factory, 156
Znak kachestva, see Mark of Quality
Zond space probes, 499, 510, 512

LIBRARY OF DAVIDSON COLLEGE